Nematodes as Biocontrol Agents

Nematodes as Biocontrol Agents

Edited by

Parwinder S. Grewal

Department of Entomology
Ohio State University, Wooster, Ohio
USA

Ralf-Udo Ehlers
Department of Biotechnology and Biological Control
Institute for Phytopathology
Christian-Albrechts-University Kiel, Raisdorf
Germany

David I. Shapiro-Ilan
United States Department of Agriculture
Agriculture Research Service
Southeastern Fruit and Tree Nut Research Laboratory, Byron, Georgia
USA

CABI Publishing

CABI Publishing is a division of CAB International

CABI Publishing
CAB International
Wallingford
Oxfordshire OX10 8DE
UK

Tel: +44 (0)1491 832111
Fax: +44 (0)1491 833508
E-mail: cabi@cabi.org
Web site: www.cabi-publishing.org

CABI Publishing
875 Massachusetts Avenue
7th Floor
Cambridge, MA 02139
USA

Tel: +1 617 395 4056
Fax: +1 617 354 6875
E-mail: cabi-nao@cabi.org

A catalogue record for this book is available from the British Library, London, UK.

Library of Congress Cataloging-in-Publication Data

Nematodes as biocontrol agents / edited by Parwinder S. Grewal, Ralf-Udo Ehlers, David I. Shapiro-Ilan.
p. cm.
Includes bibliographical references and index.
ISBN 0-85199-017-7 (alk. paper)
1. Nematoda as biological pest control agents. I. Grewal, Parwinder S. II. Ehlers, Ralf-Udo. III. Shaprio-Ilan, David I.
SB976.N46N46 2005
632′.96–dc22

2004030022

ISBN 0 85199 0177

Typeset by SPI Publisher Services, Pondicherry, India
Printed and bound in the UK by Biddles Ltd., King's Lynn

This volume is dedicated to Dr Harry K. Kaya in recognition of his numerous contributions to insect pathology. His early contributions to the ecology of entomopathogenic nematodes provided a solid foundation for the development of application strategies for use of nematodes in insect biocontrol. Dr Kaya has co-edited four books on entomopathogenic nematodes and insect pathology/biological control and co-authored a book on insect pathology. He has published more than 230 research papers on the ecology and application of insect nematodes and other pathogens and is one of the co-founding editors of the journal *Biological Control*. His outstanding leadership and scholarly accomplishments, spanning nearly three decades, have played a key role in expanding research and application in insect nematology to laboratories and industries around the world.

Contents

Contributors

Mary Barbercheck, Department of Entomology, Pennsylvania State University, University Park, PA 16802, USA (email: meb34@psu.edu)

Robin A. Bedding, Division of Entomology, CSIRO, GPO Box 1700, Canberra, ACT 2601, Australia (email: robin.bedding@csiro.au)

Guy Bélair, Horticultural Research and Development Centre, Agriculture and Agri-Food Canada, St-Jean-sur-Richelieu, Quebec, Canada J3B 3E6 (email: belairg@agr.gc.ca)

Anwar L. Bilgrami, Department of Entomology, Rutgers University, NJ 08901, USA (email: anwarbil@rci.rutgers.edu)

Noel E. Boemare, Laboratoire EMIP, UMR, INRA, Universite Montpellier II, Montpellier, Cedex 5, France (email: boemare@ensam.inra.fr)

Christopher Brey, Department of Entomology, Rutgers University, NJ 08901, USA (email: brey@rci.rutgers.edu)

H. Enrique Cabanillas, USDA, ARS, Kika de la Garza SARC, Beneficial Insects Research Unit, 2413 E. Hwy 83, Weslaco, TX 78596, USA (email: ecabanillas@westlaco.ars.usda.gov)

Ho Y. Choo, Department of Applied Biology and Environmental Science, Gyeongsang National University, Jinju, Gyeongnam 660–701, Republic of Korea (email: hychoo@nongae.gsnu.ac.kr)

Richard Cowles, Connecticut Agricultural Experiment Station, Valley Laboratory, Windsor, CT 06095, USA (email: richard.cowles@po.state.ct.us)

Giovanna Curto, Servizio Fitosanitario Regione emilia-Romagna, Bologna, Italy (email: gcurto@regione.emilia-romagna.it)

Fernando Garcia del Pino, Department of Animal Biology, Faculty of Science, Universitat Autònoma de Barcelona, 08193, Bellaterra, Barcelona, Spain (email: Fernando.Garcia@uab.es)

Larry W. Duncan, University of Florida, CREC, Lake Alfred, FL 33850, USA (email: lwdn@lal.ufl.edu)

Ralf-Udo Ehlers, Department for Biotechnology and Biological Control, Institute for Phytopathology, Christian-Albrechts-University Kiel, 24223, Raisdorf, Germany (email: ehlers@biotec.uni-kiel.de)

Albert Ester, Applied Plant Research Ltd, Business Unit Arable Farming and Field Production of Vegetables, PO Box 430, 8200 AK Lelystad, The Netherlands (email: Albert.Ester@wur.nl)

Jane Patterson Fife, Aerosol and Process Technologies, Battelle Memorial Institute, 505 King Avenue, Columbus, OH 43201, USA (email: fifej@battelle.org)

Joe E. Funderburk, North Florida Research and Education Center, University of Florida, Quincy, 155 Research Road, FL 32351, USA (email: jefunderburk@mail.ifas.ufl.edu)

Itamar Glazer, Department of Nematology, ARO, Volcani Center, Bet-Dagan, 50250, Israel (email: glazer@netvision.net.il)

Dawn H. Gouge, Department of Entomology, University of Arizona, Maricopa, AZ 85239, USA (email: dhgouge@ag.arizona.edu)

Parwinder S. Grewal, Department of Entomology, Ohio State University, Wooster, OH 44691, USA (email: grewal.4@osu.edu)

Christine T. Griffin, Department of Biology and Institute of Bioengineering and Agroecology, National University of Ireland, Maynooth, Ireland (email: christine.griffin@may.ie)

Richou Han, Guangdong Entomological Institute, Guangzhou 510260, China (email: richou-han@163.net)

Stuart Heritage, Entomology Branch, Forest Research, Northern Research Station, Roslin, Midlothian EH25 9SY, UK (email; stuart.heritage@forestry.gsi.gov.uk)

Casey W. Hoy, Department of Entomology, Ohio State University, Wooster, OH 44691, USA (email: hoy.1@osu.edu)

David J. Hunt, CABI Bioscience (UK Centre), Egham, Surrey TW20 9TY, UK (email: d.hunt@cabi.org)

Edson T. Iede, EMBRAPA Florestas, Brazil (email: iedeet@cnpf.embrapa.br)

Nobuyoshi Ishibashi, Saga University, 1090-3, Kinryu-Chifu, Saga, 849-0905, Japan (email: ishibasn@cc.saga-u.ac.jp)

Ganpati Jagdale, Department of Entomology, Ohio State University, Wooster, OH 44691, USA (email: jagdale.1@osu.edu)

Stephen Jess, Applied Plant Science Division, Department of Agriculture and Rural Development for Northern Ireland, Newforge Lane, Belfast BT9 5PX, UK (email: stephen.Jess@dardni.gov.uk)

Mairead Kilpatrick, Applied Plant Science Division, Department of Agriculture and Rural Development for Northern Ireland, NIHPBS Loughgall, County Armagh, BT61 8JB, UK (email: mairead.kilpatrick@dardni.gov.uk)

Albrecht M. Koppenhöfer, Department of Entomology, Rutgers University, New Brunswick, NJ 08901, USA (email: koppenhöfer@aesop.rutgers.edu)

Lawrence A. Lacey, USDA-ARS, Yakima Agricultural Research Laboratory, Wapato, WA 98951, USA (email: llacey@yarl.ars.usda.gov)

Kelly Sims Latsha, North Florida Research and Education Center, University of Florida, Quincy, FL 32351, USA (email: simskell@yahoo.com)

Edwin E. Lewis, Department of Nematology and Department of Entomology, 1 Shields Avenue, University of California Davis, Davis, CA 95616, USA (email: eelewis@ucdavis.edu)

Bradley A. Mullens, Department of Entomology, University of California, Riverside, CA 92521, USA (email: mullens@mail.ucr.edu)

Arne Peters, e-nema GmbH, Klausdorfer Str. 28–36, D-24223 Raisdorf, Germany (email: a.peters@ e-nema.de)

Simon Piggott, Becker Underwood, Littlehampton, West Sussex, BN17 7AU, UK (email: simon.piggott@icr.ac.uk)

Edward G. Platzer, Department of Nematology, University of California, Riverside, CA 92521, USA (email: edward.platzer@ucr.edu)

Sridhar Polavarapu, Department of Entomology, Rutgers University, Blueberry and Cranberry Research Center, Chatsworth, NJ 08019, USA (deceased)

Michael J. Raupp, Department of Entomology, University of Maryland, College Park, MD 20742, USA (email: mraupp@umd.edu)

Michael Samish, Kimron Veterinary Institute, Bet-Dagan, PO Box 12, 50250, Israel (email: samishm@int.gov.il)

Sibylle Schroer, Department for Biotechnology and Biological Control, Institute for Phytopathology, Christian-Albrechts-University Kiel, 24223, Raisdorf, Germany (email: schroer@biotec.uni-kiel.de)

Heinrich Schweizer, Department of Applied Plant Science, Queen's University of Belfast, Newforge Lane, Belfast BT9 5PX, UK (email: heinrich.hdh@tiscali.ch)

Muhammed M. Shamseldean, Department of Agricultural Zoology and Nematology, Faculty of Agriculture, Cairo University, Giza 12311, Egypt (email: mshamseldean@hotmail.com)

David I. Shapiro-Ilan, United States Department of Agriculture, Agriculture Research Service, Southeastern Fruit and Tree Nut Research Laboratory, Byron, GA 31008, USA (email: dshapiro@saa.ars.usda.gov)

S. Patricia Stock, Division of Plant Pathology and Microbiology, Department of Plant Sciences, University of Arizona, Tucson, AZ 85721, USA (email: spstock@ag.arizona.edu)

Arne Thies, e-nema France, Le Columbie, 46350 Payrac, France (email: silke@wanadoo.fr)

Marek Tomalak, Department of Biological Pest Control and Quarantine, Institute of Plant Protection, Miczurina 20, 60–318 Poznan, Poland (email: m.tomalak@ior.poznan.pl)

Peter Torr, School of Biological Sciences, University of Aberdeen, Cruickshank Building, St Machar Drive, Aberdeenshire, AB24 3UU, UK (email: p.torr@abdn.ac.uk)

Rob W.H.M. van Tol, Plant Research International, Wageningen-UR, PO Box 16, 6700 AA Wageningen, The Netherlands (email: r.w.h.m.vantol@plant.wag-ur.nl)

Raja V. Vyas, Department of Nematology, Gujarat Agricultural University, Anand 388 110, India (email: rvvyas@gauanand.com)

Roger N. Williams, Department of Entomology, Ohio State University, Wooster, OH 44691, USA (email: Williams.14@osu.edu)

Michael J. Wilson, School of Biological Sciences, University of Aberdeen, Cruickshank Building, St Machar Drive, Aberdeenshire, AB24 3UU, UK (email: m.j.wilson@abdn.ac.uk)

Denis J. Wright, Department of Biological Sciences, Imperial College London, Silwood Park Campus, Ascot, Berkshire SL5 7PY, UK (email: d.wright@ic.ac.uk)

Robert J. Wright, Department of Entomology, 202 Plant Industry Building, University of Nebraska-Lincoln, Lincoln, NE 68583, USA (email: rwright2@unl.edu)

Preface

The interest in the use of nematodes as biological pest control agents has increased exponentially over the past two decades. Thousands of researchers and practitioners worldwide are now exploring the potential of nematodes to manage noxious insects, molluscs, plant nematodes and even soil-borne plant pathogens. The entomopathogenic nematodes (EPNs) (*Steinernema* and *Heterorhabditis*) and slug-parasitic nematodes (*Phasmarhabditis*) have proven particularly successful and are now commercially mass-produced in six of the seven continents to treat pest problems in agriculture, horticulture and veterinary and human husbandry. The ease of mass production and exemption from registration requirements are the two major reasons for early interest in the commercial developments of nematodes. However, demonstrations of practical use, particularly in Europe and North America and subsequently in Japan, China and Australia, spurred developments across the world that have led to the availability of nematodes against pests that were once thought impossible to control.

In this volume 54 experts from 18 countries contribute authoritative chapters that comprehensively illustrate the remarkable developments in the use of nematodes for biocontrol of a diverse array of pests in diverse ecosystems. This volume captures the full breadth of basic and applied information on all groups of nematodes that are used or have potential as biocontrol agents of pest invertebrates and soil-borne plant pathogens. The actual application of nematodes in different cropping systems of the world is described and the huge amount of recent efficacy data on numerous target pests is summarized. We have attempted to integrate the vast amount of information for the development of novel and practical approaches for nematode application and to explain test failures that frustrated early efforts. EPNs in the families Heterorhabditidae and Steinernematidae are by far the most widely tested group. Due to a mutualistic association with bacteria in the genera *Photorhabdus* (for Heterorhabditidae) and *Xenorhabdus* (for Steinernematidae), EPNs are able to kill a diverse array of insects. The slug-parasitic nematodes, particularly *Phasmarhabditis hermaphrodita* (Rhabditidae), have shown tremendous potential for the management of mollusc pests, and recent research has shown that slug-parasitic nematodes also partner with bacteria to kill their hosts. Although the symbiotic bacteria *Photorhabdus* and *Xenorhabdus* have emerged as a source of a diverse array of toxins and antibiotics with a potential for stand-alone biocontrol agents, this aspect was considered to be beyond the scope of this book. Remarkable successes with entomopathogenic and slug-parasitic nematodes have increased interest in the development of entomophilic nematodes such as *Thripinema* for insect control, predatory nematodes for plant-parasitic nematode control and fungal-feeding nematodes for the control of soil-borne plant pathogens. All these fascinating developments are described in this volume.

As accurate definitions and usage of terminology are critical to effective communication, we begin by providing a glossary of some of the commonly used terms in insect nematology. This volume is divided into seven parts: morphology and taxonomy of all nematode groups used as biocontrol agents; EPNs; entomophilic nematodes; slug-parasitic nematodes; predatory nematodes; fungal-feeding nematodes; and conclusions. In Part II, there are five chapters devoted to biology, mass production, formulation and quality control, application technology and safety. Subsequent chapters focus on the efficacy of nematodes against target pests in different cropping systems, including turfgrass and pastures, glasshouse production, nurseries and trees, mushrooms, orchards, soft fruits, vegetable and tuber crops, cereal, fibre, medicinal and oilseed crops, forestry, veterinary and human husbandry and social insects. We separated these chapters based on cropping systems as there are vast differences in the ecology of these systems that have a profound effect on the efficacy of nematodes. Each chapter begins with a general introduction to the cropping system and target pests, followed by a critical review of the information on the application and efficacy of nematodes against specific pests. Tables to summarize efficacy data and comments on the essential components of application strategy are some of the key features of these chapters. Each chapter identifies factors in the success and failure of nematodes and is concluded with specific application recommendations and future research needs. Three additional chapters provide information on the compatibility and interactions of EPNs with agricultural chemicals, the potential of EPNs to suppress plant-parasitic nematodes and the development of a conservation approach.

There are three chapters in Part III: one providing an update on the use of *Deladenus* for the control of sirex wood wasp, the second on *Thripinema* and the third on mermithid nematodes. Part IV has two chapters: one on biology, mass production and formulation and the other on field application. Part V has one chapter covering the potential of predacious nematodes to control plant-parasitic nematodes, Part VI describes the latest research on the use of fungal-feeding nematodes, particularly *Aphelenchus avenae*, to control soil-borne fungal pathogens. Part VII provides an overall synthesis of the field and identifies critical issues and research needs for further expansion of the potential and use of nematodes in biocontrol.

This volume is dedicated to Dr Harry K. Kaya as an acknowledgement of his numerous contributions to the ecology of EPNs and for his leadership of insect nematology for nearly three decades. We thank all the contributors who made this book possible. Finally, we express gratitude to our wives, Sukhbir Grewal, Karen Ehlers and Laura Lucy-Ilan from whom we stole time for this endeavour.

Parwinder S. Grewal, Ralf-Udo Ehlers and David I. Shapiro-Ilan

August 2004

Glossary of terms

Axenic: Free from associated organisms.

Biocontrol: The introduction of natural enemies (parasites, parasitoids, predators, or pathogens) to suppress pest populations; some include certain byproducts of natural enemies in the definition.

Commensalism: A symbiotic relationship between two species in which one of the organisms benefits and the other is not apparently affected.

Dauer stage or dauer larva: A developmentally arrested dispersal stage in certain nematodes; in entomopathogenic nematodes it is the only free-living stage (also known as infective juvenile).

Entomogenous: Refers to organisms growing in or on the bodies of insects; denotes a parasitic or other intimate symbiotic relationship.

Entomoparasitic: Parasitic to insects; a relationship between an organism (e.g. nematode) and an insect, in which the organism benefits at the insect host's expense; host mortality is not necessarily a requirement for the parasite's development; nematode examples include Mermithidae, Allantonematidae, Parasitylenchidae, Phaenopsitylenchidae, Iotonchidae, Acugutturidae, Parasitaphelenchidae, Entaphelenchidae and Thelastomatidae.

Entomopathogenic: A microorganism or nematode capable of causing disease in insects; in insect nematology, the term is specifically used to refer to parasitic nematodes that are mutualistically associated with bacterial symbionts; all life stages of the nematode, except for the free-living third stage infective juvenile or dauer stage, are found inside the insect host; examples are Steinernematidae and Heterorhabditidae.

Entomophilic: Having an affinity for insects ('insect loving'); for nematodes, can refer to any association with insects (parasitic or non-parasitic).

Epizootic: An outbreak of disease in which there is an unusually large number of cases.

Incidence: The number of new cases of a particular disease within a given time period.

Infectivity: The ability of an organism to enter a susceptible host, resulting in presence of the organism within the host (whether or not this causes detectable pathological effects); the ability to produce infection.

In vitro: Outside the living organism, in an artificial environment.

In vivo: In the living organism.

Mutualism: A symbiotic relationship between two different species in which both jointly benefit.

Patent infection: An overt infection with distinct signs and symptoms of the disease.

Pathogenicity: The quality or state of being pathogenic, the potential ability to produce disease (an 'all-or-none' concept).

Phoretic: Refers to a symbiotic relationship in which one organism associates with another in order to obtain transportation, and causing little or no detectable pathology to the host; examples of nematodes having a phoretic association with insects include certain members of Rhabditidae, Diplogastridae and Aphelenchidae.

Prevalence: The total number of cases of a particular disease at a given moment of time.

Sign: An objective manifestation of disease indicated by alteration in structure.

Symbiosis: The living together of individuals of two different species, particularly the living together of two dissimilar species in an intimate association (e.g. mutualism, commensalism, parasitism).

Symptom: Any objective aberration in behaviour or function indicating disease.

Virulence: The disease-producing power of an organism, the degree of pathogenicity within a group or species.

Sources

Lacey, L.A. and Brooks, W.M. (1997) Initial handling and diagnosis of diseased insects. In: Lacey, L.A. (ed.) *Manual of Techniques in Insect Pathology.* Academic Press, San Diego, California, pp. 1–15.

Poinar, G.O., Jr (1975) *Entomogenous Nematodes: A Manual and Host List of Insect–Nematode Associations.* E.J. Brill, Leiden, The Netherlands.

Steinhaus, E.A. and Martignoni, M.E. (1970) *An Abridged Glossary of Terms Used in Invertebrate Pathology*, 2nd edn, USDA Forest Service, PNW Forest and Range Experiment Station.

Stock, S.P. (2002) Glossary of terms used in insect nematology. *The Society of Nematology Newsletter* 2002, Issue No. 3, p. 17.

Part I

Nematode Morphology and Taxonomy

1 Morphology and Systematics of Nematodes Used in Biocontrol

S.P. Stock[1] and D.J. Hunt[2]

[1]*Department of Plant Pathology, University of Arizona, Tucson, AZ, 85721-0036, USA;* [2]*CABI Bioscience (UK Centre), Bakeham Lane, Egham, Surrey TW20 9TY, UK*

1.1. Introduction

One of the first and most important needs in biocontrol programmes, is the accurate identification of the pest and any beneficial organisms with biocontrol potential. This aspect has a direct impact not only in determining the geographic range of a pest but also in the acquisition of permits necessary for release of control agents (Schauff and LaSalle, 1998). Moreover, this basic but indispensable information eventually impacts directly on their success as biocontrol agents (Lacey *et al.*, 2001).

Among the numerous beneficial organisms considered in biocontrol are nematodes. Many nematodes are associated with insects, mites and molluscs of potential importance in agriculture, forestry or health (Poinar, 1983; Petersen, 1985; Gaugler and Kaya, 1990; Bedding, 1993; Wilson *et al.*,

 Nematodes as Biocontrol Agents (eds P.S. Grewal, R.-U. Ehlers and D.I. Shapiro-Ilan)

1993, 1994; Wilson and Gaugler, 2000; Grewal *et al.*, 2003). These nematode–invertebrate associations range from 'casual' (i.e. phoretic, commensal) to obligate parasitism and pathogenesis. The number of newly discovered nematode species/isolates with biocontrol potential has significantly increased over the past decade. Accurate and prompt identification/diagnosis of these taxa requires the implementation of appropriate taxonomic tools. To meet these expectations nematode systematists have incorporated new technologies into their traditional morphological approaches including several molecular techniques.

This chapter summarizes the latest information regarding the taxonomic status of nematode groups considered as biocontrol agents of economically important pests. Morphological diagnoses to genera and/or species are provided and keys where feasible. A summary of molecular methods and markers currently used in the systematics of these groups is also presented.

1.2. Classification

More than 30 nematode families are known to host taxa that parasitize or are associated with insects (Nickle, 1972; Poinar, 1975, 1983, 1990; Maggenti, 1981; Kaya and Stock, 1997). However, because of their biocontrol potential, research has concentrated on seven families: Mermithidae, Allantonematidae, Neotylenchidae, Sphaerularidae, Rhabditidae, Steinernematidae and Heterorhabditidae, the latter two currently receiving the most attention as control agents of soil insect pests (Lacey *et al.*, 2001).

The biocontrol potential of nematodes is not restricted to insects. *Phasmarhabditis hermaphrodita* (Schneider), a member of the family Rhabditidae, is known to suppress several slug species, and has recently been developed as a biological molluscicide (Wilson *et al.*, 1993; Glen and Wilson, 1997; Wilson and Gaugler, 2000). Moreover, several predatory mononchids, dorylaimids, nygolaimids, diplogasterids and the fungal-feeding nematode (*Aphelenchus avenae* Bastian) have also been studied as potential biocontrol agents of plant-parasitic nematodes and plant pathogens (Kasab and Abdel-Kader, 1996; Lootsma and Scholte, 1997; Choudhury and Sivakumar, 2000; Matsunaga *et al.*, 1997).

In this chapter, we have adopted the new classification scheme suggested by De Ley and Blaxter (2002) to list those groups with biocontrol potential. This classification is rooted on a phylogenetic interpretation of a preliminary evolutionary tree based on 18S ribosomal DNA (rDNA) proposed by Blaxter *et al.* (1998). This molecular framework does not support the common division of Nematoda into Adenophorea and Secernentea. Instead, it recognizes the presence of three basal clades: dorylaimids, enoplids and chromadorids. Relationships between these clades have not been fully resolved, but available data support sister taxon status of dorylaims and enoplids (De Ley and Blaxter, 2002). In this new taxonomic system, dorylaims and enoplids are encompassed within the class Enoplea Inglis, 1983. The Chromadorea Inglis, 1983 comprise the majority of taxa within Nematoda, including all the former Secernentea.

In this classification system, 7 out of 11 nematode families currently considered in biocontrol are grouped within the Chromadorea; the remaining, Mononchidae, Mermithidae, Dorylaimidae and Nygolaimidae, are members of the Enoplea (Table 1.1).

1.3. Diagnosis of Major Groups

1.3.1. Family Steinernematidae Chitwood and Chitwood, 1937 (Fig. 1.1)

1.3.1.1. Diagnostic characters

Adults with truncated to slightly rounded head. Six fused lips, but tips distinct, and with one labial papilla each. Four cephalic papillae present. Amphids small. Stoma reduced, short and wide, with inconspicuous sclerotized walls. Oesophagus rhabditoid, set off from intestine. Nerve ring usually surrounding isthmus or anterior part of basal bulb. Excretory pore opening distinct.

Table 1.1. Major groups in the phylum Nematoda with biocontrol potential (classification based on De Ley and Blaxter, 2002).

CLASS CHROMADOREA INGLIS, 1983

Subclass Chromadoria Pearse, 1942
ORDER RHABDITIDA CHITWOOD, 1933
Suborder Tylenchina Thorne, 1949
Infraorder Panagrolaimomorpha De Ley and Blaxter, 2002
Superfamily Strongyloidoidea Chitwood and McIntosh, 1934
Family Steinernematidae Chitwood and Chitwood, 1937
Superfamily Aphelenchoidea Fuchs, 1937
Family Aphelenchidae Fuchs, 1937
Infraorder Tylenchomorpha De Ley and Blaxter, 2002
Superfamily Sphaerularoidea Lubbock, 1861[a]
Family Allantonematidae Pereira, 1931
Family Neotylenchidae Thorne, 1941
Suborder Rhabditina Chitwood, 1933
Infraorder Rhabditomorpha De Ley and Blaxter, 2002
Superfamily Rhabditoidea Örley, 1880
Family Rhabditidae Örley, 1880
Superfamily Strongyloidea Baird, 1853
Family Heterorhabditidae Poinar, 1975
Infraorder Diplogasteromorpha De Ley and Blaxter, 2002
Superfamily Diplogasteroidea Micoletzky, 1922
Family Diplogasteridae Micoletzky, 1922

CLASS ENOPLEA INGLIS, 1983

Subclass Dorylaimia Inglis, 1983
ORDER DORYLAIMIDA PEARSE, 1942
Suborder Dorylaimia Pearse, 1942
Superfamily Dorylaimoidea de Man, 1876
Family Dorylaimidae de Man, 1876
Suborder Nygolaimia Thorne, 1935
Superfamily Nygolaimoidea Thorne, 1935
Family Nygolaimidae Thorne, 1935
ORDER MONONCHIDA JAIRAJPURI, 1969
Suborder Mononchina Kirjanova and Krall, 1969
Superfamily Mononchoidea Chitwood, 1937
Family Mononchidae Chitwood, 1937
ORDER MERMITHIDA HYMAN, 1951
Suborder Mermithina, Andrássy, 1974
Superfamily Mermithoidea Braun, 1883
Family Mermithidae Braun, 1883

[a]Families within Sphaerularoidea are listed based on the classification proposed by Siddiqi (2000) which recognizes three families within the Sphaerularoidea: Sphaerulariidae, Lubbock, 1861; Allantonematidae, Pereira, 1931; and Neotylenchidae Thorne, 1941.

Females with paired opposed ovaries. Vagina short, muscular. Vulva located near middle of body, with or without protruding lips. Epiptygma present or absent. Males monorchic, testis reflexed. Spicules paired, symmetrical. Gubernaculum present. One single midventral and 10–14 pairs of genital papillae present of which 7–10 pairs are precloacal. Tail rounded, digitated or mucronated. Third-stage infective juvenile (IJ) with collapsed stoma. Cuticle annulated, lateral field with 6–8 ridges in middle of body. Oesophagus and intestine collapsed. Specialized bacterial pouch located

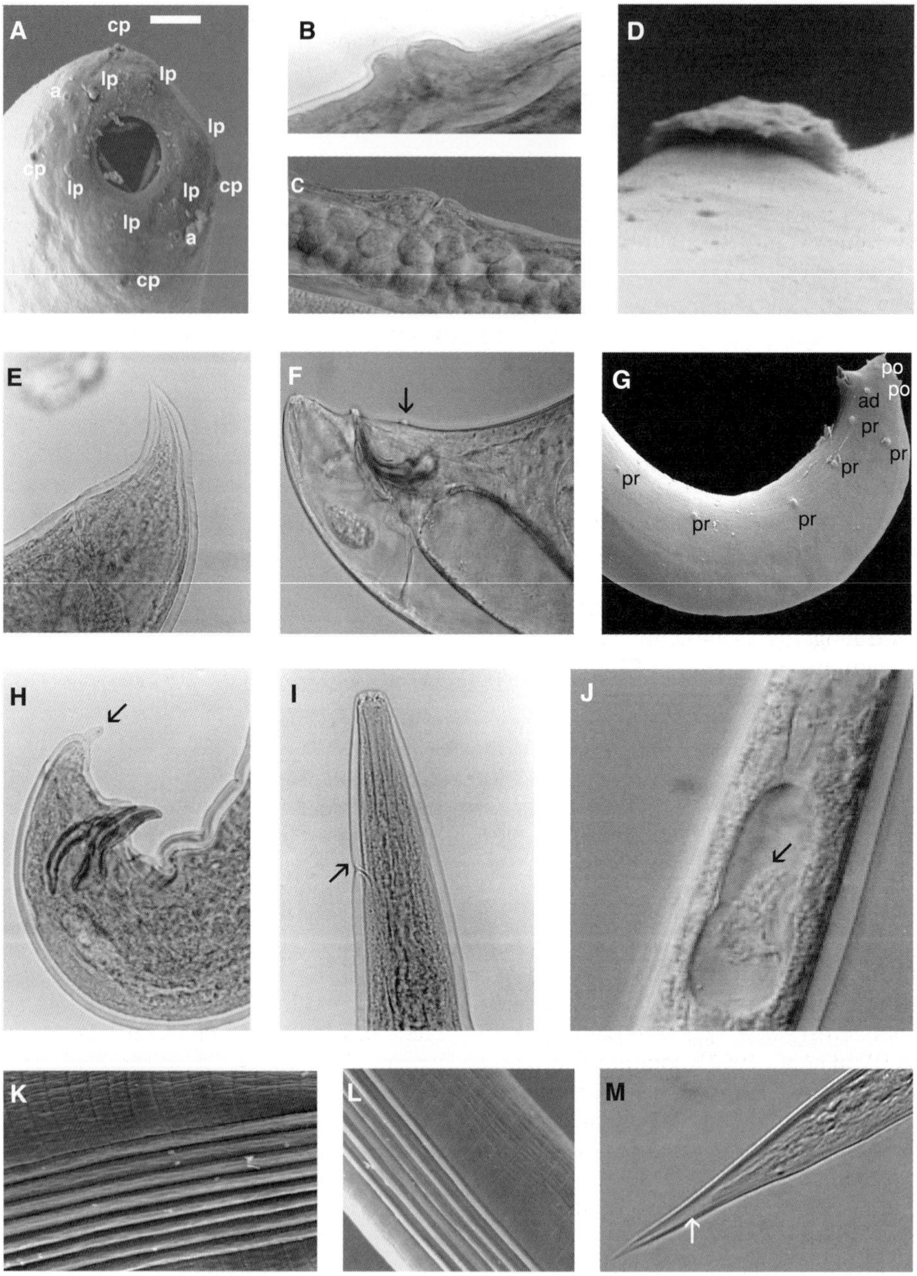

Fig. 1.1. Family Steinernematidae. A–D. First generation female: A, scanning electron micrograph (SEM) showing stomatal opening, labial and cephalic papillae; B, protruding vulval lips (lateral view); C, slightly protruding vulval lips (lateral view); D, Epiptygma. E, tail (lateral view). F–H. First generation male: F, tail (lateral view) showing single ventral papilla (arrow); G, SEM of tail showing precloacal, adcloacal and postcloacal papillae (lateral view); H, tail (lateral view) showing mucro (arrow). I–M. Third-stage infective juvenile (IJ): I, anterior end showing excretory pore (arrow); J, bacterial pouch (lateral view) showing clump of bacterial cells (arrow); K and L, SEMs of lateral field pattern with (K) eight and (L) six ridges; M, tail (lateral view) showing hyaline portion (arrow). (Scale bars: A, L = 5.5 μm; B, C, E, F = 25 μm; D = 35 μm; G = 40 μm; H = 23.5 μm; I, J = 16 μm; K = 4 μm; M = 10 μm.)

at beginning of intestine is of variable shape. Excretory pore distinct, anterior to nerve ring. Tail conoid or filiform, with variable hyaline portion. Phasmids present, prominent or inconspicuous.

The Steinernematidae currently comprise two genera, *Steinernema* Travassos, 1927 with more than 30 species and *Neosteinernema* Nguyen and Smart, 1994 with only one species (*N. longicurvicauda*) (Tables 1.2 and 1.3).

1.3.1.2. Bionomics

Steinernematids are obligate pathogens in nature and are characterized by their mutualistic association with bacteria of the genus *Xenorhabdus*. Of all nematodes studied for biocontrol of insects, the Steinernematidae together with the Heterorhabditidae have received the most attention because they possess many of the attributes of effective biocontrol agents. Details on the biology of this group are discussed in Chapter 2, this volume.

1.3.1.3. Phylogenetic relationships

The first explicit hypotheses for evolutionary relationships among *Steinernema* spp. were proposed by Reid (1994) based on phylogenetic analysis of genetic distances calculated from rDNA restriction sites for 12 species. Additional investigations were based on restriction fragment length polymorphic (RFLP) pattern analysis of the internal transcribed spacer (ITS) region of rDNA (Reid *et al.*, 1997), combined analyses of morphological data and randomly amplified polymorphic DNA (RAPD) markers (Liu and Berry, 1996), and partial small subunit (SSU; 18S) rDNA sequence analysis (Liu *et al.*, 1997). Unfortunately, the evolutionary hypotheses so obtained are of limited utility due to several factors, including an insufficient number of phylogenetically informative characters, uncertainties in character homology and, in certain cases, the use of data (e.g. RAPD markers) or tree-building methods (e.g. unweighted pair group method analysis (UPGMA) phenograms) that are inappropriate for inferring evolutionary history (Stock *et al.*, 2001). In addition, although different isolates of individual species have been included in some of these studies, less than half of the described *Steinernema* spp. were studied.

More recently, DNA sequence analysis of mitochondrial genes, i.e. cytochrome oxidase II (COII-16S) (Szalanski *et al.*, 2000), and nuclear genes, i.e. internal transcribed spacer-1 (ITS-1) region of rDNA (Nguyen *et al.*, 2001), and the large subunit (LSU; 28S) of rDNA (Stock *et al.*, 2001) have been used to assess evolutionary relationships among *Steinernema* spp. Taxon sampling, i.e. inclusion of all available *Steinernema* spp., is one of the challenges for accomplishing a robust interpretation of phylogenetic relationships of species in this genus. This will probably be a difficult task, particularly in view of the large number of newly described species in the past few years, but is essential to robustly test methods used to infer evolutionary relationships.

In this respect, the study conducted by Stock *et al.* (2001) has incorporated the most extensive list of *Steinernema* spp. to date. Results from this study were in part consistent with some traditional morphological expectations and previous phylogenetic studies. The hypotheses inferred from molecular evidence and those from combined analysis of morphological and sequence data provided the first comprehensive testable hypothesis of phylogenetic relationships for species in *Steinernema*. Following this study, the incorporation of some newly described species has not only provided a better resolution of several clades (reflected by higher bootstrap values) than the previous analysis, but has also reinforced previous considerations of the value of 28S rDNA sequences in assessing evolutionary history in *Steinernema* (Stock and Koppenhöfer, 2003) (Fig. 1.2).

1.3.2. Family Aphelenchidae Fuchs, 1937

1.3.2.1. Diagnostic characters

Labial cap distinct and often set off by a constriction. Hollow axial protrusible spear with slight basal thickenings.

Table 1.2. Taxonomic summary of the family Steinernematidae. Family Steinernematidae Chitwood and Chitwood, 1937 Syn. Neoaplectanidae Sobolev, 1953.

Taxa	Biogeography[a]	GenBank sequence data (accession number)
Type genus:		
Steinernema Travassos, 1927		
Type species:		
Steinernema kraussei (Steiner, 1923) Travassos, 1927	Europe (Germany), North America	28S (AF331896)
Other species:		
S. abbasi Elawad, Ahmad and Reid, 1997	Asia (Oman)	18S (AY035764), 28S (AF331890), ITS-1,-2 (AY248749)
S. affine (Bovien, 1937) Wouts, Mrácek, Gerdin and Bedding, 1982	Europe (Denmark)	18S (AY035765), 28S (AF331899), ITS-1,-2 (AF331912)
S. anatoliense Hazir, Stock and Keskin, 2003	Asia (Turkey)	28S (AY841761)
S. arenarium (Artyukhovsky, 1967) Wouts, Mrácek, Gerdin and Bedding, 1982	Asia (Central Russia)	18S (U70639), 28S (AF331892), ITS-1 (AF192985), COII-16S (AF192992)
S. asiaticum Anis, Shahina, Reid and Rowe, 2002	Asia (Pakistan)	NA
S. bicornutum Tallosi, Peters and Ehlers, 1995	Europe (Yugoslavia)	28S (AF331904), ITS-1,-2 (AF121048),
S. carpocapsae (Weiser, 1955) Wouts, Mrácek, Gerdin and Bedding, 1982	Asia, Europe (Czechoslovakia), North America, South America	18S (U70633, AF36604), 28S (AF331900), ITS-1 (AF192987, AF036947), ITS-1,-2 (AF331913, AF121049), COII-16S (AF192995), SAT (U12680)
S. caudatum Xu, Wang and Li, 1991	Asia (China)	NA
S. ceratophorum Jian, Reid and Hunt, 1997	Asia (China)	28S (AF331888), ITS-1,-2, (AF440765)
S. cubanum Mrácek, Hernandez and Boemare, 1994	Central America (Cuba)	28S (AF331889)
S. diaprepesi Nguyen and Duncan, 2002	North America (USA)	ITS-1,-2 (AF440764)
S. feltiae (Filipjev, 1934) Wouts, Mrácek, Gerdin and Bedding, 1982	Europe (Denmark), North America, South America	18S (U70634, AY035766), 28S (AF331906), ITS-1 (AF92983, AF92982), ITS-1,-2 (AF121050), mRNA-GSY-1 (AF241845), COII-16S (AF192991, AF192990)
S. glaseri (Steiner, 1929) Wouts, Mrácek, Gerdin and Bedding, 1982	Asia, Europe, North America (USA), South America	18S (U70640), 28S (AF331908), ITS-1 (AF192986), ITS-1,-2 (AF122015), COII-16S (AF192993), SAT (U19929)
S. intermedium (Poinar, 1985) Mamiya, 1988	North America (USA), Europe	18S (U70636), 28S (AF331909), ITS-1 (AF192989), ITS-1,-2 (AF33916, AF122016)

S. karii Waturu, Hunt and Reid, 1997	Africa (Kenya)	18S (AJ417021), 28S (AF331902)
S. kushidai Mamiya, 1988	Asia (Japan)	28S (AF331897), ITS-1,-2 (AF192984),
S. loci Phan, Nguyen and Moens, 2001	Asia (Vietnam)	ITS-1,-2 (AY355443)
S. longicaudum Shen and Wang, 1992	Asia (China), North America	18S (AY035767), 28S (AF331894)
S. monticolum Stock, Choo and Kaya, 1997	Asia (Korea)	28S (AF331895), ITS-1,-2 (AF331914, AF122017)
S. neocurtillae Nguyen and Smart, 1992	North America (USA)	ITS-1,-2 (AF122018)
S. oregonense Liu and Berry, 1996	North America (USA)	18S (U70637), 28S (AF331891), ITS-1,-2 (AF122019)
S. pakistanense Shahina, Anis, Reid, Rowe and Maqbool, 2001	Asia (Pakistan)	NA
S. puertoricense Román and Figueroa, 1994	Central America (Puerto Rico)	28S (AF331903)
S. rarum (de Doucet, 1986) Mamiya, 1988	South America (Argentina), North America (USA)	28S (AY253296, AF331905)
S. riobrave Cabanillas, Poinar and Raulston, 1994	North America (USA)	18S (U70635), 28S (AF331893), COII-16S (AF192994)
S. ritteri de Doucet and Doucet, 1990	South America (Argentina)	NA
S. sangi Phan, Nguyen and Moens, 2001	Asia (Vietnam)	ITS-1,-2, (AY355441)
S. scapterisci Nguyen and Smart, 1990	South America (Uruguay)	28S (AF331898), ITS-1,-2 (AF122020, AF331915)
S. scarabaei Stock and Koppenhöfer, 2003	North America (USA)	28S (AY172023)
S. serratum Liu, 1992[b]	Asia (China)	18S (U70638)
S. siamkayai Stock, Somsook and Kaya, 1998	Asia (Thailand)	28S (AF331907), ITS-1,-2 (AF331917)
S. tami Van Luc, Nguyen, Spiridonov and Reid, 2000	Asia (Vietnam)	18S (AY035768)
S. thanhi Phan, Nguyen and Moens, 2001	Asia (Vietnam)	ITS-1,-2 (AY355444)
S. websteri Cutler and Stock, 2003	Asia (China)	28S (AY841762)
S. thermophilum Ganguly and Singh, 2000	Asia (India)	NA
Genus: *Neosteinernema* Nguyen and Smart, 1994		
Type and only species: *Neosteinernema longicurvicauda* Nguyen and Smart, 1994	North America (USA)	

[a]Country of original isolation in parentheses.
[b]Species *inquirenda*.
NA = no sequences available.

Table 1.3. Polytomous key for Steinernematidae.

Neosteinernema

Key diagnostic features: adults and third-stage infective juveniles (IJs) with very conspicuous amphids. Males with ventrally arcuate spicules with a very prominent manubrium. IJs with very long (as long as oesophagus length) and filiform tail.

	IJs							First generation adults						
								Male				Female		
Species	TBL	MBW	EP	TL	D%	E%	LF	SpL	GuL	SW	D%	M	EPI	VL
longicurvicauda	920	24	68	167	41	41	8	61	59	1.03	NA	A	A	V
	789–1084	20–31	61–76	141–190	38–46	37–48		52–67	52–66	0.8–1.15				

Steinernema

Key diagnostic features: adults and third-stage infective juveniles (IJs) with phasmids not visible. Shape of spicules variable but not with a manubrium shape as in *Neosteinernema*. IJs with conoid tail (variable in size).

	IJs							First generation adults						
								Male				Female		
Species	TBL	MBW	EP	TL	D%	E%	LF	SpL	GuL	SW	D%	M	EPI	VL
					***carpocapsae*-group (IJ average size < 600 μm)**									
asiaticum	425[a]	23	32	NA	32[a]	78[a]	6	68[a]	53[a]	2.0[a]	44	P	P	SP
	360–450	20–25	28–34		30–36	60–90		61–74	46–62	1.6–2.5	35–57			
siamkayai	446	21	35	36	37	96	6–8	77.5	54	1.7	42	P	P	PR
	398–495	18–24	29–38	31–41	31–43	95–112		75–80	47–65	1.4–2.2	35–49			
ritteri	510	22	43	49	46	88	6	69	44	1.56	47	A	A	PR
	470–590	19–24	40–46	44–54	44–50	79–97		8–75	33–50	1.44–1.57	44–50			
rarum	511	23	38	51	35	72	6	47	34	0.94	50	P	A	PR
	443–573	18–26	32–40	4–56	30–39	63–80		42–52	23–38	0.91–1.05	44–51			
tami	530	23	36	50	31	73	6–8	77	48	2.0	44	P	A	NP
	400–600	19–29	34–41	42–57	28–34	67–86		71–84	38–55	1.4–3.0	30–60			
abbasi	541	29	48	56	53	86	8	65	45	1.56	60	A	P	PR

	496–579	27–30	46–51	52–61	51–58	79–94		57–74	33–50	1.07–1.87	51–68			
anatoliense	545	24.5	37	52	35	72	6	74	47	1.75	48.5	P	A	SP
	507–580	21–28	36–39	46–58	31.5–39	68–81.5		68–84	42–59	1.6–1.9	46.5–55			
thermophilum	555	21	40	45	46	96	8	61	36	1.7	63	A	P	PR
	510–620	21–23	37–46	40–52	42–53	81–102		44–72	30–42	1.2–2.8	50–87			
carpocapsae	558	25	38	53	26	60	6	66	47	1.72	41	P	A	PR
	438–650	20–30	30–60	46–61	23–28	54–66		58–77	39–55	1.40–2.00	27–55			
scapterisci	572	24	39	54	31	73	6	83	65	2.52	38	P	P	SP
	517–609	18–30	36–38	48–60	27–40	60–80		72–92	59–75	2.04–2.8	32–44			
websteri	584	21	36	47	31	77	6	68	49	1.8	40	P	A	NP
	553–631	17–25	29–40	37–56	24–34	62–102		64–72	42–56	1.6–2.1	30–50			
kushidai	589	26	46	50	41	92	8	63	44	1.5	51	A	A	PR
	524–662	22–31	42–50	44–59	38–44	84–95		NA	NA		NA			
***intermedium*-group** (IJ average size between 600 and 800 μm)														
riobrave	622	28	56	54	49	105	NA	67	51	1.14	71	A	A	SP
	561–701	26–30	51–64	46–59	45–55	93–111		62.5–75	47.5–56		60–80			
intermedium	671	29	65	66	51	96	6–8	91	64	1.24	67	A	A	SP
	608–800	25–32	59–69	53–74	48–58	89–108		84–100	56–75	NA	58–76			
pakistanense	683	27	54	58	47	91	8	68	41	1.8	60	P	P	SP
	649–716	24–29	49–58	53–62	42–53	87–102		62–73	36–45	1.0–2.2	50–60			
affine	693	30	62	66	49	94	8	70	46	1.17	61	P	A	PR
	608–880	28–34	51–69	64–74	43–53	74–108		67–86	37–56	NA	NA			
ceratophorum	706	27	55	66	45	84	6–8	71	40	1.4	51	A	A	SP
	591–800	23–34	47–70	56–74	40–56	74–96		54–90	25–45	1.0–2.0	33–65			
monticolum	706	37	58	77	47	76	8[b]	70	45	1.4	55	P	A	NP
	612–821	32–46	54–62	71–95	44–50	63–86		61–80	35–54	1.2–1.5	49–61			
sangi	753	35	51	81	40	62	8	63	40	1.5	49	P	A	PR
	704–784	30–40	46–54	76–89	36–44	56–70		58–80	34–46	1.2–1.6	42–63			
bicornutum	769	29	61	72	50	84	8	62	48	2.22	52	A	A	NP
	648–873	25–33	53–65	63–78	40–60	80–100		53–70	38–50	2.18–2.26	50–60			

continued

Table 1.3. *Continued.* Polytomous key for Steinernematidae.

Species	IJs							First generation adults						
								Male				Female		
	TBL	MBW	EP	TL	D%	E%	LF	SpL	GuL	SW	D%	M	EPI	VL
***feltiae*-group (IJ average size between 800 and 1000 μm)**														
feltiae	849	26	62	81	45	78	8	70	41	1.13	60	P	P	PR
	736–950	22–29	53–67	70–92	42–51	69–86		65–77	34–47	0.99–1.3	NA			
thanhi	851	31	75	63	58	119	8	72	49	1.8	73	A	A	PR
	720–960	27–39	68–84	52–72	52–67	101–138		67–78	40–56	1.5–2.1	64–82			
neocurtillae	885	34	18	80	12	23	6	58	52	1.43	19	P	P	V
	741–988	28–42	14–22	64–97	10–15	18–30		52–64	44–59	1.18–1.64	13.26			
scarabaei	918	31	77	76	60	100	8	75	44	1.7	66	P	A	SP
	890–959	25–37	72–81.5	71–80	50–75	90–110		67–83	36–50	1.5–2.0	53–77			
karii	932	33	74	74	57	96	8	83	57	NA	66	A	P	SP
	876–982	31–35	68–80	64–80	NA	NA		73–91	42–64		57–78			
kraussei	957	33	63	79	47	80	8	55	33	1.10	53	P	A	SP
	797–1102	30–36	50–6	63–86	NA	NA		52–57	23–38	NA	NA			
oregonense	980	34	66	70	50	100	6–8	71	56	1.51	73	A	A	SP
	820–1110	28–38	60–72	64–78	40–60	90–110		65–73	52–59	NA	64–75			
loci	986	37	80	75	57	107	8	71	46	1.9	73	A	A	PR
	896–1072	30–45	71–86	66–83	52–63	94–120		60–80	40–52	1.7–2.1	61–80			

					***glaseri*-group (IJ average size > 1000 μm)**									
longicaudum	1063	40	81	95	56	85	8	77	48	1.60	62	A	A	PR
	NA	NA	NA	NA	NA	NA		NA	NA		NA			
caudatum	1106	36	82	88	52	94	8	75	52	2.22	71	A	A	V
	933–1269	34–41	76–89	80–100	NA	87–100		NA	NA		NA			
glaseri	1130	43	102	78	65	131	8	77	55	2.1	70	A	A	PR
	864–1448	31–50	87–110	62–87	58–71	122–138		64–90	44–59	1.6–2.4	60–80			
puertoricense	1171	51	95	94	66	101	8	78	40	1.52	77	A	P	PR
	1057–1238	47–54	90–102	88–107	62–74	88–108		71–88	36–45					
cubanum	1283	37	106	67	70	160	8	58	39	1.41	70	A	A	PR
	1149–1508	33–46	101–114	61–77	NA	NA		50–67	37–42					

[a]Morphometric values of type isolate have incongruent and/or erroneous data in tables and text in original publication.
[b]After Stock, unpublished data.
E% = EP/TL × 100; EP = excretory pore; EPI = epiptygma; D% = EP/oesophagus length × 100; GuL = gubernaculum length; LF = number of ridges of lateral field at midbody level; M = mucro; MBW = maximum body width; SpL = spicule length; SW = SpL/cloacal body width; TBL = total body length; TL = tail length; VL = vulval lips; A = absent; NA = not available; P = present; V = variable; PR = protruding; NP = not protruding; SP = slightly protruding.
Note: All data from original descriptions unless otherwise specified. Morphometrics are given in microns.

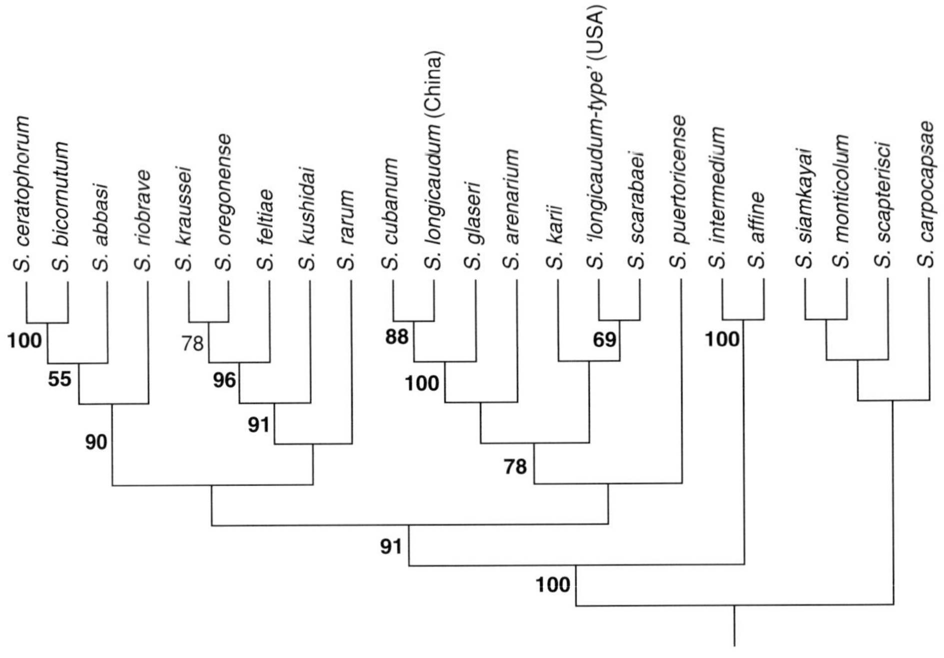

Fig. 1.2. Phylogenetic relationships among *Steinernema* spp. Single, most parsimonious tree inferred by maximum parsimony analysis of 28S rDNA sequences. Numbers represent bootstrap frequencies (1000 replicates) (Stock and Koppenhöfer, 2003).

Oesophagus with a large metacorpus (median bulb). Dorsal oesophageal gland opening into metacorpus. Oesophageal glands either forming a lobe or abutting intestine. Male bursa supported by four pairs of caudal papillae (rays). Spicules ventrally arcuate and slender. Gubernaculum present.

1.3.2.2. Bionomics

Mycophagous nematodes are found in decaying plant tissues feeding on various fungal hyphae. *A. avenae* has been studied as a biocontrol alternative to suppress fungal pathogens of plants (see Chapter 27, this volume).

1.3.2.3. Aphelenchus *Bastian, 1865* (Fig. 1.3)

DIAGNOSTIC CHARACTERS. Cuticle with transverse striae except for head region. Lateral field with about 6–14 incisures. Head slightly offset. Stylet lacking basal knobs. Oesophagus with a cylindrical procorpus; ovoid median bulb offset from procorpus and with prominent valve. Gland lobe overlapping intestine. Nerve ring circumoesophageal; located just posterior to bulb. Excretory pore at nerve-ring level. Females with posterior vulva; ovary outstretched, prodelphic. Postvulval sac present. Tail short, cylindroid with a bluntly rounded terminus. Male bursa supported by one precloacal and three postcloacal pairs of papillae. Spicules paired, slender, slightly ventrally arcuate and proximally cephalated. Gubernaculum about one-third the length of spicules.

Type Species: *A. avenae* Bastian, 1865.

1.3.3. Family Allantonematidae Pereira 1931

1.3.3.1. Diagnostic characters

Preparasitic females and free-living males with small stylet (less than 15 μm long) with or without knobs. Oesophageal glands elongated, lobe-like; subventral glands extending

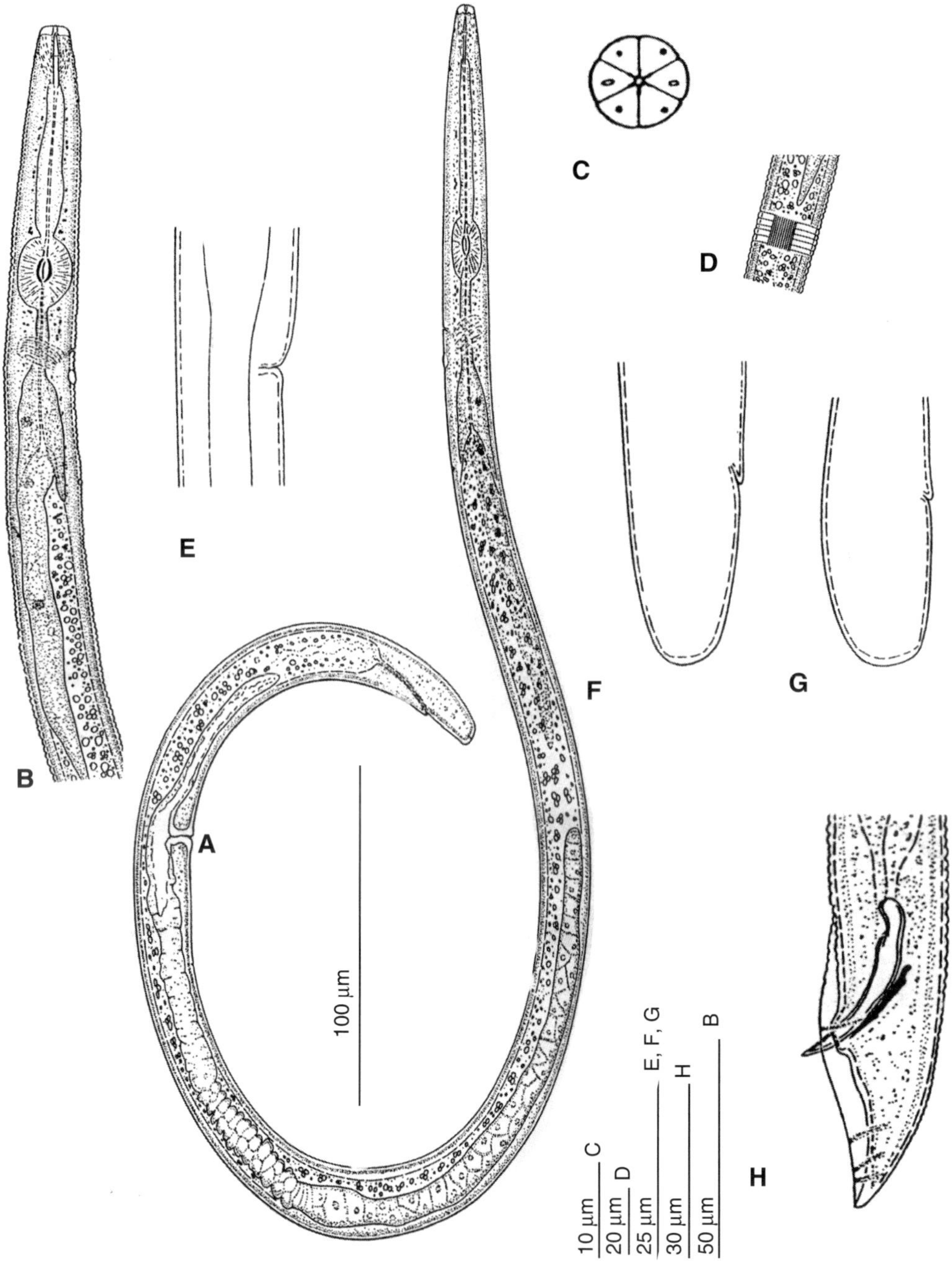

Fig. 1.3. *Aphelenchus avenae* Bastian. A, female; B, anterior region of female (lateral view); C, head (en face view); D, lateral field pattern; E, vulva (lateral view); F and G, female tails (lateral view); H, male tail (lateral view).

past dorsal lobe. Tail conoid or subcylindrical. Preparasitic females with small vulva and short vagina. Postvulval sac short or absent. Uterus elongated. Parasitic females obese, sac-like, elongate or spindle-shaped. Reproductive organs filling body cavity. Uterus not everted. Vulva a small transverse slit or indistinct. Males monorchic, testis outstretched. Spicules arcuate, pointed, usually less than 25 μm long. Gubernaculum usually present. Bursa present or absent.

1.3.3.2. Bionomics

Allantonematids have a single heterosexual cycle. Adult females are parasites of the haemocoel of mites and insects. Within this family, members of *Thripinema* Siddiqi, 1986 are known to parasitize thrips (Thysanoptera: Thripidae). A free-living stage occurs in flowers, buds and leaf galls of plants that attacks thrips. See Chapter 22, this volume, for additional information.

1.3.3.3. Thripinema *Siddiqi, 1986* (Fig. 1.4)

DIAGNOSTIC CHARACTERS (modified from Siddiqi, 1986). Infective females with straight or slightly ventrally curved body when relaxed. Cuticle finely striated. Lip region moderately sclerotized. Stylet strong, without basal knobs (except *Thripinema khrustalevi*). Orifices of dorsal and subventral oesophageal glands at 2.6–3 and 3–3.6 stylet lengths from anterior end, respectively. Oesophagus fusiform; glands elongated, extending for two-thirds of body length. Vulva inconspicuous. Ovary anteriorly outstretched. Parasitic females with small oval or elliptical body. Stylet without basal knobs, indistinct in mature females. Oesophagus atrophied. Vulva terminal or subterminal. Ovary long and convoluted occupying most of body cavity, with two to four flexures. Uterus large, usually containing one or two eggs. Males with straight or arcuate body. Stylet absent or present. Oesophagus degenerate. Monorchic, testis extending to oesophageal region. Tail subcylindroid-subclavate, about three cloacal body widths long. Spicules paired, arcuate, pointed and 14–16 μm long. Gubernaculum present but weakly developed, about one-third the length of spicules. Bursa prominent, adanal or almost terminal (Table 1.4).

1.3.4. Family Neotylenchidae Thorne, 1941

1.3.4.1. Diagnostic characters (modified from Siddiqi, 2000)

Free-living stages with smooth or finely striated cuticle. Stylet well developed, less than 20 μm long, basal knobs may be bifid. Oesophagus fusiform, basal bulb absent. Oesophageal glands free in body cavity, extending over intestine. Orifice of dorsal gland close to stylet base. Nerve ring generally circumintestinal, posterior to, or at level of, oesophago–intestinal junction. Excretory pore anterior or posterior to nerve ring. Females monodelphic or prodelphic. Vulva in posterior region, postvulval sac present or absent. Tail conoid, subcylindroid or cylindroid. Males monorchic, testis outstretched. Bursa present or absent. Spicules paired, small, cephalated or arcuate, distally pointed. Gubernaculum present or absent. Pre-adult females (free-living) with hypertrophied stylet and oesophagus. Ovary immature. Uterus long. Mature parasitic females obese, sausage-shaped or elongate tuboid. Stylet and oesophagus non-functional. Uterus hypertrophied but not everted.

1.3.4.2. Bionomics

Members of this family have a free-living generation alternating with an insect-parasitic generation. *Beddingia* Thorne, 1941 currently comprises 17 nominal species with *Beddingia siricidicola* Bedding, 1968, a parasite of the wood wasp *Sirex noctilio*, being the only taxon currently used in biocontrol. Additional reading on this matter can be found in Chapter 20, this volume.

1.3.4.3. Beddingia *Blinova and Korenchenko, 1986* (Fig. 1.5)

DIAGNOSTIC CHARACTERS (modified after Siddiqi, 2000). Free-living adult stages

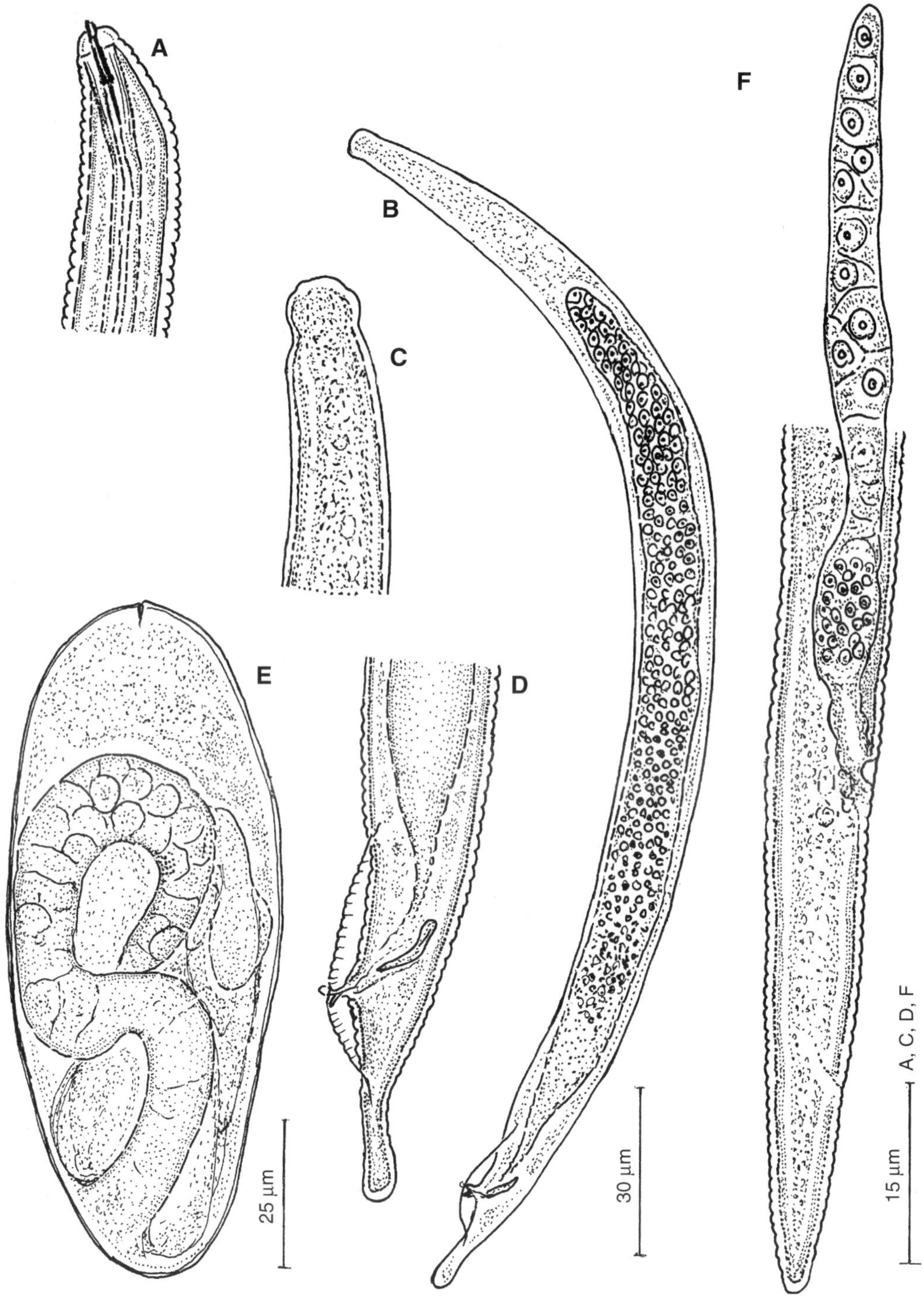

Fig. 1.4. *Thripinema reniraoi* Siddiqi. A and F, (A) anterior and (F) posterior region of partially free-living impregnated female; B, male; C and D, (C) anterior and (D) posterior regions of male; E, entomoparasitic female from haemocoel of *Megaluriothrips* sp. (After Siddiqi, 1986.)

Table 1.4. Key diagnostic features of *Thripinema* spp.

Diagnostic features	*T. aptini* (Sharga, 1932)	*T. fuscum* Tipping and Nguyen, 1998	*T. khrustalevi* Chizhov, Subbotin and Zakharenkova, 1995	*T. nicklewoodi* Siddiqi, 1986	*T. reniraoi* Siddiqi, 1986[a]
Body shape (parasitic female)	Oval, elliptical	Oval, elliptical	Oval, spherical	Oval, elliptical, bean-shaped	Oval
Vulva position	Terminal	Terminal	Terminal	*ca* 85%	Terminal
Body shape (male)	Ventrally curved	Dorsally curved	Ventrally curved	Ventrally curved	Ventrally curved
Stylet (male)	Absent	Present	Absent	Absent	Absent
Bursa (male)	Adanal	Adanal	Subterminal	Subterminal	Adanal

[a]Type species.

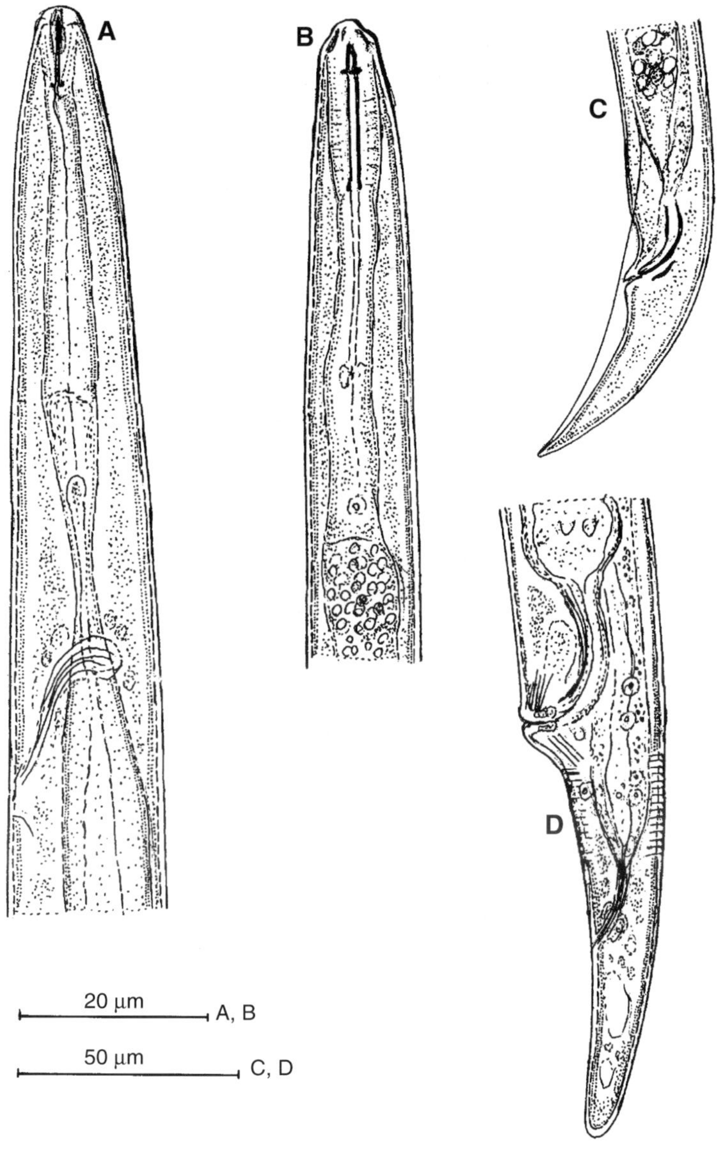

Fig. 1.5. *Beddingia siricidicola* Bedding. A, oesophageal region of fungus feeding female; B, oesophageal region of entomoparasitic pre-adult female (*Beddingia* sp.); C, male tail region; D, posterior region of fungus-feeding female. (After Siddiqi, 2000.)

(mycetophagous) straight or slightly ventrally curved. Body cylindrical, tapering anteriorly and posteriorly to vulva; slender in young females but obese or swollen in mature females. Cuticle with fine transverse striae. Stylet small, basal knobs weak to moderately developed and rounded. Oesophagus cylindroid. Oesophago-intestinal junction at, or anterior to, nerve ring. Dorsal gland large, subventral glands reduced. Nerve ring surrounding isthmus. Excretory pore location variable. Hemizonid anterior or posterior to excretory pore. Female reproductive system monovarial, amphidelphic.

Ovary outstretched and flexed. Spermatheca elongate. Vulva protuberant or not and extremely posterior. Vulval sac present or absent. Males monorchic, testis outstretched. Spicules paired, moderately robust and arcuate. Gubernaculum present. Tail conical or elongate conoid. Bursa present. Parasitic females obese, with body elongate. Cephalic region overgrown by body enlargement. Stylet present, hypertrophied, stout. Oesophagus and oesophageal glands hypertrophied in young females but degenerate in mature females. Vulva a transverse slit, lips not protuberant. Short postvulval sac secondarily formed in impregnated young females.

1.3.5. Family Rhabditidae Örley, 1880

1.3.5.1. Diagnostic characters

Stoma commonly cylindrical without distinct separation of cheilo-, gymno- and stegostom. Stoma two or more times as long as wide. Usually with six distinct lips, each with one cephalic papilla. Amphids pore-like. Oesophagus clearly divided into corpus (procorpus and metacorpus) and postcorpus (isthmus and valvated muscular portion). Male spicules separate or fused distally. Gubernaculum present. Bursa mostly well developed, peloderan or leptoderan, occasionally small or rudimentary. Nine or ten pairs of genital papillae (bursal rays). Females with one or two ovaries.

1.3.5.2. Bionomics

Most members of this family are free-living bacterivores although two species of *Phasmarhabditis*, *Phasmarhabditis hermaphrodita* (Schneider, 1859) and *P. neopapillosa* (Schneider, 1866), have parasitic associations with terrestrial slugs and snails. *P. hermaphrodita* is capable of killing several slugs, snails and slug pests, and is the only species currently used as a biocontrol agent and is mass-produced and commercialized as a molluscicide (Wilson *et al.*, 1994; Glen and Wilson, 1997) (see Chapters 24 and 25, this volume).

1.3.5.3. Phasmarhabditis *Andrássy, 1976* (Fig. 1.6)

DIAGNOSTIC CHARACTERS. Body almost straight when heat-killed, robust, elongate and tapering gradually to bluntly rounded head end. Cuticle with fine transverse and longitudinal striations. Lips rounded, arranged in three pairs each bearing a prominent labial papilla. Stoma rounded, triangular in cross-section. Stegostom well developed and with minute tubercules. Oesophageal collar present. Oesophagus with well-developed, cylindrical corpus. Basal bulb with prominent valve plates. Excretory pore usually anterior to basal bulb. Nerve ring surrounding isthmus. Deirids prominent. Females didelphic, amphidelphic. Vulva located at mid-body level. Males (when present) monorchic. Spicules separate. Bursa peloderan, open, with nine pairs of genital papillae. Tail conical, spicate or cupola-shaped. Phasmids prominent and sometimes protruding (Table 1.5).

1.3.6. Family Heterorhabditidae Poinar, 1976 (Fig. 1.7)

1.3.6.1. Diagnostic characters

Adults with six distinct protruding pointed lips surrounding oral aperture. Each lip bearing one labial papilla. Stoma short and wide. Oesophagus rhabditoid. Corpus cylindrical, metacorpus not differentiated. Isthmus short. Basal bulb pyriform with reduced valve. Excretory pore usually located at level of basal bulb. Hermaphrodite (first generation) with an ovotestis. Vulva located near middle of body. Post-anal swelling present or absent. Tail terminus blunt, with or without a mucro. Females (second generation) amphidelphic, ovaries with reflexed portions often extending past vulval opening. Vulva located near middle of body, with or without protruding lips. Tail conoid; post-anal swelling present or absent. Males (second generation) monorchic. Spicules paired, symmetrical, straight or arcuate, with pointed tips. Gubernaculum slender, about half the length of spicules. Bursa open, peloderan, attended by a complement of nine pairs of bursal rays (papillae). IJ ensheathed in cuticle of second-stage juvenile. Cuticle of second-stage juvenile with longitudinal ridges throughout most of body length, and a tessellate pattern in anteriormost region. Lateral field with two ridges. Prominent cuticular dorsal tooth present. Excretory pore located posterior to basal bulb. Tail short, conoid, tapering to a small spike-like tip.

1.3.6.2. Bionomics

Heterorhabditids have a similar life cycle to steinernematids, but adults resulting from IJs are hermaphroditic. Eggs laid by the hermaphrodites produce juveniles that develop into males and females or IJs. The males and females mate and produce eggs that develop into IJs. Additional reading on this matter can be found in Chapter 2, this volume.

Heterorhabditidae consist of one genus, *Heterorhabditis* Poinar, 1976, with

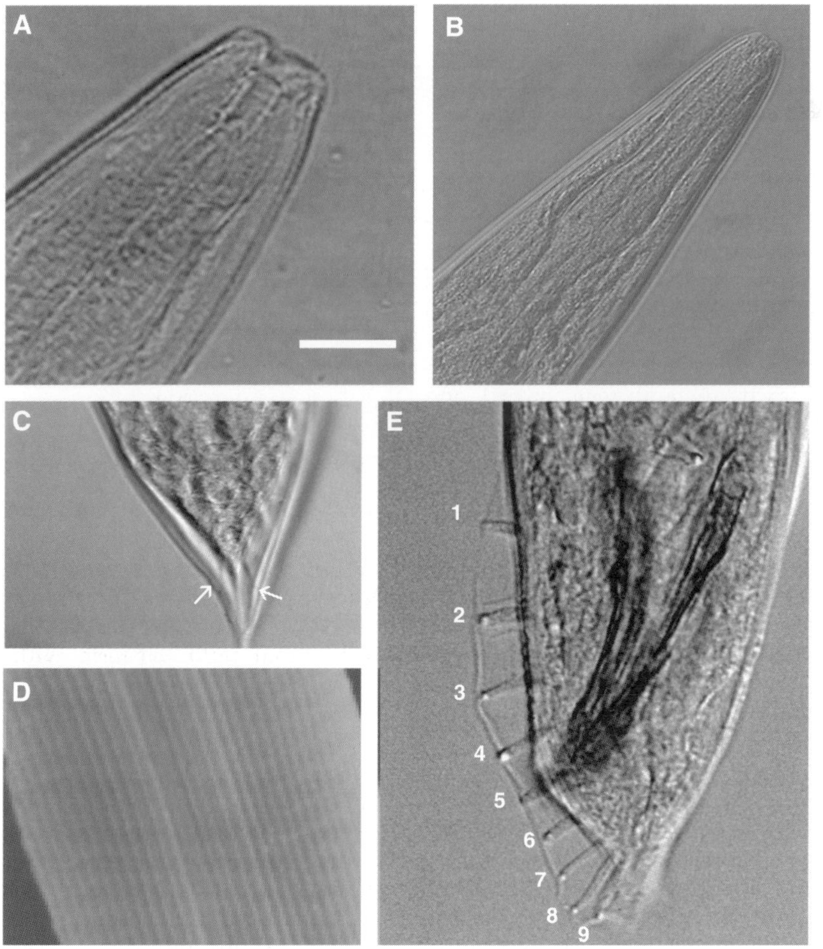

Fig. 1.6. *Phasmarhabditis* Andrássy. A, female stoma (dorsal view) of *P. hermaphrodita*; B, oesophageal region (lateral view) of *P. hermaphrodita*; C, female tail of *P. neopapillosa* showing phasmids (arrows); D, lateral field of *P. hermaphrodita*; E, male tail of *P. neopapillosa*. (Scale bars: A, E = 10 μm; B, C = 25 μm; D = 12 μm.)

Table 1.5. Key diagnostic features of *Phasmarhabditis* spp.

	Female			Male		
		Tail				
Species	TBL	Shape	Length	TBL	Bursa shape	SpL
hermaphrodita[a] (Schneider,1859) Andrássy, 1983	1799 1509–2372	Elongate, conoid	3–4 anal body widths long		Males are extraordinarily rare	
neopapillosa[a] (Mengert *in* Osche, 1952) Andrássy, 1983	2227 1817–2449	Elongate, conoid	3–4 anal body widths long	1585 1432–1771	Well-developed	1.5 times as long tail
nidrosiensis[b] (Allgén, 1933) Andrássy, 1983	1000–1750	Cupola-shaped w/pointed tip	1.5–2 anal body widths long	900–1720	Small and narrow	Twice as long as tail
papillosa[b,c] (Schneider, 1866) Andrássy, 1976	1600–3400	Cupola-shaped w/pointed tip	1.5–2 anal body widths long	1200–2400	Well-developed	1–1.5 times as long as tail
valida[b] (Sudhaus, 1974) Andrássy, 1983	NA	Cupola-shaped w/pointed tip	1.5–2 anal body widths long	NA	Well-developed	NA

[a]After Hooper *et al.*, 1999.
[b]After Andrássy, 1983.
[c]Type species.
NA = not available; SpL = spicule length; TBL = total body length.
Note: All measurements are in microns.

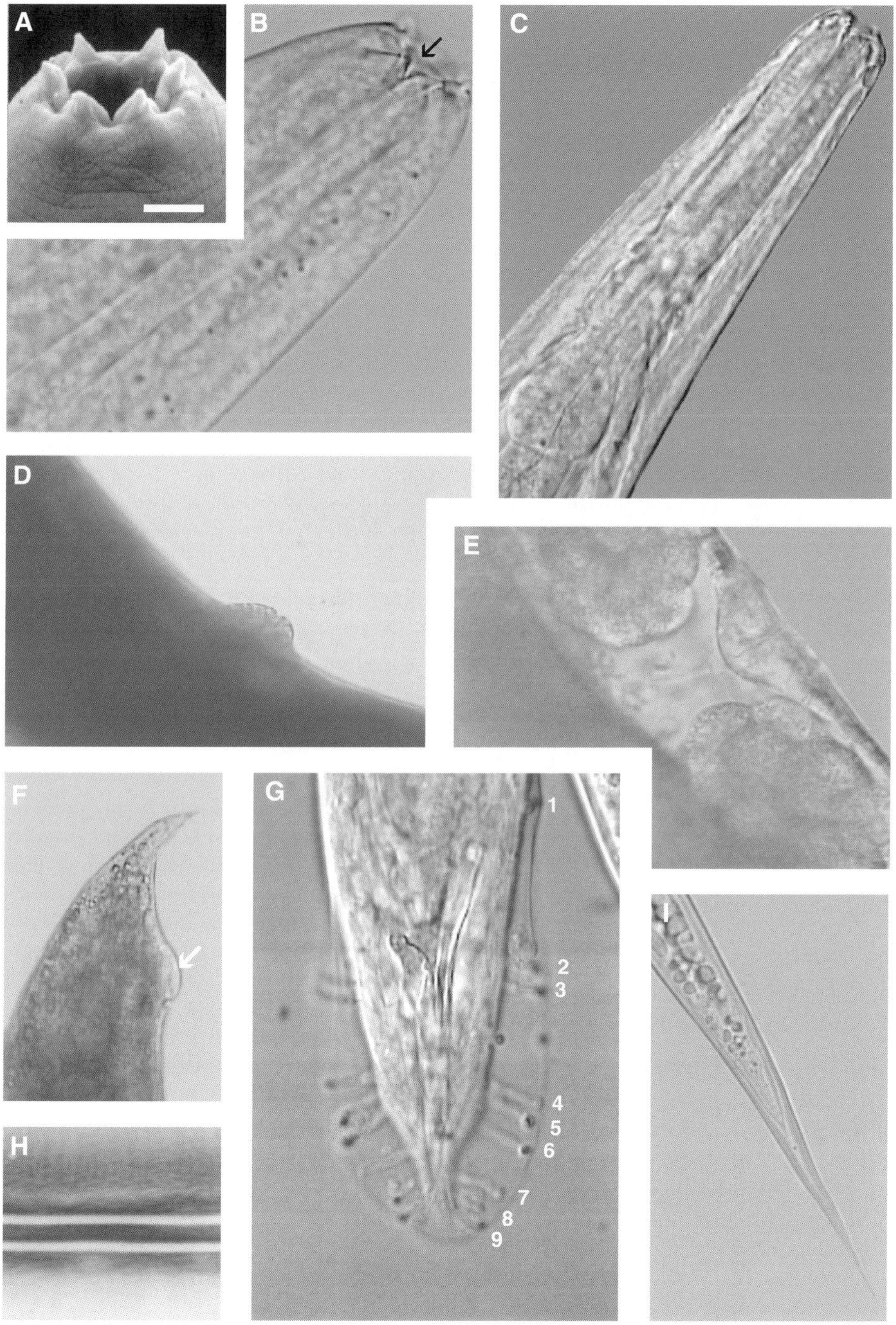

Fig. 1.7. Family Heterorhabditidae. A—E. First generation hermaphrodite: A, scanning electron micrograph (SEM) of anterior end; B, anterior end (lateral view) showing stoma (arrow); C, oesophagus (lateral view); D, protruding vulval lips (lateral view); E, non-protruding vulva (lateral view). F, tail (lateral view) showing post-anal swelling (arrow). G, tail of second generation male (ventral view) showing arrangement of genital papillae. H–I. Third-stage infective juvenile (IJ): H, tail (lateral view); I, lateral field pattern. (Scale bars: A = 4.5 μm; B = 12 μm; C, D = 25 μm; E = 20 μm; F = 15 μm; G, I = 6.5 μm; H = 3.5 μm.)

Heterorhabditis bacteriophora as the described type species and nine other described species (Tables 1.6 and 1.7).

1.3.6.3. Phylogenetic relationships (Fig. 1.8)

Evolutionary relationships among *Heterorhabditis* spp. have been explored using nucleotide sequences from nuclear (28S, 18S and ITS-1) and mitochondrial (ND4) genes (Curran and Driver, 1994; Reid, 1994; Liu *et al.*, 1997, Adams *et al.*, 1998; Liu *et al.*, 1999). Curran and Driver (1994) presented the first hypothesis of evolutionary relationships in the genus using a combination of RFLP analysis and partial sequences of 28S rDNA. Their study, although preliminary, recognized species identity on the basis of morphological and cross-hybridization tests, but did not contribute to an understanding of their phylogenetic relationships, mainly because of their limited taxon sampling. Reid (1994) also used RFLP analysis of ITS rDNA to assess evolutionary relationships among several *Steinernema* spp. and *Heterorhabditis* spp. With respect to *Heterorhabditis*, his study demonstrated a close relationship between the type isolate of *H. megidis* and the *Heterorhabditis* spp. of the NW European group, now considered to be conspecific. His study also indicated that *Heterorhabditis* spp. were more closely related to one another than were *Steinernema* spp.

Liu *et al.* (1997) inferred phylogenetic relationships for both families of entomopathogenic nematodes (EPNs) using partial 18S rDNA sequences, concluding that this region was too conserved to resolve relationships among *Heterorhabditis* spp.

More recently, evolutionary relationships among *Heterorhabditis* spp. have been inferred using sequences of the ITS-1 region of the tandem repeat unit of rDNA (Adams *et al.*, 1998). In this study, relationships between closely related 'species' (i.e. *H. indica* and *H. hawaiiensis*; *H. bacteriophora* and *H. argentinensis*) were well established. However, relationships among more distantly related species, i.e. *H. zealandica* in relation to *H. megidis* and *H. marelata*, could not be resolved. A more extensive study at the population level might contribute to a better resolution and/or interpretation of the relatedness between these species. More recently, Phan *et al.* (2003) showed that the tropical and subtropical *Heterorhabditis* spp., *H. indica* and *H. baujardi*, formed one clade separated from those species known mainly from temperate regions.

Mitochondrial genes have also been explored to study the evolutionary history of *Heterorhabditis* spp. (Liu *et al.*, 1999), the results broadly agreeing with those of Adams *et al.* (1998). Although Liu *et al.* did not study all species (*H. zealandica* and *H. downesi* were not included), their study also indicated poor support for nodes involving *H. megidis* and *H. marelata*.

1.3.7. Family Diplogasteridae Micoletzky, 1922 (Fig. 1.9)

1.3.7.1. Diagnostic characters

Lip region, never set off by a constriction, usually composed of six distinct lips or six fused lips. Amphids pore-like. Stylet absent. Stoma variable, usually broad and short with stegostom containing denticles, warts or teeth. Oesophagus with a median valvated bulb and a basal valveless bulb. Female gonad usually paired. Males with paired spicules and gubernaculum. Bursa usually small or absent. Male tail often with nine pairs of genital papillae and a pair of phasmids. Three pairs of genital papillae located pre-anally.

1.3.7.2. Bionomics

Diplogasterids are usually predators or omnivores but can also be bacterial feeders. Only a few genera (i.e. *Butlerius*, *Fictor* and *Mononchoides*) have been studied as biocontrol agents of plant-parasitic nematodes (see Chapter 26, this volume).

Table 1.6. Taxonomic summary of described *Heterorhabditis* spp.

Taxa	Biogeography[a]	GenBank sequence data (accession number)
Type and only genus:		
Heterorhabditis Poinar, 1976		
Syn. *Chromonema* Khan, Brooks and Hirschmann, 1976		
Type species:	Africa, Asia, Australia,	18S (AF036593), 5.8S (U65497),
Heterorhabditis bacteriophora Poinar, 1976	Central America, Europe,	ITS-1 (AF029708, AF029706),
Syn. *Chromonema heliothidis* Khan, Brooks and Hirschmann, 1976	North America (USA),	28S (D3) (U47560),
H. heliothidis (Khan, Brooks and Hirschmann, 1976) Poinar, Thomas and Hess, 1977	South America	ND4 (AF066890, AF066888), SAT (U19928)
H. argentinensis Stock, 1993[b]		
Other species:		
H. baujardi Phan, Subbotin, Nguyen and Moens, 2003	Asia (Vietnam)	ITS-1 (AF548768)
H. brevicaudis Liu, 1994	Asia (China)	ITS-1,-2 (AF548768)
H. downesi Stock, Burnell and Griffin, 2002	Europe (Ireland)	ITS-1 (AF029713)
H. indica Poinar, Karunakar and David, 1992	Asia (India), Central America,	18S (U70628), ITS-1 (AF029710, AF029707),
Syn. *H. hawaiiensis* Gardner, Stock and Kaya, 1994[b]	North America	ND4 (AF066879, AF066878), SAT (U68112)
H. marelata Liu and Berry, 1996	North America (USA)	18S (AF083004, U70630), ITS-1 (AF029713,
Syn. *H. hepialius* Stock, Strong and Gardner, 1996		AF029709) ND4 (AF06881, AF066880)
H. megidis Poinar, Jackson and Klein, 1987	North America (USA), Europe	18S (AF70631), ITS-1 (AF029711), ITS-1,-2 (AY293284), ND4 (AF066885)
H. poinari Kakulia and Mikaia, 1997[c]	Europe (Georgia)	
H. taysearae Shamseldean, Abou El-Sooud, Abd-Elgawad and Saleh, 1996	Asia (Egypt)	
H. zealandica Poinar, 1990	Australia (New Zealand)	ITS-1 (AF029705)

[a]Country of original isolation in parentheses.
[b]As proposed by Stock (in press).
[c]Species *inquirenda*.

Table 1.7. Polytomus key to *Heterorhabditis* spp.

	IJs							Adults						
								Hermaphrodite	Male			Female		
Species	TBL	MBW	EP	TL	RF	D%	E%	T shape	SpL	GuL	TREF	D%	BR	PAS
***indica*-group (IJ average size < 550 μm)**														
poinari	NA	NA	NA	NA	NP	NA	NA	Conoid	NA	NA	NA	NA	NA	NA
	350–410	18–22							43–55	24–32				
taysaerae	418	19	90	55	NP	82	180	Conoid	39	18	122	NA	7, 8 do not reach	P
	332–499	17–23	74–113	44–70		71–96	110–230		30–42	14–21	100–146		the bursal rim	
indica	528	20	98	101	NP	84	94	Conoid	43	21	106	122	1 may be	V
	479–573	19–22	88–107	93–109		79–90	83–103		35–48	18–23	78–132	NA	out of bursa	
													4, 7 outwards	
***bacteriophora*-group (IJ average size 550–700 μm)**														
bacteriophora	588	23	103	98	NP	84	112	Conoid	40	20	76	117	4, 7 outwards	P
	512–671	18–31	87–110	83–112		76–92	103–130		36–44	18–25	61–89	NA		
baujardi	551	20	97	90	NP	84	108	Conoid	40	20	91	NA	NA	P
	497–595	18–22	91–103	83–97		78–88	98–114		33–45	18–22	63–106			
brevicaudis	572	22	111	76	NP	90	147	Conoid	47	22	194	88	NA	P
	528–632	20–24	104–116	68–80		NA	NA		44–48	20–24	162–240	NA		
zealandica	685	27	112	102	NP	80	108	Conoid	51	22	132	118	4, 7 outwards	V
	570–740	22–30	94–123	87–119		70–84	103–109		48–55	19–25	88–173	NA		
marelata	654	28	102	107	NP	77	96	Pipette-shaped	45	19	91 + 67–136	113	4, 7 outwards	P
	588–700	24–32	81–113	99–117		60–86	89–110		42–50	18–22		NA	8 does not touch bursal rim	
***megidis*-group (IJ average size > 700 μm)**														
megidis	768	29	131	119	NP	85	110	Conoid	49	21	128	122	4, 7 outwards	P
	736–800	27–32	123–142	112–128		81–91	103–120		46–54	17–24		NA	2, 3 fused	
downesi	879	39	97	33	P	83	169	Blunt and	46	23	NA	NA	4, 7 outwards	P
	669–1066	33–55	64–107	28–42		77–92	129–216	mucronated	40–53	40–53			8 does not touch bursal rim	

Abbreviations: BR = bursal rays; D% = EP/oesophagus length × 100; E% = EP/TL × 100; EP = excretory pore; MBW = maximum body width; NA = information not available; PAS = post-anal swelling; RF = tail refractile spine; T = tail; TBL = total body length; TL = tail length; TREF = testis reflexion; V = variable; NP = not present; P = present; SpL = spicule length; GuL = gubernaculum length.

Note: All data from original descriptions unless otherwise specified. Morphometrics are given in microns.

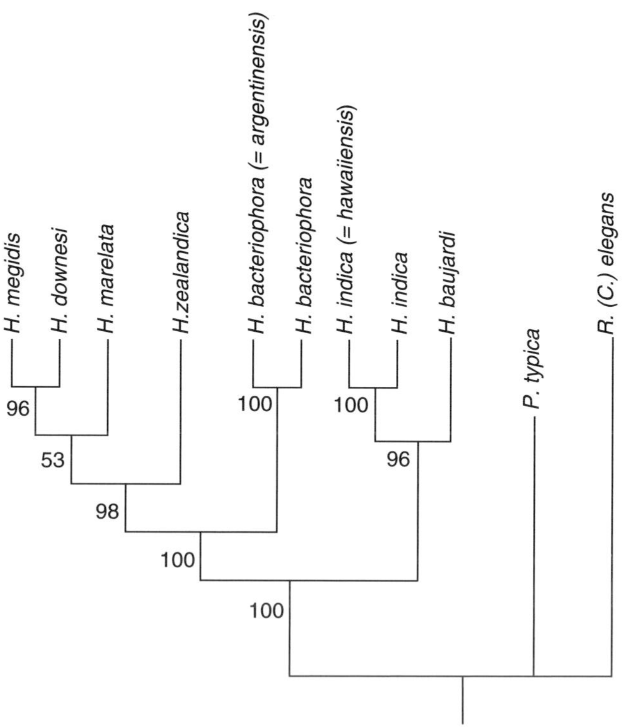

Fig. 1.8. Phylogenetic relationships among *Heterorhabditis* spp. (modified from Adams *et al.*, 1998). Single, most parsimonious tree inferred by maximum parsimony analysis of ITS-1 rDNA. Bootstraps frequencies (100 replicates) are from Phan *et al.*, 2003.

1.3.8. Family Mononchidae Chitwood, 1937 (Fig. 1.10)

1.3.8.1. Diagnostic characters

Generally large, stout nematodes. Cuticle usually appearing non-striated and smooth. Lateral field usually not differentiated. Head not distinctly offset, composed of six or fewer confluent lips, each carrying at least two papillae. Amphids small, cup-shaped. Stylet absent. Stoma forming a small to large barrel-shaped cuticularized chamber bearing an immovable dorsal tooth. Subventral teeth and/or rows of denticles or ridges may also be present. Oesophagus stout, muscular, glandular and almost cylindrical with some posterior swelling. Oesophago–intestinal junction tuberculate or non-tuberculate. Excretory pore usually absent. Females usually with paired ovaries, opposite and reflexed. Males with paired opposed testes leading to a common vas deferens. Spicules paired. Gubernaculum present. Lateral guiding piece often present. Mid-ventral row of precloacal papillae always present on males. Tail variable in form. Bursa absent (Table 1.8).

1.3.8.2. Bionomics

Mononchids are predominantly predaceous nematodes feeding on small invertebrates (including other nematodes) in soil and fresh water. Many genera have been proposed, but only *Mylonchulus* (Cobb, 1916), *Mononchus* Bastian and *Iotonchus* (Cobb, 1916) have been explored as biocontol agents. A few taxa have been used against plant-parasitic nematode species such as juveniles of *Meloidogyne incognita* and *Rotylenchulus reniformis* (Choudhury and Sivakumar, 2000) (see Chapter 26, this volume).

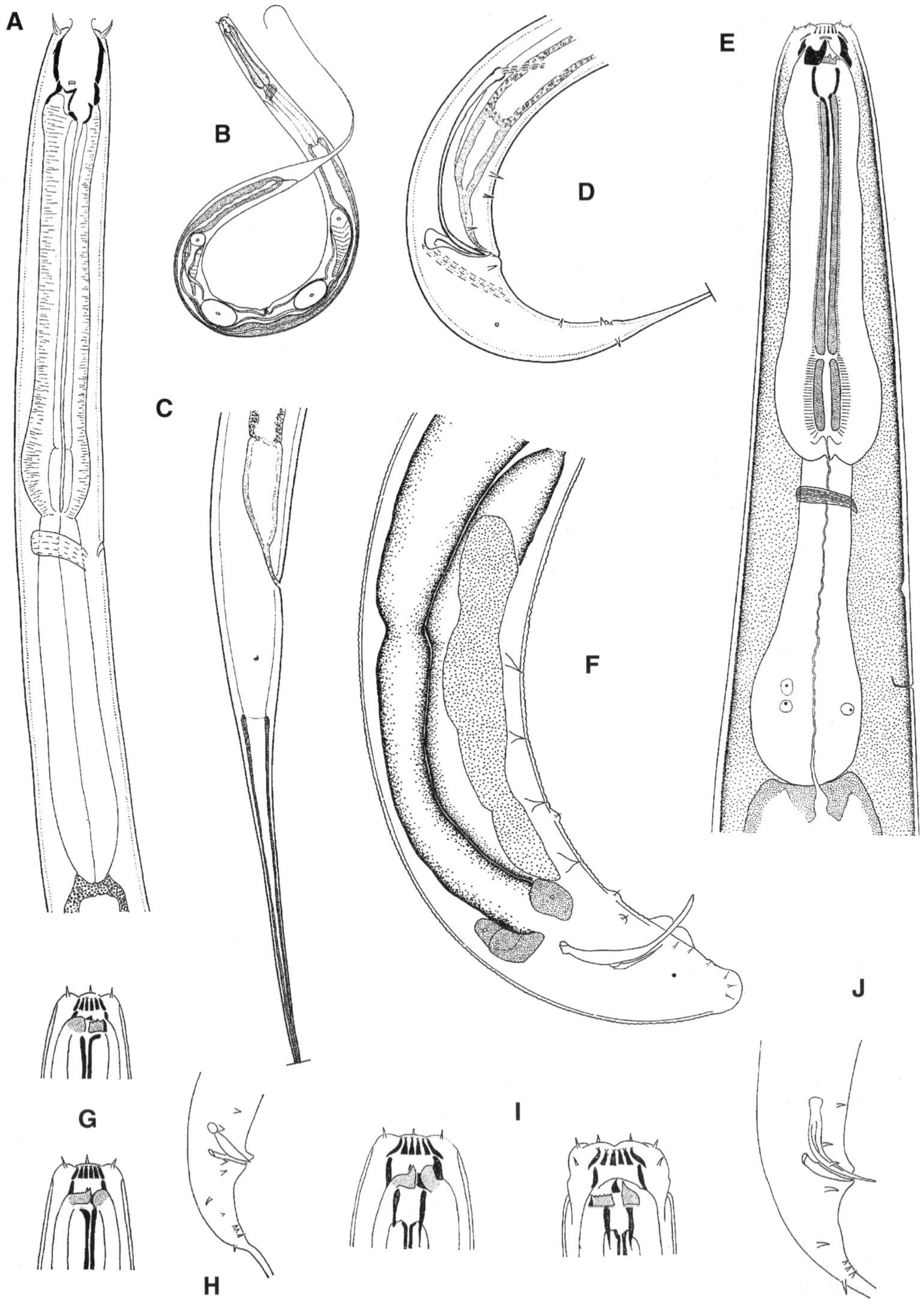

Fig. 1.9. Family Diplogasteridae. A–D. *Butlerius*: A, pharyngeal region; B, entire female; C, female tail region; D, male tail region. E and F. *Diplenteron*: E, pharyngeal region; F, male tail region. G and H. *Fictor*: G, stoma region; H, male tail region. I and J. *Mononchoides*: I, stoma region; J, male tail region. (A, C, D after Hunt, 1980, courtesy *Revue de Nématologie*; E, F after Yeates, 1984, courtesy *Nematologica*; B, G–J after Goodey, 1963, *Soil and freshwater nematodes*; various scales.)

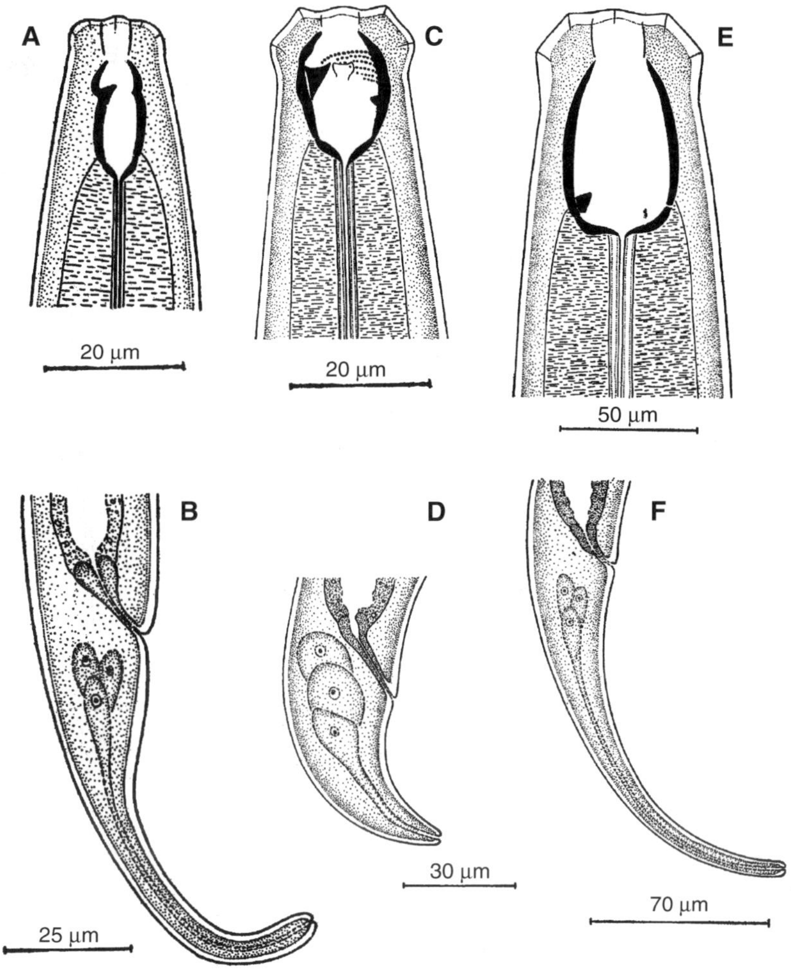

Fig. 1.10. Family Mononchidae. A and B. *Mononchus* sp.: A, anterior region; B, female tail region. C and D. *Mylonchulus minor*: C, anterior region; D, female tail region. E and F. *Iotonchus* sp.: E, anterior region; F, female tail region. (After Jairajpuri and Khan, 1982, courtesy Associated Publishing Company.)

1.3.9. Family Mermithidae Braun, 1883 (Fig. 1.11)

1.3.9.1. Diagnostic characters

Long slender nematodes sometimes reaching a length of 50 cm, but usually between 1 cm and 10 cm. Cuticle smooth or with criss-cross fibres. Anterior end containing two, four or six cephalic papillae and rarely a pair of lateral mouth papillae. Amphids tube-like or modified pouch-like. Oesophagus modified into a slender tube surrounded posteriorly by stichosomal tissue. Intestine modified into a trophosome or food-storage organ forming a blind sac soon after the nematodes enter a host. Preparasitic juveniles with a functional stylet and a pair of penetration glands that degenerate after host invasion. Ovaries paired; muscular vagina straight or curved. Males with a single fused or paired spicules. Gubernaculum and bursa absent. Several rows of genital papillae usually present (Table 1.9).

Table 1.8. Key diagnostic features of mononchid genera considered in biocontrol. (Modified from Goodey, 1963.)

Diagnostic features	*Mononchus* Bastian, 1865	*Mylonchulus* Cobb, 1916	*Iotonchus* Cobb, 1916
Oesophago-intestinal junction	Non-tuberculate	Non-tuberculate	Tuberculate
Position/direction of dorsal tooth	Anterior half and forward	Anterior half and forward	Posterior half and forward
Subventral teeth or denticles	Absent	Small pair of teeth usually opposite to base of dorsal tooth. Walls with 2–13 transverse rows of minute denticles.	Absent

1.3.9.2. Bionomics

There are numerous described genera, many of which are poorly characterized by contemporary standards. The group is in urgent need of revision before a workable key can be constructed. All known species are obligate parasites of terrestrial and aquatic arthropods and other invertebrates. Mermithids parasitize many different insect groups, including Orthoptera, Dermaptera, Hemiptera, Lepidoptera, Diptera, Coleoptera and Hymenoptera. Mermithids with significant biocontrol potential include *Romanomermis culicivorax* (a parasite of mosquito larvae) (Petersen, 1985), *Oesophagomermis* (= *Filipjevimermis*) *leipsandra* (a parasite of larval banded cucumber beetle *Diabrotica balteata*) (Creighton and Fassuliotis, 1983), *Mermis nigrescens* (a parasite of grasshoppers) (Webster and Thong, 1984) and *Agamermis unka* (a parasite of white and brown planthoppers) (Choo *et al.*, 1989, Choo and Kaya, 1994) (see Chapter 23, this volume).

1.3.10. Family Dorylaimidae de Man, 1876 (Fig. 1.12)

1.3.10.1. Diagnostic characters

Generally large and robust nematodes. Stoma with an axial odontostyle, the aperture of which is located dorsally. Oesophagus cylindrical and divided into two parts: anterior portion usually slender, sometimes with small muscular swellings, followed by an expanded posterior portion. Excretory pore rudimentary or absent. Females with one or two ovaries. Males diorchic. Spicules robust and separated. Gubernaculum usually absent, but lateral guiding pieces present. Bursa absent. Setae and caudal glands absent.

1.3.10.2. Bionomics

The feeding habits of many members are not known, although some are acknowledged as being predaceous on other nematodes and invertebrates. See Chapter 26, this volume, for additional information.

1.3.11. Family Nygolaimidae Thorne, 1935

1.3.11.1. Diagnostic characters

Stoma armed with mural tooth of variable shape. Dorylaimoid oesophagus with posterior portion enclosed in a sheath. Three large cardiac glands at oesophago–intestinal junction. Ovaries paired, opposed and reflexed. Spicules arcuate. Gubernaculum and lateral guiding pieces present in some males.

1.3.11.2. Bionomics

Nygolaimids are predaceous, some taxa (i.e. *Sectonema* spp.) have been studied for their biocontrol potential of plant-parasitic nematodes (see Chapter 26, this volume).

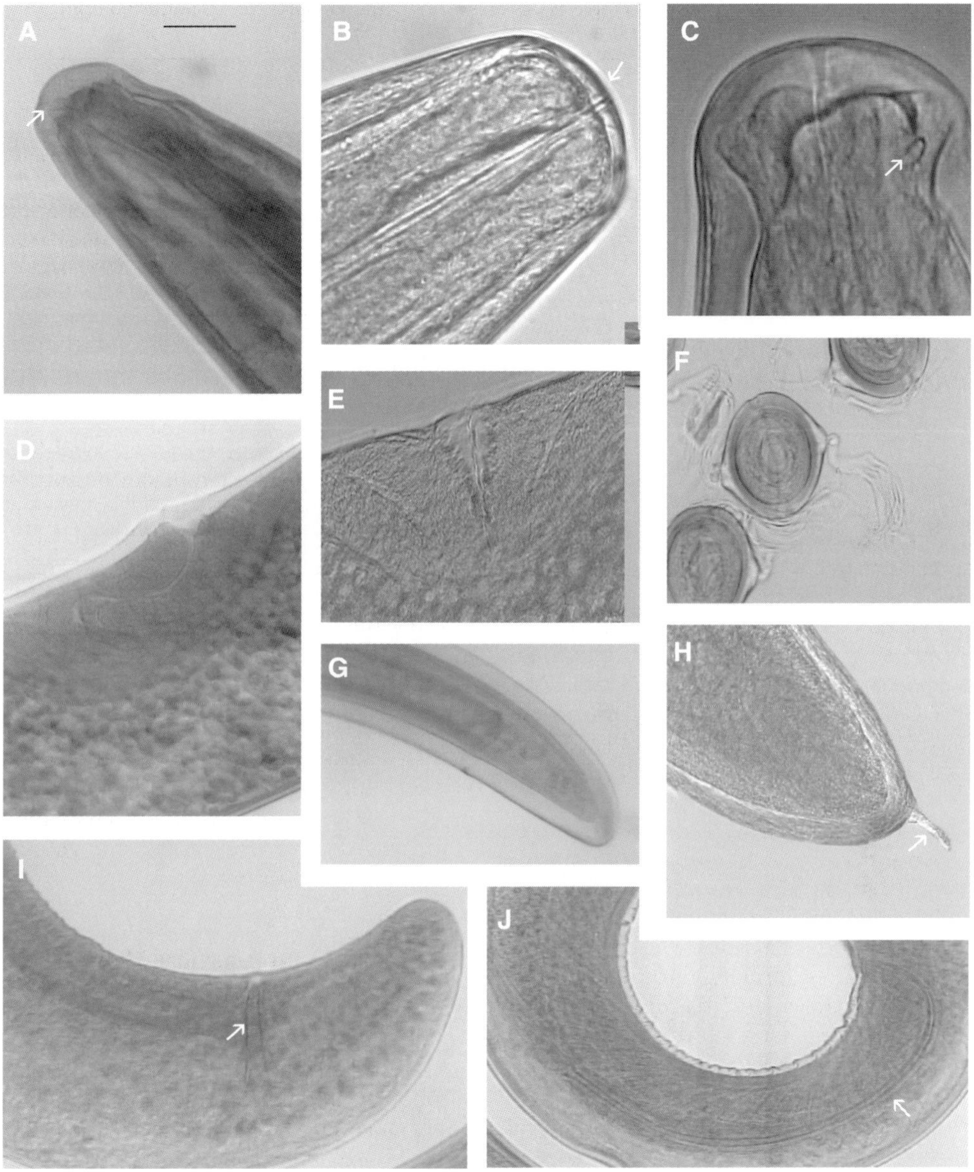

Fig. 1.11. Family Mermithidae. A, anterior end of female showing slightly shifted oral aperture (arrow); B, anterior end of postparasitic juvenile (dorsal view) showing terminal oral aperture (arrow); C, anterior end (dorsolateral view) showing amphid position (arrow); D, S-shaped vagina; E, pear-shaped vagina (lateral view); F, eggs with byssi; G and H. Tail of postparasitic juvenile (G) without digitate appendage and (H) with digitate appendage (arrow); I and J. Male tail (lateral view) showing (I) short spicules (arrow) and (J) long curved spicules (arrow). (Scale bars: A = 18 μm; B, C = 12 μm; D, E = 45 μm; F, I = 40 μm; G, J = 20 μm; H = 25 μm.)

Table 1.9. Key diagnostic features of mermithid genera considered in biocontrol.

Diagnostic features	*Agamermis* Cobb, Steiner and Christie, 1923	*Mermis* Dujardin, 1842	*Oesophagomermis* Artyukhovsky, 1969	*Romanomermis* Coman, 1961	*Strelkovimermis* Rubzov, 1969
Cephalic papillae	6	4	6	6	6
Labial papillae	Absent	Present (2)	Absent	Absent	Absent
Oral opening	Terminal	Absent	Terminal or slightly shifted to ventral side	Terminal	Terminal or slightly shifted to ventral side
Hypodermal cords	6	6	6	8	6
Vagina shape	S-shaped	S-shaped	S-shaped	Pear-shaped	S-shaped
Bursal sleeve	Absent	Absent	Absent	Absent	May be present
Parasitic and post-parasitic tail	With crater-like terminus	With tail appendage	With small tail appendage	With tail appendage	With tail appendage
Eggs	Without byssi	With byssi	Without byssi	Without byssi	Without byssi

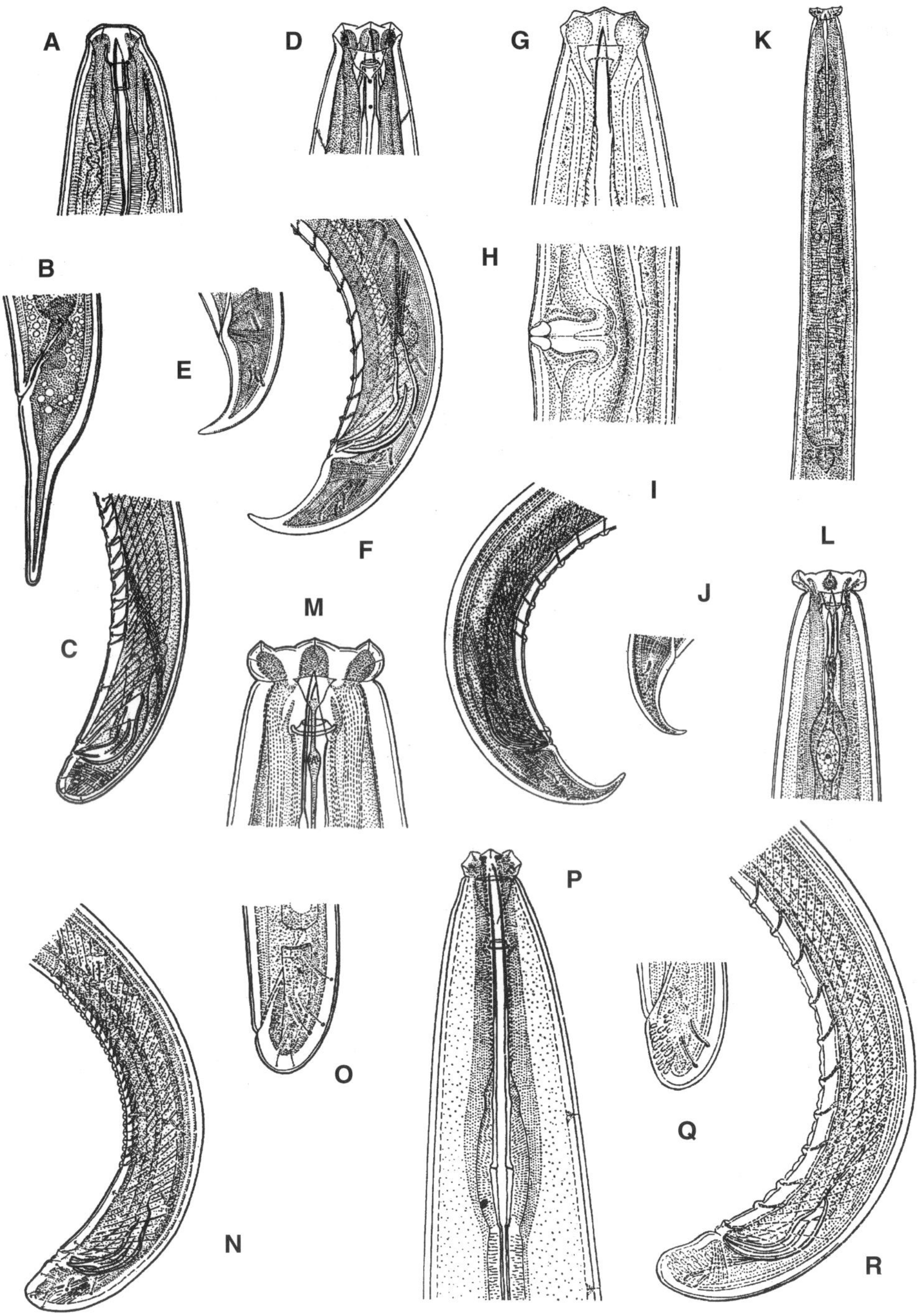

Fig. 1.12. Family Dorylaimidae. A–C. *Mesodorylaimus*: A, head region; B, female tail; C, male tail. D–F. *Allodorylaimus*: D, head region; E, female tail; F, male tail. G–J. *Eudorylaimus*: G, head region; H, vulval region; I, male tail, J, female tail. K and L. *Discolaimus*: K, pharyngeal region; L, head region. M–O. *Labronema*: M, head region; N, male tail region; O, female tail. P–R. *Pungentus*: P, head region; Q, female tail; R, male tail region. (After Jairajpuri and Ahmad, 1992, *Dorylaimida. Free-living, Predaceous and Plant-parasitic Nematodes*; various scales.)

1.4. Molecular Approaches and their Application in Nematode Taxonomy

The relative paucity of morphological traits and their limited utility in identification and/or diagnosis of many nematode groups has resulted in the exploration of alternative tools such as biochemical and molecular methods. During the past 15 years, several molecular techniques have been considered in nematode systematics. Many of these approaches have provided interesting and important insights into biodiversity and evolution, particularly for parasitic nematodes such as Steinernematidae and Heterorhabditidae (Akhurst, 1987; Reid and Hominick, 1992; Gardner *et al.*, 1994; Liu and Berry, 1995; Liu *et al.*, 1997; Reid *et al.*, 1997; Adams *et al.*, 1998; Nguyen *et al.*, 2001; Stock *et al.*, 2001).

This section reviews the most widely used molecular techniques and markers that have been applied to the groups covered by this book. Rather than promoting the latest technique, we believe it is more important that the reader understand which technique(s), gene(s) or molecular marker(s) are best suited for a particular problem and should be applied. Additional information on this subject can be found in Hussey (1981), Curran (1991), Curran and Robinson (1993), Avise (1994), Powers and Fleming (1998) and Stock and Reid (2003).

1.4.1. Molecular tools

A wide range of molecular approaches has been used and/or adopted for diagnostics/identification of nematodes with biocontrol potential. However, three methods (RAPD, RFLP and DNA sequencing) are being used most extensively.

1.4.1.1. Randomly amplified polymorphic DNA (RAPD)

The RAPD-PCR approach has been applied to the Heterorhabditidae, Steinernematidae and Aphelenchidae. RAPD-PCR was first used as a complementary tool in the identification of *Heterorhabditis* spp. and *Steinernema* spp. (Gardner *et al.*, 1994; Liu and Berry, 1996), but it has also been employed to measure genetic variability among *Heterorhabditis* and *Steinernema* isolates (Liu and Berry, 1995, 1996; Hashmi *et al.*, 1996), and to assess phylogenetic relationships among these taxa (Liu and Berry, 1996).

In the Aphelenchidae, RAPD-PCR was used to analyse the genetic diversity of *A. avenae* isolates from different locations in Japan and to correlate their geographical distribution with their host fungi preference (Ali *et al.*, 1999).

In spite of these efforts, the use of RAPDs has been discouraged, mainly because of the recognition that reproducibility of results can be affected by many factors such as the quality and concentration of DNA, PCR cycling conditions (including type of PCR machine used), etc. It can also be difficult to draw the line between inter- and intraspecific variability when using RAPD markers, leading to possible misdiagnosis.

1.4.1.2. Restriction fragment length polymorphism (RFLP)

Restriction enzymes and PCR-RFLP have been demonstrated as good diagnostic tools for the Steinernematidae and Heterorhabditidae (Reid and Hominick, 1992; Reid *et al.*, 1997; Anis *et al.*, 2000, Hussaini *et al.*, 2001; Phan *et al.*, 2001). Table 1.10 summarizes the RFLP profiles from 17 restriction enzymes that have been used to diagnose *Steinernema* spp. This method has also been applied as a diagnostic tool (Joyce *et al.*, 1994; Nasmith *et al.*, 1996; Stack *et al.*, 2000) and to complement morphological characterization of undescribed *Steinernema* spp. (Stock *et al.*, 1998; Luc *et al.*, 2000; Phan *et al.*, 2001). In addition, the PCR-RFLP approach has been used to interpret evolutionary relationships among EPNs (Reid, 1994; Reid *et al.*, 1997). However, care must be exercised when using this approach as a diagnostic tool and/or for phylogenetic history inference, since it has been recognized that even for large sequences or entire genomes, restriction enzymes vary in their

Table 1.10. Restriction fragment length polymorphism (RFLP) patterns generated by digestion of the ITS region of rDNA for *Steinernema* spp. with 17 restriction enzymes. (Modified from Reid *et al*., 1997.)

Species	*Alu*I	*Bst*O/ *Mva*I	*Dde*I	*Eco*RI	*Hae*III	*Hha*I/ *Cfo*I	*Hind*III	*Hinf*I	*Hpa*II/ *Msp*I	*Kpn*I	*Pst*I	*Pvu*II	*Rsa*I	*Sal*I	*Sau*3AI/ *Nde*II	*Sau*96/ *Bsiz*I	*Xba*I
S. feltiae	1	1	1	1	1	1	1	1	1	1	1	1	1	1	1	1	1
S. abbasi	2	1	2	1	2	17	1	9	4	1	2	4	19	1	10	3	1
S. affine	3	2	3	1	4	2	2	3	1	2	2	2	3	1	2	2	1
S. anatoliense	4	1	4	1	2	15	3	11	3	3	2	1	4	1	6	3	2
S. arenarium	5	1	5	1	2	7	4	6	1	1	2	5	9	3	7	3	1
S. asiaticum	6	1	6	1	2	21	1	14	1	1	2	1	15	1	9	3	4
S. bicornutum	7	1	7	1	2	5	1	5	1	1	2	3	5	1	16	3	1
S. carpocapsae	8	1	8	1	2	4	2	4	1	1	2	1	7	1	5	3	1
S. caudatum	+	+	+	+	+	+	+	+	+	+	+	+	+	+	+	+	+
S. ceratophorum	9	1	9	1	2	11	1	7	1	1	2	9	16	1	3	3	1
S. cubanum	10	1	10	1	2	9	2	6	2	1	2	6	14	4	7	3	1
S. diaprepesi	+	+	+	+	+	+	+	+	+	+	+	+	+	+	+	+	+
S. glaseri	11	1	10	1	2	8	1	6	2	1	2	6	10	4	8	3	1
S. intermedium	12	2	11	1	4	3	1	6	1	2	2	1	3	1	17	2	1
S. karii	13	1	12	1	2	6	1	6	1	1	2	11	8	6	7	3	1
S. kraussei	14	1	10	2	2	18	1	2	1	1	2	1	18	1	7	3	1
S. kushidai	15	1	13	2	3	12	1	8	5	2	2	1	18	1	4	3	1

continued

Table 1.10. *Continued*. Restriction fragment length polymorphism (RFLP) patterns generated by digestion of the ITS region of rDNA for *Steinernema* spp. with 17 restriction enzymes. (Modified from Reid *et al*., 1997.)

S. loci	16	1	16	1	2	13	1	20	1	1	3	10	11	5	22	3	3
S. longicaudum	17	1	14	3	2	14	1	10	1	1	2	7	19	1	8	3	6
S. monticolum	18	1	15	1	1	10	1	18	1	1	5	1	13	6	13	3	1
S. neocurtillae	+	+	+	+	+	+	+	+	+	+	+	+	+	+	+13	+	+
S. oregonense	19	1	15	4	2	18	1	19	1	1	2	1	20	1	18	3	1
S. pakistanense	20	1	17	1	2	10	1	9	1	1	4	1	15	1	14	3	1
S. puertoricense	21	1	13	1	2	10	5	6	2	2	2	12	21	1	7	3	5
S. rarum	22	1	18	1	5	10	1	15	1	1	3	1	17	1	19	4	1
S. riobrave	23	1	19	1	2	12	2	9	1	1	2	1	18	1	20	3	1
S. ritteri	+	+	+	+	+	+	+	+	+	+	+	+	+	+	+	+	+
S. sangi	24	1	20	5	2	1	1	2	6	1	2	1	2	1	11	3	1
S. scapterisci	25	1	21	1	2	4	1	16	1	1	2	4	6	1	5	3	1
S. scarabaei	26	1	10	1	2	7	1	3	7	1	2	1	9	1	15	3	1
S serratum	+	+	+	+	+	+	+	+	+	+	+	+	+	+	+	+	+
S. siamkayai	27	1	6	1	2	20	1	17	1	1	2	1	15	1	21	3	4
S. tami	28	1	6	1	2	19	1	12	1	1	2	1	15	1	21	3	1
S. thanhi	29	1	22	1	2	16	1	6	1	1	2	8	12	1	22	3	3
S. thermophilum	+	+	+	+	+	+	+	+	+	+	+	+	+	+	+	+	+
S. websteri	30	1	8	1	2	4	2	13	1	1	2	1	7	1	12	3	1

+ = no RFLP profile available.

Note: Numbers indicate different RFLP profiles yielded by a single enzyme. Species with the same number indicate identical patterns and species with different numbers indicate different patterns.

efficiency for generating RFLPs (Whitkus *et al.*, 1994). Moreover, without restriction site maps, fragment patterns cannot reliably produce homologous characters required to infer phylogenetic relationships or delimit species. Without *a priori* knowledge of cleavage site homology, interpretation of fragment patterns can be complicated or misleading (Hillis *et al.*, 1996).

1.4.1.3. DNA sequence analysis

DNA sequence analysis has recently been incorporated into nematode systematics and has been demonstrated to yield more information about variation within and between nematode species than the RFLP approach (Powers *et al.*, 1997). In addition, DNA sequence analysis has proved to be a more suitable method in assessing phylogenetic relationships at different taxonomic levels (Powers *et al.*, 1994; Hyman and Azevedo, 1996; Adams *et al.*, 1998; Blaxter *et al.*, 1998; Iwahori, 1998; Szalanski *et al.*, 2000; Stock *et al.*, 2001; Perlman *et al.*, 2003) and a useful method for species delimitation (Adams *et al.*, 1998; Nguyen *et al.*, 2001; Stock *et al.*, 2001; Stock and Koppenhöfer, 2003).

1.4.2. Target regions

1.4.2.1. Nuclear genes

Nuclear rDNA is a useful source for markers involved in delimitation of nematodes at different taxonomic levels (e.g. Curran and Driver, 1994; Blaxter *et al.*, 1998; Nadler and Hudspeth, 1998, 2000).

18S OR SMALL SUBUNIT (SSU) GENE OF rDNA. Phylogenetic interpretation of 18S sequence data for Steinernematidae and Heterorhabditidae revealed that these two families represent distinct, unrelated, lineages (Blaxter *et al.*, 1998). However, at the species level the region was demonstrated to be too conserved in resolving relationships among *Heterorhabditis* (Liu *et al.*, 1997) or *Steinernema* (Stock *et al.*, 2001).

INTERNAL TRANSCRIBED SPACER (ITS) REGION AND 5.8S GENE OF rDNA. ITS has been used in EPN systematics. This variable region has revealed numerous diagnostic markers. In the Heterorhabditidae, ITS-1 region has sufficient genetic variation for differentiating *Heterorhabditis* spp. and has proved valuable for delimitation and interpretation of evolutionary relationships between species (Adams *et al.*, 1998). ITS-1 and -2 regions, including the 5.8S gene of rDNA, have also been used to assess phylogenetic relationships and delimit species with a limited number of *Steinernema* spp. (Nguyen *et al.*, 2001). Because of its conserved nature, the 5.8S gene was uninformative in resolving phylogenetic relationships and delimitation of terminal taxa in *Steinernema* (Nguyen *et al.*, 2001).

With respect to the ITS region, a more extensive taxon sampling is necessary to prove its value in interpreting evolutionary relationships among species in this genus and to adequately address the nature of variability within and among individuals and populations of *Steinernema*. The ITS region might only be useful for resolving relationships among closely related *Steinernema* spp. (see Stock *et al.*, 2001), but is perhaps too variable to reliably infer relationships among all species in this genus.

28S OR LARGE SUBUNIT (LSU) OF rDNA. LSU sequence data has been used to assess phylogenetic relationships among *Steinernema* spp. (Stock *et al.*, 2001). In the study by Stock *et al.* (2001), 28S rDNA proved to be a suitable and informative region for interpreting evolutionary relationships among *Steinernema* spp. (see Section 1.3, this chapter). This region is also considered to be an effective and reliable approach for delimitation of terminal taxa in *Steinernema* as well as for diagnostic purposes (Stock *et al.*, 2001; Stock and Koppenhöfer, 2003).

1.4.2.2. Mitochondrial genes

At present, a few mitochondrial genes have been considered in studies of genetic variation within and among nematodes with potential as biocontrol agents. Powers *et al.* (1986) studied the molecular structure of nematode mitochondrial DNA (mtDNA) using the mermithid *R. culicivorax*. In a later study, Powers *et al.* (1994) compared several mtDNA genes (e.g. NADH dehydrogenase subunit 3 (ND3), large rRNA, and cytochrome b genes) to measure the genetic divergence from several nematode species, including *R. culicivorax*. More recently, Blouin *et al.* (1999) and Liu *et al.* (1999) studied the genetic variation among several *Heterorhabditis marelata* populations using the ND4 gene of mtDNA and found limited intraspecific variation. Other mtDNA genes studied include COXII and 16S rDNA (Szalanski *et al.*, 2000). These loci showed variation at the species level and proved useful for discrimination between a selection of *Steinernema* spp. However, they failed to show variation at the intraspecific level when tested with several *Steinernema feltiae* populations.

1.5. Origin of Invertebrate Parasitism

According to Poinar (1983), invertebrate parasitism arose in four major groups of nematodes. Based on his proposal, the most primitive group, the Rhabditida, gave rise to members of the Oxyurida (*c.* 420 million years ago) as well as to the Steinernematidae and Heterorhabditidae (375 million years ago) (Fig. 1.13). Poinar (1993) also speculated that morphological and life history similarities between these two groups were the result of convergent evolution.

Based on similarities of the buccal capsule and male tail morphology, Poinar (1993) suggested that steinernematids have evolved from a terrestrial 'proto-*Rhabditonema*' ancestor, while heterorhabditids arose from a '*Pellioiditis*-like' ancestor in a sandy marine environment. The notion that heterorhabditids and steinernematids do not share an exclusive common ancestor has also been proposed by other studies based on cladistic interpretation of morphological traits (Sudhaus, 1993) and of molecular data (Adams *et al.*, 1998), and a combination of both approaches (Liu *et al.*, 1997).

Poinar (1983) also suggested that plant-parasitic tylenchids gave rise to the Allantonematidae (300 million years ago) and Sphaerulariidae (223 million years ago) and that the insect-parasitic Entaphephelenchidae probably arose from an aphelenchoid ancestor approximately 300 million years ago (Fig. 1.13). The fourth group of invertebrate parasites for which Poinar (1983) suggested an evolutionary hypothesis was the Mermithida, where he suggested predaceous dorylaimids as their closest ancestors (185 million years ago) (Fig. 1.13).

A recent evolutionary framework of the Nematoda based on 18S rDNA (Blaxter

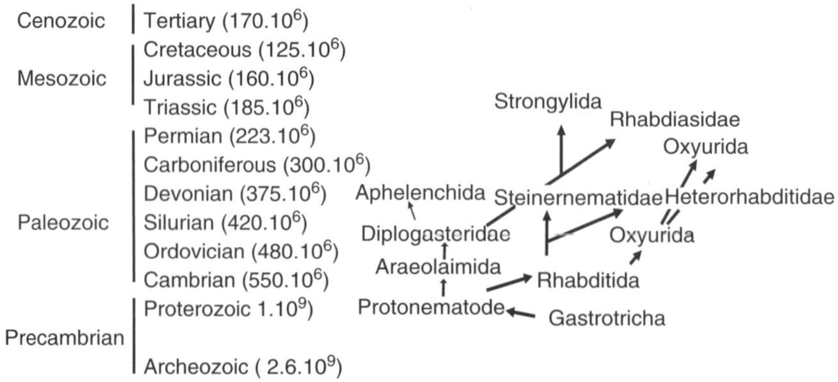

Fig. 1.13. Schematic representation of the evolution of invertebrate parasitism in Nematoda according to Poinar (1983). (Modified after Poinar, 1983.)

et al., 1998) supported Poinar's (1993) hypothesis regarding the independent origins of steinernematids and heterorhabditids. This tree depicted *Steinernema* as being most closely related to the Panagrolaimidae (mostly free-living nematodes, with some genera considered as insect associates) and *Strongyloides* (vertebrate parasites). These three groups (Steinernematidae, Panagrolaimidae and Strongyloididae) are members of a larger clade that comprises plant-parasitic, fungal-feeding and bacterivorous taxa of the order Tylenchida, Aphelenchida and Cephalobida (Fig. 1.14). The same tree also depicted *Heterorhabditis* as being most closely related to the Strongylida (vertebrate parasites), both clades sharing the rhabditoid *Pellioditis* (Rhabditida) as their most recent common ancestor (Fig. 1.14).

Blaxter *et al.*'s tree depicted the mermithids as being most closely related to the free-living mononchids, and as a member of a larger clade that included the vertebrate-parasitic trichocephalids and the plant-parasitic dorylaimids (Fig. 1.14). These results are consistent with Poinar's hypothesis of a predatory dorylaimid as the closest relative to the mermithids.

Three orders in Nematoda have representatives of mollusc-parasitic or associated

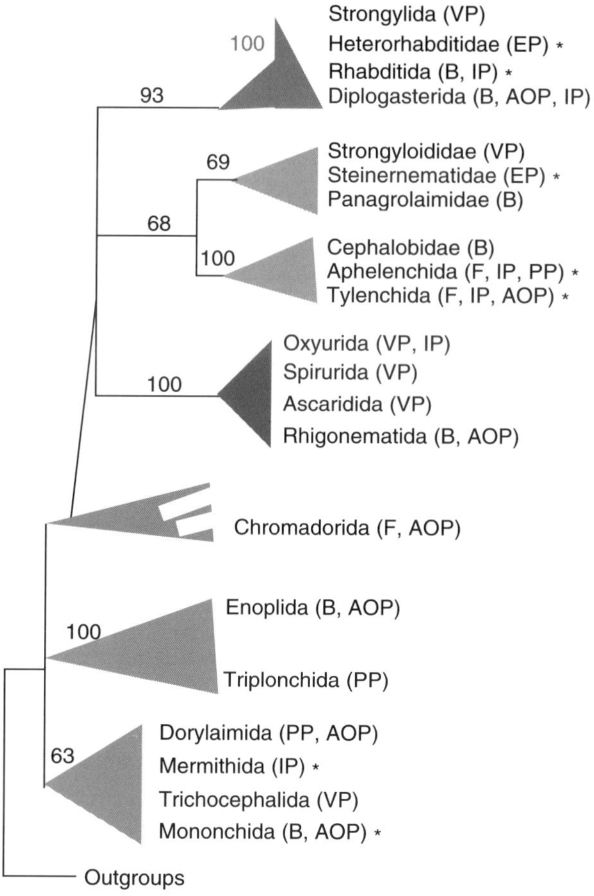

Fig. 1.14. Schematic representation of the evolution of invertebrate parasitism in Nematoda. (Modified from Blaxter *et al.*, 2000.) AOP = algivore-omnivore-predator; B = bacterivore; EP = entomopathogen; F = fungivore; IP = invertebrate parasite; PP = plant parasite; VP = vertebrate parasite, * = used or with potential as biocontrol agents.

taxa: Strongylida, Rhabditida and Aereolaimida (Grewal *et al.*, 2003). Based on interpretation of Blaxter *et al.*'s (1998, 2000) phylogeny, parasitism of molluscs seems to have arisen up to three times in Nematoda. Such a distribution suggests that utilization of molluscan hosts could be extremely lucrative for nematodes, and that nematodes display extreme adaptive plasticity (Grewal *et al.*, 2003).

In conclusion, Blaxter *et al.*'s analysis also suggested that invertebrate parasitism arose independently at least four times in the evolution of Nematoda (Fig. 1.14). These data also indicate an association between invertebrate and vertebrate parasitism, with invertebrate-pathogenic and -parasitic clades lying basal to major vertebrate-parasitic ones (Blaxter *et al.*, 1998, 2000).

Acknowledgements

We thank Alex P. Reid for providing data for RFLP profiles of *Steinernema* spp.

References

Adams, B.J., Burnell, A.M. and Powers, T.O. (1998) A phylogenetic analysis of *Heterorhabditis* (Nematoda: Rhabditidae) based on internal transcribed spacer 1 DNA sequence data. *Journal of Nematology* 30, 22–39.

Akhurst, R.J. (1987) Use of starch gel electrophoresis in the taxonomy of the genus *Heterorhabditis* (Nematoda: Heterorhabditidae). *Nematologica* 33, 1–9.

Albertson, D.G. and Thomson, J.N. (1976) The pharynx of *Caenorhabditis elegans*. *Philosophical Transactions of the Royal Society of London, Series B, Biological Sciences* 275, 299–325.

Ali, M.R., Yamaguchi, Y. and Ishibashi, N. (1999) RAPD and PCR-RFLP analysis on genetic diversity of *Aphelenchus avenae* isolates collected from Kyushu and some other districts of Japan. *Japanese Journal of Nematology* 29, 24–34.

Andrássy, I. (1983) *A Taxonomic Review of the Suborder Rhabditina (Nematoda: Secernentia).* ORSTROM, Paris, 241 pp.

Anis, M., Shahina, F., Reid, A.P. and Maqbool, M.A. (2000) Redescription of *Heterorhabditis indica,* Poinar *et al.*, 1992 (Rhabditida: Heterorhabditidae) from Pakistan. *Pakistan Journal of Nematology* 18, 11–27.

Anis, M., Shahina, F., Reid, A.P. and Rowe, J. (2002) *Steinernema asiaticum* sp. n. (Rhabditida: Steinernematidae) from Pakistan. *International Journal of Nematology* 12, 220–231.

Avise, J.C. (1994) *Molecular Markers, Natural History and Evolution.* Chapman and Hall, London, 511 pp.

Bedding, R.A. (1993) Biological control of *Sirex noctilio* using the nematode *Beddingia siricidicola.* In: Bedding, R.A., Akhurst, R. and Kaya, H.K. (eds) *Nematodes and the Biological Control of Insect Pests.* CSIRO Publications, East Melbourne, Australia, pp. 11–20.

Blaxter, M.L., De Ley, P., Garey, J.R., Liu, L.X., Scheldeman, P., Vierstraete, A., Vanfleteren J.R., Mackey, L.Y., Dorris, M., Frisse, L.M., Vida, J.T. and Thomas, W.K. (1998) A molecular evolutionary framework for the phylum Nematoda. *Nature* 392, 71–75.

Blaxter, M., Dorris, M. and De Ley, P. (2000) Patterns and processes in the evolution of animal parasitic nematodes. *Nematology* 2, 43–55.

Blouin, M.S., Liu, J. and Berry, R.E. (1999) Life cycle variation and the genetic structure of nematode populations. *Heredity* 83, 253–259.

Choo, H.Y. and Kaya, H.K. (1994) Biological control of the brown plant hopper by a mermithid nematode. *Korean Journal of Applied Entomology* 33, 207–215.

Choo, H.Y., Kaya, H.K. and Kim, J.B. (1989) *Agamermis unka* (Mermithidae) parasitism of *Nilaparvata lugens* in rice fields in Korea. *Journal of Nematology* 21, 254–259.

Choudhury, B.N. and Sivakumar, C.V. (2000) Biocontrol potential of *Mylonchulus minor* against some plant-parasitic nematodes. *Annals of Plant Protection Sciences* 8, 53–57.

Creighton, C.S. and Fassuliotis, G. (1983) Infectivity and suppression of the banded cucumber beetle (Coleoptera: Chrysomelidae) by the mermithid nematode *Filipjevimermis leipsandra* (Mermithida: Mermithidae). *Journal of Economic Entomology* 76, 615–618.

Curran, J. (1991) Application of DNA analysis to nematode taxonomy. In: Nickle, W.R. (ed.) *Manual of Agricultural Nematology.* Marcel Dekker, New York, pp. 125–143.

Curran, J. and Driver, F. (1994) Molecular taxonomy of *Heterorhabditis.* In: *Genetics of Entomopathogenic Nematode Bacterium-Complexes,* EUR 15681, St Patrick's College, Maynooth, Kildare, Ireland, pp. 41–48.

Curran, J. and Robinson, M.P. (1993) Molecular aids to nematode diagnosis. In: Evans, K., Trudgill,

D.L. and Webster, J.M. (eds) *Plant Parasitic Nematodes in Temperate Agriculture*. CAB International, Wallingford, UK, pp. 545–564.

De Ley, P. and Blaxter, M. (2002) Systematic position and phylogeny. In: Lee, D.L. (ed.). *The Biology of Nematodes*. Taylor and Francis, New York, pp. 1–30.

Gardner, S.L., Stock, S.P. and Kaya, H.K. (1994) A new species of *Heterorhabditis* from the Hawaiian islands. *Journal of Parasitology* 80, 100–106.

Gaugler, R. and Kaya, H.K. (1990) *Entomopathogenic Nematodes in Biological Control*. CRC Press, Boca Raton, Florida, 365 pp.

Glen, D.M. and Wilson, M. (1997) Slug-parasitic nematodes as biocontrol agents for slugs. *Agro Food Industry Hi Technology* 8, 23–27.

Goodey, J.B. (1963) *Soil and Freshwater Nematodes*. Methuen, London, 544 pp.

Grewal, P.S., Grewal, S.K. and Adams, B.J. (2003) Parasitism of molluscs by nematodes: types of associations and evolutionary trends. *Journal of Nematology* 35, 145–156.

Hashmi, G., Glazer, I. and Gaugler, R. (1996) Molecular comparisons of entomopathogenic nematodes using randomly amplified polymorphic DNA (RAPD) markers. *Fundamental and Applied Nematology* 19, 399–406.

Hillis, D.M., Moritz, C. and Mable, B.K. (1996) *Molecular Systematics*. 2nd edn. Sinauer Associates Press, Sunderland, Massachusetts, 655 pp.

Hooper, D.J., Wilson, M.J., Rowe, J.A. and Glen, D.M. (1999) Some observations on the morphology and protein profiles of the slug-parasitic nematodes *Phasmarhabditis hermaphrodita* and *P. neopapillosa* (Nematoda: Rhabditidae). *Nematology* 1, 173–182.

Hussaini, S.S., Ansari, M.A., Ahmad, W. and Subbotin, S.A. (2001) Identification of some Indian populations of *Steinernema* species (Nematoda) by RFLP analysis of the ITS region of rDNA. *International Journal of Nematology* 11, 73–76.

Hussey, R.S. (1981) Biochemical systematics of nematodes: a review. *Helminthological Abstracts (Series B)* 48, 141–148.

Hyman, B.C. and Azevedo, J.L.B. (1996) Similar evolutionary patterning among repeated and single copy nematode mitochondrial genes. *Molecular Biology and Evolution* 13, 221–232.

Iwahori, H., Tsuda, K., Kanzaki, N., Izui, K. and Futai, K. (1998) PCR-RFLP and sequencing analysis of ribosomal DNA of *Bursaphelenchus* nematodes related to pine wilt disease. *Fundamental and Applied Nematology* 21, 655–666.

Joyce, S.A., Burnell, A.M. and Powers, T.O. (1994) Characterization of *Heterorhabditis* isolates by PCR amplification of sequences of mDNA and rDNA genes. *Journal of Nematology* 26, 260–270.

Kakulia, G. and Mikaia, N. (1997) New species of the nematode *Heterorhabditis poinari* sp. n. (Rhabditida: Heterorhabditidae). *Bulletin of the Georgian Academy of Sciences* 155, 457–459.

Kasab, A.S. and Abel-Kader, K. (1996) Suppressing root-knot, citrus and reniform nematodes by use of less irrigation frequency and predacious nematode in olive. *Annals of Agricultural Science of Cairo* 41, 511–520.

Kaya, H.K. and Stock, S.P. (1997) Techniques in insect nematology. In: Lacey, L.A. (ed.) *Manual of Techniques in Insect Pathology*. Biological Techniques Series, Academic Press, San Diego, California, pp. 281–324.

Lacey, L.A., Frutos, R., Kaya, H.K. and Vails, P. (2001) Insect pathogens as biological control agents: do they have a future? *Biological Control* 21, 230–248.

Liu, J. and Berry, R.E. (1995) Differentiation of isolates in the genus *Steinernema* (Nematoda: Steinernematidae) by random amplified polymorphic DNA fragments and morphological characters. *Parasitology* 111, 119–125.

Liu, J. and Berry, R.E. (1996) Phylogenetic analysis of the genus *Steinernema* by morphological characters and randomly amplified polymorphic DNA fragments. *Fundamental and Applied Nematology* 19, 463–469.

Liu, J., Berry, R.E. and Moldenke, A.F. (1997) Phylogenetic relationships of entomopathogenic nematodes (Heterorhabditdae and Steinernematidae) inferred from partial 18S rRNA gene sequences. *Journal of Invertebrate Pathology* 69, 246–252.

Liu, J., Berry, R.E. and Blouin, M.S. (1999) Molecular differentiation and phylogeny of entomopathogenic nematodes (Rhabditida: Heterorhabditidae) based on ND4 gene sequences of mitochondrial DNA. *Journal of Parasitology* 85, 709–715.

Lootsma, M. and Scholte, K. (1997) Effects of the springtail *Folsomia fimetaria* and the nematode *Aphelenchus avenae* on *Rhizoctonia solani* stem infection of potato at temperatures of 10 and 15°C. *Plant Pathology* 46, 203–208.

Luc, V.P., Nguyen, K.B., Reid, A.P. and Spiridonov, S.E. (2000) *Steinernema tami* sp. n. (Rhabditida: Steinernematidae) from Cat Tien Forest, Vietnam. *Russian Journal of Nematology* 8, 33–43.

Maggenti, A.R. (1981) *General Nematology*. Springer-Verlag, New York, 372 pp.

Matsunaga, T., Wang, X.D. and Ishibashi, N. (1997) Interference by entomopathogenic and fungivorous nematodes of root invasion by plant-parasitic nematodes. *Japanese Journal of Nematology* 26, 1–11.

Nadler, S.A. and Hudspeth, D.S.S. (1998) Ribosomal DNA and phylogeny of the Ascaridoidea (Nematoda: Secernentea): implications for morphological evolution and classification. *Molecular Phylogenetics and Evolution* 10, 221–236.

Nadler, S.A. and Hudspeth, D.S.S. (2000) Phylogeny of the Ascaridoidea (Nematoda: Ascaridida) based on three genes and morphology: hypotheses of structural and sequence evolution. *Journal of Parasitology* 86, 380–393.

Nasmith, C.G., Speranzini, D., Jeng, R. and Hubbes, M. (1996) RFLP analysis of PCR amplified ITS and 26S ribosomal RNA genes of selected entomopathogenic nematodes (Steinernematidae and Heterorhabditidae). *Journal of Nematology* 28, 15–25.

Nguyen, K.B., Maruniak, J. and Adams, J.B. (2001) Diagnostic and phylogenetic utility of the rDNA internal transcribed spacer sequences of *Steinernema. Journal of Nematology* 33, 73–82.

Nickle, W.R. (1972) A contribution to our knowledge of the Mermithidae (Nematoda). *Journal of Nematology* 4, 113–146.

Perlman, S.J., Spicer, G.S., Shoemaker, D. and Jaenike, J. (2003) Associations between mycophagous *Drosophila* and their *Howardula* nematode parasites: a worldwide phylogenetic shuffle. *Molecular Biology* 12, 237–249.

Petersen, J.J. (1985) Nematode parasites. In: Chapman, H.C. (ed.) *Biological Control of Mosquitoes. American Mosquito Control Association Bulletin* 6, pp. 110–122.

Phan, K.L., Nguyen, N.C. and Moens, M. (2001) *Steinernema loci* sp. n. and *Steinernema thanhi* sp. n. (Rhabditida: Steinernematidae) from Vietnam. *Nematology* 3, 503–514.

Phan, K.L., Subbotin, S.A., Nguyen, N.C. and Moens, M. (2003) *Heterorhabditis baujardi* sp. n. (Rhabditida: Heterorhabditidae) from Vietnam and morphometric data for *H. indica* populations. *Nematology* 5, 367–382.

Poinar, G.O. Jr (1975) *Entomogenous Nematodes.* E.J. Brill, Leiden, The Netherlands, 317 pp.

Poinar, G.O. Jr (1983) *The Natural History of Nematodes.* Prentice-Hall, Englewood Cliffs, New Jersey, 323 pp.

Poinar, G.O. Jr (1990) Entomopathogenic nematodes in biological control. In: Gaugler, R. and Kaya, K.H. (eds) *Taxonomy and Biology of Steinernematidae and Heterorhabditidae.* CRC Press, Boca Raton, Florida, pp. 23–74.

Poinar, G.O. Jr (1993) Origins and phylogenetic relationships of the entomophilic rhabditids, *Heterorhabditis* and *Steinernema. Fundamental and Applied Nematology* 16, 333–338.

Powers, T.O. and Fleming, C.C. (1998) Biochemical and molecular characterization. In: Perry, R.N. and Wright, D.J. (eds) *The Physiology and Biochemistry of Free-Living and Plant-Parasitic Nematodes.* CAB International, Wallingford, UK, pp. 355–380.

Powers, T.O., Platzer, E.G. and Hyman, B.C. (1986) Large mitochondrial genome and mitochondrial DNA size polymorphism in the mosquito parasite, *Romanomermis culicivorax. Current Genetics* 11, 71–77.

Powers, T.O., Harris, T.S. and Hyman, B.C. (1994) Mitochondrial DNA sequence divergence among *Meloidogyne incognita, Romanomermis culicivorax, Ascaris suum* and *Caenorhabditis elegans. Journal of Nematology* 25, 564–572.

Powers, T.O., Todd, T.C., Burnell, A.M., Muray, P.C.B., Fleming, C.C., Szlanski, A.L., Adams, B.A. and Harris, T.S. (1997) The rDNA internal transcribed spacer region as a taxonomic marker for nematodes. *Journal of Nematology* 29, 441–450.

Reid, A.P. (1994) Molecular taxonomy of *Steinernema.* In: Burnell, A.M., Ehlers, R.-U. and Mason, J.P. (eds) *Genetics of Entomopathogenic Nematode-Bacterium Complexes.* European Commission, Luxembourg City, Luxembourg, pp. 49–58.

Reid, A.P and Hominick, W.M. (1992) Restriction fragment length polymorphisms within ribosomal DNA repeat unit of British entomopathogenic nematodes (Rhabditida: Steinernematidae). *Parasitology* 105, 317–323.

Reid, A.P., Hominick, W.M. and Briscoe, B.R. (1997) Molecular taxonomy and phylogeny of entomopathogenic nematode species (Rhabditida: Steinernematidae) by RFLP analysis of the ITS region of rDNA repeat unit. *Systematic Parasitology* 37, 187–193.

Schauff, M.E. and LaSalle, J. (1998) The relevance of systematics to biological control: protecting the investment in research. In: Zaluki, M.P., Drew, R.A.I. and White, G.G. (eds) *Pest Management – Future Challenges. Proceedings of the Sixth Australasian Applied Entomological Research Conference.* Brisbane, pp. 425–436.

Siddiqi, M.R. (1986) *Tylenchida Parasites of Plants and Insects.* Commonwealth Agricultural Bureaux, Farnham Royal, Slough, UK, 645 pp.

Siddiqi, M.R. (2000) *Tylenchida Parasites of Plants and Insects.* CAB International, Wallingford, UK, pp. 633–751.

Stack, C.M., Easwaramoorthy, S.G., Metha, U.K., Downes, M.J., Griffin, C.T. and Burnell, A.M. (2000) Molecular characterisation of *Heterorhabditis indica* isolates from India, Kenya, Indonesia and Cuba. *Nematology* 2, 477–487.

Stock, S.P. (1993) A new species of the genus *Heterorhabditis* Poinar, 1976 (Nematoda: Heterorhabditidae) parasitizing *Graphognathus* sp. larvae (Coleoptera: Curculionidae) from Argentina. *Research and Reviews in Parasitology* 53, 103–107.

Stock, S.P. and Koppenhöfer, A.M. (2003) *Steinernema scarabaei* n. sp. (Rhabditida: Steinernematidae), a natural pathogen of scarab beetle larvae (Coleoptera: Scarabaeidae) from New Jersey, USA. *Nematology* 5, 191–204.

Stock, S.P. and Reid, A.P. (2003) Biosystematics of entomopathogenic nematodes (Steinernematidae, Heterorhabditidae): current status and future directions. In: Cook, R. and Hunt, D.J. (eds) *Proceedings of the Fourth International Congress of Nematology.* Tenerife, Spain 8–13 June, 2002. Nematology Monographs and Perspectives 2, 435–446.

Stock, S.P., Somsook, V. and Reid, A.P. (1998) *Steinernema siamkayai* n. sp. (Rhabditida: Steinernematidae), an entomopathogenic nematode from Thailand. *Systematic Parasitology* 41, 105–113.

Stock, S.P., Campbell, J.F. and Nadler, S.A. (2001) Phylogeny of *Steinernema* Travassos, 1927 (Cephalobina: Steinernematidae) inferred from ribosomal DNA sequences and morphological characters. *Journal of Parasitology* 87, 877–889.

Sudhaus, W. (1993) Die mittels symbiontischer Bakterien entomopathogenen Nematoden-Gattungen *Heterorhabditis* und *Steinernema* sind keine Schwestertaxa. Verhand. *Der Deutschen Zool. Gesell.* 86, 146.

Szalanski, A.L., Taylor, D.B. and Mullin, P.G. (2000) Assessing nuclear and mitochondrial DNA sequence variation within *Steinernema* (Rhabditida: Steinernematidae). *Journal of Nematology* 32, 229–233.

Webster, J.M. and Thong, C.H.S. (1984) Nematode parasites of orthopterans. In: Nickle, W.R. (ed.) *Plant and Insect Nematodes.* Marcel Dekker, New York, pp. 697–726.

Whitkus, R., Doebley, J. and Wendel, J.F. (1994) Nuclear DNA markers in systematics and evolution. In: Phillips, L. and Vasil, I.K. (eds) *DNA-Based Markers in Plants.* Kluwer Academic Publishers, Dordrecht, The Netherlands. pp. 116–141.

Wilson, M.J. and Gaugler, R. (2000) Terrestrial mollusc pests. In: Lacey, L. and Kaya, H.K (eds) *Field Manual of Techniques in Invertebrate Pathology.* Kluwer Academic Publishers, Dordrecht, The Netherlands, pp. 787–804.

Wilson, M.J., Glen, D.M. and George, S.K. (1993) The rhabditid nematode *Phasmarhabditis hermaphrodita* as a potential biological control agent for slugs. *Biocontrol Science and Technology* 3, 503–511.

Wilson, M.J., Glen, D.M., George, S.K., Pearce, J.D. and Wiltshire, C.W. (1994) Biological control of slugs in winter wheat using the rhabditid nematode *Phasmarhabditis hermaphrodita. Annals of Applied Biology* 125, 377–390.

Part II

Entomopathogenic Nematodes

2 Biology and Behaviour

C.T. Griffin,[1] N.E. Boemare[2] and E.E. Lewis[3]
[1]Department of Biology and Institute of Bioengineering and Agroecology, National University of Ireland, Maynooth, Co. Kildare, Ireland; [2]Laboratoire EMIP (Ecologie Microbienne des Insectes et Interactions Hôte-Pathogène), C.C. 54, UMR INRA-UMII no 1133 – IFR 122, Université Montpellier II, Place Eugène Bataillon, 34095 Montpellier Cedex 5, France; [3]Department of Nematology and Department of Entomology, 1 Shields Avenue, University of California Davis, Davis, CA 95616, USA

2.1. Introduction

Entomopathogenic nematodes (EPNs) of the families Steinernematidae and Heterorhabditidae are lethal pathogens of insects. These pathogens contribute to the regulation of natural populations of insects, but the main interest in them is as an inundatively applied biocontrol agent. Their success in this role can be attributed to the unique partnership between a host-seeking nematode and a lethal insect-pathogenic bacterium. Because of their biocontrol potential, considerable attention has been directed over the past few decades to *Heterorhabditis* and *Steinernema* and their respective bacterial partners, *Photorhabdus* and *Xenorhabdus*. Landmark publications reviewing the biology and use of EPNs are Gaugler and Kaya (1990) and Gaugler (2002). A third genus of EPN, *Neosteinernema*, has received almost no attention since the first report of its association with termites by Nguyen and Smart (1994).

Although heterorhabditids and steinernematids are not closely related (Blaxter *et al.*, 1998; see also Chapter 1, this volume), they have many features in common. These similarities, including their association with insect-pathogenic bacteria, are presumed to have arisen through convergent evolution (Poinar, 1993). In both *Steinernema* and *Heterorhabditis* there is a single free-living stage, the infective juvenile (IJ), that carries in its gut bacteria of the genus *Xenorhabdus* and *Photorhabdus*, respectively (Boemare *et al.*, 1993). On encountering a suitable insect, the IJ enters through the mouth, anus or spiracles and makes its way to the haemocoel. Some species may also penetrate through the intersegmental membranes of the insect cuticle (Bedding and Molyneux, 1982; Peters and Ehlers, 1994). In *Heterorhabditis* spp. this is facilitated by the possession of an anterior tooth (Bedding and Molyneux, 1982).

In the haemocoel, the IJ releases cells of its bacterial symbiont from its intestine. The bacteria proliferate in the nutrient-rich insect haemolymph. Death of the insect ensues, normally within 24–48 h. The IJs recover from their arrested state and feed on the proliferating bacteria and digested host tissues. The nematodes develop through the fourth to the fifth (adult) stage, and then reproduce. One or more generations may occur within the host cadaver, depending on available resources.

Steinernematids and heterorhabditids differ in their mode of reproduction. In heterorhabditids, the first generation consists of self-fertile hermaphrodites, while males, females and hermaphrodites are produced in subsequent generations (Dix *et al.*, 1992). In steinernematids, all generations reproduce by amphimixis (cross-fertilization involving males and females) (Poinar, 1990). Recently, a *Steinernema* sp. was found to depart from the norm; in that species, the majority of individuals are self-fertile hermaphrodites, while a small proportion of the population in each generation are males (Griffin *et al.*, 2001). Thus, heterorhabditids and at least one *Steinernema* sp. can develop in a host when a single IJ invades, while most steinernematids require at least two individuals to colonize the host before multiplication can occur.

Initially, eggs are laid into the host medium. In older females or hermaphrodites, eggs hatch in the uterus, and the developing juveniles consume the parental tissues – a process known as 'endotokia matricida' (Johnigk and Ehlers, 1999). This use of the parental tissues results in rather efficient conversion of insect biomass to IJ biomass. Juveniles developing with adequate food supply mature to adults, while those developing in crowded conditions with limited food resources arrest as IJs. Hundreds of thousands of IJs may be produced in larger hosts. These emerge from the insect cadaver over a period of days or weeks, to begin the search for new hosts (Fig. 2.1).

Newly emerged IJs retain the moulted second-stage cuticle as a sheath. Particularly in *Heterorhabditis* spp., the sheath may help in protection against desiccation, freezing, and fungal pathogens (Timper and Kaya, 1989; Campbell and Gaugler, 1991a; Wharton and Surrey, 1994). The loose-fitting sheath of steinernematids is soon lost as the nematode moves through soil, while the tighter-fitting heterorhabditid sheath is not so easily lost (Campbell and Gaugler, 1991b; Dempsey and Griffin, 2003).

2.2. Nematode–Bacterial Symbiosis

Knowledge of the nematode–bacterial symbiosis is essential to understanding the pathogenicity of the complex for target insects, and is fundamental for successful mass production. Both partners benefit from the association: the bacteria are largely responsible for the rapid death of the insect, they provide a suitable nutritive medium for nematode growth and reproduction, and suppress competing organisms by the production of antibiotics. The nematode

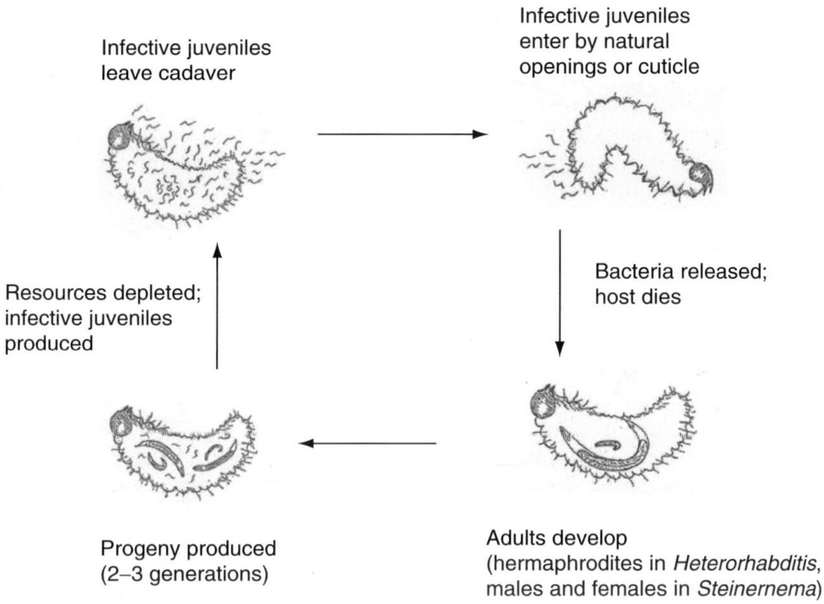

Fig. 2.1. Simplified life cycle of entomopathogenic nematodes (EPNs) (*Steinernema* spp. and *Heterorhabditis* spp.).

protects the bacteria in the external environment, vectors them into the insect haemocoel and, in some associations, inhibits the insect immune response.

The nematode–bacterial interaction is not obligate: each partner can be cultured separately, but when combined they present a high degree of specificity. The paradox of 'apparent independence and high specificity' is one of the fascinating aspects of the relationship. The symbionts occupy two different ecological niches or states in the life cycle, and thus interact with the nematode at two levels. The first is a phoretic state where the bacteria are retained in, and interact with, the intestine of the non-feeding IJ, apparently without any significant multiplication. *Xenorhabdus* occur in a special intestinal vesicle of *Steinernema* IJs (Bird and Akhurst, 1983), while *Photorhabdus* are mainly located in the anterior part of the intestine in *Heterorhabditis* (Boemare *et al.*, 1996). The second state is a vegetative one, when the bacteria overcome the insect host's defence system, allowing them to multiply unrestrained inside the infected insects.

2.2.1. Bacterial taxonomy and co-speciation with nematodes

Xenorhabdus and *Photorhabdus* are members of the γ-subclass of Proteobacteria and belong to the family Enterobacteriaceae (Boemare, 2002). Since their original description, they have been considered to be Gram-negative, facultatively anaerobic rods, as are all the Enterobacteriaceae. However, both genera are negative for nitrate reductase, and *Xenorhabdus* are negative for catalase: two major positive characters of this family. Moreover, recent results seem to indicate that some groups are strictly aerobic. These recent data, which are incompatible with the classical bacteriological canons, may result in a revision of the description of both genera (Pagès and Boemare, 2003, unpublished data).

There is a close relationship between the taxonomy of the symbiont species and of their nematode hosts. In general, for each species of nematode there is a specific association with a species or subspecies of bacteria (Fischer-Le Saux *et al.*, 1998; Boemare and Akhurst, 2001, 2003; Akhurst and Boemare, 2003). However, some nematode species share the same species of bacterium. For example, *Xenorhabdus bovienii* is associated with four species of *Steinernema*, and *X. poinarii* is associated with two (Table 2.1). More rarely, some bacterial species share the same nematode species; for example, *Photorhabdus luminescens* and *P. temperata* are both associated with the *H. bacteriophora* group (Table 2.1). The specificity of the nematode–bacterial association can be considered to be the result of partial co-speciation, together with some recent acquisitions.

2.2.2. Phenotypic variation

Phenotypic or phase variation occurs for every strain of symbiont known so far. The

Table 2.1. Correspondence between taxonomy of the bacteria and of the nematodes.

Xenorhabdus spp.	Genotype[a]	*Steinernema* spp.[b]
X. nematophila	No 1, 2 and 3	*S. carpocapsae*
X. japonica	No 18	*S. kushidai*
X. beddingii	No 4	*Steinernema* sp.
X. bovienii	No 5 and 7	*S. feltiae*
	No 5 and 7	*S. affine*
	No 7 and 8	*S. kraussei*
	No 6	*S. intermedium*
X. poinarii	No 17	*S. cubanum*
		S. glaseri
Xenorhabdus spp.	No 9	*S. karii*
		S. monticolum
	No 10	*S. serratum*
	No 10 and 11	*S. longicaudum*
	No 12	*S. siamkayai*
	No 13	*S. ceratophorum*
	No 15	*S. arenarium* (syn.: *S. anomalae*)
	No 20	*S. rarum*
	No 21	*S. puertoricense*
	No 23	*S. abbasi*
	No 24	*S. scapterisci*
	No 25	*S. riobrave*
Photorhabdus spp.	Genotype[c]	*Heterorhabditis* spp.
P. luminescens luminescens	No 10	*H. bacteriophora* group Brecon[d]
P. luminescens laumondii	No 13 and 28	*H. bacteriophora* group HP88[d]
P. luminescens akhurstii	No 12 and 27	*H. indica*
P. luminescens	No 11	*Heterorhabditis* sp.
P. temperata temperata	No 14	*H. megidis* Palaearctic group
P. temperata	No 14b	*H. downesi*
P. temperata	No 15	*H. megidis* Nearctic group
	No 16	*H. bacteriophora* group NC[d]
	No 17	*H. zealandica*

[a]New numbering using the PCR-RFLP of 16S rRNA genes methodology of Fischer Le Saux *et al.* (1998) but updated to take account of new genotypes in course of identification (Pagès, Brunel and Boemare, Montpellier, France, unpublished data).
[b]N. Boemare and P. Stock, unpublished.
[c]Numbering of the genotype follows that of Fischer-Le Saux *et al.* (1998), except for symbionts of the Irish strains of *H. downesi* that have the provisional No 14b.
[d]According to Boemare (2002), the NC strain of a nematode identified in the past as *H. bacteriophora* harbours *P. temperata* and not a subspecies of *P. luminescens* as other symbionts of *H. bacteriophora*. The re-isolation of this group in nature is required to control for possible confusion in the previous sampling.

initial isolate from the wild nematode, termed the Phase I variant, possesses two major properties: dye adsorption and antibiotic production (Akhurst, 1980). After *in vitro* subculture, there appears a variable proportion of clones, called Phase II variants, that not only have lost these two properties but are also affected in a range of other phenotypic characters, including colony and cell morphology, motility, endo- and exo-enzymatic activity, respiratory enzymes and secondary metabolites (Boemare and Akhurst, 1988; Smigielski *et al.*, 1994; Givaudan *et al.*, 1995). For every character that can be evaluated the difference between phase variants is quantitative (e.g. the emitted luminescence of the *Photorhabdus* Phase II variant is about 1% that of the Phase I variant) and is probably under the control of a genetic regulatory mechanism that is not yet understood (Forst *et al.*, 1997; Forst and Clarke, 2002). For the purposes of numerical taxonomy, any character that is recorded as positive for any variant should be considered as a positive character of that strain.

What is the ecological role of Phase II? Although such variants may also kill the insect host and are capable of colonizing the IJs, they have never been found associated with naturally occurring nematodes (Akhurst and Boemare, 1990). Moreover, some *Photorhabdus* Phase II variants may be deleterious for their original *Heterorhabditis* (Ehlers *et al.*, 1990). So far, there is no consistent ecological explanation of the significance of Phase II variants, though it has been suggested that they represent a survival form (Smigielski *et al.*, 1994).

2.2.3. Pathogenicity

The pathogenic process depends on characteristics of each of the three partners of the interaction: the insect, nematode and bacteria. It is influenced by insect resistance (including humoral and cellular defences) and by virulence factors of the bacteria and of the nematode acting separately or together to overcome the defence system (reviewed by Dowds and Peters, 2002). Pathogenicity, as evaluated by injection into the insect haemocoel, varies between insects. Differences in pathogenicity among bacterial species have also been recorded, principally in larvae of the wax moth *Galleria mellonella*. Thus, most species of *Xenorhabdus* are highly pathogenic, with LD_{50} of less than 20 cells (Akhurst and Dunphy, 1993). In contrast, *X. poinarii* and the symbiont of *Steinernema scapterisci* have very little pathogenicity for *G. mellonella* when injected alone ($LD_{50} > 5000$ cells), and their axenic nematode hosts, *S. glaseri* and *S. scapterisci*, are also not pathogenic when injected alone. Re-combination of both partners re-establishes the pathogenicity towards *G. mellonella* (Akhurst, 1986; Bonifassi *et al.*, 1999), illustrating the need for cooperation between both partners to kill the insect. Most *Photorhabdus* strains examined to date have been reported to be entomopathogenic, the LD_{50} usually being < 100 cells (Akhurst and Boemare, 1990). However, some non-pathogenic strains of *Photorhabdus temperata* have been found recently (Pagès, Gaudriault, 2003, unpublished data).

The recent discovery of some strains of *Photorhabdus* that are pathogenic to insects by ingestion (ffrench-Constant and Bowen, 1999) has resulted in an enhanced level of interest in these bacteria. Although development of the bacteria in the insect gut has not yet been reported, some symbionts produce a toxin that is active on the intestinal epithelium from both sides (gut lumen as well as the haemocoel) (Blackburn *et al.*, 1998). *P. luminescens* possesses toxins, called Tc or toxin complex, that are orally active against Coleoptera and Lepidoptera (ffrench-Constant and Bowen, 2000). Such toxins have also been identified during the sequencing of the genome of another strain of *Photorhabdus* (Duchaud *et al.*, 2003), and in *Serratia entomophila* (Hurst *et al.*, 2000). Several other virulence factors participate in the pathogenicity of *Photorhabdus* and *Xenorhabdus* (Dowds and Peters, 2002; Forst and Clarke, 2002), including motility (Givaudan *et al.*, 1995, 1996; Givaudan and Lanois, 2000) and haemolysins (Brillard *et al.*, 2001, 2002, 2003).

2.2.4. Isolation of symbionts and maintenance of monoxeny

Only one natural symbiont species has been found in the gut of the IJs of any one nematode species, and this is true for all the species of *Steinernema* and *Heterorhabditis* collected throughout the world over the last 30 years with the exception of the *Heterorhabditis bacteriophora* group, strains of which are associated with two *Photorhabdus* spp. Some nematode species carry fewer bacterial cells, and carry them in only a proportion of the IJs. For instance, *Steinernema scapterisci* carries significantly less symbionts than *S. riobrave* and *S. carpocapsae* (Sicard *et al.*, 2003). Therefore, to be sure of isolating symbiont clones in good condition, the nematode sample from which they are isolated should contain a reasonable number of IJs (*c.* 100–1000).

Sometimes bacterial strains other than the symbionts have been found associated with *Steinernema* (Aguillera *et al.*, 1993) or with *Heterorhabditis* (Jackson *et al.*, 1995; Babic *et al.*, 2000), mainly following prolonged maintenance in laboratories. It was shown that they were mostly contaminants of the cuticle (Bonifassi *et al.*, 1999) and there is no definitive evidence that any are inhabitants of the intestine. Recently, sporangia of *Paenibacillus* spp. have been noted adhering to the cuticle of *Heterorhabditis* spp. IJs, and it is suggested that the bacteria exploit the nematode as a phoretic host (Enright *et al.*, 2003).

Mechanisms involved in the specificity of the association between the nematode and its symbiont operate both in the cadaver and in the IJ. Large amounts of antimicrobial organic compounds are produced during *in vivo* multiplication of *Xenorhabdus* spp. and *Photorhabdus* spp. (Webster *et al.*, 2002), preventing global microbial contamination. Bacteriocins active against closely related bacteria such as other species of *Xenorhabdus*, *Photorhabdus* and the nearest genus, *Proteus*, are also produced (Boemare *et al.*, 1992; Thaler *et al.*, 1995). So antimicrobial barriers may play an important role in protecting the specificity of the symbiosis by eliminating microbial competitors, though some bacteria, such as the *Paenibacillus* spp. mentioned above, appear to be resistant to these antimicrobials (Enright and Griffin, 2004). Additionally, the symbiotic bacteria must be retained in the monoxenic nematodes by an active recognition process, as illustrated by the fact that aposymbiotic (without symbiont) *Steinernema* did not retain any non-symbiotic bacteria, and rejected any symbiont that was not their natural partner (Sicard *et al.*, 2003). The nature of this recognition process has yet to be discovered, but an important step towards understanding the molecular mechanism of the association was obtained by disrupting the *rpo*S gene of *X. nematophila* (Vivas and Goodrich-Blair, 2001). This gene encodes the sigma S factor that controls interactions with hosts in other Gram-negative bacteria. Vivas and Goodrich-Blair (2001) obtained a mutant that was able to induce pathogenesis in insects, but was unable to mutualistically colonize nematode intestines, and such a mutant should prove to be a useful tool for further studies.

2.2.5. Importance of the bacterial symbiont

Recently, Sicard *et al.* (2003) undertook gnotobiological experiments demonstrating the importance of the symbiont for the nematode. Aposymbiotic nematodes inoculated into insect hosts had reduced fitness relative to symbiotic nematodes, showing the importance of the bacteria for efficient reproduction of their corresponding nematode host. This was demonstrated for three species (*S. carpocapsae*, *S. scapterisci* and *S. riobrave*); the most extreme results were those with *S. riobrave*, which did not reproduce without its symbiotic bacteria (Sicard *et al.*, 2003). These results, together with previous ones, such as those showing that combination of *S. scapterisci* and its symbiont re-established the pathogenicity of the complex towards *G. mellonella* and gave the best yields of IJs when produced in this insect or *in vitro* on artificial diet

(Bonifassi *et al.*, 1999), demonstrate the importance of the symbiont for the nematode host. In addition, although development of non-infective stages of *S. scapterisci* occurred on all *Xenorhabdus* spp., the development of IJs to the fourth stage ('dauer recovery') was significantly delayed with *Xenorhabdus* other than the natural symbiont. This development was restored when the culture medium was supplemented with cell-free filtrates from the *Xenorhabdus* native strain (Grewal *et al.*, 1997).

Thus, apart from their pathogenicity for insects, the role played by the bacteria is possibly a nutritional one or the production of a food signal (hormonal). This signal is apparently essential for nematode development, as the experiments of Grewal *et al.* (1997) suggest. This is also indirectly demonstrated by the fact that the symbiotic bacteria are required for successful production of nematodes in bioreactors (see Chapter 3, this volume). Like many soil-dwelling rhabditids, *Steinernema* and *Heterorhabditis* are microbivorous grazers. Nevertheless, the specific requirements provided by their specific bacteria are still unknown.

2.3. Infective Juvenile (IJ) Behaviour

The IJ is morphologically, physiologically and behaviourally adapted to its role in transmission – and hence to its acquired role as the active ingredient of a biological pesticide. A thorough understanding of the materials used is essential for predicting efficacy of any pest management product. As EPNs are active organisms that move, seek their hosts and prefer some hosts to others, a treatment of their behaviour, as it relates to efficacy, follows. IJs have a pair of sensory organs, the amphids, at their anterior end, which are used in detecting cues potentially associated with hosts, and a behavioural repertoire appropriate to their role in host-finding. Their behaviours are divided into four categories that are not mutually exclusive: dispersal, foraging strategies, host discrimination and infection.

2.3.1. Dispersal

Among the many behavioural characters that impact the biocontrol potential, the location of the IJ within the soil profile is one of the most important (Lewis, 2002). To provide control, the parasite and the host must be in the same place at the same time. The location of an IJ is dictated by how it disperses after application and by the method of application. Since application technology is covered elsewhere, we will concentrate on how the IJs disperse. The dispersal behaviours and capabilities of EPNs vary among species, strains and even among individuals emerging from the same infection (Lewis, 2002).

EPNs disperse horizontally and vertically after application. The studies that have been conducted on dispersal phenomena can be grouped into laboratory studies that measured EPN movement through various media, field studies that recorded the distribution of native EPN populations that make inferences about dispersal and field studies that re-isolated EPNs after they were applied. Different kinds of information are provided by each of these types of studies.

Laboratory studies are the easiest to conduct and have been carried out on the widest variety of species and strains; yet one must take care in extrapolating these results to field populations. Interspecies variation has been measured in several studies. *S. carpocapsae* IJs move upwards in soil columns (Georgis and Poinar, 1983; Schroeder and Beavers, 1987), whereas *S. glaseri* and *H. bacteriophora* move downwards, but they also disperse throughout the soil column. Studies of movement through soil arenas have shown that *Heterorhabditis* spp. tended to migrate farther than did *Steinernema* spp. (Westerman, 1995; Downes and Griffin, 1996). Koppenhöfer and Kaya (1996) suggested that differential distribution patterns may allow some species, such as *S. glaseri* and *S. carpocapsae*, to coexist since they would not compete for the same hosts.

While laboratory studies may be limited in their ability to predict behaviour in the field, there are aspects of dispersal behaviour that are best addressed in a small controlled environment. Variation with age, variation among IJs emerging from the same cadaver and the impact of harvesting IJs in water are three examples. Lewis *et al.* (1995) compared changes in several aspects of IJ behaviour as they aged in water and found that the behaviours of *H. bacteriophora*, including locomotory rate on agar plates, degraded at a faster rate than those of *S. carpocapsae* or *S. glaseri*. They also found that the nictation rate of *S. carpocapsae* declined with age. Differences among individuals emerging from the same cadaver represent a source of variation usually not considered. IJs emerge from host cadavers for up to 3 weeks in some species, and several differences among those emerging first versus last have been shown. In *S. glaseri* male IJs emerge before females, and those males emerging first are more responsive to host cues than are females (Lewis and Gaugler, 1994). This is not the case for *S. carpocapsae* or *S. feltiae*, where males did not emerge first (Lewis, 2002). Male IJs of some EPN species are more responsive to host cues (Grewal *et al.*, 1993c) and disperse quicker (Lewis and Gaugler, 1994) than females. These findings gave rise to the 'male colonization hypothesis', which suggests that males establish infections before females. In the only direct test of this hypothesis to date, however, Stuart *et al.* (1998) found no evidence of earlier invasion by male than female IJs of *S. glaseri*, despite the documented behavioural differences. *H. megidis* IJs that emerged early differed in their behaviour, but also differed in their tolerance of temperature extremes and desiccation from those that emerged later (O'Leary *et al.*, 1998). Ryder and Griffin (2003) showed that the infectivity of *H. megidis* IJs produced in the first and second generation differed, and that infectivity of juveniles was further affected by the extent of crowding in the insect cadaver in which they developed. Shapiro and Glazer (1996) compared the dispersal of EPNs emerging from their host cadaver into sand with nematodes applied in water and found that *H. bacteriophora* and *S. carpocapsae* directly moving from their host cadaver to the soil had greater movement. How these findings relate to nematodes applied as products is impossible to know, but these findings may allow development of production technologies to favour particular characteristics.

Several field studies describe the distribution of EPNs. In the vertical plane, natural populations of *S. carpocapsae* were found in the upper 1–2 cm of soil, whereas *H. bacteriophora* was distributed throughout the upper 8 cm of soil (Campbell *et al.*, 1995). Ferguson *et al.* (1995) compared the vertical distributions of three species after application. *S. carpocapsae* and an undescribed *Steinernema* sp. remained near the soil surface, while *H. bacteriophora* strains moved to greater depths. Horizontal distribution studies on natural populations show that EPNs are patchily distributed, with a variable degree of patchiness among species (Stuart and Gaugler, 1994; Campbell *et al.*, 1995; Strong *et al.*, 1996). In general, *H. bacteriophora* populations are patchier than either *S. carpocapsae* or *S. feltiae* populations (Campbell *et al.*, 1998). Host distribution, nematode behaviour and soil factors will all contribute to the spatial distribution of the nematodes.

Populations of *H. bacteriophora*, which were applied in a homogeneous layer, had a patchy distribution that mirrored native populations within 2 months of application (Campbell *et al.*, 1998), but the mechanism – whether due to recycling in patchily distributed hosts or redistribution of the applied nematodes – was unknown. Wilson *et al.* (2002), while studying the possibility of using different spatial application patterns to lengthen nematode persistence, showed that *H. bacteriophora* can move up to 3 m from their point of application.

2.3.2. Foraging strategies

Understanding foraging behaviour is essential to accurate prediction of efficacy for

EPNs because foraging mode predicts where the nematodes will be located and what hosts they are likely to contact (Gaugler *et al.*, 1997). EPN foraging strategies vary along a continuum from ambush to cruise foraging (Lewis *et al.*, 1992; Grewal *et al.*, 1994a; Campbell and Gaugler, 1997). The variation in foraging behaviour among species is considerable.

The way nematodes search for hosts has a direct impact on efficacy because mobile nematodes tend to find sedentary hosts and vice versa. Ambushing nematodes nictate during foraging by raising nearly all of their bodies off the substrate (Fig. 2.2) (Campbell and Gaugler, 1993). Of the commercially available EPN species, *S. carpocapsae* and *S. scapterisci* are the most extreme ambushers and may nictate for hours at a time (Campbell and Gaugler, 1993). Ambushing nematode species are usually associated with highly mobile, surface-dwelling hosts. Cruising nematodes never nictate and probably spend most of the IJ stage moving through the soil. Commercially available cruise foraging species include the *Heterorhabditis* spp. and *S. glaseri* (Lewis, 2002). These species are usually effective against relatively sedentary hosts located throughout the soil column. Some EPN species, e.g. *S. riobrave* and *S. feltiae*, adopt an intermediate foraging strategy (Table 2.2) and have been effective against pests with a range of habits from mobile to sedentary.

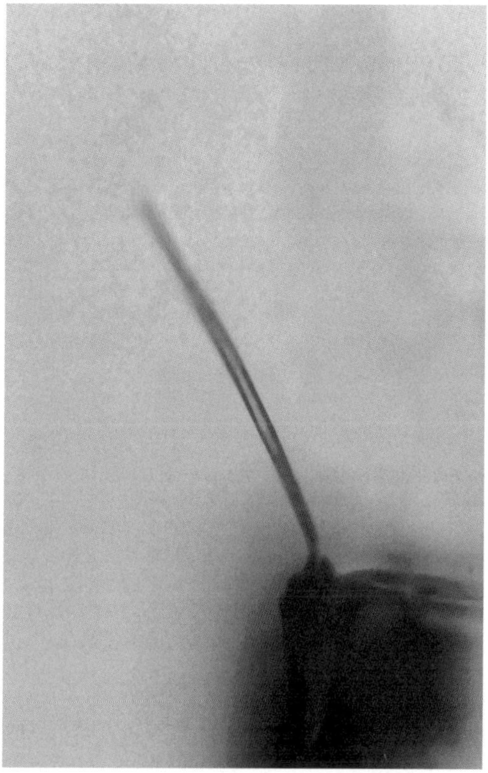

Fig. 2.2. Nictating infective juvenile (IJ) of *Steinernema carpocapsae*. The nematode stands on its tail and waves from side to side. (Photo: Jim Campbell, USDA ARS GMPRC, Kansas, USA.)

2.3.3. Host discrimination

Dispersal and foraging strategy constrain the host range of EPN species indirectly. The IJs themselves discriminate directly among potential hosts. Knowledge of natural host ranges of EPNs could help predict which nematodes would be effective against a particular insect pest. When an EPN is isolated from soil, we are essentially ignorant of its natural host range because of the use of *G. mellonella* as a bait (Bedding and Akhurst, 1975). Current knowledge of natural EPN host ranges is limited to anecdotal accounts of native populations found infecting a host in the field (Peters, 1996). There is also information on potential host range to be gleaned from field trials that test EPN species against particular hosts (Chapters 7–17).

Host recognition behaviour has been studied in a few species of EPNs, and has been measured by recording changes in several behaviours in response to host-related materials. Responses of *H. bacteriophora*, *S. glaseri*, *S. carpocapsae* and *S. scapterisci* to gut contents of four host species suggested consistent host affiliations: infectivity of nematode species to hosts was correlated with their behavioural responses to those hosts (Grewal *et al.*, 1993a). Grewal *et al.* (1993b) also suggested that these EPN species respond differently to excretory

Table 2.2. Foraging strategy and summary of behavioural tests for four species of *Steinernema* (J.F. Campbell, unpublished data).

Steinernema spp.	Foraging strategy[a]	Nictation	Jumping	Dispersal decreased by sand	Ranging to localized search by host contact	Attraction increased by host contact
S. carpocapsae	Ambusher	Yes	Yes	Yes	No	Yes
S. feltiae	Intermediate	No	No	No	No	No
S. riobrave	Intermediate	No	Yes	No	No	No
S. glaseri	Cruiser	No	No	No	Yes	No

[a]Based on attachment to mobile versus immobile host.
Note: For a more complete treatment of IJ foraging behaviour see Lewis (2002) and Campbell *et al.* (2003).

products of various natural and experimental hosts. Lewis *et al.* (1996) studied the behavioural recognition response of *S. carpocapsae* IJs by measuring their response to volatiles from *G. mellonella* larvae following exposure to contact with the cuticle of nine candidate host species. Again, the level of recognition response to different hosts was correlated with the infectivity of the nematodes for those hosts, and also with IJ production per gram of host tissue. Measures of host recognition might be useful in the characterization of new isolates from the field, and a standard testing procedure for assessment of host range could be developed.

2.3.4. Infection behaviours

Once an IJ has located a host and found it acceptable, penetration into the host haemocoel is the next step. Different species use different routes of entry into hosts: via the natural openings (mouth, anus, spiracles) or by penetration through the external cuticle. Wang and Gaugler (1999) compared the penetration behaviour of *S. glaseri* and *H. bacteriophora* into *Popillia japonica* larvae and found that *S. glaseri* penetrated primarily through the gut. *H. bacteriophora* was not efficient at penetrating the gut, presumably because of the thick peritrophic membrane, but penetrated through the intersegmental membranes of the cuticle. Cui *et al.* (1993) found that *S. glaseri* IJs would penetrate through existing holes in the gut made by previous nematodes. Renn (1999) found that *S. feltiae* IJs also followed established routes of penetration in larval houseflies.

Fan and Hominick (1991) suggested that in the 'phased infectivity hypothesis' less than 40% of *S. feltiae* IJs that emerged from a host were infectious at any time, regardless of host availability. Nematodes were assumed to be either infectious or non-infectious, and to convert from one state to the other. Bohan and Hominick (1996, 1997) described short- and long-term interactions between a cohort of IJs and potential hosts that support this idea. However, Campbell *et al.* (1999) found that *S. feltiae* IJs will infect hosts when enough are available, but they also collected data for *H. bacteriophora* that support the phased infectivity hypothesis for this species. Infectivity of *H. megidis* shows an initial increase from time of emergence from the host cadaver, before eventually declining (Griffin, 1996; Dempsey and Griffin, 2002; Ryder and Griffin, 2003), and Griffin (1996) proposed that individual infectious nematodes may have variable levels of infectivity (tendency to infect), as an alternative to the dichotomous (infectious versus non-infectious) phased infectivity hypothesis.

2.4. Ecology

Field studies show that numbers of EPNs recovered from soil decline sharply in a short period following application (Selvan *et al.*, 1993a; Gaugler *et al.*, 1997). Although soil is a relatively buffered environment, IJs may experience stressful conditions such as desiccation and high temperatures, especially at the soil surface immediately after application, while waterlogged soils may develop anoxic conditions. Nematodes in soil also face a variety of diseases and predators. If they are not killed by antagonists or lethal levels of abiotic factors, IJs can survive for months in the soil, and have evolved a suite of adaptations such as high levels of energy reserves and a protective sheath that allow them to persist in this sometimes hostile environment. Consideration of the survival mechanisms of IJs is important for formulation also.

2.4.1. Energy reserves and starvation

The IJ does not feed, but relies on stored energy reserves. Lipids (especially triglycerides) constitute up to 40% of the body weight (Selvan *et al.*, 1993b; Fitters *et al.*, 1999) and are the most important energy reserve, though proteins and the carbohydrates, glycogen and trehalose, also yield energy (Qiu and Bedding, 2000). It is probable that, unless subjected to other mortality factors, IJs will starve to death. Thus, the lifespan is largely determined by the quantity and quality of reserves that it has built up during its prior feeding phase and by the rate at which the reserves are depleted (Qiu and Bedding, 2000). Both the rate of activity and basal metabolic rate – and hence the rate at which reserves are utilized – are affected by ambient conditions, most notably temperature. IJs survive longer at low temperatures, with optimal temperature for survival of most species typically between 5°C and 15°C (Georgis, 1990), though 20°C is optimal for storage of certain tropical strains. The tendency of IJs to become inactive in the absence of stimulation, even when temperature and other conditions permit movement, also favours energy conservation. Foraging strategies have been related to several life history characters that have an impact on survival. Lewis *et al.* (1995) found that *S. carpocapsae*, an ambush forager, had a lower metabolic rate than *H. bacteriophora*. We also find that the products with the longest shelf-life tend to comprise ambush foragers. Foraging strategy also affects the choice of appropriate formulation for species of EPNs. For example, formulation in water-dispersible granules is very successful with the ambush forager *S. carpocapsae*, while the cruise foraging *S. feltiae* and *S. riobrave* rapidly migrate out of the granules (Grewal, 2002). Before starvation reaches critical lethal levels, motility and infectivity of the IJ may have declined (Lewis *et al.*, 1995; Patel *et al.*, 1997b), with the result that viability is not the only indicator of nematode quality.

2.4.2. Abiotic stress

Desiccation and temperature extremes are the most important abiotic factors affecting survival of EPNs (reviewed by Glazer, 2002). Nematodes require free water for movement, and as it disappears they necessarily become inactive. As the environment dries further, water is lost from the nematode body. *Steinernema* and *Heterorhabditis* have relatively limited tolerance of desiccation, and are classed as partial anhydrobiotes. Even partially anhydrobiotic nematodes have lowered energy consumption and increased tolerance to temperature extremes, making induction into this state the Holy Grail of formulation technology (see Chapter 4, this volume). Most studies have concentrated on *S. carpocapsae*, which is noted as one of the more desiccation-tolerant species (Patel *et al.*, 1997a), perhaps related to its tendency to remain near the soil surface, waiting to ambush passing hosts.

Exposure to extremes of temperature is damaging for nematodes, but the extent and nature of damage depends on the duration of exposure. Steinernematids and heterorhabditids tolerate exposure to sub-zero temperatures for several days (Wharton and Surrey, 1994) and, with suitable preconditioning, IJs may be stored indefinitely in liquid nitrogen (Popiel and Vasquez, 1991). This is an important property, allowing the maintenance of genetic stock without the need for repeated subculture and the attendant risk of inadvertant selection (Wang and Grewal, 2002). Temperatures above 30°C inhibit infection and reproduction of several species of EPNs, though others such as *S. riobrave* reproduce at 32°C and infect at up to 39°C (Grewal *et al.*, 1994b). In laboratory assays, IJs of *S. carpocapsae* are killed by short periods (hours) at 40°C (Somasekhar *et al.*, 2002), but an Arkansas isolate of *S. carpocapsae* survived for 2 weeks at 40°C in soil (Gray and Johnson, 1983). Indeed, the limited ability of EPNs to tolerate ultraviolet light, desiccation and high temperature undoubtedly reflects their soil-dwelling evolutionary history.

2.4.3. Biotic stress

In soil, IJs are subject to attack by a variety of microbial and invertebrate antagonists (reviewed by Kaya, 2002). The main natural enemies with the potential to affect the survival of EPNs in soil are predatory mites and collembolans (e.g. Epsky *et al.*, 1988), nematode-trapping fungi (e.g. Poinar and Jansson, 1986) and parasitic fungi that produce adhesive spores (Timper *et al.*, 1991). Little is known about the impact of such organisms on natural or applied populations of EPNs. Indirect evidence for an effect of naturally occurring antagonists on nematode survival comes from the observation that nematodes survived longer when applied to sterilized soil (Ishibashi and Kondo, 1986). Developmental stages of EPNs are also at risk from scavengers attacking the cadavers (Baur *et al.*, 1998), and the fact that some cadavers deter predation by ants (Zhou *et al.*, 2002) suggests that such predation may exert selective pressure.

2.5. Geographical Distribution of Natural Populations

EPNs are very common in cultivated and uncultivated soils, and numerous surveys have documented their occurrence throughout the world (reviewed by Hominick *et al.*, 1996; Hominick, 2002). The level of effort that has been applied to the recovery of EPNs varies, with Europe being the most intensively studied continent. Amongst the species recovered are those with a global distribution: *S. carpocapsae* and *S. feltiae* are widely distributed in temperate regions; *H. bacteriophora* is common in regions with continental and Mediterranean climates; and *H. indica* is found throughout the tropics and subtropics. For some species, the known distribution is much more restricted, e.g. *S. cubanum* and *S. kushidai* are so far known only from Cuba and Japan, respectively.

The distribution of EPNs on a global scale, like that of other taxa, is probably strongly influenced by climate and chance dispersal events, including those associated with human activities. Soil texture, vegetation and availability of suitable hosts are amongst the factors that have been implicated in affecting local distribution patterns. There is growing evidence of preferences of nematode species for certain habitats. For example, *S. affine* is found largely in arable lands and grasslands, and is virtually absent in forests, while *S. kraussei* is commonly found in forests (Hominick, 2002). It is likely that such habitat preferences are at least partly due to host preferences, and the fact that associations with habitat are rather weak probably reflects the lack of strict host specificity in most EPN species (Peters, 1996). More striking is the association of some species with soil of a particular texture, in particular sand. *H. megidis* and *H. indica* are almost exclusively found in sandy soils, resulting in a mainly coastal distribution (Hara *et al.*, 1991; Amarasinghe *et al.*, 1994; Griffin *et al.*, 1994, 2000), and there is some evidence of a similar association for tropical steinernematids (Amarasinghe *et al.*, 1994; Griffin *et al.*, 2000). While laboratory assays are useful in predicting the effect of ecological factors on the potential of inundatively applied nematodes to survive and infect, predictions of whether such applied nematodes will establish as self-renewing populations are best informed by knowledge of the factors affecting the prevalence of natural populations (see Chapter 18, this volume). For example, from the known association of *H. megidis* with sandy soils, it could be predicted that this species is highly unlikely to persist long term in peat or clay soils.

While a grower with little knowledge of the biology of EPNs can apply them in line with the supplier's instructions, even a small amount of knowledge will increase the likelihood of his or her success. Continuing advances by researchers in understanding the complex requirements and strategies of these organisms in their natural environment will lead to the much more efficient targeting and expanded use of EPNs in the future.

References

Aguillera, M.M., Hodge, N.C., Stall, R.E. and Smart, G.C. Jr (1993) Bacterial symbionts of *Steinernema scapterisci*. *Journal of Invertebrate Pathology* 62, 68–72.

Akhurst, R.J. (1980) Morphological and functional dimorphism in *Xenorhabdus* spp., bacteria symbiotically associated with the insect pathogenic nematodes *Neoaplectana* and *Heterorhabditis*. *Journal of General Microbiology* 121, 303–309.

Akhurst, R.J. (1986) *Xenorhabdus nematophilus* subsp. *poinarii*: its interaction with insect pathogenic nematodes. *Systematic and Applied Microbiology* 8, 142–147.

Akhurst, R.J. and Boemare, N.E. (1990) Biology and taxonomy of *Xenorhabdus*. In: Gaugler, R. and Kaya, H.K. (eds) *Entomopathogenic Nematodes in Biological Control*. CRC Press, Boca Raton, Florida, pp. 75–90.

Akhurst, R.J. and Boemare, N.E. (2005) The *Xenorhabdus* genus. In: Krieg, N.R., Staley, J.T. and Brenner, D.J. (eds) *Bergey's Manual of Systematic Bacteriology*. Williams & Wilkins, Baltimore, Maryland.

Akhurst, R.J. and Dunphy, G.B. (1993) Tripartite interactions between symbiotically associated entomopathogenic bacteria, nematodes, and their insect hosts. In: Beckage, N.E., Thompson, S.N. and Federici, B. (eds) *Parasites and Pathogens of Insects*. Academic Press, New York, pp. 1–23.

Amarasinghe, L.D., Hominick, W.M., Briscoe, B.R. and Reid, A.P. (1994) Occurrence and distribution of entomopathogenic nematodes in Sri Lanka. *Journal of Helminthology* 68, 277–286.

Babic, I., Fischer-Le Saux, M., Giraud, E. and Boemare, N.E. (2000) Occurrence of natural dixenic associations between the symbiont *Photorhabdus luminescens* and bacteria related to *Ochrobactrum* spp. in tropical entomopathogenic *Heterorhabditis* spp. (Nematoda: Rhabditida). *Microbiology* 146, 709–718.

Baur, M.E., Kaya, H.K. and Strong, D.R. (1998) Foraging ants as scavengers on entomopathogenic nematode-killed insects. *Biological Control* 12, 231–236.

Bedding, R.A. and Akhurst, R.J. (1975) A simple technique for the detection of insect parasitic rhabditid nematodes in soil. *Nematologica* 21, 109–110.

Bedding, R.A. and Molyneux, A.S. (1982) Penetration of insect cuticle by infective juveniles of *Heterorhabditis* spp. (Heterorhabditidae: Nematoda). *Nematologica* 28, 354–359.

Bird, A.F. and Akhurst, R.J. (1983) The nature of the intestinal vesicle in nematodes of the family Steinernematidae. *International Journal of Parasitology* 13, 599–606.

Blackburn, M., Golubeva, E., Bowen, D. and ffrench-Constant, R.H. (1998) A novel insecticidal toxin from *Photorhabdus luminescens* toxin complex a (Tca), and its histopathological effects on the midgut of *Manduca sexta*. *Applied and Environmental Microbiology* 64, 3036–3041.

Blaxter, M.L., De Ley, P., Garey, J.R., Liu, L.X., Scheldeman, P., Vierstraete, A., Vanfleteren, J.R., Mackey, L.Y., Dorris, M., Frisse, L.M., Vida, J.T. and Thomas, W.K. (1998) A molecular evolutionary framework for the phylum Nematoda. *Nature* 392, 71–75.

Boemare, N.E. (2002) Biology, taxonomy and systematics of *Photorhabdus* and *Xenorhabdus*. In: Gaugler, R. (ed.) *Entomopathogenic Nematology*. CAB International, Wallingford, UK, pp. 35–56.

Boemare, N.E. and Akhurst, R.J. (1988) Biochemical and physiological characterization of colony form variants in *Xenorhabdus* spp. (Enterobacteriaceae). *Journal of General Microbiology* 134, 751–761.

Boemare, N.E. and Akhurst, R.J. (2001) The genera *Photorhabdus* and *Xenorhabdus*. In: Dworkin, M., Falkow, S., Rosenberg, E., Schleifer, K.-H. and Stackebrandt, E. (eds) *The Prokaryotes: An Evolving Electronic Resource for the Microbiological Community*. Springer-Verlag, New York. Available at: http://www.prokaryotes.com

Boemare, N.E. and Akhurst, R.J. (2005) The *Photorhabdus* genus. In: Krieg, N.R., Staley, J.T. and Brenner, D.J. (eds) *Bergey's Manual of Systematic Bacteriology*. Williams & Wilkins, Baltimore, Maryland.

Boemare, N.E., Boyer-Giglio, M.-H., Thaler, J.-O., Akhurst, R.J. and Brehélin, M. (1992) Lysogeny and bacteriocinogeny in *Xenorhabdus nematophilus*, and other *Xenorhabdus* spp. *Applied and Environmental Microbiology* 58, 3032–3037.

Boemare, N.E., Akhurst, R.J. and Mourant, R.G. (1993) DNA relatedness between *Xenorhabdus* spp. (Enterobacteriaceae), symbiotic bacteria of entomopathogenic nematodes, and a proposal to transfer *Xenorhabdus luminescens* to a new genus, *Photorhabdus* gen. nov. *International Journal of Systematic Bacteriology* 43, 249–255.

Boemare, N.E., Laumond, C. and Mauléon, H. (1996) The nematode-bacterium complexes: biology, life cycle, and vertebrate safety. *Biocontrol Science and Technology* 6, 333–345.

Bohan, D.A. and Hominick, W.M. (1996) Investigations on the presence of an infectious proportion amongst populations of *Steinernema feltiae* (Site 76 Strain) infective stages. *Parasitology* 112, 113–118.

Bohan, D.A. and Hominick, W.M. (1997) Long-term dynamics of infectiousness within the infective-stage pool of the entomopathogenic nematode *Steinernema feltiae* (site 76 strain) Filipjev. *Parasitology* 114, 301–308.

Bonifassi, E., Fischer-Le Saux, M., Boemare, N., Lanois, A., Laumond, C. and Smart, G. (1999) Gnotobiological study of infective juveniles and symbionts of *Steinernema scapterisci*: a model to clarify the concept of the natural occurrence of monoxenic associations in entomopathogenic nematodes. *Journal of Invertebrate Pathology* 74, 164–172.

Brillard, J., Ribeiro, C., Boemare, N., Brehélin, M. and Givaudan, A. (2001) Two distinct hemolytic activities in *Xenorhabdus nematophila* are active against immunocompetent insect cells. *Applied and Environmental Microbiology* 67, 2515–2525.

Brillard, J., Duchaud, E., Boemare, N., Kunst, F. and Givaudan, A. (2002) The PhlA hemolysin from the entomopathogenic bacterium *Photorhabdus luminescens* belongs to the two-partner secretion family of hemolysins. *Journal of Bacteriology* 184, 3871–3878.

Brillard, J., Boyer-Giglio, M.-H., Boemare, N. and Givaudan, A. (2003) Holin locus characterisation from lysogenic *Xenorhabdus nematophila* and its involvement in *Escherichia coli* SheA haemolytic phenotype. *FEMS Microbiology Letters* 218, 107–113.

Campbell, L.R. and Gaugler, R. (1991a) Mechanisms for exsheathment of entomopathogenic nematodes. *International Journal of Parasitology* 21, 219–224.

Campbell, L.R. and Gaugler, R. (1991b) Role of the sheath in desiccation tolerance of two entomopathogenic nematodes. *Nematologica* 37, 324–332.

Campbell, J.F. and Gaugler, R. (1993) Nictation behaviour and its ecological implications in the host search strategies of entomopathogenic nematodes (Heterorhabditidae and Steinernematidae). *Behaviour* 126, 155–169.

Campbell, J.F. and Gaugler, R. (1997) Inter-specific variation in entomopathogenic nematode foraging strategy: dichotomy or variation along a continuum? *Fundamental and Applied Nematology* 20, 393–398.

Campbell, J.F., Lewis, E.E., Yoder, F. and Gaugler, R. (1995) Entomopathogenic nematode (Heterorhabditidae and Steinernematidae) seasonal population dynamics and impact on insect populations in turfgrass. *Biological Control* 5, 598–606.

Campbell, J.F., Orza, G., Yoder, F., Lewis, E.E. and Gaugler, R. (1998) Entomopathogenic nematode distribution in turfgrass: variation among sites, correlation with *Popillia japonica* larvae and edaphic factors, and influence of inoculative releases. *Entomologia Experimentalis et Applicata* 86, 1–11.

Campbell, J.F., Koppenhöfer, A.M., Kaya, H.K. and Chinnasri, B. (1999) Are there temporarily non-infectious dauer stages in entomopathogenic nematode populations: a test of the phased infectivity hypothesis. *Parasitology* 118, 499–508.

Campbell, J.F., Lewis, E.E., Stock, S.P., Nadler, S. and Kaya, H.K. (2003) Evolution of host search strategies in entomopathogenic nematodes (Nematoda: Steinernematidae). *Journal of Nematology* 35, 142–145.

Cui, L., Gaugler, R. and Wang, Y. (1993) Penetration of steinernematid nematodes (Nematoda: Steinernematidae) into Japanese beetle larvae, *Popillia japonica* (Coleoptera: Scarabaeidae). *Journal of Invertebrate Pathology* 62, 73–78.

Dempsey, C.M. and Griffin, C.T. (2002) Phased activity in *Heterorhabditis megidis* infective juveniles. *Parasitology* 124, 605–613.

Dempsey, C.M. and Griffin, C.T. (2003) The infectivity and behaviour of exsheathed and ensheathed *Heterorhabditis megidis* infective juveniles. *Nematology* 5, 49–53.

Dix, I., Burnell, A.M., Griffin, C.T., Joyce, S.A. and Nugent, J.M. (1992) The identification of biological species in the genus *Heterorhabditis* (Nematoda: Heterorhabditidae) by cross breeding second-generation amphimictic adults. *Parasitology* 104, 509–518.

Dowds, B.C.A. and Peters, A. (2002) Virulence mechanisms. In: Gaugler, R. (ed.) *Entomopathogenic Nematology*. CAB International, Wallingford, UK, pp. 79–98.

Downes, M.J. and Griffin, C.T. (1996) Dispersal behaviour and transmission strategies of the entomopathogenic nematodes *Heterorhabditis* and *Steinernema*. *Biocontrol Science and Technology* 6, 347–356.

Duchaud, E., Rusniok, C., Frangeul, L., Buchrieser, C., Givaudan, A., Taourit, S., Bocs, S., Boursaux-Eude, C., Chandler, M., Charles, J.-F., Dassa, E., Derose, R., Derzelle, S., Freyssinet, G., Gaudriault, S., Médigue, C., Lanois, A., Powell, K., Siguier, P., Vincent, R., Wingate, V., Zouine, M., Glaser, P., Boemare, N., Danchin, A. and Kunst, F. (2003) The genome sequence of the entomopathogenic bacterium *Photorhabdus luminescens*. *Nature Biotechnology* 21, 1307–1313.

Ehlers, R.-U., Stoessel, S. and Wyss, U. (1990) The influence of phase variants of *Xenorhabdus* spp. and *Escherichia coli* (Enterobacteriaceae) on the propagation of entomopathogenic nematodes of the genera *Steinernema* and *Heterorhabditis*. *Revue de Nématologie* 13, 417–424.

Enright, M.R. and Griffin, C.T. (2004) Specificity of association between *Paenibacillus* spp. and the entomopathogenic nematodes, *Heterorhabditis* spp. *Microbial Ecology* 48, 412–421.

Enright, M.R., McInerney, J.O. and Griffin, C.T. (2003) Characterization of endospore-forming bacteria associated with entomopathogenic nematodes, *Heterorhabditis* spp., and description of *Paenibacillus nematophilus* sp. nov. *International Journal of Systematic and Evolutionary Microbiology* 53, 435–441.

Epsky, N.D., Walter, D.E. and Capinera, J.L. (1988) Potential role of nematophagus microarthropods as biotic mortality factors of entomogenous nematodes (Rhabditida: Steinernematidae, Heterorhabditidae). *Journal of Economic Entomology* 81, 821–825.

Fan, X. and Hominick, W.M. (1991) Effects of low storage temperature on survival and infectivity of two *Steinernema* species (Nematoda: Steinernematidae). *Revue de Nématologie* 14, 407–412.

Ferguson, C.S., Schroeder, P.C. and Shields, E.J. (1995) Vertical distribution, persistence, and activity of entomopathogenic nematodes (Nematoda: Heterorhabditidae and Steinernematidae) in alfalfa snout beetle (Coleoptera: Curculionidae) infested fields. *Environmental Entomology* 24, 149–158.

ffrench-Constant, R.H. and Bowen, D. (1999) *Photorhabdus* toxins: novel biological insecticides. *Current Opinion in Microbiology* 2, 284–288.

ffrench-Constant, R.H. and Bowen, D.J. (2000) Novel insecticidal toxins from nematode-symbiotic bacteria. *Cellular and Molecular Life Sciences 57*, 828–833.

Fischer-Le Saux, M., Mauleon, H., Constant, P., Brunel, B. and Boemare, N. (1998) PCR-ribotyping of *Xenorhabdus* and *Photorhabdus* isolates from the Caribbean region in relation to the taxonomy and geographic distribution of their nematode hosts. *Applied and Environmental Microbiology* 64, 4246–4254.

Fitters, P.F.L., Patel, M.N., Griffin, C.T. and Wright, D.J. (1999) Fatty acid composition of *Heterorhabditis* sp. during storage. *Comparative Biochemistry and Physiology B* 124, 81–88.

Forst, S. and Clarke, D.J. (2002) Bacteria-nematode symbiosis. In: Gaugler, R. (ed.) *Entomopathogenic Nematology*. CAB International, Wallingford, UK, pp. 57–77.

Forst, S., Dowds, B., Boemare, N. and Stackebrandt, E. (1997) *Xenorhabdus* spp. and *Photorhabdus* spp.: bugs that kill bugs. *Annual Review of Microbiology* 51, 47–72.

Gaugler, R., (ed.) (2002) *Entomopathogenic Nematology*. CAB International, Wallingford, UK, 388 pp.

Gaugler, R. and Kaya, H.K, (eds) (1990) *Entomopathogenic Nematodes in Biological Control.* CRC Press, Boca Raton, Florida, 365 pp.

Gaugler, R., Lewis, E.E. and Stuart, R.J (1997) Ecology in the service of biological control: the case of entomopathogenic nematodes. *Oecologia* 109, 483–489.

Georgis, R. (1990) Formulation and application technology. In: Gaugler, R. and Kaya, H.K. (eds) *Entomopathogenic Nematodes in Biological Control.* CRC Press, Boca Raton, Florida, pp. 173–191.

Georgis, R. and Poinar, G.O. Jr (1983) Effect of soil texture on the distribution and infectivity of *Neoaplectana carpocapsae* (Nematoda: Steinernematidae). *Journal of Nematology* 15, 308–311.

Givaudan, A. and Lanois, A. (2000) *flhDC*, the flagellar master operon of *Xenorhabdus nematophilus*: requirement for motility, lipolysis, extracellular hemolysis, and full virulence in insects. *Journal of Bacteriology* 182, 107–115.

Givaudan, A., Baghdiguian, S., Lanois, A. and Boemare, N. (1995) Swarming and swimming changes concomitant with phase variation in *Xenorhabdus nematophilus. Applied and Environmental Microbiology* 61, 1408–1413.

Givaudan, A., Lanois, A. and Boemare, N. (1996) Cloning and nucleotide sequence of a flagellin encoding genetic locus from *Xenorhabdus nematophilus*: phase variation leads to differential transcription of two flagellar genes (*FliCD*). *Gene* 183, 243–253.

Glazer, I. (2002) Survival biology. In: Gaugler, R. (ed.) *Entomopathogenic Nematology.* CAB International, Wallingford, UK, pp. 169–187.

Gray, P.A. and Johnson, D.T. (1983) Survival of the nematode *Neoaplectana carpocapsae* in relation to soil temperature, moisture and time. *Journal of the Georgia Entomological Society* 18, 454–460.

Grewal, P.S. (2002) Formulation and application technology. In: Gaugler, R. (ed.) *Entomopathogenic Nematology.* CAB International, Wallingford, UK, pp. 265–287.

Grewal, P.S., Gaugler, R. and Lewis, E.E. (1993a) Host recognition behavior by entomopathogenic nematodes during contact with insect gut contents. *Journal of Parasitology* 79, 495–503.

Grewal, P.S., Gaugler, R. and Selvan, S. (1993b) Host recognition by entomopathogenic nematodes: behavioral response to contact with host feces. *Journal of Chemical Ecology* 19, 1219–1231.

Grewal, P.S., Selvan, S., Lewis, E.E. and Gaugler, R. (1993c) Male insect-parasitic nematodes: a colonizing sex. *Experientia* 49, 605–608.

Grewal, P.S., Lewis, E.E., Campbell, J.F. and Gaugler, R. (1994a) Host finding behavior as a predictor of foraging strategy for entomopathogenic nematodes. *Parasitology* 108, 207–215.

Grewal, P.S., Selvan, S. and Gaugler, R. (1994b) Thermal adaptation of entomopathogenic nematodes: niche breadth for infection, establishment, and reproduction. *Journal of Thermal Biology* 19, 245–253.

Grewal, P.S., Matsuura, M. and Converse, V. (1997) Mechanisms of specificity of association between the nematode *Steinernema scapterisci* and its symbiotic bacterium. *Parasitology* 114, 483–488.

Griffin, C.T. (1996) Effects of prior storage conditions on the infectivity of *Heterorhabditis* sp. (Nematoda: Heterorhabditidae). *Fundamental and Applied Nematology* 19, 95–102.

Griffin, C.T., Joyce, S.A., Dix, I., Burnell, A.M. and Downes, M.J. (1994) Characterisation of the entomopathogenic nematode *Heterorhabditis* (Nematoda: Heterorhabditidae) from Ireland and Britain by molecular and cross-breeding techniques, and the occurrence of the genus in these islands. *Fundamental and Applied Nematology* 17, 245–253.

Griffin, C.T., Chaerani, R., Fallon, D., Reid, A.P. and Downes, M.J. (2000) Occurrence and distribution of the entomopathogenic nematodes *Steinernema* spp. and *Heterorhabditis indica* in Indonesia. *Journal of Helminthology* 74, 143–150.

Griffin, C.T., O'Callaghan, K. and Dix, I. (2001) A self-fertile species of *Steinernema* from Indonesia: further evidence of convergent evolution amongst entomopathogenic nematodes? *Parasitology* 122, 181–186.

Hara, A.H., Gaugler, R., Kaya, H.K. and Lebeck, L.M. (1991) Natural populations of entomopathogenic nematodes (Rhabditida: Heterorhabditidae, Steinernematidae) from the Hawaiian islands. *Environmental Entomology* 20, 211–216.

Hominick, W.M. (2002) Biogeography. In: Gaugler, R. (ed.) *Entomopathogenic Nematology.* CAB International, Wallingford, UK, pp. 115–143.

Hominick, W.M., Reid, A.P., Bohan, D.A. and Briscoe, B.R. (1996) Entomopathogenic nematodes: biodiversity, geographical distribution and the convention on biological diversity. *Biocontrol Science and Technology* 6, 317–331.

Hurst, M.R.H., Glare, T.R., Jackson, T.A. and Ronson, C.W. (2000) Plasmid-located pathogenicity determinants of *Serratia entomophila,* the causal agent of amber disease of grass grub, show similarity to the insecticidal toxins of *Photorhabdus luminescens. Journal of Bacteriology* 182, 5127–5138.

Ishibashi, N. and Kondo, E. (1986) A possible quiescence of the applied entomogenous nematode, *Steinernema feltiae,* in soil. *Japanese Journal of Nematology* 16, 66–67.

Jackson, T.J., Wang, H., Nugent, M.J., Griffin, C.T., Burnell, A.M. and Dowds, B.C.A. (1995) Isolation of insect pathogenic bacteria, *Providencia rettgeri,* from *Heterorhabditis* spp. *Journal of Applied Bacteriology* 78, 237–244.

Johnigk, S.-A. and Ehlers, R.-U. (1999) *Endotokia matricida* in hermaphrodites of *Heterorhabditis* spp. and the effect of the food supply. *Nematology* 1, 717–726.

Kaya, H. (2002) Natural enemies and other antagonists. In Gaugler, R. (ed.) *Entomopathogenic Nematology.* CAB International, Wallingford, UK, pp. 189–203.

Koppenhöfer, A.M. and Kaya, H.K. (1996) Coexistence of two steinernematid nematode species (Rhabditida: Steinernematidae) in the presence of two host species. *Applied Soil Ecology* 4, 221–230.

Lewis, E. (2002) Behavioural ecology. In: Gaugler, R. (ed.) *Entomopathogenic Nematology.* CAB International, Wallingford, UK, pp 205–223.

Lewis, E.E. and Gaugler. R. (1994) Entomopathogenic nematode sex ratio relates to foraging strategy. *Journal of Invertebrate Pathology* 64, 238–242.

Lewis, E.E., Gaugler, R. and Harrison, R. (1992) Entomopathogenic nematode host finding: response to host contact cues by cruise and ambush foragers. *Parasitology* 105, 309–319.

Lewis, E.E., Selvan, S., Campbell, J.F. and Gaugler, R. (1995) Changes in foraging behaviour during the infective juvenile stage of entomopathogenic nematodes. *Parasitology* 110, 583–590.

Lewis, E.E., Ricci, M. and Gaugler, R. (1996) Host recognition behavior reflects host suitability for the entomopathogenic nematode, *Steinernema carpocapsae*. *Parasitology* 113, 573–579.

Nguyen, K.B. and Smart, G.C. (1994) *Neosteinernema longicurvicauda* n. gen., n. sp. (Rhabditida: Steinernematidae), a parasite of the termite *Reticulitermes flavipes* (Koller). *Journal of Nematology* 26, 162–174.

O'Leary, S.A., Stack, C.M., Chubb, M. and Burnell, A.M. (1998) The effect of day of emergence from the insect cadaver on the behavior and environmental tolerances of infective juveniles of the entomopathogenic nematode *Heterorhabditis megidis* (strain UK211). *Journal of Parasitology* 84, 665–672.

Patel, M.N., Perry, R.N. and Wright, D.J. (1997a) Desiccation survival and water contents of entomopathogenic nematodes, *Steinernema* spp. (Rhabditida: Steinernematidae). *International Journal of Parasitology* 27, 61–70.

Patel, M.N., Stolinski, M. and Wright, D.J. (1997b) Neutral lipids and the assessment of infectivity in entomopathogenic nematodes: observations on four *Steinernema* species. *Parasitology* 114, 489–496.

Peters, A. (1996) The natural host range of *Steinernema* and *Heterorhabditis* spp. and their impact on insect populations. *Biocontrol Science and Technology* 6, 389–402.

Peters, A. and Ehlers, R.-U. (1994) Susceptibility of leather jackets (*Tipula paludosa* and *Tipula oleracea*; Tipulidae: Nematocera) to the entomopathogenic nematode *Steinernema feltiae*. *Journal of Invertebrate Pathology* 63, 163–171.

Poinar, G.O. Jr (1990) Biology and taxonomy of Steinernematidae and Heterorhabditidae. In: Gaugler, R. and Kaya, H.K. (eds) *Entomopathogenic nematodes in biological control.* CRC Press, Boca Raton, Florida, pp. 23–61.

Poinar, G.O. Jr (1993) Origins and phylogenetic relationships of the entomophilic rhabditids, *Heterorhabditis* and *Steinernema*. *Fundamental and Applied Nematology* 16, 333–338.

Poinar, G.O. and Jansson, H.B. (1986) Infection of *Neoaplectana* and *Heterorhabditis* (Rhabditida: Nematoda) with the predatory fungi, *Monacrosporium ellipsosporum* and *Arthrobotrys oligospora* (Moniliales: Deuteromycetes). *Revue de Nématologie* 9, 241–244.

Popiel, I. and Vasquez, E.M. (1991) Cryopreservation of *Steinernema carpocapsae* and *Heterorhabditis bacteriophora*. *Journal of Nematology* 23, 432– 437.

Qiu, L. and Bedding, R. (2000) Energy metabolism and its relation to survival and infectivity of infective juveniles of *Steinernema carpocapsae* under aerobic conditions. *Nematology* 2, 551–559.

Renn, N. (1999) Routes of penetration of the entomopathogenic nematode *Steinernema feltiae* attacking larval and adult houseflies (*Musca domestica*). *Journal of Invertebrate Pathology* 72, 281–287.

Ryder, J.J. and Griffin, C.T. (2003) Phased infectivity in *Heterorhabditis megidis*: the effects of infection density in the parental host and filial generation. *International Journal of Parasitology* 33, 1013–1018.

Schroeder, W.J. and Beavers, J.B. (1987) Movement of the entomogenous nematodes of the families Heterorhabditidae and Steinernematidae in soil. *Journal of Nematology* 19, 257–259.

Selvan, S., Campbell, J.F. and Gaugler, R. (1993a) Density dependent effects on entomopathogenic nematodes (Heterorhabditidae and Steinernematidae) within an insect host. *Journal of Invertebrate Pathology* 62, 278–284.

Selvan, S., Gaugler, R. and Lewis, E.E. (1993b) Biochemical energy reserves of entomopathogenic nematodes. *Journal of Parasitology* 79, 167–172.

Shapiro, D.I. and Glazer, I. (1996) Comparison of entomopathogenic nematode dispersal from infected hosts versus aqueous suspension. *Environmental Entomology* 25, 1455–1461.

Sicard, M., Le Brun, N., Pagès, S., Godelle, B., Boemare, N. and Moulia, C. (2003) Effect of native *Xenorhabdus* on the fitness of their *Steinernema* hosts: contrasting types of interactions. *Parasitology Research* 91, 520–524.

Smigielski, A.J., Akhurst, R.J. and Boemare, N.E. (1994) Phase variation in *Xenorhabdus nematophilus* and *Photorhabdus luminescens*: differences in respiratory activity and membrane energization. *Applied and Environmental Microbiology* 60, 120–125.

Somasekhar, N., Grewal, P.S. and Klein, M.G. (2002) Genetic variability in stress tolerance and fitness among natural populations of *Steinernema carpocapsae*. *Biological Control* 23, 303–310.

Strong, D.R., Kaya, H.K., Whipple, A.V., Child, A.L., Kraig, S., Bondonno, M., Dyer, K. and Maron, J.L. (1996) Entomopathogenic nematodes: natural enemies of root-feeding caterpillars on bush lupine. *Oecologia* 108, 167–173.

Stuart, R.J. and Gaugler, R. (1994) Patchiness in populations of entomopathogenic nematodes. *Journal of Invertebrate Pathology* 64, 39–45.

Stuart, R.J., Abu Hatab, M. and Gaugler, R. (1998) Sex ratio and the infection process in entomopathogenic nematodes: Are males the colonizing sex? *Journal of Invertebrate Pathology* 72, 288–295.

Thaler, J.-O., Baghdiguian, S. and Boemare, N. (1995) Purification and characterization of xenorhabdicin, a phage tail-like bacteriocin, from the lysogenic strain F1 of *Xenorhabdus nematophilus*. *Applied and Environmental Microbiology* 61, 2049–2052.

Timper, P. and Kaya, H.K. (1989) Role of the second stage cuticle of entomogenous nematodes in preventing infection by nematophagous fungi. *Journal of Invertebrate Pathology* 54, 314–321.

Timper, P., Kaya, H.K. and Jaffee, B.A. (1991) Survival of entomogenous nematodes in soil infested with the nematode-parasitic fungus *Hirsutella rhossiliensis* (Deuteromycotina: Hyphomycetes). *Biological Control* 1, 42–50.

Vivas, E.I. and Goodrich-Blair, H. (2001) *Xenorhabdus nematophilus* as a model for host-bacterium interactions: *rpoS* is necessary for mutualism with nematodes. *Journal of Bacteriology* 183, 4687–4693.

Wang, Y. and Gaugler, R. (1999) Host and penetration site location by entomopathogenic nematodes against Japanese beetle larvae. *Journal of Invertebrate Pathology* 72, 313–318.

Wang, X.D. and Grewal, P.S. (2002) Rapid genetic deterioration of environmental tolerance and reproductive potential of an entomopathogenic nematode during laboratory maintenance. *Biological Control* 23, 71–78.

Webster, J.M., Chen, G., Hu, K. and Li, J. (2002) Bacterial metabolites. In: Gaugler, R. (ed.) *Entomopathogenic Nematology*. CAB International, Wallingford, UK, pp. 99–114.

Westerman, P.R. (1995) Comparative vertical migration of twenty one isolates of the insect parasitic nematode *Heterorhabditis* spp. in sand at 20°C. *Fundamental and Applied Nematology* 18, 149–158.

Wharton, D.A. and Surrey, M.R. (1994) Cold tolerance mechanisms of the infective larvae of the insect parasitic nematode, *Heterorhabditis zelandica,* Poinar. *Cryo Letters* 15, 353–360.

Wilson, M.J., Lewis, E.E., Yoder, F. and Gaugler, R. (2002) The influence of application pattern on persistence of *Heterorhabditis bacteriophora. Biological Control* 26, 180–188.

Zhou, X.S., Kaya, H.K., Heungens, K. and Goodrich-Blair, H. (2002) Response of ants to a deterrent factor(s) produced by the symbiotic bacteria of entomopathogenic nematodes. *Applied and Environmental Microbiology* 68, 6202–6209.

3 Mass Production

R.-U. Ehlers[1] and D.I. Shapiro-Ilan[2]

[1]*Department for Biotechnology and Biological Control, Institute for Phytopathology, Christian-Albrechts-University Kiel, Klausdorfer Str. 28–36, 24223 Raisdorf, Germany;* [2]*USDA-ARS, SE Fruit and Tree Nut Research Lab, Byron, GA 31008, USA*

3.1. Introduction

For laboratory use and small-scale field-testing, *in vivo* production of entomopathogenic nematodes (EPN) is the appropriate method. *In vivo* production is also appropriate for niche markets, grower cooperatives and other commercial arenas where a lack of capital outlay, scientific expertise or infrastructure cannot justify large investments into *in vitro* culture technology (e.g. bioreactors, downstream equipment and installations). When it comes to commercial use of EPN at a larger scale for international markets, *in vitro* production is currently the only economically reasonable means to supply EPN at high quality and at reasonable costs. This chapter summarizes production technology.

3.2. *In vivo* Production

Production methods for culturing EPN in insect hosts have been reported by various authors (Dutky *et al.*, 1964; Poinar, 1979; Woodring and Kaya, 1988; Lindegren *et al.*, 1993; Flanders *et al.*, 1996; Kaya and Stock, 1997). These references essentially describe systems based on the White trap (White, 1927) (Fig. 3.1), which takes advantage of the infective juvenile's (IJ) natural migration away from the host cadaver upon emergence. The methods described consist of

Fig. 3.1. Modified White trap. Insect larvae (*Galleria mellonella*) infected with entomopathogenic nematodes (EPNs) (*Heterorhabditis bacteriophora*) placed on moist filter paper in an inverted Petri dish lid (60 mm). As infective juvenile (IJ) nematodes emerge from the insect cadavers they migrate into water, which is held in a larger Petri dish (100 mm), and surrounds the central dish.

inoculation, harvest, concentration and (if necessary) decontamination (Shapiro-Ilan and Gaugler, 2002). Mass production is accomplished in a two-dimensional system of trays and shelves. Insects are inoculated with nematodes on dishes or trays lined with an absorbent substrate. After 2–5 days, infected insects are transferred to the White traps (i.e. harvest dishes). Following harvest, concentration of nematodes can be accomplished by gravity settling (Dutky *et al.*, 1964) and/or vacuum filtration (Lindegren *et al.*, 1993). Centrifugation is also feasible (Kaya and Stock, 1997), but, for commercial *in vivo* operations, the capital outlay for a centrifuge of sufficient capacity may be unwarranted.

Yield is affected by choice of nematode and host species. Among nematode species, yield is generally inversely proportional to size (see Grewal *et al.*, 1994 and Hominick *et al.*, 1997). The most common insect host used for *in vivo* laboratory and commercial EPN production is the last instar of the greater wax moth, *Galleria mellonella*, because of its high susceptibility to most nematodes, ease in rearing, wide availability and ability to produce high yields (Woodring and Kaya, 1988). Only a few EPNs (i.e. *S. kushidai*, *S. scarabaei* and *S. scapterisci*) exhibit relatively poor reproduction in *G. mellonella* due to extremes in host specificity (Mamiya, 1989; Nguyen and Smart, 1990; Kaya and Stock, 1997; Grewal *et al.*, 1999; Koppenhöfer and Fuzy, 2003). Various other Lepidoptera and Coleoptera have been studied as hosts during *in vivo* nematode production (Shapiro-Ilan and Gaugler, 2002).

In general, nematode yield is proportional to insect host size (Blinova and Ivanova, 1987; Flanders *et al.*, 1996), but yield per milligram insect (within host species) and susceptibility to infection is usually inversely proportional to host size or age (Dutky *et al.*, 1964; Blinova and Ivanova, 1987; Shapiro *et al.*, 1999). In addition to yield, ease of culture and infection are important factors when choosing a host (Blinova and Ivanova, 1987; Shapiro-Ilan and Gaugler, 2002). Ultimately, the choice of host species and nematode for *in vivo* production should rest on nematode yield per cost of insect, and the suitability of the nematode for the target pest (Blinova and Ivanova, 1987; Shapiro-Ilan *et al.*, 2002).

However, nematode quality may also need to be considered in choosing a host because nematodes reared on various hosts may differ in quality (Abu Hatab *et al.*, 1998), and nematodes can adapt to the host they are reared on (Stuart and Gaugler, 1996).

Other factors affecting *in vivo* production yields include inoculation and environmental parameters. Successful infection and yield have been reported to be optimum with increasing dosage (Shapiro-Ilan *et al.*, 2002) or at intermediate dosages (Boff *et al.*, 2000). Increased host density per unit area tends to decrease infection efficiency (Shapiro-Ilan *et al.*, 2002). Rearing temperature is critical as it affects both yield and life cycle duration (time to emergence) (Grewal *et al.*, 1994). Grewal *et al.* (1994) determined the optimum rearing temperature and time to emergence in *G. mellonella* for 12 species and strains of EPNs; optimum temperatures varied from 18°C to 28°C. In addition to appropriate temperatures, adequate aeration (Burman and Pye, 1980; Friedman, 1990) and high levels of humidity are important environmental factors that must be maintained throughout the production cycle (Woodring and Kaya, 1988).

In vivo production of EPN offers several advantages and disadvantages relative to *in vitro* culture. *In vivo* production requires the least capital outlay and technical expertise (Friedman, 1990; Gaugler and Han, 2002). Some studies indicated that the quality of EPN produced *in vivo* could be higher than that of EPN produced *in vitro* (Gaugler and Georgis, 1991; Yang *et al.*, 1997). However, the lower quality in EPN produced *in vitro* observed in these studies was likely the result of poor understanding of *in vitro* production technology. Indeed, several subsequent studies detected no difference between *in vitro* and *in vivo* culture methods (Shapiro and McCoy, 2000; R.-U. Ehlers unpublished). The major disadvantage of *in vivo* production is cost of labour and insects, which tends to make *in vivo* culture the least cost-efficient approach.

Despite limitations in cost efficiency and scale, *in vivo* production has managed to sustain itself as a cottage industry (Gaugler *et al.*, 2000, Gaugler and Han, 2002). *In vivo* production is likely to continue as small business ventures for niche markets or in developing countries where labour is inexpensive. Further innovations to improve efficiency and scalability will enable *in vivo* production to play an expanded role in pest management programmes in niche markets and developing countries. For example, a recently developed scalable *in vivo* system 'LOTEK' promises to increase cost efficiency by decreasing labour and space requirements relative to the White trap approach (Gaugler *et al.*, 2002). Another method for improving *in vivo* production efficiency and field efficacy may be through production and application of EPNs in infected hosts (Shapiro and Glazer, 1996; Shapiro and Lewis, 1999; Shapiro-Ilan *et al.*, 2001, 2003). Using this approach, infected host cadavers are applied to the target site, and pest suppression is subsequently achieved by the emerging IJ progenies.

3.3. *In vitro* Production

3.3.1. Dauer juvenile (DJ) and recovery

Producing EPN *in vitro* requires knowledge on the biology and behaviour of the nematode species produced. The only stage that can be commercially used is the so-called dauer (German for enduring) juvenile (DJ), a morphologically distinct juvenile, formed as a response to depleting food sources and adverse environmental conditions. The DJ carries between 200 and 2000 cells of its symbiont in the anterior part of its intestine (Endo and Nickle, 1994). After invasion of its host, the DJ exits from this stage as a response to yet unknown signals encountered in the haemolymph of the insect (Strauch and Ehlers, 1998). Pharynx, digestive tract and excretory metabolism are activated. Analogous to *Caenorhabditis elegans*, this process is called 'recovery' and 'food signal' (the recovery-inducing signal) (Riddle *et al.*, 1997). During recovery, the DJ releases the symbiont cells into the insect's haemocoel.

3.3.2. Phase variation

The biology of the symbiotic bacterium needs particular attention. Typical for symbionts of both genera, *Xenorhabdus* and *Photorhabdus*, is the phenomenon of phase variation, the two extremes of which are the primary and the secondary phase (Akhurst, 1980). Intermediate phases have been reported (Gerritsen and Smits, 1997). The primary phase is isolated from the DJ or infected insects, whereas the secondary phase occurs after *in vitro* and also *in vivo* subculturing, when the nematodes emigrate from the cadaver (Grunder, 1997). The secondary phase is not retained by the DJ of *Heterorhabditis bacteriophora* (Han and Ehlers, 2001). Krasomil-Osterfeld (1995) induced secondary phase by cultivating primary forms under stress conditions, for instance in media with low osmotic strength. When the bacteria were subcultured at standard conditions, they reverted to the primary phase. Prolonged subculture at stress conditions produced stable secondary phase cultures. The major drawback related to phase shift is the detrimental effect of secondary phase on nematode development and yields, particularly in liquid culture (Ehlers *et al.*, 1990; Völgyi *et al.*, 1998; Han and Ehlers, 2001). All measures should therefore be taken during production to avoid the occurrence of phase variation. In general, the phase shift can be prevented by carefully reducing stress (lack of oxygen, high temperature, deviation from optimum osmotic strength of medium) during bacterial inoculum production, inoculation and the preculture. The mechanisms causing the phase transition are as yet unresolved. Genetic variation was excluded (LeClerc and Boemare, 1991; Akhurst *et al.*, 1992a,b; Wang and Dowds, 1993).

3.3.3. Solid-phase production

When used for the first time in history to control larvae of the Japanese beetle (*Popillia japonica*) in the USA, Glaser (1931) had already tried to mass-produce *Steinernema glaseri in vitro* on solid media. EPN can be grown on Petri dishes using different agar media (House *et al.*, 1965; Wouts, 1981; Dunphy and Webster, 1989). A major breakthrough in mass production was achieved when Bedding (1981) published his results on the growth of *Steinernema* spp. on a three-dimensional medium in flasks, using polyether–polyurethane sponge as an inert medium carrier. Autoclavable plastic bags aerated with aquarium pumps and inoculated with approximately 2000 DJ/g medium can be used to scale up this method (Bedding, 1984), and currently the companies Andermatt (Switzerland), Bionema (Sweden), Oviplant (Poland) and Biologic (USA) use this system. Bedding *et al.* (1991) developed a culture vessel comprising a tray with side walls and overlapping lids that allowed gas exchange through a layer of polyether–polyurethane foam. These trays are particularly well suited for developing countries as forced aeration is not necessary, making this system independent from cuts in the power supply. Nematodes can be extracted from solid media with centrifugal sifters, or by washing nematodes out of the sponge in simple washing machines and then separating the DJ by sedimentation or migration.

Solid-state production has several advantages. The effect of phase variation on the yields is less than in liquid cultures (Han and Ehlers, 2001). Little investment in biotechnology equipment is necessary and the risk for process failure is partitioned over several smaller production units. In developing countries this system is still superior to liquid culture technology (Bedding, 1990; Ehlers *et al.*, 2000). Solid-state production was later transferred to large stainless-steel blenders used to produce mushroom spawn (Gaugler and Han, 2002). As costs of these vessels can even surpass those of conventional bioreactors, this approach is only feasible if the blenders are not always used for spawn production. When it comes to large-scale production, the disadvantages of solid media are overwhelming. The solid-state culture is labour-intensive, vulnerable to contamination during up- and downstream processing and can hardly be monitored online. The uneven distribution of the nematodes in the medium prevents systematic sampling and thus improves the technique. An exploitation of the potential of EPN for plant protection required the development of liquid culture technology.

3.3.4. Liquid culture

EPNs were first cultured axenically in liquid media by Stoll (1952) using raw liver extract in shaken flasks. The first attempt to use bioreactors was described by Pace *et al.* (1986). They cultured nematodes in a standard 10-l bioreactor (Braun Biostat E) and showed that shear from a flat-blade impeller, expressed as impeller tip velocity of 1 m/s or greater, leads to the disruption of adult females. They therefore recommended shear to be less than 0.3 m/s for maximum yields. Using a kidney homogenate-yeast extract medium, they inoculated a culture of *Xenorhabdus nematophilus* 24 h before the inoculation of 2000 DJ/ml of the nematode *S. carpocapsae*. When the nematodes were inoculated, the temperature was reduced from 28°C to 23°C and the velocity of the impeller set at 180 rpm to maintain 20% oxygen saturation. After 10 days the culture yielded 40,000 DJ/ml. In order to increase yields and reduce losses obtained by shear stress, they exchanged the conventional flat-blade impeller with a paddle stirrer. The first commercial application of the liquid culture technology was made by the company Biosys, Palo Alto, California. The company was incorporated in 1987 and soon started to produce in liquid culture. In 1992 large-scale production of *S. carpocapsae* began and was scaled up to volumes of 80,000 l. Today the majority of EPN products result from liquid culture and are produced by the European companies E-Nema GmbH (www.e-nema.de), Koppert B.V. (www.koppert.nl) and Becker Underwood (www.beckerunderwood.com) and by the US-based company Certis (www. certisusa.com).

3.3.5. Liquid culture process technology

Due to the even distribution of fluids and organisms obtained through the mixing of liquids in bioreactors and the long process time, EPN cultures are particularly vulnerable to contamination. The presence of any non-symbiotic microorganisms will reduce nematode yields and prevent the subsequent scale-up. As a nematode process can last up to 3 weeks, maintenance of sterile conditions is a challenge for process engineers. The monoxenicity of the cultures must be ensured from the onset of inoculum production. The symbiotic bacteria can easily be isolated from nematode-infected insect larvae. Stock cultures are mixed with glycerol at 15% (v/v), and aliquots are frozen at −80°C. Details on the determination of the symbiotic bacteria are provided by Boemare and Akhurst (1988). More laborious is the establishment of bacteria-free nematodes. Surface-sterilized DJ should not be used because this procedure cannot exclude the presence of contaminants (Lunar *et al.*, 1993). The preparation of nematode inoculum is preferably done with nematode eggs obtained from gravid female stages. Detailed descriptions about the production of monoxenic nematode inoculum are provided by Lunau *et al.* (1993) and Han and Ehlers (1998). Monoxenic cultures can be stored on shakers at 20 rpm and 4°C for several months until they are inoculated into the bioreactor. Strain collections of nematodes can be kept in liquid nitrogen (Popiel and Vasquez, 1991).

Owing to the potentiality of *Xenorhabdus* and *Photorhabdus* spp. to metabolize almost every kind of protein-rich medium, the selection of appropriate culture media for EPN production can largely follow economic aspects. A standard medium should start with a carbon source (e.g. glucose or glycerol), a variety of proteins of animal and plant origin, yeast extract and lipids of animal or plant origin (e.g. Pace *et al.*, 1986; Friedman *et al.*, 1989; Han *et al.*, 1995; Surrey and Davies, 1996; Ehlers *et al.*, 1998). The osmotic strength of the medium must not surpass 600 milliosmol/kg. Improvements of the medium and adaptation to requirements of different species are feasible approaches to increase yields (Ehlers, 2001). Essential amino acid requirements have only been defined for *S. glaseri* (Jackson, 1973). Nematodes have nutritive demands for sterols, but they can metabolize necessary sterols from a variety of steroid sources (Ritter, 1988) that are provided through the addition of lipids of animal or plant origin. In general, *S. carpocapsae* requires proteins of animal origin (Yang *et al.*, 1997) and is unable to reproduce without the addition of lipid sources to the medium, whereas *H. bacteriophora* produces offspring in a liquid medium without the addition of lipids (Han and Ehlers, 2001). *Photorhabdus luminescens* provides or metabolizes necessary lipids; however, lipids should always be added to increase total DJ fat content. The lipid composition of the medium has an effect on the fatty acid composition of the bacteria and DJ (Abu Hatab *et al.*, 1998), and low fat content of DJ can reduce efficacy (Patel *et al.*, 1997a,b).

Conventional equipment used in biotechnology (e.g. conventional bioreactors), stirred with flat-blade impellers, bubble columns, airlift and internal loop bioreactors, have been successfully tested (Pace *et al.*, 1986; Surrey and Davies, 1996; Ehlers *et al.*, 1998). In a direct comparison with flat-blade impeller-stirred tanks (R.-U. Ehlers, unpublished) or airlift bioreactors (A. Peters, unpublished), internal loop bioreactors always yielded higher DJ concentrations. Figure 3.2 provides an overview of the production process. Cultures are always pre-incubated for 24–36 h with the specific symbiont bacterium before DJs are inoculated. The inoculum density for the symbiotic bacterium is between 0.5% and 1% of the culture volume. A specific (universal) nematode inoculation rate cannot be given because the optimum temperature varies depending on species and media composition. However, an optimum number of adults per millilitre can be calculated, which is defined by the percentage of DJ bound to recover (see section on nematode population dynamics). Usually the nematode inoculum is between 5% and 10% of the culture volume. Process parameters favouring the growth and reproduction of the nematode–bacterium complex have not yet been studied systematically and only a few results have been published. The optimum growth temperature for the symbiont of *H. indica* was investigated under continuous culture conditions (Ehlers *et al.*, 2000). Optimum growth was recorded between 35°C and 37°C. Optimum culture temperature should always be defined before mass production of a new isolate. Any deviation surpassing the optimum temperature can induce the formation of the secondary phase, which impedes nematode reproduction. The culture medium should be between pH 5.5 and 7.0 when the culture is started. Attempts to control the pH at 7.0 always had a negative influence on nematode yields (R.-U. Ehlers, unpublished). The pH appears to be well regulated by the organisms themselves. Oxygen supply must be maintained at approximately 30% saturation, also to prevent the bacteria from shifting to the secondary phase. An important parameter is the aeration rate. Strauch and Ehlers (2000) compared the yields of *H. megidis* in 10-l bioreactor cultures aerated at 0.3 vvm and 0.7 vvm, and obtained a significantly higher number of adult nematodes 8 days after DJ inoculation and higher number of DJ final yields in cultures aerated at 0.7 vvm. Increasing the aeration rate often increases foaming. The addition of silicon oil usually prevents foaming; however, it should be used carefully because higher concentrations can be detrimental to the nematodes. Long-chain fatty acids tested to control foaming had negative effects on *H. bacteriophora* (R.-U. Ehlers, unpublished data). Data on final DJ yields from liquid culture have been reported by many authors (Pace *et al.*, 1986; Bedding *et al.*, 1993; Han, 1996; Surrey and Davies, 1996; Ehlers *et al.*, 1998; Strauch and Ehlers, 2000). Maximum yields of > 500,000 DJ/ml were recorded by Ehlers *et al.* (2000) for *H. indica*. Yields show a negative correlation with the body length of the DJ, which is genetically defined and, although being quite stable within a species, differs according to strain and culture

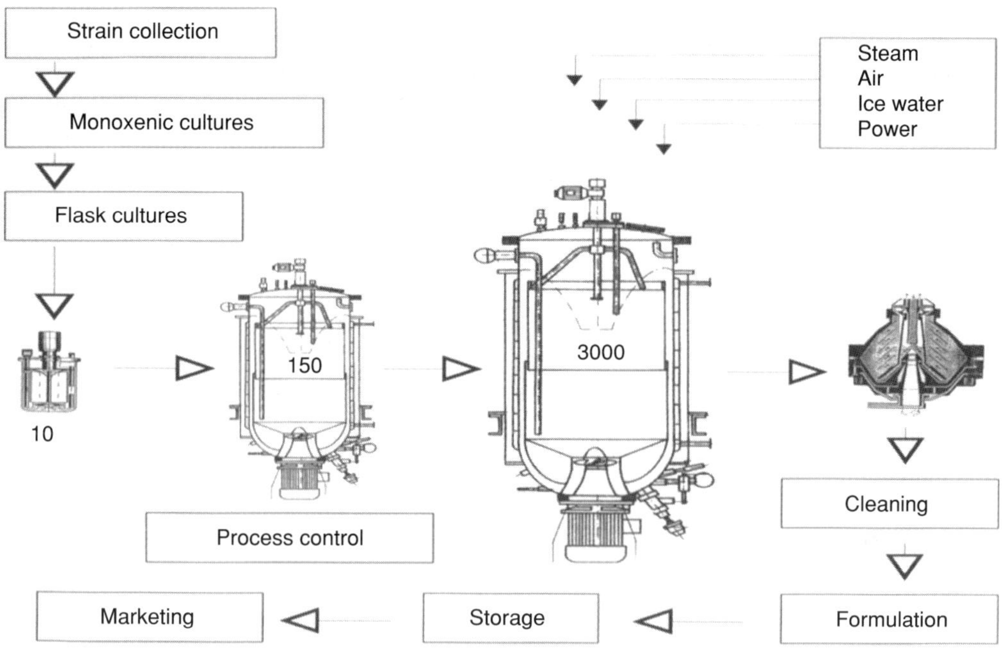

Fig. 3.2. Flow chart of nematode production process. After monoxenic cultures are established they are scaled up to a 3000 l internal loop bioreactor. After 12 days the dauer juveniles (DJs) are harvested with a separator. The nematode paste is then cleaned by passing through centrifugal sifters and formulated.

conditions. If a species with a small DJ has the same biocontrol potential as a species with a long DJ, the former species will always be cheaper to produce.

3.3.6. Developmental biology in liquid media

The environment in liquid culture is not what EPNs encounter in nature. Whereas insect cadavers infested with *Steinernema* spp. are often quite liquid, the cadavers occupied by *Heterorhabditis* are viscous. In a bioreactor the nematodes are driven around by impellers or air bubbles. This environment has consequences for nematode development, feeding and copulation, which need to be considered when adapting culture methods.

Success in liquid culture is dependent on the ability to accurately manage nematode population dynamics. In order to understand the critical phases during the process, the nematode developmental biology needs to be explained in more detail. Figure 3.3 presents the life cycle of *Heterorhabditis* spp., including alternative pathways and developmental steps, indicated by numbers. In principle, the development is driven by the availability of food. Low food concentration induces DJ formation, whereas high food concentration induces the development of additional adult generations or the recovery of the DJ. As the DJ (upper left corner Fig. 3.3) is developmentally arrested, it can be stored until needed for process inoculation. Once inoculated into the culture of the symbiont, the DJ recovers development (step 1A). The resulting IJs develop through the fourth-stage juvenile into hermaphrodites (step 3), which are automictic (self-fertilizing). The final yield can be predicted from the density of the hermaphrodites and their length (S.-A. Johnigk and R.-U. Ehlers, unpublished results). The length of the hermaphrodites as well as the number of eggs that will be laid are positively correlated with

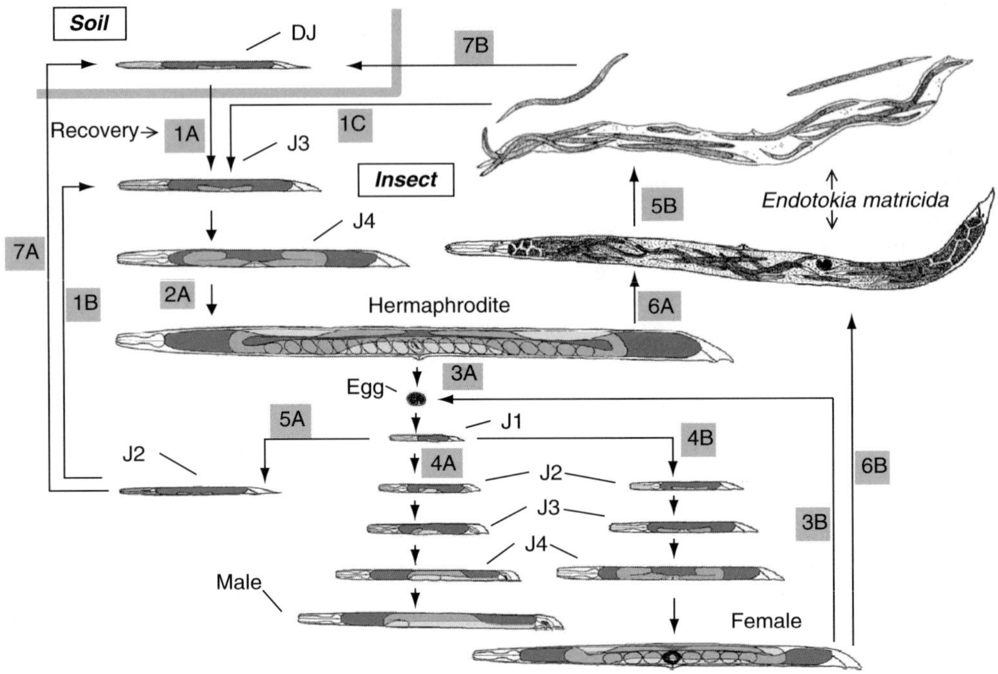

Fig. 3.3. Detailed life cycle of a *Heterorhabditis* sp. with alternative developmental pathways. Numbers indicate the critical developmental steps during the process. 1: Recovery of dauer juvenile (DJ) from free-living stage (1A), pre-dauer stage second-stage juvenile (J2) originating from laid eggs (1B) or from *endotokia matricida* (1C). 2: Development of hermaphrodite (2A). 3: Egg-laying by automictic hermaphrodite (3A) or amphimictic female (3B). 4: Development through third (J3) and fourth (J4) juvenile stage into amphimictic male (4A) and female (4B). 5: DJ formation of first-stage juvenile (J1) originating from laid eggs (5A) or from *endotokia matricida* (5B). 6: *Endotokia matricida* of hermaphrodite (6A) or amphimictic female (6B). 7: DJ emigration of DJ originating from laid eggs (7A) or from *endotokia matricida* (7B). Further explanations are given in text.

food supply. At first, the hermaphrodites lay eggs into the surrounding medium (step 3A). After 12 h of the hatching of the first-stage juvenile, male phenotypes can be identified (step 4A). After another 12 h female phenotypes are distinguishable (step 4B) (Johnigk and Ehlers, 1999a). In the insect or on solid media, the amphimictic adults copulate and produce another generation (egg-laying females, step 3B). In liquid media, however, the male is unable to attach itself to the female for insemination (Strauch *et al.*, 1994). Consequently, the development ends at this point and females only contain unfertilized eggs identified by the enlarged nucleus. Only automictic offspring can continue the life cycle in liquid media, which are a result of DJ formation (steps 5A and B). The DJ is always bound to become an automictic hermaphrodite. The decision for the development into amphimictic adults or into DJ occurs during the first stage. High concentrations of food induce the development of amphimictic adults (step 4), whereas low concentrations induce DJ formation (step 5A) (Strauch *et al.*, 1994). This mechanism is valid for nematodes of both genera. If the DJs do not yet emigrate from the infected insect (step 7A), the late second-stage juveniles recover and continue their development into the hermaphrodite (step 1B) to produce another generation of offspring.

After egg-laying of the parental hermaphrodites ceases, the juveniles hatch inside the uterus and *endotokia matricida* (intra-uterine birth causing maternal death) starts (step 6A). High food concentrations delay the beginning of the *endotokia matricida* and consequently enhance the number of eggs laid (Johnigk and Ehlers, 1999b). The length of the hermaphrodite defines the number of offspring in the uterus. The first hatched first-stage juveniles immediately feed on sperm, non-fertilized eggs and oogonia, so once *endotokia matricida* has started, no further offspring can develop. In the uterus the DJ formation (step 5B) is induced due to high nematode density and low food sources. A rapid change in food supply occurs when the juveniles have destroyed the uterus and intestinal tissue. They then have access to the body content of the adult and to cells of the symbiotic bacteria, which they retain in their intestine. Food provided by the body content of the hermaphrodite is well tuned to feed the defined number of offspring in the uterus. The resulting DJs accumulate a maximum of fat reserves and are of excellent quality (Johnigk and Ehlers, 1999b). Only in insects and solid cultures *endotokia matricida* is also observed in amphimictic females (step 6B). Emigrating DJs either result from DJs that have developed from laid eggs (step 7A) or from *endotokia matricida* (step 7B). *Steinernema* spp. have a similar life cycle, except that amphimictic adults develop from DJs, which must copulate in order to produce offspring. Males of *Heterorhabditis* spp. have a fan endowed with sensory receptors, the 'bursa copulatrix', which enables them to attach to the female at the vulval region and copulate, forming a lambda or 'y' with the female. As males of *Steinernema* spp. lack this structure they wind around the female's body forming a spiral. This copulation behaviour is less impeded in liquid culture.

One would expect that a certain medium can provide nutrients for fixed number of nematodes. However, yields in the same medium can vary considerably (Ehlers *et al.*, 1998; Strauch and Ehlers, 2000). The reason why the population dynamics are so important becomes apparent when data obtained from the commercial production are analysed. Up to a hermaphrodite density of 4000/ml at day 3 of the process the DJ yields are positively correlated to hermaphrodite density. Consequently, an inoculation of > 4000 DJ/ml is enough to obtain maximum yields. This hermaphrodite density, however, cannot be obtained by defining the DJ inoculation density, as DJ recovery is highly variable in liquid culture. Whereas almost 100% of the DJs recover within a day after having entered the haemocoel of an insect, liquid media lack any kind of food signal that could trigger recovery. Fortunately, the symbiotic bacteria produce such food signals, and they therefore enable the production of EPN *in vitro* through preculturing of the symbiotic bacteria. However, the levels of recovery caused by bacterial food signals are variable (18% and 90% within a period of several days) (Strauch and Ehlers, 1998).

The main reason for unstable DJ yields in *in vitro* culture is unpredictable, unsynchronized and low DJ recovery. It prevents population management that is required to maximize yields and to shorten the process time, and it makes necessary additional scale-up steps. Low recovery results in a low hermaphrodite density. At a low density, the abundance of food causes the hermaphrodites to lay many eggs from which the majority develop into amphimictic adults instead of DJs. This is, although prolonging the process time, acceptable when culturing steinernematids, as the amphimictic adults can copulate in liquid culture and produce an F2 offspring generation (Strauch *et al.*, 1994). It usually results in process failure in heterorhabditid cultures, as the F1 amphimictic adults cannot produce offspring. Furthermore, when reproductive F1 generation hermaphrodites have developed from second-stage juveniles (J2s) (step 5A) or from *endotokia matricida* (step 5B), amphimictic adults have already consumed much of the bacterial culture. Offspring production of the F1 hermaphrodites is low, and those that remained in the DJ stage (steps 7A and B) are of low quality as they have already consumed part of their fat reserves at the moment of harvest. In some cases high yields might even be achieved at low hermaphrodite density. This is due to the potential of the hermaphrodites to adapt to variable nematode density and respond with increasing body length and higher number of offspring. But this is only observed in cases of synchronous DJ recovery. With increasing numbers of hermaphrodites (> 2000/ml), their feeding activity reduces the bacterial concentration. Less offspring develop into amphimictic adults, but many develop into DJs and remain in this stage. The yield increases until a point is reached where the hermaphrodites hardly lay any eggs, and almost all offspring originate from *endotokia matricida*. This composition of the nematode population results in high yields of high-quality DJs within a minimum process time. Competition for food reduces the number of DJs per hermaphrodite. When the number of hermaphrodites is too high, the resources go into the basic maintenance of the adult instead of DJ production and the yields decline. Observations from flask cultures have shown that the body length of the hermaphrodites also decreases.

3.3.7. How to increase recovery in liquid culture

Production of *Steinernema* is less vulnerable to reduced recovery and DJs usually respond well to the food signal supplied by *Xenorhabdus* spp. However, the key for the industrial scale production of *Heterorhabditis* spp. is a synchronized, reproducible and high DJ recovery in order to reach an optimum number of parental hermaphrodites. To increase DJ recovery, several process parameters were investigated (Table 3.1). Recovery can already be influenced during the bacterial preculture. The higher the bacterial density, the higher the food signal concentration. Nematodes should therefore not be inoculated too early as the food signal concentration increases until the stationary growth phase is reached (Strauch and Ehlers, 1998). The moment when the conditions become favourable coincides with a significant drop of the respiration coefficient and a drop of the pH (F. Ecke, S.-A. Johnigk and R.-U. Ehlers, unpublished data). Fed-batch cultivation, adding glucose at the end of the exponential growth, is a possible measure to increase bacterial density (Jeffke *et al.*, 2000) and to enhance food signal production. Glucose fed-batch can thus be used to increase DJ recovery (unpublished data).

Jessen *et al.* (2000) found that increasing the CO_2 concentration in the medium enhanced DJ recovery. The influence of decreasing pH caused by the CO_2 concentration was excluded. A pH below 6.5 significantly reduces DJ recovery. The positive effect of CO_2 could be confirmed by comparing two parallel bioreactor runs: one at standard conditions and the other with a CO_2 concentration at 5%. Cultures were inoculated with DJs of the same origin at 12,000/ml.

Table 3.1. Parameters influencing dauer juvenile (DJ) recovery (Strauch and Ehlers, 1998; Jessen *et al.*, 2000; F. Ecke, S.-A. Johnigk, U. Böttcher, R.-U. Ehlers, unpublished data).

Process parameter/culture condition	Effect on DJ recovery
Food signal insect haemolymph	++
Food signal symbiotic bacterium	+
Compounds of artificial media	−
High bacterial density	+
Symbiont culture in stationary phase	+
pH within 6.5–9.0	+
pH < 6.5	−
Increasing CO_2 concentration	+
DJ originate from laid eggs	+
DJ originate from *endotokia matricida*	−
Age of DJ	Variable
DJ fat reserves	Variable

The artificial increase of the CO_2 resulted in a higher percentage of DJ recovery and caused the DJ to recover earlier. The yields were more than doubled (Ehlers, 2001).

When the response of different DJ batches is compared at standard conditions, it becomes obvious that a major source of variability is the DJs themselves (Strauch and Ehlers, 1998; Jessen *et al.*, 2000). The response to the food signal differs considerably from batch to batch. This difference may be due to variable fat reserves of the DJ. The lower the energy reserves, the higher would be the predisposition of the DJ to recover. Several experiments that tested the influence of DJ ageing (loss of fat reserves) did not support this hypothesis. An insignificant increase is usually recorded after 1 week of DJ inoculum storage; however, higher DJ recovery is often hampered by increasing DJ mortality during storage. The only significant difference was recorded for DJs originating from *endotokia matricida* or from laid eggs. The latter had a significantly better predisposition to recover (R.-U. Ehlers, unpublished data).

3.4. Conclusions

EPNs are no longer just used in niche markets or glasshouses, but have taken the step to outdoor environments (citrus, turf, strawberries). In various crops (such as vegetables and fruits) there are many pests that can be controlled by EPNs. However, many of these new potential markets will demand nematode products only when a cost-competitive price is available. *In vivo* producers and production on solid media will likely be limited in their ability to meet these cost demands. Even with technical improvements these systems will probably never reach the scale-up potential of liquid culture technology. Although the cost of nematode products has halved since the introduction of liquid culture technology, the prices are still prohibitively high for application in low-value crops. The continuous scale-up of bioreactor volumes will bring along further reduction of production costs. Other factors are strengthening process stability and downstream processing, increasing EPN shelf-life, improving transport logistics and marketing (a major limiting factor). If progress is made in these areas also, EPNs will further substitute insecticides and contribute to stabilize agriculture environments and crop yields.

References

Abu Hatab, M., Gaugler, R. and Ehlers, R.-U. (1998) Influence of culture method on *Steinernema glaseri* lipids. *Journal of Parasitology* 84, 215–221.

Akhurst, R.J. (1980) Morphological and functional dimorphism in *Xenorhabdus* spp., bacteria symbiotically associated with insect pathogenic nematodes *Neoaplectana* and *Heterorhabditis*. *Journal of Genetic Microbiology* 121, 303–309.

Akhurst, R.J., Bedding, R.A., Bull, R.M. and Smith, R.J. (1992a) An epizootic of *Heterorhabditis* spp. (Heterorhabditidae: Nematoda) in sugar cane *scarabaeida* (Coleoptera). *Fundamental and Applied Nematology* 15, 71–73.

Akhurst, R.J., Smigielski, A.J., Mari, J., Boemare, N. and Mourant, R.G. (1992b) Restriction analysis of phase variation in *Xenorhabdus* spp. (Enterobacteriaceae), entomopathogenic bacteria associated with nematodes. *Systematic and Applied Microbiology* 15, 469–473.

Bedding, R.A. (1981) Low cost *in vitro* mass production of *Neoaplectana* and *Heterorhabditis* species (Nematoda) for field control of insect pests. *Nematologica* 27, 109–114.

Bedding, R.A. (1984) Large scale production, storage and transport of the insect-parasitic nematodes *Neoaplectana* spp. and *Heterorhabditis* spp. *Annals of Applied Biology* 104, 117–120.

Bedding, R.A. (1990) Logistics and strategies for introducing entomopathogenic nematode technology into developing countries. In: Gaugler, R. and Kaya, H.K. (eds) *Entomopathogenic Nematodes in Biological Control.* CRC Press, Boca Raton, Florida, pp. 233–248.

Bedding, R.A., Stanfield, M.A. and Crompton, G. (1991) Apparatus and method for rearing nematodes, fungi, tissue cultures and the like, and for harvesting nematodes. *International Patent Publication* WO 91/15569.

Bedding, R.A., Akhurst, R.J. and Kaya, H.K. (1993) Future prospects of entomogenous and entomopathogenic nematodes. In: Bedding,

R.A., Akhurst, R.J. and Kaya, H.K. (eds) *Nematodes and the Biological Control of Insect Pests.* CSIRO Publications, East Melbourne, Australia, pp. 157–170.

Bedding, R.A., Akhurst, R.J. and Kaya, H.K. (1993) *Nematodes and the Biological Control of Insect Pests.* CSIRO Publications, East Melbourne, Australia.

Blinova, S.L. and Ivanova, E.S. (1987) Culturing the nematode-bacterial complex of *Neoaplectana carpocapsae* in insects. In: Sonin, M.D. (ed.) *Helminths of Insects.* Amerind Publishing, New Delhi, pp. 22–26.

Boemare, N.E. and Akhurst, R.J. (1988) Biochemical and physiological characterization of colony form variants in *Xenorhabdus* spp. (Enterobacteriaceae). *Journal of Genetic Microbiology* 134, 751–761.

Boff, M.I.C., Wiegers, G.L., Gerritsen, L.J.M. and Smits, P.H. (2000) Development of the entomopathogenic nematode *Heterorhabditis megidis* strain NLH-E 87,3 in *Galleria mellonella. Nematology* 2, 303–308.

Burman, M. and Pye, A.E. (1980) *Neoaplectana carpocapsae*: respiration of infective juveniles. *Nematologica* 26, 214–219.

Dunphy, G.B. and Webster, J.M. (1989) The monoxenic culture of *Neoaplectana carpocapsae* DD-136 and *Heterorhabditis heliothidis. Revue de Nématologie* 12, 113–123.

Dutky, S.R., Thompson, J.V. and Cantwell, G.E. (1964) A technique for the mass propagation of the DD-136 nematode. *Journal of Insect Pathology* 6, 417–422.

Ehlers, R.-U. (2001) Mass production of entomopathogenic nematodes for plant protection. *Applied Microbiology and Biotechnology* 56, 623–633.

Ehlers, R.-U., Stoessel, S. and Wyss, U. (1990) The influence of phase variants of *Xenorhabdus* spp. and *Escherichia coli* (Enterobacteriaceae) on the propagation of entomopathogenic nematodes of the genera *Steinernema* and *Heterorhabditis. Revue de Nématologie* 13, 417–424.

Ehlers, R.-U., Lunau, S., Krasomil-Osterfeld, K. and Osterfeld, K.H. (1998) Liquid culture of the entomopathogenic nematode-bacterium complex *Heterorhabditis megidis/Photorhabdus luminescens. Biocontrol* 43, 77–86.

Ehlers, R.-U., Niemann, I., Hollmer, S., Strauch, O., Jende, D., Shanmugasundaram, M., Mehta, U.K., Easwaramoorthy, S.K. and Burnell, A. (2000) Mass production potential of the bacterio-helminthic biocontrol complex *Heterorhabditis indica–Photorhabdus luminescens. Biocontrol Science and Technology* 10, 607–616.

Endo, B.Y. and Nickle, W.R. (1994) Ultrastructure of the buccal cavity region and oesophagus of the insect parasitic nematode, *Heterorhabditis bacteriophora. Nematologica* 40, 379–398.

Flanders, K.L., Miller, J.M. and Shields, E.J. (1996) *In vivo* production of *Heterorhabditis bacteriophora* 'Oswego' (Rhabditida: Heterorhabditidae), a potential biological control agent for soil-inhabiting insects in temperate regions. *Journal of Economic Entomology* 89, 373–380.

Friedman, M.J. (1990) Commercial production and development. In: Gaugler, R. and Kaya, H.K. (eds) *Entomopathogenic Nematodes in Biological Control.* CRC Press, Boca Raton, Florida, pp. 153–172.

Friedman, M.J., Langston, S. and Pollitt, S. (1989) Mass production in liquid culture of insect-killing nematodes. *International Patent Publication* WO 89/04602, *US Patent* 5,023,183.

Gaugler, R. and Georgis, R. (1991) Culture method and efficacy of entomopathogenic nematodes (Rhabditida: Steinernematidae and Heterorhabditidae). *Biological Control* 1, 269–274.

Gaugler, R. and Han, R. (2002) Production technology. In: Gaugler, R. (ed.) *Entomopathogenic Nematology.* CAB International, Wallingford, UK, pp. 289–310.

Gaugler, R., Grewal, P., Kaya, H.K. and Smith-Fiola, D. (2000) Quality assessment of commercially produced entomopathogenic nematodes. *Biological Control* 17, 100–109.

Gaugler, R., Brown, I., Shapiro-Ilan, D. and Atwa, A. (2002) Automated technology for *in vivo* mass production of entomopathogenic nematodes. *Biological Control* 24, 199–206.

Gerritsen, L.J.M. and Smits, P.H. (1997) The influence of *Photorhabdus luminescens* strains and form variants on the reproduction and bacterial retention of *Heterorhabditis megidis. Fundamental and Applied Nematology* 20, 317–322.

Glaser, R.W. (1931) The cultivation of a nematode parasite of an insect. *Science* 73, 614–615.

Grewal, P.S., Selvan, S. and Gaugler, R. (1994) Thermal adaptation of entomopathogenic nematodes: niche breadth for infection, establishment, and reproduction. *Journal of Thermal Biology* 19, 245–253.

Grewal, P.S., Converse, V. and Georgis, R. (1999) Influence of production and bioassay methods on infectivity of two ambush foragers (Nematoda: Steinernematidae). *Journal of Invertebrate Pathology* 73, 40–44.

Grunder, J.M. (1997) *Photorhabdus luminescens* als Symbiont insektenpathogener Nematoden. PhD thesis. ETH Zürich, Switzerland.

Han, R.C. (1996) The effects of inoculum size on yield of *Steinernema carpocapsae* and *Heterorhabditis bacteriophora* in liquid culture. *Nematologica* 42, 546–553.

Han, R.C. and Ehlers, R.-U. (1998) Cultivation of axenic *Heterorhabditis* spp. dauer juveniles and their response to non-specific *Photorhabdus luminescens* food signals. *Nematologica* 44, 425–435.

Han, R.C. and Ehlers, R.-U. (2001) Effect of *Photorhabdus luminescens* phase variants on the *in vivo* and *in vitro* development and reproduction of *Heterorhabditis bacteriophora* and *Steinernema carpocapsae. FEMS Microbiology/Ecology* 35, 239–247.

Han, R.C., Pang, X. and Li, L. (1995) Optimization of medium components for the solid culture of entomopathogenic *Steinernema* and *Heterorhabditis* nematodes. *Natural Enemies of Insects* 17, 153–164.

Hominick, W.M., Briscoe, B.R., del Pino, F.G., Heng, J., Hunt, D.J., Kozodoi, E., Mracek, Z., Nguyen, K.B., Reid, A.P., Spiridonov, S., Stock, P., Sturhan, D., Waturu, C. and Yoshida, M. (1997) Biosystematics of entomopathogenic nematodes: current status, protocols, and definitions. *Journal of Helminthology* 71, 271–298.

House, H.L., Welch, H.E. and Cleugh, T.R. (1965) Food medium of prepared dog biscuit for the mass-production of the nematode DD-136 (Nematoda; Steinernematidae). *Nature* 206, 847.

Jackson, G.J. (1973) *Neoaplectana glaseri*: essential amino acids. *Experimental Parasitology* 34, 111–114.

Jeffke, T., Jende, D., Mätje, C., Ehlers, R.-U. and Berthe-Corti, L. (2000) Growth of *Photorhabdus luminescens* in batch and fed-batch culture. *Applied Microbiology and Biotechnology* 54, 326–330.

Jessen, P., Strauch, O., Wyss, U., Luttmann, R. and Ehlers, R.-U. (2000) Carbon dioxide triggers dauer juvenile recovery of entomopathogenic nematodes (*Heterorhabditis* spp.). *Nematology* 2, 319–324.

Johnigk, S.-A. and Ehlers, R.-U. (1999a) Juvenile development and life cycle of *Heterorhabditis bacteriophora* and *H. indica* (Nematoda: Heterorhabditidae). *Nematology* 1, 251–260.

Johnigk, S.-A. and Ehlers, R.-U. (1999b) *Endotokia matricida* in hermaphrodites of *Heterorhabditis* spp. and the effect of the food supply. *Nematology* 1, 717–726.

Kaya, H.K. and Stock, P. (1997) Techniques in insect nematology. In: Lacey, L.A. (ed.) *Manual of Techniques in Insect Pathology*. Academic Press, San Diego, California, pp. 282–324.

Koppenhöfer, A.M. and Fuzy, E.M. (2003) Ecological characterization of *Steinernema scarabaei*, a scarab-adapted entomopathogenic nematode from New Jersey. *Journal of Invertebrate Pathology* 83, 139–148.

Krasomil-Osterfeld, K.C. (1995) Influence of osmolarity on phase shift in *Photorhabdus luminescens. Applied Environmental Microbiology* 61, 3748–3749.

LeClerc, M.C. and Boemare, N.E. (1991) Plasmids and phase variation in *Xenorhabdus* spp. *Applied Environmental Microbiology* 57, 2597–2601.

Lindegren, J.E., Valero, K.A. and Mackey, B.E. (1993) Simple *in vivo* production and storage methods for *Steinernema carpocapsae* infective juvenile. *Journal of Nematology* 25, 193–197.

Lunau, S., Stoessel, S., Schmidt-Peisker, A.J. and Ehlers, R.-U. (1993) Establishment of monoxenic inocula for scaling up *in vitro* cultures of the entomopathogenic nematodes *Steinernema* spp. and *Heterorhabditis* spp. *Nematologica* 39, 385–399.

Mamiya, Y. (1989) Comparison of the infectivity of *Steinernema kushidai* (Nematode: Steinernematidae) and other steinernematid and heterorhabditid nematodes for three different insects. *Applied Entomology and Zoology* 24, 302–308.

Nguyen, K.B. and Smart, G.C. Jr (1990) *Steinernema scapterisci* n. sp. (Rhabditida: Steinernematidae). *Journal of Nematology* 22, 187–199.

Pace, G.W., Grote, W., Pitt, D.E. and Pitt, J.M. (1986) Liquid culture of nematodes. *International Patent Publication* WO 86/01074.

Patel, M.N., Perry, R.N. and Wright, D.J. (1997a) Desiccation survival and water contents of entomopathogenic nematodes, *Steinernema* spp. (Rhabditida: Steinernematidae). *International Journal of Parasitology* 27, 61–70.

Patel, M.N., Stolinski, M. and Wright, D.J. (1997b) Neutral lipids and the assessment of infectivity in entomopathogenic nematodes: observations on four *Steinernema* species. *Parasitology* 114, 489–496.

Poinar, G.O. Jr (ed.) (1979) *Nematodes for Biological Control of Insects.* CRC Press, Boca Raton, Florida.

Popiel, I. and Vasquez, E.M. (1991) Cryopreservation of *Steinernema carpocapsae* and *Heterorhabditis bacteriophora. Journal of Nematology* 23, 432–437.

Riddle, D.L., Blumenthal, T., Meyer, B.J. and Priess, J.R. (eds) (1997) *C. elegans II.* Cold Spring Habor Press, Cold Spring Habor, New York, 1222 pp.

Ritter, K.S. (1988) *Steinernema feltiae* (= *Neoaplectana carpocapsae*): effect of sterols and hypolipidemic agents on development. *Experimental Parasitology* 67, 257–267.

Shapiro, D.I. and Glazer, I. (1996) Comparison of entomopathogenic nematode dispersal from infected hosts versus aqueous suspension. *Environmental Entomology* 25, 1455–1461.

Shapiro, D.I. and Lewis, E.E. (1999) Comparison of entomopathogenic nematode infectivity from infected hosts versus aqueous suspension. *Environmental Entomology* 28, 907–911.

Shapiro, D.I. and McCoy, C.W. (2000) Effect of culture method and formulation on the virulence of *Steinernema riobrave* (Rhabditida: Steinernematidae) to *Diaprepes abbreviatus* (Curculionidae). *Journal of Nematology* 32, 281–288.

Shapiro, D.I., Cate, J.R., Pena, J., Hunsberger, A. and McCoy, C.W. (1999) Effects of temperature and host range on suppression of *Diaprepes abbreviatus* (Coleoptera: Curculionidae) by entomopathogenic nematodes. *Journal of Economic Entomology* 92, 1086–1092.

Shapiro-Ilan, D.I. and Gaugler, R. (2002) Production technology for entomopathogenic nematodes and their bacterial symbionts. *Journal of Industrial Microbiology and Biotechnology* 28, 137–146.

Shapiro-Ilan, D.I., Lewis, E.E., Behle, R.W. and McGuire, M.R. (2001) Formulation of entomopathogenic nematode-infected cadavers. *Journal of Invertebrate Pathology* 78, 17–23.

Shapiro-Ilan, D.I., Gaugler, R., Tedders, W.L., Brown, I. and Lewis, E.E. (2002) Optimization of inoculation for *in vivo* production of entomopathogenic nematodes. *Journal of Nematology* 34, 343–350.

Shapiro-Ilan, D.I, Lewis, E.E., Tedders, W.L. and Son, Y. (2003) Superior efficacy observed in entomopathogenic nematodes applied in infected-host cadavers compared with application in aqueous suspension. *Journal of Invertebrate Pathology* 83, 270–272.

Stoll, N.R. (1952) Axenic cultivation of the parasitic nematode, *Neoaplectana glaseri*, in a fluid medium containing raw liver extract. *Journal of Parasitology* 39, 422–444.

Strauch, O. and Ehlers, R.-U. (1998) Food signal production of *Photorhabdus luminescens* inducing the recovery of entomopathogenic nematodes *Heterorhabditis* spp. in liquid culture. *Applied Microbiology and Biotechnology* 50, 369–374.

Strauch, O. and Ehlers, R.-U. (2000) Influence of the aeration rate on yields of the biocontrol nematodes *Heterorhabditis megidis* in monoxenic liquid cultures. *Applied Microbiology and Biotechnology* 54, 9–13.

Strauch, O., Stoessel, S. and Ehlers, R.-U. (1994) Culture conditions define automictic or amphimictic reproduction in entomopathogenic rhabditid nematodes of the genus *Heterorhabditis. Fundamental and Applied Nematology* 17, 575–582.

Stuart, R.J. and Gaugler, R. (1996) Genetic adaptation and founder effect in laboratory populations of the entomopathogenic nematode *Steinernema glaseri. Canadian Journal of Zoology* 74, 164–170.

Surrey, M.R. and Davies, R.J. (1996) Pilot-scale liquid culture and harvesting of an entomopathogenic nematode, *Heterorhabditis bacteriophora. Journal of Invertebrate Pathology* 67, 92–99.

Völgyi, A., Fodor, A., Szentirmai, A. and Forst, S. (1998) Phase variation in *Xenorhabdus nematophilus. Applied Environmental Microbiology* 64, 1188–1193.

Wang, H. and Dowds, C.A. (1993) Phase variation in *Xenorhabdus luminescens*: cloning and sequencing of the lipase gene and analysis of its expression in primary and secondary phases of the bacterium. *Journal of Bacteriology* 175, 1665–1673.

White, G.F. (1927) A method for obtaining infective nematode larvae from cultures. *Science* 66, 302–303.

Woodring, J.L. and Kaya, H.K. (1988) *Steinernematid and Heterorhabditid Nematodes: A Handbook of Biology and Techniques.* Southern Cooperative Bulletin 331, Arkansas Agricultural Experiment Station, Fayettville, Arkansas, 30 pp.

Wouts, W.M. (1981) Mass production of the entomogenous nematode *Heterorhabditis heliothidis* (Nematoda: Heterorhabditidae) on artificial media. *Journal of Nematology* 13, 467–469.

Yang, H., Jian, H., Zhang, S. and Zhang, G. (1997) Quality of the entomopathogenic nematode *Steinernema carpocapsae* produced on different media. *Biological Control* 10, 193–198.

4 Formulation and Quality

P.S. Grewal[1] and A. Peters[2]
[1]*Department of Entomology, Ohio State University, Wooster, OH 44691, USA;*
[2]*E-nema GmbH, Klausdorfer Straße 28–36, D-24223 Raisdorf, Germany*

4.1. Introduction

Formulation and quality control are two of the most important aspects in the commercialization of nematodes as biocontrol agents. Mass-produced nematodes are formulated for ease of storage, transport and application. In addition, formulations also provide a means to enhance nematode storage stability and field efficacy. The topic of nematode formulations has been reviewed recently (see Grewal, 2002), and is therefore treated briefly except for the recent developments. However, the aspects of nematode quality are discussed in more detail to develop a broader view and consensus on the methods of quality assessment.

4.2. Formulation

Although entomopathogenic nematode (EPN) infective juveniles (IJs) can be stored in water for several months in refrigerated bubbled tanks, high cost and difficulties of maintaining quality preclude the routine use of this method. Settling of nematodes, high oxygen demand, sensitivity of some species to low temperature, susceptibility to microbial contamination and the effect of antimicrobial agents on nematode longevity are some of the major factors influencing nematode quality during storage in water. Therefore, nematodes are formulated to improve their storage stability. Formulation refers to the preparation of a product from an ingredient by the addition of certain active (functional) and non-active (inert) substances. Formulation is usually intended to improve activity, absorption, delivery, and ease of use or storage stability of an active ingredient. Typical examples of pesticide formulation ingredients (additives) include absorbents, adsorbents, anticaking agents, antimicrobial agents, antioxidants, binders, carriers, dispersants, humectants, preservatives, solvents, surfactants, thickeners and ultraviolet (UV) absorbers. Although the overall concept of nematode formulations is similar to traditional pesticide formulations, nematodes present unique challenges. High oxygen and moisture requirements of concentrated nematodes, sensitivity to temperature extremes and behaviour of IJs limit the choice of the formulation method and ingredients. Major goals of developing nematode formulations include maintenance of quality, enhancement of storage stability, improvement in ease of transport and use, reduction of transport costs, and enhancement of nematode survival during and after application. Formulations to improve nematode application and post-application survival are discussed in Chapter 5.

4.2.1. Principles of nematode formulation

Two distinct approaches have been used to formulate nematodes for storage and transport. In one approach the nematodes are placed in inert carriers that allow free gas exchange and movement of nematodes, while in the other approach functional ingredients are added to reduce nematode activity and metabolism. Although the placement of nematodes in inert carriers such as sponge or vermiculite provides a convenient means to ship small quantities of nematodes, the high activity of nematodes rapidly depletes their stored energy reserves. Sometimes the nematodes even escape from the inert carriers and dry out. Therefore, formulations have been developed in which the mobility/metabolism of nematodes is minimized through physical trapping, inclusion of metabolic inhibitors or via the induction of partial anhydrobiosis (i.e. life without water). Formulations and expected shelf-life of commercially produced EPNs are given in Table 4.1.

4.2.2. Inert carrier formulations

Inert carriers such as polyether–polyurethane sponge and vermiculite are widely used for storage and transport of small quantities of nematodes throughout the nematode industry. These formulations are easy and less expensive to make, but require constant refrigeration as the nematodes remain active, freely moving in, or on, the substrates. Shelf-life of these formulations under refrigeration (2–10°C) varies from 1 month to 3–4 months depending upon the nematode species (Table 4.1). The strict refrigeration requirement even during transport makes these formulations very expensive to the end-user.

4.2.3. Active carrier formulations

The active carrier formulations include functional ingredients that either physically trap nematodes to reduce their movement, use metabolic inhibitors or reduce nematode activity and metabolism through the induction of partial anhydrobiosis. The nematodes are physically trapped in alginate and flowable gel formulations that contain sufficient moisture to prevent induction of nematode anhydrobiosis. In one formulation, sheets of calcium alginate spread over plastic screens have been used to trap nematodes (Georgis, 1990). Trapping of nematodes in alginate gels allows storage at room temperatures. For example, in one alginate gel formulation, *Steinernema carpocapsae* can be stored for 3–4 months at 25°C and *S. feltiae* for 2–4 weeks (Grewal, 2002). In another formulation, the nematodes are mixed in a viscous flowable gel or paste to reduce activity (Georgis, 1990); however, room temperature storage stability is lower than the alginate formulation.

Table 4.1. Formulations and expected shelf-life of commercially produced *Steinernema* and *Heterorhabditis* spp.

			Shelf-life (months)	
Nematode species	Strain	Formulation	22–25°C	2–10°C
S. carpocapsae	All	Sponge	0.03–0.1	2.0–3.0
	All	Vermiculite	0.1–0.2	5.0–6.0
	All	Liquid concentrate	0.16–0.2	0.4–0.5
	All	Wettable powder	2.0–3.5	6.0–8.0
	All	Water-dispersible granule (WG)	4.0–5.0	9.0–12.0
S. feltiae	SN	Vermiculite	0.03–0.1	4.0–5.0
	UK	Wettable powder	2.5–3.0	5.0–6.0
	EN02	Wettable powder	0.5–1	3–4
	SN	WG	1.5–2.0	5.0–7.0
	Umeå	Nemagel(TM)	12[a]	12
S. glaseri	NJ43	Wettable powder	0.03–0.06	1.0–1.5
S. riobrave	RGV	Liquid concentrate	0.1–0.13	0.23–0.3
S. scapterisci	Uruguay	Wettable powder	1.0–1.5	3.0–4.0
H. bacteriophora	HP88	Sponge	0	1.0–2.0
	Hybrid	Sponge	0	0.75–1.5
	EN01	Wettable powder	0.5–1.0	2–3
H. indica	LN2	Wettable powder	0.25–0.50	0
	LN2	Sponge	0.25	0
H. marelata	Oregon	Sponge	0	1.0–2.0
H. megidis	UK	Wettable powder	2.0–3.0	4.0–5.0
H. zealandica	X1	Wettable powder	1.0–2.0	0

[a]Only for small nematode concentrations (2 million in 25 ml).

Nematodes have also been formulated in various heteropolysaccharides (agarose, carbopol, carrageenan, dextran, guar gum or gellan gum) surrounded by a paste of hydrogenated oil. Up to 35 days storage of *S. carpocapsae* at room temperature has been reported for this hydrogenated oil formulation (Chang and Gehert, 1995). Recently, a liquid concentrate was developed for the transport of nematodes in bulk tanks that contained a proprietary metabolic inhibitor to reduce nematode oxygen demand (Grewal, 1998).

The induction of anhydrobiosis reduces nematode metabolism and makes them more tolerant of both warm and cold temperatures (Glazer and Salame, 2000; Grewal and Jagdale, 2002). Partial anhydrobiosis can be induced in steinernematid and heterorhabditid nematodes by controlling water activity (A_W) of the substrate through the composition of formulation ingredients (Bedding, 1988; Silver *et al.*, 1995; Grewal, 2000a,b). Water activity is a measure of how tightly water is bound, structurally or chemically, to the substrate. As opposed to water content, A_W is influenced by bonding of water molecules to the surfaces, as well as osmosis. A_W equals the relative humidity of air, in equilibrium with a nematode sample in a sealed container. The formulations containing partially anhydrobiotic nematodes include gels, powders and granules. Bedding and Butler (1994) developed a formulation in which nematode slurry was mixed in anhydrous polyacrylamide, so that the resulting gel attained a water activity between 0.800 and 0.995. The nematodes were partially desiccated, but survival at room temperature was low. A composition of 2–3 g of polyacrylate with proprietary additives (Nemagel(TM)) to 250 ml of nematode slurry containing 40 million *S. feltiae* resulted in a 2-year survival at 4°C (Hokkanen and Menzler-Hokkanen, 2002). At room temperature, 1-year survival was recorded in 25-ml bags with 2 million *S. feltiae*. The A_W in this formulation was much higher

(> 0.995). Bedding (1988) described another formulation in which nematodes were mixed in clay to remove excess surface moisture and to induce partial anhydrobiosis (Bedding, 1988). The formulation, termed 'sandwich', consisted of a layer of nematodes between two layers of clay. In a slightly different formulation where nematode slurry (concentrated nematodes) was mixed in attapulgite or bentonite clay, Strauch *et al.* (2000) reported that *Heterorhabditis bacteriophora* (hybrid strain) and *H. indica* (LN2 strain) only survived for 2 weeks and 1 week respectively, at 25°C. At 5°C, the survival of *H. bacteriophora* was superior in sponge than in clay, but that of *H. indica* was superior in clay than in sponge at 15°C.

Granular formulations have also been developed for storage and transport of nematodes. Capinera and Hibbard (1987) described a formulation in which nematodes were partially encapsulated in lucerne meal and wheat flour. Later, Connick *et al.* (1993) described an extruded or formed granule in which nematodes were distributed throughout a wheat gluten matrix. This 'Pesta' formulation included a filler and a humectant to enhance nematode survival. The process involved drying of granules to low moisture to prevent nematode migration and reduce risk of contamination. However, granules rapidly dry out during storage resulting in poor nematode survival. A water-dispersible granule (WG) formulation has been developed in which IJs are encased in 10–20-mm diameter granules consisting of mixtures of various types of silica, clays, cellulose, lignin and starches (Georgis *et al.*, 1995; Silver *et al.*, 1995). The granular matrix allows access of oxygen to nematodes during storage and transport. At optimum temperature, the nematodes enter into a partial anhydrobiotic state due to the slow removal of body water by the substrate. The induction of partial anhydrobiosis is usually evident within 4–7 days by a three- to fourfold reduction in oxygen consumption of the nematodes following an initial increase (Grewal, 2000a,b). WG formulation offers several advantages over other formulations. This is the first commercial formulation that enabled storage of *S. carpocapsae* for over 6 months at 25°C at a nematode concentration of over 300,000/g (Grewal, 2000a). This shelf-life represents an extension of IJ longevity by 3 months as compared to the nematodes stored in water (Grewal, 2000a,b). The WGs also enhanced nematode tolerance to temperature extremes enabling easier and less expensive transport, improved ease of use of nematodes by eliminating time-consuming and labour-intensive preparation steps, decreased container size and coverage ratio, and reduced disposal material (i.e. screens and containers). However, this WG formulation is prone to microbial contamination when stored at room temperature. Therefore, antimicrobial and antifungal agents are often added to suppress the growth of contaminating microbes. A detailed discussion of the factors affecting the survival of EPNs in formulations can be found in Grewal (2002).

Nematodes can also be applied in the form of infected insect cadavers for small-scale applications (Shapiro-Ilan *et al.*, 2001, 2003). Cadavers can be coated with a protective formulation (e.g. starch and clay mixture) to prevent rupturing during storage and shipping (Shapiro-Ilan *et al.*, 2001).

4.3. Quality Control and Standardization

EPNs do not require registration in many countries. Therefore, the quality of commercially produced nematodes is essentially self-regulated. However, the University Extension and government advisory services can play a role in quality control of commercial nematode products (see Gaugler *et al.*, 2000). When nematodes are mass-produced in small companies, resources are often limited for the development of quality control methods and routine assessment of quality. Quality assessment also requires training of employees and a strong commitment from the management. Below we describe the various aspects of quality in relation to nematode products.

4.3.1. Defining nematode quality

The dictionary meaning of quality is the degree of excellence of a product, and quality control is a system of maintaining standards in manufactured products by testing a sample of the output against the specification. In this regard, nematode quality and nematode product quality are distinct parameters and are measured differently. EPN quality entails accuracy of the species identity, total number of live nematodes, ratio of live and dead nematodes, matching of host-finding behaviour to the target pest, pathogenicity and reproduction (recycling) ability in the target pest, age of the nematodes, storability, heat tolerance, and cold or warm temperature activity. The product quality includes the size and sturdiness of packaging, clarity and accuracy of instructions for the consumers, dispersibility, ease of transport and application, absence of contaminants, product cost, availability, and field efficacy.

4.3.2. Maintaining nematode quality during production, formulation and storage

Maintenance of high viability and virulence during production, formulation and storage forms the backbone of an effective quality control strategy. Viability refers to the percentage of living IJs (compared with dead and non-infective stages) whereas total viable nematodes are the total numbers of living IJs in a suspension. This distinction is important as dead nematodes dissolve over time and viability alone may be misleading. Also, some nematode species adopt quiescent postures that may be easily confused with dead nematodes. Therefore, motionless nematodes should be either probed or agitated by adding a drop of hydrogen peroxide to facilitate assessments. Overpacking is a method of ensuring minimum total viable nematodes in a product.

Nematode viability and virulence can be influenced by many factors during mass production, formulation and storage (Table 4.2). These may include the source and genetic diversity of the master stock, quality of the host or media, exposure to environmental extremes (temperature, aeration, sheer), contamination, and toxicity of antifoaming and antimicrobial agents. In addition, factors such as moisture content and the rate of water loss from the formulations, thermal cycling (temperature shifts) during storage and relative humidity may impact the quality of the nematodes. Also, the optimum levels of various factors may differ with nematode species and therefore close attention should be paid to monitor each factor. For instance, the optimum storage temperature differs with nematode species. Although low temperatures (2–5°C) generally reduce nematode metabolic activity and can therefore enhance their shelf-life, some warm-adapted species such as *H. indica* and *S. riobrave* do not store well at temperatures below 10°C (Strauch *et al.*, 2000; Grewal, 2002).

As the product ages, the depletion in stored energy reserves may reduce virulence (Patel *et al.*, 1997b; Wright *et al.*, 1997), nictation ability (Lewis *et al.*, 1995) and environmental tolerance (Selvan *et al.*, 1993a,b; Patel *et al.*, 1997a) of IJs. Therefore, time from production to formulation, formulation to packaging, and packaging to shipping is usually controlled. Batch codes and expiration dating are useful methods of tracking and controlling the inventory life (refrigerated storage time before application). Assessment of microbial contamination is also an integral part of nematode product quality assessment. Physical characteristics such as product colour and weight, granule size distribution, formulation dispersibility, product temperature and packaging are also monitored to reduce batch-to-batch variability and maintain consistency.

Nematode production batches can also differ in quality. For example, batches of *S. carpocapsae* produced in liquid culture were found to differ in lipid content (the major energy reserve) of the IJs (Grewal and Georgis, 1998). There are various optical and biochemical methods available to measure the lipid and glycogen content of IJs (Fitters *et al.*, 1997; Patel and Wright, 1997). Similarly, differences in the virulence of nematode batches are quite common. There is also the risk of genetic

Table 4.2. Quality control during mass production and formulation of entomopathogenic nematodes (EPNs).

Process	Quality control parameter
Master stock	Source
	Maintenance of genetic diversity
	Prevention of genetic deterioration
	Contamination avoidance
Mass production	Quality of the host (*in vivo* rearing) or media (*in vitro* production)
	Temperature
	Aeration
	Sheer stress
	Contamination avoidance
	Toxicity of antifoaming agents
Harvesting and bulk storage	Temperature
	Aeration
	Sheer stress
	Contamination avoidance
	Toxicity of detergents, antifoaming agents and antimicrobial agents
	Length of storage period
Formulation	Temperature
	Aeration
	Moisture content and rate of water loss
	Contamination avoidance
	Toxicity of antimicrobial agents
Product storage (inventory)	Temperature
	Aeration
	Relative humidity
	Contamination avoidance
	Toxicity of antimicrobial agents
	Length of storage period

deterioration via genetic drift or inadvertent selection during repeated subculturing of nematodes. In this regard, some nematode species may be more prone to rapid deterioration than others. For example, Wang and Grewal (2002) observed decline in the environmental stress tolerance of *H. bacteriophora* within three to six passages through *Galleria mellonella* in the laboratory. They also demonstrated that the best method to prevent this genetic deterioration is through storage of the master stock in liquid nitrogen.

4.3.3. Maintaining nematode quality during transport and application

Both extremes and fluctuations in temperature (thermal cycling) may reduce nematode quality during transport. Nematodes respond physiologically to changes in temperature and would thus expend a considerable amount of stored energy reserves to acclimate to external temperature conditions. Changes in temperature during transport can be measured by including temperature monitors in the product containers.

All nematode products should be applied as soon as they are received by the end-user. If the product cannot be used immediately, it should be refrigerated at 2–10°C or according to the label instructions. Nematode products should never be frozen, as freezing is detrimental to all commercially available nematodes. Likewise nematode-containing products should never be exposed to hot sun or stored in warm places.

Most nematode products are applied as aqueous suspension. Therefore, the nematodes need to be mixed in water for application. The WGs, wettable powders and vermiculite formulations can be directly mixed in water, but the nematodes have to be squeezed out of sponges. The nematodes contained in the alginate gel matrix are released by dissolving the gel in water with the aid of sodium citrate (Georgis, 1990). As nematodes settle out of suspension they need to be constantly mixed during preparation of the suspension and application. Nematodes require oxygen that can be simply provided via mixing. The temperature of the water used for preparation of nematode suspension and application should not exceed 30°C. The choice of the application equipment for nematodes is described in Chapter 5.

4.3.4. Philosophies of nematode quality assessment

An assessment of nematode quality should provide information on whether or not a nematode will control the target insect in the field, given that there are no environmental constraints like drought, or high or low temperatures. Suggested assays can be classified into holistic and reductionistic. A compilation of quality assessment methods can be found on the Internet (www.cost850.ch) and in Glazer and Lewis (2000). The most holistic approach would be a quality assessment using the target insect under field conditions. For most target insects this is costly and time-consuming and therefore not practical. Moreover, the field conditions add variation to the test result and hence reduce its predictive value. On the other hand, it has been suggested to test every single trait suspected to impact nematode quality like the content of stored energy reserves, the proportion of nematodes retaining bacteria and the number of bacteria per nematode, the agility of the IJs responding to temperature gradients and their sensitivity to chemical host cues. The problem with this reductionistic approach is the poor understanding of the contribution of each individual trait to overall nematode performance and the lack of insight into the interaction among the traits. Reductionistic assays are, however, indispensable for detecting the sources of variation in nematode quality.

The most commonly used assays compromise between the holistic and reductionistic approaches. Model insects are challenged with a well-defined number of nematodes in an artificial arena that reduces variability compared to field conditions. A good assessment of nematode quality should be designed to include as many events of the infection process (see Table 4.3) as possible in one test. On the other hand, variance should be minimized and the assay should be reproducible. Cost, and especially time efficiency, are other requirements of quality assessment methods since information is needed before the product is released and the product's shelf-life is limited.

Virulence of EPNs, i.e. the ability to search, recognize, penetrate and kill insect hosts, can be measured by several different methods, including one-on-one bioassays (Converse and Miller, 1999; Grewal *et al.*, 1999), LC_{50} bioassays (Georgis, 1990), establishment efficiency (Hominick and Reid, 1990; Epsky and Capinera, 1994) or penetration efficiency (invasion rate) bioassays (Glazer, 1992). However, bioassays using multiple nematodes against single or multiple hosts are considered inappropriate for quality control purposes due to host–parasite interactions. The invasion into pre-infected hosts has been shown to be more likely than into non-infected hosts (Grewal *et al.*, 1993; Hay and Fenlon, 1995), whereas other studies indicate a repellence of IJs from infected cadavers (Glazer, 1997; Grewal *et al.*, 1997). If grouped into one arena, infected cadavers would affect infection of further insects. The goal of nematode quality assessment must be to expose all defective IJs (the smallest infectious unit). Thus, one-on-one bioassays should be the most sensitive to 'impaired' nematodes compared to the assays using multiple IJs, because multiple nematode bioassays have the potential to hide the defective individuals.

One-on-one bioassays have been developed and are routinely used to assess quality of commercially produced EPNs in some companies. The choice of the insect host used, however, depends upon its susceptibility and availability. The original one-on-one bioassays were developed using the wax moth *G. mellonella* larvae due to their high susceptibility to EPNs and commercial availability. These bioassays used filter papers placed in 24-well plates on which individual last instar larvae were exposed to single nematode IJs (Converse and Miller, 1999). These methods work well for *S. carpocapsae, S. feltiae* and *S. riobrave* (Converse and Miller, 1999; Grewal, 2002), the nematode species that cause around 50% larval mortality at one IJ per larva. However, some nematode species such as *H. bacteriophora* and *S. scapterisci* do not cause significant mortality of *G. mellonella* on filter papers, and rates of 5–50 IJs/larva, respectively, are required to obtain around 50% mortality. In an effort to reduce the rate of nematodes used in these quality assessment assays, a new bioassay was developed in which filter paper was replaced with playsand in the 24-well plates (Grewal *et al.*, 1999). This new 'sandwell' bioassay resulted in substantial reduction in the rate of IJs required to cause significant mortality of *G. mellonella* larvae. For example, in the sandwell bioassay, *S. scapterisci* caused 30–70% mortality at 15 IJs per larva as opposed to 50 IJs required to cause the same level of mortality in the filter paper assay (Grewal *et al.*, 1999).

Filter paper arenas are generally more suitable for host-finding by ambushing nematodes whereas sand columns are optimal for cruisers (Grewal *et al.*, 1994). However, ambushers and cruisers performed equally well in the sandwell bioassay, which facilitates both ambushing and cruising behaviours by IJs (Grewal *et al.*, 1999). Recent tests demonstrate that the sandwell bioassay can be used for quality assessment of almost all the species of *Steinernema* and *Heterorhabditis*, at the rate of 1 IJ/larva (Table 4.4; P.S. Grewal and S.K. Grewal, unpublished data), except for *S. scapterisci* (Grewal *et al.*, 1999). The sandwell bioassay is easy to set up and is closer to field conditions than the filter paper bioassay. Therefore, it has been proposed to adopt the sandwell bioassay as a standard quality assessment tool for EPNs. The stepwise

Table 4.3. Events in the infection process of entomopathogenic nematodes (EPNs) and the traits of nematodes or symbiotic bacteria that determine infection success and should be incorporated into quality control protocols.

Infection event	Traits determining success
Host-finding	Host-sensing (chemotaxis, thermotaxis, thigmotaxis, etc.)
	Host-finding behaviour (ambushing or cruising)
	Host-finding efficiency (distance and time)
	Attack strategy (mass attack, leaders versus followers)
Host recognition	Specificity to the target host
	Responsiveness to host cues
Host penetration	Route of penetration
	Penetration efficiency (invasion rate)
Host establishment	Evasion from non-self-recognition
	Suppression of immune reactions
	Production of anti-immune factors (e.g. cecropins)
Bacterial release	Quantity and frequency of bacteria carried
	Bacterial release efficiency
Host mortality	Bacterial defence against host immune response
	Rate of bacterial proliferation
	Expression of bacterial virulence factors

set-up of the sandwell bioassay can be found in Grewal (2002).

One criticism of the use of *G. mellonella* in quality assessment has been the fact that it is too susceptible to EPNs and thus may not be sensitive to impaired nematodes. This may be true when multiple IJs are used per larva, but differences in the quality of production batches and ages of nematodes have been detected with the use of single IJ per larva in the sandwell bioassay. For example, the *G. mellonella* larval mortality caused by the 7-week-old IJs of *H. zealandica* and *H. indica* was significantly lower than that caused by the 3-week-old nematodes (Fig. 4.1; P.S. Grewal and S.K. Grewal, unpublished data). Another commercially available insect host, the mealworm *Tenebrio molitor*, is used in the quality assessment of EPNs particularly in Europe. Currently, a group of 40 mealworms in sand-filled arenas are exposed to 5, 10 or 20 IJ nematodes per larva for *S. carpocapsae, S. feltiae* and *H. bacteriophora*, respectively. Mortality is recorded after 7 days. In order to determine the feasibility of using mealworm larvae in the sandwell bioassay we tested rates of 1, 2, 5, 10 and 20 IJs of three EPN species against single mealworm larvae in the 24-well plates. We found an excellent dose response for all three nematode species (P.S. Grewal, unpublished data) and, more importantly, single IJs of all three species caused 31–45% mortality (Table 4.4). The IJs used in this test were 1 month old. These preliminary results suggest that even mealworm larvae can be used in the one-on-one sandwell bioassay to assess the quality of EPNs.

A good assay must be able to detect differences between various nematode batches or age groups of a nematode species. The resolution of an assay may be described by the *F*-statistic calculated during analysis of variance. When using a dose of 30 *H. bacteriophora* per mealworm the resolution with grouped insects was superior to the assay with isolated insects (Peters, 2000). Further research is required to compare the resolution of the one-on-one bioassay using *G. mellonella* or *Tenebrio molitor* with multiple nematode and multiple insect bioassays.

For other biopesticides, such as *Bacillus thuringiensis*, a standard is always included in infectivity bioassays and relative efficiency is measured. Such standards for

Table 4.4. Mean percentage mortality (± SE) of last instar *Galleria mellonella* or *Tenebrio molitor* in the 1:1 sandwell bioassay after exposure of each larva to one infective juvenile (IJ) of different species of *Steinernema* or *Heterorhabditis* at 25°C.

Nematode species	Strain	*G. mellonella*[a]	*T. molitor*[b]
S. carpocapsae	All	79 (4.17)	41 (9.53)
S. feltiae	SN	72 (5.56)	45 (6.88)
S. glaseri	NJ	66 (4.28)	–
S. intermedium	NC	33 (4.16)	–
S. karii	Kenya	86 (2.78)	–
S. oregonensis	Oregon	25 (3.75)	–
S. rarum	Argentina	47 (3.36)	–
H. bacteriophora	HP88	63 (5.37)	–
H. bacteriophora	GPS11	31 (5.22)	31 (2.69)
H. indica	LN2	29 (2.15)	–
H. marelata	Oregon	38 (5.87)	–
H. megidis	UK	42 (3.48)	–
H. zealandica	X1	40 (4.89)	–

[a]Larval mortality after 72 h (P.S. Grewal and S.K. Grewal, unpublished data).
[b]Larval mortality after 96 h (P.S. Grewal, unpublished data).

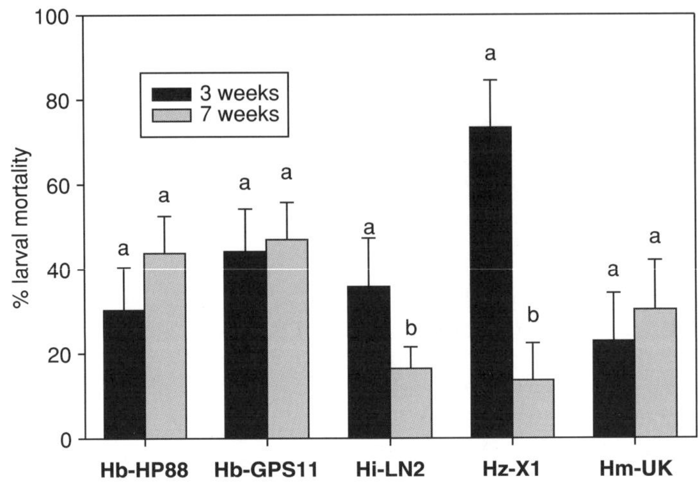

Fig. 4.1. Mean percentage *Galleria mellonella* mortality (± standard deviation) in the one-on-one sandwell bioassay by the 3- and 7-week-old infective juveniles (IJs) of entomopathogenic nematodes (EPNs). Hb = *Heterorhabditis bacteriophora;* Hi = *H. indica;* Hz = *H. zealandica;* Hm = *H. megidis.* The letters and numbers followed by species names represent strain. Same letter on the two bars for the same species indicate no significant difference at $P < 0.05$.

EPNs are not feasible due to the limited shelf-life of the IJs. This problem has been resolved by establishing a 'standard' based on the results from several bioassay runs under standard laboratory conditions. To establish a standard for a particular mass-production process of a nematode species, 30–40 bioassay runs are conducted and the larval mortality data are tested for normal distribution. The lower cut-off points (i.e. minimum larval mortality required for a 'pass') are then established for each nematode species for the rejection of an inferior production batch. This standard cut-off point will, of course, vary for different strains of the same species and for a particular mass-production process.

Besides good resolution, quality assessment methods should produce similar results if performed by different laboratories. The reproducibility of a method for counting nematode numbers in commercial packages and multiple nematodes–multiple mealworm bioassays was investigated in a 2-year project between nematode producers and retailers in Germany. While the counting method was highly reproducible, the absolute values for nematode infectivity varied considerably between laboratories. In comparisons of differently treated nematode packages, however, all laboratories came out with the same ranking. Similar results were obtained in a joint project between two nematode-producing companies (Peters, 2000).

4.3.5. Assessing the quality of commercially produced nematodes

Gaugler *et al.* (2000) assessed the quality of commercially produced nematodes aimed at a mail-order market in the USA. They found that most companies were accessible, and they reliably shipped pure populations of the correct species on time, in sturdy containers, often with superb accompanying instructions. Nematodes were received in satisfactory condition with acceptable levels of viability. Consistency, however, was a problem, with each supplier having one or more weak spots to bolster. Most shipments did not contain the expected nematode quality, and one shipment had no nematodes. Pathogenicity of several products against *G. mellonella* larvae was not equivalent to laboratory standards. *H. bacteriophora* was not always available when ordered. A few products contained mixed populations of *S. carpocapsae* and *H. bacteriophora*. Application rate recommendations provided by several suppliers were unsound. They concluded that the cottage industry lacks rigorous quality control, self-regulation is problematic without feedback and consumers are rarely able to provide this feedback. Improved reliability by the nematode industry will most likely be achieved via industry-generated agreement on standards for quality. Along these lines, the association of suppliers of biocontrol organisms in Germany (Verein der Nützlingsanbieter Deutschlands) has developed standards for packing, cooling and transport durations for nematode products. Moreover, they have proactively organized workshops for retailers, extension services and interested end-users to teach them how to assess the quality of nematode products.

References

Bedding, R.A. (1988) Storage of insecticidal nematodes. *World Patent No.* WO 88/08668.

Bedding, R.A. and Butler, K.L. (1994) Method for storage of insecticidal nematodes. *World Patent No.* WO 94/05150.

Capinera, J.L. and Hibbard, B.E. (1987) Bait formulations of chemical and microbial insecticides for suppression of crop feeding grasshoppers. *Journal of Agricultural Entomology* 4, 337–344.

Chang, F.N. and Gehret, M.J. (1995) Stabilized insect nematode compositions. *US Patent No.* 5,401,506.

Connick, W.J. Jr, Nickle, W.R. and Vinyard, B.J. (1993) 'Pesta': new granular formulations for *Steinernema carpocapsae*. *Journal of Nematology* 25, 198–203.

Converse, V. and Miller, R.W. (1999) Development of the one-on-one quality assessment assay for entomopathogenic nematodes. *Journal of Invertebrate Pathology* 74, 143–148.

Epsky, N.D. and Capinera, J.L. (1994) Invasion efficiency as a measure of efficacy of the entomopathogenic nematode *Steinernema carpocapsae* (Rhabditida: Steinernematidae). *Journal of Economic Entomology* 87, 366–370.

Fitters, P.F.L., Meijer, E.M.J., Wright, D.J. and Griffin, C.T. (1997) Estimation of lipid reserves in unstained living and dead nematodes by image analysis. *Journal of Nematology* 29, 160–167.

Gaugler, R., Grewal, P.S., Kaya, H.K. and Smith-Fiola, D. (2000) Quality assessment of commercially produced entomopathogenic nematodes. *Biological Control* 17, 100–109.

Glazer, I. (1992) Invasion rate as a measure of infectivity of steinernematid and heterorhabditid nematodes to insects. *Journal of Invertebrate Pathology* 59, 90–94.

Glazer, I. (1997) Effects of infected insects on secondary invasion of steinernematid entomopathogenic nematodes. *Parasitology* 114, 597–604.

Glazer, I. and Lewis, E.E. (2000) Bioassays for entomopathogenic nematodes. In: Navon, A. and Ascher, K.R.S. (eds) *Bioassays of Entomopathogenic Microbes and Nematodes.* CAB International, Wallingford, UK, pp. 229–247.

Glazer, I. and Salame, L. (2000) Osmotic survival of the entomopathogenic nematode *Steinernema carpocapsae*. *Biological Control* 18, 251–257.

Georgis, R. (1990) Formulation and application technology. In: Gaugler, R. and Kaya, H.K. (eds) *Entomopathogenic Nematodes in Biological Control.* CRC Press, Boca Raton, Florida, pp. 173–191.

Georgis, R., Dunlop, D.B. and Grewal, P.S. (1995) Formulation of entomopathogenic nematodes. In: Hall, F.R. and Barry, J.W. (eds) *Biorational Pest Control Agents: Formulation and Delivery.* American Chemical Society, Maryland, pp. 197–205.

Grewal, P.S. (1998) Formulations of entomopathogenic nematodes for storage and application. *Japanese Journal of Nematology* 28, 68–74.

Grewal, P.S. (2000a) Enhanced ambient storage stability of an entomopathogenic nematode through anhydrobiosis. *Pest Management Science* 56, 401–406.

Grewal, P.S. (2000b) Anhydrobiotic potential and long-term storage stability of entomopathogenic nematodes (Rhabditida: Steinernematidae). *International Journal of Parasitology* 30, 995–1000.

Grewal, P.S. (2002) Formulation and application technology. In: Gaugler, R. (ed.) *Entomopathogenic Nematology.* CAB International, Wallingford, UK, pp. 265–287.

Grewal, P.S. and Georgis, R. (1998) Entomopathogenic nematodes. In: Hall, F.R. and Menn, J. (eds) *Methods in Biotechnology: Biopesticides: Use and Delivery.* Humana Press Inc, Totowa, New Jersey, pp. 271–299.

Grewal, P.S. and Jagdale, G.B. (2002) Enhanced trehalose accumulation and desiccation survival of entomopathogenic nematodes through cold preacclimation. *Biocontrol Science and Technology* 12, 533–545.

Grewal, P.S., Selvan, S., Lewis, E.E. and Gaugler, R. (1993) Male insect-parasitic nematodes: a colonizing sex. *Experientia* 49, 605–608.

Grewal, P.S., Lewis, E.E., Gaugler, R. and Campbell, J.F. (1994) Host finding behavior as a predictor of foraging strategy of entomopathogenic nematodes. *Parasitology* 108, 207–215.

Grewal, P.S., Lewis, E.E. and Gaugler, R. (1997) Response of infective stage parasites (Nematoda: Steinernematidae) to volatile cues from infected hosts. *Journal of Chemical Ecology* 23, 503–515.

Grewal, P.S., Converse, V. and Georgis, R. (1999) Influence of production and bioassay methods on infectivity of two ambush foragers (Nematoda: Steinernematidae). *Journal of Invertebrate Pathology* 73, 40–44.

Hay, D.B. and Fenlon, J.S. (1995) A modified binomial model that describes the infection dynamics of the entomopathogenic nematode *Steinernema feltiae* (Steinernematidae: Nematoda). *Parasitology* 111, 627–633.

Hokkanen, H.M.T. and Menzler-Hokkanen, I. (2002) Formulation of the nematodes *Steinernema feltiae* for a shelf-life up to one year. *VIII International Colloquim on Invertebrate Pathology and Microbial Control*, Foz do Iguassu, Brazil, Poster abstract MC25.

Hominick, W.M. and Reid, A.P. (1990) Perspectives on entomopathogenic nematology. In: Gaugler, R. and Kaya, H.K. (eds) *Entomopathogenic Nematodes in Biological Control.* CRC Press, Boca Raton, Florida, pp. 327–345.

Lewis, E.E., Selvan, S., Campbell, J.F. and Gaugler, R. (1995) Changes in foraging behavior during the infective stage of entomopathogenic nematodes. *Parasitology* 105, 583–590.

Patel, M.N. and Wright, D.J. (1997) Phospholipids and their fatty acids in infective juveniles of entomopathogenic steinernematid nematodes. *Comparative Biochemistry and Physiology* 118B, 649–657.

Patel, M.N., Perry, R.N. and Wright, D.J. (1997a) Desiccation survival and water contents of entomopathogenic nematodes, *Steinernema* spp. (Rhabditida: Steinernematidae). *International Journal of Parasitology* 27, 61–70.

Patel, M.N., Stolinski, M. and Wright, D.J. (1997b) Neutral lipids and the assessment of infectivity of entomopathogenic nematodes: observations on four *Steinernema* species. *Parasitology* 114, 489–496.

Peters, A. (2000) Insect based assays for entomopathogenic nematode infectiousness: definitions, guidelines, problems. *IOBC WPRS Bulletin* 23, 109–114.

Selvan, S., Gaugler, R. and Grewal, P.S. (1993a) Water content and fatty acid composition of infective juvenile entomopathogenic nematodes during storage. *Journal of Parasitology* 79, 510–516.

Selvan, S., Gaugler, R. and Lewis, E.E. (1993b) Biochemical energy reserves of entomopathogenic nematodes. *Journal of Parasitology* 79, 167–172.

Shapiro-Ilan, D.I., Lewis, E.E., Behle, R.W. and McGuire, M.R. (2001) Formulation of entomopathogenic nematode-infected cadavers. *Journal of Invertebrate Pathology* 78, 17–23.

Shapiro-Ilan, D.I, Lewis, E.E., Tedders, W.L. and Son, Y. (2003) Superior efficacy observed in entomopathogenic nematodes applied in infected-host cadavers compared with application in aqueous suspension. *Journal of Invertebrate Pathology* 83, 270–272.

Silver, S.C., Dunlop, D.B. and Grove, D.I. (1995) Granular formulations of biological entities with improved storage stability. *World Patent No.* WO 95/0577.

Strauch, O., Niemann, I., Neumann, A., Schmidt, A.J., Peters, A. and Ehlers, R.-U. (2000) Storage and formulation of the entomopathogenic nematodes *Heterorhabditis indica* and *H. bacteriophora. Biocontrol* 54, 483–500.

Wang, X. and Grewal, P.S. (2002) Rapid deterioration of environmental tolerance and reproductive potential of an entomopathogenic nematode during laboratory maintenance. *Biological Control* 23, 71–78.

Wright, D.J., Grewal, P.S. and Stolinski, M. (1997) Relative importance of neutral lipids and glycogen as energy stores in dauer larvae of two entomopathogenic nematodes, *Steinernema carpocapsae* and *Steinernema feltiae. Comparative Biochemistry and Physiology* 118B, 269–273.

5 Application Technology

D.J. Wright,[1] A. Peters,[2] S. Schroer[2] and J.P. Fife[3]
[1]*Department of Biological Sciences, Imperial College London, Silwood Park Campus, Ascot, Berkshire SL5 7PY, UK;* [2]*Institute for Phytopathology, Christian-Albrechts-University Kiel, Klausdorfer Str. 28–36, D-24223 Raisdorf, Germany;* [3]*Food, Agricultural, and Biological Engineering, The Ohio State University, 1680 Madison Ave, Wooster, OH 44691, USA*

5.1. Introduction

Application technology for entomopathogenic nematodes (EPNs) has been a relatively neglected area, partly because spray equipment for chemical pesticides and standard irrigation systems can be used to apply nematode infective juveniles (IJs) without major modifications (Georgis, 1990). EPNs are, however, some of the most expensive active ingredients (a.i.) used for insect control. They are also, like other biopesticides, particulate and can have different optimal application requirements to chemicals (Matthews, 2000). Improvements in application technology for EPNs that fundamentally aim at minimizing losses during the transfer of an a.i. from the mixing tank to the target insect are therefore badly needed.

Application is thus one of the most important barriers to the more widespread adoption of EPNs in insect pest management. Improvements to the application systems currently in use will aid in the more efficient transfer from chemical regimes. EPNs are most commonly used for the treatment of soil-borne insects, where good control is often obtained, although improvements in application technology are still required to make their use more reliable for growers. Targeted application methods, such as baits or infection stations, could also widen the economically exploitable host range of EPN.

The host range of EPN also covers a large number of serious foliar feeding insects; hence foliar application is an interesting option to explore. Control of foliar pests has been attempted, but effective control has been limited to specific, more protected environments (Lacey *et al.*, 1993; Bennison *et al.*, 1998); control in more exposed conditions has proved much more variable (Mason and Wright, 1997; Williams and Walters, 2000; Unruh and Lacey, 2001).

5.2. Nematode-specific problems

Nematodes are usually applied in aqueous suspensions. The water used should not be too hot (4–30°C) and it should not be heavily chlorinated. Black irrigation hoses can heat up considerably unless buried and most nematodes will not withstand temperatures > 35°C for more than 30 min. Moreover, the solubility of oxygen decreases dramatically with increasing temperature and low oxygen concentrations will inactivate nematodes. Care should be taken for nematode compatibility with chemical pesticides. Although nematodes are fairly resistant to fungicides and herbicides, they can be very susceptible to insecticides (Patel and Wright, 1996). The most comprehensive list of pesticide side effects on nematodes is provided in Chapter 20. More research is currently being carried out to explore possible tank-mixing with pesticides following standardized IOBC guidelines (Peters, 2003).

With a density of about 1.05 g/cm^3, IJs are heavier than water and they will settle in a spray tank. Settling velocities of some nematodes are given in Table 5.1; larger IJs appear to sediment faster than smaller ones. Sedimentation will result in an unequal distribution over time and can cause substantial problems when applying nematodes using irrigation systems; some sedimentation may also occur in spray tanks. Sedimentation can be mitigated by increasing the viscosity of the water. Figure 5.1 shows the effect of adding carboxymethylcellulose (CMC) on the sedimentation speed of *Steinernema feltiae* in aqueous suspensions. Adding 0.1% (w/v) CMC decreases the sedimentation speed of *S. feltiae* by about 83% (Peters and Backes, 2003).

With most application methods IJs will be exposed to shear forces, which occur in the pumps, when they pass through filters or nozzles and when they hit the canopy. For example, high volume (> 10 m^3/ha/h) overhead irrigation equipment requires high pressures, and the shear forces involved might be detrimental for nematodes.

Extensive recirculation of the tank mix can also be detrimental to EPNs. Nilsson and Gripwall (1999) reported that the survival of *S. feltiae* decreased by approximately 10% during a 20-min pumping period, using a piston pump. They suggested that the reason for the decreased viability was probably mechanical stress from the pump and nozzles, but may also have been due to the rise of temperature in the liquid. Other work showed that nematode viability is not influenced by passage

Table 5.1. Settling velocities of biocontrol nematodes in water.

Nematode species	Settling velocity (mm/min)	References
Steinernema carpocapsae	3.6	Schroer *et al.*, 2005
S. feltiae	14	Young *et al.*, 1998
S. feltiae	5.8	Peters and Backes, 2003
Heterorhabditis megidis	6	Young *et al.*, 1998
H. bacteriophora	1	G. Marini and R.-U. Ehlers, unpublished data
Phasmarhabditis hermaphrodita	8	Young *et al.*, 1998

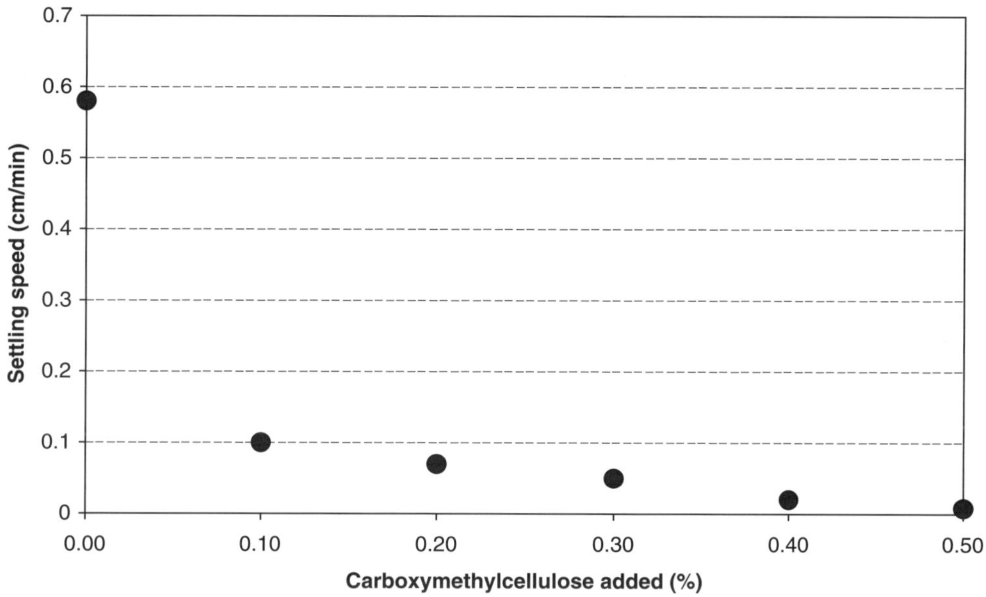

Fig. 5.1. Settling speed of *Steinernema feltiae* infective juveniles (IJ) in water at 20°C with different concentrations of carboxymethylcellulose.

through different pumps (centrifugal, diaphragm, roller, piston) operated at standard pressures (Klein and Georgis, 1994; Fife, 2003), which suggests that reductions in viability are likely the result of temperature influences rather than mechanical stress.

Liquid temperature within a spray tank increases during pump recirculation, and can produce conditions that are incompatible with EPNs. The general recommendation is to avoid temperatures exceeding 30°C within the pump, tank and nozzles (Grewal, 2002). Lower-capacity pumps, such as a diaphragm or roller pump, are better suited for use with EPNs compared with a high-capacity centrifugal pump, which can contribute significant heat to the spray system (Fig. 5.2). Additionally, liquid volume within the spray system is important because the smaller the volume

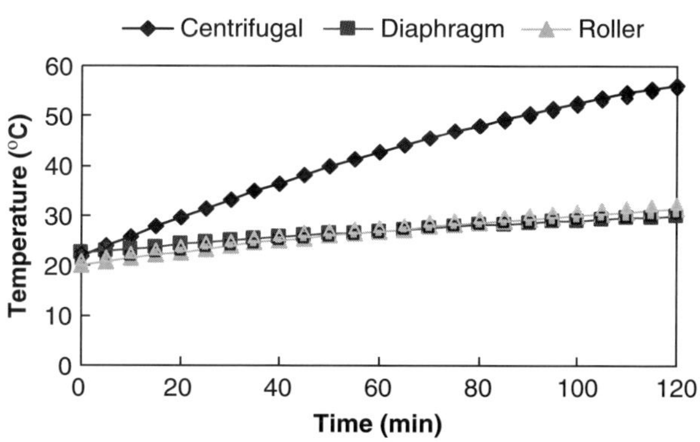

Fig. 5.2. Temperature during recirculation of 45.4 l of water at a volumetric flow rate of 15.1 l/min (4 gpm) using centrifugal, diaphragm and roller pumps.

of liquid in the tank, the more times the liquid will pass through the pump, causing the temperature to increase at a greater rate.

When applying EPNs, filters and sieves should be at least 300 μm wide (= 50 mesh) or they should be removed before application (Klein and Georgis, 1994). Nozzle apertures > 500 μm are recommended for nematode applications. Nematode species can differ in shear sensitivity; IJs of *S. carpocapsae* are able to withstand greater pressure differentials (Fig. 5.3) (Fife *et al.*, 2003) and more intensive hydrodynamic conditions (Fife *et al.*, 2004) than *Heterorhabditis bacteriophora* or *H. megidis*. Consequently, EPN species is an important factor to consider when defining spray-operating conditions. Operating pressures within a spray system should not exceed 20 bar (2000 kPa; 295 p.s.i.) for *S. carpocapsae* and *H. bacteriophora*, and 13.8 bar for *H. megidis*. Other EPN species may require lower pressures. For example, *P. hermaphrodita* appears to be particularly susceptible compared with *S. feltiae* and *H. megidis* (Young *et al.*, 1998).

Strongest shear forces will most likely occur at the nozzles. The shear forces depend on the nozzle geometry, material of the nozzles and the velocity at which the nematodes pass the nozzle. This in turn is dependent on the pressure. In North America it is usually advised that pressures up to 20.7 bar (2068 kPa; 300 p.s.i.) can be used, whereas 5 bar is usually the limit stated by European nematode distributors. Who is right? The absolute pressure nematodes can tolerate is certainly much higher than 20.7 bar because they lack gas-filled body compartments. What matters is the shear forces involved in spraying suspensions at higher pressures; but they are very dependent on the geometry of the tubing and nozzles.

Fife (2003) evaluated the distinct differences in the flow characteristics of fan- and cone-type nozzles (Fig. 5.4) with respect to EPN damage. The internal shape of the fan nozzle causes liquid from a single direction to curve inwards so the two streams of liquid meet at the elliptical exit orifice, producing the characteristic fan pattern. Within a cone nozzle, the liquid is forced through tangential slits into a swirl chamber giving the liquid a high rotational velocity, producing the cone pattern at the circular exit orifice. It was found that the reduced flow area of the narrow, elliptic exit orifice

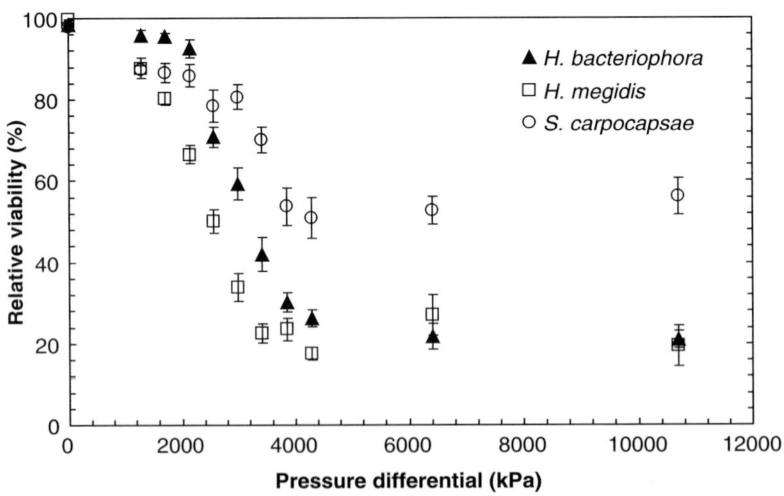

Fig. 5.3. Relative viability of *Heterorhabditis bacteriophora*, *H. megidis* and *Steinernema carpocapsae* after pressure differential treatments. Error bars represent ± SE ($n = 6$ for *H. bacteriophora* and *H. megidis*, and $n = 9$ for *S. carpocapsae*).

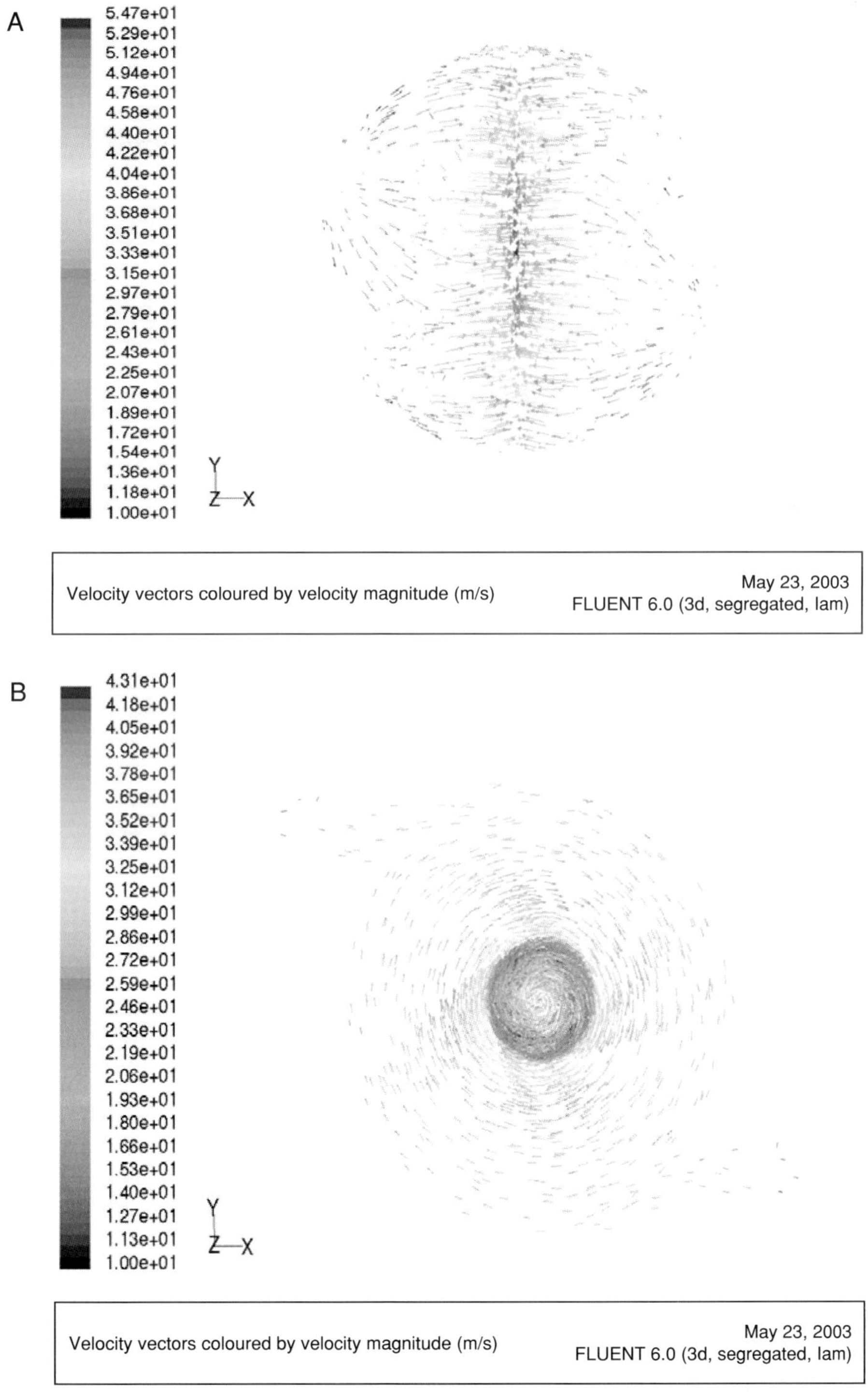

Fig. 5.4. Numerical simulation of (A) a Spraying Systems XR8001VS flat fan nozzle and (B) a Spraying Systems TXA8001VK hollow cone nozzle using FLUENT, a commercial computational fluid dynamics (CFD) program. Simulation results show the velocity (m/s) within the exit orifices of each nozzle.

of a flat fan nozzle generates an extensional flow regime, where tensile stresses developed are large enough to cause nematode damage. The high rotational flow component within a cone nozzle does not produce hydrodynamic conditions conducive to causing EPN damage. However, it is important that the cone nozzle exit orifice is sufficiently larger than the EPN length to avoid any damage. Overall, it was found that common 01-type hydraulic nozzles are acceptable for spray application of EPNs when following the manufacturer recommendations. Larger-capacity hydraulic nozzles are recommended, particularly for soil-applied treatments where a high volume of water is required. Particles in the spray suspension, which partly block the nozzle orifice, can considerably reduce the viability of the nematodes passing through the nozzle (Gwynn *et al.*, 1999).

5.3 Soil Application

5.3.1. Conventional sprayers

Most EPN are probably applied as a drench with a high volume of water. The recommended water volume varies considerably but is always much higher than for chemical insecticides, which are applied to cover the leaf area. In greenhouses, the recommended water volume can quite easily be applied with hand-held showers or by incorporating nematode application in the daily irrigation regime.

Spray equipment used in the open field is usually built for maximum volumes of 500–600 l/ha, and it is unreasonable to expect that more than 1000 l/ha will be applied to broad acre crops. On golf courses in Europe, the recommended volume for applying nematodes is 1200 l/ha but application post-irrigation is recommended. Such large water volumes require appropriate spray nozzles. The international code for nozzles gives the angle of the spray swath and the flow rate in US gallon/min at 2.81 bar (e.g. 120–08 for 120° spray swath and 0.8 US gallon/min flow rate). For nematodes, nozzles with the highest flow rate should be chosen. Logically, these nozzles will also have the largest orifice and create relatively the lowest shear stress. The maximum flow rate found in the most commonly used nozzle type, the flat-fan nozzle, is 0.8 gallon/min at 5 bar, which transfers to 1200 l/ha at 5 km/h. Tongue-nozzles for applying soil herbicides are made for up to 1850 l/ha at 6 km/h. The optimum nozzle type for applying nematodes has not been fully resolved (Section 5.4.3). Nor has it been elucidated whether an even coverage of the soil is superior to an application using nozzles or hoses hanging down from the spray rig that apply the whole volume in lines 10–50 cm apart.

EPNs only fit into relatively large spray droplets that are not prone to drift (Lello *et al.*, 1996). Spray technology for chemical application is usually aimed at covering the highest possible proportion of the above-ground parts of crops or weeds. They are not optimized to transport material into the soil. Increasing pressure or using advanced nozzles to lower droplet volume is useless for nematodes. Any droplet that does not contain nematodes and does not hit the soil is a waste of water and spray adjuvants. Therefore, the technology for soil fertilizers or irrigation rather than chemical pesticides is probably better suited to apply EPNs to the soil.

Controlling grubs in turf is one of the most promising applications of EPNs, but turf also poses large challenges to the application technology (see Chapter 7, this volume). The thatch layer, a layer of densely packed dead plant material, is a sink for nematodes. Zimmerman and Cranshaw (1991) recorded only 10–17% penetration of *H. bacteriophora* and *S. feltiae* through the thatch, even after three irrigation treatments of 0.64 cm over 48 h. Turf tends to develop dry patches with low water permeability (Ritsema and Dekker, 2003). Any liquid applied to the soil will run off the surface from these patches and penetrate only the interpatch areas. Anionic and non-ionic products, such as sulfonated carbonic acids (e.g. Kick™; Compo, Germany), ethylene-oxide and propylene-oxide copolymers (e.g. Foresight®; Famura; UK) or alkylpolyglycosides and fatty acids (Magic Wet; Cognis; Germany) are available to make these dry patches permeable. These substances can and should be tank-mixed with nematodes applied to turf. A pretreatment with these substances during regular turf irrigation will help to transfer nematodes into the soil as well.

Above ground, EPNs are exposed to ultraviolet (UV) radiation and desiccation, and should therefore enter the soil as quickly as possible. In wheat, incorporation of spray-applied *P. hermaphrodita* by a spring-tine cultivator was shown to significantly increase the infection of slugs (Wilson *et al.*, 1996). Numerous studies have shown that post-application irrigation increases nematode performance dramatically (e.g. Curran, 1992; Downing, 1994; Boselli *et al.*, 1997). In turf, a minimum of 0.64 cm irrigation is recommended within 24 h of nematode application. In field trials, irrigation frequency proved to be of major importance on the efficacy of *H. bacteriophora* against the Japanese beetle (Georgis and Gaugler, 1991). Besides the transport function, irrigation keeps the water tension in the soil at a level allowing nematode activity. At low water tension, EPNs tend to remain inside infected cadavers rather than emerging and infecting new hosts (Koppenhöfer *et al.*, 1997). It is therefore crucial to also irrigate 2–4 weeks post application to enhance secondary cycling of nematode infections.

5.3.2. Irrigation systems

Various nematode-specific problems relating to application were discussed in Section 5.2. Leakages in drip irrigation hoses can also result in substantial losses of nematodes and this will decrease pressure and flow velocity in the remaining part of the hose. The flow velocity in irrigation hoses decreases in any case after every exit hole, and at low velocities nematodes can get trapped into the hose due to sedimentation (Section 5.2). Reed *et al.* (1986) recovered only 37–59% of the nematodes injected into a trickle irrigation system, and Conner *et al.* (1998) demonstrated that such losses were due to EPNs settling in tubing further away from the injection point. Increasing the pressure and especially increasing the viscosity of the irrigation solution (Section 5.2) can mitigate this problem.

In field experiments, EPNs have been successfully applied with centre-pivot irrigators in maize (Wright *et al.*, 1993), furrow irrigation in maize (Cabanillas and Raulston, 1996a,b) and cotton (Jech and Henneberry, 1997), and trickle irrigation systems (Reed *et al.*, 1986; Curran and Patel, 1988; Gouge *et al.*, 1997; Kakouli-Duarte *et al.*, 1997; Wennemann *et al.*, 2003). When compared to conventional spraying, delivering nematodes by irrigation was generally more successful (Cabanillas and Raulston, 1996a,b). Ellsbury *et al.* (1996) applied *S. carpocapsae* to maize by a lateral-move irrigation system and observed a threefold greater concentration of EPNs at the base of the maize plants by stem flow compared with the overall ground level.

If done properly, excellent application of EPNs through irrigation systems can be achieved. EPN rates can be substantially reduced, for example, from 5 to 2 billion IJs/ha in strawberries (Kramer and Grunder, 1998). In the grower's field, however, there is considerable variation in the technical standard of irrigation equipment and this can severely affect the distribution of IJs. Education of growers is indispensable to make such systems work.

5.3.3. Other techniques

Given the limitations of spray technology for applying EPNs to the soil, other equipment has been tried, especially since placing nematodes beneath grass roots by top application is difficult, even if ample water is used (Section 5.3.1). Subsurface application with an adapted seed-driller has been found to improve the delivery of *S. glaseri* to turfgrass by fourfold compared with application with a boom sprayer (Smits, 1999). When using a subsurface applicator

(Toro Co., USA) for EPN application on a golf course against the garden chafer (*Phyllopertha horticola*), excellent efficacy was achieved with one-third the dose used with boom sprayers (e-nema, unpublished data). Shetlar *et al.* (1993) recorded good control of billbug (*Sphenophorus parvulus*) larvae using a similar subsurface injector to apply 2.6 billion *S. feltiae*/ha at a depth of 2 cm at 1200 l/ha. Likewise, soil injectors have been used to treat strawberry plants under plastic mulch (see Chapter 12, this volume).

With any crop planted or sown in rows, a large quantity of EPNs is wasted between the plants. A more targeted application to the root system, by dipping plants into a nematode suspension, can give excellent results (Pye and Pye, 1985; Klingler, 1988). When problems with the efficiency of *H. bacteriophora* against *Otiorhynchus sulcatus* were recorded in German tree nurseries, growers dipped cuttings into a nematode solution before transplanting into the field. This method gave improved control and reduced the number of EPNs applied per hectare by 60%. Thickeners (e.g. 0.5% CMC) can be used to increase the amount of nematode solution retained by plant roots following dipping.

Nematodes can also be applied during sowing in granular formulations. This could be an efficient way, for example, to control maize rootworm (*Diabrotica* spp.) or sugar beet weevil (*Temnorhinus mendicus*). However, sowing or transplanting may not be the optimum time to control the target insect and slow-release granules are required to improve nematode persistence. Substantial progress has been made in developing such granules for the EPN bacteria *Serratia entomophila* (Johnson and Pearson, 2002). In oilseed rape, nematodes were applied in tea bags containing superabsorbant gel (Menzler-Hokkanen and Hokkanen, 2003) and persistence was good. Similarly, superadsorbant gel has been added to the soil to prolong persistence of *S. carpocapsae* against the citrus root weevil (*Diaprepes abbreviatus*) in Florida (Georgis, 1990). Infected insects can also serve as slow-release systems for EPNs (Shapiro-Ilan *et al.*, 2003), although they would be difficult to apply with conventional machinery, and rearing insects is only commercially viable where labour is a cheap resource. Technology for overcoming these limitations is being investigated (D. Shapiro-Ilan 2004, personal communication). Nematodes are expensive products and enclosing them into baits or infection (autodissemination) stations can reduce costs. Such methods may also open up new areas of application. Wheat bran and alginate beads containing *S. carpocapsae* have been successfully applied to control black cutworm (*Agrotis ipsilon*) larvae on maize; however, this treatment was not superior to a spray application (Capinera *et al.*, 1988). Bait stations with actively nictating *S. carpocapsae* outperformed standard insecticide-containing baits in field trials with cockroaches (Appel *et al.*, 1993), and a commercial version is available in the USA (Pye *et al.*, 2001). For houseflies in pigsties, a bait station with *H. megidis* or *S. feltiae* gave significantly better control than methomyl baits (Renn, 1998). Unlike in most other applications, adult insects are targeted in infection stations, rather than larvae. Dissemination of nematodes by infecting and releasing adult mole crickets in sound traps was reported by Parkman and Frank (1993). The use of adult Japanese beetles to disseminate *S. glaseri* in the field was first reported by Lacey *et al.* (1993).

5.4. Above-ground Application and Formulation Technology

5.4.1. Treatment of stem borers

Thanks to the moist and sunlight-protected environment inside trunks, EPNs can actively move to stem-boring insects. For controlling currant borer moth (*Synanthedon tipuliformis*) in blackcurrant, lateral spraying devices have been used; blackcurrant cuttings have also been treated with a hand-held sprayer followed by overnight incubation under a plastic cover to maintain high humidity (Miller and Bedding,

1982). Cotton flocs have been used to manually apply nematodes into the holes left by stem-boring *Zeuzera pyrina* (Deseö and Rovesti, 1992). In China, hundreds of hectares have been treated this way to control the peach fruit moth *Carposina nipponensis* (Wang, 1990) and over 100,000 shade trees have been injected with nematodes to control the cossid *Holcocerus insularis* (Bedding, 1990). In Italy, larvae of the cerambicid beetle (*Saperda carcharias*) were successfully controlled with EPNs by injecting nematode suspensions into the holes made by the larvae (Barani *et al.*, 2000).

5.4.2. Foliar application

Promising results with EPNs have been achieved under commercial conditions on protected ornamental and vegetable crops. Glasshouse trials have shown that *S. feltiae* can give effective control (up to 80%) of the agromyzid leafminers (*Liriomyza huidobrensis*, *Liriomyza bryoniae* and *Chromatomyia syngensiae*) on vegetables (lettuce, tomato) and leafminers and thrips on ornamentals (Hara *et al.*, 1993; Williams and MacDonald, 1995; Bennison *et al.*, 1998; Williams and Walters, 2000). Nematodes have also shown potential for controlling various other insects on foliage, including *Liriomyza trifolii* (Broadbent and Olthof, 1995) and *Bemisia tabaci* (Cuthbertson *et al.*, 2003). A common feature of these and other reports is that high relative humidities (80–90% or greater) were required for optimum control.

Under more exposed field conditions the results have been more variable (Begley, 1990; Glazer *et al.*, 1992; Baur *et al.*, 1998) although the potential of EPNs against early season apple pests has been reported (Belair *et al.*, 1998). Nematodes have also been suggested as possible components of integrated pest management (IPM) programmes for the diamondback moth (DBM; *Plutella xylostella*) on cruciferous vegetable crops (Baur *et al.*, 1998).

In most trials on foliar application of EPNs, standard hydraulic application equipment has been used. Mistblowers (Matthews, 2000) have also been used to spray EPNs against thrips and agromyzid leafminers on ornamentals in commercial greenhouses (L.R. Wardlow and S.J. Piggott, 2003, personal communication).

5.4.3. Spray equipment

In most cases the objectives for spraying EPNs or chemical insecticides on foliage are the same: to obtain the optimum cover and placement on the leaf surface in order to optimize contact with the target insect. For example, in leafminer control the aim is to maximize the density and distribution of EPNs on leaf surfaces to enable as many IJs as possible to locate a mine entrance (see Chapter 13, this volume). Cover and placement is usually more critical for EPNs, since their residual infectivity is generally only a few hours, and there are very limited possibilities for redistribution of IJs on the plant to compensate for suboptimal placement.

Standard spray systems that are designed for chemical application do not perform very efficiently when applying particulate materials such as nematode IJs (Lello *et al.*, 1996; Mason *et al.*, 1998a, 1999) or fungal spores (Matthews, 2000). Hydraulic nozzles (flat-fan and full-cone) produce a wide range of droplet sizes, many of which are too small to carry an IJ and therefore have a high water-to-nematode ratio. Higher-output (flow rate) nozzles give the best coverage or deposition of nematodes (IJ/cm^2 of leaf) and, in laboratory studies, greater insect control (Fig. 5.5) (Lello *et al.*, 1996). An ultra-low-volume spinning disc applicator (Ulva+, Micron Sprayers Ltd, Hereford, UK) gave lower deposition rates and poorer insect control compared with hydraulic nozzles (Fig. 5.5), but since it used 90% less nematodes such systems were thought to have greater potential if their use could be modified (Lello *et al.*, 1996).

Conventional spinning discs have a large number of narrow grooves and 'zero issue points' (teeth) that are designed to produce very small droplets, most of which are too

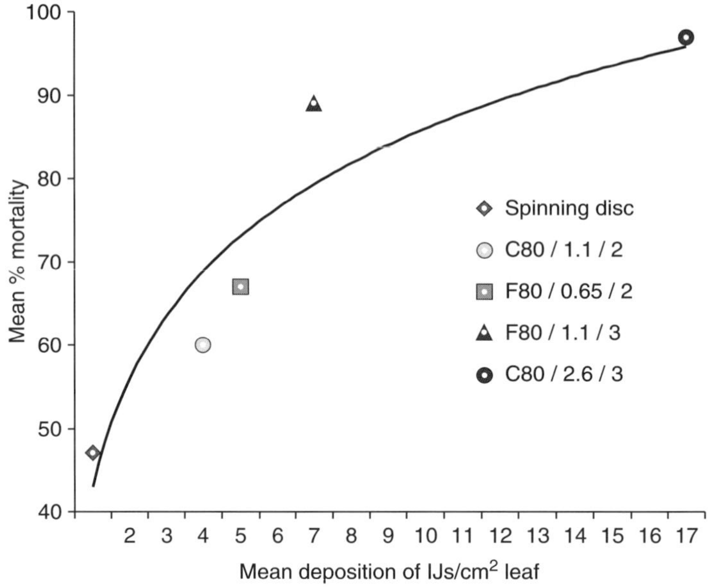

Fig. 5.5. Relationship between mean number of *S. carpocapsae* infective juvenile (IJ) spray deposited per square centimetre on Chinese cabbage leaf discs and mortality (24 h) of fourth instar larvae of *P. xylostella*. Fan (F) and full-cone (C) hydraulic nozzles (Lurmark Ltd, Cambridge, UK) with a 800 spray angle and flow rates of 0.65–2.6 l/min operated at 2 or 3 bar (200 or 300 kPa). Ulva+ spinning disc operated at 3 V (3000 rpm) fitted with red restrictor. Nematodes (2500 IJ/ml) were applied using a linear track sprayer at 0.1 m/s (spinning disc) or 1.0 m/s (hydraulic). (Adapted from Lello *et al.*, 1996.)

small to carry nematode IJs. Studies on two commercially available systems (Ulva+ and Herbaflex; Micron Sprayers Ltd) showed that deposition of nematodes was generally greater at slower rotational disc speeds (low operating voltages) since these produced larger droplets with a greater carrying capacity for IJs (Mason *et al.*, 1998a, 1999), although the equipment was not optimally designed to operate at such low speeds. Mason *et al.* (1998a) found that increasing the flow rate (application rate), and the initial concentration of IJs in the spray reservoir, proportionally increased the rate of deposition of IJs on leaf surfaces (per cm^2) for both the Ulva+ and Herbaflex. However, these spinning discs still failed to produce a droplet spectrum that carried sufficient IJs to compete with hydraulic systems. In addition, Piggott *et al.* (2003) showed that with the Ulva+ some IJs are separated from the carrier liquid on the disc surface by centrifugation, leading to aggregations of IJs in the disc grooves (Fig. 5.6) and their emission from the disc in semi-dry clumps to beyond the normal swath width.

A prototype spinning disc with an improved efficiency of application for EPNs was developed by Piggott *et al.* (2003). This disc is flatter than a standard disc, with fewer, larger grooves and has no teeth. These modifications give increased liquid flow over the disc surface, eliminate clumping of IJs and increase droplet size, resulting in improved deposition rates of IJs compared with conventional discs. However, the prototype disc tended to form clusters of IJs in larger, more dispersed droplets when compared with the Ulva+, which could reduce their effectiveness against target insects.

Even if such design problems can be overcome, it is uncertain whether novel application systems for biopesticides are commercially viable, since growers may be unwilling on economic or other grounds to

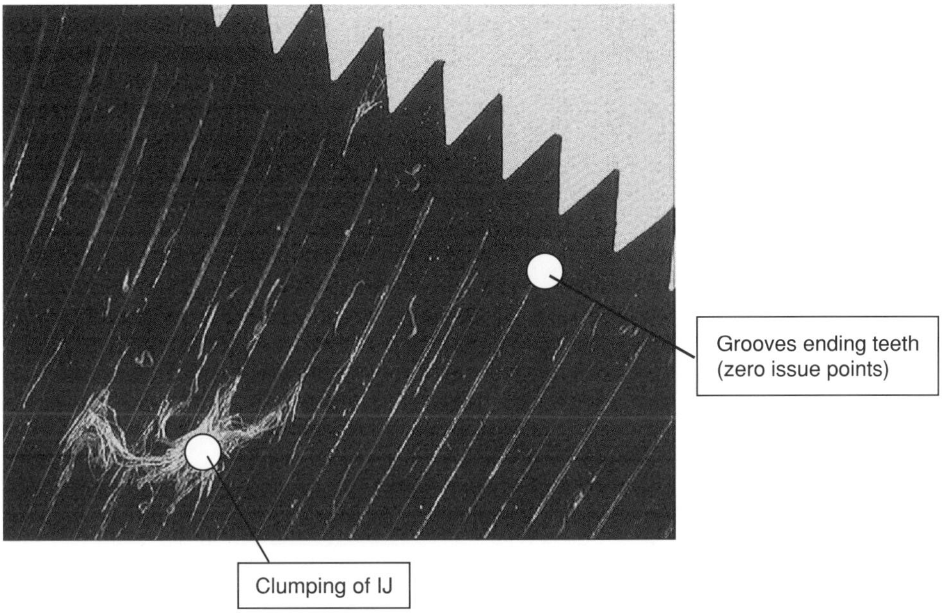

Fig. 5.6. Clumps of infective juvenile (IJ) (*Steinernema* sp.) formed in the grooves of an Ulva+ spinning disc. (Plate by M.N. Patel; adapted from Piggott *et al.*, 2003.)

replace their existing systems. Chapple *et al.* (1996) discuss such constraints and describe a double nozzle system designed to reduce the amount of biopesticide required and thus reduce a major limiting factor for such products – their relatively high cost compared with chemical alternatives.

5.4.4. Formulation technology

Large spray droplets, such as those containing IJs (Section 5.4.3), are particularly vulnerable to runoff from leaf surfaces by 'bouncing' because of their high kinetic energy and/or because of the contact angle of water on some (waxy) leaf surfaces (Matthews, 2000). The addition of some surfactants can enhance droplet retention on foliage by reducing the surface tension, although their effects can vary depending upon the nature of the leaf surface (Matthews, 2000). Mason *et al.* (1998b) showed that the addition of several glycerol or oil-based anti-desiccants, or non-ionic surfactants, significantly increased the deposition rate of nematode IJs applied by spinning disc (Section 5.4.3) onto Chinese cabbage leaf discs. The evidence suggested that this was due to a change in the swath pattern rather than an effect on the spray droplet spectrum or total spray output.

Under field conditions, crops with waxy, densely packed leaves, such as many varieties of cabbage, represent a particular challenge for applying EPNs against pests such as the DBM (*P. xylostella*) (Baur *et al.*, 1997, 1998; Mason and Wright, 1997; Mason *et al.*, 1999). The DBM is the most important pest on crucifer plants worldwide. In functional ecosystems a wide range of antagonists will be found that are able to control up to 80% of this pest. The immense use of insecticides decreases the potential of naturally occurring antagonists, while the DBM has developed resistance against every insecticide applied on crucifer crops. The DIABOLO project (2001–2004; EC INCO Programme) aimed to develop integrated control programmes for *P. xylostella* on crucifer crops by conserving natural enemies and developing a set of biocontrol agents: parasitoids against eggs

(*Trichogramma* spp.), *Bacillus thuringiensis* and viruses for the first instars, and EPNs against third and fourth instars. Four universities in China, Indonesia, Ireland and Germany were involved in this project. To enhance EPN efficacy on the leaf, research on genetic improvement of desiccation tolerance and on appropriate formulation adjuvants is going on. Nematodes get entrapped in droplets if applied with water. Due to the waxy surface of cabbage plants EPNs are lost in water drops due to runoff. Adjuvants that lower the surface tension and enhance binding properties to decrease EPN runoff were evaluated in the laboratory. The formulation of a surfactant suitable to emulsify heavy plant oil (e.g. Rimulgan® Themmen, Germany) and a polymer with the feature to increase the viscosity at low concentration (xanthan gum) raises EPN efficacy significantly and decreases EPN runoff. Currently, the recommended concentration for both components is 0.3%. The formulation supports EPN movement on foliage, while decreasing motility of the pest, resulting in faster EPN infection. In the laboratory, efficacy was improved by 50% with this formulation; however, EPNs persisted for less than 10 h. Other ingredients, like polyacrylamides, silicate and alginate, did not significantly improve EPN persistence (Schroer and Ehlers, 2005). For maximum efficacy, the DBM larvae, which tend to hide underneath the leaf or inside the leaf, should be covered with the EPN formulation. Detailed instructions for the mode and timing of spraying need to be elaborated to hit the susceptible stages optimally with the respective biocontrol agents.

There are a number of environmental factors that can lead to reduced efficacy of EPNs on foliage. The most critical factor is usually desiccation (Baur *et al.*, 1995; Mason and Wright, 1997; Grewal, 2002), although its significance is reduced at high ambient relative humidities (Section 5.4.2). Other interlinked factors that can be important are high temperatures and UV radiation (Grewal, 2002). All these factors are generally more acute in field crops, which represent a much greater challenge for the foliar application of EPNs compared with protected crops (Section 5.4.2). Spraying of EPNs in the late afternoon or early evening can be one practical way of reducing all of the above problems and prolonging nematode infectivity (Lello *et al.*, 1996).

Nematode survival and efficacy on foliage has also been shown to be enhanced to varying degrees by the addition of various adjuvants to the spray mixture, which have antidesiccant (e.g. glycerol, various polymers) or UV-protective (brighteners) actions (MacVean *et al.*, 1982; Glazer *et al.*, 1992; Nickle and Shapiro, 1994; Broadbent and Olthof, 1995; Baur *et al.*, 1997; Mason *et al.*, 1998b; Grewal, 2002; Navon *et al.*, 2002), although more needs to be done to enhance post-application survival. A polymer-based formulation of *S. feltiae*, Nemasys F® (Becker Underwood Ltd., Littlehampton, UK), has been reported to give improved control of leafminers and thrips (Section 5.4.2) on ornamentals in commercial greenhouses (S.J. Piggott, personal communication).

5.5. Conclusions

While some progress has been made in developing application technologies for IJs against soil and above-ground pests, it is clear that further improvements are required to give the levels of reliability and efficiency for EPNs to compete more effectively with insecticides outside their current niche markets. The withdrawal of approvals for agrochemicals on many horticultural food crops in Europe, North America and elsewhere is likely to represent an increasing market opportunity for biopesticide products, but the application of EPNs will need to be both cost-effective and robust if their usage is to be maximized, especially against pests of field crops.

Progress should be possible in all areas, including formulation of IJs, optimization of existing application equipment (e.g. nozzle choice and operating pressure, use of irrigation systems), the development of novel systems (especially inexpensive adaptations to commonly used equipment) and the development of optimal application strategies (e.g. use of split doses timed to coincide with peak numbers of susceptible pest stages; Fenton *et al.*, 2002). In all these areas, the particular requirements of the EPN species used, and the target pest and crop, need to be taken into account.

Foliar application of EPNs is still a relatively new area and very little is known, for example, on how droplets containing IJs behave on foliage and how their distribution on plants can therefore be optimized in relation to the target pest. The greatest potential for using EPNs against foliar pests is almost certainly in IPM programmes, in conjunction with other biocontrol agents (e.g. Sher and Parella, 1999) or selective chemicals (Rovesti and Deseo, 1990; Baur *et al.*, 1998; Head *et al.*, 2000).

References

Appel, A.G., Benson, E.P., Ellenberger, J.M. and Manweiler, S.A. (1993) Laboratory and field evaluations of an entomogenous nematode (Nematoda, Steinernematidae) for German cockroach (Dictyoptera, Blatellidae) control. *Journal of Economic Entomology* 86, 777–784.

Barani, A., Vezzadini, S., Ferrari, M., Vai, N., Mazzini, F. and Scaravelli, P. (2000) La difesa da *Saperda carcharias,* xilofago del pioppo. *L'Informatore Agrario* 56, 6–73.

Baur, M.E., Kaya, H.K. and Thurston, G.S. (1995) Factors affecting entomopathogenic nematode infection of *Plutella xylostella* on a leaf surface. *Entomologia Experimentalis et Applicata* 77, 239–250.

Baur, M.E., Kaya, H.K., Gaugler, R. and Tabashnik, B.E. (1997) Effects of adjuvants on entomopathogenic nematode persistence and efficacy against *Plutella xylostella. Biocontrol Science and Technology* 7, 513–525.

Baur, M.E., Kaya, H.K., Tabashnik, B.E. and Chilcutt, C.F. (1998) Suppression of diamondback moth (Lepidoptera: Plutellidae) with an entomopathogenic nematode (Rhabditidae: Steinernematidae) and *Bacillus thuringiensis* Berliner. *Journal of Economic Entomology* 91, 1089–1095.

Bedding, R.A. (1990) Logistics and strategies for introducing entomopathogenic nematode technology into developing countries. In: Gaugler, R. and Kaya, H.K. (eds) *Entomopathogenic Nematodes in Biological Control.* CRC Press, Boca Raton, Florida, pp. 233–248.

Begley, J.W. (1990) Efficacy against insects in habitats other than soil. In: Gaugler, R. and Kaya, H.K. (eds) *Entomopathogenic Nematodes in Biological Control.* CRC Press, Boca Raton, Florida, pp. 215–231.

Belair, G., Vincent, C. and Chouuinard, G. (1998) Foliar sprays with *Steinernema carpocapsae* against early season apple pests. *Journal of Nematology* 30, 599–606.

Bennison, J.A., Maulden, K.A., Wardlow, L.R., Pow, E.M. and Wadhams, L.J. (1998) Novel strategies for improving biological control of western flower thrips on protected ornamentals – potential new biological control agents. *Proceedings of the Brighton Crop Protection Conference, Pests and Disease* Vol. 1, pp. 193–198.

Boselli, M., Curto, G.M. and Tacconi, R. (1997) Field efficacy of entomopathogenic nematodes against the sugar-beet weevil *Temnorhinus* (= *Conorrhynchus*) *mendicus* Gyll. (Coleoptera: Curculionidae). *Biocontrol Science and Technology* 7, 231–238.

Broadbent, A.B. and Olthof, T.H.A. (1995) Foliar application of *Steinernema carpocapsae* (Rhabditida, Steinernematidae) to control *Liriomyza trifolii* (Diptera, Agromyzidae) larvae on chrysanthemums. *Environmental Entomology* 24, 431–435.

Cabanillas, H.E. and Raulston, J.R. (1996a) Evaluation of *Steinernema riobravis, Steinernema carpocapsae* and irrigation timing for the control of corn earworm, *Helicoverpa zea. Journal of Nematology* 28, 75–82.

Cabanillas, H.E. and Raulston, J.R. (1996b) Effects of furrow irrigation on the distribution and infectivity of *Steinernema riobravis* against corn earworm in corn. *Fundamental and Applied Nematology* 19, 273–281.

Capinera, J.L., Pelissier, D., Menout, G.S. and Epsky, N.D. (1988) Control of black cutworm, *Agrotis ipsilon* (Lepidoptera, Noctuidae) with entomogenous nematodes (Nematoda, Steinernematidae, Heterorhabditidae). *Journal of Invertebrate Pathology* 52, 427–435.

Chapple, A.C., Downer, R.A., Wolf, T.M., Taylor, R.A.J. and Hall, F.R. (1996) The application of biological pesticides: limitations and a practical solution. *Entomophaga* 41, 465–474.

Conner, J.M., McSorley, R., Stansly, P.A. and Pitts, D.J. (1998) Delivery of *Steinernema riobravis* through a drip irrigation system. *Nematropica* 28, 95–100.

Curran, J. (1992) Influence of application method and pest population size on the field efficacy of entomopathogenic nematodes. *Journal of Nematology* 24, 631–636.

Curran, J. and Patel, V. (1988) Use of trickle irrigation system to distribute entomopathogenic nematodes (Nematoda, Heterorhabditidae) for the control of weevil pests (Coleoptera, Curculionidae) of strawberries. *Australian Journal of Experimental Agriculture* 28, 639–643.

Cuthbertson, A.G.S., Head, J., Walters, K.F.A. and Gregory, S.A. (2003) The efficacy of the entomopathogenic nematode, *Steinernema feltiae*, against immature stages of *Bemisia tabaci*. *Journal of Invertebrate Pathology* 83, 267–269.

Deseö, K.V. and Rovesti, L. (1992) *Lotta microbiologica contro i fitofagi*, Vol. 1. Edagricole, Bologna, Italy.

Downing, A.S. (1994) Effect of irrigation and spray volume on the efficacy of entomopathogenic nematodes (Rhabditida, Heterorabditidae) against white grubs (Coleoptera, Scarabaeidae). *Journal of Economic Entomology* 87, 643–646.

Ellsbury, M.M., Jackson, J.J., Woodson, W.D., Beck, D.L. and Stange, K.A. (1996) Efficacy, application distribution, and concentration by stemflow of *Steinernema carpocapsae* (Rhabditida: Steinernematidae) suspensions applied with a lateral-move irrigation system for corn rootworm (Coleoptera: Chrysomelidae) control in maize. *Journal of Economic Entomology* 89, 74–81.

Fenton, A., Gwynn, R.L., Gupta, A., Norman, R., Fairbairn, J.P. and Hudson, P.J. (2002) Optimal application strategies for entomopathogenic nematodes: integrating theoretical and empirical approaches. *Journal of Applied Ecology* 39, 481–492.

Fife, J.P. (2003) Investigation of the effect of agricultural spray application equipment on damage to entomopathogenic nematodes – a biological pest control agent. PhD thesis, The Ohio State University, Columbus, Ohio.

Fife, J.P., Derksen, R.C., Ozkan, H.E. and Grewal, P.S. (2003) Effects of pressure differentials on the viability and infectivity of entomopathogenic nematodes. *Biological Control* 27, 65–72.

Fife, J.P., Derksen, R.C., Ozkan, H.E., Grewal, P.S., Chalmers, J.J. and Krause, C.R. (2004) Evaluation of a contraction flow field on hydrodynamic damage to entomopathogenic nematodes – a biological pest control agent. *Biotechnology and Bioengineering* 86, 96–107.

Georgis, R. (1990) Formulation and application technology. In: Gaugler, R. and Kaya, H.K. (eds) *Entomopathogenic Nematodes in Biological Control*. CRC Press, Boca Raton, Florida, pp. 173–191.

Georgis, R. and Gaugler, R. (1991) Predictability in biological control using entomopathogenic nematodes. *Journal of Economic Entomology* 84, 713–720.

Glazer, I., Klein, M., Navon, A. and Nakache, Y. (1992) Comparison of efficacy of entomopathogenic nematodes combined with antidesiccants applied by canopy sprays against 3 cotton pests (Lepidoptera, Noctuidae). *Journal of Economic Entomology* 85, 1636–1641.

Gouge, D.H., Lee, L.L., Berkum, J.R.V., Henneberry, T.J., Smith, K.A., Payne, C., Ortega, D. and Van Berkum, J.R. (1997) Control of pink bollworm, *Pectinophora gossypiella*, (Saunders) (Lepidoptera: Gelechiidae) with biocontrol and biorational agents. *Proceedings of the Beltwide Cotton Conferences*, Los Angeles, USA, January 6–10, 1997, pp. 1066–1072.

Grewal, P.S. (2002) Formulation and application technology. In: Gaugler, R. (ed.) *Entomopathogenic Nematology*. CAB International, Wallingford, UK, pp. 265–287.

Gwynn, R.L., Chapple, A.C. and Smits, P.H. (1999) Post-application persistence of entomopathogenic nematodes. In: Gwynn, R.L., Smits, P.H., Griffin, C., Ehlers, R.-U., Boemare, N. and Masson, J.-P. (eds) *Cost 819 Entomopathogenic Nematodes: Application and Persistence of Entomopathogenic Nematodes*. European Communities, Luxembourg, pp. 89–94.

Hara, A.H., Kaya, H.K., Gaugler, R., Lebeck, L.M. and Mello, C.L. (1993) Entomopathogenic nematodes for biological control of the leafminer, *Liriomyza trifoliii* (Dipt., Agromyzidae). *Entomophaga* 38, 359–369.

Head, J., Walters, K.F.A. and Langton, S. (2000) The compatibility of the entomopathogenic nematode, *Steinernema feltiae*, and chemical insecticides for the control of the South American leafminer, *Liriomyza huidobrensis*. *Biocontrol* 45, 345–353.

Jech, L.F. and Henneberry, T.J. (1997) *Pink Bollworm Larval Mortality Following Application of* Steinernema riobravis *Entomopathogenic Nematodes in Cotton Furrow Irrigation*. National Cotton Council, Memphis, Tennessee, pp. 1192–1194.

Johnson, W. and Pearson, J.F. (2002) A release composition and method of preparation. *Patent No.* WO0215703.

Kakouli-Duarte, T., Labuschagne, T. and Hague, N.G.M. (1997) Biological control of the black vine weevil, *Otiorhynchus sulcatus* (Coleoptera: Curculionidae) with entomopathogenic nematodes (Nematoda: Rhabditida). *Annals of Applied Biology* 131, 11–27.

Klein, M.G. and Georgis, R. (1994) Application techniques for entomopathogenic nematodes. In: *Proceedings of the VI International Colloquium* on *Invertebrate Pathology and Microbial Control*, Montpellier, France, pp. 483–484.

Klingler, J. (1988) Investigations on the parasitism of *Otiorhynchus salicicola* and *Otiorhynchus sulcatus* (Coleoptera, Curculionidae) by *Heterohabditis* sp. (Nematoda). *Entomophaga* 33, 325–331.

Koppenhöfer, A.M., Baur, M.E., Stock, S.P., Choo, S.Y., Chinnasri, B. and Kaya, H.K. (1997) Survival of entomopathogenic nematodes within host cadavers in dry soil. *Applied Soil Ecology* 6, 231–240.

Kramer, I. and Grunder, J. (1998) Efficacy of EPNs application by irrigation systems on black vine weevil larvae control. *IOBC WPRS Bulletin* 21, 179–182.

Lacey, L.A., Bettencourt, R., Garrett, F.J., Simões, N.J. and Gaugler, R.H. (1993) Factors influencing parasitism of adult Japanese beetles, *Popillia japonica* (Coleoptera: Scarabeidae) by entomopathogenic nematodes. *Entomophagia* 38, 502–507.

Lello, E.R., Patel, M.N., Matthews, G.A. and Wright, D.J. (1996) Application technology for entomopathogenic nematodes against foliar pests. *Crop Protection* 15, 567–574.

MacVean, C.M., Brewer, J.W. and Capinera, J.L. (1982) Field tests of antidesiccants to extend the infection period of an entomogenous nematode, *Neoaplectana carpocapsae*, against Colorado potato beetle. *Journal of Economic Entomology* 75, 97–101.

Mason, J.M. and Wright, D.J. (1997) Potential for the control of *Plutella xylostella* larvae with entomopathogenic nematodes. *Journal of Invertebrate Pathology* 70, 234–242.

Mason, J.M., Matthews, G.A. and Wright, D.J. (1998a) Appraisal of spinning disc technology for the application of entomopathogenic nematodes. *Crop Protection* 17, 453–461.

Mason, J.M., Matthews, G.A. and Wright, D.J. (1998b) Screening and selection of adjuvants for the spray application of entomopathogenic nematodes against a foliar pest. *Crop Protection* 17, 463–467.

Mason, J.M., Matthews, G.A. and Wright, D.J. (1999) Evaluation of spinning disc technology for the application of entomopathogenic nematodes against a foliar pest. *Journal of Invertebrate Pathology* 73, 282–288.

Matthews, G.A. (2000) *Pesticide Application Methods*, 3rd edn. Blackwell Science, Oxford, UK, 432 pp.

Menzler-Hokkanen, I. and Hokkanen, H.M.T. (2003) Developing entomopathogenic nematode delivery systems for biological control of oilseed rape pests. In: *Proceedings of the European Meeting of the IOBC/WPRS Working Group Insect Pathogens and Insect Parasitic Nematodes*, Solzau, Germany, May 23–29, 2003, p. 63.

Miller, L.A. and Bedding, R.A. (1982) Field testing of the insect parasitic nematode, *Neoaplectana bibionis* (Nematoda, Steinernematidae) against currant borer moth, *Synanthedon tipuliformis* (Lepidoptera, Sesiidae) in blackcurrants. *Entomophaga* 27, 109–114.

Navon, A., Nagalakshmi, V.K., Levski, S., Salame, L. and Glazer, I. (2002) Effectiveness of entomopathogenic nematodes in an alginate gel formulation against lepidopterous pests. *Biocontrol Science and Technology* 12, 737–746.

Nickle, W.R. and Shapiro, M. (1994) Effects of eight brighteners as solar radiation protectants for *Steinernema carpocapsae*, all strain. *Journal of Nematology* 26 (Suppl.), 784–786.

Nilsson, U. and Gripwall, E. (1999) Influence of application technique on the viability of biological control agents *Verticillium lecanii* and *Steinernema feltiae*. *Crop Protection* 18, 53–59.

Parkman, J.P. and Frank, J.H. (1993) Use of a sound trap to inoculate *Steinernema scapterisci* (Rhabditida, Steinernematidae) into pest mole cricket populations (Orthoptera, Gryllotalpidae). *Florida Entomologist* 76, 75–82.

Patel, M.N. and Wright, D.J. (1996) The influence of neuroactive pesticides on the behaviour of entomopathogenic nematodes. *Journal of Helminthology* 70, 53–61.

Peters, A. (2003) Pesticides and entomopathogenic nematodes – current status and future work. *IOBC/WPRS Bulletin* 26, 107–110.

Peters, A. and Backes, J. (2003) Impact of substrate conditions and application method on the efficacy of *Steinernema feltiae*. *IOBC/WPRS Bulletin* 26, 151–158.

Piggott, S.J., Clayton, R., Matthews, G.A. and Wright, D.J. (2003) Development of a new application apparatus for entomopathogenic nematodes. *Pest Management Science* 59, 1344–1348.

Pye, A.E. and Pye, N.L. (1985) Different applications of the insect parasitic nematode *Neoaplectana carpocapsae* to control the large pine weevil, *Hylobius abietis*. *Nematologica* 31, 109–116.

Pye, N., Holbrook, G. and Pye, A. (2001) New bait station with *Steinernema carpocapsae* nematodes kills cockroaches. In: *Proceedings of the 34th Annual Meeting of the Society for Invertebrate Pathology*, Cordoba, Spain, pp. 62.

Reed, D.K., Reed, G. and Creighton, C.S. (1986) Introduction of entomogenous nematodes into trickle irrigation systems to control striped cucumber beetle (Coleoptera, Chrysomelidae). *Journal of Economic Entomology* 79, 1330–1333.

Renn, N. (1998) The efficacy of entomopathogenic nematodes for controlling housefly infestations of intensive pig units. *Medical and Veterinary Entomology* 12, 46–51.

Ritsema, C.J. and Dekker, L.W. (2003) *Soil Water Repellency: Occurrence, Consequences and Amelioration.* Elsevier Science, The Netherlands, 352 pp.

Rovesti, L. and Deseö, K.V. (1990) Compatibility of chemical pesticides with the entomopathogenic nematodes, *Steinernema carpocapsae* Weiser and *Steinernema felitae* Fiipjev (Nematoda, Steinernematidae). *Nematologica* 36, 237–245.

Schroer, S. and Ehlers, R.-U. (2005) Foliar application of the entomopathogenic nematode *Steinernema carpocapsae* for biological control of diamondback moth larvae (*Plutella xylostella*). *Biological Control* (in press).

Schroer, S., Ziermann, D. and Ehlers, R.-U. (2005) Mode of action of a surfactant-polymer-formulation to support performance of the entomopathogenic nematode *Steinernema carpocapsae* for control of diamondback moth (*Plutella xylostella*). *Biocontrol Science and Technology* (in press).

Shapiro-Ilan, D.I., Lewis, E.E., and Tedders, W.L. (2003) Superior efficacy observed in entomopathogenic nematodes applied in infected host cadavers compared with application in aqueous suspension. *Journal of Invertebrate Pathology* 83, 270–272.

Sher, R.B. and Parella, M.P. (1999) Biological control of the leafminer, *Liriomyza trifolii*, in chrysanthemums: implications for intraguild predation between *Diglyphus begini* and *Steinernema carpocapsae. Bulletin of the International Organization for Biological and Integrated Control of Noxious Animals and Plants: Integrated Control in Glasshouses* 22, 221–224.

Shetlar, D.J., Power, K.T. and Niemczyk, H.D. (1993) Effect of subsurface application of entomopathogenic nematodes on non-target arthropods in turfgrass. In: Carrow, R.N., Christians, N.E. and Shearman, R.C. (eds) *International Turfgrass Society Research Journal Volume* 7 – 1993, 385–390.

Smits, P.H. (1999) Persistence of entomopathogenic nematodes. In: Gwynn, R.L., Smits, P.H., Griffin, C., Ehlers, R.-U., Boemare, N. and Masson, J.-P. (eds) *Cost 819 Entomopathogenic Nematodes: Application and Persistence of Entomopathogenic Nematodes.* European Communities, Luxembourg, pp. 77–88.

Unruh, T.R. and Lacey, L.A. (2001) Control of codling moth, *Cydia pomonella* (Lepidoptera: Tortricidae), with *Steinernema carpocapsae*: effects of supplemental wetting and pupation site on infection rate. *Biological Control* 20, 48–56.

Wang, J. (1990) Use of the nematodes *Steinernema carpocapsae* to control the major apple pest *Carposina nipponensis* in China. *Proceedings of the Vth International Colloquium Invertebrate Pathology and Microbial Control,* Adelaide, Australia, pp. 392.

Wennemann, L., Cone, W.W., Wright, L.C., Perez, J., Conant, M.M. (2003) Distribution patterns of entomopathogenic nematodes applied through drip irrigation systems. *Journal of Economic Entomology* 96, 287–291.

Williams, E.C. and MacDonald, O.C. (1995) Critical factors required by the nematode *Steinernema feltiae* for the control of the leafminers *Liriomyza huidobrensis, Liriomyza bryoniae* and *Chromatomyia syngenesiae. Annals of Applied Biology* 127, 329–341.

Williams, E.C. and Walters, K.F.A. (2000) Foliar application of the entomopathogenic nematode *Steinernema feltiae* against leafminers on vegetables. *Biocontrol Science and Technology* 10, 61–70.

Wilson, M.J., Huges, L.A., Hamacher, G.M., Barahona, L.D. and Glen, D.M. (1996) Effects of soil incorporation on the efficacy of the rhabditoid nematode, *Phasmarhabditis hermaphrodita,* as a biological control agent for slugs. *Annals of Applied Biology* 128, 117–126.

Wright, L.C., Witkowski, J.F., Echtenkamp, G. and Georgis, R. (1993) Efficacy and persistence of *Steinernema carpocapsae* (Rhabditida, Steinernematidae) applied through a center-pivot irrigation system against larval corn rootworms (Coleoptera, Chrysomelidae). *Journal of Economic Entomology* 86, 1348–1354.

Young, J.M., Dunnill, P. and Pearce, J.D. (1998) Physical properties of liquid nematode cultures and the design of recovery operations. *Bioprocess Engineering* 19, 121–127.

Zimmerman, R.J. and Cranshaw, W.S. (1991) Short-term movement of *Neoaplectana* spp. (Rhabditida, Steinernematidae) and *Heterorhabditis* spp. (Rhabditida, Heterohabditidae) through turfgrass thatch. *Journal of Economic Entomology* 84, 875–878.

6 Forum on Safety and Regulation

R.-U. Ehlers
Department for Biotechnology and Biological Control, Institute for Phytopathology, Christian-Albrechts-University Kiel, 24223 Raisdorf, Germany

6.1 Safety and Potential Effects on Non-target Organisms (NTOs)

Entomopathogenic nematodes (EPNs) are exceptionally safe biocontrol agents. Biocontrol nematodes are certainly more specific and are less of a threat to the environment than chemical insecticides (Ehlers and Peters, 1995). Since the first use of the EPN *Steinernema glaseri* against the white grub *Popillia japonica* in New Jersey (Glaser and Farrell, 1935), not even minor damages or hazards caused by the use of EPNs to the environment have been recorded. Application of EPNs is safe to the user. EPNs and their associated bacteria cause no detrimental effect to mammals or plants (Poinar *et al.*, 1982; Bathon, 1996; Boemare *et al.*, 1996; Akhurst and Smith, 2002). A joint workshop supported by the European Co-operation in the Field of Scientific and Technical Research (EU COST) Action 819, 'Entomopathogenic Nematodes', and the Organisation for Economic Co-operation and Development (OECD) Research Programme, 'Biological Resource Management for Sustainable Agriculture Systems', which met in 1995 to discuss potential risks related with the use of EPNs in biocontrol, concluded that EPNs are safe to production and application personnel and to the consumers of agriculture products treated with EPNs (Ehlers and Hokkanen, 1996). The expert group could not identify any risk for the general public related to the use of EPNs.

No reports exist that document any effect on humans by the symbiotic bacteria. A related non-symbiotic species, *Photorhabdus asymbiotica*, was reported five times from humans in the USA (Farmer *et al.*, 1989). Another group of non-symbiotic *Photorhabdus* was reported from five patients in Australia (Peel *et al.*, 1999). From most of the patients, other human-pathogenic bacteria were also recorded, thus the *Photorhabdus* spp. were considered opportunistic. The route of the infections was not established. Three infections might have been related to spider bites. Both clinical groups lack symbiotic relations with nematodes, and strains within each group have a high level of within-group relatedness but do not cluster in groups containing the nematode symbionts (Szállás *et al.*, 1997; Akhurst and

Smith, 2002). The existence of bacterial species with and without pathogenic effects on humans within one genus is common (e.g. *Bacillus*). No action is therefore required and no conclusions should be drawn from the reports of pathogenic effects on humans by non-symbiotic *Photorhabdus* spp. about the potential risks related to the use of EPNs and their symbiotic bacteria.

The COST–OECD expert group evaluated possible risks to the environment. Long-term effects on non-target organisms (NTOs) or other environmental impacts following the application of indigenous or exotic EPNs have not been reported. Even after release of an exotic nematode species, no detrimental effects were observed (Parkman and Smart, 1996). The possible short-term environmental risks of using EPNs are effects on predators and parasitoids of the target pest and effects on NTOs in the soil or cryptic environments. These risks were classified as remote to moderate and temporary (Ehlers and Hokkanen, 1996).

Much scientific information on the safety and possible impacts of EPNs on NTOs and the environment is available. Significant effects on foliage-inhabiting NTOs can be excluded as EPNs cannot survive for long above the soil (Glazer, 2002). Bathon (1996) summarized available results on non-target effects on soil-inhabiting insects and concluded that mortality caused by released EPNs among non-target arthropod populations can occur, but will only be temporary, will be spatially restricted and will affect only part of the population. The potential wide host range of 200 species recorded from laboratory assays (Poinar, 1986) could not be supported in field trials (Georgis and Gaugler, 1991; Buck and Bathon, 1993; Koch and Bathon, 1993, Bathon 1996). Bathon (1996) summarized results of extensive field studies performed over a period of 3 years with several 100 m^2 plots in different environments. A total of approximately 400,000 specimens were evaluated. EPN application never resulted in the extinction of any local population. The density of a few species was reduced (some increased) after EPN application; however, the reduction was temporary and spatially restricted. In general, the impact on the non-target populations was negligible.

Commercial applications of EPNs have also been found to be safe to soil nematode and microbial communities. Somasekhar *et al.* (2002) reported that EPNs significantly reduced the abundance, species richness, diversity and maturity of the nematode community by reducing the number of genera and abundance of plant-parasitic nematodes, but not free-living nematodes (also see Chapter 18, this volume). Bacterivorous, fungivorous and omnivorous nematodes are unaffected by EPN application to the soil (Jagdale *et al.*, 2002; Somasekhar *et al.*, 2002). Also, no negative impact of EPN application on microbial biomass, respiration and nitrogen pools in microcosms has been detected (E.A.B. De Nardo, P.S. Grewal, D. McCartney and B.R. Stinner, unpublished data).

Effects nematodes can have on NTOs are transient. Several environmental factors limit survival of EPNs in the soil (Glazer, 1996). The half-life of EPNs is between a few days and 1 month (Strong, 2002). After inundative release with 0.5 million nematodes/m^2, EPN population density rapidly declines, followed by a period of about 2 weeks with lower rates of decline, after which the population reaches background levels of about 10,000 m^2 (Smits, 1996). Consequently, EPNs need to reproduce in order to establish and have long-term effects on an insect population. Their population density is always correlated with the occurrence and density of potential host insect populations, which, on the other hand, is a result of available food resources supporting these host insect populations (Strong, 2002). Density and distribution of EPNs in a field thus depends on recycling in hosts and is a consequence of the distribution of host insects. Like the distribution of host insect populations, EPN populations are typically patchy and aggregated (Stuart and Gaugler, 1994; Spiridonov and Voronov, 1995). The polyphagous nature of EPN antagonists in the soil (Kaya and Koppenhöfer, 1996) is another factor limiting EPN population density and dispersal. Considering the low overall density, the high patchiness and a reduced mobility of nematodes, the risk for large impacts on NTO populations is negligible.

A high risk was rated by the experts for the possible 'biological pollution' with exotic EPN species. Although one could also argue that it is beneficial to the agroecosystem when an additional antagonist has been successfully established, others think that the original species structure should not be disturbed. Barbercheck and Millar (2000) introduced exotic *S. riobrave* from Texas on plots in North Carolina with an endemic population of *S. carpocapsae* and *Heterorhabditis bacteriophora*. The introduction resulted in a reduction of insect mortality caused by the endemic species when soil samples were baited with *Galleria mellonella*. Data suggest that coexistence of the three nematode species in the field was possible and that the risk for local extinction of the native nematodes was minimal. However, the results indicate that the application of the exotic species can cause a reduction of endemic species populations.

Coexistence is facilitated by highly aggregated populations. The relatively low mobility of EPNs is likely to result in fragmented populations. The highly aggregated distribution (Taylor *et al.*, 1998) will ensure that parts of the population survive while others might become transiently extinct by introduction of exotic populations. Survivors can later recolonize locally extinct populations. These metapopulation dynamics are of major importance for the survival and coexistence of species (Harrison and Taylor, 1997).

Naturally occurring nematode populations cause sustainable reduction of pest populations (Ehlers, 1998). However, these effects have not been very well exploited because of the limited understanding of the EPN population dynamics, although possibilities to enhance EPN populations by cultural practices have been reported (Fischer and Führer, 1990; Brust, 1991; also see Chapter 18, this volume). Until now, the economic benefits of these sustainable effects have not been determined. The economic effect of introducing an exotic species is easier to assess. In the case of a pest population surpassing the economic threshold, the use of an exotic nematode might be economically reasonable. It is often argued that before the release of exotic species it should be tested whether an endemic population might solve the problem. However, the naturally occurring species, even if superior in its control potential, might not be commercially available. Waiting until the endemic population has increased and reached an even distribution to significantly reduce the pest population will result in economic losses. The benefit from introducing the exotic species will overwhelm the damage caused by a reduction of the population of the endemic EPN species. Should the exotic species persist, we have a case of 'biological pollution'. However, the question needs to be asked whether this potential 'damage' to the agroecosystem is outweighed by the benefit to the farmer. As exotic species have not been recorded to eliminate the endemic EPN species, no real hazard has yet been identified with the introduction of the exotic species and the 'biological pollution'.

6.2 Registration

In biocontrol science, EPNs are assigned to the group of beneficial invertebrate parasites and predators. However, they are also classified as pathogens or microbial control agents because of their mutualistic relation with their symbiotic bacteria. In regard to registration policy, EPNs are usually covered within the macroorganisms together with beneficial arthropods. For that reason they have been exempted from registration in many countries. There are strong arguments why nematodes should be considered macroorganisms and, if necessary, be registered as such. Users of EPN products do not get into contact with the symbiotic bacteria, as the bacterial cells are embedded in the intestine of the infective juvenile (IJ). On the other hand, the number of bacteria, is relatively small (200—2000/IJ). Should EPNs be registered as microbial agents due to their symbiotic relation with *Xenorhabdus* and *Photorhabdus* spp., decision makers in regulatory offices will face a serious problem. They would have to also consider endosymbionts of insects, e.g. *Wolbachia* spp. or virus symbionts, which often contribute to insect death. Harwood and Beckage (1994), for instance, identified a polydnavirus associated with eggs of the parasitoid *Cotesia congregata*. During deposition of the parasitoid egg, the virus is also injected into the haemocoel of the lepidopteran host. The virus suppresses the immune response of the host *Manduca sexta*, which otherwise would encapsulate the eggs of the parasitoid in the haemolymph. Should *Cotesia* spp. now be considered microbial control agents and be registered as such? Besides, all beneficial arthropods are grown under non-sterile conditions and hence carry a large variety of microorganisms in the intestine.

In most countries EPNs are exempted from registration requirements (Table 6.1). Only a few countries have developed requirements for registration, which are usually not comparable with the data needed for the registration of chemical compounds or microbial agents. Safety data files (and associated costs, i.e. > $200,000) that SDS Biotech had to file in Japan for registration of *S. carpocapsae* and *S. glaseri* were

Table 6.1. Requirements for registration of entomopathogenic nematodes (EPNs) in different countries.

Country	Registration necessary
Australia	No; importation and release of exotic species requires permits from a series of authorities (see Bedding *et al.*, 1996)
Austria	Yes; although following the requirements for chemical pesticides, the time-consuming procedure for EPNs is limited to data that are in a reasonable context with biocontrol agent
Belgium	Yes; required only for new EPN species not marketed yet
Brazil	Yes; required for field testing of all indigenous and non-indigenous species
Canada	No; but guidelines for registration are being developed
Czech Republic	Yes; requirements include efficacy data from field trials
European Union	No; Directive 91/414/EEC distinguishes between chemical pesticides and microorganisms and viruses. Nematodes and macroorganisms are not mentioned; EU tries to avoid implementation of registration for low-risk products
Germany	No; but guidelines have been developed
Hungary	Yes; requirements include efficacy data from field trials
Ireland	Yes; new law recently implemented
Japan	Yes; data requirements are not different from those for chemical compounds; costs are enormously high
Netherlands	Required for new EPN species not marketed yet
New Zealand	Yes; although other macroorganisms do not require registration, nematodes must be registered (see Bedding *et al.*, 1996)
Norway	Yes; requirements follow recommendations of the OECD guidelines, except that the assessment of the environmental risk is not necessary
Poland	Yes; efficacy data from field trials in Poland requested
Sweden	Yes; EPNs must be approved under the Act on Preliminary Examination of Biological Pesticides, limited data requirement
Switzerland	Yes; but rarely more than a paperwork exercise
United Kingdom	No; indigenous EPNs do not need registration, but the introduction of non-indigenous species or strains is controlled through the Wildlife and Countryside Act (see Richardson, 1996).
United States	No; but any import of living material must be accompanied by shipment permits; release of exotic species is regulated by the Animal and Plant Health Inspection Service (APHIS) and other federal organizations (see Rizvi *et al.*, 1996; Akhurst and Smith, 2002)

Denmark, Finland, France, Greece, Italy, Portugal and Spain: no registration required.

comparable to data files and cost requirements for chemical registration (Satoshi Yamanaka, personal communication).

In most European countries no registration is required. The exemption from registration requirement aided the commercial development of EPN-based products. Those countries that require registration usually ask for information that is freely available in the scientific literature. In Switzerland, for instance, all biocontrol agents need to be registered; however, the requirements are not comparable with those for chemical compounds. Even microbial agents undergo a reduced procedure in Switzerland that is not comparable to EU requirements. The registration of EPNs is based on published efficacy data and safety information, accompanied by descriptions of the production and quality control procedures. In Austria, Sweden and Norway the requirements are similar. Eastern European countries ask for data of each new product from field trials performed within their borders (Poland, Czech Republic and Hungary). A complete file is required for every new product. Even if other EPN products containing the same species of strain exist in these markets, authorities go through the whole bureaucratic process again for every new product. This practice causes high costs and loss of time as the registration process lasts for at least 2 years until a product can be marketed. Many small and medium-sized enterprises would not have been able to start commercializing their EPN products if registrations were required in all EU countries and the USA.

Attempts to control the use of invertebrate biocontrol agents are underway. The Netherlands and Belgium implemented a registration procedure recently for all nematode-based products that are not yet marketed. Germany wants to implement a similar procedure to avoid uncontrolled release of exotic species. Products that are already in the market will be covered on a positive list and will not need registration. The Pesticide Steering Committee of the OECD produced guidelines for the regulation of invertebrate biocontrol agents. This document exaggerated the risks involved with the use of biocontrol organisms, and therefore implementation of the requirements would result in severe negative impacts on the development and marketing of EPN-based products. It is most unfortunate that the OECD Steering Committee spent much time in producing this recommendation instead of working on a consensus document including a positive list of invertebrate biocontrol agents that have a history of safe use. This approach was taken by the European and Mediterranean Plant Protection Organisation (EPPO), which has produced the document PM 6/3(2), containing a positive list (EPPO, 2002). The EPPO states:

> There is extensive previous knowledge and experience of the use of introduced biological control agents in a number of countries in the EPPO region, sufficient to indicate the absence of significant risks, or the availability of reliable risk management measures, for many individual organisms. This list accordingly specifies indigenous, introduced and established biological control agents, which are recognized by the EPPO Panel on Safe Use of Biological Control to have been widely used in several EPPO countries. Other EPPO countries may therefore presume with some confidence that these agents can be introduced and used safely.

The list includes five nematode species used in biocontrol.

6.3 Should Entomopathogenic Nematodes (EPNs) be Regulated?

In risk analysis the major hazard is the loss of human lives. Never in the past has there been a loss of human lives related to the use of EPN, and the environmental damage caused by biocontrol agents is of much less magnitude than hazards related to the use of chemical pesticides. A particular problem is the conception that products or activities are either 'safe' or 'unsafe'. But the real world is not a risk-free existence. Biocontrol agents are not necessarily hazard-free. However, the risks associated with biocontrol agents are much less compared with those associated with alternative control measures, and biocontrol risks can be accepted by users and consumers. We must be aware that regulation of EPNs poses risks as well. For instance, regulation of EPNs might keep older, riskier chemical pesticides in use. If, as a consequence of regulation, chemical insecticides have to be used, farmers can be harmed, particularly in glasshouse environments where they are highly exposed to chemical compounds.

Governments should attempt to use effective and inexpensive tools to regulate EPNs. If we take costly steps to address all risks, however improbable they may be, we will quickly impoverish ourselves. The search for cheaper and more effective tools to achieve the basic goal is of major importance and might produce creative solutions for risk assessment. Trade-offs of regulation must be considered and evaluated. Weighing the costs related with the assessment of risks of EPNs and adding the costs related with countervailing risks, our societies should search for more effective possibilities to regulate risks related with the use of EPNs, rather than implementing registration procedures following the rules used to register chemical compounds and microbials. Biocontrol currently needs less regulation instead of more bureaucratic hurdles. Therefore, as a first principle, any kind of regulation of indigenous EPNs should be avoided. Regulating the use of indigenous EPNs is overregulation without valid foundation concerning ecological risks (Blum *et al.*, 2003). If our baseline concept for cost-effective regulation of EPNs is driven by the fact that EPNs have a long history of safe use, we can waive any kind of regulations for those agents that have already been used for many years without any problems, including exotic EPN species.

The COST–OECD expert committee concluded that the use of exotic EPNs, which have never been used in biocontrol in an ecosystem or country, needs some regulation. Species should be accurately identified and specimens should be deposited. Expert opinions based on available information on the origin, natural distribution, biology, host range and safety for the user are desirable to assess possible risks related with the release of exotic species (Ehlers and Hokkanen, 1996). These data should be evaluated by expert committees, with the final goal of listing the exotic species on a positive list if no major risks can be identified related with the use of the exotic species. This committee should also consider costs related with the risk assessment and perform a risk/benefit trade-off analysis. If further risk assessments are necessary before the experts can make a decision, these should be supported by the public. In order to reduce the costs for risk assessments, public–private partnerships are one possibility to gather necessary information on potential risks. Unfortunately, many countries adopt the precaution principle 'better safe than sorry' and do not allow the use of exotic species at all (e.g. Norway). The consequence is that fewer biocontrol products are on the market.

Any regulation of the use of EPNs in biocontrol should consider the tremendous benefits to the environment resulting from the use of EPNs. Biocontrol nematodes are exceptionally safe for users and the environment, and the benefits outweigh potential risks to NTOs.

References

Akhurst, R. and Smith, K. (2002) Regulation and safety. In: Gaugler, R. (ed.) *Entomopathogenic Nematology*. CAB International, Wallingford, UK, pp. 311–332.

Barbercheck, M.E. and Millar, L.C. (2000) Environmental impacts of entomopathogenic nematodes used for biological control in soil. In: Follett, P.A. and Duan, J.J. (eds) *Non-target Effects of Biological Control*. Kluwer Academic Publishers, Dordrecht, The Netherlands, pp. 287–308.

Bathon, H. (1996) Impact of entomopathogenic nematodes on non-target hosts. *Biocontrol Science and Technology* 6, 421–434.

Bedding, R.A., Tyler, S. and Rochester, N. (1996) Legislation on the introduction of exotic entomopathogenic nematodes into Australia and New Zealand. *Biocontrol Science and Technology* 6, 465–475.

Blum, B., Ehlers, R.-U., Haukeland-Salinas, S., Hokkanen, H., Jung, K., Kuhlmann, U., Menzler-Hokkanen, I., Ravensberg, W., Strasser, H., Warrior, P. and Wilson, M. (2003) Biological control agents: safety and regulatory policy. *Biocontrol* 48, 477–484.

Boemare, N.E., Laumond, C. and Mauleon, H. (1996) The entomopathogenic nematode-bacterium complex: biology, life cycle and vertebrate safety. *Biocontrol Science and Technology* 6, 333–346.

Brust, G.E. (1991) Augmentation of an endemic entomogenous nematode by agroecosystem manipulation for the control of a soil pest. *Agriculture Ecosystem and Environment* 36, 175–184.

Buck, M. and Bathon, H. (1993) Auswirkungen des Einsatzes entomopathogener Nematoden *Heterorhabditis* sp. im Freiland auf die Nichtzielfauna. 2. Teil: Diptera. *Anzeiger für Schädlingskunde, Pflanzenschutz, Umweltschutz* 66, 84–88.

Ehlers, R.-U. (1998) Entomopathogenic nematodes – save biocontrol agents for sustainable systems. *Phytoprotection* 79, 94–102.

Ehlers, R.-U. and Hokkanen, H.M.T. (1996) Insect biocontrol with non-endemic entomopathogenic nematodes *Steinernema* and *Heterorhabditis* spp.: conclusions and recommendations of a combined OECD and COST workshop on scientific and regulatory policy issues. *Biocontrol Science and Technology* 6, 295–302.

Ehlers, R.-U. and Peters, A. (1995) Entomopathogenic nematodes in biological control: feasibility, perspectives and possible risks. In: Hokkanen, H.M.T. and Lynch, J.M. (eds) *Biological Control: Benefits and Risks.* Cambridge University Press, UK, pp. 119–136.

EPPO (2002) List of biological control agents widely used in the EPPO region. Available at: http://www.eppo.org/quarantine/biocontrol/bio_list.html

Farmer, J.J., III Jörgensen, J.H., Grimont, P.A.D., Akhurst, R.J., Poinar, G.O. Jr, Ageron, E., Pierce, G.V., Smith, J.A., Carter, G.P., Wilson, K.L. and Hickman-Brenner, F.W. (1989) *Xenorhabdus luminescens* DNA hybridization group 5 from human clinical specimens. *Journal of Clinical Microbiology* 27, 1594–1600.

Fischer, P. and Führer, E. (1990) Effect soil acidity on the entomophilic nematode *Steinernema kraussei* Steiner. *Biology and Fertility of Soils* 9, 174–177.

Georgis, R. and Gaugler, R. (1991) Predictability in biological control using entomopathogenic nematodes. *Journal of Economic Entomology* 84, 713–720.

Glaser, R.W. and Farrell, C.C. (1935) Field experiments with the Japanese beetle and its nematode parasite. *Journal of the New York Entomological Society* 43, 345.

Glazer, I. (1996) Survival mechanisms of entomopathogenic nematodes. *Biocontrol Science and Technology* 6, 373–378.

Glazer, I. (2002) Survival biology. In: Gaugler, R. (ed.) *Entomopathogenic Nematology.* CAB International, Wallingford, UK, pp. 169–187.

Harrison, S. and Taylor, A.D. (1997) Empirical evidence for metapopulation dynamics. In: Hanski, I. and Gilpin, M.E. (eds) *Metapopulation Biology, Ecology, Genetics and Evolution.* Academic Press, San Diego, California, pp. 27–42.

Harwood, S.H. and Beckage, N.E. (1994) Purification and characterization of an early-expressed polyDNAvirus-induced protein from the hemolymph of *Manduca sexta* larvae parasitized by *Cotesia congregata. Insect Biochemical Molecular Biology* 24, 685–698.

Jagdale, G.B., Somasekhar, N., Grewal, P.S. and Klein, M.G. (2002) Suppression of plant-parasitic nematodes with application of live and dead entomopathogenic nematodes on boxwood (*Buxus* spp.). *Biological Control* 24, 42–49.

Kaya, H.K. and Koppenhöfer, A.M. (1996) Effects of microbial and other antagonistic organisms and competition on entomopathogenic nematodes. *Biocontrol Science and Technology* 6, 357–372.

Koch, U. and Bathon, H. (1993) Results of the outdoor application of entomopathogenic nematodes on nonobjective fauna – I Coleoptera. *Anzeiger für Schädlingskunde, Pflanzenschutz und Umweltschutz* 66, 65–68.

Parkman, J.P. and Smart, G.C. Jr (1996) Entomopathogenic nematodes, a case study: introduction of *S. scapterisci* in Florida. *Biocontrol Science and Technology* 6, 413–419.

Peel, M.M., Alfredson, D.A., Gerrad, J.G., Davis, J.M., Robson, J.M., McDougall, R.J., Scullie, B.L. and Akhurst, R.J. (1999) Isolation, identification and molecular characterization of strains of *Photorhabdus luminescens* from infected humans in Australia. *Journal of Clinical Microbiology* 37, 3647–3653.

Poinar, G.O. Jr (1986) Entomophagous nematodes. In: Franz, B.D. (ed.) *Biological Plant and Health Protection.* Gustav Fischer Verlag, Stuttgart, Germany, pp. 95–121.

Poinar, G.O. Jr, Thomas, G.M., Presser, S.B. and Hardy, J.L. (1982) Inoculation of entomogenous nematodes, *Neoaplectana* and *Heterorhabditis* and their associated bacteria, *Xenorhabdus* spp., into chicks and mice. *Environmental Entomology* 11, 137–138.

Richardson, P.N. (1996) British and European legislation regulating rhabditid nematodes. *Biocontrol Science and Technology* 6, 449–463.

Rizvi, S.A., Hennessey, R. and Knott, D. (1996) Legislation on the introduction of exotic nematodes in the US. *Biocontrol Science and Technology* 6, 477–480.

Smits, P.H. (1996) Post-application persistence of entomopathogenic nematodes. *Biocontrol Science and Technology* 6, 379–387.

Somasekhar, N., Grewal, P.S., De Nardo, E.A.B. and Stinner, B.R. (2002) Non-target effects of entomopathogenic nematodes on the soil nematode community. *Journal of Applied Ecology* 39, 735–744.

Spiridonov, S.E. and Voronov, D.A. (1995) Small scale distribution of *Steinernema feltiae* juveniles in cultivated soil. In: Griffin, C.T., Gwynn, R.L. and Masson, J.-P. (eds) *COST 819 Biotechnology – Ecology and Transmission Strategies of Entomopathogenic Nematodes.* Official Publications of the EC, Luxembourg, EN 16269 EN, pp. 36–41.

Strong, D.R. (2002) Population of entomopathogenic nematodes in foodwebs. In: Gaugler, R. (ed.) *Entomopathogenic Nematology.* CAB International, Wallingford, UK, pp. 225–240.

Stuart, R.J. and Gaugler, R. (1994) Patchiness in populations of entomopathogenic nematodes. *Journal of Invertebrate Pathology* 64, 39–45.

Szállás, E., Koch, C., Fodor, A., Burghardt, J., Buss, O., Szentirmai, A., Nealson, K.H. and Stackebrandt, E. (1997) Phylogenetic evidence for the taxonomic heterogeneity of *Photorhabdus luminescens. International Journal of Systematic Bacteriology* 47, 402–407.

Taylor, D.B., Szalanski, A.L., Adams, B.J. and Peterson II, R.D. (1998) Susceptibility of house fly (Diptera: Muscidae) larvae to entomopathogenic nematodes (Rhabditida: Heterorhabditidae, Steinernematidae). *Environmental Entomology* 27, 1524–1529.

7 Lawn, Turfgrass and Pasture Applications

P.S. Grewal,[1] A.M. Koppenhöfer[2] and H.Y. Choo[3]
[1]*Department of Entomology, Ohio State University, Wooster, Ohio 44691, USA;*
[2]*Department of Entomology, Rutgers University, New Brunswick, NJ 08901, USA;*
[3]*Department of Applied Biology and Environmental Science, Gyeongsang National University, Jinju, Gyeongnam 660–701, Republic of Korea*

7.1. Introduction

Grasses are the dominant vegetation in many environments that vary in size and composition, from the great prairies to manicured golf courses, bowling greens and home lawns. Natural grasslands cover millions of hectares throughout the world, providing sustenance for vast numbers of wildlife. Grasslands, improved by sowing and managing desirable species, support livestock industries around the world. Wear-tolerant grass species are used to create recreational spaces in the urban environment. Such amenity turfgrasses occupy > 12 million ha in the USA alone, comprising over 50 million lawns, 14,500 golf courses, many parks, athletic fields, cemeteries and sod farms (Potter, 1998). Besides its recreational and livestock uses, grass sequesters carbon, controls soil erosion, captures and cleans runoff water from urban areas, provides soil improvement and restoration, moderates temperature, reduces glare and noise, reduces pests, reduces pollen and human disease exposure, creates good wildlife habitats, and improves physical and mental health of urban populations (Beard and Green, 1994).

7.2. Major Pests

Permanent turf provides a habitat for many invertebrate species, most of which feed on vegetation and detritus without causing obvious damage or loss of productivity. Spectacular outbreaks of grasshoppers, armyworms or white grubs can occur over large areas of natural grasslands, but such attacks are rare (Klein *et al.*, 2000). More intense management of grasslands by sowing palatable species and increasing fertility has provided greater energy resources for some herbivorous species that have become key pests of forage systems. Amenity turf is under constant critical scrutiny from the public, and its high cosmetic value and low damage thresholds have led to a large number of insect species being regarded as pests. In the USA, more than 24.5 million people spend over 2.4 billion h on golf courses each year, and about 56 million take part in their lawn care. Between golf courses and professional and homeowner lawn care, turf maintenance has become a $45 billion per year industry. A substantial amount of this budget and time is spent on insect and mite management (Danneberger, 1993).

Insect pests of turfgrass vary in their behaviour and feeding location. While white grubs, larvae of the Scarabaeidae (Coleoptera), usually feed on the grass roots, webworms and hepialids (Lepidoptera) create burrows in the soil from which they emerge at night to feed on the growing grass shoots. Armyworms (Lepidoptera) live on the surface, feeding on the foliage of grass plants, whereas some weevils, billbugs (Coleoptera) and fly larvae (Diptera) may bore into the stem, killing the tillers. Pests of grasslands and turf have been reviewed by Tashiro (1987), Delfosse (1993), Watschke *et al.* (1995), Potter (1998) and Vittum *et al.* (1999). While grasses support a wide variety of living organisms, less than 1% of these organisms acquire pest status requiring control. Major pests and the part of the plants they attack are listed in Table 7.1. Root-feeding white grubs, stem- and crown-feeding weevils, and foliage- and

Table 7.1. Major lawn, turfgrass, pasture pests, part of the plants they attack and geographic problem areas.

Plant part attacked	Pests	Pest life stage	Geographic location
Roots	White grubs	Larva	Worldwide
	Mole crickets	Adult and nymph	South-eastern USA, Korea
Stem/crown	Annual bluegrass weevil	Larva	North-eastern USA
	Billbugs	Larva	USA, Japan, New Zealand, Australia
	Crane flies	Larva	Europe, north-western USA, south-western Canada
Leaf/stem	Armyworms	Larva	Worldwide
	Cutworms	Larva	Worldwide
	Sod webworms	Larva	USA
	Chinchbugs	Adult and nymph	Central and eastern USA, south-eastern Canada, Japan
	Greenbug aphids	Adult and nymph	USA
	Mites	Adult and nymph	USA
	Spittlebugs	Adult and nymph	Eastern USA, Brazil
	Scales	Adult and nymph	Southern USA, Japan
	Mealybugs	Adult and nymph	Southern USA, New Zealand

stem-feeding Lepidoptera are pests worldwide, but other groups have a more limited distribution. The following sections will concentrate on those pest species that have received the most attention as targets for entomopathogenic nematodes (EPNs).

7.3. White Grubs

Root-feeding larvae of scarabaeid beetles are among the most damaging pests of lawns, turf and pastures in different parts of the world (Jackson, 1992). In some species, the adults can also cause extensive damage by feeding on foliage or flowers of ornamentals and fruit trees. Important endemic scarab pests include *Cyclocephala* spp. and *Phyllophaga* spp. in many parts of the Americas, *Holotrichia* spp. and *Heteronychus* spp. throughout Asia and Africa, *Melolontha* spp., *Amphimallon* spp. and *Phyllopertha* spp. in Europe, and *Anomala* spp. in Japan and Korea. Exotic species that have invaded new regions include the Japanese beetle, *Popillia japonica*; the oriental beetle, *Anomala* (= *Exomala*) *orientalis*; the European chafer, *Rhizotrogus majalis*; the Asiatic garden beetle, *Maladera castanea* in North America; and the South African beetle, *Heteronychus arator*, in New Zealand and Australian pastures.

The most important grub species have annual life cycles with adults emerging in summer (Potter, 1998). The females lay eggs in the soil below the grass. The grubs feed on the roots, which at high larval densities and under warm, dry conditions can lead to wilting of plants, gradual thinning of the turf and death of large turf areas. In addition, foraging skunks, raccoons, crows or other animals often cause further disruption of the turf surface by digging for the grubs (Watschke *et al.*, 1995). For most North American annual white grub species, most grubs reach the third instar by the middle of September but they may continue feeding well into October. Larvae move downwards into the soil for overwintering before the soil surface freezes. After overwintering in the soil, the grubs resume feeding in the spring before they pupate and emerge as adults in the summer. Some grub species, such as *Melolontha*, *Amphimallon* and some *Phyllophaga* have 2- or 3-year life cycles, and damage is dependent on the larval stage and species present. The typical life cycle of an annual white grub is shown in Fig. 7.1.

7.3.1. Nematodes for white grub control

White grubs are parasitized by a large number of nematode species (Poinar, 1975, 1992). EPNs are by far the most extensively studied parasites of white grubs. At least five species of EPNs, *Steinernema anomali, S. glaseri, S. kushidai, S. scarabaei* and *Heterorhabditis megidis*, were originally collected and described from naturally infected white grubs, and many more species have been documented as using

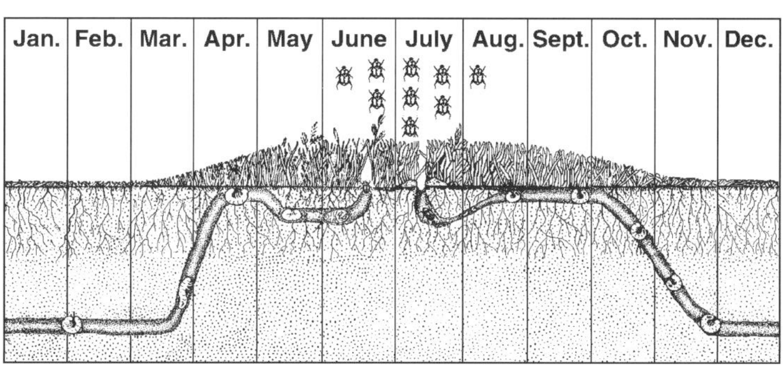

Fig. 7.1. Generalized life cycle of an annual white grub in turfgrass.

white grubs as natural hosts (Poinar, 1975, 1990, 1992; Peters, 1996; Stock and Koppenhöfer, 2003). Much of the research in the last two decades has focused on the potential use of nematodes for inundative application against white grubs. Four species, *H. bacteriophora*, *H. zealandica*, *H. marelata* and *S. glaseri*, are currently available commercially for grub control in the world.

7.3.2. Nematode field efficacy

Attempts to use nematodes in inundative control of white grubs began in the 1980s, when nematodes were first commercially mass-produced in liquid culture. Generally, *S. glaseri* and *H. bacteriophora* were found to be more effective than *S. carpocapsae* and *S. feltiae*. However, even with the more effective nematode species, early results were often variable (Kard *et al.*, 1988; Shetlar *et al.*, 1988; Villani and Wright, 1988; Wright *et al.*, 1988; Georgis and Poinar, 1989; Klein, 1990, 1993). Georgis and Gaugler (1991) analysed data from 82 field trials and concluded that most control failures against *Popillia japonica* could be explained on the basis of unsuitable nematode strain or environmental conditions. Much of the work since then has focused on discovery and evaluation of new species and strains, elucidation of factors affecting nematode efficacy and determination of the interactions between nematodes and other control agents. The available field data on the efficacy of nematodes against different white grub species is presented in Table 7.2. However, interpretations need to be made with caution as application rates, evaluation timing, post-application irrigation regimes and nematode quality may have differed between experiments. Below, we summarize the results from more recent field trials.

P. japonica has been studied extensively as a target for the field application of nematodes (Table 7.2). Multiple trials conducted between 2001 and 2003 have demonstrated the superiority of three species of EPNs, *H. bacteriophora* GPS11 (83–96% control) and TF (65–92% control) strains, *H. zealandica* X1 strain (96–98% control), and *S. scarabaei* AMK001 (100% control) over all the other nematode species tested (Cappaert and Koppenhöfer, 2003; Koppenhöfer and Fuzy, 2003a; Grewal *et al.*, 2004). Strains of *S. glaseri* (0–82%), *S. kushidai* (37–73%) and *S. carpocapsae* (38–66%) have been less effective and some others have shown very little efficacy. Against *Cyclocephala borealis*, *H. zealandica* X1 (72–96%) may be the most effective species followed by *S. scarabaei* AMK001 (58–84%) and *H. bacteriophora* GPS11 (47–83%) (Koppenhöfer and Fuzy, 2003a; Grewal *et al.*, 2004). Other *H. bacteriophora* strains and *S. kraussei* have provided some control whereas *S. glaseri* is ineffective. Against *Cyclocephala hirta* and *C. pasadenae*, none of the nematodes tested has provided useful levels of control, but the newer species/strains have not yet been evaluated against these scarabs in the field.

Against *A. orientalis*, *S. scarabaei* AMK001 has been the most effective species (60–96% at 21 DAT and 100% at 35 DAT) among the nematode species and strains evaluated in the field (Table 7.2). Other nematodes including *S. kushidai*, *H. bacteriophora* GPS11 and *H. zealandica* may be similarly effective; however, no field data are yet available. In a greenhouse trial, *S. kushidai* provided 88–94% control of *A. orientalis* (Table 7.3). Other nematodes that provided some *A. orientalis* control included *S. longicaudum* (41–56%), *S. glaseri* (0–70%), *S. carpocapsae* (56%), *Heterorhabditis* sp. Gyeongsan (67%). For other scarab species, only limited field and/or greenhouse data are available. Against *Phyllopertha horticola*, *H. bacteriophora* has provided better control than *H. megidis*. Against a mixture of three *Phyllophaga* spp. (*anxia*, *comes*, *fusca*), *H. bacteriophora* and *S. carpocapsae* strains provided variable results and showed no dosage effects, but overall *S. carpocapsae* All strain provided the highest control (75%). Against *R. majalis*, only *S. scarabaei* has provided good control (75–89%), whereas *S. glaseri* and *H. bacteriophora* TF

Table 7.2. Field efficacy of *Steinernema* and *Heterorhabditis* against white grubs in turfgrass (late summer/early autumn applications only).[a]

Grub species	Nematode species	Strain	Rate ($\times 10^9$ IJs/ha)	Mean % control	Duration	References
Anomala orientalis	*H. bacteriophora*	TF	1.0	11–40	21–39	Grewal *et al.*, 2004
			1.25	40	21	Koppenhöfer *et al.*, 2002
			2.5	0–52	21–39	Koppenhöfer and Fuzy, 2003a,b,c; Koppenhöfer *et al.*, 2002
	Heterorhabditis sp.	Gyeongsan	0.5	54	28	Lee *et al.*, 2002
			1.0	67	28	Lee *et al.*, 2002
	S. carpocapsae	Pocheon	1.0	56	28	Lee *et al.*, 2002
	S. glaseri	Dongrae	1.0	50	28	Lee *et al.*, 2002
			2.5	49	21	Koppenhöfer *et al.*, 1999
	S. glaseri	Mungyeong	1.0	50	28	Lee *et al.*, 2002
	S. glaseri	Biosys #326	1.24	0	28–44	Yeh and Alm, 1995
			2.47	21–70	28–44	Yeh and Alm, 1995
			4.9	54–68	28–44	Yeh and Alm, 1995
	S. longicaudum	Gongju	1.0	56	28	Lee *et al.*, 2002
	S. longicaudum	Nonsan	1.0	41–55	28	Lee *et al.*, 2002
	S. scarabaei	—	0.4	43/63–100	21/39	Koppenhöfer and Fuzy, 2003a; A.M. Koppenhöfer, unpublished data
			1.0	60–89/100	21/39	Koppenhöfer and Fuzy, 2003a; A.M. Koppenhöfer, unpublished data
			2.5	87–96/100	21/39	Koppenhöfer and Fuzy, 2003a; A.M. Koppenhöfer, unpublished data
Aphodius contaminatus	*H. bacteriophora*	EN0043	5.0	55	29	Sulistyanto and Ehlers, 1996
	H. megidis	HSH2	5.0	40	42	Sulistyanto and Ehlers, 1996
Ataenius spretulus	*S. carpocapsae*	All	4.9	46	15	Alm *et al.*, 1992
	S. glaseri	Biosys #2	4.9	14	15	Alm *et al.*, 1992
Cyclocephala borealis	*H. bacteriophora*	GPS11	2.5	47–83	28–35	Grewal *et al.*, 2004
		TF	1.0	6	21	Koppenhöfer and Fuzy, 2003a
			2.5	20	21	Koppenhöfer and Fuzy, 2003a
	H. zealandica	X1	2.5	72–96	28–35	Grewal *et al.*, 2004
	S. glaseri	MB	2.5	0	28	Grewal *et al.*, 2004

continued

Table 7.2. *Continued.* Field efficacy of *Steinernema* and *Heterorhabditis* against white grubs in turfgrass (late summer/early autumn applications only).[a]

Grub species	Nematode species	Strain	Rate ($\times 10^9$ IJs/ha)	Mean % control	Duration	Reference
	S. kraussei	UK	2.5	50	21	Grewal *et al.*, 2004
	S. scarabaei	—	1.0	58	21	Koppenhöfer and Fuzy, 2003a
			2.5	84	21	Koppenhöfer and Fuzy, 2003a
Cyclocephala hirta	*H. bacteriophora*	NC1	2.5	16/34–48	18/26	Koppenhöfer *et al.*, 1999, Koppenhöfer *et al.*, 2000a
			5.0	13	18	Koppenhöfer *et al.*, 2000b
	S. glaseri	NC	2.5	9	20	Koppenhöfer *et al.*, 2000a
	S. kushidai	—	5.0	33	18	Koppenhöfer *et al.*, 2000b
Cyclocephala pasadenae	*H. bacteriophora*	NC1	2.5	8	18	Koppenhöfer *et al.*, 1999
Maladera castanea	*H. bacteriophora*	TF	2.5	12–33	14–21	Koppenhöfer and Fuzy, 2003b
	S. scarabaei	—	1.0	51–60	14–21	Koppenhöfer and Fuzy, 2003b
			2.5	71–86	14–21	Koppenhöfer and Fuzy, 2003b
Phyllopertha horticola	*H. bacteriophora*	EN0043	5.0	55–74	21–42	Ehlers and Peters, 1998; Sulistyanto and Ehlers, 1996
	H. megidis	HSH2	5.0	40	42	Ehlers and Peters, 1998; Sulistyanto and Ehlers, 1996
Phyllophaga spp.	*H. bacteriophora*	?	1.35	61	35–42	Kard *et al.*, 1988
(*anxia*, *fusca*, *comes*)	(= *H. heliothidis*)		2.69	0	35–42	Kard *et al.*, 1988
			5.38	44	35–42	Kard *et al.*, 1988
	S. carpocapsae	DD-136	1.35	67	35–42	Kard *et al.*, 1988
	(= *S. feltiae*)		2.69	48	35–42	Kard *et al.*, 1988
			5.38	22	35–42	Kard *et al.*, 1988
	S. carpocapsae	Mexican	1.08	50	35–42	Kard *et al.*, 1988
	(= *S. feltiae*)		2.69	68	35–42	Kard *et al.*, 1988
			5.38	40	35–42	Kard *et al.*, 1988
	S. carpocapsae	All	1.08	87	35–42	Kard *et al.*, 1988
	(= *S. feltiae*)		2.69	77	35–42	Kard *et al.*, 1988
			5.38	61	35–42	Kard *et al.*, 1988
Popillia japonica	*H. bacteriophora*	GPS11	2.5	34–97	22–35	Grewal *et al.*, 2004
		HP88	2.5	52–74	22–35	Georgis and Gaugler, 1991; Grewal *et al.*, 2003
			5.0	51	21	Selvan *et al.*, 1993
			7.5	67	28–35	Georgis and Gaugler, 1991
	H. bacteriophora	NC	2.5	57	28–35	Georgis and Gaugler, 1991
			7.5	62	28–35	Georgis and Gaugler, 1991

	H. bacteriophora	NC1	1.0	40	25	Koppenhöfer *et al.*, 2000a
			2.0	85	25	Koppenhöfer *et al.*, 2000a
	H. bacteriophora	NJ2	5.0	70	21	Selvan *et al.*, 1993
	H. bacteriophora	TF	1.25	58	22	Koppenhöfer *et al.*, 2002
			2.5	65–92	21–22	Koppenhöfer and Fuzy, 2003a,c; Koppenhöfer *et al.*, 2000a, 2002
			5.0	51–63	18	Koppenhöfer *et al.*, 2000b
	H. indica	man 16	2.5	9	21	Koppenhöfer *et al.*, 2000a
	H. zealandica	X1	2.5	73–98	22–35	Grewal *et al.*, 2004
	S. arenarium	Ryazan	5.0	40	21	Selvan *et al.*, 1994
	S. carpocapsae	All	2.5	38	28–35	Georgis and Gaugler, 1991
			4.9	66	20	Alm *et al.*, 1992
			7.5	45	28–35	Georgis and Gaugler, 1991
	S. feltiae	Biosys #27	4.9	10–15	28–42	Alm *et al.*, 1992
	S. feltiae	Biosys #980	4.9	13–19	42	Alm *et al.*, 1992
	S. glaseri	Biosys #2	4.9	39	20	Alm *et al.*, 1992
	S. glaseri	Biosys #326	1.24	0	21	Yeh and Alm, 1995
			2.47	0–47	21–25	Yeh and Alm, 1995
			4.9	55–82	21–25	Yeh and Alm, 1995
	S. glaseri	NC	2.5	62	21	Koppenhöfer *et al.*, 2000a
			5.0	49–52	21	Selvan *et al.*, 1993, 1994
	S. glaseri	NJ	2.5	20	28	Grewal *et al.*, 2004
	S. glaseri	NJ43	5.0	70–72	21	Selvan *et al.*, 1993, 1994
	S. glaseri	MB	2.5	41–58	28	Grewal *et al.*, 2004
	S. glaseri	SI-12	5.0	72	21	Selvan *et al.*, 1994
	S. kraussei	UK	2.5	30	21	Grewal *et al.*, 2004
	S. kushidai	—	5.0	37–73	18	Koppenhöfer *et al.*, 2000b
	S. riobrave	RGV	5.0	32	21	Selvan *et al.*, 1994
	S. scarabaei	—	1.0	100	14	Koppenhöfer and Fuzy, 2003a
			2.5	100	14	Koppenhöfer and Fuzy, 2003a
Rhizotrogus majalis	*H. bacteriophora*	TF	2.5	38	21	Cappaert and Koppenhöfer, 2003
	S. scarabei	—	1.0	75	21	Cappaert and Koppenhöfer, 2003
			2.5	89	21	Cappaert and Koppenhöfer, 2003

[a]Data are shown from tests that were conducted only under conducive conditions (e.g. sufficiently high soil temperature, post-treatment irrigation, etc.) where data are separated by white grub species if more than one species was present, and only rates $\leq 7.5 \times 10^9$ IJs/ha.
?, Strain unknown.

Table 7.3. Efficacy of *Steinernema* and *Heterorhabditis* against white grubs in turfgrass greenhouse/pot experiments.[a]

Grub species	Nematode species	Strain	Rate ($\times 10^9$ IJs/ha)	Mean % control	References
Anomala orientalis	*H. bacteriophora*	NC1	0.5	6	Koppenhöfer *et al.*, 2000a
		TF	1.25	2440	Koppenhöfer and Fuzy, 2003; Koppenhöfer *et al.*, 2002
		CT	1.25	35	Koppenhöfer and Fuzy, 2003
		O	1.25	44	Koppenhöfer and Fuzy, 2003
	H. megidis	IN	1.25	11	Koppenhöfer *et al.*, 2002
	H. megidis	UK211	1.25	28	Koppenhöfer *et al.*, 2002
	S. feltiae	SN	1.25	0	Koppenhöfer *et al.*, 2002
	S. glaseri	NC	0.63	555	Koppenhöfer *et al.*, 2000a,b
			1.25	3072	Koppenhöfer *et al.*, 2002; Koppenhöfer and Fuzy, 2003
		38	1.25	36	Koppenhöfer and Fuzy, 2003
	S. kushidai	—	0.63	88	Koppenhöfer *et al.*, 2000b
			1.25	94	Koppenhöfer *et al.*, 2000b
	S. scarabaei	AMK001	0.16, 0.31, 0.63, 1.25	63, 73, 91, 96	Koppenhöfer and Fuzy, 2003a
Cyclocephala borealis	*H. bacteriophora*	NC1	0.5	0	Koppenhöfer *et al.*, 2000a
		TF	1.25	48	Koppenhöfer and Fuzy, 2003a
			2.5	46	Koppenhöfer and Fuzy, 2003a
		O	1.25	59	Koppenhöfer and Fuzy, 2003a
	S. glaseri	NC	0.6	0	Koppenhöfer *et al.*, 2000a
			1.25	13	Koppenhöfer and Fuzy, 2003a
		38	1.25	12	Koppenhöfer and Fuzy, 2003a
	S. scarabaei	AMK001	0.31	42	Koppenhöfer and Fuzy, 2003a
			0.63	55	Koppenhöfer and Fuzy, 2003a
			1.25	68	Koppenhöfer and Fuzy, 2003a
Cyclocephala hirta	*H. bacteriophora*	NC1	0.4	1353	Koppenhöfer and Kaya, 1998; Koppenhöfer *et al.*, 1999, 2000a
			0.6	20	Koppenhöfer *et al.*, 2000b
			1.2	2229	Koppenhöfer *et al.*, 2000a,b
	S. glaseri	NC	0.4	13	Koppenhöfer *et al.*, 2000a
			1.2	3739	Koppenhöfer *et al.*, 2000a,b
	S. kushidai	—	0.4	48	Koppenhöfer *et al.*, 2000a
			0.6	4063	Koppenhöfer *et al.*, 2000b
			1.2	5255	Koppenhöfer *et al.*, 2000a,b

Cyclocephala pasadenae	*H. bacteriophora*	NC1	0.5	20	Koppenhöfer *et al.*, 2000a
			1.0	1424	Koppenhöfer and Kaya, 1998; Koppenhöfer *et al.*, 1999
			2.0	12	Koppenhöfer *et al.*, 1999
	S. glaseri	NC	0.5	22	Koppenhöfer *et al.*, 2000a
	S. kushidai	—	0.5	39	Koppenhöfer *et al.*, 2000a
Hoplia philanthus	*H. megidis*	VBM30	2.5	12	Ansari *et al.*, 2003
			5.0	32	Ansari *et al.*, 2003
			7.5	37	Ansari *et al.*, 2003
	S. feltiae	MA40	2.5	12	Ansari *et al.*, 2003
			5.0	6	Ansari *et al.*, 2003
			7.5	14	Ansari *et al.*, 2003
	S. glaseri	NC	2.5	35	Ansari *et al.*, 2003
			5.0	49	Ansari *et al.*, 2003
			7.5	54	Ansari *et al.*, 2003
Maladera castanae	*H. bacteriophora*	TF	1.25	13	Koppenhöfer and Fuzy, 2003b
	S. glaseri	NC	1.25	17	Koppenhöfer and Fuzy, 2003b
	S. scarabaei	—	1.25	71	Koppenhöfer and Fuzy, 2003b
			2.5	94	Koppenhöfer and Fuzy, 2003b
Popillia japonica	*H. bacteriophora*	TF	0.31	7791	Koppenhöfer and Fuzy, 2003a
			0.5	2025	Koppenhöfer *et al.*, 2000a
			1.25	81	Koppenhöfer and Fuzy, 2003a
	S. glaseri	NC	0.5	321	Koppenhöfer *et al.*, 2000a
			1.25	81	Koppenhöfer and Fuzy, 2003a
	S. kushidai	—	0.5	36	Koppenhöfer *et al.*, 2000a
	S. scarabaei	—	0.16	67	Koppenhöfer and Fuzy, 2003a
			0.32	88	Koppenhöfer and Fuzy, 2003a
			0.63	90	Koppenhöfer and Fuzy, 2003a
			1.25	96	Koppenhöfer and Fuzy, 2003a
Rhizotrogus majalis	*H. bacteriophora*	TF	1.25	27	Cappaert and Koppenhöfer, 2003
	S. glaseri	NC	1.25	38	Cappaert and Koppenhöfer, 2003
	S. scarabaei	—	0.16	87	Cappaert and Koppenhöfer, 2003
			0.31	91	Cappaert and Koppenhöfer, 2003
			0.63	91	Cappaert and Koppenhöfer, 2003
			1.25	98	Cappaert and Koppenhöfer, 2003

[a]All data shown are control rates observed at 14 DAT.

strain are ineffective. *M. castanea* follows the same pattern as *R. majalis* but appears to be somewhat less susceptible to all nematodes tested both in the field and greenhouse (Tables 7.2 and 7.3). For *Hoplia philanthus* (subfamily: Melolonthinae), only greenhouse data are published, which suggest that *S. glaseri* and *H. megidis* may provide acceptable control, whereas *S. feltiae* is ineffective.

7.3.3. Factors affecting nematode efficacy

Major factors affecting the infection and field efficacy of EPNs against white grubs are shown in Fig. 7.2. In general, white grubs are less susceptible to EPNs than most lepidopteran larvae. This low susceptibility is due to a series of ecological, behavioural, morphological and physiological barriers to infection against EPNs. First, the location of white grub larvae in the soil profile precludes infection by the nematode species that utilize ambush-type foraging behaviour (Gaugler *et al.*, 1997). The detection of a potential host may be made more difficult through the white grubs' tendency to release CO_2 in bursts rather than continuously. CO_2 is an important volatile host cue for EPNs (Lewis *et al.*, 1993). Nematodes that have successfully located a white grub and attached to its cuticle can be effectively eliminated by the grub's aggressive grooming behaviours. These behaviours include rubbing with an abrasive raster situated on the ventral end of the abdomen or brushing with legs or mouth parts (Gaugler *et al.*, 1994). In addition, white grubs evade nematode attack by moving away from the

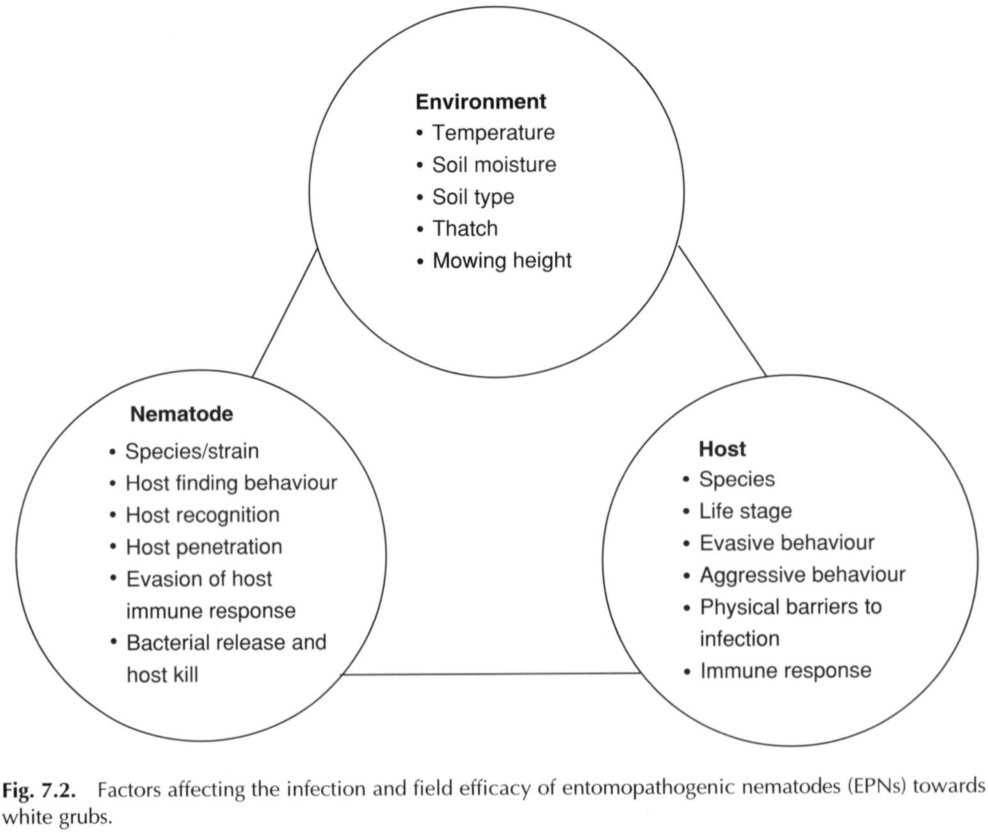

Fig. 7.2. Factors affecting the infection and field efficacy of entomopathogenic nematodes (EPNs) towards white grubs.

nematodes (Schroeder *et al.*, 1993; Gaugler *et al.*, 1994). Both aggressive and evasive behaviours have been demonstrated for *P. japonica* larvae.

Nematode penetration into a host can occur (i) through the thin regions of the cuticle (only common in *Heterorhabditis* spp.); (ii) through tracheae via the spiracles; or (iii) through the midgut epithelium via mouth or anus, depending on which routes are accessible and the specific stage of the insect (Eidt and Thurston, 1995). In white grubs, the spiracles are covered with sieve plates that are impenetrable to nematodes (Hinton, 1967; Galbreath, 1976; Forschler and Gardner, 1991). Nematode penetration through the midgut epithelium is delayed by a dense peritrophic membrane (Forschler and Gardner, 1991). This delay increases the chances of nematode inactivation by gut fluids (Wang *et al.*, 1995) and/or removal by food passage from the alimentary tract. In *P. japonica* larvae, *S. glaseri* possess superior gut penetration ability and do not seem to penetrate through the cuticle (Wang and Gaugler, 1998). In contrast, *H. bacteriophora* are more quickly deactivated by gut juices but possess remarkable cuticular penetration ability, especially at membranous areas such as leg and maxilla joints (Wang and Gaugler, 1998).

Nematodes that have penetrated into the grubs' haemocoel may still have to face a strong immune response that results in melanotic encapsulation (Wang *et al.*, 1994, 1995). *H. bacteriophora* elicit a strong immune response in *P. japonica* larvae but release their symbiotic bacteria before the nematodes are killed. The bacteria produce insecticidal toxins that rapidly kill the host and allow later invading *H. bacteriophora* infective juveniles (IJs) to escape encapsulation (Wang *et al.*, 1994, 1995). *S. glaseri*, although initially encapsulated in *P. japonica* larvae, escape from the capsules (Wang *et al.*, 1995) because their surface coat proteins suppress the immune response in *P. japonica* larvae and destroy haemocytes (Wang and Gaugler, 1999). Differences in the encapsulation of *H. bacteriophora* strains by *P. japonica* and *C. borealis* grubs have also been reported (Grewal *et al.*, 2002).

7.3.3.1. Grub species and nematode species/strain

Large variation exists in the virulence of nematode species and strains against white grub species. In general, *H. bacteriophora*, *H. zealandica*, *H. megidis*, *S. glaseri*, *S. kushidai* and *S. scarabaei* are more virulent against white grubs than *H. indica*, *H. marelata*, *S. anomali*, *S. carpocapsae*, *S. feltiae* and *S. riobrave* (Table 7.4). In addition, different strains of the same nematode species may vary considerably in their virulence to different white grub species (Grewal *et al.*, 2002). Grewal *et al.* (2002) determined the virulence of 16 species and strains of EPNs against *P. japonica* and *C. borealis* in 30-ml cups containing 20 g of sand, and found that *H. zealandica* X1 strain and *H. bacteriophora* GPS11 strain were significantly more virulent than other strains and species towards both *P. japonica* and *C. borealis*. Although it is often suggested that local strains may be more virulent than exotic strains, Grewal *et al.* (2002) found no significant differences in the virulence of nematode species and strains isolated from within and outside the geographic ranges of *P. japonica* and *C. borealis*. Differences in the virulence of nematode species have been attributed to differences in foraging behaviour (Kaya and Gaugler, 1993), penetration efficiency (Grewal *et al.*, 2002), ability to escape from the host immune response (Wang *et al.*, 1995; Grewal *et al.*, 2002), number of cells of the symbiotic bacteria carried (Selvan *et al.*, 1993) and virulence of the symbiotic bacteria (Yeh and Alm, 1992; Ansari *et al.*, 2003). Grewal *et al.* (2002) compared the penetration efficiency of *H. bacteriophora*, strains GPS11 and HP88, and *H. zealandica*, strain X1, into *P. japonica* and *C. borealis* grubs. *H. zealandica* X1 had the highest penetration, followed by *H. bacteriophora* GPS11 and *H. bacteriophora* HP88 in both scarab species. They also found that a significantly lower percentage of penetrated *H. zealandica* X1 and *H. bacteriophora* GPS11 were melanized and killed due to encapsulation than *H. bacteriophora* HP88.

Table 7.4. Relative virulence of entomopathogenic nematode (EPN) species and strains to white grub species in laboratory bioassays. Third instars were tested unless otherwise stated.

Grub/nematode species	Strain	Nematodes/grub	Mean % mortality	References
Anomala cuprea **(Coleoptera: Scarabaeidae: Rutelinae)**				
Steinernema kushidai	?	1000	96.7	Fujie *et al.*, 1993
Anomala orientalis **(Coleoptera: Scarabaeidae: Rutelinae)**				
Heterorhabditis bacteriophora	GPS11	100, 1000	50, 75	Grewal *et al.*, 2002
	HP88	100, 1000	35, 45	Grewal *et al.*, 2002
	TF	400	18–42	Koppenhöfer and Fuzy, 2003a
		400	29	Koppenhöfer *et al.*, 2004
	CT	400	20	Koppenhöfer and Fuzy, 2003a
	O	400	30–60	Koppenhöfer and Fuzy, 2003a
H. megidis	UK	1000	58	Grewal *et al.*, 2002
H. zealandica	X1	100, 1000	50, 65	Grewal *et al.*, 2002
Heterorhabditis sp.	Gyeongsang	300	39	Lee *et al.*, 2002
	Korea	400	22–58	Koppenhöfer and Fuzy, 2003a
S. carpocapsae	Pocheon	300	20	Lee *et al.*, 2002
S. glaseri	Dongrae	300	28	Lee *et al.*, 2002
	Dongrae	46	18–36	Koppenhöfer *et al.*, 1999
S. glaseri	Mungyeong	300	18	Lee *et al.*, 2002
	NC	400	42–62	Koppenhöfer and Fuzy, 2003a
	NC	400	53	Koppenhöfer *et al.*, 2004
S. longicaudum	Gongju	300	28	Lee *et al.*, 2002
	Nonsan	300	16	Lee *et al.*, 2002
S. scarabaei	AMK001	13, 20, 25, 50, 100, 200, 400	30, 53, 70, 96, 96, 94, 96	Koppenhöfer and Fuzy, 2003a
	AMK001	50, 400	94, 96	Koppenhöfer *et al.*, 2004
Popillia japonica **(Coleoptera: Scarabaeidae: Rutelinae)**				
H. bacteriophora	HP88	969	87	Yeh and Alm, 1995
	HP88	100	47	Selvan *et al.*, 1994
	HP88	100, 1000	60, 100	Simard *et al.*, 2001
	HP88	100, 200, 1000	20, 30, 25	Grewal *et al.*, 2002
	Acows	200	50	Grewal *et al.*, 2002
	Oswego	200	20	Grewal *et al.*, 2002
	NC1	200	30	Grewal *et al.*, 2002
	Lewiston	200	5	Grewal *et al.*, 2002

	KMD10	200	50	Grewal *et al.*, 2002
	KM 19	200	45	Grewal *et al.*, 2002
	GPS1	200	25	Grewal *et al.*, 2002
	GPS2	200	4	Grewal *et al.*, 2002
	GPS3	200	50	Grewal *et al.*, 2002
	GPS5	200	20	Grewal *et al.*, 2002
	GPS11	100, 200, 1000	45, 65, 75	Grewal *et al.*, 2002
	NJ2	100	77	Selvan *et al.*, 1994
	C1	969	64	Yeh and Alm, 1995
	TF	400	93	Koppenhöfer and Fuzy, 2003a
	TF	400	91	Koppenhöfer *et al.*, 2004
	CT	400	80	Koppenhöfer and Fuzy, 2003a
	O	400	100	Koppenhöfer and Fuzy, 2003a
	TF	80	40	Cappaert and Koppenhöfer, 2003
	Cruiser(TM)	1000	98	Amaral, 1996
	Azorean	1000	50	Amaral, 1996
H. indica	LN2	200	10	Grewal *et al.*, 2002
H. marelata	Oregon	200	18	Grewal *et al.*, 2002
H. marelata	IN	159, 318	100, 100	Maninon *et al.*, 2000
H. megidis	UK	100, 200, 1000	15, 35, 55	Grewal *et al.*, 2002
H. zealandica	X1	100, 200, 1000	55, 70, 95	Grewal *et al.*, 2002
Heterorhabditis sp.	Korea	400	90	Koppenhöfer and Fuzy, 2003a
S. anomali	Ryazan	500	34	Simard *et al.*, 2001
S. carpocapsae	All	100, 1000	20, 35	Wang *et al.*, 1994
	All	969	29	Yeh and Alm, 1995
	Mexican	969	18	Yeh and Alm, 1995
S. feltiae	Biosys N27	969	33	Yeh and Alm, 1995
	Hyl	1000	77	Amaral, 1996
S. glaseri	Biosys N-2	969	86	Yeh and Alm, 1995
	NC	100	40	Selvan *et al.*, 1994
	NC	100, 1000	80, 100	Wang *et al.*, 1994
	NC	500	45	Simard *et al.*, 2001
	NC	400	89	Koppenhöfer and Fuzy, 2003a
	NC	400	87	Koppenhöfer *et al.*, 2004
	SI-12	500	90	Simard *et al.*, 2001
	NJ 43	100	83	Selvan *et al.*, 1994

continued

Table 7.4. *Continued.* Relative virulence of entomopathogenic nematode (EPN) species and strains to white grub species in laboratory bioassays. Third instars were tested unless otherwise stated.

Grub/nematode species	Strain	Nematodes/grub	Mean % mortality	References
	NJ 43	500	88	Simard *et al.*, 2001
	Biosys N326	1000	98	Amaral, 1996
S. riobrave	RGV	500	44	Selvan *et al.*, 1994
S. scapterisci	Uruguay	1000	0	Townsend *et al.*, 1998
S. scarabaei	AMK001	6, 13, 25, 50, 100, 200, 400	22, 45, 90, 100, 98, 98, 100	Koppenhöfer and Fuzy, 2003a
	AMK001	50, 400	100, 98	Koppenhöfer *et al.*, 2004
	AMK001	16, 32, 64, 80	78, 80, 100, 90	Cappaert and Koppenhöfer, 2003
Costelytra zealandica **(Coleoptera: Scarabaeidae: Dynastinae)**				
S. glaseri	?	4500	100	Kain *et al.*, 2003
H. bacteriophora	?	4500	95	Kain *et al.*, 2003
Cyclocephala borealis **(Coleoptera: Scarabaeidae: Dynastinae)**				
H. bacteriophora	Oswego	200	30	Grewal *et al.*, 2002
	NC1	200	52	Grewal *et al.*, 2002
	Lewiston	200	40	Grewal *et al.*, 2002
	KMD10	200	54	Grewal *et al.*, 2002
	KMD19	200	35	Grewal *et al.*, 2002
	GPS1	200	18	Grewal *et al.*, 2002
	GPS2	200	4	Grewal *et al.*, 2002
	GPS3	200	4	Grewal *et al.*, 2002
	GPS5	200	5	Grewal *et al.*, 2002
	GPS11	100, 200, 1000	15, 50, 8	Grewal *et al.*, 2002
	HP88	100, 200, 1000	15, 20, 40	Grewal *et al.*, 2002
	Acows	200	45	Grewal *et al.*, 2002
	TF	400	48–58	Koppenhöfer and Fuzy, 2003a
	TF	400	50	Koppenhöfer *et al.*, 2004
	CT	400	60	Koppenhöfer and Fuzy, 2003a
	O	400	60–82	Koppenhöfer and Fuzy, 2003a
H. indica	LN2	200	15	Grewal *et al.*, 2002
H. marelata	Oregon	200	20	Grewal *et al.*, 2002
H. megidis	UK	100, 200, 1000	30, 42, 70	Grewal *et al.*, 2002
H. zealandica	X1	100, 200, 1000	35, 58, 88	Grewal *et al.*, 2002
Heterorhabditis sp.	Korea	400	58–62	Koppenhöfer and Fuzy, 2003a

S. glaseri	NC	400	18–20	Koppenhöfer and Fuzy, 2003a
S. scarabaei	AMK001	400	43–54	Koppenhöfer and Fuzy, 2003a
Cyclocephala hirta **(Coleoptera: Scarabaeidae: Dynastinae)**				
H. bacteriophora	HP88	125	65	Converse and Grewal, 1998
	NC1	400	70	Koppenhöfer *et al.*, 2000b
	Ecogen	400	48	Koppenhöfer *et al.*, 2000b
H. megidis	UK	125	60	Converse and Grewal, 1998
Heterorhabditis sp.	Chino Hill	125	84	Converse and Grewal, 1998
	Merced	125	84	Converse and Grewal, 1998
	Nebraska	125	78	Converse and Grewal, 1998
S. carpocapsae	All	125	0	Converse and Grewal, 1998
	Mexican	125	0	Converse and Grewal, 1998
S. feltiae	Argentina	125	15	Converse and Grewal, 1998
S. glaseri	NC	125	62	Converse and Grewal, 1998
	NJ21	125	100	Converse and Grewal, 1998
	NJ29	125	100	Converse and Grewal, 1998
	NJ32	125	95	Converse and Grewal, 1998
	NJ40	125	85	Converse and Grewal, 1998
	NJ41	125	80	Converse and Grewal, 1998
	NJ42	125	100	Converse and Grewal, 1998
	NJ43	125	68	Converse and Grewal, 1998
	NJ63	125	72	Converse and Grewal, 1998
	NJ65	125	100	Converse and Grewal, 1998
S. kushidai	Hamakita	125	52	Converse and Grewal, 1998
	Kubota	40	67–78	Koppenhöfer *et al.*, 2000
S. riobrave	RGV	125	0	Converse and Grewal, 1998
S. scarabaei	Colon	125	0	Converse and Grewal, 1998
Cyclocephala lurida **(Coleoptera: Scarabaeidae: Dynastinae)**				
H. bacteriophora	TF	400	52	Koppenhöfer *et al.*, 2004
S. glaseri	NC	400	36	Koppenhöfer *et al.*, 2004
S. scarabaei	AMK001	50, 400	33, 50	Koppenhöfer *et al.*, 2004
Cyclocephala pasadenae **(Coleoptera: Scarabaeidae: Dynastinae)**				
H. bacteriophora	TF	400	8	Koppenhöfer *et al.*, 2004
S. glaseri	NC	400	25	Koppenhöfer *et al.*, 2004
S. scarabaei	AMK001	50, 400	11, 30	Koppenhöfer *et al.*, 2004

continued

Table 7.4. *Continued.* Relative virulence of entomopathogenic nematode (EPN) species and strains to white grub species in laboratory bioassays. Third instars were tested unless otherwise stated.

Grub/nematode species	Strain	Nematodes/grub	Mean % mortality	References
***Amphimallon solstitiale* (Coleoptera: Scarabaeidae: Melonthinae)**				
S. glaseri	Morocco	1000	—	Peters *et al.*, 2002
***Hoplia philanthus* (Coleoptera: Scarabaeidae: Melolonthinae)**				
H. megidis	VBM30	10,000	100	Ansari *et al.*, 2003
S. feltiae	MA40	10,000	38	Ansari *et al.*, 2003
S. glaseri	NC	10,000	100	Ansari *et al.*, 2003
***Maladera castanea* (Coleoptera: Scarabaeidae: Melolonthinae)**				
H. bacteriophora	TF	100, 200, 400	10, 16, 25–30,	Koppenhöfer and Fuzy, 2003a
	TF	400	5	Koppenhöfer *et al.*, 2004
	CT	400	45	Koppenhöfer and Fuzy, 2003a
	O	400	10	Koppenhöfer and Fuzy, 2003a
Heterorhabditis sp.	Korea	400	50	Koppenhöfer and Fuzy, 2003a
S. glaseri	NC	400	30	Koppenhöfer and Fuzy, 2003a
	NC	400	23	Koppenhöfer *et al.*, 2004
S. scarabaei	AMK001	50, 100, 200, 400	65, 76, 90, 98	Koppenhöfer and Fuzy, 2003a
	AMK001	50, 400	78, 98	Koppenhöfer *et al.*, 2004
***Melolontha melolontha* (Coleoptera: Scarabaeidae: Melolonthinae)**				
H. bacteriophora	HI 191, HI 127, HI 23, HI 273, NJ	1500, 5000	—	Peters, 2000
	HK3	1000	30	Berner and Schnetter, 2001
	HH-Bp 1202	1000	100 (2nd)	Selvan *et al.*, 1993
	AZ32	1000	60 (2nd)	Selvan *et al.*, 1993
H. marelata	?	1000	—	Peters and Keller, 2000
	?	1000	20	Berner and Schnetter, 2001
H. megidis	HSH-2	1500, 5000	—	Peters, 2000
		1000	—	Peters and Keller, 2000
S. arenaria	?	1000	10	Berner and Schnetter, 2001
S. feltiae	Neud	1000	0	Berner and Schnetter, 2001
	?	1000	100 (2nd)	Selvan *et al.*, 1993
S. glaseri	NC	1000	90-(2nd)	Selvan *et al.*, 1993
	NC	1000	40	Berner and Schnetter, 2001

	RS92	1000	60	Berner and Schnetter, 2001
	RS92	1500, 5000	—	Peters, 2000
	Morocco	—	—	Peters *et al.*, 2002
S. riobrave	Biosys N355	1000	100 (2nd)	Selvan *et al.*, 1993
Steinernema sp.	Morocco	1000	60	Berner and Schnetter, 2001
Phyllophaga crinita **(Coleoptera: Scarabaeidae: Melolonthinae)**				
H. bacteriophora	TF	400	9	Koppenhöfer *et al.*, 2004
S. glaseri	NC	400	0	Koppenhöfer *et al.*, 2004
S. scarabaei	AMK001	50, 400	67, 94	Koppenhöfer *et al.*, 2004
Phyllophaga congrua **(Coleoptera: Scarabaeidae: Melolonthinae)**				
H. bacteriophora	TF	400	13	Koppenhöfer *et al.*, 2004
S. glaseri	NC	400	18	Koppenhöfer *et al.*, 2004
S. scarabaei	AMK001	50, 400	64, 89	Koppenhöfer *et al.*, 2004
Phyllophaga georgiana **(Coleoptera: Scarabaeidae: Melolonthinae)**				
H. bacteriophora	TF	400	57	Koppenhöfer *et al.*, 2004
S. glaseri	NC	400	8	Koppenhöfer *et al.*, 2004
S. scarabaei	AMK001	50, 400	35, 90	Koppenhöfer *et al.*, 2004
Rhizotrogus majalis **(Coleoptera: Scarabaeidae: Melolonthinae)**				
H. bacteriophora	GPS11	100, 1000	0, 0	Grewal *et al.*, 2002
	HP88	100, 1000	0, 8	Grewal *et al.*, 2002
	TF	400	25	Koppenhöfer and Fuzy, 2003a
	TF	400	23	Koppenhöfer *et al.*, 2004
	TF	80	10	Cappaert and Koppenhöfer, 2003
	CT	400	58	Koppenhöfer and Fuzy, 2003a
	?	1000, 5000	20, 30	Townsend *et al.*, 1994
	Heliothidis	9300	94	Yeh and Alm, 1995
H. megidis	UK	100, 1000	20, 18	Grewal *et al.*, 2002
H. zealandica	X1	100, 1000	5, 18	Grewal *et al.*, 2002
S. carpocapsae	?	100, 1000	5, 12	Townsend *et al.*, 1994
	All	9300	35	Yeh and Alm, 1995
S. feltiae	?	1000, 5000	0, 0	Townsend *et al.*, 1994
S. glaseri	NC	1000, 5000	78, 90	Townsend *et al.*, 1994
	NC	400	42	Koppenhöfer and Fuzy, 2003a
	NC	400	43	Koppenhöfer *et al.*, 2004

continued

Table 7.4. *Continued.* Relative virulence of entomopathogenic nematode (EPN) species and strains to white grub species in laboratory bioassays. Third instars were tested unless otherwise stated.

Grub/nematode species	Strain	Nematodes/grub	Mean % mortality	References
S. scarabaei	AMK001	400	100	Koppenhöfer and Fuzy, 2003a
	AMK001	50, 400	96, 100	Koppenhöfer *et al.*, 2004
	AMK001	10, 20, 40, 80	74, 83, 96, 100	Cappaert and Koppenhöfer, 2003
Ataenius spretulus **(Coleoptera: Scarabaeidae: Melolonthinae)**				
H. bacteriophora	TF	400	80	Koppenhöfer *et al.*, 2004
S. glaseri	NC	400	50	Koppenhöfer *et al.*, 2004
S. scarabaei	AMK001	50, 400	20, 50	Koppenhöfer *et al.*, 2004
Cotinus nitida **(Coleoptera: Scarabaeidae: Cetoninae)**				
H. bacteriophora	?	1000-peroral	63	Townsend *et al.*, 1998
	?	1000	34	Wang *et al.*, 1994
	TF	400	10	Koppenhöfer *et al.*, 2004
S. carpocapsae	All	1000-peroral	65	Townsend *et al.*, 1998
	All	1000	12	Wang *et al.*, 1994
S. feltiae	NC	1000-peroral	45	Townsend *et al.*, 1998
	NC	1000	18	Wang *et al.*, 1994
S. glaseri	?	1000-peroral	65	Townsend *et al.*, 1998
	?	1000	22	Wang *et al.*, 1994
	NC	400	5	Koppenhöfer *et al.*, 2004
S. scarabaei	AMK001	400	22	Koppenhöfer *et al.*, 2004
	AMK001	50	9	Koppenhöfer *et al.*, 2004

?, Strain unknown.

Differences in the susceptibility of different white grub species is another factor influencing the efficacy of nematodes (Table 7.2). Grewal *et al.* (2002) compared the susceptibilities of *P. japonica*, *A. orientalis*, *C. borealis* and *R. majalis* to *H. bacteriophora* strains HP88 and GPS11, *H. megidis* UK and *H. zealandica* X1 in laboratory bioassays. They found that *R. majalis* was the least susceptible of the white grub species to all four nematode strains, with grub mortality never exceeding 20%. *P. japonica* and *A. orientalis* (both Rutelinae) and *C. borealis* (Dynastinae) were generally equally susceptible to all four nematode strains (Grewal *et al.*, 2002). Koppenhöfer *et al.* (2004) compared the pathogenicity of three nematode species (*H. bacteriophora* TF strain, *S. glaseri* NC strain, *S. scarabaei* AMK001) to 12 white grub species in the laboratory. Generally, *P. japonica* was the most susceptible species, being highly susceptible to all three nematodes; *Ataenius orientalis*, *R. majalis*, *M. castanea* and three *Phyllophaga* spp. were highly susceptible to *S. scarabaei*, but showed mediocre to very low susceptibility to *S. glaseri* and *H. bacteriophora*; *A. spretulus* (subfamily: Aphodiinae) showed high susceptibility to *H. bacteriophora* but lower susceptibility to *S. glaseri* and *S. scarabaei*; and three *Cyclocephala* spp. and *Cotinus nitida* showed mediocre to very low susceptibility to all three nematodes (Table 7.4).

In summary, nematode efficacy can vary considerably with nematode species and strains and white grub species. Overall, members of the subfamily Melolonthinae have generally shown very low susceptibility to *H. bacteriophora* and *S. glaseri* but are susceptible to *S. scarabaei*. Given the relatively limited number of nematode species–grub species combinations studied at this point, extrapolations on nematode efficacy against other white grub species should be made with care.

7.3.3.2. Larval stage

Susceptibility to EPNs also varies with white grub larval stage. We have found (K.T. Power and P.S. Grewal, 2003, unpublished data) that the third instar *P. japonica* is relatively less susceptible to *H. bacteriophora* GPS11 than the first or second instars, both in the laboratory and field tests. Similarly, second instar *A. orientalis* were more susceptible than third instars to EPNs (Lee *et al.*, 2002). Koppenhöfer and Fuzy (2004) observed the same trend in *P. japonica* for *H. bacteriophora* TF strain, but for *S. scarabaei* observed no difference between second and third instars. In *A. orientalis*, first and second instars were more susceptible than third instars to *H. bacteriophora* TF strain, but there was no difference between second and third instars with *S. scarabaei* and *S. glaseri*. In addition, young third instars (< 100 mg) were more susceptible than more mature third instars (> 175 mg) to *H. bacteriophora* TF strain, with a similar but weaker effect observed for *S. scarabaei* (Koppenhöfer and Fuzy, in press). In *M. melolontha,* first and early second instars were the most susceptible stages to *S. glaseri* and a strain of *Heterorhabditis* sp. (Deseö *et al.*, 1990). In *M. castanea*, third instars were more susceptible than second instars to *S. scarabaei* (Koppenhöfer and Fuzy, in press) and in *M. matrida* (Glazer and Gol'berg, 1989, 1993), the third instars were more susceptible to *H. bacteriophora* than first and second instars. In *P. horticola*, instar susceptibility increased significantly for *S. glaseri*, *H. downesi* and *H. bacteriophora*, but not for *S. arenarium* (= *anomali*) and *S. carpocapsae* (Smits *et al.*, 1994). Similarly, the third instars of *A. cuprea* were more susceptible than the second and first instars to *S. kushidai* (Fujiie *et al.*, 1993). In summary, instar susceptibility varies with white grub and nematode species, and the nematode applications targeted against the most susceptible instars may be more effective than those targeted exclusively against the third instars. Also, applications made against earlier grub instars may allow enough time for nematodes to recycle in the grub populations, leading to the possibility of secondary infections and enhanced grub control during the season.

7.3.3.3. Nematode application rate

Application rates between 0.5 and 12.5×10^9 IJs/ha have been tested in various studies. Overall, a rate of 2.5×10^9 IJs/ha has been recommended (Georgis and Gaugler, 1991). Although grub control with two out of the three nematode species increased with an increase in application rate from 2.5×10^9 to 5×10^9 IJs/ha, differences were not significant (Grewal *et al.*, 2004). Nematode rates in excess of 2.5×10^9 IJs/ha are not economical at this time (Grewal and Georgis, 1998). However, lower nematode application rates need to be re-evaluated, especially against the more susceptible *P. japonica*, when all other biotic and abiotic factors are optimum for nematode activity. Additionally, new species such as *S. scarabaei* may be effective at lower rates against scarab species such as *R. majalis*, *P. japonica* and *A. orientalis*.

7.3.3.4. Thatch depth and mowing height

Thatch, a layer of tightly intermingled living and dead roots, crowns, rhizomes, stolons and organic debris, which sometimes accumulates on the soil surface in turfgrass systems due to a low rate of decomposition, may present an impenetrable barrier to the nematodes (Zimmerman and Cranshaw, 1991), reducing their efficacy (Georgis and Gaugler, 1991). Hydrophobicity of thatch reduces water intake, resulting in nematode runoff in water during application. Anionic and non-ionic wetting agents, such as sulfonated carbonic acids (e.g. Kick™; Compo, Germany), ethylene oxide and propylene oxide copolymers (e.g. Foresight®; Famura; UK), or alkylpolyglycosides and fatty acids (e.g. Magic Wet™; Cognis; Germany) can enhance permeability through the thatch. A pretreatment with these substances during regular turf irrigation will help transfer nematodes into the soil. Also these substances can be tank-mixed with nematodes applied to turf (R.-U. Ehlers, personal communication). The practice of aeration, i.e. making holes in the ground by removing thin soil cores, is often used to enhance the movement of air, water and nutrients into the soil where thatch becomes a problem. This practice can also improve nematode penetration into the soil. Aeration increased the mortality of *A. orientalis*, caused by *Heterorhabditis* sp. Gyeongsan strain, from 71% to 85% and that by *S. carpocapsae* Pocheon strain from 35% to 80% (H.Y. Choo, 2003, unpublished data).

High mowing height can also reduce nematode efficacy by restricting nematode contact with soil. Mortality of third instar *A. orientalis* was 89% and 94% by *S. glaseri* Dongrae strain and *S. longicaudum* Gongju strain, respectively, at 5 mm turf height but only 52% and 64%, respectively, at 14 mm (H.Y. Choo, 2003, unpublished data). Thus, it is recommended to mow the turfgrass to the lowest height acceptable before nematode application.

7.3.3.5. Soil moisture and the amount of post-application irrigation

Optimum soil moisture is extremely important for nematode activity and survival. Shetlar *et al.* (1988) reported that at least 0.74 cm of post-application irrigation is required for the activity and establishment of nematodes in turfgrass. They also suggested that moderate soil moisture should be maintained after nematode application. Georgis and Gaugler (1991) reported that an irrigation frequency of 1–4-day interval was optimum for grub control with nematodes. Grewal *et al.* (2004) found that a total of 10 cm of post-application irrigation plus rainfall over a period of 4–5 weeks after application was optimum, at which the *H. bacteriophora* GPS11 and *H. zealandica* X1 strains produced 83–97% and 96–98% control of the two white grub species, respectively. Ehlers and Peters (1998) have indicated that optimum soil moisture is critical for obtaining sustainable effects of nematode applications.

Timing of post-application irrigation is also very important when applying nematodes to turfgrass. As nematodes that get stuck to grass may be rapidly inactivated by heat and ultraviolet (UV) radiation, they must be rinsed off as soon as possible. Post-application rinse irrigation applied immediately after nematodes significantly increased the efficacy of *S. glaseri* and *H. bacteriophora* against *P. japonica* (Selvan *et al.*, 1993). In this regard, the turfgrass should be mowed to the lowest acceptable height before nematode application to enhance nematode contact with the soil.

7.3.3.6. Soil temperature

Georgis and Gaugler (1991) reported that the nematode applications made in late summer/early autumn were more effective against white grubs than those made in the spring, and advocated that soil temperature should be above 20°C for maximum white grub control with nematodes. However, the influence of temperature on the efficacy depends upon the nematode species (Grewal *et al.*, 1994, 2004). In bioassays conducted at 23°C, *S. glaseri* and *H. bacteriophora* caused 100% *P. japonica* larval mortality and *S. carpocapsae* caused 56% mortality, but at temperatures below 15°C only *S. glaseri* remained effective (Simões *et al.*, 1993).

7.3.3.7. Nematode trait stability

Circumstantial evidence suggests that the virulence against white grubs and other desired traits of nematode strains may deteriorate over time. Selvan *et al.* (1994) attributed the poor performance of NC strain of *S. glaseri* against *P. japonica* to its prolonged laboratory culture. Similarly, Lee *et al.* (2002) reported rapid decline in the performance of Dongrae strain of *S. glaseri* against *A. orientalis*. Rapid genetic deterioration in environmental stress tolerance has been reported for heterorhabditids (Shapiro *et al.*, 1996; Wang and Grewal, 2002). Wang and Grewal (2002) demonstrated that genetic deterioration in traits of EPNs can be prevented/reduced through cryopreservation of the master stock in liquid nitrogen or storage at low temperature coupled with less frequent culturing. Repeated or frequent culturing of the master stock in white grubs may also maintain or even enhance virulence of nematodes towards white grubs (Selvan *et al.*, 1994). Additionally, beneficial traits can be stabilized in selected inbred lines (Bai *et al.*, 2004, unpublished data).

7.3.4. Current status and analysis

Chemical insecticides have been the primary tools in the management of white grubs. In the USA, insecticides are usually applied in late July or August after oviposition ends and the bulk of the population is in the first or early second instar. This is usually well before damage becomes apparent. The efficacy of most insecticides declines when larvae reach the third instar. Four insecticides, trichlorfon (Dylox or Proxol), chlorpyrifos (Dursban), carbaryl (Sevin) and diazinon were frequently applied for curative grub control. However, all these insecticides are under scrutiny by the United States Environmental Protection Agency due to the implementation of the Food Quality Protection Act, and both diazinon and chlorpyrifos have been removed from usage. Local ordinances and public opinion have also restrained the use of the remaining products. Turfgrass managers have few options for curative control of existing populations of white grubs. An increase in the preventive use of products such as imidacloprid and halofenozide applied at or before egg-laying has been seen (Niemczyk and Shetlar, 2000).

It should be noted that there is no single nematode species that provides the best control of all white grub species. Although acceptable control of *P. japonica* can be obtained using several different nematode species and strains, the control of other grub species will require the use of specific nematode species. For example, the most effective control of Melolonthine species (e.g. *R. majalis*, *Phyllophaga* spp., *Amphimallon* spp. and *M. melolontha*) can be obtained with *S. scarabaei*, but *H. bacteriophora* GPS11 strain and *H. zealandica* should provide more effective control of Dynastinae (*Cyclocephala* spp.).

Although the lack of predictability is the most often-cited liability of biocontrol agents (Gaugler *et al.*, 1997; Grewal, 1999; Shapiro-Ilan *et al.*, 2002; Shetlar, 2002), the new species and strains of EPNs provide consistency in grub control that exceeds the standard chemical insecticides (Grewal *et al.*, 2004). A combined analysis of the results from eight trials containing 46 treatments revealed that the grub control provided by trichlorfon varied between 0% and 92%, but that by *H. zealandica* X1 and *H. bacteriophora* GPS11 strains varied only between 48% and 98%, and 34% and 97%, respectively (Grewal *et al.*, 2004).

At present only the GPS11 and HP88 strains of *H. bacteriophora* are commercially available in the USA, and *H. zealandica* X1 strain is available in Australia. Fortunately, *H. zealandica* has been recently found to naturally occur in Florida (B. Adams, 2002, personal communication), thus potentially clearing regulatory hurdles for the registration of this species in the USA. Unfortunately, *S. scarabaei* have proven to be extremely difficult to mass-produce in liquid culture, and the commercial development of *S. glaseri* and *S. kushidai* has been halted due to technical and financial constraints.

7.4. Mole Crickets

Mole crickets were accidentally introduced into Florida from South America around 1990 and have since become the most destructive pest of turf and pastures in south-eastern USA. The tawny mole cricket (*Scapteriscus vicinus*) and the southern mole cricket (*S. borellii*) are the two most destructive crickets and are distributed throughout the coastal plain region of south-eastern USA. Overwintering occurs primarily in the nymphal stage (*S. borellii*) or adult stage (*S. vicinus*). Another species, *Gryllotalpha orientalis*, is an occasional pest but sometimes severely damages turfgrass in Korea. Heavily infested turf has virtually no root system and is very susceptible to damage from foot traffic or golf carts. Adult and nymphal mole crickets cause damage by feeding on grass roots and shoots, and by tunnelling through the ground. A single mole cricket can create 10–20 feet of tunnel in just one night, drying out the soil and causing serious damage to plant roots. Annual costs of controlling mole crickets are estimated to exceed US $50 million in Florida alone.

7.4.1. Nematodes for mole cricket control

Nematodes have been successful in reducing damage to turfgrass by mole crickets. *S. scapterisci*, which was originally isolated from infected mole crickets in Uruguay (Nguyen and Smart, 1990), showed 75–100% infection of adult mole crickets under laboratory conditions (Nguyen and Smart, 1991). In an inoculative release effort, *S. scapterisci* was introduced into pastures during the summer of 1985 (Hudson *et al.*, 1988). Based on the evaluation of field-collected mole crickets over a 5-year period, the nematodes were established at all the sites, with the mean number of adults infected being 11% for the entire period (Parkman *et al.*, 1993, 1994, 1996; Parleman and Smart, 1996).

Another nematode species, *S. riobrave*, has been used in biocontrol of mole crickets. In one test, 66–86% reduction in turf injury was observed with a single application of 2.5×10^9 *S. riobrave*/ha in South Carolina (Gorsuch, 1995). *S. carpocapsae* has also been examined as a control agent of mole crickets and was the focus of early investigations; field trials using *S. carpocapsae* resulted in an average of 58% control (Georgis and Poinar, 1994).

7.4.2. Factors affecting nematode efficacy

The efficacy of *S. scapterisci* was affected by mole cricket species and developmental stage (Hudson and Nguyen, 1989a,b; Nguyen and Smart, 1991; Parkman and Frank, 1992). The short-winged mole cricket, *S. abbreviatus*, is less susceptible than *S. vicinus* and *S. borellii* in laboratory studies. In addition, *S. borellii* was more susceptible than *S. vicinus* in field studies, probably because the greater activity arising out of its predatory behaviour increases its chances of contact with the ambusher *S. scapterisci*. Nymphal mole crickets were substantially less susceptible to *S. scapterisci* than adults, and small nymphs were not infected. *S. riobrave* is also ineffective against mole cricket nymphs, and does not recycle in infected mole crickets (K. Smith, personal communication).

7.4.3. Current status and analysis

A commercial product (Vector MC®) containing *S. riobrave* was marketed by Lesco, Inc. for the control of mole crickets in turf during 1994, but was later discontinued following the sale of Biosys Inc. A *S. scapterisci*-based product became available in 1993 but failed due to the lack of a consistent mass-production technique. Becker Underwood Ltd has recently acquired a licence for *S. scapterisci* from the University of Florida and a product (Nematac S®) became available on the market in 2003. *S. scapterisci* is an ideal control agent for pastures and turfgrass areas that can tolerate some mole cricket damage. In pastures, the potentially biggest market, nematodes are applied using slit injectors in strips covering 12.5% of the area. The nematodes then spread throughout the pasture over a period of several years. This approach reduces the cost to US$62/ha, considerably lower than chemical insecticides that provide only short-term suppression. In the turf market, *S. scapterisci* is applied to low-profile and environmentally sensitive areas on golf courses, sod farms and recreational areas at a rate of 2.5×10^9/ha (cost US$500/ha). In more damage-prone areas, *S. scapterisci* use is likely to remain limited due to the competition from the more effective but expensive insecticide fipronil (US$550/ha). The nematodes have to be applied in spring or autumn when adults are present, while control measures are typically necessary in summer against nymphs.

7.5. Weevils

Billbugs, *Sphenophorus* spp., are important turfgrass pests throughout much of the USA and Japan. The younger larvae feed inside the stem and crown and older larvae feed externally on the below-ground parts of the plant. Seasonal life cycles vary depending on species and latitude. No detailed studies on billbug–nematode interaction have been published. The bluegrass billbug, *S. parvulus*, is one of the most important pests of Kentucky bluegrass and perennial ryegrass but also attacks other cool-season grasses. The EPNs, *S. carpocapsae* and *H. bacteriophora*, have been shown to control billbug larvae and adults very effectively (Georgis and Poinar, 1989; Klein, 1990; Watschke *et al.*, 1995). Field tests in Ohio indicated that *S. parvulus* can be controlled with *S. carpocapsae* (average 78%) or *H. bacteriophora* (average 74%) (Georgis and Poinar, 1994; Smith, 1994). The hunting billbug, *S. venatus vestitus*, causes damage to warm-season turfgrasses including bermudagrass and zoysiagrass. In Japan, *S. carpocapsae* has been more effective for control of *S. venatus vestitus* than standard insecticides (average 84% versus 69% control), (Smith, 1994; Kinoshita and Yamanaka, 1998). However, *S. carpocapsae* sales for billbug control have significantly declined since the recent registration of imidacloprid for turfgrass uses in Japan.

The annual bluegrass weevil or hyperodes weevil is an important pest of *Poa annua* and annual bluegrass on golf courses in north-eastern USA. *S. carpocapsae* and *H. bacteriophora* have shown good results as a rescue treatment for weevil larvae in May in turfgrass (P. Vittum, personal communication).

7.6. Cutworms, Webworms and Armyworms

Lepidopterous larvae primarily cause defoliation but some feed on roots as well. The primary foliage feeders are species of cutworm and sod webworm. The cutworms, which are semi-subterranean pests, also burrow into the ground or thatch and damage the roots. They emerge at night to chew grass blades and shoots. The black cutworm, *Agrotis ipsilon*, is a cosmopolitan pest of short-cut bentgrass on golf courses, and the cutworm, *A. segetum*, is an important pest in Korea. On golf courses, sporadic damage by cutworm occurs from early spring to autumn. Another cutworm, the Japanese lawn cutworm, *Spodoptera depravata*, may occur at outbreak levels three or four times a year. *A. ipsilon* is found throughout North America and is a perennial problem on bentgrass turf of golf course greens, tees and fairways, but rarely damages lawns. The bronzed, variegated and glassy cutworms are pests of homelawn turf. Cutworms are semi-subterranean pests and usually dig a burrow into the ground or thatch and emerge at night to clip off grass blades and shoots. *S. carpocapsae* can be used effectively to manage all cutworm species. Black cutworm larvae can be controlled on golf course greens by applying nematodes at a rate of 2.5×10^9/ha (Georgis and Poinar, 1989; Watschke *et al.*, 1995).

The common armyworm, fall armyworm and yellow-striped armyworm most commonly damage home lawns, and only occasionally are pests on golf course turf. Armyworms also damage other ornamentals and vegetables in the gardens. Armyworms are very susceptible to nematodes, as all larval stages and the pupae are infected. Rosa and Simões (2004) evaluated 28 isolates of *H. bacteriophora* against the armyworm *Pseudaletia unipuncta* and found large variation in their virulence. The mortality of the sixth-instar larvae of *P. unipuncta* varied from 33% to 100% after 96 h exposure to nematodes in Petri dishes. Based on the LC_{50} and LT_{50} values, Rosa and Simões (2004) identified Az29 isolate to be the most virulent to the armyworm larvae. The field evaluations indicated that Az29 isolate was more effective to control *P. unipunctata* larvae than *S. carpocapsae* Az20 and *H. bacteriophora* Az32 isolates.

The bluegrass, larger, western, striped, elegant and vagabond sod webworms, along with the closely related cranberry girdler, sometimes damage cool season grasses. The tropical sod webworm is the most damaging pest of warm-season grasses. Both *S. carpocapsae* and *H. bacteriophora* are effective against sod webworms in turfgrass.

7.7. Crane Flies

The larvae of two crane fly species, *Tipula paludosa* (also called the European crane fly) and *T. oleracea*, are important turfgrass pests in the northern Palaearctic region in Europe and in parts of North America, including Nova Scotia, British Columbia, Oregon and Washington. In North America, *T. oleracea* was only recently recognized but seems to have a similar distribution and importance as *T. paludosa*. Larvae of both species are susceptible to heterorhabditid nematodes, and particularly to *S. feltiae* (Ehlers and Gerwien, 1993). In both species susceptibility to *S. feltiae* decreases with larval development (Peters and Ehlers, 1994).

7.8. Miscellaneous Pests

Other minor pests of turfgrass include the homopterous pests *Balanococcus takahashii*, *Aspidiella phragmis*, *Margarodes* and *Nephrotoma* sp. Ants (*Camponotus japonicus*, *Formica japonica* and *Lasius neoniger*) cause serious problems in golf courses by making nests or biting golfers in the USA (Watschke *et al.*, 1995) and in Korea (Choo *et al.*, 2000). Although fleas are not pests of turfgrass, their larvae feed on dead organic matter and develop in the grass. The use of nematodes for the control of fleas and ants is described in Chapters 16 and 17, respectively.

7.9. Conservation of Entomopathogenic Nematodes (EPNs) in Turfgrass

Although conservation of EPNs may be difficult to achieve in agroecosystems due to tillage disturbance, it may be easier in no-till systems, natural systems (e.g. forestry) and grassland systems including golf courses, pastures and lawns. In surveys for endemic populations of EPNs in golf courses in Ohio, over 40% of the golf course fairways and over 60% of the golf course rough areas were positive (A. Alumai and P.S. Grewal, unpublished data). Kaya (1990) proposed the following set of conditions for inoculative control to be effective: the soil pest or complex of pests is present throughout the year; pests have a high economic threshold and are moderately susceptible to nematodes; and soil conditions are favourable for nematode persistence. Thus, the turfgrass ecosystem is ideally suited for both inoculative and conservation approaches with the nematodes. Between 1939 and 1942, Glaser and his co-workers mass-produced and inoculatively released *S. glaseri* into the fields against *P. japonica* (see Gaugler *et al.*, 1992). Their colonization efforts were unsuccessful, probably due to the lack of knowledge about the symbiotic bacteria at that time, as the released nematodes were mass-produced without their symbiotic bacterium (Gaugler *et al.*, 1992). However, Akhurst *et al.* (1992) reported that two *Heterorhabditis* spp. caused an epizootic that extended over 5 ha among four species of white grubs feeding on sugarcane roots. Campbell *et al.* (1999) reported that the occurrence of *H. bacteriophora* in turfgrass was correlated with reduced numbers of *P. japonica*. The persistence of EPNs beyond a season following their application against third-instar white grubs has been reported (Sexton and Williams, 1981; Poinar *et al.*, 1987; Klein and Georgis, 1992), thus suggesting the potential impact of EPNs on multiple generations of white grubs. Obviously, more research is needed to build a sound conservation approach for using EPNs in turfgrass.

7.10. General Recommendations and Conclusions

Application of chemical insecticides has been the main method of defence against damage by turfgrass pests. However, many of the chemical insecticides used for turfgrass pest control are under scrutiny by the United States Environmental Protection Agency due to the proposed implementation of Food Quality Protection Act (FQPA), and several have already been removed from usage. Local ordinances and public opinion have further restrained the use of the remaining products in various parts of the world. Turfgrass managers have few options for curative control of pest populations. EPNs are effective biocontrol agents of most turfgrass pests. The lack of consistency in pest control has been the major hurdle in the adoption of nematodes by golf course superintendents and lawn care companies. Tremendous progress has been made in the past few years in the identification of more virulent nematode strains, particularly for white grubs. These new nematode strains, *H. bacteriophora* GPS11 and TF, *H. zealandica* X1 and *S scarabaei* AMK001, have shown increased consistency in white grub control. These strains provide equal or better curative grub control than the most commonly used chemical insecticides. Two strains, *H. bacteriophora* GPS11 and *H. zealandica* X1, have already become commercially available in the USA and Australia, respectively. Unfortunately, *S. scarabaei* has proven difficult to mass-produce with established nematode mass-production technology (R.-U. Ehlers, personal communication).

EPNs are currently used for the control of white grubs, crane fly and flea larvae, billbugs and mole crickets on home lawns in the USA and Canada. Small lawn care companies, particularly those that provide organic or natural lawn care, have begun to use nematodes to manage white grubs and billbugs. In Australia, the nematodes are used for white grub control in public properties such as urban parks. In Japan, the nematodes are applied for the control of billbugs and white grubs on golf courses, and in Europe, the nematodes are used mainly for white grub control on golf courses. Further expansion in the nematode use will require the availability of large quantities of good-quality products and several companies are expanding production capacity (R.-U. Ehlers, personal communication).

Appropriate application strategy is the key to obtain successful control of turfgrass insects with EPNs. Selecting the best nematode species or strain for each target pest is important, as there are large differences in the virulence of nematode species and strains against different species of pests. Targeting the most susceptible stage of the pest for nematode applications cannot be overemphasized. Therefore, timing of nematode applications to match susceptible stages of the pest is important and can be achieved by close monitoring of pest life cycles. There are also special requirements for nematode applications in turfgrass ecosystems. The thick ground cover, composed of numerous grass stems and leaves and a layer of thatch (dead, non-decomposed plant material), on the surface of the soil can restrict nematode penetration into the soil under turfgrass. Hydrophobicity of grass leaves can also reduce movement of nematodes that get trapped in droplets of water. Therefore, turfgrass sites should be prepared for nematode application by mowing the grass to the lowest acceptable height to ensure good contact of the nematodes with the soil. Also, the soil must be moist before nematode application. A pre-application irrigation may be necessary. The nematodes must be applied when soil temperatures are optimum for nematode activity (20–28°C) and UV radiation is minimum. Thus, the best time to apply the nematodes is in the late evening, which allows nematodes to enter the soil before the sun comes out the next morning. Alternatively, the nematodes may be applied under a cloud cover. The value of post-application irrigation (which can also moderate temperature) and maintenance of optimum soil moisture for up to 2–3 weeks after nematode application should not be underestimated. The actual amount and frequency of irrigation will depend upon the site, soil type and the amount of rain. Soil aeration (mechanical removal of soil cores), which is often used to reduce soil compaction, can improve nematode movement in the soil profile, thus enhancing insect control. The use of wetting agents for enhancing the penetration of nematodes into the soil should also be considered, especially for sites with thatch problems.

Nematode application technology also needs to be addressed in the context of different sectors of the turfgrass industry. The selection of equipment for the application of nematodes in agriculture discussed in Chapter 5 is also appropriate for golf courses and pastures. However, the current pesticide application equipment used by the commercial lawn care industry is not adequate for handling nematodes. Although a hydraulic spray application system can be used for safe delivery of nematodes, there are limitations to the equipment and operating conditions. A hydraulic spray application system usually consists of a tank, pump, valves, spray hose and nozzle(s). The nematodes will settle out of suspension within a short period of time, so there must be agitation in the tank either through recirculation of a portion of the spray liquid or mechanical mixing. Some pumps have moving parts in direct contact with the nematodes that could mechanically tear them apart. During recirculation of the tank mix through the pump system, the liquid temperature can rise considerably, which may be harmful to the nematodes.

The current recommendations for spray application of nematodes are to use nozzles with openings larger than 500 μm, operating pressures less than 2070 kPa (300 psi), and to remove all mesh screens from the system unless they have orifices larger than 300 μm or 50 mesh (Grewal, 2002). These recommendations are based on observations of *S. carpocapsae*, the most widely used and robust insecticidal nematode species, and might not be representative for all species. Fife *et al.* (2003) found that with increases in pressure change, *S. carpocapsae* had significantly higher viability compared to *H. bacteriophora* and *H. megidis*. To maintain viability above 85%, they recommended that operating pressures be kept below 2000 kPa (290 psi) for *S. carpocapsae* and *H. bacteriophora*, and less than 1380 kPa (200 psi) for *H. megidis*. Fife *et al.* (2004) evaluated the effect of three different nozzle types (flat-fan, hollow-cone and full-cone) on four different nematode species (*S. carpocapsae*, *S. glaseri*, *H. bacterio-*

phora and *H. megidis*). Results indicated that the flat-fan nozzle (the smallest size commercially available) caused higher levels of nematode damage compared with a similar capacity hollow-cone nozzle. Larger-sized flat-fan nozzles did not cause damage. They recommend using cone-type nozzles or large flat-fan nozzles for spray application. The effect of different pump types on nematode damage is currently under investigation.

With commercial lawn care, the operating environment commonly encountered during a workday offers additional challenges for safe delivery of insecticidal nematodes (Grewal, 2002). Generally, the spray system consists of a hand-held spray boom that is connected to the tanker truck by long lengths of hose. The tank and hoses are often exposed to the sun for several hours during the workday. It is anticipated that the temperature of the liquid inside could reach levels that are lethal to the nematodes (i.e. > 30°C), and because of the large size of the tank and the warm temperatures encountered, oxygen deprivation of the nematodes could also occur. When travelling from one property to another, the hose must be reeled each time and the nematodes would tend to settle to the bottom of the hose loops causing inconsistent spray distribution at the next property. In addition, other chemicals in the tank or residuals from previous tank mixes may be lethal to the nematodes.

In conclusion, the EPNs have proven very useful in the management of important turfgrass pests including white grubs, mole crickets and billbugs in home lawns, pastures and golf course situations. One reason for this success is that substantial research with nematodes has been conducted on these pests. EPNs also possess potential for the management of other important pests including armyworms, cutworms, webworms, cranefly and flea larvae, and ants, but more research is needed to identify the most effective nematode species and strains for these pests. Also, the development of more effective application technology and strategy is extremely important for the acceptance of nematodes by turfgrass managers. Finally, researchers need to focus on the development of a conservation approach for using nematodes, as turfgrass systems are ideally suited for such an approach.

References

Akhurst, R.J., Bedding, R.A., Bull, R.M. and Smith, D.R.J. (1992) An epizootic of *Heterorhabditis* spp. (Heterorhabditidae: Nematoda) in sugar cane scarabaeids (Coleoptera). *Fundamental and Applied Nematology* 15, 71–73.

Alm, S.R., Yeh, T., Hanula, J.L. and Georgis, R. (1992) Biological control of Japanese, oriental, and black turfgrass ataenius beetle (Coleoptera: Scarabaeidae) larvae with entomopathogenic nematodes (Nematoda: Steinernematidae, Heterorhabditidae). *Journal of Economic Entomology* 85, 1660–1665.

Amaral, J.J.S. (1996) Pathogenicity of Azorean isolates of *Steinernema* spp. and *Heterorhabditis* spp. against larvae of *Popillia japonica* Newman (Coleoptera: Scarabaeidae) at three temperatures. In: *Proceedings of the Cost 819 Workshop on the Pathogenicity of Entomopathogenic Nematodes versus Insect Defense Mechanisms, and Impact of Selection of Virulent Strains.* Universidade dos Acores, Ponta Delgada, Acores, Portugal, p. 269.

Ansari, M.A., Tirry, L. and Moens, M. (2003) Entomopathogenic nematodes and their symbiotic bacteria for the biological control of *Hoplia philanthus* (Coleoptera: Scarabaeidae). *Biological Control* 28, 111–117.

Beard, J.B. and Green, R.L. (1994) The role of turfgrass in environmental protection and their benefits to humans. *Journal of Environmental Quality* 23, 452–460.

Berg, G.N., Williams, P., Bedding, R.A. and Akhurst, R.J. (1987) A commercial method of application of entomopathogenic nematodes pasture for controlling subterranean insect pests. *Plant Protection Quarterly* 2, 174–177.

Berner, M. and Schnetter, W. (2001) Wirksamkeit entomopathgener Nematoden gegen Engerlinge der Maikäfer *Melolontha melolontha* und *M. hippocastani. Mitteilungen der Deutschen Gesellschaft für allgemeine und angewandte. Entomologie* 13, 165–167.

Campbell, J.F., Koppenhöfer, A.M., Kaya, H.K. and Chinnasri, B. (1999) Are there temporarily non-infectious dauer stages in entomopathogenic nematode populations: a test of the phased infectivity hypothesis. *Parasitology* 118, 499–508.

Cappaert, D.C. and Koppenhöfer, A.M. (2003) *Steinernema scarabaei*, an entomopathogenic nematode for control of the European chafer. *Biological Control* 28, 379–386.

Choo, H.Y., Lee, D.W., Lee, S.M., Lee, T.W., Choi, W.G., Chung, Y.K. and Sung, Y.T. (2000) Turfgrass insect pests and natural enemies in golf courses. *Korean Journal of Applied Entomology* 39, 171–179.

Converse, V. and Grewal, P.S. (1998) Virulence of entomopathogenic nematodes to the Western masked chafer *Cyclocephala hirta* (Coleoptera: Scarabaeidae). *Journal of Economic Entomology* 91, 428–432.

Danneberger, T.K. (1993) *Turfgrass Ecology and Management.* Franzak and Foster G.I.E., Cleveland, Ohio.

Delfosse, E.S. (1993) *Pests of Pastures. Weed, Invertebrate and Disease Pests of Australian Sheep Pastures.* CSIRO Entomology, Canberra, Australia.

Deseö, K.V., Bartocci, R., Tartaglia, A. and Rovesti, L. (1990) Entomopathogeneous nematodes for contol of scarab larvae. *IOBC/WPRS Bulletin* 14, 57–58.

Ehlers, R.-U. and Gerwien, A. (1993) Selection of entomopathogenic nematodes (Steinernematidae and Heterorhabditidae, Nematoda) for the biological control of cranefly larvae *Tipula paludosa* (*Tipulidae, Diptera*). *Zeitschrift für Pflanzenkrankheiten und Pflanzenschutz* 100, 343–353.

Ehlers, R.-U. and Peters, A. (1998) Bekämpfung von Engerlingen auf Sportsrasen. *Rasen Turf Gazon* 3, 1–8.

Eidt, D.C. and Thurston, G.S. (1995) Physical deterrents to infection by entomopathogenic nematodes in wireworms (Coleoptera: Elateridae) and other soil insects. *The Canadian Entomologist* 127, 423–429.

Fife, J.P., Derksen, R.C., Ozkan, H.E. and Grewal, P.S. (2003) Effects of pressure differentials on the viability and infectivity of entomopathogenic nematodes. *Biological Control* 27, 65–72.

Fife, J.P., Derksen, R.C., Ozkan, H.E., Grewal, P.S., Chalmers, J.J. and Krause, R. (2004) Evaluation of a contraction flow field on hydrodynamic damage to entomopathogenic nematodes – a biological pest control agent. *Biotechnology and Bioengineering* 86, 96–107.

Forschler, B. and Gardner, W. (1991) Parasitism of *Phyllophaga hirticula* (Coleoptera: Scarabaeidae) by *Heterorhabditis heliothidis* and *Steinernema carpocapsae. Journal of Invertebrate Pathology* 58, 396–407.

Fujiie, A., Yokomoto, T., Fujikata, M., Sawada, M. and Hasegawa, M. (1993) Pathogenicity of an entomopathogenous nematode, *Steinernema kushidai* Mamiya (Nematoda: Steinernematidae), on *Anomala cuprea* (Coleoptera: Scarabaeidae). *Japanese Journal of Applied Entomology and Zoology* 37, 53–60.

Galbreath, R.A. (1976) Spiracular structure and function in *Costelytra zealandica* larvae (Coleoptera: Scarabaeidae). *New Zealand Journal of Zoology* 3, 333–337.

Gaugler, R., Campbell, J.F., Selvan, M.S. and Lewis, E.E. (1992) Large-scale inoculative releases of the entomopathogenic nematodes *Steinernema glaseri*: assessment of establishment 50 years later. *Biological Control* 2, 181–187.

Gaugler, R., Wang, Y. and Campbell, J.F. (1994) Aggressive and evasive behaviors in *Popillia japonica* (Coleoptera: Scarabaeidae) larvae: defense against entomopathogenic nematode attack. *Journal of Invertebrate Pathology* 64, 193–199.

Gaugler, R., Lewis, E.E. and Stuart, R.J. (1997) Ecology in the service of biological control: the case of entomopathogenic nematodes. *Oecologia* 109, 483–489.

Georgis, R. and Gaugler, R. (1991) Predictability in biological control using entomopathogenic nematodes. *Journal of Economic Entomology* 84, 713–720.

Georgis, R. and Poinar, G.O. Jr (1989) Field effectiveness of entomophilic nematodes *Neoaplectana* and *Heterorhabditis.* In: Leslie, A.R. and Metcalf, R.L. (eds) *Integrated Pest Management for Turfgrass and Ornamentals*, pp. 213–224.

Georgis, R. and Poinar, G.O. Jr (1994) Nematodes as bioinsecticides in turf and ornamentals. In: Leslie, A. (ed.) *Integrated Pest Management for Turf and Ornamentals.* CRC Press, Boca Raton, Florida, pp. 477–489.

Glazer, I. and Gol'berg, A. (1989) Laboratory evaluation of steinernematid and heterorhabditid nematodes for control of the beetle *Maladera matrida* Argaman (Coleoptera: Scarabaeidae). *Phytoparasitica* 17, 3–11.

Glazer, I. and Gol'berg, A. (1993) Field efficacy of entomopathogenic nematodes against the beetle *Maladera matrida* (Coleoptera: Scarabaeidae). *Biocontrol Science and Technology* 3, 367–376.

Gorsuch, C.S. (1995) Biological control of mole crickets. *Carolinas Green* 31, 18–19.

Grewal, P.S. (1999) Factors in the success and failure of microbial control in turfgrass. *Integrated Pest Management Reviews* 4, 287–294.

Grewal, P.S. (2002) Formulation and application technology. In: Gaugler, R. (ed.) *Entomopathogenic Nematology*, CAB International, Wallingford, UK, pp. 265–287.

Grewal, P.S. and Georgis, R. (1998) Entomopathogenic nematodes. In: Hall, F.R. and Menn, J.J. (eds) *Biopesticides: Use and Delivery*, Humana Press, Totowa, New Jersey, pp. 271–299.

Grewal, P.S., Selvan, S. and Gaugler, R. (1994) Thermal adaptation of entomopathogenic nematodes: niche breadth for infection, establishment, and reproduction. *Journal of Thermal Biology* 19, 245–253.

Grewal, P.S., Grewal, S.K., Malik, V.S. and Klein, M.G. (2002) Differences in the susceptibility of introduced and native white grub species to entomopathogenic nematodes from various geographic localities. *Biological Control* 24, 230–237.

Grewal, P.S., Power, K.T., Grewal, S.K., Suggars, A. and Haupricht, S. (2004) Enhanced consistency in biological control of white grubs (Coleoptera: Scarabaeidae) with new strains of entomopathogenic nematodes. *Biological Control* 30, 73–82.

Hinton, H.E. (1967) Structure and ecdysial process of the larval spiracles of the Scarabaeoidea, with special reference to those *Lepidoderma. Australian Journal of Zoology* 17, 1185–1199.

Hudson, W.G. and Nguyen, K.B. (1989a) Effect of soil moisture, exposure time, nematode age, and nematode density on infection of *Scapteriscus vicinus* and *S. acletus* (Orthoptera: Gryllotalpidae) by a Uruguayan *Neoaplectana* sp. (Rhabditida: Steinernematidae). *Environmental Entomology* 18, 719–722.

Hudson, W.G. and Nguyen, K.B. (1989b) Infection of *Scapteriscus vicinus* (Orthoptera: Gryllotalpidae) nymphs by *Neoaplectana* sp. (Rhabditida: Steinernematidae). *Florida Entomologist* 72, 383–384.

Hudson, W.G., Frank, J.H. and Castner, J.L. (1988) Biological control of *Scapteriscus* spp. mole crickets (Orthoptera: Gryllotalpidae) in Florida. *Bulletin of Entomological Society of America* 34, 192–198.

Jackson, T.A. (1992) Scarabs – pests of the past or future? In: Jackson, T.A. and Glare, T.R. (eds) *Use of Pathogens in Scarab Pest Management*. AgResearch, Lincoln, New Zealand, pp. 1–6.

Kain, W.M., Bedding, R.A. and Van der Mespel, C.J. (1982) Preliminary evaluations of parasitic nematodes for grass grub (*Costelytra zealandica* (White)) control in central Hawke's Bay of New Zealand. *New Zealand Journal of Experimental Agriculture* 10, 447–450.

Kard, B.M.R., Hain, F.P. and Brooks, W.M. (1988) Field suppression of three white grub species (Coleoptera: Scarabaeidae) by the entomogenous nematodes *Steinernema feltiae* and *Heterorhabditis heliothidis. Journal of Economic Entomology* 81, 1033–1039.

Kaya, H.K. (1990a) Entomopathogenic nematodes in biological control of insects. In: Baker, R.P. and Dunn, P.E. (eds) *New Directions in Biological Control*. Liss, New York, pp. 189–198.

Kaya, H.K. (1990b) Soil ecology. In: Gaugler, R. and Kaya, H.K. (eds) *Entomopathogenic Nematodes in Biological Control*. CRC Press, Boca Raton, Florida, pp. 93–116.

Kaya, H.K. and Gaugler, R. (1993) Entomopathogenic nematodes. *Annual Review of Entomology* 38, 181–206.

Kinoshita, M. and Yamanaka, S. (1998) Development and prevalence of entomopathogenic nematodes in Japan. *Japanese Journal of Nematology* 28, 42–45.

Klein, M.G. (1990) Efficacy against soil-inhabiting insect pests. In: Gaugler, R. and Kaya, H.K. (eds) *Entomopathogenic Nematodes in Biological Control*. CRC Press, Boca Raton, Florida, pp. 195–214.

Klein, M.G. (1993) Biological control of scarabs with entomopathogenic nematodes. In: Bedding, R., Akhurst, R. and Kaya, H.K. (eds) *Nematodes and the Biological Control of Insect Pests*. CSIRO Publications, East Melbourne, Australia, pp. 49–58.

Klein, M.G. and Georgis, R. (1992) Persistence of control of Japanese beetle (Coleoptera: Scarabaeidae) larvae with steinernematid and heterorhabditid nematodes. *Journal of Economic Entomology* 85, 727–730.

Klein, M.G., Grewal, P.S. and Jackson, T.A. (2000) Lawn, turf and grassland pests. In: Lacey, L.A. and Kaya, H.K. (eds) *Field Manual of Techniques in Invertebrate Pathology*. Kluwer Academic Publishers, The Netherlands, pp. 681–706.

Koppenhöfer, A.M. and Fuzy, E.M. (2003a) *Steinernema scarabaei* for the control of white grubs. *Biological Control* 28, 47–59.

Koppenhöfer, A.M. and Fuzy, E.M. (2003b) Biological and chemical control of the Asiatic garden beetle, *Maladera castanea* (Coleoptera: Scarabaeidae). *Journal of Economic Entomology* 96, 1076–1082.

Koppenhöfer, A.M. and Fuzy, E.M. (2003c) Effects of turfgrass endophytes (Clavicipitaceae: Ascomycetes) on white grub (Coleoptera: Scarabaeidae) control by the entomopathogenic nematode *Heterorhabditis bacteriophora* (Rhabditida: Heterorhabditidae). *Environmental Entomology* 32, 392–396.

Koppenhöfer, A.M. and Fuzy, E.M. (2004) Effect of white grub developmental stage on susceptibility to entomopathogenic nematodes. *Journal of Economic Entomology* 97, 1842–1849.

Koppenhöfer, A.M. and Kaya, H.K. (1998) Synergism of imidacloprid and an entomopathogenic nematode: a novel approach to white grub (Coleoptera: Scarabaeidae) control in turfgrass. *Journal of Economic Entomology* 91, 618–623.

Koppenhöfer, A.M., Choo, H.Y., Kaya, H.K., Lee, D.W. and Gelernter, W.D. (1999) Increased field and greenhouse efficacy against scarab grubs with a combination of an entomopathogenic nematode and *Bacillus thuringiensis*. *Biological Control* 14, 37–44.

Koppenhöfer, A.M., Brown, I.M., Gaugler, R., Grewal, P.S., Kaya, H.K. and Klein, M.G. (2000a) Synergism of entomopathogenic nematodes and imidacloprid against white grubs: greenhouse and field evaluation. *Biological Control* 19, 245–251.

Koppenhöfer, A.M., Wilson, M.G., Brown, I., Kaya, H.K. and Gaugler, R. (2000b) Biological control agents for white grubs (Coleoptera: Scarabaeidae) in anticipation of the establishment of the Japanese beetle in California. *Journal of Economic Entomology* 93, 71–80.

Koppenhöfer, A.M., Cowles, R.S., Cowles, E.A., Fuzy, E.M. and Baumgartner, L. (2002) Comparison of neonicotinoid insecticides as synergists for entomopathogenic nematodes. *Biological Control* 24, 90–97.

Koppenhöfer, A.M., Fuzy, E.M., Crocker, R., Gelernter, W. and Polavarapu, S. (2004) Pathogenicity of *Steinernema scarabaei*, *Heterorhabditis bacteriophora* and *S. glaseri* to twelve white grub species. *Biocontrol Science and Technology* 14, 87–92.

Lee, W., Choo, H.Y., Kaya, H.K., Lee, S.M., Smitley, D.R., Shin, H.K. and Park, C.G. (2002) Laboratory and field evaluation of Korean entomopathogenic nematode isolates against the oriental beetle *Exomala orientalis* (Coleoptera: Scarabaeidae). *Journal of Economic Entomology* 95, 918–926.

Lewis, E.E., Gaugler, R. and Harrison, R. (1993) Response of cruiser and ambusher entomopathogenic nematodes (Steinernematidae) to host volatile cues. *Canadian Journal of Zoology* 71, 765–769.

Mannion, C.M., Winkler, H.E., Shapiro, D.I. and Gibb, T. (2000) Interaction between halofenozide and entomopathogenic nematode *Heterorhabditis marelatus* for control of Japanese beetle (Coleoptera: Scarabaeidae) larvae. *Journal of Economic Entomology* 93, 48–53.

Nguyen, K.B. and Smart, G.C. Jr (1990) *Steinernema scapterisci* n. sp. (Rhabditida: Steinernematidae). *Journal of Nematology* 22, 187–199.

Nguyen, K.B. and Smart, G.C. Jr (1991) Pathogenicity of *Steinernema scapterisci* to selected invertebrates. *Journal of Nematology* 23, 7–11.

Niemczyk, H.D. and Shetlar, D.J. (2000) *Destructive Turf Insects*, 2nd edn, HDN Books, Wooster, Ohio.

Parkman, J.P. and Frank, J.H (1992) Infection of sound-trapped mole crickets, *Scapteriscus* spp., by *Steinernema scapterisci*. *Florida Entomologist* 75, 163–165.

Parkman, J.P. and Smart, G.C. Jr (1996) Entomopathogenic nematodes, a case study: introduction of *Steinernema scapterisci* in Florida. *Biocontrol Science and Technology* 6, 413–419.

Parkman, J.P., Hudson, W.G., Frank, J.H., Nguyen, K.B. and Smart, G.C. Jr (1993) Establishment and persistence of *Steinernema scapterisci* (Rhabditida: Steinernematidae) in field populations of *Scapteriscus* spp. mole crickets (Orthoptera: Gryllotalpidae). *Journal of Entomological Science* 28, 182–190.

Parkman, J.P., Frank, J.H., Nguyen, K.B. and Smart, G.C. Jr (1994) Inoculative release of *Steinernema scapterisci* (Rhabditida: Steinernematidae) to suppress pest mole crickets (Orthoptera: Gryllotalpidae) on golf course. *Environmental Entomology* 23, 1331–1337.

Parkman, J.P., Frank, J.H., Walker, T.J., and Schuster, D.J. (1996) Classical biological control of *Scapterisci* spp. (Orthoptera: Gryllotalpidae) in Florida. *Environmental Entomology* 25(6), 1415–1420.

Peters, A. (1996) The natural host range of *Steinernema* and *Heterorhabditis* spp. and their impact on insect populations. *Biocontrol Science and Technology* 6, 389–402.

Peters, A. (2000) Susceptibility of *Melolontha melolontha* to *Heterorhabditis bacteriophora*, *H. megidis* and *Steinernema glaseri*. *IOBC/WPRS Bulletin* 23(8), 39–46.

Peters, A. and Ehlers, R-U. (1994) Susceptibility of leatherjackets (*Tipula paludosa* and *T. oleracea*; Tipulidae: Nematocera) to the entomopathogenic nematode *Steinernema feltiae*. *Journal of Invertebrate Pathology* 63, 163–171.

Peters, A. and Keller, S. (2000) Susceptibility of *Melolontha melolontha* to *Heterorhabditis bacteriophora*, *H. megidis*, and *Steinernema glaseri*. Integrated control of the soil pest '*Melolontha*' Sion, Switzerland, 19–21 October 1998. *Bulletin-OILB-SROP* 23(8), 39–45.

Peters, A., Galarza, M. and Keller, S. (2002) Evaluation of a new *Steinernema* sp. of *Melolontha melolontha* and *Amphimallon solstittale*. Proceedings of the IOBC-WPRS Working Group Meeting on Integrated Control of Soil Pest

'Melolontha', Aosta, Italy, 24–26 September 2001. *Bulletin-OILB-SROP* 25(7), 23–28.

Poinar, G.O. Jr (1975) *Entomogenous Nematodes.* E.J. Brill, Leiden, The Netherlands, 317 pp.

Poinar, G.O. Jr (1990) Taxonomy and biology of Steinernematidae and Heterorhabditidae. In: Gaugler, R. and Kaya, H.K. (eds) *Entomopathogenic Nematodes in Biological Control.* CRC Press, Boca Raton, Florida, pp. 23–61.

Poinar, G.O. Jr (1992) Nematodes associated with Scarabaeidae. In: Jackson, T.A. and Glare, T.R. (eds) *Use of Pathogens in Scarab Pest Management.* AgResearch, Lincoln, New Zealand, pp. 93–109.

Poinar, G.O. Jr, Jackson, T.A. and Klein, M.G. (1987) *Heterorhabditis megidis* sp. n. (Heterorhabdtidae: Rhabdida), parasitic in the Japanese beetle, *Popillia japonica* (Scarabaediae: Coleoptera), in Ohio. *Proceedings of the Helminthological Society of Washington* 54, 53–58.

Potter, D.A. (1998) *Destructive Turfgrass Insects: Biology, Diagnosis, and Control.* Ann Arbor Press, Michigan, 344 pp.

Rosa, J.S. and Simões, N. (2004) Evaluation of twenty-eight strains of *Heterorhabditis bacteriophora* isolated in Azores for biological control of the armyworm, *Pseudaletia unipunctata* (Lepidoptera: Noctuidae). *Biological Control* 29, 409–417.

Schroeder, P.C., Villani, M.G., Ferguson, C.S., Nyrop, J.P. and Shields, E.J. (1993) Behavioral interactions between Japanese beetle (Coleoptera: Scarabaeidae) grubs and an entomopathogenic nematode (Nematoda: Heterorhabditidae) within turf microcosms. *Environmental Entomology* 22, 595–600.

Selvan, S., Gaugler, R. and Campbell, J.F. (1993) Efficacy of entomopathogenic nematode strains against *Popillia japonica* (Coleoptera: Scarabaeidae) larvae. *Journal of Economic Entomology* 86, 353–359.

Selvan, S., Grewal, P.S., Gaugler, R., Tomalak, M. (1994) Evaluation of steinernematid nematodes against *Popillia japonica* (Coleoptera: Scarabaeidae) larvae: species, strains, and rinse after application. *Journal of Economic Entomology* 87, 605–609.

Sexton, S. and Williams, P. (1981) A natural occurrence of parasitism of *Graphognathus leucoloma* (Boheman) by the nematode *Heterorhabditis* sp. *Journal of Australian Entomological Society* 20, 253–255.

Shapiro, D.I., Glazer, I. and Segal, D. (1996) Trait stability and fitness of the heat tolerant entomopathogenic nematode *Heterorhabditis bacteriophora* IS5 strain. *Biological Control* 6, 238–244.

Shapiro-Ilan, D.I., Gauge, D.H. and Koppenhöfer, A.M. (2002) Factors affecting commercial success: case studies in cotton, turf and citrus. In: Gaugler, R. (ed.) *Entomopathogenic Nematology.* CAB International, Wallingford, UK, pp. 333–355.

Shetlar, D.J. (2002) The 2000–2001 turfgrass, insect and mite update. *OTF TurfNews* 63, 18–19.

Shetlar, D.J., Suleman, P.E. and Georgis, R. (1988) Irrigation and use of entomogenous nematodes, *Neoaplectana* spp. and *Heterorhabditis heliothidis* (Rhabditida: Steinernematidae and Heterorhabditidae), for control of Japanese beetle (Coleoptera: Scarabaeidae) grubs in turfgrass. *Journal of Economic Entomology* 81, 1318–1322.

Simard, L., Belair, G. and Brodeur, J. (2001) Susceptibility of the European chafer (Coleoptera: Scarabaeidae) to entomopathogenic nematodes (Rhabditida: Steinernematidae, Heterorhabditidae). *Supplement to the Journal of Nematology* 33, 297–301.

Simões, N., Laumond, C. and Bonifassi, E. (1993) Effectiveness of *Steinernema* spp. and *Heterorhabditis bacteriophora* against *Popillia japonica* in the Azores. *Journal of Nematology* 25, 480–485.

Smith, K.A. (1994) Control of weevils with entomopathogenic nematodes. In: Smith, K.A. and Hatsukade, M. (eds) *Control of Insect Pests with Entomopathogenic Nematodes.* Food and Fertilizer Technology Center, Republic of China in Taiwan, pp. 1–13.

Smits, P.H., Wiegers, G.L. and Vlug, H.J. (1994) Selection of insect parasitic nematodes for biological control of the garden chafer, *Phyllopertha horticola. Entomologia Experimentalis et Applicata* 70, 77–82.

Stock, S.P. and Koppenhöfer, A.M. (2003) *Steinernema scarabaei* n. sp. (Rhabditida: Steinernematidae), a natural pathogen of scarab beetle larvae (Coleoptera: Scarabaeidae) from New Jersey, USA. *Nematology* 5, 191–204.

Sulistyanto, D. and Ehlers, R.-U. (1996) Efficacy of the entomopathogenic nematodes *Heterorhabditis megidis* and *Heterorhabditis bacteriophora* for control of grubs (*Phyllopertha horticola* and *Aphodius contaminatus*) in golf course turf. *Biocontrol Science and Technology* 6, 247–250.

Tashiro, H. (1987) *Turfgrass insects of the United States and Canada.* Cornell University Press, Ithaca, New York, 391 pp.

Townsend, M.L., Steinkraus, D.C. and Johnson, D.T. (1994) Mortality response of green June beetle (Coleoptera: Scarabaeidae) to four species of entomopathogenic nematodes. *Journal of Entomological Science* 29, 268–275.

Townsend, M.L., Johnson, D.T. and Steinkraus, D.C. (1998) Laboratory studies of the interactions of environmental conditions on the susceptibility of green June beetle (Coleoptera: Scarabaeidae) grubs to entomopathogenic nematodes. *Journal of Entomological Science* 33, 40–48.

Villani, M.G. and Wright, R.J. (1988) Entomogenous nematodes as biological control agents of European chafer and Japanese beetle (Coleoptera: Scarabaeidae) larvae infesting turfgrass. *Journal of Economic Entomology* 81, 484–487.

Vittum, P.J., Villani, M.G. and Tashiro, H. (1999) *Turfgrass Insects of the United States and Canada.* Cornell University Press, Ithaca, New York.

Wang, Y. and Gaugler, R. (1998) Host and penetration site location by entomopathogenic nematodes against Japanese beetle larvae. *Journal of Invertebrate Pathology* 72, 313–318.

Wang, Y. and Gaugler, R. (1999) *Steinernema glaseri* surface coat protein suppresses the immune response of *Popillia japonica* (Coleoptera: Scarabaeidae) larvae. *Biological Control* 14, 45–50.

Wang, X. and Grewal, P.S. (2002) Rapid deterioration of environmental tolerance and reproductive potential of an entomopathogenic nematode during laboratory maintenance. *Biological Control* 23, 71–78.

Wang, Y., Gaugler, R. and Cui, L. (1994) Variations in immune response of *Popillia japonica* and *Acheta domesticus* to *Heterorhabditis bacteriophora* and *Steinernema* species. *Journal of Nematology* 26, 11–18.

Wang, Y., Campbell, J.F. and Gaugler, R. (1995) Infection of entomopathogenic nematodes *Steinernema glaseri* and *Heterorhabditis bacteriophora* against *Popillia japonica* (Coleoptera: Scarabaeidae) larvae. *Journal of Invertebrate Pathology* 66, 178–184.

Watschke, T.L., Dernoeden, P.H. and Shetlar, D.J. (1995). *Managing Turfgrass Pests.* Lewis Publishers, Boca Raton, Florida, pp. 171–343.

Wright, R.J., Villani, M.G. and Agudelo-Silva, F. (1988) Steinernematid and heterorhabditid nematodes for control of larval European chafers and Japanese beetles (Coleoptera: Scarabaeidae) in potted yew. *Journal of Economic Entomology* 81, 152–157.

Yeh, T. and Alm, S.R. (1992) Effects of entomopathogenic nematode species, rate, soil moisture, and bacteria on control of Japanese beetle (Coleoptera: Scarabaeidae) larvae in the laboratory. *Journal of Economic Entomology* 85, 2144–2148.

Yeh, T. and Alm, S.R. (1995) Evaluation of *Steinernema glaseri* (Nematoda: Steinernematidae) for biological control of Japanese and oriental beetles (Coleoptera: Scarabaeidae). *Journal of Economic Entomology* 88, 1251–1255.

Zimmerman, R.J. and Cranshaw, W.S. (1991) Short-term movement of *Neoaplectana* spp. (Rhabditida: Steinernematidae) and *Heterorhabditis* 'HP88' strain (Rhabditida: Heterorhabditidae) through turfgrass thatch. *Journal of Economic Entomology* 84, 875–878.

8 Glasshouse Applications

M. Tomalak,[1] S. Piggott[2] and G.B. Jagdale[3]
[1]*Department of Biological Pest Control and Quarantine, Institute of Plant Protection, Miczurina 20, 60-318 Poznan, Poland;* [2]*Becker Underwood, Littlehampton, West Sussex, BN17 7AU, UK;* [3]*Department of Entomology, Ohio State University, OARDC, 1680 Madison Avenue, Wooster, OH 44691, USA*

8.1. Introduction

Worldwide glasshouse industry, generally referred to as the greenhouse industry in North America, is the most rapidly growing segment of agriculture with more than 300,000 ha of land under cultivation of vegetables (65% area) and ornamentals (35% area) worth billions of dollars in annual sales (Albajes *et al.*, 1999; Parella *et al.*, 1999; Jerardo, 2004). Of the total land under glasshouse cultivation, 250,000 ha are under plastic cover and 50,000 ha under glass cover. Glasshouses are regularly used for propagation, overwintering and full production cycle for many plant species, which are generally grown on a variety of organic and mineral substrates. Covered houses allow a degree of control over many abiotic environmental conditions required for survival and proper growth of plants. Unfortunately, these conditions favour rapid growth and multiplication of many economically important pests and diseases. Arrays of pests that threaten glasshouse industry include insects, mites, nematodes, slugs and snails. The presence of even a few of these pests, dead or live, and their damage in the glasshouses or in the ship-

 Nematodes as Biocontrol Agents (eds P.S. Grewal, R.-U. Ehlers and D.I. Shapiro-Ilan)

ments can cause losses of millions of dollars to the greenhouse industry because of rejected and returned shipments by the retailers and wholesalers. For example, nurseries in the Pacific Northwest spend more than US$1 million per year scouting for black vine weevils, and relinquish over US$500,000 per year on shipments of plant stocks returned due to the presence of the weevils.

In the glasshouses, although integrated pest management (IPM) is a common practice increasingly adopted by many growers, its development and implementation is rather difficult because of the complexity of plant species being grown on a variety of media, the presence and immigration of a wide range of insect pests, and incompatibility of many biocontrol agents with pesticides. Most growers have relied on cultural practices (plant hygiene and light/sticky traps) and/or chemical pesticides to manage insect pests of glasshouse crops (Lindquist *et al.*, 1985), but use of pesticides has been increasingly restricted because of the development of resistance, environmental pollution, human health concerns (van Lenteren, 2003) and statutory reductions in the availability of effective pesticides (Nielsen, 2003). In addition, pesticides have posed a serious threat to beneficial organisms that are frequently used in pollination and plant protection programmes in the glasshouses. Currently, over 30 biocontrol agents including parasitoids, predatory insects and mites, and pathogens including bacteria, fungi and entomopathogenic nematodes (EPNs) are commercially available for the control of several glasshouse pests (van Lenteren, 2003). So far the efficacy of only a few nematode species has been evaluated; some of those tested have proven effective against key target pests in the glasshouses (Table 8.1). This chapter focuses on the application of EPNs in the glasshouses for the management of root- and foliage-feeding insect pests.

8.2. Glasshouse Environment

Controlled glasshouse environment generally favours excellent growth and production of many plant species, development and rapid multiplication of several pests and diseases, and survival and effectiveness of beneficial organisms including pollinators and biocontrol agents (Hussey and Scopes, 1985; Albajes *et al.*, 1999; Parrella *et al.*, 1999; van Lenteren, 2003). Year-round warm temperatures can help maintain a high level of pest population, a main food source for biocontrol agents, but its fluctuation (too high or too low temperatures) could affect activity and efficacy of biocontrol agents including EPNs (Grewal *et al.*, 1994).

Glasshouse crops are generally grown on a broad range of soils, soil mixtures and non-soil media with different chemical and physical properties. The most commonly used rooting and plant-growing media consist of soil mixtures, which are prepared by mixing peat, vermiculite, perlite, composted bark, composted wastes or sewage sludge with soil to modify its texture and structure to that required for the proper growth and development of plants (Adams and Fonteno, 2003). However, certain media can serve as excellent substrates for the development and reproduction of insect pests (Lindquist *et al.*, 1985; Olson *et al.*, 2002; Jagdale *et al.*, 2004). For example, the nursery mix (hardwood bark) was the most conducive medium to fungus gnat colonization when compared with ball mix, metro-mix, pro-mix and pine bark (Lindquist *et al.*, 1985; Jagdale *et al.*, 2004). Peat-based mixes have also provided favourable conditions for survival and development of both fungus gnats (Olson *et al.*, 2002) and black vine weevils (Moorhouse *et al.*, 1992). In contrast, the rock wool-based non-soil media, which have become very popular in cultivation of vegetables, slow development and reduce reproduction of both thrips and leafminers when compared with soil. Specific conditions (pH or moisture) and ingredients (sandy soil, soils with high organic content, peat moss, composted bark, rock wool, or mixtures of all of them with perlite, vermiculite, etc.) prevailing in these media can also affect the survival, recycling, persistence and efficacy of biocontrol agents including EPNs against target pests (Oetting

Table 8.1. List of different species/strains of entomopathogenic nematodes (EPNs) used for the control of insect pests in the glasshouses.

Insect pest	Target stages and plant parts infested	Nematode species/strains	References
Black vine weevil, *Otiorhynchus sulcatus*	Larvae Roots and crowns	*Heterorhabditis bacteriophora* (= *H. heliothidis*) *H. megidis* *Steinernema carpocapsae* All *S. feltiae* *S. glaseri*	Bedding and Miller, 1981; Georgis and Poinar, 1984; Kakouli *et al.*, 1997; Simons, 1981; Stimmann *et al.*, 1985
Fungus gnat, *Bradysia* spp.	Larvae Roots and stems	*S. feltiae* SN *S. carpocapsae* All *S. anomali* (= *S. arenarium*) *S. riobrave* *H. bacteriophora* *H. indica* *H. zealandica*	Jagdale *et al.*, 2004; Harris *et al.*, 1995; Kim *et al.*, 2004; Lindquist *et al.*, 1994; G. B. Jagdale and P.S. Grewal, unpublished data; M. Tomalak, unpublished data
Leafminer, *Liriomyza* spp. Shore fly, *Scatella stagnalis*	Larvae Leaves Larvae Tender plant parts	*S. feltiae* *Heterorhabditis* sp. *S. feltiae* *S. carpocapsae* *S. arenarium* *H. megidis*	Head and Walters, 2003; Williams and MacDonald, 1995; Gouge, 1994; Morton and García del Pino, 2003
Western flower thrip, *Frankliniella occidentalis*	Adults and nymphs Stems, leaves; vectors of viral diseases	*S. feltiae* *H. bacteriophora* *S. abassi* *S. arenarium* *S. bicornutum* *S. carpocapsae* *H. indica* *H. marelatus* *Heterorhabditis* sp.	Ebssa *et al.*, 2001a,b; 2004

and Latimer, 1991; Gouge and Hague, 1994; 1995a; Jagdale *et al.*, 2004).

In the IPM approach, broad-spectrum chemical insecticides are important components used for suppressing insect pests of various crops, and compatibility with these chemicals is essential for the survival of biocontrol agents during and after their applications. Since many studies have shown that EPNs are relatively compatible with many chemical pesticides used in plant protection (see Chapter 18, this volume), no special limitations are imposed on their use in the routine integrated pest control programmes. However, as a safe practice, it is generally recommended that nematodes should be applied separately because of the potential deleterious effects of osmotic pressure on infective juveniles (IJs) if mixed with chemical pesticides or fertilizers.

8.3. Soil Application

8.3.1. Fungus gnats, *Bradysia* spp.

Fungus gnats, *Bradysia* spp. (Diptera: Sciaridae), are relatively small (3–4 mm) flies commonly associated with compost and soils with high organic contents and are one of the most common pests of production nurseries and glasshouse crops (Harris *et al.*, 1995). Although several species of fungus gnats are present in the glasshouse environment, *Bradysia paupera* and *B. coprophila* are economically the most important species reported on many crops in Europe and North America, respectively (Harris *et al.*, 1995). Adult females often lay about 200 eggs in small batches and as many as 1000 eggs in a lifetime on the media or soil surface (Nielsen, 2003). Eggs hatch within 4–6 days; maggots develop through four instars within 12–14 days, pupate in the soil for 3–4 days and then emerge as adults. Thus, egg-to-egg life cycle can be completed within 20–25 days at 20–25°C (Wilkinson and Daugherty, 1970; Nielsen, 2003).

Fungus gnat maggots primarily feed on fungi and organic matter (Freeman, 1983), but they can also cause serious damage to the roots of many ornamentals including African violets, carnations, chrysanthemums, cyclamen, lilies, geraniums, impatiens and poinsettias. Commercially these plants are propagated using stem cuttings, and feeding by maggots on these fresh cuttings can prevent callus development and root formation. In already rooted plants, maggots often feed on the roots and stems by chewing/stripping and tunnelling, respectively (Binns, 1973). Severely injured plants generally lose their healthy appearance, turn off-colour and eventually dry. In addition, direct injuries caused by maggots to the roots can become the major route of entry for many soil-borne pathogens, *Fusarium*, *Phoma*, *Pythium* and *Verticillium*, which are generally responsible for root and stem rots. Thus, maggots are capable of transmitting fungal pathogens during feeding (Ludwig and Oetting, 2001), whereas adult flies are a nuisance to people and also disseminate fungal spores from plant to plant when they migrate through the glasshouse (Gillespie and Menzies, 1993). Since seedlings, rooted stock material or young plants (shortly after transplantation) are most sensitive to fungus gnat damage, the greatest economic losses are generally observed in the nurseries.

Continuous and overlapping generations of fungus gnats in the glasshouses have made most control strategies ineffective. Chemical insecticides such as diazinon and oxamyl are not very effective for the control of fungus gnats and can also be phytotoxic to seedlings and young plants. In addition, the application of pyrethroids against adult flies is ineffective because of continuous immigration and emergence of new generation adults from the plant-growing substrates. First attempts to use EPNs as biocontrol agents to control fungus gnats in the glasshouses were undertaken in the late 1980s (Bedding and Miller, 1981; Simons, 1981; Nedstam and Burman, 1990). Several species of nematodes including *Steinernema feltiae*, *S. carpocapsae*, *S. arenarium* (= *S. anomali*), *S. riobrave*, *S. glaseri* and *Heterorhabditis bacteriophora* were evaluated, but only *S. feltiae* proved to be as effective as chemical insecticides in controlling fungus gnats (Harris *et al.*, 1995). According to Gouge and Hague (1994), *S. feltiae* usually enters the fungus gnat larva through both the anus and mouth, and once inside, it kills the larva within 20 h. These authors and Tomalak (1994a) noted that due to the small size of fungus gnat larva, nematodes completed only one generation inside the cadaver and produced about 1000 IJs/cadaver (Fig. 8.1) within 6–7 days of infection. Currently, the use of commercially produced EPNs, especially *S. feltiae* in Europe, has become a common practice to control fungus gnats in greenhouse productions, but in the USA commercial success has been limited.

8.3.1.1. Nematode application rate

Determining an appropriate concentration of EPNs is a crucial step in the cost-effective control of fungus gnats in greenhouse pro-

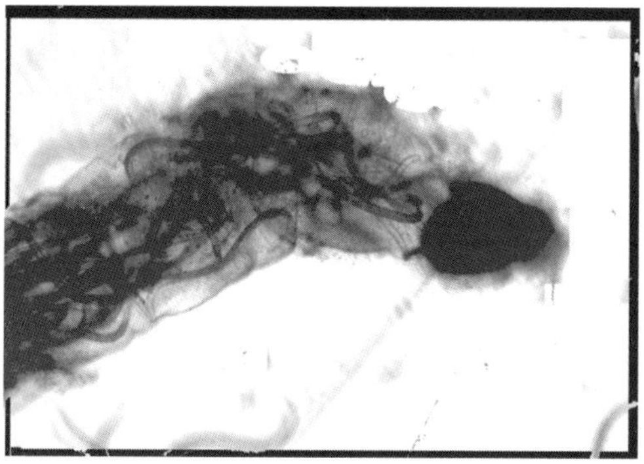

Fig. 8.1. Fungus gnat larva infected with entomopathogenic nematode (EPN) *Steinernema feltiae*.

duction. Gouge and Hague (1995a) and Lindquist and Piatkowski (1993) used relatively high concentrations of 7.8×10^5 IJs and 8.86×10^5 IJs of *S. feltiae*/m^2, respectively, and obtained up to 92% control of *B. paupera*. In contrast, Jagdale *et al.* (2004) applied only 2.5×10^5 IJs of *S. feltiae*/m^2 and obtained up to 100% control of *B. coprophila*, and Harris *et al.* (1995), using the same rate of *S. feltiae*, obtained about 80% control.

8.3.1.2. Method of application

Although nematodes need to be applied in water suspensions to the surface of plant-growing substrate, top spraying is convenient and commonly used for control of fungus gnats infesting small seedlings and compost-filled trays before transplanting. Since there is a potential problem of nematode retention on the surface of leaves, washing off nematodes with water spray or flood irrigation for larger plants is most often useful to treat soil surface under plant canopy. Dripping of nematodes with the aid of central capillary system seems to be less feasible because of aggregations of IJs caused by a slow flow of the suspension inside the tubes. Short distance of nematode movement in the pots makes precise spraying and even distribution of IJs a general requirement for good control of insects, although adult fungus gnats infected with *S. feltiae* can occasionally help in dispersion of nematodes to nematode-free compost (Gouge and Hague, 1995a).

8.3.1.3. Life stage

Synchronization of nematode application with the most susceptible developmental stage of the target pest is important, especially when persistence of the nematodes is expected to be low. The second and fourth larval instars of the fungus gnat *B. coprophila* were significantly more susceptible to *S. feltiae* than the pupae (Harris *et al.*, 1995). The third and fourth instars of another fungus gnat species, *B. agrestis*, were highly susceptible to *S. carpocapsae* Pocheon strain (Kim *et al.*, 2004). Therefore, in the greenhouses, targeting cohorts of second, third or fourth instars is a vital step for suppressing fungus gnat populations below economic threshold level. In a growth chamber study in which mature *B. coprophila* adults were used for inoculation, Jagdale *et al.* (2004) reported that nematodes applied after 16 days of transplanting when fourth instars were expected significantly suppressed *B. coprophila* population but not when they were applied after 0, 4 and 8 days.

8.3.1.4. Potting media

In the greenhouse production, potting medium can be a very important factor for the survival and infectivity of EPNs (Oetting and Latimer, 1991). The IJs of *S. feltiae* actively searched for the sciarid larvae and persisted in the media over 60 days when they were applied to the soil or compost surface (Gouge and Hague, 1994, 1995b). The application of *S. feltiae* in the ball mix (pinewood bark mix), metro-mix (30–40% coconut coir pith, 20–30% vermiculite, 20–30% compost pine bark, 10–20% horticultural perlite) and pro-mix (75–85% of Canadian sphagnum peat and 15–25% of perlite, vermiculite, limestone) equally reduced the overall population of *B. coprophila* over the control by 40%, 50% and 56%, respectively (Jagdale *et al.*, 2004). However, nematode application to nursery mix (pinewood 3: hardwood 1: peat 1) only produced 27% reduction in fungus gnat population. According to Hoitink (1989), continued decomposition of hardwood bark (nursery mix) during the growing season increases water-holding capacity and decreases air porosity, which in turn increases fungus gnat populations and suppresses plant-parasitic nematode populations. Therefore, the low efficacy of nematodes against fungus gnats in the nursery mix obtained by Jagdale *et al.* (2004) may be due to the unfavourable environment for *S. feltiae*. Also, the addition of perlite in the growing substrates or use of perlite alone as medium can adversely affect the efficacy of EPNs (A. Peters, personal communication, 2004). This may be due to the sharp edges of the perlite particles that injure and kill moving IJs (M. Tomalak, unpublished data).

8.3.1.5. Host plant

It has been demonstrated that the host plant could influence fungus gnat colonization and the efficacy of *S. feltiae*. Jagdale *et al.* (2004) found that poinsettia supported significantly higher numbers of fungus gnats than impatiens. They also found that nematode efficacy against fungus gnats was higher in impatiens than poinsettia. Also, the efficacy of different nematode concentrations against fungus gnats in the greenhouse was host plant dependent. Nematodes applied at 1.25×10^5 IJs/m^2 caused 55% reduction in fungus gnats in impatiens and only 18% in poinsettia 30 days after treatment. When applied at 2.5×10^5 IJs/m^2, the reduction was 41% in impatiens and only 20% in poinsettia 12 days after treatment. Since many plant species can affect both chemical and physical properties of potting medium (Argo, 1998), the efficacy of nematodes in poinsettia may be low due to unfavourable conditions in the rhizosphere. Obviously, this area needs further investigation.

8.3.1.6. Temperature

In the greenhouse, the efficacy of EPNs against fungus gnats is generally temperature- and species-dependent. *S. feltiae* is a cold-adapted nematode species with infection occurring between 8°C and 30°C and reproduction between 10°C and 25°C (Grewal *et al.*, 1994). Poor persistence and lack of reproduction of *S. feltiae* at warm temperatures poses a serious constraint for the use of this species in greenhouses where temperatures often exceed 30°C during the summer. Gouge and Hague (1994) reported that the efficacy of the cold-adapted *S. feltiae* against fungus gnats was reduced if soil temperatures in the greenhouse remained above 25°C for prolonged periods of time, and they suggested that *S. feltiae* should be used against sciarids at temperatures between 15°C and 26°C for most satisfactory results. In a subsequent study these researchers demonstrated that the warm-adapted *Heterorhabditis* spp., *S. anomali* and *S. riobrave* provided better control of sciarids than *S. feltiae* at 30°C (Gouge and Hauge, 1995b). Jagdale *et al.* (2004) demonstrated that *S. feltiae* produced significantly higher fungus gnat control at cooler temperature (22 ± 1°C) in the growth chamber (73–80%) than at warmer temperature (25 ± 5°C) in the glasshouse (34–41%). In an effort to find a suitable warm-adapted species that can be used to control sciarids in glasshouses in the USA, where temperatures may often exceed 30°C during the summer, G.B. Jagdale and P.S. Grewal (2003; unpublished data) compared the efficacies of *H. bacteriophora* (GPS 11 strain), *H. indica*, *H. zealandica* and *S. carpocapsae* against fungus gnats with *S. feltiae* in a growth chamber at fluctuating temperatures from 22°C to 29°C. Of the four warm-adapted species, *H. bacteriophora* (GPS 11 strain) and *H. indica* were significantly more effective than *S. feltiae* in controlling *B. coprophila* infesting poinsettia.

8.3.2. Black vine weevil, *Otiorhynchus sulcatus*

The black vine weevil, *O. sulcatus* (Coleoptera: Curculionidae), is a common pest of many glasshouse and field-cultivated plants. Over 150 plant species, including *Azalea*, *Cyclamen*, *Euonymus*, *Fragaria*, *Fuxia*, *Gerbera*, *Primula*, *Rosa*, *Rhododendron*, strawberries and *Taxus* can support development and reproduction of the black vine weevil (Moorhouse *et al.*, 1992). Larvae of *O. sulcatus* damage roots and corms of plants whereas flightless adults feed on leaves and flowers, leaving small characteristic notches along the leaf margins/petals, thus causing significant economic and aesthetic losses. Since weevils do not fly, they are usually brought into the glasshouses together with compost or wood bark used as a growing media. In the glasshouse, the parthenogenic female lays about 50–1000 eggs in the soil near roots; after hatching, larvae usually feed on the roots and enter diapause in the autumn. In the laboratory, females remain active and continue oviposition throughout the year without any interruption of diapause (Sol, 1991). The weevil larvae (previous-year generation) overwinter deep in the soil with pupation occurring in the spring (May). Most of the adults die in autumn but some may overwinter. Adult beetles are mostly nocturnal and they hide in the soil or compost underneath the host plants or under leaf litter during the day. Although *O. sulcatus* produces only one generation per year, soil temperatures and host plants influence its development from egg to adult. In the glasshouse, *O. sulcatus* takes about 84 days to complete its life cycle on rhododendron at 18–22°C, but on outdoor-grown rhododendrons it takes 211 days (La Lone and Clarke, 1981).

During the last two decades, the EPNs, *Steinernema* spp. and *Heterorhabditis* spp., have been tested and found to be effective alternatives to chemical pesticides in controlling black vine weevil larvae in glasshouse production (Bedding and Miller, 1981; Simons, 1981; Georgis and Poinar, 1984; Sitmmann *et al.*, 1985; Kakouli *et al.*, 1997). However, it has been found that the different nematode species/strains, their dosage rates, host insect stages, host plants, application methods, and temperature can influence the efficacy of these nematodes against weevils.

8.3.2.1. Nematode species/strains

Susceptibility of *O. sulcatus* larvae varies among different nematode species and strains. The application of *H. bacteriophora* (= *H. heliothidis*) and *S. feltiae* at the same dosage rate caused 89% and 78% black vine weevil larval mortality, respectively (Stimmann *et al.*, 1985). The IJs of *S. carpocapsae* and *H. megidis*, when applied at the rate of 5000–20,000/pot (10 cm^2), produced 52–93% and 84–100% mortality of weevils, respectively (Kakouli *et al.*, 1997). According to Bedding and Miller (1981), application of *H. bacteriophora* and *S. feltiae* (= *S. bibionis*) at the same rate produced 60% and 100% weevil mortality in potted grapes, respectively. These researchers also reported obvious efficacy differences between T310 and T327 strains of *H. bacteriophora* against *O. sulcatus* larvae. When compared, T327 strain showed higher mortality (93%) of *O. sulcatus* larvae than T310 strain (87%). In contrast, no significant differences were observed in the efficacies of three nematode species, *H. bacteriophora*, *S. glaseri* and *S. carpocapsae*, against *O. sulcatus* larvae when applied to the soil surface (Georgis and Poinar, 1984). At identical application rates between 1000 and 30,000 IJs/pot (250 ml capacity), *H. megidis* HF85 strain gave better control of *O. sulcatus* than *S. carpocapsae* N25 strain (Miduturi *et al.*, 1994).

8.3.2.2. Nematode application rate

Several inconsistencies have been reported in the effectiveness of nematode dosages against black vine weevil in the greenhouses. Stimmann *et al.* (1985) applied between 15,000 and 30,000 IJs of two nematode species, *H. bacteriophora* (= *H. heliothidis*) and *S. feltiae*, per pot (unknown size) and obtained 71–90% weevil mortality. In contrast, Bedding and Miller (1981) applied 5000–60,000 IJs of *H. bacteriophora* or 60,000 *S. bibionis* (= *S. feltiae*) per pot (15–20 cm diameter) and obtained 40–100% and 60% weevil mortality, respectively. Georgis and Poinar (1984) used 15,000 IJs of *H. bacteriophora* per pot (15 cm diameter) and obtained

50–97% *O. sulcatus* larval mortality, whereas Simons (1981) applied either 50 or 100 IJs of unknown *H. megidis* per square centimetre of soil area and obtained 90–97% larval mortality of weevils. Recently, P.S. Grewal and K.T. Power (2003; unpublished data) applied 9000 IJs of *H. bacteriophora* GPS11 and HP88 strains per pot (15 cm diameter) and obtained 100% larval mortality of *O. sulcatus*.

8.3.2.3. Method of application

The method of application and host-finding ability of EPNs is important for targeting black vine weevil larvae, which are generally found at various depths in the pots. Georgis and Poinar (1984) found that the IJs of *H. bacteriophora*, *S. glaseri* and *S. carpocapsae*, when applied on the soil surface to target *O. sulcatus* late instars, which were placed at 5 cm, 10 cm and 20 cm soil depth, were equally effective in killing 70–93% *O. sulcatus* located at 5 cm and 10 cm soil depth, but at 20 cm soil depth only *S. glaseri* was effective in killing 70% of the larvae. These researchers also demonstrated that the IJs of all three species, if injected at 5 cm depth, which is near to host larvae, can effectively kill 70–80% of host larvae located at different soil depths (5–20 cm), but this practice could be costly and time-consuming for large-scale glasshouse productions. Shapiro-Ilan *et al.* (2003) observed greater suppression of *O. sulcatus* in the glasshouse when *H. bacteriophora* (Oswego) was applied in nematode-infected host cadavers compared with application of nematodes in aqueous suspension.

8.3.2.4. Life stage

Developmental stages of black vine weevil can also influence the effectiveness of EPNs. First and second instars of *O. sulcatus* were significantly more susceptible to *H. bacteriophora* (80% mortality) than to *S. glaseri* and *S. carpocapsae* (40% and 20% mortality), but third and fourth instars of *O. sulcatus* were equally susceptible to all three nematode species (Georgis and Poinar, 1984). When compared at a given rate against third or fourth instars of *O. sulcatus*, *H. megidis* consistently produced higher mortality than *S. carpocapsae* in both bags and pots containing strawberry plants (Kakouli *et al.*, 1997). In addition, sixth-instar larvae and pupae of *O. sulcatus* showed higher susceptibility to *H. bacteriophora* (85–95% mortality) than to *S. feltiae* (71–78% mortality) (Stimmann *et al.*, 1985).

8.3.2.5. Temperature

Although the discrepancies between thermal requirements of *O. sulcatus* and EPNs for their survival, development and reproduction imposes some limitations on field control of weevils using nematodes, especially during cold seasons, the glasshouse environment provides excellent thermal and moisture conditions for both organisms. In a Petri dish bioassay, Schirocki and Hague (1997) reported that the temperature required for successful infection in *O. sulcatus* was between 10°C and 28°C for *H. megidis*, between 15°C and 33°C for *Heterorhabditis* spp. and between 13°C and 33°C for *S. carpocapsae*. These authors also reported that the uninfected larvae of *O. sulcatus* were killed when exposed to temperatures above 30°C. In addition, two strains (T327 and T310) of *H. bacteriophora* (= *H. heliothidis*) behaved differently at low temperatures. At 12°C, two strains, T327 and T310, caused 100% and 60% larval mortality of black vine weevil, respectively, whereas at 10°C, only strain T327 was effective in killing 86% larvae (Bedding and Miller, 1981). In the glasshouse, it has been demonstrated that both *H. megidis* and *S. carpocapsae* are able to infect and recycle successfully in black vine weevil larvae at 20°C (Kakouli, 1995).

8.3.3. Western flower thrips (WFT), *Frankliniella occidentalis*

The WFT, *F. occidentalis* (Thysanoptera: Thripidae), is considered one of the major pests of many field- and glasshouse-grown vegetables and ornamentals throughout the world (Yudin *et al.*, 1986; Mantel and van de Vrie, 1988; Tommasini and Maini, 1995). WFT are capable of producing large populations within a few generations, and their typical life cycle contains eggs, two larval, prepupal and pupal, and adult stages. Adults lay eggs in the parenchyma tissue; after hatching, first-instar larvae begin to feed by piercing cell contents of aerial plant parts. Second-instar larvae feed voraciously until they move to the soil or to cryptic habitats for pupation, which lasts for 2–5 days. Newly emerged adults leave the soil, immediately disperse in the glasshouse and begin feeding on leaves and flowers. In the glasshouse, WFT generally take about 10–20 days to complete egg-to-egg life cycle at 25–30°C. The direct feeding of adults by piercing and scraping of the tissues of leaves, stems, flowers and fruits leads to discoloration and drying of the wounded area, which give a flecked and malformed appearance to the infested plant parts. In addition to direct feeding, WFT are also capable of transmitting viral plant diseases (Ullman *et al.*, 1997), which lead to substantial losses in the horticultural industry.

High resistance to chemical insecticides and cryptic behaviour makes *F. occidentalis* control very difficult. Moreover, the use of many chemical compounds must be limited due to concerns over the safety of beneficial organisms, which are concurrently released for pollination and control of glasshouse pests. The biological agents including mites (*Amblyseius* spp.) and bugs (*Orius* spp.) are presently available and could provide a potential alternative to chemical pesticides, but they are only partially effective in controlling *F. occidentalis*. Currently, several species of EPNs belonging to genera *Steinernema* and *Heterorhabditis* have been used as potential biocontrol agents and they have shown great promise for controlling soil-dwelling late second instar nymphs, prepupal and pupal stages of thrips (Tomalak, 1994b; Helyer *et al.*, 1995; Chyzik *et al.*, 1996; Ebssa *et al.*, 2001a,b, 2004; Premachandra *et al.*, 2003a,b). It has been reported that the IJs of *S. feltiae* infected and killed both the prepupal and pupal stages of the WFT within only 2–4 h after infection. The actual death of WFT stages occurred because of the direct body damage resulting from the vigorously moving nematodes, and the small body size of pupae prevented further development and reproduction of nematodes (Tomalak, 1994b).

8.3.3.1. Nematode species/strains

In general, the efficacies of different nematode species and strains vary against soil-dwelling stages of WFT, and heterorhabditid nematodes tend to be more infectious than the steinernematids (Ebssa *et al.*, 2001a,b; 2004; Premachandra *et al.*, 2003a,b). Ebssa *et al.* (2004) screened six species of *Steinernema* (*S. abassi, S. arenarium, S. bicornutum, S. carpocapsae, S. feltiae* and *Steinernema* sp.) and three species of *Heterorhabditis* (*H. bacteriophora, H. indica* and *H. marelata*) against mixed stages of WFT in the soil, and demonstrated that *Heterorhabditis* spp. caused overall higher WFT mortality (24–60%) than *Steinernema* spp. (2.6–54%). According to Ebssa *et al.* (2004), of the four strains of each of *S. carpocapsae* (A1 B5; S.N2, DD136 and S.S2) and *H. bacteriophora* (PS8 hybrid, PALH04, PALH05 and HK3), only DD136, S.S2, PALH04, PALH05 and HK3 showed significant mortality of mixed stages of WFT in the soil. In a laboratory bioassay, a similar trend was observed by Premchandra *et al.* (2003a), who demonstrated that two strains of *H. bacteriophora*, HD01 and HK3, caused significantly higher mortality of prepupal and pupal stages of WFT than the HBN strain. In addition, IS5 strain of *H. bacteriophora* was less effective in controlling WFT than HP88 strain of *H. bacteriophora* and two *Steinernema* spp. (Chyzik *et al.*, 1996).

Although all the available data suggest that various species of EPNs can be effective against WFT in the soil, relatively high nematode dosages are still needed to obtain satisfactory control level. It is probable that further improvement of the nematode performance and optimization of nematode dosages against WFT could be obtained by finding new more effective strains of EPNs. Another possible approach could be genetic selection/improvement of nematodes. Intraspecific hybridization and selection of *S. feltiae* resulted in development of a series of improved strains with much greater ability to control the pest populations in the glasshouse (Tomalak, 1994b). Application of the best strains at the rate of 100 IJs/cm^2 of soil surface resulted in 44–76% WFT mortality, and this increased efficacy was apparently related to the increased proportion of small IJs (less than 600 μm long) present in the nematode populations resulting from intraspecific hybridization and selection. Unfortunately, in spite of a strong selection pressure, the high proportion of individuals with 'small' phenotypes could not be maintained after relaxation of selection pressure (M. Tomalak, unpublished data). Therefore, further research is needed to produce stable populations of improved strains, perhaps through mutagenesis.

8.3.3.2. Nematode application rate

In order to achieve economically feasible control of WFT, the appropriate concentration of nematodes needs to be determined. Application of *S. feltiae* at a rate of 100 IJs/cm^2 of soil surface caused only a 10–18% reduction in the emerging WFT adults. In contrast, application of *S. carpocapsae* at a rate of 25 × 10^4 IJs/l of compost produced better results (76.6% control) in controlling both the prepupal and pupal stages of the WFT (Helyer *et al.*, 1995), but this rate seems to be very high and uneconomical for recommendation in the glasshouses. Recently, several researchers tested different concentrations ranging from 100 IJs to 1000 IJs of several different nematode species per square centimetre of soil/medium and concluded that the concentration of 400 IJs/cm^2 of soil surface was the best rate for achieving more than 50% control of various soil-dwelling stages of WFT (Chyzik *et al.*, 1996; Ebssa *et al.*, 2001a,b, 2004; Premachandra *et al.*, 2003a,b).

8.3.3.3. Life stage

Although nematodes showed a high efficacy against mixed stages of WFT in the soil, the nematode efficacies were different between the WFT stages. The efficacy of *H. bacteriophora* HK3 strain was highest against both larvae and prepupae but lowest against pupae. Similarly, of the three strains of *S. feltiae* (CR, OBSIII and Sylt), OBSIII and Sylt caused higher larval mortality, OBSIII caused highest mortality of prepupae and Sylt caused highest mortality of pupae. Between the species, both *H. bacteriophora* HK3 strain and *S. feltiae* Sylt strain caused overall highest mortality of all the three soil-dwelling stages including larvae, prepupae and pupae (Ebssa *et al.*, 2001a). According to Premchandra *et al.* (2003a) both *Steinernema* and *Heterorhabditis* spp. showed higher efficacy against larval stages than pupal stages, but according to Ebssa *et al.* (2001b) they were more effective against prepupal and pupal stages than late larval stages.

8.3.3.4. Temperature and moisture

The OBSIII strain of *S. feltiae* was more effective against late larval and prepupal stages under high soil moisture but less effective against pupal stages under comparatively dry conditions (Ebssa *et al.*, 2001a,b). It has been reported that the cold-adapted species, *S. bicornutum*, was more effective in reducing populations of WFT at lower (≤25°C) than at higher temperatures. Similarly, warm-adapted *H. indica* LN2 strain was more effective against WFT at higher (25–30°C) than at lower (20°C) temperatures (Ebssa *et al.*, 2004).

8.3.4. Shore flies, *Scatella stagnalis*

Shore flies, *S. stagnalis*, are frequent and occasionally numerous insects present in glasshouse cultures. Shore fly larvae feed on blue-green algae developing on the surface of a variety of organic and mineral plant-growing substrates in moist habitats.

Adult insects can reduce the aesthetic value of ornamentals by leaving faecal spots on leaves and flowers (Foote, 1977; Zack and Foote, 1978) whereas larvae damage the tender crop tissues causing them to dry and affecting the overall productivity and quality of the plants (Ciampolini and Suss, 1994). Moreover, the larvae are capable of ingesting spores of fungal plant pathogens, *Pythium* spp. and *Fusarium* spp., which survive in their intestine and are transmitted to new hosts by adult flies (Goldberg and Stanghellini, 1990; Corbaz and Fischer, 1994). Although chemical methods are primarily focused on reduction of growth of blue-green algae (Vänninen and Koskula, 1998), they give limited control of shore flies (Lindqiust *et al.*, 1994). Since treatments with hydrogen peroxide reduce the growth of young plants (Vänninen *et al.*, 1996), safer and more effective methods are needed to reduce the glasshouse populations of shore flies.

In laboratory trials, *S. stagnalis* showed a very high susceptibility to IJs of *S. feltiae* (Gouge, 1994; Morton and García del Pino, 2003), *S. carpocapsae* (Gouge, 1994), *S. arenarium* and *H. megidis* (Morton and García del Pino, 2003). According to Morton and García del Pino (2003), all three nematode species, *S. feltiae*, *S. arenarium* and *H. megidis*, caused between 65% and 100% *S. stagnalis* larval mortality, which was dose-dependent. Nematode concentrations of as low as 3 IJs/cm^2 of soil surface were sufficient to obtain 87% insect mortality. The most effective nematode species, *S. feltiae*, when applied at the rate of 50 IJs/cm^2, caused 100% mortality within 2 days of treatment. Similarly, *H. megidis* and *S. arenarium*, when applied at the same rate, also caused over 96% control. The rate of penetration of IJs into the insect haemocoel increased with increased dose of all the nematode species except *H. bacteriophora*, and the greatest penetration was observed in the case of IJs of *S. arenarium*.

Surprisingly, glasshouse trials conducted by several researchers in Europe and the USA provided rather poor results of nematode efficacy against shore flies. No satisfactory control of adult flies was obtained on cucumber seedlings when *S. feltiae* was applied at economically feasible concentrations in Finland (Vänninen *et al.*, 1996) or at a high rate of 2.5×10^6 IJs/m^2 on compost over the 10-week period of the experiment in the USA (Lindquist *et al.*, 1994). Furthermore, the applications of either *S. feltiae* (ScP strain) or *H. megidis* at the rate of 5×10^6 IJs/m^2 in commercial gerbera culture showed no visible effects on the population of shore flies in Poland (M. Tomalak, 1996, 1997, unpublished data). Although these failed attempts have clearly prevented EPNs from being recommended against shore flies in commercial glasshouse cultures, recent studies by Morton and Garcia del Pino (2003), who reported 65–100% larval mortality, suggest that more research is needed to develop alternative application methods to achieve cost-effective control of shore flies with nematodes.

8.4. Foliar Application

Several attempts have been made to use EPNs as biocontrol agents against target pests located on the crop foliage but the early results were not encouraging both in the glasshouses (Hara *et al.*, 1993) and fields (Kaya *et al.*, 1981; Hara *et al.*, 1993; Mason and Wright, 1997; Bélair *et al.*, 1998; Grewal and Georgis, 1998). However, recent development of more effective application techniques/tools for nematodes promises to improve the efficacy of nematodes against foliar pests, particularly under glasshouse conditions.

8.4.1. WFT, *F. occidentalis*

Foliar feeding stages of WFT tend to cause damage to the rapidly growing points of plants. In the early stages of flower development, a small amount of damage can lead to excessive aesthetic damage upon the opening of flower buds. This is also true for many other field crops including food crops, where young shoot damage by WFT can often lead to a production of malformed and unmarketable products. Although the majority of researchers have considered soil application of nematodes as a best strategy for the control of WFT, the use of foliar applications of nematodes may have been considered more likely after reported successes in controlling leafminers. One of the first positive results using EPNs against foliar feeding stages of WFT was reported by Bennison *et al.* (1998), who achieved significant control of larval stages on the leaves both in the laboratory and glasshouse tests. This research led to the first successful use of commercially produced nematodes for controlling WFT by foliar applications (Wardlow *et al.*, 2001). Thereafter, a number of sites in the UK were treated with *S. feltiae* using medium volumes of sprays (approximately 1000 l/ha), which not only led to successful control of the WFT populations, but in a number of cases their population reduction was better than what was achieved by the standard chemical treatments. This successful control was attainable by the prudent use of adjuvants to enable suitable targeting of the nematodes to the WFT on the foliage. Currently, throughout Europe and North America, use of commercially produced nematodes is being seen as a viable solution to control WFT when chemical compounds are withdrawn due to current legislation.

8.4.2. Leafminers, *Liriomyza* spp.

Serpentine leafminers (*Liriomyza* spp.) are reported as economically important polyphagous pests in many countries. *Liriomyza* spp. commonly infest indoor vegetables including aubergines, beans, beet, carrots, celery, cucumbers, lettuce, melons, onions, peas, peppers, potatoes, squash and tomatoes. They also infest flowering plants including chrysanthemum, gerbera and gypsophila. Adult leafminers measure less than 2 mm in length and live for about 13–18 days depending on temperature. Females lay eggs in punctures just beneath the epidermis on either the under or upper side of the leaf depending on the species. The eggs hatch within 4–8 days; larvae immediately begin feeding on the mesophyll and undergo three moults within 7–13 days. Mature larvae eventually cut through the leaf epidermis, move to the soil for pupation – or may overwinter provided they mature in autumn – and adults emerge within 3 weeks of pupation in summer (Parrella, 1987).

Major damage is often in the form of punctures caused by females during feeding, mining and oviposition, which result in the destruction of leaf mesophyll, and stippled appearance on foliage, especially at the leaf tip and along the leaf margins. Noticeable damage occurs 3–4 days after oviposition and it increases with increasing size of larvae and mines. Both extensive mining and stippling on the leaves can greatly decrease the level of photosynthesis, which results in premature leaf drop (Parrella *et al.*, 1985). Excessive leaf drop can reduce the amount of shade, which in turn causes sun-scalding of fruits. Wounding of the foliage also allows entry of bacterial and fungal disease-causing pathogens. Floricultural crops are generally less tolerant to leafminer damage than vegetable crops such as tomatoes, which are quite resilient and capable of withstanding considerable leaf damage.

In many countries, *Liriomyza* spp. are statutory pests and there is a general requirement for eradication of this pest on imported crops and in outbreak situations, but due to the protection of immature stages from conventional pesticide sprays, and their ability to develop resistance to many insecticides, leafminers are difficult pests to control (Mason *et al.*, 1987; Parrella *et al.*, 1999). Therefore, biocontrol agents including parasites and EPNs are being used for the control of leafminers in glasshouses as an alternative to chemical pesticides (Harris *et al.*, 1990; Olthof and Broadbent, 1992; Williams and Walters, 1994; Parella *et al.*, 1999).

8.4.2.1. Nematode species and application rate

Recently, the Central Science Laboratories in the UK conducted several glasshouse trials involving insecticide-resistant leafminer populations infesting lettuce and demonstrated that the EPN *S. feltiae* was effective in causing over 85% larval mortality, which was exceeding the mortality achieved by chemical treatments (Head and Walters, 2003). Similarly, in the glasshouse, the foliar application of *S. feltiae* caused 82% leafminer (*Liriomyza huidobrensis*) mortality, which was significantly higher than the chemical treatment, heptenophos (Williams and Walters, 2000). These findings are in agreement with the findings of Harris *et al.* (1990), who demonstrated that *S. carpocapsae*, when applied at the rate of 5×10^8 IJs/ha, was equally effective as the insecticide abamectin (0.17 a.i./ha) in causing mortality of *L. trifolii* infesting chrysanthemum foliage in the greenhouse.

The efficacy of nematodes against leafminers was species/strains-specific. Both *S. carpocapsae* and *S. feltiae*, when applied at same concentration (9×10^9 IJs/ha), produced higher mortality (> 63%) of *L. trifolii* on the bean leaves than *Heterorhabditis* sp. (< 33%) in the foghouse (Hara *et al.*, 1993). However, when tested against another species of leafminers (*L. huidobrensis*), both *S. feltiae* and *Heterorhabditis* sp. were equally effective in reducing pupae formation (Williams and MacDonald, 1995). In addition, *S. feltiae* significantly and equally reduced the pupal production of three different species of leafminers, *L. huidobrensis*, *L. bryoniae* and *Chromatomyia syngenesiae* (Williams and Walters, 2000). In a laboratory test, strains of *S. feltiae* (SN, MG-24, MG-14 and MG25R3), *Heterorhabditis* sp. (H-3, MB-7, OK-3), and *S. carpocapsae* (S.20) were equally effective causing 67–80% mortality of *L. trifolii* (Hara *et al.*, 1993).

8.4.2.2. Life stage

Since the efficacy of nematodes is varied among the developing stages of leafminers, timing of nematode application is important for targeting susceptible stages of this pest. Harris *et al.* (1990) reported that the prepupal and pupal stages of leafminers were not susceptible to *S. carpocapsae* (Mexican strain). LeBeck *et al.* (1993) reported that all the larval and prepupal stages, except > 1-h-old pupal stage, were susceptible to *S. carpocapsae* All strain. Among the different larval stages of *L. trifolii*, second-stage larvae were most susceptible to *S. carpocapsae*, which caused 93% larval mortality (LeBeck *et al.*, 1993). Foliar application of *S. feltiae* also caused high mortality of all three larval stages of another leafminer, *L. huidobrensis*, second-stage being more susceptible than first- and third-stage larvae (Williams and MacDonald, 1995). Since most of the studies demonstrated that the second and early third stages of leafminers are susceptible to nematodes, repeat nematode applications could target these stages (Williams and Walters, 2000; Head and Walters, 2003). Using a model that simulates pest populations, a single commercial application of nematodes led to successful control (75%) of *L. huidobrensis* infesting Chinese brassica, but three repeated applications with the same nematode dosage caused comparatively more (99%) larval mortality (Head and Walters, 2003).

Some species of leafminers specifically lay their eggs on the underside of the foliage where they complete their life cycle and escape location and infection by the nematodes. Improved application technology is needed to target leafminer stages underneath the leaf.

8.4.2.3. Temperature and relative humidity

Temperature combined with relative humidity is an important factor limiting the success of EPNs applied to the foliage for the control of leafminers. Williams and MacDonald (1995) reported that the foliar application of *S. feltiae* and *Heterorhabditis* sp. was effective in killing second-stage larvae of *L. huidobrensis* at 20°C and > 80% relative humidity. It was also observed that *S. feltiae* was equally effective at temperatures between 10°C and 30°C against second-stage larvae of *L. huidobrensis* but > 90% relative humidity was the best for their efficacy (Williams and MacDonald, 1995). The efficacy of *S. feltiae* against pupal production of *L. huidobrensis* was also highest at 10°C and > 90% relative humidity compared to the control. The efficacy of *S. feltiae* against pupal production of *L. huidobrensis* at 20°C in three different ranges of relative humidity was compared; pupal production was 57%, 68% and 88% at 50–65%, 75–85% and > 90% relative humidity, respectively (Williams and MacDonald, 1995). Recently, Williams and Walters (2000) demonstrated that *S. feltiae* was also highly effective against larval stages of three different species of leafminers (*L. huidobrensis*, *L. bryoniae* and *C. syngenesiae*) at 20°C and > 90% relative humidity. Another nematode species, *S. carpocapsae* All strain, also produced over 50% leafminer (*L. trifolii*) mortality at 22°C and 95% relative humidity (Olthof and Broadbent, 1992). According to Hara *et al.* (1993), the All strain of *S. carpocapsae* also caused the highest (67%) mortality of *L. trifolii* at 23°C and 92% relative humidity. These studies suggest that maintaining 20°C temperature and > 90% relative humidity in the glasshouses would be ideal for obtaining effective control of leafminers.

Use of enough water is of prime importance because it enables nematodes to move on the leaf surface freely, locate and enter the mine through the oviposition hole, and easily infect leafminer larvae. Post-application conditions such as high relative humidity also enable improved control of leafminers due to the reduced mortality of nematodes in the open environment (Olthof and Broadbent, 1992; Hara *et al.*, 1993; Williams and Walters, 1994). Nematodes applied in the evening have a better chance of reaching their targets, as the leaf surface remains moist for a longer period in the evenings than in the mornings during a sunny day. Hara *et al.* (1993) showed that there is high variability between nematode strains and also between the humidity levels in a glasshouse. Research has highlighted the requirement of enough water in the application to enable suitable targeting of the miner, but it has been noted that too much water can cause runoff of the nematode suspension and loss of efficacy. Improved formulations to reduce the rate of drying of nematodes on foliage may improve this situation, especially the use of antidesiccants that will improve water retention on the leaf surface and may also aid in the host-seeking ability of the nematodes (Glazer *et al.*, 1992).

8.4.3. General restrictions for foliar applications of nematodes in protected crops

The utilization of a biological agent for the control of a pest can often be achieved by simply the addition of a natural enemy at greater numbers than that found in natural predator–prey demographic cycles. However, the process of application utilizes mechanization processes, which may damage the nematodes or apply them in an unnatural homogeneity, both of which may reduce the effect of otherwise natural occurrences. It is important to understand the effects of such application technologies upon the organism since there are often ways to lessen any deleterious effects.

The major area requiring improvement to allow greater use of nematodes as biocontrol agents is application apparatus (Curran, 1992), particularly as the apparatus employed in many situations comes directly from chemical application technology. However, a full understanding of nematode behaviour within the spray has yet to be attained (Lello *et al.*, 1996; Mason *et al.*, 1998a,b; Piggott *et al.*, 2003). The use of EPNs against foliar targets is particularly problematic due to the harsh nature of the environment, but the process of applying nematodes to the target in the first instance also creates a particular quandary.

It has been noted that nematodes tend to fall out of suspension when held in a container and this has the tendency to block gravity-fed applicators, such as spinning disc atomizers, during application (J. Mason, personal communication). Such occurrences are compounded when the nematodes are held within solutions that are less viscous. However, it has been shown that carefully controlled conditions and the use of correct apparatus for new formulations can provide good targeting and control of pest insects on foliage (Glazer *et al.*, 1992; Lello *et al.*, 1996; Navon *et al.*, 1998). Other workers have identified certain nematode attributes, such as anhydrobiosis and cold tolerance (Womersley and Ching, 1989; Brown and Gaugler, 1996; Grewal, 2000), which indicate that there may be potential for survival under foliar conditions.

Independent of such environmental factors is also the desire by the grower for the use of as few products as possible, whether chemical or biological. For foliar application against WFT, for example, it was noted that chemical products could have a side effect on WFT, therefore precluding the need for biologicals. It is often the case that growers who are able to move most of their pest control to biocontrol are able to use nematodes in an IPM system. This is not solely due to their biological bias but predominantly due to the fact that within such IPM systems there is often little overlap and nematodes therefore have their own foliar niche.

8.5. Potential New Target Pests and Cultures

The progress made in recent years on nematode application against glasshouse pests holds promise that still other insect species and crops can be targeted. New *Steinernema* and *Heterorhabditis* species and strains are continuously being discovered, which potentially carry new characteristics of practical value. Better sensitivity to host-specific cues, better desiccation tolerance and better adaptation to high moisture conditions in the soil are only some of the characters that would be particularly welcome in new nematode products. Research on further improvements of application techniques and equipment, including better timing as well as better formulation for foliar treatments is still underway (see Wright *et al.*, Chapter 5, this volume).

As the technology of glasshouse production develops further, new environmental conditions may lead to new problems for pest control. One such problem has recently been identified in tomato and cucumber cultures grown on rock wool. It has been observed that many populations of sciarids (*B. paupera*) have adapted to environmental conditions prevailing in the rock wool and occasionally become serious pests and plant pathogen vectors in the glasshouse. Although *S. feltiae* has proved to be very effective against this pest, direct application of nematodes to the surface of the growing substrate is difficult due to the polyethylene lining of the rock wool blocks. A series of laboratory and glasshouse experiments revealed that nematode infectivity to sciarid larvae exposed in the rock wool was not significantly different from that in horticultural compost (M. Tomalak, unpublished data). However, effective nematode application techniques had to be developed. The standard rock wool block is 1 m long with three evenly spaced cubes, with individual plants inserted through slots made on the top of the polyethylene lining. The nematodes applied at a single dose of 5×10^4 IJs/plant to the top of each plant cube were able to colonize the whole rock wool block within only a few days. In 27% of the examined blocks, nematodes were present throughout the length 2 days after application. On day 6 almost all blocks were colonized completely. In the glasshouse experiment, nematode efficacy against sciarids was unaffected by the type of mineral wool (i.e. Flormin® or Grodan®). The respective pest control level was 73% and 77% at a nematode dose of 0.5×10^6 IJs/m^2, and 85% and 86% at a dose of 10^6 IJs/m^2. Both nematode rates significantly differed from the control, and between each other.

8.6. Conclusions

Within a relatively short period EPNs have become widely accepted as economic biological agents for the control of a number of soil-dwelling and foliage pests of many crops in the glasshouses. They can be safely used with most of the chemical pesticides and all the other beneficial organisms routinely employed in this environment.

Control of fungus gnats and the black vine weevil with EPNs has already become a standard procedure in glasshouse crops. New target pests and new niches for nematode application are continuously identified and positively verified by small- and large-scale glasshouse trials. Particularly promising is the recent commercial success of *S. feltiae*-based products in the control of WFT and leafminers on plant foliage. Results of the research on the control of other pests, such as shore flies, and soil-dwelling developmental stages of WFT are also promising. Nevertheless, further research on the development of new nematode strains, application methods and equipment is needed to meet the growing demands of modern glasshouse production.

References

Adams, R. and Fonteno, R. (2003) Media. In: Hamrick, D. (ed.) *Ball Redbook Crop Production.* Ball Publishing, Batavia, Illinois, pp. 19–27.

Albajes, R., Lodovica Gullino, M., Van Lenteren, J.C. and Elad, Y. (eds) (1999) *Integrated Pest and Disease Management in Greenhouse Crops.* Kluwer Academic Publishers, Dordrecht, The Netherlands, 545 pp.

Argo, W.R. (1998) Root medium chemical properties. Available at: http://www.imok.ufl.edu/veghort/docs/media_072602b.pdf

Bedding, R.A. and Miller, L.A. (1981) Use of a nematode, *Heterorhabditis heliothidis,* to control black vine weevil, *Otiorhynchus sulcatus,* in potted plants. *Annals of Applied Biology* 99, 211–216.

Bélair, G., Vincent, C. and Chouinard, G. (1998) Foliar sprays with *Steinernema carpocapsae* against early-season apple pests. *Journal of Nematology* 30, 599–606.

Bennison, J.A., Maulden, K.A., Wardlow, L.R., Pow, E.M. and Wadhams, L.J. (1998) Novel strategies for improving biological control of western flower thrips on protected ornamentals – potential new biological control agents. *Brighton Crop Protection Conference – Pests and Disease* 1, 193–198.

Binns, E.S. (1973) Fungus gnats (Diptera: Mycetophilidae, Sciaridae) and the role of mycophagy in soil: a review. *Review of Ecological Biology and Soil* 18, 77–90.

Brown, I.M. and Gaugler, R. (1996) Cold tolerance of steinernematid and heterorhabditid nematodes. *Journal of Thermal Biology* 21, 115–121.

Chyzik, R., Glazer, I. and Klein, M. (1996) Virulence and efficacy of different entomopathogenic nematode species against western flower thrips (*Frankliniella occidentalis*). *Phytoparasitica* 24, 103–110.

Ciampolini, M. and Suss, L. (1994) *Scatella stagnalis* Fallen (Diptera: Ephydridae) dannosa allo stadio adulto a piante ortive e floreali in serra. *Bollettino di Zoologia Agraria e di Bachicoltura* 26, 115–126.

Corbaz, R. and Fischer, S. (1994) La mouche *Scatella stagnalis* Fall. *(Diptera, Ephydridae)* responsable de la dispersion de *Fusarium oxysporum* f. sp. *lycopersici* dans les cultures de tomates hors sol. *Revue Suisse de Viticulture, d'Arboriculture et d'Horticulture* 26, 383–385.

Curran, J. (1992) Influence of application method and pest population size on the field efficacy of entomopathogenic nematodes. *Journal of Nematology* 24, 631–636.

Ebssa, L., Borgemeister, C., Berndt, O. and Poehling, H.M. (2001a) Efficacy of entomopathogenic nematodes against soil-dwelling life stages of western flower thrips, *Frankliniella occidentalis* (Thysanoptera: Thripidae). *Journal of Invertebrate Pathology* 78, 119–127.

Ebssa, L., Borgemeister, C., Berndt, O. and Poehling, H.M. (2001b) Impact of entomopathogenic nematodes on different soil-dwelling stages of western flower thrips, *Frankliniella occidentalis* (Thysanoptera: Thripidae), in the laboratory and under semi-field conditions. *Biocontrol Science and Technology* 11, 515–525.

Ebssa, L., Borgemeister, C. and Poehling, H.M. (2004) Effectiveness of different species/strains of entomopathogenic nematodes for control of western flower thrips (*Frankliniella occidentalis*) at various concentrations, host densities and temperatures. *Biological Control* 29, 145–154.

Foote, B.A. (1977) Utilization of blue-green algae by larvae of shore flies. *Environmental Entomology* 6, 812–814.

Freeman, P. (1983) Sciarid flies, Diptera; Sciaridae. *Handbooks for the Identification of British Insects 9, Part 6.* Royal Entomology Society, London, pp. 68.

Georgis, R. and Poinar, G.O. Jr (1984) Greenhouse control of the black vine weevil, *Otiorhynchus sulcatus* (Coleoptera: Curculionidae) by heterorhabditid and steinernematid nematodes. *Environmental Entomology* 13, 1138–1140.

Glazer, I., Klein, M., Navon, A., and Nakache, Y. (1992) Comparison of efficacy of entomopathogenic nematodes combined with antidesiccants applied by canopy sprays against three cotton pests. *Journal of Economic Entomology* 85, 1636–1640.

Gillespie, D.R. and Menzies, J.G. (1993) Fungus gnat vector *Fusarium oxysporum* f. sp. *radicislycopersici*. *Annals of Applied Biology* 123, 539–544.

Goldberg, N.P. and Stanghellini, M.E. (1990) Ingestion-egestion and aerial transmission of *Pythium aphanidermatum* by shore flies (*Ephydrinae: Scatella stagnalis*). *Phytopathology* 80, 1244–1246.

Gouge, D.H. (1994) Biological control of sciarid and other flies with entomopathogenic nematodes. PhD thesis, University of Reading, Reading, UK.

Gouge, D.H. and Hague, N.G.M. (1994) Development of *Steinernema feltiae* (Steinernematidae: Nematoda) in *Bradysia paupera* (Sciaridae: Diptera). *IOBC / WPRS Bulletin* 17, 132–135.

Gouge, D.H. and Hague, N.G.M. (1995a) The susceptibility of different species of sciarid flies to entomopathogenic nematodes. *Journal of Helminthology* 69, 313–318.

Gouge, D.H. and Hague, N.G.M. (1995b) Glasshouse control of fungus gnats, *Bradysia paupera*, on fuchsias by *Steinernema feltiae*. *Fundamental and Applied Nematology* 18, 77–80.

Grewal, P.S. (2000) Enhanced ambient storage stability of an entomopathogenic nematode through anhydrobiosis. *Pest Management Science* 56, 401–406.

Grewal, P.S. and Georgis, R. (1998) Entomopathogenic nematodes. In: Hall, F.R. and Menn, J.J. (eds) *Biopesticides: Use and Delivery*. Humana Press, Totowa, New Jersey, pp. 271–299.

Grewal, P.S., Selvan, S. and Gaugler, R. (1994) Thermal adaptation of entomopathogenic nematodes: niche breadth for infection, establishment, and reproduction. *Journal of Thermal Biology* 19, 245–253.

Hara, A.H., Kaya, H.K., Gaugler, R., Lebeck, L.M. and Mello, C.L. (1993) Entomopathogenic nematodes for biological control of the leafminer, *Liriomyza trifolii* (Diptera: Agromyzidae). *Entomophaga* 38, 359–369.

Harris, M.A., Begley, J.W. and Warkentin, D.L. (1990) *Liriomyza trifolii* (Diptera: Agromyzidae) suppression with foliar application *Steinernema carpocapsae* (Rhabditida: Steinernematidae) and abamectin. *Journal of Economic Entomology* 83, 2380–2384.

Harris, M.A., Oetting, R.D. and Gardner, W.A. (1995) Use of entomopathogenic nematodes and a new monitoring technique for control of fungus gnats, *Bradysia coprophila* (Diptera: Sciaridae), in floriculture. *Biological Control* 5, 412–418.

Head, J. and Walters, K.F.A. (2003) Augmentation biological control utilising the entomopathogenic nematode, *Steinernema feltiae*, against the South American leafminer, *Liriomyza huidobrensis*. *Proceedings of the 1st International Symposium on Biological Control* (Hawaii, USA, 13–18 January 2002). USDA Forest Service, FHTET-03-05, 136–140.

Helyer, N.L., Brobyn, P.J., Richardson, P.N. and Edmondson, R.N. (1995) Control of western flower thrips (*Frankliniella occidentalis* Pergande) pupae in compost. *Annals of Applied Biology* 127, 405–412.

Hoitink, H.A.J. (1989) Integrated control of *Pythium, Rhizoctonia* and *Fusarium* diseases. In: *Proceedings of the Fifth Conference on Insect and Disease Management on Ornamentals*. 26–28 February. Society of American Florists, Alexandria, Virginia, pp. 116–121.

Hussey, N.W. and Scopes, N. (eds) (1985) *Biological Pest Control: The Glasshouse Experience*. Blandford Press, Poole, UK, 240 pp.

Jagdale, G.B., Casey, M.L., Grewal, P.S. and Lindquist, R.K. (2004) Effects of application rate and timing, potting medium, and host plant on efficacy of *Steinernema feltiae* against fungus gnat, *Bradysia coprophila*, in floriculture. *Biological Control* 29, 296–305.

Jerardo, A. (2004) *Floriculture and nursery crops outlook*. Electronic Outlook Report from the Economic Research Service. Available at http://www.ers.usda.gov/publications

Kakouli, T. (1995) Biological control of the black vine weevil, *Otiorhynchus sulcatus* (*F.*) (Coleoptera: Curculionidae), with entomopathogenic nematodes (Nematoda: Rhabditida). PhD thesis, University of Reading, Reading, UK.

Kakouli, T., Labuschange, L. and Hague, N.G.M. (1997) Biological control of the black vine weevil, *Otiorhynchus sulcatus* (*F.*) (Coleoptera: Curculionidae), with entomopathogenic nematodes (Nematoda: Rhabditida). *Annals of Applied Biology* 131, 11–27.

Kaya, H.K., Hara, A.H. and Reardon, R.C. (1981) Laboratory and field evaluation of *Neoplectana carpocapsae* (Rhabditida: Steinernematidae) against the elm leaf beetle (Coleoptera: Chrysomelidae) and the western spruce budworm (Lepidoptera: Tortricidae). *Canadian Entomologist* 113, 787.

Kim, H.H., Choo, H.Y., Kaya, H.K., Lee, D.W., Lee, S.M. and Jeon, H.Y. (2004) *Steinernema carpocapsae* (Rhabditida: Steinernematidae) as a biological control agent against fungus gnat *Bradysia agrestis* (Diptera: Sciaridae) in propagation houses. *Biocontrol Science and Technology* 14, 171– 183.

La Lone, R.S. and Clarke, R.G. (1981) Larval development of *Otiorhynchus sulcatus* (Coleoptera: Curculionidae) and effects of larval density on larval mortality and injury to rhododendron. *Environmental Entomology* 10, 190–191.

LeBeck, L.M., Gaugler, R., Kaya, H.K., Hara, A.H. and Johnson, M.W. (1993) Host stage suitability of the leafminer *Liriomyza trifolii* (Diptera: Agromyzidae) to the entomopathogenic nematode *Steinernema carpocapsae* (Rhabditida: Steinernemattidae). *Journal of Invertebrate Pathology* 62, 58–83.

Lello, E.R., Patel, M.N., Matthews, G.A. and Wright, D.J. (1996) Application technology for entomopathogenic nematodes against foliar pests. *Crop Protection* 15, 567–574.

Lindquist, R.K. and Piatkowski, J. (1993) Evaluation of entomopathogenic nematodes control of fungus gnat larvae. *IOBC/WPRS Bulletin* 16, 97–100.

Lindquist, R.K., Faber, W.R. and Casey, M.L. (1985) Effect of various soilless root media and insecticides on fungus gnats. *HortScience* 20, 358–360.

Lindquist, R., Buxton, J. and Piatkowski, J. (1994) Biological control of sciarid flies and shore flies in glasshouses. *Brighton Crop Protection Conference: Pests and Diseases,* BCPC Publications 3, 1067–1072.

Ludwig, S.W. and Oetting, R.D. (2001) Evaluation of medium treatments for management of *Frankliniella occidentalis* (Thripidae: Thysanoptera) fungus gnats, *Bradysia coprophila* (Diptera: Sciaridae). *Pest Management Science* 57, 1114–1118.

Mantel, W.P. and van de Vrie, M. (1988) De Californische trios, *Frankliniella occidentalis,* een nieuwe Schadelijke tripssoort in de tuinbouw onder glas in Nederland. *Entomologische Berichten* 48, 140–144.

Mason, J.M. and Wright, D.J. (1997) Potential of the control of *Plutella xylostella* larvae with entomopathogenic nematodes. *Journal of Invertebrate Pathology* 70, 234–242.

Mason, G.A., Tabashnik, B.E. and Johnson, M.W. (1987) Susceptibility of *Liriomyza sativae* and *L. trifolii* (Diptera: Agromyzidae) to permethrin and fenvalerate. *Journal of Economic Entomology* 80, 1262–1266.

Mason, J.M., Matthews, G.A. and Wright, D.J. (1998a) Appraisal of spinning disc technology for the application of entomopathogenic nematodes. *Crop Protection* 17, 453–461.

Mason, J.M., Matthews, G.A. and Wright, D.J. (1998b) Screening and selection of adjuvants for the spray application of entomopathogenic nematodes against foliar pests. *Crop Protection* 17, 463–470.

Miduturi, J.S., De Clercq, R. and De Grisse, A. (1994) Greenhouse and field control of black vine weevil, *Otiorhynchus sulcatus F.* with *Steinernema carpocapsae* and *Heterorhabditis* sp. In: Burnell, A.M., Ehlers, R.-U. and Masson, J.P (eds) *Genetics of Entomopathogenic Nematode – Bacterium Complexes.* EUR 15681 EN, ECSC-EC EAEC, Brussels, Luxembourg, 221 pp.

Moorhouse, E.R., Charnley, A.K. and Gillespie, A.T. (1992) A review of the biology and control of the vine weevil, *Otiorhynchus sulcatus* (Coleoptera: Curculionidae). *Annals of Applied Biology* 121, 431–454.

Morton, A. and Garcia del Pino, F. (2003) Potential of entomopathogenic nematodes for the control of shore flies *(Scatella stagnalis*). *Growing Biocontrol Markets Challenge Research and Development. 9th European Meeting IOBC/WPRS Working Group 'Insect Pathogens and Entomopathogenic Nematodes',* Abstracts, 67.

Navon, A., Keren, S., Salame, L. and Glazer, I. (1998) An edible-to-insects calcium alginate gel as a carrier for entomopathogenic nematodes. *Biocontrol Science and Technology* 8, 429–437.

Nedstam, B. and Burman, M. (1990) The use of nematodes against sciarids in Swedish greenhouses. *SPOR/WPRS Bulletin* 13, 147–148.

Nielsen, G.R. (2003) Fungus gnats. Available at: http://www.uvm.edu/extension/publications/el/el50.htm

Oetting, R.D. and Latimer, J.G. (1991) An entomogenous nematode *Steinernema carpocapsae* is compatible with potting media environments created by horticultural practices. *Journal of Entomological Science* 26, 390–394.

Olson, D.L., Oetting, R.D. and van Lersel, M.W. (2002) Effect of soilless media and water management on development of fungus gnats (Diptera: Sciaridae) and plant growth. *HortScience* 37, 919–923.

Olthof, T.H.A. and Broadbent, A.B. (1992) Evaluation of steinernematid nematodes for control of a leafminer, *Liriomyza trifolii,* in greenhouse chrysanthemums. *Journal of Nematology* 24, 612.

Parrella, M.P. (1987) Biology of *Liriomyza. Annual Review of Entomology* 32, 201–224.

Parrella, M.P., Jones, V.P., Youngman, R.R. and LeBeck, L.M. (1985) Effect of leaf mining and leaf stippling of *Liriomyza* spp. on photosynthetic rates of Chrysanthemum. *Annals of Entomological Society of America* 78, 90–93.

Parrella, M.P., Hansen, L.S., and van Lenteren, J. (1999) Glasshouse environments. In: Bellows,

T.S. and Fisher, T.W. (eds) *Handbook of Biological Control.* Academic Press, San Diego, California, pp. 819–839.

Piggott, S.J., Clayton, R., Matthews, G.A. and Wright, D.J. (2003) Development of a new application apparatus for entomopathogenic nematodes. *Pest Management Science* 50, 1344–1348.

Premachandra, W.T.S.D., Borgemeister, C., Berndt, O., Ehlers, R.-U. and Poehling, H.-M. (2003a) Laboratory bioassays of virulence of entomopathogenic nematodes against soil-inhabiting stages of *Frankliniella occidentalis* Pergand (Thysanoptera: Thripidae). *Nematology* 5, 539–547.

Premachandra, W.T.S.D., Borgemeister, C., Berndt, O., Ehlers, R.-U. and Poehling, H.-M. (2003b) Combined release of entomopathogenic nematodes and the predatory mite *Hypoaspis aculeifer* to control soil-dwelling stages of western flower thrips *Frankliniella occidentalis. Biocontrol* 48, 529–541.

Schirocki, A. and Hague, N.G.M. (1997) Evaluation of UK Heterorhabditids and Steinernematids for the control of the black vine weevil (*Otiorhynchus sulcatus*). *Annals of Applied Biology* 130 (Suppl.), 46–47.

Shapiro-Ilan, D.I., Lewis, E.E., Son, Y. and Tedders, W.L. (2003) Superior efficacy observed in entomopathogenic nematodes applied in infected-host cadavers compared with application in aqueous suspension. *Journal of Invertebrate Pathology* 83, 270–272.

Simons, W.R. (1981) Biological control of *Otiorhynchus sulcatus* with Heterorhabditid nematodes in the glasshouse. *Netherlands Journal of Plant Pathology* 87, 149–158.

Sol, R. (1991) De eierproduktie en de levensduur van de gegroefde lapsnuitkever (taxuskever) (*Otiorhynchus sulcatus* F.) (Coleoptera, Curculionidae) onder laboratoriumomstandigheden. *Mededelingen van de Faculteit Landbouwwetenschappen, Rijksuniversiteit Gent* 56, 1041–1046.

Stimmann, M.W., Kaya, H.K., Burlando, T.M. and Studdert, J.P. (1985) Black vine weevil management in nursery plants. *California Agriculture* 39, 25–26.

Tomalak, M. (1994a) Selective breeding of *Steinernema feltiae* (Filipjev) (Nematoda: Steinernematidae) for improved efficacy in control of a mushroom fly, *Lycoriella solani* Winnertz (Diptera: Sciaridae). *Biocontrol Science and Technology* 4, 187–198.

Tomalak, M. (1994b) Genetic improvement of *Steinernema feltiae* for integrated control of the western flower thrips, *Frankliniella occidentalis. IOBC/WPRS Bulletin* 17, 17–20.

Tommasini, M. and Maini, S. (1995) *Frankliniella occidentalis* and other thrips harmful to vegetables and ornamental crops in Europe. In: *Biological Control of Thrip Pests.* Wageningen Agriculture University Papers 95, 1–42.

Ullman, D.E., Sherwood, J.L. and German, T.L. (1997) Thrips as vectors of plant pathogens. In: Lewis, T. (ed.) *Thrips as Crop Pests.* CAB International, Wallingford, UK, pp. 539–565.

van Lenteren, J.C. (2003) Commercial availability of biological control agents. In: van Lenteren, J.C. (ed.) *Quality Control and Production of Biological Control Agents: Theory and Testing Procedures.* CAB International, Wallingford, UK, pp. 167–179.

Vänninen, I. and Koskula, H. (1998) Effect of hydrogen peroxide on algal growth, cucumber seedlings and the reproduction of shore flies (*Scatella stagnalis*) in rockwool. *Crop Protection* 17, 547–553.

Vänninen, I., Koskula, H. and Van Lenteren, J.C. (1996) Biology and management of shore flies (*Scatella stagnalis*) in a cucumber seedling crop grown in rockwool. *IOBC/WPRS Bulletin* 19, 187–190.

Wardlow, L.R., Piggott, S. and Goldsworthy, R. (2001) Foliar application of *Steinernema feltiae* for the control of flower thrips. *Medeelingen Faculteit Landbouwkundige en Toegepaste Biologische Wetenschappen Universiteit Gent* 66, 285–291.

Wilkinson, J.D. and Daugherty, D.M. (1970) Comparative development of *Bradysia impatiens* (Diptera: Sciaridae) under constant and variable temperatures. *Annals of the Entomological Society of America* 63, 1079–1083.

Williams, E.C. and MacDonald, O.C. (1995) Critical factors required by the nematode *Steinernema feltiae* for the control of the leafminers *Liriomyza huidobrensis, Liriomyza bryoniae* and *Chromatomyia syngenesiae. Annals of Applied Biology* 127, 329–341.

Williams, E.C. and Walters, K.F.A. (1994) Nematode control of leafminers: efficacy, temperature and timing. *Brighton Crop Protection Conference – Pests and Disease* 1079–1084.

Williams, E.C. and Walters, K.F.A. (2000) Foliar application of the entomopathogenic nematode *Steinernema feltiae* against leafminers on vegetables. *Biocontrol Science and Technology* 10, 61–70.

Womersley, C.Z. and Ching, C. (1989) Natural dehydration regimes as a prerequisite for the successful induction of anhydrobiosis in the nematode *Rotylenchus reniformis. Journal of Experimental Biology* 143, 359–372.

Yudin, L.S., Cho, J.J. and Mitchell, W.C. (1986) Host range of western flower thrips, *Frankliniella occidentalis* (Thysanoptera: Thripidae) with special references to *Leucaena glauca*. *Environmental Entomology* 15, 1292–1295.

Zack, R.S. and Foote, B.A. (1978) Utilization of algal monoculture by larvae of *Scatella stagnalis*. *Environmental Entomology* 7, 509–511.

9 Nursery and Tree Application

R.W.H.M. van Tol[1] and M.J. Raupp[2]
[1]*Plant Research International, Wageningen-UR, PO Box 16, 6700 AA Wageningen, The Netherlands;* [2]*Department of Entomology, University of Maryland, College Park, MD 20742, USA*

9.1. Introduction

Production of ornamental plants in nurseries and greenhouses is one of the fastest-growing areas of agriculture in the USA and Europe. Total annual plant sales for the greenhouse and nursery industry in the USA are estimated at over US$6.2 billion in 1998 (USDA fact sheets 1998, http://www.nass.usda.gov). Approximately US$2.7 billion of this production accounts to nursery crops. Nursery production includes the production of woody ornamental trees and shrubs, woody and herbaceous ground covers and propagation materials. Hardy nursery stock in the Netherlands and the UK – having the largest production areas in Europe – had an annual plant value of US$1.1 billion in 2002. The total ornamental plant value in Europe is estimated to be close to the value of the total nursery production in the USA.

Ornamental production is unique because of its enormous variety of individual species and cultivars. It is common to find dozens of different plant species grown in a single nursery. This diversity creates very complex pest- and plant-management problems. The nursery industry relies heavily on pesticides to control all these pests. In contrast to greenhouse production, there are only limited biocontrol alternatives available for nurseries. One of the positive exceptions is the biocontrol of soil-borne pests. Entomopathogenic nematodes (EPNs) have become an increasingly successful means to control several of these pest problems (mainly grubs). The grubs of several weevil and beetle species cause serious economic damage to many nursery plants. The root weevil complex causes damage to hardy ornamentals equal to *c.* 25% of their market value in western and eastern USA (http://pestdata.ncsu.edu/cropprofiles). The black vine weevil (*Otiorhynchus sulcatus*), one of the more important weevil species in the USA and Canada, threatens large areas of cranberry (annual crop value of US$386 million, of which 20% are heavily threatened and 50% are moderately threatened by this pest), strawberry (annual crop value of US$940 million, of which > 30% are threatened) and red raspberries (annual crop value of US$53 million, of which > 60% are threatened), and is one of the more important pests in landscaping. Based on USDA data, on average US$25–70 million is spent in the USA and Canada annually for the control of this pest. In the Netherlands approximately US$0.5–2 million is spent yearly in hardy ornamental production for control of the vine weevil. The total annual costs for growers in the Netherlands accountable to this pest are estimated to be US$10–17 million.

This review was performed to determine the current status of the biocontrol of pest problems in the nursery industry with EPNs. Our specific objectives were to (i) review the results of pot and field tests; (ii) determine the key factors of success and failure of EPNs in practice; (iii) give recommendations for practical application; and (iv) identify research necessary to solve the limiting factors in control of pests with EPNs in nursery and tree applications. We thereby looked at aspects of effectiveness, information availability, management complexity, labour requirements and costs of EPNs. By determining the essential components for successful application of EPNs we come to recommendations for growers and extension personnel as well as scientists, on how to make biocontrol with EPNs more effective and acceptable in the nursery industry.

9.2. Growers' Perception

Although the use of alternatives for chemical insecticides has increased in the nursery industry, often as part of an integrated pest management (IPM) programme, chemical pesticides continue to be the primary pest control method. A national study on insect control methods used by the greenhouse and nursery industry in the USA (Hudson *et al.*, 1996) revealed that the majority of the respondents used several different types of alternatives like *Bacillus thuringiensis*, insect growth regulators, horticultural oils and insecticidal soaps. According to Hudson *et al.* (1996), the extensive use of pesticide-like products may be the result of easy integration into a system accustomed to applying chemical pesticides. Further, cultural methods and monitoring/scouting were incorporated and considered as effective for pest control by most nurserymen. Biological methods were the least used alternative control measure. The available EPN products at that time (Biosafe and Exhibit), however, were used by almost 50% of the respondents and of these 75% reported them effective.

One of the aspects that provides interesting information considering success and failure of EPN use in nursery industry is the perception of nurserymen towards factors that may limit the adoption of alternative pest-management practices (Hudson *et al.*, 1996). The effectiveness was clearly the most important in adoption of non-chemical measures. Another key limiting factor was the lack of information (protocols, extension programme support), followed by management complexity. Surprisingly, costs of the alternative means were considered to be the least limiting factor. Only a few reports in Europe deal with the perception of growers towards the use of EPNs and other biological means in nurseries. A survey of German nurserymen (Von Reibnitz and Backhaus, 1994) indicated that the perception of growers was quite similar to that of growers in the USA. It was shown that, although more than 90%

of the nurseries performed treatments to control vine weevil, only 17% applied EPNs and 78% applied chemicals. More than 80% of the nurseries were familiar with the existence of EPNs for control of vine weevil larvae. In 2003, more than 90% of the nurserymen in Germany used EPNs to control vine weevil according to R.-U. Ehlers (personal communication). Information about the use in other countries was at the time unavailable. Chemical control of adult weevils and larvae were considered to be highly effective by only 49% and 35%, respectively, of the interviewed nurserymen in the study of Von Reibnitz and Backhaus (1994). In contrast, more than 50% of the growers who applied EPNs considered biocontrol of the larvae very effective. Like the Americans, the German nurserymen did not consider costs to be the limiting factor to use of EPNs, but rather believed that information about efficacy, availability of the product, information about where to get EPNs and proper protocols for application were important limiting factors.

9.3. Economics

A study in the Netherlands on the feasibility of vine weevil control on hardy ornamental nurseries as part of an IPM programme revealed that the main obstacle to efficient EPN use was the high cost of labour related to monitoring the pest (Vander Horst and van Tol, 1995; van Tol, 1996a). The main cost of pest control was in the use of EPNs, but the total costs for pest control did not increase due to the high level of natural control of other pests. The local and limited use of (persistent) insecticides stimulated the build-up of high populations of natural enemies that kept several other pests below the economic injury level. Before introduction of IPM on this field nursery the costs of crop protection in plant production were estimated at 1.5 cents per dollar, of which 30% were assigned to pest control. After 2 years of IPM, the costs increased to 2.6 cents per dollar due to high labour costs of monitoring. The contribution of EPN application to these costs accounted for only 0.4%. The costs of crop protection in plant production would therefore rise from 1.5 to 1.9 cents per dollar (20% increase) if monitoring costs were absent.

A study in the USA (Maryland) revealed that the implementation of IPM programmes for commercial nurseries reduced costs associated with pest management by 52–72% (Cornell, 1994). Pesticide costs for foliage nursery industry in 1991 ranged from 1.5 cents per dollar for large firms to 2.7 cents per dollar for small firms (Hodges and Hull, 1991). Similar costs were reported for foliage plants and woody ornamental nurseries (Hodges, 1991). A survey of the American Association of Nurserymen Pest Management Committee resulted in an estimate that chemical pest control costs averaged 8% of the total cost of production (Thomas, 1996). They provided no data on the increase in costs if EPNs were applied but they did estimate the rise in costs if certain chemicals were no longer available to nurserymen and no acceptable alternatives were available. The loss of acephate, for example, increases these costs from 8.5% to 8.7% of the production costs. Acephate is used to control several important weevil and other beetle pests, as well as many other insect pests. Therefore, only part of this 8.5% can be accounted to pests that can be controlled with EPNs.

9.4. Factors Affecting Efficacy

9.4.1. Coleoptera – vine weevil (*Otiorhynchus sulcatus*)

9.4.1.1. Nematode species, strains and products

In trials with *Primula*, Fitters *et al.* (2001) showed that the efficacy of a commercial EPN product compared to the same EPN strain reared in the laboratory can give dramatically different results, especially when soil temperatures are low. Neubauer (1997) also found that under controlled temperature conditions the efficacy of nematode products strongly differs between the years. In this chapter we will not further discuss the quality aspect of EPN products in relation to efficacy for control, but we do realize that this may explain only part of the variation in results between the different field tests.

In this study we compared field results for EPN products, and if no product was available, we used the laboratory-reared EPN results. In Tables 9.1 and 9.2 results of vine weevil control with EPNs are summarized where soil temperature (> 12°C) and dosage of nematodes applied (> 0.5×10^6 EPNs/1-l pot and > 1.0×10^6 EPNs/m^2 open field) were not limiting EPN efficacy. For *Heterorhabditis megidis*, the Nl-H-F85 strain seems to be superior in controlling vine weevil larvae in both field and pot trials while the UK-H-211 strain gave generally 20% lower control. For *H. bacteriophora* we did not find large differences in efficacy between the different products when applied in autumn (van Tol, 1993a,b, 1998). Results in field and pot trials varied between 60% and 70% control. Only Gill *et al.* (2001) found relatively high control (97%). Excellent control with all the *Heterorhabditis* sp. when applied in spring (Table 9.2) is related to

Table 9.1. Efficacy of entomopathogenic nematode (EPN) species and strains applied in autumn for control of vine weevil larvae (*Otiorhynchus sulcatus*).

EPN species (strain, product)	% control in pots	*N*	% control in the field	*N*	References
Heterorhabditis megidis (UK-H-211, Nemasys H)	63	13	56	17	Fitters, 2001; van Tol, 1993a,b, 1996b, 1998; van Tol *et al.*, 1998
Metarhizium anisopliae (DSM 3884)	92	4	57	2	van Tol, 1993a,b, 1996b, 1998
M.anisopliae + *H. megidis* (DSM3884 + UK-H-211)	91	1	86	1	van Tol, 1996b, 1998
H. megidis (NI-H-F85, Larvanem)	89	4	77	6	van Tol, 1993a,b, 1996b, 1998; van Tol *et al.*, 1998
H. megidis (NI-H-E87.3)	88	1	11	1	van Tol, 1998
H. megidis (D-H-SH, Optimaaltje)	90	3	17	1	van Tol, 1993a,b, 1994, 1998
H. bacteriophora (D-H-D, Optimaaltje)	29	1	—	1	van Tol, 1993a, 1998
H. heliothidis	61	1	—	—	Stimmann *et al.*, 1985
Heterorhabditis sp.	55	1	—	—	Heungens and Buysse, 1992
H. bacteriophora (Au-H-?, Otinem)	63	1	59	1	van Tol, 1993a, 1998
H. bacteriophora (I-H-?, Bioerre)	73	2	60	1	van Tol, 1993a,b, 1998
H. bacteriophora (HP88, Cruiser)	97	1	—	—	Gill *et al.*, 2001
Steinernema carpocapsae (US-S-25, Exhibit)	54	3	0	1	van Tol, 1993b, 1996b, 1998
S. carpocapsae (UK-S-9387)	—	—	19	1	van Tol, 1998
S. feltiae (NI-S-OBS3)	39	3	—	—	van Tol, 1993b, 1998
S. feltiae (NZ-S-CA)	76	2	—	—	van Tol, 1998
S. feltiae	67	1	—	—	Stimmann *et al.*, 1985
S. kraussei (CZ-S-Mraçek)	48	2	—	—	van Tol, 1998

Note: Results present average values from field tests with natural infestation or egg inoculation, soil temperatures equal or higher than 12°C and the dosage of applied EPNs in pot and field respectively higher than 0.5×10^6 and 1.0×10^6 nematodes/m^2.
Abbreviation: N = number of tests included.

Table 9.2. Efficacy of entomopathogenic nematode (EPN) species and strains applied in spring for control of vine weevil larvae (*Otiorhynchus sulcatus*).

EPN species (strain, product)	% control in pots[a,b,c]	*N*	% control in the field[a]	*N*	References
Heterorhabditis sp. (NZ-type)	—	—	80[a]	1	Backhaus, 1994
Heterorhabditis bacteriophora (C1)	—	—	66[a]	1	Hanula, 1993
H. bacteriophora (HP88, Cruiser)	—	—	63[a]	1	Hanula, 1993
Steinernema feltiae (S27, Biosys)	—	—	64[a]	1	Hanula, 1993
S. carpocapsae (US-S-25, Exhibit)	45[b]	2	73[a]	1	Hanula, 1993; Neubauer, 1997
Heterorhabditis sp. (HL-type)	96[b]	1	—	—	Backhaus, 1994
H. megidis (UK-H-211, Nemasys H)	74[b]	4	—	—	Neubauer, 1997
H. bacteriophora (Nematop)	87[b]	2	—	—	Neubauer, 1997
H. bacteriophora (Heteromask)	100[b]	1	—	—	Gill *et al.*, 2001
H. bacteriophora (HP88, Cruiser)	93[a]	1	—	—	Gill *et al.*, 2001
H. heliothidis	87[a]	2	—	—	Stimmann *et al.*, 1985
H. heliothidis	79[b,c]	2	—	—	Georgis and Poinar, 1984
S. carpocapsae	45[b,c]	2	—	—	Georgis and Poinar, 1984
S. glaseri	54[b,c]	2	—	—	Georgis and Poinar, 1984
S. feltiae	77[a]	2	—	—	Stimmann *et al.*, 1985

[a]Natural infestation.
[b]Artificial inoculation with larvae before EPN application.
[c]Average result of test with L1/L2 and L3/L4 larvae.
Abbreviation: *N* = number of tests included.
Note: Results present average values from field tests with soil temperatures equal or higher than 12°C and the dosage of applied EPNs in pot and field respectively higher than 0.5×10^6 and 1.0×10^6 nematodes/m^2.

higher soil temperatures (15–28°C) and the presence of predominantly large larvae in the soil, as discussed later in this section. Of the tested *Steinernema* products, *Steinernema carpocapsae* strains generally give no control to very poor control when applied in autumn. Positive results (Table 9.2) are related to spring applications (Hanula, 1993), high soil temperatures (Georgis and Poinar, 1984; Hanula, 1993) and unrealistic imitation of field situations where identical late instar larvae were released (Georgis and Poinar, 1984; Miduturi *et al.*, 1994; Neubauer, 1997). For *S. feltiae*, large differences in efficacy seem to exist between the different strains. In some, but not in all cases (New Zealand strain; van Tol, 1998), positive results are related to trial set-up, larval stages and high soil temperatures. *S. kraussei* is considered to be a promising new species for vine weevil control (Willmott *et al.*, 2002) but field trials with ornamentals are essential to confirm this promise before introduction in practice. van Tol (1998) tested another promising strain of *S. kraussei* in a soil pre-mix application as well as an autumn application in pots, and found no control in the pre-mix test and only moderate control in the autumn application (average 47.5% control after 2 years of testing). The new strain of *S. kraussei* gave 75% control in the open field when applied in October at 0.5×10^6 nematodes/m^2 while *H. megidis* (product Larvanem) failed to do so. Application of both strains in March gave no control, indicating that the lower soil temperatures at that time of the year and the poor persistence and activity in the field are limiting the successful winter or early spring application (R.W.H.M. van Tol, unpublished).

Furthermore, a combined application of the entomopathogenic fungus *Metarhizium anisopliae* (product BIO1020) applied as a pre-mix in the field soil (spring) and an EPN application in autumn (*H. megidis*, strain UK-H-211) is more effective than application of the fungus or EPNs alone (van Tol, 1998). The weevil larvae and eggs escaping from infection by the fungus during spring and summer are infected by the EPNs applied in autumn. Where a higher dosage of nematodes applied does not give better control in the field, higher control can obviously be achieved by combining these two different biocontrol agents. Better control obtained with the combined application was not related to the larval stages (van Tol, 1998), despite the fact that the fungus is capable of infecting eggs and small larvae.

9.4.1.2. Nematode application rate

For maximum control of vine weevil larvae a recommended dosage of 0.5×10^6 nematodes/1 l pot and 1.0×10^6 nematodes/m^2 of field surface area is currently advised. This recommendation is based on many field tests performed in the past. As the results summarized in Fig. 9.1 show, these dosages are in general correct. In the open field, dosages lower than 1.0×10^6 nematodes/m^2 give lower or variable results. Doubling the field concentration to 2.0×10^6 nematodes/m^2 does not give better control (Hanula, 1993). With *S. carpocapsae* (US-S-25, product Exhibit) no control at any dosage was achieved in the open field when applied in autumn. In contrast, the same strain tested by Hanula (1993) (Table 9.2) provided excellent control of the weevil larvae. Spring application of this product in relation to high soil temperatures, larger larvae and location of these larvae near the soil surface are the reasons for this good control. In pot trials an identical result is found for all the tested EPN species and strains. Lowering the applied concentration below 0.5×10^6 nematodes/m^2 reduces control and higher concentrations do not provide any improvement of control.

9.4.1.3. Soil temperature

For a long time soil temperature appeared to be the most limiting factor for successful control of this pest. In fact, the life cycle of the vine weevil dictates the timing of the EPN application. An application in summer does not control the freshly hatched larvae and eggs that are still laid by the weevils, and an application in autumn has only limited success because of lower soil temperatures. Application in late spring would be most effective, but for many nursery plants damage occurring during autumn and winter is unacceptable. The first EPN products that became available for growers were giving no reliable control because the minimum soil temperature for effective control was too high. In the last 10–15 years EPN products became available that are effectively controlling the larvae at temperatures as low as 12°C. The increased activity at low temperature makes application in autumn no longer a problem, although a further selection of EPNs at even lower temperature activity would give growers more time to apply EPNs. Currently more limiting for control is the rapid reduction in efficacy of the EPNs after application. Several weeks after application no more effective control can be expected. For this reason EPNs have to be applied as late as possible in the season. Early application (May) of EPNs (*S. carpocapsae* – product Exhibit, *S. feltiae* – product Nemasys S and strain Nl-S-OBS3, *S. kraussei* – strain CZ-S-Mraçek) in pots and the field have so far revealed no control of vine weevil larvae in a natural setting (van Tol, 1998). If EPNs would remain effective for months after application in summer, this would definitely control the vine weevil better and perhaps, more importantly, the acceptance by growers to use EPNs would increase.

In Fig. 9.2 we summarize the results of field and pot tests where soil temperature was determined and EPNs were applied in autumn. Results shown are based on tests with egg inoculation of the vine weevil and where dosage of applied EPNs is optimal for control ($> 0.5 \times 10^6$ EPNs/1 l pot and $> 1.0 \times 10^6$ EPNs/m^2 open field). Fitters *et al.* (2001), van Tol (1993a,b, 1994, 1996b, 1998) and van Tol *et al.* (1998) measured soil temperature continuously in relation to EPN efficacy in pots and field, thereby providing very detailed information about the effect of temperature on control. More than 6 years of field and pot trials with two strains of *H. megidis* (UK-H-211 and Nl-H-F85) and some other EPNs enables us to draw reliable conclusions

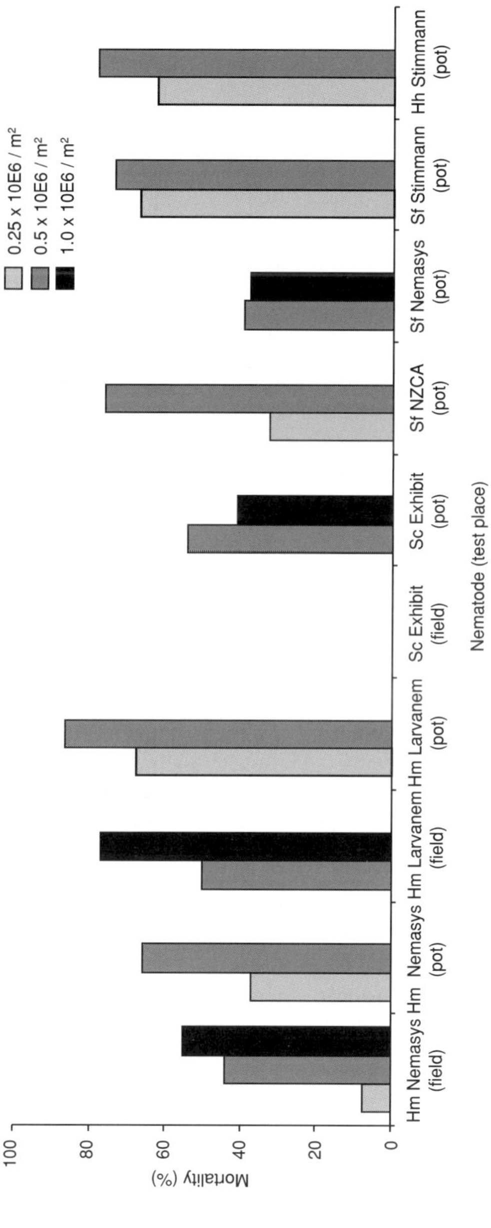

Fig. 9.1. Influence of entomopathogenic nematode (EPN) dosage on vine weevil control in autumn. Values present average results from field tests with natural infestation or egg inoculation of the vine weevil and where soil temperature is equal or higher than 12°C. Hm = *Heterorhabditis megidis;* Hh = *H. heliothidis;* Sf = *Steinernema feltiae;* Sc = *S. carpocapsae.* (Data used from Stimman *et al.*, 1985; van Tol, 1998 and van Tol *et al.*, 1998.)

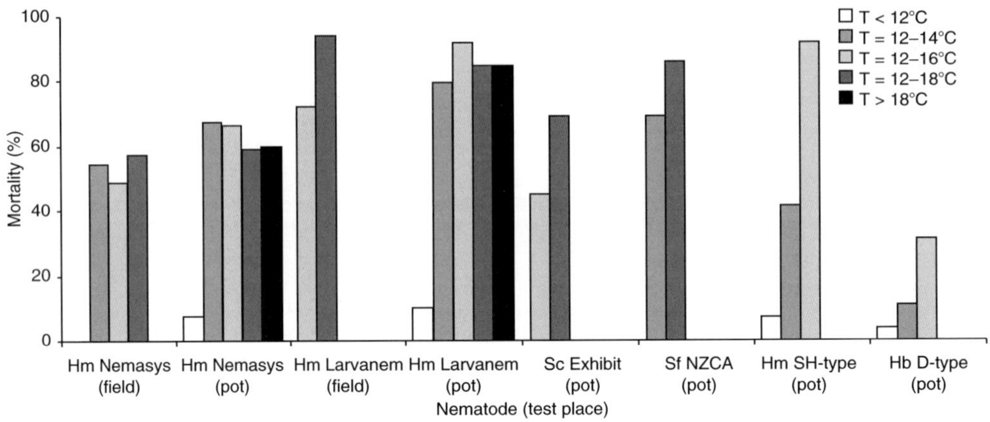

Fig. 9.2. Influence of soil temperature on vine weevil control in pot and field tests with entomopathogenic nematodes (EPNs) applied in autumn. Values present average results from field tests and tests in temperature-controlled rooms with egg inoculation of the vine weevil and a dosage of EPNs applied in pot and field respectively higher than 0.5×10^6 and 1.0×10^6nematodes/m^2. Hm = *Heterorhabditis megidis;* Hb = *H. bacteriophora;* Sf = *Steinernema feltiae;* Sc = *S. carpocapsae.* (Data used from van Tol, 1993a,b, 1994, 1996b, 1998; van Tol *et al.*,1998; Fitters *et al.*, 2001.)

about the effect of soil temperature on efficacy in a natural setting. The results show that a soil temperature of 12°C is sufficient for most tested EPN species and strains. Results of temperature measurements in some years (van Tol, 1998; Fitters *et al.*, 2001) and detailed studies in temperature controlled rooms (van Tol, 1993a,b) revealed that only a few hours of 12°C was enough to get maximum control in pots, while in the open field a longer period of this temperature was needed. In the open field a soil temperature of 12–13°C for 48 h was too short for effective control (*H. megidis*, strain UK-H-211: 9% reduction), while a total of 144 h at this temperature range gave 60–70% control (van Tol, 1996b, 1998; Fitters *et al.*, 2001). Based on these results extension personnel and growers are advised to apply EPNs only if soil temperatures are expected to be higher than 12°C for at least 1 week. For the EPN strain UK-H-211 of *H. megidis* the control does not improve with increasing soil temperatures in both pot and field, but for the strain Nl-H-F85 of *H. megidis* the control in the field improves by a further increase in temperature above 12°C. Also, for *S. carpocapsae* (US-S-25), *S. feltiae* (NZ-S-CA), *H. megidis* (D-H-SH) and *H. bacteriophora* (D-H-D) in pots, a further increase in efficacy with increasing soil temperatures above 12°C is evident. An autumn application of *S. carpocapsae* (US-S-25) in the open field with soil temperatures remaining in the range of 12–15°C (in total 264 h at 12–14°C and 24 h at 14–15°C) gave no control (van Tol, 1996b, 1998), while a spring application at temperatures over 19°C gave 73% control with the same strain (Hanula, 1993). As further discussed in Section 9.4.1.4 these differences in control are more likely related to differences in ratio of larval stages and the location of the larvae in the soil in autumn and spring.

9.4.1.4. Life stages

Many field and pot trials are performed by inoculation of plants with identical larval stages several days before application of the EPNs. This set-up, in most cases, cannot imitate a realistic field situation. Usually there is a large variation in larval stages present at naturally infested places, and application of EPNs in autumn to control vine weevil larvae involves all larval stages between freshly hatched eggs and third larval stage. The late instars and pupae are only found in spring. Results of trials with larval inoculation may therefore at best predict the efficacy that can be expected in practice with a spring application and not with an autumn application. R.W.H.M. van Tol and R.L. Gwynn (unpublished data) analysed these problems related to trial set-up for several pests in ornamentals and developed protocols for field/pot efficacy testing of vine weevil, scarabs, sciarids and slugs. In greenhouses a mix of all larval stages can be present during spring, summer and winter depending on favourable temperature and light conditions for the adult weevils to oviposit. Control with EPNs in these circumstances necessitates a chemical eradication of the adult weevil population and repeated application of EPNs to control both larvae and still-hatching larvae from the eggs that escape the first EPN application.

To determine whether larval stages have an impact on control we compared all field trials with information about the larval stages (Fig. 9.3A and B). The results of Stimmann *et al.* (1985) and van Tol (1996b, 1998) in Fig. 9.3A are based on field tests with egg inoculation to the plants several months before application of EPNs in autumn. These results indicate that there is an influence of the larval stage on control in the field but not in pots for the two tested strains of *H. megidis* (UK-H-211 and Nl-H-F85). Especially the high number of small larvae (L2) we recovered in spring indicates that some of these larvae had escaped infection in autumn. The absence of larval stage effects in pots may be related to the small soil volume (easier for nematodes to find their host) but also to the faster development of the larvae in pots. When EPNs are applied in autumn there are less eggs and/or small larvae present in pots than in the open field.

Results of spring application of EPNs are presented in Fig. 9.3B. The results of Hanula (1993) and Stimmann *et al.* (1985) are based on natural infestation with vine weevils. In these tests EPNs (*S. feltiae*, product Biosys; *S.carpocapsae* All strain, product Exhibit or Biosafe; *H. bacteriophora* C1 and HP88 strains; *H. heliothidis* and *S. feltiae*, undescribed strains) were applied in spring, when larvae were already large and/or in the pupal stage and soil temperatures were higher than 19°C. All the tested EPNs gave 60–80% control of these larval stages indicating that these circumstances avoid most problems related to larval stage. The excellent control by *S. carpocapsae* seems to be in contrast to the results of van Tol (1993b, 1996b) with the same strain/product in pots, but this is strongly related to the influence of soil temperature (see Fig. 9.2) and larval stage (Fig. 9.3B). Georgis and Poinar (1984) tested different larval stages for infection by releasing small (L1+L2) or larger (L3+L4) larvae in pots before EPN application. They found that for *S. carpocapsae* (Exhibit) very small larvae were only poorly infected (17%) while control of larger larvae was 72% (Fig. 9.3B). The same was found for *S. glaseri* but not for *H. bacteriophora* (= *H. heliothidis*).

9.4.1.5. Plant species

According to Stock *et al.* (1999) coniferous forests harbour the largest biodiversity of EPNs. Association with conifers is ancestral for the weevil species and was likely formed in the Jurassic period or earlier. When angiosperm plants appeared, multiple shifts to angiosperm host plants occurred, associated with increases in species diversity (Marvaldi *et al.*, 2002). The trophic relation between weevils and EPNs may thus have evolved in ancient times in relation to the specific host plants. A study of van Tol *et al.* (2001) showed that conifer roots (*Thuja occidentalis*), when damaged by vine weevil larvae, release compounds that attract *H. megidis* to the plant roots. Weevil-damaged roots were preferred to mechanically damaged roots indicating that these roots release very specific signals in response to weevil damage, thereby alerting EPNs. In contrast, Boff *et al.* (2001) found that mechanically damaged strawberry roots were preferred to weevil-damaged roots indicating no such relation. In this study the strain Nl-H-E87.3 of

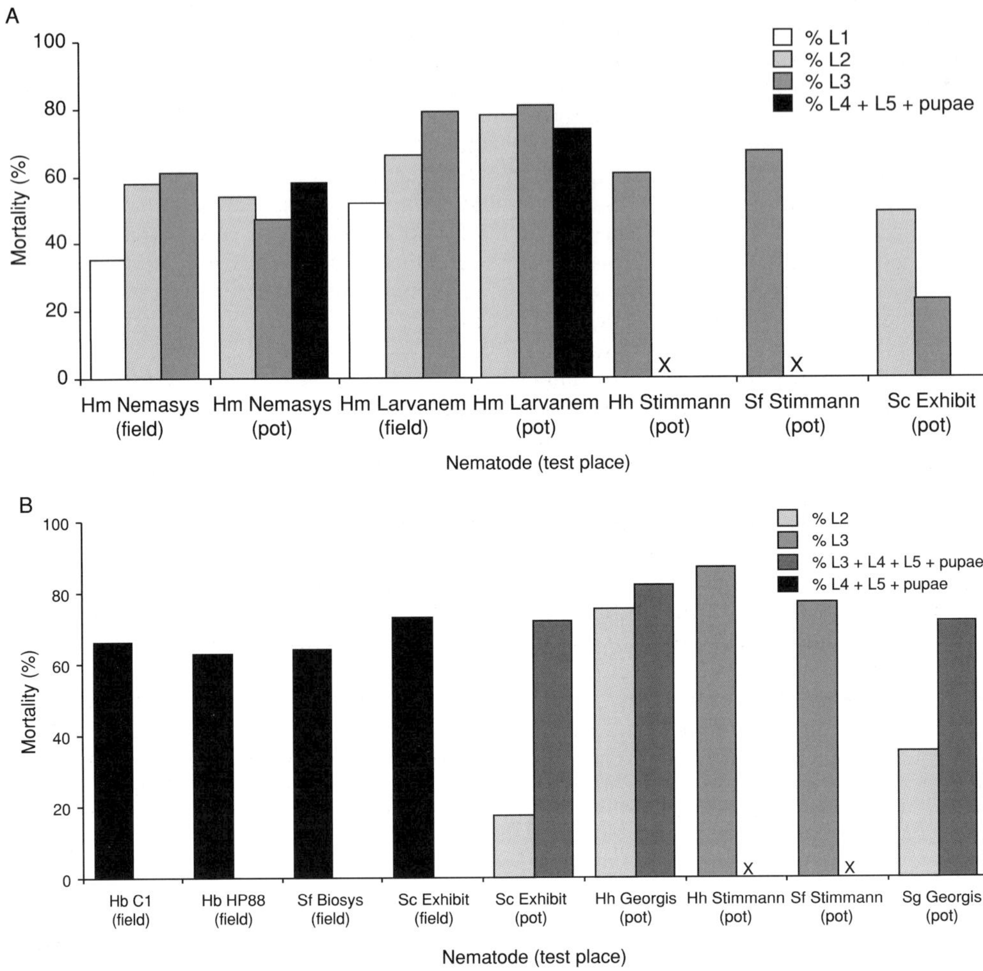

Fig. 9.3. Control of different larval stages of the vine weevil in pot and field tests with entomopathogenic nematodes (EPNs) applied in (A) autumn and (B) spring. Values present average results from tests with natural infestation or egg inoculation of the vine weevil (except for Sc Exhibit, Hh Georgis and Sg Georgis in pots in B, where either L1+L2 larvae or L3+L4 larvae are released before EPN application), and where soil temperature is equal or higher than 12°C and the dosage of EPNs applied in pot and field higher than 0.5×10^6 and 1.0×10^6nematodes/m^2 respectively. Hm = *Heterorhabditis megidis*; Hb = *H. bacteriophora*; Hh = *H. heliothidis*; Sf = *Steinernema feltiae*; Sc = *S. carpocapsae*; Sg = *S. glaseri*; X = no older larval stages present in control pots. (Data from Georgis and Poinar, 1984; Stimmann *et al.*, 1985; Hanula, 1993; van Tol, 1996b, 1998.)

H. megidis was used while van Tol *et al.* (2001) used the strain UK-H-211 of *H. megidis*. The pot and field results for both strains show that the UK-H-211 strain performed well in pot and field, while the Nl-H-E87.3 strain of *H. megidis* is only effective in pots and not in the field (Table 9.1; van Tol, 1996b, 1998). This indicates that the Nl-H-E87.3 strain is a random searcher and/or unsuccessful in olfactory orientation to host and plant-root odours and this may explain the contrasting results by Boff *et al.* (2001).

An additional explanation may come from another study of van Tol *et al.* (2004), which indicates that the commonly used strawberry cultivar (*Fragaria* × *ananassa*) is not a primary host plant of the vine weevil, and that the relation between the vine weevil and this American crossbreed of strawberry is very recent (around 1900). If EPN species and/or strains vary in their ability to find the host insect, and if they are better host finders in relation to the primary host plants that release specific alerting compounds, this may have consequences for the efficacy of the applied EPNs in different agricultural systems. To determine whether plant species and nematode species and/or strains have influence on the control we compared 17 different field studies performed with seven different plant species, two nematode species and two strains of *H. megidis* in both pot and field tests. We only compared results of tests with a trial set-up similar to the natural situation of the pest (egg inoculation timed according to the natural life cycle of the weevil), optimal soil temperature (> 12°C), nematode application rate (> 0.5×10^6 EPNs/1 l pot and > 1.0×10^6 EPNs/m^2 open field) and the same time of year (autumn). The results indicate that a plant–EPN interaction has an effect on weevil control (Fig. 9.4; van Tol, 1993a,b, 1994, 1996b, 1998; van Tol *et al.*, 1998; Fitters *et al.*, 2001). For the tested strain Nl-II-F85 of *H. megidis* no such effect was found, but for the UK-H-211 strain of *H. megidis* clearly plant effects were found in both pot and field tests. In the field, *Rhododendron* and *Cornus* experienced generally 20% lower control than *Fragaria*, *Thuja* or *Taxus*. Pot trials with *Primula* and *Waldsteinia* experienced 25–30% lower control of vine weevil larvae when compared with *Thuja* as a test plant. The field results were confirmed by a large-scale test on a field nursery with a heavy vine weevil infestation in large *Rhododendron* and *Taxus* plants. No control was achieved by application of *H. megidis* (UK-H-211) in the *Rhododendron* plants but 75% control was achieved in the *Taxus* field (van Tol, 1998). Results with *S. carpocapsae* (US-S-25) also revealed a plant effect. Control was more than 20% lower in potted *Waldsteinia* when compared with control in potted *Thuja*. A study by Gill *et al.* (2001) revealed that *H. bacteriophora* gave almost complete control in the perennials *Heuchera* and *Epimedium*. Comparing these results with the demonstrated results in Fig. 9.4 is, however, difficult because soil temperature, larval stages and application timing of the EPNs differed considerably. Probably the interaction is more complex than we suggest since plant species also have a

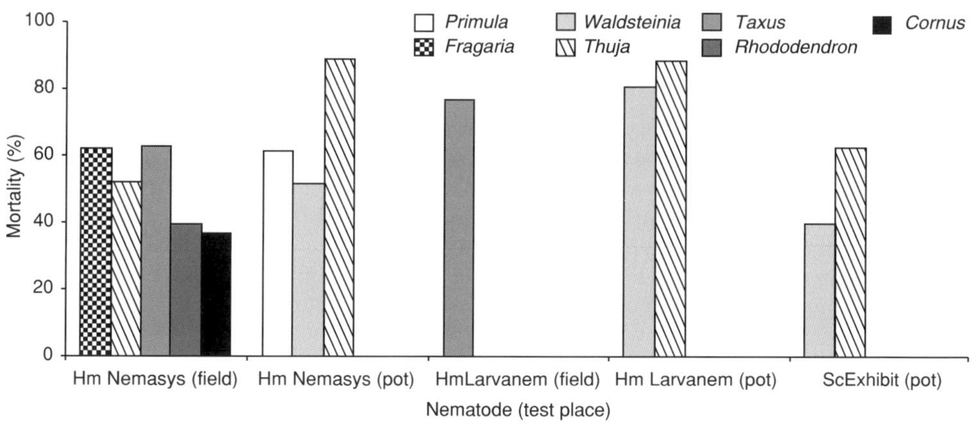

Fig. 9.4. Influence of plant species on vine weevil control with entomopathogenic nematodes (EPNs) applied in autumn. Values present average results from field tests with egg inoculation of the vine weevil and where soil temperature is equal or higher than 12°C and the dosage of EPNs applied in pot and field respectively higher than 0.5×10^6 and 1.0×10^6 nematodes/m^2. Hm = *Heterorhabditis megidis*; Sc = *Steinernema carpocapsae*. (Data used from van Tol, 1993a,b, 1994, 1996b, 1998 ; van Tol *et al.*, 1998; Fitters *et al.*, 2001.)

large influence on the development and natural mortality of the larvae (LaLone and Clarke, 1981; van Tol, 1996b, 1998; R.W.H.M. van Tol and R.L. Gwynn, unpublished data). Field studies and a study on a broad range of plant species in pots revealed that *Taxus* and *Thuja* have low larval mortality and faster larval development than in many other host plants like *Rhododendron* (van Tol, 1998; van Tol *et al.*, 2004). It is important for growers and advisers to consider these effects. In the case of *Rhododendron*, it is advisable to apply EPNs in spring when larvae are large enough to be infected. It is also important that larger instars and pupae are closer to the surface of the soil and, thereby, easier and faster to locate by the nematodes. A practical problem is, however, that spring application may be too late to prevent serious damage to the plants during autumn and winter.

9.4.2. Coleoptera – root-feeding white grubs

Results from various field and pot trials with different root-feeding white grubs (Coleoptera: Scarabaeidae) are summarized in Table 9.3. Mannion *et al.* (2001) evaluated *H. bacteriophora* (HP88) and *H. marelata* and found that both provided poor to moderate control of Japanese beetle grubs (*Popillia japonica*) when applied at rates of 5 billion/ha to the soil beneath several species of woody trees growing in nurseries. Wright *et al.* (1988) investigated the use of *S. feltiae*, *S. glaseri*, *H. bacteriophora* (= *H. heliothidis*) and *Heterorhabditis* sp. 'Holland strain' to control Japanese beetle and European chafer (*Rhizotrogus majalis*) grubs in potted Japanese yew (*Taxus cuspidata*). Control of Japanese beetle grubs with both *H. bacteriophora* and *Heterorhabditis* sp. 'Holland strain' ranged from about 60% to 90%. Control of Japanese beetle grubs with *S. feltiae* and *S. glaseri* was lower (0–86%) and more variable. All four nematode species provided poor to moderate (0–58%) control of European chafer grubs. Nielsen and Cowles (1998) evaluated *H. bacteriophora* for control of Japanese beetle, European chafer and Oriental beetle grubs (*Exomala orientalis*) in potted *Cotoneaster* and found that *H. bacteriophora* failed to control any of the three species. The relatively high control of beetle species with EPNs by Wright *et al.* (1988) may be related to the release of larvae before EPN application, while Mannion *et al.* (2001) and Nielsen and Cowles (1998) performed a more realistic field test with infestation of plants with beetle eggs several months before application of EPNs.

Mannion *et al.* (2001) found that doubling the dose of *H. marelata* from 2.5 to 5 billion/ha roughly doubled the mortality of Japanese beetle grubs (20–53%). For Japanese beetle grubs, increasing the rates of several tested nematode species from 46 to 385 nematodes/cm^2 generally increased levels of grub mortality (Wright *et al.*, 1988). For *Steinernema* spp. the highest levels of mortality occurred at the highest rates of nematodes. Control of Japanese beetle grubs with both *H. heliothidis* and *Heterorhabditis* sp. 'Holland strain' was generally greater than that observed with *S. feltiae* and *S. glaseri*. There was little change in grub mortality when the rate of nematodes applied exceeded 192 nematodes/cm^2. With respect to European chafer grubs, control by *S. feltiae* was poor and the rate of nematodes applied did not affect efficacy. *S. glaseri*, *H. heliothidis* and *Heterorhabditis* sp. 'Holland strain' all exhibited slightly greater levels of mortality when the rate of nematodes was increased from 46 to 92 nematodes/cm^2.

Control of beetle species with EPNs has improved in the last few years due to the discovery of more effective strains of *Heterorhabditis* and *Steinernema* (Grewal *et al.*, 2002, 2004; Lee *et al.*, 2002; Koppenhöfer and Fuzy, 2003). Many of these new strains have been tested in turfgrass (see Chapter 7, this volume) but their efficacy in nursery plants still needs to be demonstrated, and products for the growers have yet to be developed. The recent results with EPNs indicate that biocontrol of many beetle species may become as effective as for the vine weevil in the next few years.

Table 9.3. Efficacy of entomopathogenic nematode (EPN) species and strains applied for control of beetle larvae in ornamental trees, shrubs and perennials.

Insect species[a,b]	Plant species	EPN species (strain, product)	% control in pots (N)	Dosage	% control in the field (N)	Dosage	References
Root-feeding grubs							
Popillia japonica[a]	*Tilia*	*Heterorhabditis bacteriophora*	0 (1)	3×10^4 (/6.2 l pot)	0–65 (3)	$0.5 \times 10^6/m^2$	Nielsen and Cowles, 1998; Mannion *et al.*, 2001
	Gleditsia						
	Fraxinus	(HP88-Cruiser)					
	Prunus						
	Cercis						
	Cotoneaster						
P. japonica[a]	*Malus*	*H. marelata*	—	—	53 (1)	$0.5 \times 10^6/m^2$	Mannion *et al.*, 2001
P. japonica[b]	*Taxus*	*Steinernema feltiae*	28 (1)	$385/cm^2$	—	—	Wright *et al.*, 1988
P. japonica[b]	*Taxus*	*S. glaseri*	86 (1)	$385/cm^2$	—	—	Wright *et al.*, 1988
P. japonica[b]	*Taxus*	*H. heliothidis* sp. (Holland)	92 (1)	$192/cm^2$	—	—	Wright *et al.*, 1988
P. japonica[b]	*Taxus*	*Heterorhabditis* sp. (Holland)	92 (1)	$192/cm^2$	—	—	Wright *et al.*, 1988
Rhizotrogus majalis[b]	*Taxus*	*S. feltiae*	0 (1)	$385/cm^2$	—	—	Wright *et al.*, 1988
R. majalis[b]	*Taxus*	*S. glaseri*	58 (1)	$385/cm^2$	—	—	Wright *et al.*, 1988
R. majalis[b]	*Taxus*	*H. heliothidis*	45 (1)	$192/cm^2$	—	—	Wright *et al.*, 1988
R. majalis[b]	*Taxus*	*Heterorhabditis* sp. (Holland)	54 (1)	$192/cm^2$	—	—	Wright *et al.*, 1988
R. majalis[a]	*Cotoneaster*	*H. bacteriophora* (HP88, Cruiser)	0 (1)	3×10^4 (/6.2 l pot)	—	—	Nielsen and Cowles, 1998
Exomala orientalis[a]	*Cotoneaster*	*H. bacteriophora* (HP88, Cruiser)	0 (1)	3×10^4 (/6.2 l pot)	—	—	Nielsen and Cowles, 1998
Wood-boring beetles							
Scolytus scolytus[a]	*Ulmus* (bark)[c]	*S. carpocapsae* (DD-136)	—	—	15–40 (1)	$5 \times 10^3/m^2$	Finney and Walker, 1979
Anoplophora glabripennis[a]	*Salix* (galleries)[c]	*S. bibionis*	—	—	62 (1)	2×10^3/gallery	Qin *et al.*, 1988

continued

Table 9.3. *continued.* Efficacy of entomopathogenic nematode (EPN) species and strains applied for control of beetle larvae in ornamental trees, shrubs and perennials.

Insect species[a,b]	Plant species	EPN species (strain, product)	% control in pots (*N*)	Dosage	% control in the field (*N*)	Dosage	References
Leaf beetles							
Agelastica alni	*Alnus*	*S. feltiae*	28 (1)	$1 \times 10^5/m^2$	—	—	Tomalak, 2004
A. alni	*Alnus*	*H. megidis*	100 (1)	$1 \times 10^5/m^2$	—	—	Tomalak, 2004
Altica quercetorum	*Quercus*	*S. feltiae*	79 (1)	$1 \times 10^5/m^2$	—	—	Tomalak, 2004
A. quercetorum	*Quercus*	*H. megidis*	99 (1)	$1 \times 10^5/m^2$	—	—	Tomalak, 2004
Xanthogaleruca luteola[a]	*Ulmus* (soil)[c]	*S. carpocapsae* (All-strain)	—	—	50 (2)	8×10^5/tree	Kaya *et al.*, 1981
X. luteola[a]	*Ulmus* (leaf)[c]	*S. carpocapsae* (All-strain)	53 (2)	200/ml (until runoff)	67 (2)	8×10^3/ml (? ml/tree)	Kaya *et al.*, 1981

[a]Egg inoculation or natural infestation several months before EPN application.
[b]Inoculation with L3 larvae before EPN application.
[c]Application of EPNs on bark, into galleries, on leaves or on the soil to control the insect larvae.
Abbreviation: *N* = number of tests included.
Note: Results are from field tests with minimum dosage of nematodes applied for maximum attainable control.

9.4.3. Coleoptera – wood-boring beetles

Results from various field and pot trials with wood-boring beetles are summarized in Table 9.3. Finney and Walker (1979) evaluated *S. carpocapsae* (DD-136) applied to the bark of elm logs infested with larger European elm bark beetle (*Scolytus scolytus*) and found that emergence was reduced by 15–40%. Qin *et al.* (1988) injected *S. feltiae* into galleries of the Asian long-horned beetle (*Anoplophora glabripennis*) and found moderate levels of mortality (62%). Control of wood-boring beetles is not only ineffective with the currently available EPN strains, but there is also a need for improvement in formulations and application techniques to increase nematode efficacy and reduce the costs of labour.

9.4.4. Coleoptera – leaf beetles

Results from various field and pot trials with leaf beetles are summarized in Table 9.3. Kaya *et al.* (1981) obtained intermediate levels of control (41–74%) of larvae and pupae of the elm leaf beetle (*Xanthogaleruca luteola*) treated with *S. carpocapsae*. Nematodes were applied to leaves or soil beneath trees outdoors or to the foliage of elms infested with elm leaf beetle larvae on potted plants outdoors. In Petri dish studies adults were more susceptible to *S. carpocapsae* than larvae and pupae. On excised foliage this trend was reversed. In the field, mortality rates of larvae on foliage exceeded those of pupae in soil. In a semi-field test Tomalak (2004) revealed that *H. megidis* could eliminate the pupating leaf beetles *Altica quercetorum* (99% control) and *Agelastica alni* (100% control) in soil under the canopy of urban trees. *S. feltiae* caused only 28% control of larvae of *A. alni* but 79% control of *A. quercetorum*.

9.4.5. Lepidoptera – caterpillars boring roots, trunks and branches

Results from various trials with caterpillars boring roots, trunks and branches are summarized in Table 9.4. The larvae of the Iris moth, *Macronoctua onusta*, are important pests for several species of iris. *S. carpocapsae* applied to the soil and corms of *Iris* × *germanica* provided 100% control of this pest (Gill and Raupp, 1997). In the same study *H. bacteriophora* provided a reduction of 91%. Increasing the rate of *S. carpocapsae* applied to the soil from 77 to 154 nematodes/cm^2 did not improve the control of iris borer *M. onusta* on *Iris* × *germanica*. The lower rate of 77 nematodes/cm^2 was sufficient to provide 100% control of this pest (Gill and Raupp, 1997).

Larvae of many species of clearwing moths (Lepidoptera: Sessiidae) are wood borers and attack a wide variety of woody trees and shrubs in nurseries and landscapes. The standard approach for controlling moth borers includes application to bark of residual insecticides to kill newly hatched larvae that must chew through the insecticide barrier to reach sapwood. Application of *S. feltiae* directly into borer galleries with syringes has proven highly effective in controlling carpenterworms (*Holcocerus insularis*) in urban trees (Qin *et al.*, 1988). The same has been true for steinernematid nematodes (*S. feltiae* and *S. carpocapsae*) injected directly into galleries of clearwing borers *Synanthedon culiciformis* and *Paranthrene robiniae* in alder and birch, respectively, where control ranged from 86% to 93% (Kaya and Lindgren, 1983; Kaya and Brown, 1986). Nematodes applied to the bark of trees in these studies provided generally more variable and less effective control. *S. feltiae* applied on bark of *Alnus* for the control of *S. culiciformis* provided 81% control, while application on bark of *Platanus* to control *S. resplendens* gave only 13% control. Little is known about the relationship between doses and efficacy of nematodes for control of clearwing borers (*Sesiidae* spp.) as few studies have examined the use of multiple rates under similar conditions. In one study the lower application rate of 77.5 *S. carpocapsae*/cm^2 of bark provided numerically greater levels of control than a higher rate of 145 *S. carpocapsae*/cm^2 for the banded ash clearwing borer, *Podosesia aureocincta* (Gill *et al.*, 1994).

Table 9.4. Efficacy of entomopathogenic nematode (EPN) species and strains applied for the control of Lepidoptera and Hymenoptera species in ornamental trees, shrubs and perennials.

Insect species[a,b]	Plant species[c]	EPN species (strain, product)	% control in pots (*N*)	Dosage	% control in the field (*N*)	Dosage	References
Lepidoptera – caterpillars boring in roots, trunks and branches							
Holocerus insularis[a]	*Ligustrum* (galleries)	*Steinernema feltiae*	—	—	99 (1)	2×10^3/gallery	Qin *et al.*, 1988
Macronoctua onusta[b]	*Iris* (soil)	*Heterorhabditis bacteriophora* (Lawn patrol)	—	—	91 (1)	46/cm^2	Gill and Raupp, 1997
M. onusta[b]	*Iris* (soil)	*S. carpocapsae* (Vector)	—	—	100 (1)	77/cm^2	Gill and Raupp, 1997
Paranthrene robiniae[a]	*Betula* (galleries)	*S. carpocapsae* (All-strain)	—	—	89 (3)	7×10^5/gallery	Kaya and Lindgren, 1983
Podosesia aureocincta[a]	*Fraxinus* (bark)	*S. carpocapsae* (S25, Exhibit)	—	—	17–70 (5)	77.5/cm^2	Gill *et al.*, 1994; Smith-Fiola *et al.*, 1996
P. aureocincta[a]	*Fraxinus* (bark)	*S. feltiae* (S27)	—	—	0–74 (4)	77.5/cm^2	Gill *et al.*, 1994; Smith-Fiola *et al.*, 1996
P. aureocincta[a]	*Fraxinus* (bark)	*S. glaseri* (# 326)	—	—	54 (1)	3.9×10^3/cm^2	Smith-Fiola *et al.*, 1996
Rhyacionia frustana[a]	*Pinus* (leaf)	*S. carpocapsae* (DD-136)	—	—	15–35 (2)	4×10^3/ml (until runoff)	Nash and Fox, 1969
Synanthedon culiciformis[a]	*Alnus* (galleries)	*S. feltiae*	—	—	90 (2)	1.8×10^4/plant	Kaya and Brown, 1986
S. culiciformis[a]	*Alnus* (bark)	*S. feltiae*	—	—	81 (2)	6.5×10^6/plant	Kaya and Brown, 1986
S. exitiosa[a]	*Prunus* (bark)	*S. carpocapsae* (S25, Biosafe)	—	—	66 (1)	198/cm^2	Gill *et al.*, 1992
S. resplendens[a]	*Platanus* (bark)	*S. feltiae*	—	—	13 (1)	11.3×10^6/plant	Kaya and Brown, 1986
S. resplendens[a]	*Platanus* (bark)	*S. bibionis*	—	—	61 (1)	8.6×10^6/plant	Kaya and Brown, 1986
S. scitula[a]	*Cornus* (bark)	*S. carpocapsae* (S25, Biosafe)	—	—	85 (1)	77.5/cm^2	Davidson *et al.*, 1992
Lepidoptera – leaf-eating caterpillars							
Choristoneura occidentalis[a]	*Abies* (leaf)	*S. carpocapsae* (All-strain)	—	—	0–52 (3)	4×10^5/ml (until runoff)	Kaya and Reardon, 1986; Kaya *et al.*, 1981
Hyphantria cunea[a]	*Prunus* (leaf)	*S. feltiae*	—	—	98 (3)	1×10^4/ml	Yamanaka *et al.*, 1986

Lymantria dispar[a]	*Quercus* (refuge band)	*S. feltiae*	—	—	0–74 (4)	538/cm^2	Reardon *et al.*, 1986
L. dispar[a]	*Quercus* (refuge band)	*S. bibionis*	—	—	0–74 (2)	1072/cm^2	Reardon *et al.*, 1986
Operophtera spp.		*H. megidis*	95 (1)	1×10^5/m^2	—	—	Tomalak, 2003
Operophtera spp.		*S. feltiae*	27 (1)	1×10^5/m^2	—	—	Tomalak, 2003
Thyridopterix ephemeraeformis[b]	*Thuja* (leaf)	*S. carpocapsae*	93 (2)	200/cm^2	—	—	Gill and Raupp, 1994
T. ephemeraeformis[b]	*Cupresso-cyparis* (leaf)	*S. feltiae*	85 (2)	200/cm^2	—	—	Gill and Raupp, 1994
T. ephemeraeformis[b]	*Cupresso-cyparis* (leaf)	*S. feltiae*	49 (4)	200/cm^2	—	—	Gill and Raupp, 1994
Hymenoptera – leaf-eating sawfly larvae							
Cephalacia lariciphila[b]	*Larix* (leaf)	*S. feltiae*	—	—	22 (2)	2×10^4/branch	Georgis and Hague, 1988
C. lariciphila[a]	*Larix* (soil)	*S. feltiae*	—	—	61 (1)	200/cm^2	Georgis and Hague, 1988

[a]Natural infestation.
[b]Artificial inoculation with larvae before EPN application.
[c]Application of EPNs on bark, on refuge bands, into galleries, on leaves or on the soil to control the insect larvae.
Abbreviation: N = number of tests included.
Note: Results are from field tests with minimum dosage of nematodes applied for maximum attainable control.

Davidson *et al.* (1992) achieved 85% control of the dogwood borer, *S. scitula*, on dogwoods in a commercial nursery using *S. carpocapsae* applied to bark. The same strain of *S. carpocapsae* applied to the bark of *Prunus* gave 66% control of *S. exitiosa* (Gill *et al.*, 1992). Attempts to control the banded ash clearwing borer, *P. aureocincta*, with applications of *S. carpocapsae*, *S. feltiae* and *S. glaseri* to the bark of ash trees in landscapes have proven far more variable and generally less effective with control ranging from 0% to 74% (Gill *et al.*, 1994; Smith-Fiola *et al.*, 1996). Kaya and Brown (1986) found bark applications of *S. feltiae* to provide poor to moderate reductions (13–61%) of *S. resplendens* on plane trees (*Platanus*). Some authors suggested that wetting the bark before the application of nematodes might improve performance but this idea has not been strongly supported by the data (Kaya and Brown, 1986; Smith-Fiola *et al.*, 1996). Kaya and Brown (1986) found that different rates of application of *S. feltiae* to the bark and borer galleries did not differ greatly in controlling *S. resplendens* on plane trees (*Platanus*).

Conifer-boring caterpillars such as the Nantucket pine tip moth, *Rhyacionia frustrana*, appear to be relatively resistant to applications of *S. carpocapsae*. The best level of control using foliar applications of *S. carpocapsae* was 35% reduction in the first generation of tip moths under field conditions (Nash and Fox, 1969).

Smith-Fiola *et al.* (1996) found little difference in mortality of early or late instar larvae of the banded ash clearwing borer, *P. aureocincta*, with applications of *S. carpocapsae* and *S. feltiae*. Early summer applications directed at late instar larvae resulted in mortality ranging from 0% to 26% and autumn applications directed at early instar larvae killed 12–17% of the larvae. Kaya and Brown (1986) found that bark applications of *S. feltiae* directed at large larvae of the clearwing borer, *S. resplendens*, provided 61% control while autumn applications directed at early instar larvae provided only 10% control.

9.4.6. Lepidoptera – leaf-eating caterpillars

Results from various trials with leaf-eating caterpillars are summarized in Table 9.4. Rapid desiccation of nematodes applied to leaf surfaces is generally believed to limit their usefulness as control agents for leaf-feeding herbivores (Begley, 1990). However, under conditions of high humidity and protection against sunlight, nematodes have proven effective in controlling some leaf-eating herbivores. Too much moisture may be as big a problem as inadequate moisture. Rain may wash nematodes from leaf surfaces and rainfall near the time of application has been implicated in reduced levels of efficacy for control of spruce budworm (Kaya *et al.*, 1981), larch sawfly (Georgis and Hague, 1988) and bagworms (Gill and Raupp, 1994).

In a nursery setting, Gill and Raupp (1994) obtained moderate (41%) to very high (100%) levels of control of the bagworm, *Thyridopteryx ephemeraeformis*, with *S. carpocapsae* and *S. feltiae* applied to the foliage of potted evergreens. Control with *S. feltiae* was on average lower (49%) than with *S. carpocapsae* (85–93%). In landscape settings, Yamanaka *et al.* (1986) achieved very satisfactory levels of control (95–100% at highest rates) of fall webworm, *Hyphantria cunea*, with *S. feltiae*. Kaya and Reardon (1982) evaluated *S. carpocapsae* as a control agent for spruce budworm, *Choristoneura occidentalis*, on foliage of spruce and found levels of control to be relatively poor (0–52%). Attempts to control gypsy moth, *Lymantria dispar*, larvae under larval refuge bands with *S. feltiae* were variable and generally unsatisfactory, with control ranging from 0% to 74% (Reardon *et al.*, 1986). Larvae of the winter moths *Operophtera brumata* and *O. fagata* descending to the soil for pupation were controlled effectively (95%) by *H. megidis* while *S. feltiae* provided only 27% control (Tomalak, 2003).

Yamanaka *et al.* (1986) demonstrated dramatic improvement in mortality of fall webworm, *H. cunea*, with increased rates of *S. feltiae*. At the lowest rate of 200 nematodes/ml mortality was nil. Mortality increased to 98% when the concentration of nematodes increased to 10,000/ml. In evaluating *S. carpocapsae* as a control agent for spruce budworm, *C. occidentalis*, Kaya and Reardon (1982) found no relationship between nematode dose and mortality of budworm larvae. No dose effect was found for spruce budworm pupae.

Two trials evaluating *S. carpocapsae* as a control agent for spruce budworm, *C. occidentalis*, allow for inferences to be made regarding the effect of air temperature on nematode efficacy. In both cases Kaya *et al.* (1981) and Kaya and Reardon (1982) suggested that part of the reason for poor performance of the nematode was attributable to the fact that temperatures in the treatment areas were low. They ranged from 1°C to 10°C in one study and from 4°C to 13°C in the other. While these temperatures may have enhanced nematode survival, they likely reduced infectivity.

Gill and Raupp (1994) examined the efficacy of *S. carpocapsae* and *S. feltiae* to control different larval instars of the bagworm, *T. ephemeraeformis*. The highest levels of mortality were observed when nematodes were applied to middle instar (70–94%) compared to late instar (41–58%) larvae. Control of the leaf beetles, *O. brumata* and *O. fagata*, was successful for larvae descending to the soil for pupation, but pupal stages were not infected by any of the tested nematode species (Tomalak, 2003).

9.4.7. Hymenoptera – leaf-eating sawfly larvae

Control of the foliar-feeding web-spinning larch sawfly, *Cephalacia lariciphila*, larvae with *S. feltiae* applied to foliage or pupation sites in the soil around larch proved variable, with control ranging from 2% to 61% (Table 9.4). Control of larvae and pupae in the soil was relatively successful (61%) when compared to control of larvae on the leaves (22%). Infection rates of larvae with *S. feltiae* increased from 3% to 15% for first and second instars, and from 6% to 29% for third and fourth instars as nematode concentrations increased from 5000 to 20,000/branch (Georgis and Hague, 1988). Soil application to control prepupae of the larch sawfly showed an increase of infection rate from 9% to 61% as nematode concentrations increased from 20/cm^2 to 200/cm^2 soil surface. In the soil, sawfly pupae had infection rates of 2–17%. Georgis and Hague (1988) noted that lower soil temperatures (below 10°C) might have reduced infection rates of pupae in these studies.

9.5. Essential Components for Field Efficacy

The essential components can be grouped into three categories: growers' perception, economics and efficacy of the product. These three factors together determine the decision growers make concerning the use of pest control means and strategies. Effectiveness is considered to be the most important factor in deciding to use non-chemical control measures by growers. The cost of the EPN product is relatively high compared with chemical alternatives but this is not considered to be a key factor for non-adoption of this biocontrol agent according to surveys amongst growers. Limiting factors are the high costs of labour for monitoring the pest and applying the EPNs. The complexity of the systems approach (monitoring weevils, yes/no control of adult weevils, when/where to apply EPNs), lack of proper protocols and professional support, and absent or unsatisfactory visualization of the efficacy are some of the other factors that influence grower decisions. The increase of natural control of other pests because of reduced use of chemicals to control the weevils lowers the total costs for pest control in nurseries. Economic limits to control above-ground pests with EPNs are related to the variable efficacy mainly due to poor formulations and application techniques, as well as the high costs of labour to apply the nematodes. Increasing awareness of the environmental problems of many agrochemicals has led to the ban of more and more chemicals, which are not replaced by other chemicals. Also, limited effectiveness of the available chemicals to control several soil-borne pests favours the use of EPNs in nursery and tree growing.

Our analysis of field trials in nursery and tree growing shows that quality of the EPN products and variation in efficacy between the available nematode species and strains are important factors causing variable results under field conditions for grub control. For other pests, variable efficacy, application techniques and product formulations currently limit the introduction in practice. Detailed studies with vine weevil reveal that beside nematode species and strains, timing of EPN application causes the largest differences in field efficacy. Knowledge of the insect life cycle on location as well as the ratio of small and larger larvae shortly before application of EPNs may serve as important indicators for control that can be expected in the field. The tritrophic relation between plant species, insect larvae and EPNs has a strong influence on the control achieved. Larval development, natural mortality, threshold damage level (number of larvae per plant that cause economic growth reduction or cause the death of the plant) and nematode searching behaviour are also closely related to the plant species. The knowledge and influence of this system in relation to efficacy of the applied EPN strains are still poorly understood, but the results from field tests show that these problems are highly underestimated. Soil temperature is currently not limiting for autumn control although more cold-active nematodes would give growers more time in autumn to apply the product. Efficacy of most EPN products does not result in further rise in soil temperature above 12°C or higher rates of nematode application. More general problems are related to the low persistence of activity in soil after application and the invisibility of the efficacy after application.

Several new nematode species and strains with promising efficacy against beetle pests have been found recently. Much is yet unknown about the key factors that influence field efficacy of nematodes. The experience with the use of nematodes for the control of vine weevil may serve as a model for future use of nematodes against other pests.

Although several attempts have been made to control above-ground pest insects with EPNs, only a few have been successful, and in most cases these involve control of the soil stage of these insects or injection of nematodes in protected insect galleries in trees. Except for the control of soil stages, successful applications involve laborious treatments and regulation of humidity on leaves and bark to create circumstances that allow the nematodes to survive and infect the target insect. Control of foliar-feeding insects requires modifications of the environment in such a way that the nematodes can survive and find and infect the insect. Further, application techniques are still not practical enough for large-scale introduction of nematodes in some systems. Competing with well adapted natural enemies of these pests with little or no need to modify the environment may turn out to be a limiting factor for EPN introduction in practice. Above-ground pests with a soil life stage (e.g. thrips, iris borer, and certain gall midges, leaf beetles and lepidopterans) are pests for which effective control with EPNs may be achieved.

9.6. Recommendations and Future Research

This chapter has summarized the current status of EPN field application against pest insects in ornamental trees, shrubs and perennials. Vine weevil is one of the first pest insects where control with EPNs has become a practical solution that is even more effective than chemical control. The many years of field experience with EPNs in practice to control vine weevil should serve as a model for control of other insect pests. The analysed key factors for efficacy and perhaps, more importantly, the economics and grower perception in acceptance and integration of EPNs in the total pest management strategy indicate the importance of early involvement of growers in field testing. Testing of products in field settings should be a standard procedure to learn more about the possibilities, variability and limits of the EPN products. Development of proper protocols for application as well as a systems approach (monitoring, timing of EPN application, control of efficacy, involvement of grower in monitoring and decision making, integration in total pest control programme) are necessary to prevent failures and make growers confident with the new approach.

Applied and fundamental research should focus more on the ecology of EPNs. Understanding the behaviour of a limited number of EPN strains in relation to insect and plant species increases our knowledge about the possibilities and limits of EPNs in practice. Understanding the essential factors that influence efficacy may help provide growers with reliable solutions for control of their pests with EPNs in the future.

In conclusion, research on above-ground pest control with EPNs should consider the practicability of this control strategy. Efficacy, competition with other biocontrol agents and the labour involved with application and formulation of the nematodes may become practical hurdles that will limit the use in practice. Above-ground pests with a soil life stage have better prospects for control with EPNs than pests without a soil life stage.

The potential for EPNs to control pests in high-valued ornamental crops is excellent. Except for technical solutions and more and better adapted EPN strains, we should pay more attention to the concerns of growers, and provide a structure and instruments to support growers in using the EPNs and make them confident with the product. Growing markets for new products are usually those that have been accepted and considered reliable and easy to use by the end-users.

References

Backhaus, G.F. (1994) Biological control of *Otiorhynchus sulcatus* F. by use of entomopathogenic nematodes of the genus *Heterorhabditis*. *Acta Horticulturae* 364, 131–142.

Begley, J.W. (1990) Efficacy against insects in habitats other than soil. In: Gaugler, R. and Kaya, H.K. (eds) *Entomopathogenic Nematodes in Biological Control*. CRC Press, Boca Raton, Florida, pp. 215–231.

Boff, M.I.C., van Tol, R.W.H.M. and Smits, P.H. (2001) Behavioural response of *Heterorhabditis megidis* towards plant roots and insect larvae. *Biocontrol* 47, 67–83.

Cornell, C.F. (1994) Integrated pest management – implementation in the nursery industry. In: Heuser, C.W. Jrand Heuser, P. (eds) *Recycling and Resource Conservation – A Reference Guide for the Nursery and Landscape Industries*. Pennsylvania Nurserymans Association, Harrisburg, Pennsylvania, pp. 75–84.

Davidson, J.A., Gill, S. and Raupp, M.J. (1992) Controlling clearwing moths with entomopathogenic nematodes: the dogwood borer case study. *Journal of Arboriculture* 18, 81–84.

Finney, J.R. and Walker, C. (1979) Assessment of a field trial using the DD-136 strain of *Neoaplectana* sp. for the control of *Scolytus scolytus*. *Journal of Invertebrate Pathology* 33, 239–241.

Fitters, P.F.L., Dunne, R. and Griffin, C.T. (2001) Vine weevil control in Ireland with entomopathogenic nematodes: optimal time of application. *Irish Journal of Agricultural and Food Research* 40, 199–213.

Georgis, R. and Hague, N.G.M. (1988) Field evaluation of *Steinernema feltiae* against the web-spinning larch sawfly, *Cephalcia lariciphila*. *Journal of Nematology* 20, 317–320.

Georgis, R. and Poinar, G.O. Jr (1984) Greenhouse control of black vine weevil *Otiorhynchus sulcatus* (Coleoptera: Curculionidae) by Heterorhabditid and Steinernematid nematodes. *Environmental Entomology* 13, 1138–1140.

Gill, S.A. and Raupp, M.J. (1994) Using entomopathogenic nematodes, conventional and biorational pesticides for controlling bagworm. *Journal of Arboriculture* 20, 318–322.

Gill, S.A. and Raupp, M.J. (1997) Evaluation of biological and chemical applications for control of Iris borer. *Journal of Environmental Horticulture* 15, 108–110.

Gill, S., Davidson, J.A. and Raupp, M.J. (1992) Control of the peach tree borer, *Synanthedon exitiosa* (Lepidoptera), in a landscape setting utilizing entomopathogenic nematodes. *Journal of Arboriculture* 18, 184–187.

Gill, S., Davidson, J.A., MacLachlan, W. and Potts, W. (1994) Controlling banded ash clearwing moth borer using entomopathogenic nematodes. *Journal of Arboriculture* 20, 146–149.

Gill, S., Lutz, J., Shrewsbury, P. and Raupp, M.J. (2001) Evaluation of biological and chemical control methods for the black vine weevil, *Otiorhynchus sulcatus* (F.) (Coleoptera: Curculionidae) in container grown perennials. *Journal of Environmental Horticulture* 19, 166–170.

Grewal, P.S., Grewal, S.K., Malik, V.S. and Klein, M.G. (2002) Differences in the susceptibility of introduced and native white grub species to entomopathogenic nematodes from various geographic localities. *Biological Control* 24, 230–237.

Grewal, P.S., Power, K.T., Grewal, S.K., Suggars, A. and Haupricht, S. (2004) Enhanced consistency in biological control of white grubs (Coleoptera: Scarabaeidae) with new strains of entomopathogenic nematodes. *Biological Control* 30, 73–82.

Hanula, J.L. (1993) Vertical distribution of black vine weevil (Coleoptera: Curculionidae) immatures and infection by entomogenous nematodes in soil columns and field soil. *Journal of Economic Entomology* 86, 340–347.

Heungens, A. and Buysse, G. (1992) Bestrijding van lapsnuitkeverlarven in de azaleateelt met *Heterorhabditis*-nematoden en de insekticiden endosulfan en etrimfos. *Parasitica* 48(2), 65–72.

Hodges, A. (1991) Business analysis of woody ornamental container and field nurseries in Florida, 1989. *Florida Cooperative Extension Service* IFAS Circular 953.

Hodges, A. and Hull, D. (1991) Business analysis of foliage plant nurseries in Central and South Florida, 1989. *Florida Cooperative Extension Service* IFAS Circular 953.

Hudson, W.G., Garber, M.P., Oetting, R.D., Mizell, R.F., Chase, A.R. and Bondari, K. (1996) Pest management in the United States greenhouse and nursery industry: V. Insect and mite control. *HortTechnology* 6, 216–221.

Kaya, H.K. and Brown, L.R. (1986) Field application of entomogenous nematodes for control of clear-wing moth borers in alder and sycamore trees. *Journal of Arboriculture* 12, 150–154.

Kaya, H.K. and Lindgren, J.E. (1983) Parasitic nematode controls western poplar borer. *California Agriculture* 37, 31–32.

Kaya, H.K. and Reardon, R.C. (1982) Evaluation of *Neoaplectana carpocapsae* for biological control of the western spruce budworm, *Choristoneura occidentalis*: ineffectiveness and persistence of tank mixes. *Journal of Nematology* 14, 595–597.

Kaya, H.K., Hara, A.H. and Reardon, R.C. (1981) Laboratory and field evaluations of *Neoaplectana carpocapsae* (Rhabditida: Steinernematidae) against the elm leaf beetle (Coleoptera: Chrysomelidae) and the western spruce budworm (Lepidoptera: Tortricidae). *Canadian Entomologist* 113, 787–793.

Koppenhöfer, A.M. and Fuzy, E.M. (2003) *Steinernema scarabaei* for the control of white grubs. *Biological Control* 28, 47–59.

LaLone, R.S. and Clarke, R.G. (1981) Larval development of *Otiorhynchus sulcatus* (Coleoptera: Curculionidae) and effects of larval density on larval mortality and injury to rhododendron. *Environmental Entomology* 10, 190–191.

Lee, D.W., Choo, H.Y., Kaya, H.K., Lee, S.M., Smitley, D.R., Shin, H.K. and Park, C.G. (2002) Laboratory and field evaluation of Korean entomopathogenic nematode isolates against the Oriental beetle *Exomala orientalis* (Coleoptera: Scarabaeidae). *Journal of Economic Entomology* 95, 918–926.

Mannion, C.M., McLane, W., Klein, M.G., Moyseenko, J., Oliver, J.B. and Cowan, D. (2001) Management of early-instar Japanese beetle (Coleoptera: Scarabaeidae) in field-grown nursery crops. *Journal of Economic Entomology* 94, 1151–1161.

Marvaldi, A.E., Sequeira, A.S., O'Brien, C.W. and Farrell, B.D. (2002) Molecular and morphological phylogenetics of weevils (Coleoptera, Curculionidae): do niche shifts accompany diversification? *Systematic Biology* 51, 761–785.

Miduturi, J.S., De Clerq, R. and De Grisse, A. (1994) Greenhouse and field control of black vine weevil *Otiorhynchus sulcatus* F., with *Steinernema carpocapsae* and *Heterorhabditis* sp. *IOBC/WPRS Bulletin* 17, 148–154.

Nash, R.F. and Fox, R.C. (1969) Field control of the Nantucket pine tip moth by the nematode DD-136. *Journal of Economic Entomology* 62, 660–663.

Neubauer, C. (1997) Mit Nematoden gegen Dickmaulrüssler: Bodentemperatur und Qualität der Nützlinge beeinflussen Erfolg. *Gärtnerbörse* 15, 854–857.

Nielsen, D.G. and Cowles, R.S. (1998) Preventing white grub infestation in container-grown nursery stock. *Journal of Environmental Horticulture* 16, 202–207.

Qin, X., Kao, R., Yang, H. and Zhang, G. (1988) Study on application of entomopathogenic nematodes of *Steinernema bibionis* and *S. fel-*

tiae to control *Anoplophora glabripennis* and *Holcocerus insularis*. *Forest Research* 1, 179–185.

Reardon, R.C., Kaya, H.K., Fusco, R.A. and Lewis, F.B. (1986) Evaluation *of Steinernema feltiae* and *S. bibionis* (Rhabditida: Steinernematidae) for suppression of *Lymantria dispar* (Lepidoptera: Lymantriidae) in Pennsylvania, USA. *Agriculture, Ecosystems, and Environment* 15, 1–9.

Smith-Fiola, D.C., Gill, S.A. and Way, R.G. (1996) Evaluation of entomopathogenic nematodes as biological control against the banded ash clearwing borer. *Journal of Environmental Horticulture* 14, 67–71.

Stimmann, M.W., Kaya, H.K., Burlando, T.M. and Studdert, J.P. (1985) Black vine weevil management in nursery plants. *California Agriculture* Jan/Feb, 25–26.

Stock, S.P., Pryor, B.M. and Kaya, H.K. (1999) Distribution of entomopathogenic nematodes (Steinernematidae, Heterorhabditidae) in natural habitats in California. *Biodiversity and Conservation* 8, 339–345.

Thomas, W.A. (1996) Potential economic effects of selected pesticides. In: Garber, M.P., Hudson, W.G., Norcini, J.G., Thomas, W.A., Jones, R.K. and Bondari, K. (eds) *Biological and Economic Assessment of Pest Management in the United States Greenhouse and Nursery Industry.* NAPIAP Report, No. I-CA-96, Cooperative Extension Service, The University of Georgia, College of Agricultural and Environmental Sciences, Athens, Georgia, pp. 131.

Tomalak, M. (2003) Biocontrol potential of entomopathogenic nematodes against winter moths (*Operophtera brumata* and *O. fagata*) (Lepidoptera: Geometridae) infesting urban trees. *Biocontrol Science and Technology* 13, 517–527.

Tomalak, M. (2004) Infectivity of entomopathogenic nematodes to soil-dwelling developmental stages of the tree leaf beetles *Altica quercetorum* and *Agelastica alni*. *Entomologia Experimentalis et Applicata* 110, 125–133.

Van der Horst, M.J. and van Tol, R.W.H.M. (1995) Integrated Pest Management in nursery stock in the Netherlands. *Mededelingen Faculteit Landbouwwetenschappen Universiteit Gent* 60/3a, 759–762.

van Tol, R.W.H.M. (1993a) Control of the black vine weevil (*Otiorhynchus sulcatus*) with different isolates of *Heterorhabditis* sp. and *Metarhizium anisopliae* in nursery stock. *Proceedings Experimental and Applied Entomology, NEV Amsterdam* 4, 181–186.

van Tol, R.W.H.M. (1993b) Efficacy of control of the black vine weevil (*Otiorhynchus sulcatus*) with strains of *Heterorhabditis* sp., *Steinernema* sp. and the fungus *Metarhizium anisopliae* in nursery stock. *Mededelingen Faculteit Landbouwwetenschappen Universiteit Gent* 58/2a, 461–467.

van Tol, R.W.H.M. (1994) Influence of temperature on the control of the black vine weevil with strains of some insect-parasitic nematodes. *IOBC/WPRS Bulletin* 17, 116–119.

van Tol, R.W.H.M. (1996a) A strategy for control of black vine weevil (*Otiorhynchus sulcatus*) in an Integrated Pest Management programme in nursery stock. *Mitteilungen aus der Biologischen Bundesanstalt für Land-und Forstwirtschaft* 316, 76–80.

van Tol, R.W.H.M. (1996b) Prospects for biological control of black vine weevil (*Otiorhynchus sulcatus*) in nursery stock. *Mitteilungen aus der Biologischen Bundesanstalt für Land-und Forstwirtschaft* 316, 69–75.

van Tol, R.W.H.M. (1998) Biological and chemical control of the vine weevil (*Otiorhynchus sulcatus*) in nursery stock. *BPO Report Vine Weevil 1991–1997.*

van Tol, R.W.H.M., Van Bezooijen, J. and Ketelaars, T.A.C.M. (1998) Searching behaviour of entomopathogenic nematodes: roots and soil temperature determine success of black vine weevil (*Otiorhynchus sulcatus*) control. *IOBC/WPRS Bulletin Insect Pathogens and Insect Parasitic Nematodes* 21, 187–191.

van Tol, R.W.H.M., Van der Sommen, A.T.C., Boff, M.I.C., Van Bezooijen, J., Sabelis, M.W. and Smits, P.H. (2001) Plants protect their roots by alerting the enemies of grubs. *Ecology Letters* 4, 292–294.

van Tol, I, R.W.H.M., Van Dijk, N. and Sabelis, M.W. (2004) Host plant preference and performance of the vine weevil *Otiorhynchus sulcatus*. *Agricultural and Forest Entomology* 6, 267–278.

Von Reibnitz, C. and Backhaus, G.F. (1994) Untersuchungen zum Bedarf biologischer Pflanzenschutzverfahren gegen *Otiorhynchus* spp. im Gartenbau und zur Kostenstruktur der Produktionsverfahren für entomopathogene Nematoden (*Heterorhabditis*). *Gartenbauwissenschaft* 59, 199–206.

Willmott, D.M., Hart, A.J., Long, S.J., Edmondson, R.N. and Richardson, P.N. (2002) Use of a cold-active entomopathogenic nematode *Steinernema kraussei* to control overwintering larvae of the black vine weevil *Otiorhynchus sulcatus* (Coleoptera: Curculionidae) in outdoor strawberry plants. *Nematology* 4, 925–932.

Wright, R.J., Villani, M.G. and Agudelo-Silva, F. (1988) Steinernematid and Heterorhabditid nematodes for control of larval European chafers and Japanese beetles (Coleoptera: Scarabaeidae) in potted yew. *Journal of Economic Entomology* 81, 152–157.

Yamanaka, S., Seta, K. and Yasuda, M. (1986) Evaluation of the use of entomogenous nematode, *Steinernema feltiae* (str. Mexican) for the biological control of the fall webworm *Hyphantria cunea* (Lepidoptera: Arctiidae). *Japanese Journal of Nematology* 16, 26–31.

10 Mushroom Applications

S. Jess,[1,2] H. Schweizer[2] and M. Kilpatrick[3]

[1]*Applied Plant Science Division, Department of Agriculture and Rural Development for Northern Ireland, Newforge Lane, Belfast BT9 5PX, UK;* [2]*Department of Applied Plant Science, Queen's University of Belfast, Newforge Lane, Belfast BT9 5PX, UK;* [3]*Applied Plant Science Division, Department of Agriculture and Rural Development for Northern Ireland, NIHPBS Loughgall, County Armagh, BT61 8JB, UK*

10.1. Introduction

Mushroom production is a significant component of the horticultural industry. During 2003, the estimated output value of global mushroom production exceeded US$7 billion at the world market price (Anon, 2004). In the UK it was estimated that insect pest control, particularly targeted at two

Dipteran families, Sciaridae and Phoridae, accounted for 2% of the total farmgate value of the crop (White, 1995). Despite this expenditure, the same author estimated that a further 4% of the yield was lost to insect pests.

Agaricus bisporus (Lange) Imbach is the most commonly cultivated edible fungus, accounting for approximately 32% of world production (Chang, 1999). Considering the economic importance of *A. bisporus* and of its two most important insect pest groups, sciarids and phorids, the discussion in this chapter is limited to the potential of entomopathogenic nematodes (EPNs) in this crop pest system.

10.2. Mushroom Cultivation Processes

Fungal growth is strongly affected by environmental conditions and the requirements of the cultivated mushroom are such that a diversity of cultivation systems have been developed. This diversity is primarily influenced by geographic and socio-economic factors in the different regions in which mushrooms are produced (Gaze, 1985). Despite differences in the housing and containers used for cultivation, the principal methods associated with crop production, management and protection in growing systems are fundamentally similar and comprise a number of discrete phases.

10.2.1. Compost preparation

Most cultivated fungi are saprophytes that decompose organic matter. Historically, wheat straw and horse manure have been the predominant raw materials for the production of mushroom compost, although alternative agricultural by-products, e.g. poultry litter, pig manure, are now widely used (Fletcher *et al.*, 1989). The principal objective of composting is to transform nutrients from the raw materials into forms that are available to, and provide a selective and homogenous substrate for, the mushroom. This is achieved in a two-phase process of fermentation (Phases I and II) by microbial degradation of organic matter.

Traditionally, during Phase I, raw ingredients are mixed, wetted and formed outdoors into large stacks, which are mechanically turned in an attempt to maintain aerobic conditions. Temperature differentials arise between the centre (70–80°C) and edges (ambient) of the stacks. At this time, unwanted organisms such as flies, mites and nematodes can readily develop in these outer compost layers. Towards the end of Phase I, compost is formed into 'windrows' in covered but open-sided compost sheds, and regularly turned through a 7- to 8-day period.

Primarily to reduce emissions (odour and ammonia) and minimize anaerobic cores, understack aeration was introduced (Von Minnigerode, 1981) to improve Phase I process control. This led to a number of developments in composting technology around the world (Noble and Gaze, 1994; Gerrits *et al.*, 1995; Perrin and Macauley, 1995) that demonstrated Phase I could be shortened and more effectively controlled in positively aerated chambers. The process, known as in-vessel or bunker composting, is achieved by pumping air into the compost through air spigots, with the duration dependent on oxygen demand of the microbial population. This system has been further refined and modified to develop total indoor systems of compost production.

The Phase II process is a highly controlled, aerobic, thermophilic, solid substrate fermentation that takes place in either the ultimate growing room or specifically constructed tunnels. An integral function of this process is pasteurization, wherein compost temperature is uniformly raised to 58–59°C for 8–10 h, thus ensuring destruction of all damaging organisms that survive Phase I. For optimal selectivity, the compost is then conditioned in an ideal environment for thermophilic microflora activity (45–48°C for 4–6 days). During Phase II, easily decomposable carbohydrates are broken down and the nitrogen-rich, lignin–humus complex formed. When the compost is ammonia-free (< 5 ppm), spawning with mushroom mycelium can commence.

10.2.2. Mushroom production

Mushroom spawn is prepared by growing mushroom mycelium on sterilized cereal grains. Spawning of the compost is accomplished by mixing the mycelium-covered grains into the compost. The incubation period during which mycelium moves off the cereal grain and colonizes the compost is termed spawn-running. Traditionally carried out *in situ* on mushroom farms in bags, blocks or trays, technological developments now enable the preparation of spawn-run compost (Phase III) in bulk tunnels at composting facilities. During a 14- to 18-day incubation period, optimum environmental conditions are compost temperatures of 25°C, carbon dioxide levels in the range 0.3–1.5% and high (> 90%) relative humidity.

To induce *A. bisporus* fruitbody production it is essential to cover the colonized compost with an approximate 5-cm layer of relatively inert casing material. Typically, mixtures of sphagnum peat and calcium carbonate are used. The casing provides water to enable the growth and development of mycelium and fruitbodies and protects the compost from desiccation. During case-run, similar environmental conditions to spawn-running are maintained for *c.* 7–10 days. When the mycelia reach the casing surface, reducing air temperature to 16–18°C and carbon dioxide levels to 600–1000 ppm induces sporophore ('mushroom') formation. Sporophore production occurs in discrete 'flushes' or 'breaks', commencing 2–3 weeks after casing and continuing at approximately weekly intervals. Typically, three or four flushes are harvested from each crop.

10.3. Invertebrate Pests of Mushrooms

The pasteurization process at conclusion of Phase I reduces invertebrate populations within the compost. However, the environment used for mushroom cultivation provides ideal conditions (optimum temperature, humidity and the constant supply of nutrients within the compost) for recolonization by a diverse fauna (Fletcher *et al.*, 1989). The principal pests are Dipterans of the families Sciaridae, Phoridae and Cecidomyiidae. Other arthropod pests include mites (order Acarina) of the families Tyroglyphidae, Anoetidae, Eupodidae and Tarsonemidae. Nematodes from the orders Rhabditida and Tylenchida have also been cited as pests.

10.3.1. Family Sciaridae

Previously, several species of Sciaridae have been associated with *A. bisporus* cultivation. More recently, and subsequent to taxonomic reclassification, Menzel (1998) and Menzel and Mohrig (1999) have simplified the list of pest species. It now comprises two species of *Lycoriella* (*L. ingenua* (Dufour) = *L. mali* = *L. solani* and *L. castanescens* (Lengersdorf) (Menzel and Mohrig, 1998) = *L. auripila*) and three species of *Bradysia* (*B. brunnipes* (Meigen) (Freeman, 1983), *B. difformis* (Frey) and *B. lutaria* (Winnertz)) (White *et al.*, 2000).

The phenology of *L. ingenua* relating to the different phases of the mushroom cultivation process is shown in Fig. 10.1A. Adult sciarids, principally females, can be found in proximity to outdoor Phase I compost stacks, into which they may deposit eggs. However, it has been argued that the high temperature during pasteurization kills subsequent immature stages (Anon, 1982). Nevertheless, sciarid populations may increase on composting sites and exert considerable infestation pressure to compost after pasteurization. Consequently, imperfect elimination during pasteurization or imperfect physical exclusion of adults after pasteurization may lead to significant sciarid levels in Phase II compost being delivered to the growers (Al-Amidi, 1995). The generation time for *Lycoriella* spp., under mushroom-growing conditions (16–24°C), is approximately 21 days (Hussey and Gurney, 1968; Snetsinger, 1972; Ganney, 1973; Kielbasa and Snetsinger, 1978). Therefore, during one cropping cycle, several insect generations will occur.

Cantelo (1988) described the vertical distribution of immature *L. ingenua* in

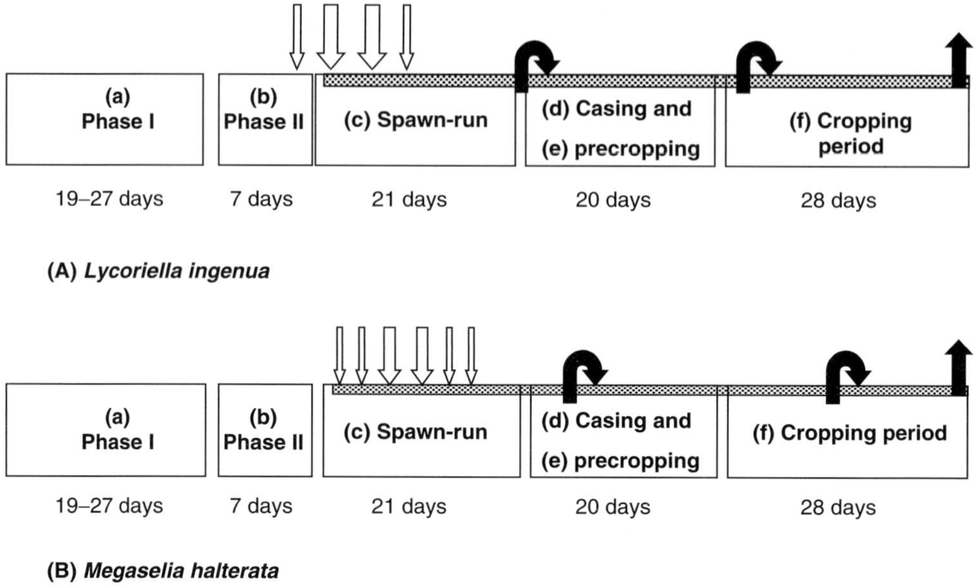

Fig. 10.1. Diagrammatic representation of the process of commercial mushroom production and the development of (A) *Lycoriella ingenua* and (B) *Megaselia halterata* under experimental conditions. (a) Phase I, initial composting of raw ingredients; (b) Phase II, a pasteurization process to produce an *Agaricus bisporus*-selective compost; (c) spawn-run, colonization of compost by *A. bisporus* mycelium; (d) casing, addition of moist peat-chalk layer required to promote fruiting; (e) precropping, mycelial growth through compost and casing layer culminating in production of primordia; (f) cropping period, production of mushrooms. ⇨ = initial oviposition; ▒ = immature stages; ➡ = adult emergence and re-infestation. (Adapted from White, 1997.)

mushroom-growing beds. The majority of eggs, approximately 70%, are found in the top 4 cm of the growing compost. As larvae hatch, feed and develop, many of them move deeper into the compost. More than 60% of third instar larvae are found below 5 cm. Late fourth instars and pupae, again, are concentrated in the upper levels, and 20 days post-oviposition more than 90% of pupae are found in the top 5 cm. If oviposition occurs before casing, very few of the larvae move into the casing layer. Sciarid larvae can develop in unspawned compost, but development is improved when small amounts of fungal mycelia are present (Binns, 1973, 1975; White, 1985). Damage may occur directly due to larval feeding on mycelia and developing sporophores. Additionally, larval excreta alter the chemical and physical properties of the compost, rendering it unsuitable for mycelial development (Hussey and Gurney, 1968). Collectively, these mechanisms may result in significant yield reduction and White (1986b) estimated that one sciarid larva per 125 g of casing caused 0.45% loss in total yield.

10.3.2. Family Phoridae

Phoridae associated with mushroom cultivation belong to the genus *Megaselia*. Two species, *Megaselia halterata* (Wood) and *M. nigra* (Meigen), are considered pests (Fletcher *et al.*, 1989). *M. nigra* females require daylight for oviposition (Hussey *et al.*, 1969) and, as this is excluded from modern mushroom production houses, *M. halterata* remains the principal phorid pest in *A. bisporus* culture in Europe and the USA (Hussey *et al.*, 1969; Rinker and Snetsinger, 1984).

Differences in the life cycles of *L. ingenua* and *M. halterata* (Fig. 10.1A and B) affect their occurrence in the mushroom cultivation process and, consequently, their relative importance as pests. In contrast to *Lycoriella* spp., adult *Megaselia* spp. are rarely found in association with Phase I compost. Oviposition by female *M. halterata* appears to be stimulated by mycelium development and has been observed to be maximal between days 7 and 12 of the spawn-run (Richardson and Hesling, 1978). Under normal mushroom cultivation conditions (compost temperature 18–20°C) the total development time for *M. halterata* is approximately 24–30 days (Hussey, 1959; Richardson and Hesling, 1978). Therefore, only two generations may be completed within the normal 10-week cropping cycle. In addition, flight activity of *M. halterata* is limited at air temperatures below 12°C. Consequently, incursion of wild populations to mushroom production sites occurs only during the warmer seasons.

Within the growing substrate, oviposition by *M. halterata* is confined to the zone with actively growing mycelia (Hussey, 1959), and successful development of immatures has only been observed in substrate with fungal mycelia (Hussey, 1959; Scheepmaker *et al.*, 1996). If oviposition occurs immediately before casing, a significant proportion of larvae migrate upwards into the casing layer as soon it becomes colonized by mycelia (Scheepmaker *et al.*, 1997a). However, in compost trays on UK farms, Hussey (1959) found that larval densities of *M. halterata* increased with substrate depth and only a few larvae were found in the casing layer. Immature stages comprise three larval instars, lasting a total of approximately 9–14 days at 17–20°C, followed by a pupal stage of 14–28 days (Hussey and Wyatt, 1962; Richardson and Hesling, 1978; Finley *et al.*, 1984).

The major economic impact of phorids probably results from the adults vectoring fungal pathogens, especially *Verticillium* spp. (Cross and Jacobs, 1968; Gandy, 1968; White, 1981). Relatively low densities (75 adults/m^2) can cause major disease outbreaks. Yield reduction directly due to larval feeding is only problematic at very high fly densities. Damage thresholds were estimated between 10^4 larvae/m^2 (Moreton and John, 1955; Rinker and Snetsinger, 1984) and 2.0×10^5 larvae/m^2 (Hussey, 1961). At such population levels, mycelial growth in the casing layer was retarded with consequent yield reduction.

10.4. Shortcomings of Established Insect Control Measures

After the proposition by Shanahan (1948) to incorporate the insecticide hexachlorocyclohexane (formerly benzene hexachloride) into the compost, pest control came to rely heavily on chemicals. Eventually, several drawbacks of this approach became evident. Many pesticides exhibit negative effects on mushroom yield (Moreton, 1955; Wyatt, 1978; Cantelo *et al.*, 1982; White, 1986a, 1999). In addition, it is broadly accepted that the use of chemical insecticides in mushroom production presents risks to the environment and to consumers (White, 1995).

Heightened public awareness of these problems generally led to increasingly restrictive legislation regarding pesticide registration. Consequently, the rate of new active ingredients becoming available to growers has declined. Reliance on a smaller number of products promoted the development of pesticide resistance. Resistance to organophosphates within sciarid populations is now considered widespread (Binns, 1976; White and Gribben, 1989; Smith and White, 1996), and more recently, resistance to diflubenzuron in UK *Lycoriella* spp. populations has been detected (Smith, 2002). In future, resistance may reduce the number of useful products further, creating a vicious circle in which reliance on fewer insecticides will further increase the rate of resistance development.

Physical exclusion of insects from the mushroom-growing substrate is another mainstay of modern pest management. At its simplest, this involves insect-proof mushroom-growing houses with fly-proof screens covering air vents. Phase I compost production inside enclosed bunkers reduces sciarid populations at the composting sites. Moreover, the transfer of the spawn-running phase from growing houses into enclosed structures at the composting sites (growers are supplied with spawn-run Phase III compost) shortens the period during which the substrate is exposed to insect pests. Complete physical exclusion of insects from growing houses is difficult to achieve, owing to essential personnel traffic during cultivation. Consequently, shorter cropping cycles in growing houses reduces the risk of insect contamination, limiting the number of insect generations and minimizing population build-up. However, physical exclusion, especially compost production in enclosed bunkers and Phase III compost preparation, requires considerable capital investment, which may be a limiting factor for small independent growers.

10.5. Nematodes for Biocontrol of Mushroom Pests

The relatively high moisture content of the mushroom-growing substrate, conducive to dipteran pest development, offers a unique potential for exploiting EPNs. Consequently, Cantelo *et al.* (1977) evaluated *Steinernema carpocapsae* (=*Neoaplectana carpocapsae*, strain DD-136). In their assay system, adult flies were allowed to oviposit into Petri dishes that contained compost with fungal mycelia. Nematodes were applied when fly larvae reached the second or third instar. No evidence of the nematode suppressing either *L. ingenua* or *M. halterata* was observed. Following these initial negative results, research concentrated on an obligate parasite of *M. halterata*, the allantonematid nematode *Howardula husseyi*, Richardson, Hesling and Riding (Richardson *et al.*, 1977). According to Riding and Hague (1974), this nematode destroys the insect fat body and follicular membrane and decreases copulation and oviposition. Additionally, the longevity of infested *M. halterata* is reduced by approximately 6 days in females and 2 days in males (Richardson and Chanter, 1979). At high infestation levels, insects may become completely sterilized (Riding and Hague, 1974). Nematodes, liberated as second instars, search and infect new hosts. Riding and Hague (1974) noted that problems with viability and storage of the free-living nematode life stage would pose a severe obstacle to direct inoculation of the mushroom substrate with the parasite. They reasoned that nematode release would have to rely on the distribution of parasitized flies. With this objective, Richardson and Chanter (1981) attempted to mass-produce parasitized *M. halterata*. However, the low yield of their rearing system in conjunction with the limited effect of the nematode on the fly population led them to conclude that further efforts towards rearing *H. husseyi* were unwarranted.

By the mid-1980s, considerable efforts were under way to develop EPNs of the genera *Steinernema* and *Heterorhabditis* as biological insecticides. Consequently, *Steinernema* spp. and *Heterorhabditis* spp. became a focus of attention for mushroom pest control, and discussion in the remainder of this chapter is restricted to these nematodes.

10.6. Factors Affecting the Efficacy of Entomopathogenic Nematodes (EPNs)

10.6.1. Nematode species and isolates

Nematode species and strains may account for a large amount of variation in infectivity to the target insect. Infection probabilities of different *S. feltiae* isolates to Sciaridae have been estimated to range from approximately 0.11 to 0.36 (Hay and Richardson, 1995). These data were generated by exposing individual *L. ingenua* larvae to single nematodes in peat-filled wells of ELISA plates. Similarly, by challenging a related sciarid species, *B. paupera*, with nematodes, Gouge and Hague (1995) found large intra- and interspecific variation in infectivity and inflicted host mortality. Among five *Steinernema* spp., *S. affinis*, *S. feltiae*, *S. kraussei* and two undescribed species, *S. feltiae* exhibited the highest infection probability against *L. ingenua* (Hay and Richardson, 1995).

Certain nematode isolates may be more effective than others against a wide range of hosts. Tested against six sciarid species, *Bradysia amonea*, *B. confinis*, *B. tritici*, *B. paupera*, *L. castanescens* and *L. ingenua*, infectivity of *S. feltiae* (NemasysM®) was consistently higher than that of *S. carpocapsae* (Gouge and Hague, 1995). Therefore, a nematode isolate, which performs well against one *Lycoriella* sp., is probably also effective against related *Lycoriella* spp.

Throughout several years, a number of different isolates have been tested against *L. castanescens* and *L. ingenua* in laboratory assays (Table 10.1). Studies, properly designed to compare different nematode species, suggest that *S. feltiae* is currently the most effective nematode at controlling *L. castanescens* and *L. ingenua* (Gouge and Hague, 1995; Hay and Richardson, 1995; Scheepmaker *et al.*, 1998c). Furthermore, within the species *S. feltiae*, the isolate sold as NemasysM® (Becker and Underwood, UK) and the strain Sus94 consistently exhibited higher infectivities against *L. ingenua* than other conspecifics (Hay and Fenlon, 1995; Hay and Richardson, 1995).

Similarly, a number of nematode isolates have been tested against the phorid *M. halterata* in the laboratory (Table 10.1). Variation in efficacy for control of *M. halterata* is less evident than in the case of *Lycoriella* spp. In a test on filter paper in Petri dishes, *S. feltiae*, *S. carpocapsae*, *Heterorhabditis megidis* and *H. bacteriophora* all resulted in significant *M. halterata* mortality (61–70%), with no clear differences between species (Scheepmaker *et al.*, 1998a). In another test, 11 *Steinernema* and *Heterorhabditis* isolates were compared. All isolates, with the possible exception of *S. feltiae* (NemasysM®), were ineffective against *M. halterata* larvae (Scheepmaker *et al.*, 1998a). This test was conducted in wells of tissue culture plates with agar compost and finely chopped compost straw. Furthermore, in a study by Long *et al.* (1998), from a total of ten isolates, three *Steinernema* spp. isolates were found to suppress *M. halterata*, and differences in LD_{50} values between these latter three isolates were insignificant.

Finally, results from larger-scale experiments involving the use of 14 l plastic buckets with compost and casing suggested that among *S. feltiae*, *S. carpocapsae*, *H. megidis* and *H. bacteriophora*, *S. carpocapsae* is slightly more effective than the others at controlling *M. halterata* (Scheepmaker *et al.*, 1998c). However, differences between the species were not very pronounced and from the reported results it is impossible to evaluate statistical significance.

10.6.2. Host-finding potential of nematodes in mushroom substrate

Several studies investigated the propensity of surface-applied nematodes to disperse into the depth of the mushroom substrate in the absence of hosts. Nickle and Cantelo (1991) applied *S. feltiae* on top of a 10-cm-deep Phase II compost layer. After 6–28 days, they divided the substrate into the upper and lower 5 cm and subjected it to Bearman funnel extraction. Over 60% of the recovered nematodes were found in the bottom 5 cm. In a study involving both *S. feltiae* and *H. megidis*, Scheepmaker (1999) found over 66% of nematodes in the casing layer (5 cm) and between 17% and 21% in the top 5 cm of the compost (total depth of compost, 20 cm). Five days after application, Jess and Bingham (2004) extracted over 90% of *S. feltiae* from the top 5 cm of a 15-cm-deep compost or casing columns.

The results of Nickle and Cantelo (1991) appear somewhat inconsistent with the other two studies. This may be due to a number of factors, e.g. substrate density, humidity, chemical and physical structure, status of mycelial growth and the amount of water used to apply the nematodes. The published reports do not allow further scrutiny of these factors. Furthermore, none of the publications referred to any efforts to verify the nematode identity. Were the extracted nematodes really the applied *S. feltiae*? Mushroom compost of inferior quality may contain saprophagous and mycophagous nematodes (Rhabditidae and Tylenchidae). If such substrate contamination was the case, one might expect a

Table 10.1. Laboratory evaluations of entomopathogenic nemotodes (EPN) against mushroom pests.

Nematode species, strain, isolate, source	Effect on host insect: *Lycoriella ingenua*	*Lycoriella castanescens*	*Megaselia halterata*	Test system	References
Heterorhabditis bacteriophora					
From H.K. Kaya, University of California, Davis, USA	—	—	++ M	Petri dish, filter paper, spawned rye grains	Scheepmaker *et al.*, 1998b
From Ecogen, Glenorchy, Tasmania, Australia	—	++ M	++ M	Buckets (14 l) with compost and casing	Scheepmaker *et al.*, 1998c
From R. Ehlers, Kiel, Germany	—	—	(+) M	Tissue culture plate well, compost agar, spawned rye grain, chopped compost straw	Scheepmaker *et al.*, 1998b
Heliothidis strain, From R. Ehlers, Kiel, Germany	—	—	(+) M		
H. heliothidis					
NZ-strain	—	++ M	++ M	Beakers with compost	Richardson, 1987
H. megidis					
NLH85	—	—	(+) M	Tissue culture plate well, compost agar, spawned rye grain, chopped compost straw	Scheepmaker *et al.*, 1998b
NLH-F85	—	+ M	+ M	Buckets (14 l) with compost and casing	Scheepmaker *et al.*, 1998c
Schleswig-Holstein			++ M	Petri dish, filter paper, spawned rye grains	Scheepmaker *et al.*, 1998b
Steinernema affinis					
Mg166	+ I	—	—	Wells of ELISA plates with peat	Hay and Richardson, 1995
S. affine					
From A. Reid, Silwood Park, UK	—	—	0 M	Tissue culture plate well, compost agar, spawned rye grain, chopped compost straw	Scheepmaker *et al.*, 1998b
S. anomale					
From R. Ehlers, Kiel, Germany	—	—	0 M	Tissue culture plate well, compost agar, spawned rye grain, chopped compost straw	Scheepmaker *et al.*, 1998b
S. carpocapsae					
DD-136	0 M	—	0 M	Petri dishes with compost	Cantelo *et al.*, 1977
"	—	—	+ M	Beakers with compost	Richardson, 1987
All, from Biosys	—	—	(+) M	Tissue culture plate well, compost agar, spawned rye grain, chopped compost straw	Scheepmaker *et al.*, 1998b
No 252, from Biosys, Palo Alto, CA, USA	+ I	+ I	—	Petri dishes with sand	Gouge and Hague, 1995
From Biosys, Palo Alto, CA, USA	—	+ M	++ M	Buckets (14 l) with compost and casing	Scheepmaker *et al.*, 1998c
Mexican strain	—	—	++ M	Petri dish, filter paper, spawned rye grains	Scheepmaker *et al.*, 1998b
S. feltiae					
Agriotos strain	—	++ M	—	Beakers with compost	Richardson, 1987

From Biosys, Palo Alto, CA, USA	++ M	—	—	Containers with 500 g compost	Nickle and Cantelo, 1991
Nemasys M®, from MicroBio, UK	++ I	++ I	—	Petri dishes with sand	Gouge and Hague, 1995
"	—	++ I	—	Wells of ELISA plates with peat	Hay and Fenlon, 1995
"	++ I	—	—	"	Hay and Richardson, 1995
"	++ M	—	++ M	Polystyrene containers, 200 ml compost	Jess and Bingham, 2004
"	—	—	(+) M	Tissue culture plate well, compost agar, spawned rye grain, chopped compost straw	Scheepmaker *et al.*, 1998b
Polish isolate	++ I	—	—	Tubes (7.5 cm diam. × 22 cm height) with compost	Tomalak and Lippa, 1991
Sus94	++ I	—	—	Wells of ELISA plates with peat	Hay and Richardson, 1995
"	—	++ I	—	"	Hay and Fenlon, 1995
Sus11	++ I	—	—	"	Hay and Richardson, 1995
"	—	++ I	—	"	Hay and Fenlon, 1995
Nor14	++ I	—	—	"	Hay and Richardson, 1995
OBS III	—	—	++ M	Petri dish, filter paper, spawned rye grains	Scheepmaker *et al.*, 1998b
Amersfoort, The Netherlands	—	—	(+) M	Tissue culture plate well, compost agar, spawned rye grain, chopped compost straw	Scheepmaker *et al.*, 1998b
From MicroBio, Agricultural Genetics Co. UK	—	++ M	++ M	Buckets (14 l) with compost and casing	Scheepmaker *et al.*, 1998c
S. intermedium					
From A. Reid, Silwood Park, UK	—	—	0 M	Tissue culture plate well, compost agar, spawned rye grain, chopped compost straw	Scheepmaker *et al.*, 1998b
S. kraussei					
War95	+ I	—	—	Wells of ELISA plates with peat	Hay and Richardson, 1995
War97	+ I	—	—	"	Hay and Richardson, 1995
M170	+ I	—	—	"	Hay and Richardson, 1995
War96	0 I	—	—	"	Hay and Richardson, 1995
Glo84	0 I	—	—	"	Hay and Richardson, 1995
Glo85	0 I	—	—	"	Hay and Richardson, 1995
Glo86	0 I	—	—	"	Hay and Richardson, 1995
From R. Ehlers, Kiel, Germany	—	—	0 M	Tissue culture plate well, compost agar, spawned rye grain, chopped compost straw	Scheepmaker *et al.*, 1998b
S. riobrave					
From Biosys, Palo Alto, CA, USA	—	—	(+) M	Tissue culture plate well, compost agar, spawned rye grain, chopped compost straw	Scheepmaker *et al.*, 1998b
Steinernema spp.					
D1 (low45)	+ I	—	—	Wells of ELISA plates with peat	Hay and Richardson, 1995
F1 (Gwe63)	++ I	—	—	"	Hay and Richardson, 1995
F1 (Pow81)	+ I	—	—	"	Hay and Richardson, 1995
F1 (Sus9)	0 I	—	—	"	Hay and Richardson, 1995

Note: ++ = significant effect ($P < 0.05$); + = tendency of effect ($P < 0.1$); (+) = some effect evident, no statistics reported; 0 = no effect; M = mortality assessed according to adult emergence in nematode treatments relative to untreated checks; I = infectivity, proportion of host larvae with nematodes present.

uniform nematode background throughout the whole depth. Hence, the conclusion by Nickle and Cantelo (1991) may be misleading and in the absence of hosts, nematodes have limited propensity to migrate into the depth of casing and compost.

However, surface-applied *S. feltiae* were able to infect *Galleria mellonella* at a depth of 22 cm in both casing and compost (Tomalak and Lippa, 1991). In contrast, *L. ingenua* were never infected at this depth in compost, whereas some infections of *L. ingenua* were observed in casing material at a depth of 22 cm. This does not contradict the above suggestion that penetration of substrate by *S. feltiae* is limited. Tomalak and Lippa (1991) did not compare infection probabilities of hosts in different depths. Scheepmaker (1999) presented *G. mellonella* larvae at 5- to 10-cm compost depth, beneath a 5-cm casing layer, and still extracted 66% and 49% of surface-applied *S. feltiae* and *H. megidis*, respectively, from the casing layer.

Infectivity of *S. feltiae* was greater in a casing soil mixture than in spawned compost for both *G. mellonella* and *L. ingenua*, (Tomalak and Lippa, 1991). Furthermore, Scheepmaker *et al.* (1998a) observed a tendency for increased *M. halterata* mortality in casing when compared with compost. The evidence from these studies suggests that EPNs locate and infect hosts more easily in casing than in compost substrate.

Additionally, substrate moisture is likely to affect nematode efficacy. The infectivity of *S. feltiae* to *L. ingenua* was maximal in casing with 60–71% gravimetric water content (tested range 0–87%) (Tomalak and Lippa, 1991). In mushroom houses, the moisture content in samples taken 4–6 h after watering was 75–80%. Therefore, it may be concluded that in typical commercial practice, the substrate could be too wet for optimal nematode performance (Tomalak and Lippa, 1991).

10.6.3. Optimum timing of nematode application

Due to the protected nature of mushroom-cropping systems, initial insect infestations are likely to occur only during certain limited time intervals. High-risk periods occur when the substrate is relatively exposed, e.g. at the end of the Phase II composting process, during compost-filling into bags, blocks or into the growing houses and during casing (see Fig. 10.1). Consequently, insect generations tend to be discrete and synchronized (Cantelo *et al.*, 1977; Scheepmaker *et al.*, 1997b; Fenton *et al.*, 2002). The combination of potentially differential susceptibility of insect life stages, and the limited persistence of nematodes in the absence of a host, indicates that application timing may be a critical factor for successful pest control. For optimum efficacy, nematodes should be applied when a majority of hosts are susceptible to nematode infection.

Two studies addressed differential susceptibility of *L. ingenua* instars to *S. feltiae*. Nickle and Cantelo (1991) applied nematodes at different times as *L. ingenua* proceeded through its life stages. Resulting mortalities suggested that second, third and early fourth instar larvae are the most susceptible, and first and late fourth instar larvae are somewhat less susceptible. Gouge and Hague (1995) challenged individual *L. ingenua* at specific life stages with 20 infective juveniles (IJs) of *S. feltiae*. Subsequently, the insects were dissected and the numbers of nematodes recorded. With the exception of the egg, all life stages from first instar larvae to adults were infected with nematodes. A mean of 3–4 nematodes per insect were found between the second larval and pupal stage, a mean of 1–2 nematodes were found in first instar larvae, and less than 1 nematode was found in adults. Thus the combined evidence suggests that *L. ingenua* is most susceptible to *S. feltiae* from the second through to the fourth larval instar. Concerning the susceptibility of *L. ingenua* life stages to *H. bacteriophora* (= *H. heliothidis*), Olthof and Rinker (1990) provided indirect evidence. In a laboratory experiment, *H. bacteriophora* was applied at 0, 3, 6, 10, 13, and 18 days post-oviposition to spawned compost in glass jars at 25°C. The highest mortality was observed following application on day 10 and, considering the environmental conditions, we assume that a majority of larvae were at the third instar.

We are unaware of any research directly addressing susceptibility of *M. halterata* life stages. However, Scheepmaker *et al.* (1998c) provided circumstantial evidence from a study in which *S. feltiae*, applied 6 and 10 days after an oviposition period of 72 h, were more effective than when applied immediately and 3 days after oviposition.

In mushroom-growing systems, applications of *S. feltiae* at the beginning of a case-run have repeatedly resulted in increased suppression of *Lycoriella* spp. when compared with applications at the beginning of spawn-run (Richardson and Grewal, 1991; Jess and Kilpatrick, 2000). Richardson and Grewal (1991) mixed the nematodes into the Phase II compost at spawning and into the casing soil before application and evaluated pest suppression with 'natural' *L. castanescens* infestations in a tray system with 15-cm compost depth. In the second study, Jess and Kilpatrick (2000) drenched the nematodes in water on to the surface of the compost at the beginning of the spawn-run, or on to the casing, 1 day post-casing. Mushrooms were grown in bags on approximately 40-cm-deep compost, and to ensure presence of pests, adult *L. ingenua* were added to the compost at the beginning of the spawn-run (immediately before the first nematode application). Fly populations were evaluated according to adult emergence from casing samples. In both experiments, applications of *S. feltiae* to the casing significantly suppressed the fly populations and no pest reduction was observed with nematode applications at the beginning of spawn-run.

Currently, growers typically apply *S. feltiae* as a surface drench shortly after casing. However, evidence suggests that pest control could be improved by applying *S. feltiae* somewhat later. Scheepmaker *et al.* (1997b), in a Dutch commercial mushroom-growing house with 'natural' fly infestations, compared: (i) an early *S. feltiae* application of $3 \times 10^6/m^2$ on to the compost before casing plus $3 \times 10^6/m^2$ on to the casing 1 day post-casing (total $6 \times 10^6/m^2$) to (ii) a late nematode application with $3 \times 10^6/m^2$ 1 week post-casing. By day 24 post-casing, both *M. halterata* and *L. castanescens* were more effectively controlled with the late nematode application at 50% of the rate used for early application. In a similar experiment, comparing applications of $3 \times 10^6/m^2$ *S. feltiae* at 1, 7 and 13 days post-casing, female *L. castanescens* were most effectively controlled with the nematode application at day 7 (Scheepmaker *et al.*, 1997b).

Finally, in a UK mushroom-cropping system, Fenton *et al.* (2002), fitted a life-stage-structured population model to adult emergence data of *Lycoriella* spp. from casing samples. Model parameters, estimated in the absence of pest control, suggested a first adult peak during the week immediately before casing. Assuming that only second to fourth instar larvae are susceptible to *S. feltiae*, they modelled and explored the outcome of several, post-casing, single and double, nematode application strategies. A single application would result in maximal fly suppression, when conducted approximately 5 days post-casing. Further, by splitting application of the dose into two, (50% at casing and 50% at 5 days post-casing), control of *Lycoriella* spp. could be improved. Alternatively, with the split application, the total dose could be reduced without compromising pest control. This latter conclusion was subsequently corroborated by experimental data (Fenton *et al.*, 2002).

10.6.4. Nematode application methods in mushroom-growing systems

The initial spatial distribution of the nematodes within the substrate (compost or casing) and the consequent initial likelihood of host encounters are dependent on the nematode application method. Virtually all application techniques are based on an aqueous suspension of IJ nematodes. Such a suspension may be applied in two different ways: (i) it can be mixed or injected into the substrate and (ii) it can be applied to the surface of the substrate. Furthermore, depending on the amount of suspension applied per unit surface area, different initial depth penetration of the nematodes may be expected (e.g. high dilution rates with a lot of water may transport a substantial proportion of nematodes into greater depth).

These issues have attracted relatively limited research. Regarding control of *Lycoriella* spp., we are unaware of any study allowing direct comparison of different application techniques. In the case of *M. halterata*, only one experiment by Scheepmaker *et al.* (1998c) is reported. Fully spawn-run compost was exposed to ovipositing females for 3 days. Subsequently, 500 g of the infested compost was placed in 14 l buckets on top of 1 kg of uninfested compost, and a layer of casing soil was applied. *S. feltiae* ($3 \times 10^6/m^2$) in 1 l water/m^2 were applied in three different ways: (i) as surface drench on to the compost; (ii) mixed into the fly infested compost (only the 500 g on top in each bucket); and (iii) as a surface drench on to the casing. Only the treatment with nematodes mixed into the compost significantly reduced *M. halterata* emergence. Nematode drenches on to compost or on to the casing did not provide any noticeable pest reduction. All nematode applications occurred within a relatively short time interval from immediately before to after casing. Therefore, it can be assumed that timing did not confound the results. However, the process of mixing the nematodes into the compost, which already contained *M. halterata* eggs (or perhaps early larval instars), may have resulted in mortality. According to the authors, mortality due to mixing had been assessed in preliminary experiments. However, no details or data from these experiments were reported.

10.6.5. Interaction of nematodes with insecticides

Interactions between EPNs and chemical insecticides have been the subject of a number of studies (for a review, see Chapter 20, this volume). In relation to mushroom pests, Sznyk-Basalyga and Bednarek (2003a,b) conducted experiments to investigate the interaction between the nematodes *S. feltiae* and *H. megidis* and the insecticide cyromazine. Experimental units comprised pots containing 100 g of damp moss, or compost, which were inoculated with *L. ingenua* or *M. halterata*, respectively. The following treatments were compared with each other: (i) the recommended rate of cyromazine (0.45 gAI/m^2 (3 g/Trigard-15WP® product/m^2, Novartis Crop Protection, Switzerland)); (ii) a low rate of cyromazine (0.045 gAI/m^2); (iii) nematodes ($0.5 \times 10^6/m^2$); and (iv) a combination of the low cyromazine rate with nematodes.

The data of Sznyk-Basalyga and Bednarek (2003a) suggest that *S. feltiae* and cyromazine act independently on the mortality of *L. ingenua*. Cyromazine and *H. megidis* act antagonistically; the mortality of *L. ingenua* in the combined treatment (70%) was less than expected (84%). However, in the case of *M. halterata*, there is evidence of synergistic mortality between cyromazine and *S. feltiae*; observed mortality in the combined treatment was 85%, compared with the expected 72% (Sznyk-Basalyga and Bednarek, 2003b).

10.6.6. Effect of selective breeding on nematode efficacy

Gaugler (1987) referred to the potential of artificial selection to enhance the efficacy of EPNs. Subsequently, Tomalak (1994) devised a breeding system to improve *S. feltiae* as a control agent of *L. ingenua*. This system comprised 7.5-cm-diameter tubes, filled with compost to a level of 18 cm and with a 5-cm casing layer on top. Third and fourth instar *L. ingenua* larvae were placed inside small copper mesh cages just above the interface between compost and casing, and IJ *S. feltiae* were applied to the surface of the casing. Gravimetric water content of the casing was maintained at approximately 78%. One day after adding the nematodes, the insect larvae were retrieved. Larvae infested with nematodes (visible inside the translucent body) were washed to remove contaminants from their surface and were incubated until nematode emergence. Nematodes were collected and stored in aerated water at 4°C for 7–14 days until used for the next breeding cycle.

Using this method, Tomalak (1994) selected a Polish strain of *S. feltiae* for a total of 34 generations, and after each generation, the infectivity of the selected strain was compared with the unselected parent strain (probably reared on *G. mellonella*). The test system for comparison of infectivity was identical to the selection system. Significant improvement of nematode infectivity to *L. ingenua* was already evident after two generations of selection. During 34 generations, the proportion of *L. ingenua* larvae infected with the selected nematodes increased from an initial 22.5% to 80–90%. The increase in infectivity was initially very steep and after approximately ten generations reached a plateau. Following 16 selection cycles, a comparison on a larger scale was conducted. Pots of 14 l with 1 kg casing material were used as experimental units. The casing originated from a mushroom culture, heavily infested with *L. ingenua*, at the conclusion of the cropping period. At an application rate of 25 nematodes/cm^2 of surface area, the selected nematode strain reduced fly emergence significantly better than the unselected parent strain.

Further studies compared the selected nematode strain with the SN-strain (Biosys, Palo Alto, CA, USA). Regarding fly control, Grewal *et al.* (1993) found only marginal (insignificant) superiority of the selected strain, whereas Tomalak (1994) found the latter resulted in significantly improved fly suppression. Conversely, persistence of the selected strain was improved in the experiment by Grewal *et al.* (1993), whereas Tomalak (1994) found no differences in persistence of the two strains. Methodological differences between the two studies possibly explain the contrasting results. Grewal *et al.* (1993) applied the nematodes to the surface immediately after casing. To assess fly emergence and nematode persistence, casing samples were collected four times at weekly intervals and pest levels were relatively low in this experiment. Tomalak (1994) applied the nematodes on to the casing surface of an old mushroom crop, which was approximately 4 weeks into sporophore production. Fly emergence and nematode persistence were evaluated from one set of casing samples collected 6 days post treatment. Fly emergence during a 27-day period resulted in the significant difference. Tomalak (1994) maintained the samples for another 24 days, during which no differences in fly emergence were noticed and nematode persistence was evaluated at the end of the 51-day period.

10.7. Implementation of Nematodes in Mushroom-growing Systems

10.7.1. Pest suppression in mushroom culture

The potential for nematodes to suppress *Lycoriella* spp. has been documented in a number of studies covering a variety of *A. bisporus*–growing systems (Table 10.2). Notable sciarid control typically resulted from nematode applications into or on to casing at around casing time, whereas earlier applications, into or on to compost during spawn-running, resulted in rather poor control (Table 10.2). In contrast, suppression of *M. halterata* with nematodes appears more difficult. Field studies that did address the issue indicate only marginal and inconsistent effects of *S. feltiae* (Table 10.2).

10.7.2. Mycotoxic effects on *Agaricus bisporus*

On several occasions reduced yields from the first mushroom flush have been observed after application of high nematode rates. However, later flushes typically compensated for this early yield loss. Richardson and Grewal (1991) observed a significant yield reduction in the first flush with compensation in the fourth and fifth flushes following incorporation of *S. feltiae*

Table 10.2. Field tests of entomopathogenic nematodes (EPNs) against mushroom pests.

	Pest		Nematode		Effect on pest	
Mushroom growing system	Infestation source time	Sampling method timing	Application method time	*Species* (source) Application rate ($10^6/m^2$)	*Species* Reduction of pest population (%) of untreated check	Reference
Wooden trays,	Artificial infestation	NR	NR	*H. bacteriophora*	*L. ingenua*	1
63 kg compost/m²	NR	NR	12 days post adult fly introduced	1.4–11.2	86–100	
	Artificial infestation	NR	NR	*S. feltiae* (Biosys #27)	*L. ingenua*	1
"	NR	NR	12 days post adult fly introduced	1.4–11.2	86–100	
	Natural fly population	Adult emergence from casing	In 0.9 l water/m² mixed into	*S. feltiae* (Ag. Genetics Co. UK)	*L. castanescens*	2
"	probably around spawning	samples week 2–7 post casing	compost at spawning	10.8	4	
			In 0.9 l water/m² mixed into casing			
"	"	"	at casing	10.8	76	
			10.8 mixed into compost at spawning			
"	"	"	10.8 mixed into casing at casing	21.6	80	
	Artificial inoculation with	Adult emergence from casing	In 0.9 l water/m² mixed into casing	*S. feltiae* (Ag. Genetics Co. UK)	*L. castanescens*	3
"	adults begin spawn run	samples week 0–8 post casing	at casing	2.7	85	
"	"	"	"	5.4	88	
"	"	"	"	10.8	94	
"	"	"	"	21.6	96	
Shelf-system in Pennsylvania	Natural fly population	Emergence traps on casing surface	In 0.6 l water/m² drenched on to casing	*S. feltiae* (SN strain, Biosys)	*L. ingenua*	4
	probably around spawning	for 14 days shortly after casing	at casing	0.5	87	
"		Adult emergence from casing samples	"	1	93	
		weekly 1 to 4 weeks post casing				
				S. feltiae (ScP, Tomalak, 1994)		
"		"	"	0.5	88	
"		"	"	1	96	
"	Natural mixed infestation of	Emergence traps on casing surface	In 0.6 l water/m² drenched on to casing	*S. feltiae* (SN strain,Biosys)	*L. ingenua/M. halterata*	4
"	L. ingenua and M. halterata	for 14 days shortly after casing	at casing	0.5	52/27	
"	probably around spawning	"	"	1	72/43	
"	during spawn run					
				S. feltiae (ScP, Tomalak, 1994)		
"	"	"	"	0.5	55/28	
"	"	"	"	1	83/26	

Tray system in Canada	Artificial inoculation with adults	Adult emergence from casing samples	In 1 l water/m^2 drenched on to casing	*H. bacteriophora*	*L. ingenua*	5
"	begin spawn run	once, 4 days post nematode appl.	at casing	0.3	52	
"	"	"	"	0.6	85	
"	"	"	"	0.8	80	
"	"	"	"	1.1	80	
"	"	"	"	1.4	85–96	
"	"	"	"	2.8	100	
"	"	"	"	5.6	96	
"	"	"	"	8.4	100	
"	"	"	"	11.2	96	
				S. feltiae (Biosys #27)	*L. ingenua*	5
"	"	"	"	0.11	39	
"	"	"	"	0.17	69	
"	"	"	"	0.22	38	
"	"	"	"	0.28	69–86	
"	"	"	"	0.6	97	
"	"	"	"	0.8	97	
"	"	"	"	1.1	100	
"	"	"	"	1.4	88–95	
"	"	"	"	2.8	90	
"	"	"	"	5.6	100	
"	"	"	"	8.4	96	
"	"	"	"	11.2	100	
Dutch growing system	Natural mixed infestation of	Adult emergence from casing samples	In 1 l water/m^2 drenched on to casing	*S. feltiae* (Ag. Genetics Co. UK)	*L. castanescens/M. halterata*	6
"	*L. castanescens* and *M. halterata*	3, 10, 24 and 38 days post casing	1 day post casing	3	22/42	
"	probably around spawning		7 days post casing	3	43/68	
	during spawn run					
	Natural mixed infestation of	Adult emergence from casing samples	In 1 l water/m^2 drenched on to casing	*S. feltiae* (Koppert BV, NL)	*L. castanescens/M. halterata*	6
"	*L. castanescens* and *M. halterata*	10, 24 and 38 days post casing	1 day post casing	1	57/0	
"	probably around spawning	"	"	2	56/0	
"	during spawn run	"	"	3	60/26	
"	"	"	7 days post casing	3	55/10	
"	"	"	13 day post casing	3	47/0	
"	"	"	1 and 13 days post casing	0.5 + 0.5	51/0	
"	"	"	1 and 13 days post casing	1.5 + 1.5	56/6	
"	"	"	3 days pre casing on to compost and 1 and 13 days post casing	1.5 + 1.5 + 1.5	74/13	

continued

Table 10.2. *Continued.* Field tests of entomopathogenic nematodes (EPNs) against mushroom pests.

	Pest		Nematode		Effect on pest	
Mushroom growing system	Infestation source time	Sampling method timing	Application method timing	Species (source) Application rate ($10^6/m^2$)	Species Reduction of pest population (%) of untreated check	References
Irish bag system approx. 22 kg compost / bag of 43 cm diameter	Artificial inoculation with adults begin spawn run	Adult emergence from compost/casing samples	Surface drench	*S. feltiae* (NemasysM, Microbio, UK)	*L. ingenua*	7
		end of spawn run, end of case run, 1st flush, and 3rd flush	begin of spawn run on to compost	3	13	
			begin of case run on to casing	3	67	

1: Olthof *et al.* (1991); 2: Richardson and Grewal (1991); 3: Grewal and Richardson (1993); 4: Grewal *et al.* (1993); 5: Rinker *et al.* (1995); 6: Scheepmaker *et al.* (1997); 7: Jess and Kilpatrick 2000

NR: Not reported. Percentages of pest suppression have been calculated by summing insect counts from all samples of the whole sample period. In case of the two sampling methods counts from both methods have been summed.

into the compost at spawning (6×10^6 nematodes/34 kg compost). In another study, Rinker *et al.* (1995) applied nematodes on to the casing surface at rates from 0 to 11.2×10^6 nematodes/m^2, and at the higher rates noticed a slight decline in the yield of the first flush, with compensation from later flushes.

However, in other studies significant yield increases were noticed after incorporation of high rates of *S. feltiae* into the casing soil (10.8×10^6 nematodes/m^2 by Richardson and Grewal (1991) and 5.4×10^6 nematodes/m^2 by Grewal and Richardson (1993)). First flush yield reductions were observed only at exceedingly high nematode rates ($21.6 \times 10^6/m^2$) (Grewal and Richardson, 1993). In all of these studies, *Lycoriella* spp. were present in the substrate, and flies were significantly suppressed by the nematodes. Therefore, yield increases may have been mediated through reduced fly populations. Early differences in *Lycoriella* spp. levels, which might have affected the yield of the first flush, cannot be compared from the reported data. In an experiment with very low pest pressure, a slight yield reduction in the first flush was observed when *S. feltiae* were incorporated into the casing mixture at the rate of 5.4×10^6 nematodes/m^2 (which previously resulted in increased yield) and, additionally, the first flush was delayed by almost 1 day (Grewal *et al.*, 1992).

Subsequently, Olthof *et al.* (1991) and Rinker *et al.* (1995) examined the effect of *H. bacteriophora* (= *H. heliothidis*) and *S. feltiae* on mycelial growth of *A. bisporus*. Nematodes were applied on to the surface immediately after casing. At pinning, the extent of mycelial coverage and mycelial penetration of the casing layer were visually examined and scored. With increasing application rates from 0 to 11.2×10^6 nematodes/m^2 of both *H. bacteriophora* and *S. feltiae* a negative correlation between nematode rate and mycelial growth was observed. At very low rates (0 to $0.28 \times 10^6/m^2$), with compost exposed to ovipositing *L. ingenua* at the beginning of spawn-run, no negative effect of *S. feltiae* on mycelial growth was evident. The effects of *H. bacteriophora* were not examined at the lower rates (Rinker *et al.*, 1995).

10.7.3. Product contamination

The saprobic nematode *Caenorhabditis elegans*, when present in the casing layer, may contaminate sporophores, result in distorted mushrooms and thus reduce marketable yield (Richardson and Grewal, 1991). Hence, *S. feltiae*, if used for insect control, might contaminate mushrooms and reduce qualitative yield.

The first investigation into this issue involved three methods of *S. feltiae* applications at the rate of $10.7 \times 10^6/m^2$, in 0.9 l water/m^2. Nematodes were: (i) mixed into the compost during spawning; (ii) mixed into the casing soil before casing; and (iii) mixed into the compost and mixed into the casing (total 21.6×10^6 nematodes/m^2) (Richardson and Grewal, 1991). The second experiment involved *S. feltiae* treatments of: (i) mixing nematodes into the casing before application; and (ii) sprinkling nematodes on to the substrate 14 days post casing at rates of 5.3 and 10.8×10^6 nematodes/m^2 (Grewal *et al.*, 1992). At harvest, sporophores were washed and nematodes were extracted from the water and identified.

Nematode applications into or on to the casing resulted in very few *S. feltiae* being retrieved from the sporophores (Richardson and Grewal, 1991; Grewal *et al.*, 1992). However, when sprinkled on to the substrate 14 days post-casing, significant numbers of *S. feltiae* were recovered from sporophores (Grewal *et al.*, 1992). This was ascribed to direct contamination of the sporophore primordia during the application. No evidence of sporophore distortion was reported. Therefore, if nematodes are not applied directly to the sporophores, no significant contamination is to be expected.

10.7.4. Economic competitiveness with established crop protection techniques

The principal method to control *Lycoriella* spp. involves drenching diflubenzuron into the casing at the beginning of case-run (Scheepmaker, 1999; Jess and Kilpatrick, 2000). *S. feltiae* can provide control levels comparable with diflubenzuron (Richardson and Grewal, 1991; Grewal and Richardson, 1993; Grewal *et al.*, 1993; Scheepmaker *et al.*, 1997b). Diflubenzuron may reduce mushroom yield owing to mycotoxicity (Kalberer and Vogel, 1978; Grewal *et al.*, 1992; Scheepmaker *et al.*, 1998b). *S. feltiae* at high application rates may reduce yield in early flushes; later flushes typically compensate these early losses (Richardson and Grewal, 1991; Grewal and Richardson, 1993; Rinker *et al.*, 1995). Hence, the total yield reduction due to *S. feltiae* appears negligible and has only been observed at application rates considerably exceeding those required for economic sciarid control. In summary, the level of sciarid control provided by *S. feltiae* is comparable with diflubenzuron and, in the absence of negative yield effects, *S. feltiae* is probably economically more advantageous compared with diflubenzuron (Scheepmaker *et al.*, 1998b).

Previously, incorporation of diazinon into the compost was the standard method for controlling *M. halterata* (Scheepmaker, 1999; Jess and Kilpatrick, 2000). Recently, however, in Europe, approval for such use of this insecticide was discontinued. At present, the only recommended chemicals for control of *M. halterata* are aerosol formulations of pyrethroids targeted at adult flies (Scheepmaker, 1999). We are unaware of any reliable economic appraisal of this approach to phorid control. Efficiency of available EPNs to control *Megaselia* is marginal at best. However, the scarcity of available alternative control measures warrants further efforts to develop EPNs against *M. halterata*.

10.8. Conclusions and Directions for Future Work

Commercial use of *S. feltiae* for the control of sciarids in mushroom cultivation began in the mid-1990s and in testimony to its efficacy and economic viability, growers are increasingly adopting this control method. The recommended strategy involves application of 3×10^6 nematodes in 1 l water/m^2 as a drench immediately after casing. However, the reliability of *S. feltiae* for sciarid control in the field remains unfavourable when compared with diflubenzuron. Poor product quality, handling, storage of the nematodes during distribution and application by retailers or growers are some of the reasons attributed to occasional failures (Staunton *et al.*, 1999). These issues can be addressed through better education of the relevant personnel. However, it must be accepted that the pest–nematode system is not yet completely understood. EPNs are likely to specifically target certain insect life stages, whereas chemicals tend to affect a broad range of life stages. Consequently, success in pest control may depend crucially on the temporal and spatial coincidence of virulent biocontrol agents and susceptible pests.

Delayed application of nematodes, 5–7 days post-casing, may provide more reliable sciarid control (Fenton *et al.*, 2002). These recommendations are based on research with Phase II compost, in which initial sciarid infestations probably occurred around spawning time. If fly infestation occurs at a different time, as has to be expected if growers use full-grown (Phase III) compost, nematode application shortly after casing might be suboptimal. Furthermore, host finding and infectivity of nematodes in compost is considerably reduced when compared with casing (Tomalak and Lippa, 1991). However, a significant proportion of susceptible second and third instar sciarid larvae reside for a considerable time period in the compost, before moving upwards into the casing layer (Cantelo, 1988). Many of the susceptible pests may remain for only a short period in the casing layer. Consequently, they are only exposed to optimum nematode infectivity for a short time and, therefore, timing of nematode application is even more critical.

In the case of *M. halterata*, control by EPNs is not commercially available at this time, and a breakthrough would seem unlikely in the near future. A few laboratory experiments recorded more than 60% mortality due to *Steinernema* or *Heterorhabditis* spp. (Scheepmaker *et al.*, 1998a; Jess and

Bingham 2004). In experiments involving mushroom-growing systems, *S. feltiae* provided only marginal and inconsistent control of *M. halterata* (Grewal *et al.*, 1993; Scheepmaker *et al.*, 1997b). Furthermore, the reasons for the relative ineffectiveness of nematodes against this pest remain largely unknown.

Scheepmaker (1999) considered that the size of natural body openings on *M. halterata* larvae may limit entry by *Steinernema* spp. into the host. This explanation is supported circumstantially by the observation that nematode applications at 6 or 10 days post-oviposition were more effective than applications immediately after oviposition (Scheepmaker *et al.*, 1998c). Later application presumably coincides with larger insects, and less constraints of small body openings. In addition, the synergism between *S. feltiae* and the chemical cyromazine suggests a limitation in nematode penetration. Cyromazine has been shown to alter mechanical properties of the insect cuticle (Kotze and Reynolds, 1993). However, *H. bacteriophora*, which can rupture the cuticle of the host with its anterior tooth (Bedding and Molineux, 1982), did not result in significantly higher *M. halterata* mortality than *Steinernema* spp. (Scheepmaker *et al.*, 1998a). In summary, the hypothesis is supported only by ambiguous and circumstantial evidence and requires further testing.

Scheepmaker *et al.* (1998c) observed that mixing *S. feltiae* directly into compost containing *M. halterata* eggs (or possibly early instar larvae) resulted in better pest control (31% reduction) than a drench application on to the compost or casing surface. A significant proportion of *M. halterata* larvae might be feeding for prolonged periods at increased compost depth (Hussey, 1959). We do not have a good understanding about differential susceptibility of the various *M. halterata* life stages. Nevertheless, circumstantial evidence suggests that late instar larvae are most susceptible. Assuming that analogous to *Lycoriella* spp., *M. halterata* eggs, pupae and adults are relatively immune, one might conclude that the third larval instar is the most susceptible life stage. Overall, the duration of the three larval instars, of which only the third may be reasonably susceptible to nematodes, is approximately 9 days. This contrasts with the duration of the immune pupal instar, which is 14 days. Consequently, we consider that two factors contribute to the poor efficacy of nematodes: (i) short duration of susceptible life stages and (ii) location of a significant proportion of susceptible hosts in the compost where nematodes are relatively ineffective.

The following directions for future research and development are based on the above reasoning:

- Development of a monitoring technique to accurately predict peaks of susceptible sciarid larvae, thus allowing more optimum timing of nematode application against this pest.
- Production of a slow release nematode formulation (e.g. slow rehydration), or nematode isolates, which are very persistent in the absence of hosts to obviate the need for optimum application timing.
- Characterization of the susceptibility of different *M. halterata* instars to nematodes.
- Clarification of the spatial distribution of susceptible *M. halterata* in compost or casing.
- Identification of nematode isolates, which are effective against a wide range of host instars (first instar larvae through to pupae).
- Identification of limiting factors (physical and chemical) of nematode efficacy in the compost.
- Identification of nematode isolates with improved host-finding and infection capability in the compost.
- Development of application techniques to inject nematodes into the compost beneath the casing, although the technical challenges and cost could be prohibitive.
- Breeding of more virulent nematode genotypes by selection of available isolates in suitable systems. The potential for this strategy has been demonstrated by Tomalak (1994).

References

Al Amidi, A.H.K. (1995) Occurrence of insects and mites in mushroom compost in Ireland. In: Elliott, T.J. (ed.) *Science and Cultivation of Edible Fungi.* Balkema 2, Rotterdam, The Netherlands, pp. 539–546.

Anon. (1982) Mushroom Pests. *Ministry of Agriculture Fisheries and Food Advisory Leaflet,* 583, 14 pp.

Anon. (2004) *FAOSTAT.* Food and Agriculture Organization of the United Nations online statistical database. Available at: http://faostat.fao.org

Bedding, R.A., and Molyneux, A.S. (1982) Penetration of insect cuticle by infective juveniles of *Heterorhabditis* spp. (Nematoda: Heterorhabditidae). *Nematologica* 28, 354–359.

Binns, E.S. (1973) Laboratory rearing, biology and chemical control of the mushroom sciarid *Lycoriella auripila* (Diptera: Lycoriidae). *Annals of Applied Biology* 73, 119–126.

Binns, E.S. (1975) Mushroom mycelium and compost substrates in relation to the survival of the larva of the sciarid *Lycoriella auripila. Annals of Applied Biology* 80, 1–15.

Binns, E.S. (1976) Mushroom Pests. *Report of the Glasshouse Crops Research Institute for 1975,* pp. 92–93.

Cantelo, W.W. (1988) Movement of *Lycoriella mali* (Diptera: Sciaridae) through mushroom-growing medium. *Journal of Economic Entomology* 81, 195–200.

Cantelo, W.W., McDaniel, J.S. and Thimijan, R.W. (1977) Research to reduce fly problems in the American mushroom industry. In: Smith, F.F. and Webb, R.E. (eds) *Pest Management in Protected Culture Crops.* 15 pp.

Cantelo, W.W., Henderson, D. and Argauer, R.J. (1982) Variation in sensitivity of mushroom strains to diazinon compost treatment. *Journal of Economic Entomology* 75, 123–125.

Chang, S.T. (1999) World production of edible and medicinal mushrooms in 1997 with emphasis on *Lentinus edodes* (Berk.) Sing. in China. *International Journal of Medicinal Mushrooms* 1, 291–301.

Cross, M. and Jacobs, L. (1968) Some observations on the biology of spores of *Verticillium malthousei. Mushroom Science* 8, 239–244.

Fenton, A., Gwynn, R.L., Gupta, A., Norman, R., Fairbairn, J.P. and Hudson, P.J. (2002) Optimal application strategies for entomopathogenic nematodes: integrating theoretical and empirical approaches. *Journal of Applied Ecology* 39, 481–492.

Finley, R.J., Wuest, P.J., Royse, D.J., Snetsinger, R., Tetrault, R. and Rinker, D.L. (1984) Mushroom flies: a review of biology and a systems approach for integrated control. *Mushroom Journal* 139, 240–247.

Fletcher, J.T., White, P.F. and Gaze, R.H. (1989) *Mushrooms: Pest and Disease Control,* 2nd edn, Intercept, Andover, UK.

Freeman, P. (1983) Revisionary notes on British Sciaridae (Diptera). *Entomologists Monthly Magazine* 119, 161–170.

Gandy, D.G. (1968) A technique for screening bacteria causing brown blotch of cultivated mushrooms. *Glasshouse Crops Research Institute Annual Report 1967,* pp. 150–154.

Ganney, G.W. (1973) Incorporation of pesticides into the casing and compost to control mushroom pests. *The Mushroom Journal* 10, 446–448.

Gaugler, R. (1987) Entomogenous nematodes and their prospect for genetic improvement. In: Maramorosch, K. (ed.) *Biotechnology in Invertebrate Pathology and Cell Culture.* Academic Press, San Diego, California, pp. 457–484.

Gaze, R.H. (1985) Cultivation systems and their evolution. In: Flegg, P.B., Spencer, D.M. and Wood, D.A. (eds) *The Biology and Technology of the Cultivated Mushroom.* John Wiley, Chichester, UK, pp. 23–41.

Gerrits, J.P.G., Amsing, J.G.M., Straatsma, G., and van Griensven, L.J.L.D. (1995) Phase I process in tunnels for the production of *Agaricus bisporus* compost with special reference to the importance of water. *Mushroom Science* 14, 203–221.

Grewal, P.S., and Richardson, P.M. (1991) Effects of *Caenorhabditis elegans* (Nematoda: Rhabditidae) on yield and quality of the cultivated mushroom *Agaricus bisporus. Annals of Applied Biology* 118, 381–394.

Grewal, P.S. and Richardson, P.N. (1993) Effects of application rates of *Steinernema feltiae* (Nematoda: Steinernematidae) on biological control of the mushroom fly *Lycoriella auripila* (Diptera: Sciaridae). *Biocontrol Science and Technology* 3, 29–40.

Grewal, P.S., Richardson, P.N., Collins, G. and Edmondson, R.N. (1992) Comparative effects of *Steinernema feltiae* (Nematoda: Steinernematidae) and insecticides on yield and cropping of the mushroom *Agaricus bisporus. Annals of Applied Biology* 121, 511–520.

Grewal, P.S., Tomalak, M., Keil, C.B.O. and Gaugler, R. (1993) Evaluation of a genetically selected strain of *Steinernema feltiae* against the mushroom sciarid *Lycoriella mali. Annals of Applied Biology* 123, 695–702.

Gouge, D.H. and Hague, N.G.M. (1995) The susceptibility of different species of sciarid flies to entomopathogenic nematodes. *Journal of Helminthology* 69, 313–318.

Hay, D.B. and Fenlon, J.S. (1995) A modified binomial model that describes the infection dynamics of the entomopathogenic nematode *Steinernema feltiae* (Nematoda: Steinernematidae). *Parasitology* 111, 627–633.

Hay, D.B. and Richardson, P.N. (1995) Inter- and intraspecific variation in infectivity of *Steinernema* spp. to larvae of the mushroom fly *Lycoriella solani. Entomologia Experimentalis et Applicata* 77, 11–15.

Hussey, N.W. (1959) Biology of mushroom phorids. *Mushroom Science* 4, 260–269.

Hussey, N.W. (1961) Recent work on the control of the Worthing phorid. *Mushroom Growers Association Bulletin* 144, 495–505.

Hussey, N.W. and Gurney, B. (1968) Biology and control of the sciarid *Lycoriella auripila* (Winnertz) (Diptera: Lycoriidae) in mushroom culture. *Annals of Applied Biology* 62, 395–402.

Hussey, N.W. and Wyatt, I.J. (1962) The interaction between mushroom mycelium and insect pest populations. *Mushroom Science* 5, 509–517.

Hussey, N.W., Read, W.H. and Hesling, J.J. (1969) Insect pests of mushrooms. *Pests of Protected Cultivation.* Arnold, London, pp. 305–337.

Jess, S. and Bingham, J.F.W. (2004) Biological control of sciarid and phorid pests of mushroom with predatory mites from the genus *Hypoapsis* (Acari: Hypoaspidae) and the entomopathogenic nematode *Steinernema feltiae. Bulletin of Entomological Research* 94, 159–167.

Jess, S. and Kilpatrick, M. (2000) An integrated approach to the control of *Lycoriella solani* (Diptera: Sciaridae) during production of the cultivated mushroom (*Agaricus bisporus*). *Pest Management Science* 56, 477–485.

Kalberer, P. and Vogel, E. (1978) Control of sciarids (*Lycoriella auripila*) in mushroom cultures (*Agaricus bisporus*) by diflubenzuron in the casing soil. *Zeitschrift fur Pflanzenkrankheiten und Pflanzenschutz* 85, 328–333.

Kielbasa, R. and Snetsinger, R. (1978) Effects of mushroom-growing temperatures and *Agaricus bisporus* strains on the biology of a mushroom-infesting sciarid, *Lycoriella mali. Melsheimer Entomological Series* 25, 16–24.

Kotze, A.C. and Reynolds, S.E. (1993) Effects of cyromazine on the mechanical properties of the larval integument of *Lucilia cuprina* (Diptera: Calliphoridae). *Bulletin of Entomological Research* 83, 389–393.

Long, S.J., Richardson, P.N., Welch, D.M. and Edmondson, R.N. (1998) Potential of entomopathogenic nematodes for biological control of the mushroom phorid fly. *Nematologica* 44, 529–530.

Menzel, F. (1998) Sciaridae. Checklists of insects of the British Isles (New Series). Part 1 Diptera. *Handbook for the Identification of British Insects* 12, 20–24.

Menzel, F. and Mohrig, W. (1998) Revision der palaarktischen Trauermucken (Diptera: Sciaridae) unter besonderer Beruckuchtigung der deutschen Fauna. *Studia Dipterologica* 3 (Suppl.), 663 pp.

Menzel, F. and Mohrig, W. (1999) Revision der palaarktischen Trauermucken (Diptera: Sciaridae), *Studia Dipterologica* 6 (Suppl.), 1–720.

Moreton, B.D. (1955) Story of the Worthing phorid fly. *The Grower* 44, 1537.

Moreton, B.D. and John, M.E. (1955) The mushroom phorid fly epidemic of 1953. *Plant Pathology* 4, 9–11.

Nickle, W.R. and Cantelo, W.W. (1991) Control of a mushroom-infesting fly, *Lycoriella mali,* with *Steinernema feltiae. Journal of Nematology* 23, 145–147.

Noble, R. and Gaze, R.H. (1994) Controlled environment composting for mushroom cultivation: substrates based on wheat and barley straw and deep litter poultry manure. *Journal of Agricultural Science* 123, 71–79.

Olthof, T.H.A. and Rinker, D.L. (1990) Efficacy of entomophilic nematodes in the control of a mushroom fly, *Lycoriella mali* (Diptera: Sciaridae). *Nematologica* 36, 379.

Olthof, T.H.A., Rinker, D.L. and Dano, J. (1991) The effect of entomophilic nematodes on a sciarid fly, *Lycoriella mali,* and on growth and yield of mushrooms, *Agaricus bisporus. Journal of Nematology* 23, 545.

Perrin, P.S. and Macauley, B.J. (1995) Positive aeration of conventional (Phase I) mushroom compost stacks for odour abatement and process control. *Mushroom Science* 14, 223–232.

Richardson, P.N. (1987) Susceptibility of mushroom pests to the insect-parasitic nematodes *Steinernema feltiae* and *Heterorhabditis heliothidis. Annals of Applied Biology* 111, 433–438.

Richardson, P.N. and Chanter, D.O. (1979) Phorid fly (Phoridae: *Megaselia halterata*) longevity and the dissemination of nematodes (Allantonematidae: *Howardula husseyi*) by parasitised females. *Annals of Applied Biology* 93, 1–11.

Richardson, P.N. and Chanter, D.O. (1981) Aspects of the laboratory production of mushroom phorid flies (*Megaselia halterata*) parasitised by the nematode *Howardula husseyi. Annals of Applied Biology* 99, 1–9.

Richardson, P.N. and Grewal, P.S. (1991) Comparative assessment of biological (Nematoda: *Steinernema feltiae*) and chemical methods of control for the mushroom fly *Lycoriella auripila* (Diptera: Sciaridae). *Biocontrol Science and Technology* 1, 217–228.

Richardson, P.N. and Hesling, J.J. (1978) Laboratory rearing of the mushroom phorid *Megaselia halterata* (Diptera: Phoridae). *Annals of Applied Biology* 88, 211–217.

Richardson, P.N., Hesling, J.J. and Riding, I.L. (1977) Life cycle and description of *Howardula husseyi* N. Sp. (Tylenchida: Allantonematidae), a nematode parasite of the mushroom phorid *Megaselia halterata* (Diptera: Phoridae). *Nematologica* 23, 217–231.

Riding, I.L. and Hague, N.G.M. (1974) Some observations on a tylenchid nematode *Howardula* sp. parasitizing the mushroom phorid *Megasalia halterata* (Diptera: Phoridae). *Annals of Applied Biology* 78, 205–211.

Rinker, D.L. and Snetsinger, R.J. (1984) Damage threshold to a commercial mushroom by a mushroom-infesting phorid (Diptera: Phoridae). *Journal of Economic Entomology* 77(2), 449–453.

Rinker, D.L., Olthof, T.H.A., Dano, J. and Alm, G. (1995) Effects of entomopathogenic nematodes on control of a mushroom-infesting sciarid fly and on mushroom production. *Biocontrol Science and Technology* 5, 109–119.

Scheepmaker, J.W.A. (1999) Biological control of the mushroom sciarid *Lycoriella auripila* and the phorid *Megaselia halterata* by entomopathogenic nematodes. PhD thesis, Catholic University of Nijmegen, Department of Microbiology, The Netherlands.

Scheepmaker, J.W.A., Geels, F.P., Van Griensven, L.J.L.D. and Smits, P.H. (1996) Substrate dependent larval development and emergence of the mushroom pests *Lycoriella auripila* and *Megaselia halterata*. *Entomologia Experimentalis et Applicata* 79, 329–334.

Scheepmaker, J.W.A., Geels, F.P., Smits, P.H. and van Griensen, L.G.L.D. (1997a) Location of immature stages of the mushroom insect pest *Megaselia halterata* in mushroom-growing medium. *Entomologia Experimentalis et Applicata* 83, 323–327.

Scheepmaker, J.W.A., Geels, F.P., Smits, P.H. and van Griensven, L.G.L.D. (1997b) Control of the mushroom pests *Lycoriella auripila* (Diptera: Sciaridae) and *Megaselia halterata* (Diptera: Phoridae) by *Steinernema feltiae* (Nematoda: Steinernematidae) in field experiments. *Annals of Applied Biology* 131, 359–368.

Scheepmaker, J.W.A., Geels, F.P., van Griensven, L.G.L.D. and Smits, P.H. (1998a) Susceptibility of larvae of the mushroom fly *Megaselia halterata* to the entomopathogenic nematode *Steinernema feltiae* in bioassays. *Biocontrol* 43(2), 201–214.

Scheepmaker, J.W.A., Geels, F.P., Smits, P.H. and van Griensven, L.G.L.D. (1998b) Influence of *Steinernema feltiae* and diflubenzuron on yield and economics of the cultivated mushroom *Agaricus bisporus* in Dutch mushroom culture. *Biocontrol Science and Technology* 8, 269–275.

Scheepmaker, J.W.A., Geels, F.P., Smits, P.H., Rutjens, A.J. and van Griensven, L.G.L.D. (1998c) Comparison of the efficacy of entomopathogenic nematodes for the biological control of the mushroom pests *Lycoriella auripila* (Sciaridae) and *Megaselia halterata* (Phoridae). *Biocontrol Science and Technology* 8, 277–288.

Shanahan, G.J. (1948) BHC and mushroom pests, a promising means of control. *Agricultural Gazette of New South Wales* 59, 184.

Smith, J.E. (2002) Dimilin resistance in mushroom sciarids. *The Mushroom Journal* 626, 15.

Smith, J.E. and White, P.F. (1996) Diazinon resistance in mushroom pests. *Project News of the Horticultural Development Council* 36, 12–15.

Snetsinger, R. (1972) Laboratory studies of mushroom-infesting arthropods. *Mushroom Science* 8, 199–208.

Staunton, L., Dunne, R.M., Cormican, T. and Donovan, M. (1999) Chemical and biological control of mushroom pests and diseases. Report: Teagasc, Kinsealy Research Centre, Dublin.

Sznyk-Basalyga, A. and Bednarek, A. (2003a) Integrated control of *Lycoriella solani* (Diptera: Sciaridae) with entomopathogenic nematodes and insecticides. *Bulletin OILB/SROP* 26, 189–192.

Sznyk-Basalyga, A. and Bednarek, A. (2003b) Effect of entomopathogenic nematodes and insecticides on *Megaselia halterata* mortality. *Bulletin OILB/SROP* 26, 147–150.

Tomalak, M. (1994) Selective breeding of *Steinernema feltiae* (Filipjev) (Nematoda: Steinernematidae) for improved efficacy in control of a mushroom fly, *Lycoriella solani* Winnertz (Diptera: Sciaridae). *Biocontrol Science and Technology* 4, 187–198.

Tomalak, M.L. and Lippa J.J. (1991) Factors affecting entomophilic activity of *Neoaplectana feltiae* in mushroom compost. *Entomologia Experimentalis et Applicata* 59, 105–110.

Von Minnigerode, H.F. (1981) A method for controlling and regulating the composting process. *Mushroom Science* XI, 265–278.

White, P.F. (1981) Spread of the mushroom disease *Verticillium fungicola* by *Megaselia halterata* (Diptera: Phoridae). *Protection Ecology* 3, 17–24.

White, P.F. (1985) Pests and Pesticides. In: Flegg, P.B., Spencer, D.M. and Wood, D.A. (eds) *The Biology and Technology of the Cultivated Mushroom.* John Wiley, Chichester, UK, pp. 279–293.

White, P.F. (1986a) Effects of bendiocarb and diflubenzuron on mushroom cropping. *Annals of Applied Biology* 108, 11–20.

White, P.F. (1986b) The effect of sciarid larvae (*Lycoriella auripila*) on cropping of the cultivated mushroom (*Agaricus bisporus*). *Annals of Applied Biology* 109, 11–17.

White, P.F. (1995) Biological control of mushroom pests: an evaluation. In: *Proceedings of the International Congress on the Science and Cultivation of Edible Fungi*, Oxford, Balkema, Rotterdam, The Netherlands 2, 475–484.

White, P.F. (1997) The use of chemicals, antagonists, repellents and physical barriers for the control of *Lycoriella auripila* (Diptera: Sciaridae), a pest of the cultivated mushroom *Agaricus bisporus. Annals of Applied Biology* 131, 29–42.

White, P.F. (1999) Comparative effects of three insect growth regulator insecticides and a dipteran-active strain of *Bacillus thuringiensis* on cropping of the cultivated mushroom *Agaricus bisporus. Annals of Applied Biology* 134, 35–43.

White, P.F. and Gribben, D.A. (1989) Variation in resistance to diazinon by the mushroom sciarid *Lycoriella auripila. Mushroom Science* 12, 851–859.

White, P.F., Smith, J.E. and Menzel, F. (2000) Distribution of Sciaridae (Dipt.) species infesting commercial mushroom farms in Britain. *Entomologist's Monthly Magazine* 136, 207–209.

Wyatt, I.J. (1978) Principles of insecticide action on mushroom cropping: incorporation into casing. *Annals of Applied Biology* 88, 89–103.

11 Orchard Applications

D.I. Shapiro-Ilan,[1] L.W. Duncan,[2] L.A. Lacey[3] and R. Han[4]
[1]*USDA-ARS, SE Fruit and Tree Nut Research Lab, Byron, GA 31008, USA;* [2]*University of Florida, CREC, Lake Alfred, FL 33850, USA;* [3]*USDA-ARS, Yakima Agricultural Research Lab, Wapato, WA 98951, USA;* [4]*Guangdong Entomological Institute, Guangzhou 510260, China*

11.1. Introduction

Orchards consist of perennial tree plantings that provide various agricultural products. A number of economically important pests occur in most types of orchards. Many orchards contain attributes, e.g. hosts available through much of the year, favourable soil conditions (moist, sandy) and shade, that are amenable to insect suppression using entomopathogenic nematodes (EPNs). Additionally, crops produced in orchards are often relatively high in value, which facilitates economic feasibility of nematode applications. As a result, a number of orchard pests have been extensively studied for their potential to be controlled by EPNs, and some have become commercial success stories. For example, root weevils attacking citrus in Florida (including *Diaprepes abbreviatus* and *Pachnaeus* spp.) have become the largest US target for commercially produced nematodes. Approximately 20,000 ha of citrus were treated with *S. riobrave* to control citrus root weevils in

1999 (M. Dimock, Certis USA, Columbia, MD, 2003, personal communication). In this chapter, we review significant research on EPN control of orchard pests indicating successes and failures, research needs and potential for the future.

11.2. Apples, Pears and Stone Fruits

11.2.1. Lepidopteran pests

11.2.1.1. Codling moth

Codling moth, *Cydia pomonella*, is a key worldwide tortricid pest of apple, pear, walnut and other fruit. The most vulnerable stages in terms of microbial control with EPNs are the full-grown larvae after they exit the fruit, and the cocooned prepupae and pupae. *C. pomonella* overwinter as full-grown cocooned larvae in cryptic habitats (under bark, within prop piles, fruit bins and leaf litter). Research results indicate good *C. pomonella* control potential with *Steinernema carpocapsae* (which was originally isolated from *C. pomonella* (Weiser, 1955)) and a number of other nematode species when adequate moisture is maintained and temperatures are above 10–15°C (Kaya *et al.*, 1984; Sledzevskaya, 1987; Nachtigall and Dickler, 1992; Lacey and Unruh, 1998; Lacey and Chauvin, 1999; Unruh and Lacey, 2001). Dosages in the range of $1–2 \times 10^6$ infective juveniles (IJs) per tree and surrounding area can provide effective control of cocooned larvae under optimum conditions of adequate moisture and temperature. Protocols for the field evaluation of EPNs against cocooned stages of *C. pomonella* are presented by Lacey *et al.* (2000). The major obstacles for successful *C. pomonella* control are low temperatures and desiccation of IJs. Ideally, sprayed trees and surrounding areas should be kept moist for 8 h or more (Lacey and Unruh, 1998; Unruh and Lacey, 2001). Recent research indicates EPNs to be compatible with other biocontrol agents (e.g. ichneumonid parasitoids) for *C. pomonella* control (Lacey *et al.*, 2003).

Fruit bins infested with cocooned *C. pomonella* can be a significant source of invading moths in mid-to late summer when they are placed in orchards for harvest. EPNs offer potential for decontaminating fruit bins when they are submerged in drop tanks (Lacey and Chauvin, 1999) or sprayed with drenchers used for treating or cooling fruit (Cossentine *et al.*, 2002).

11.2.1.2. Other lepidopteran pests

A variety of other lepidopterans are pests of apple, pear and stone fruits to varying degrees, depending on locality. EPNs offer a narrow window of opportunity for control of defoliating Lepidoptera that have soil stages. Noctuids, for example, are most vulnerable as prepupae when they search for a soil site in which they pupate. Laboratory studies demonstrated fair to good activity of several nematode species against leafrollers (torticids that construct retreats in rolled leaves or shoots), e.g. the obliquebanded leafroller, *Choristoneura rosaceana* (Poinar, 1991; Belair *et al.*, 1999). However, only limited field trials have been conducted without evidence of effective control (Belair *et al.*, 1999). A major apple pest in China, *Carposina nipponensis*, has been shown to be highly susceptible to EPN control; field trials resulted in greater than 90% larval mortality (Bedding, 1990). Additionally, substantial efficacy of *Steinernema* spp. against species of tree borers in the genus *Synanthedon* has been demonstrated in apple and stone fruit orchards (Deseö and Miller, 1985; Cossentine *et al.*, 1990; Kahounova and Mracek, 1991). In contrast, field trials to suppress another lepidopteran borer, the American plum borer, *Euzophera semifuneralis*, did not provide any significant control with *S. feltiae* or *Heterorhabditis bacteriophora* (Kain and Agnello, 1999).

11.2.2. Non-lepidopteran pests

11.2.2.1. Fruit flies

Some of the most harmful pests of cherries are fruit flies. Research demonstrates that several species of fruit flies are susceptible to EPNs (Beavers and Calkins, 1984;

Lindegren and Vail, 1986; Lindegren *et al.*, 1990; Gazit *et al.*, 2000), but most investigations have been limited to laboratory research. The western cherry fruit fly, *Rhagoletis indifferens* (a serious pest of sweet cherries in western USA), has been investigated for control using *Steinernema* spp. and *Heterorhabditis* spp. under laboratory and field conditions (Patterson Stark and Lacey, 1999; Yee and Lacey, 2003). Yee and Lacey (2003) evaluated *S. carpocapsae*, *S. feltiae* and *S. intermedium* in soil against *R. indifferens* larvae, pupae and adults in the laboratory. Larvae were the most susceptible stage, with mortality ranging from 62% to 100%. *S. carpocapsae* and *S. feltiae* were equally effective against larvae at 50 and 100 IJs/cm^2. Mortalities of *R. indifferens* larvae 0–6 days following their introduction into soil previously treated with 50 IJs/cm^2 of *S. carpocapsae* or *S. feltiae* were 78.6–77.5%. Pupae were not infected, but adult flies were infected by all three nematode species in the laboratory at a concentration of 100 IJs/cm^2. In field trials *S. carpocapsae* and *S. feltiae* were equally effective against larvae (59–85% mortality) when applied to soil under cherry trees at 50–100 IJs/cm^2. Because abandoned orchards and trees in yards of homeowners represent a threat to commercial cherry orchards by providing significant sources of invading flies, Yee and Lacey (2003) proposed the use of EPNs in these situations for the control of *R. indifferens*.

11.2.2.2. Other non-lepidopteran pests

There are a wide variety of other non-lepidopteran pests of apple, pears and peaches, but EPNs have only been evaluated against a few species. Vincent and Belair (1992) and Belair *et al.* (1998) reported control of the apple sawfly, *Hoplocampa testudinea*, with EPNs. Applications of *S. carpocapsae* every 2–3 days from early May until mid-June by Belair *et al.* (1998) reduced primary damage caused by larvae of *H. testudinea* by 98–100% in two seasons (1992–1993), but treatments were ineffective in the following year. The western flower thrips, *Frankliniella occidentalis*, attack a wide range of crops and can be a pest of several fruit varieties including apple, pear and cherry. Potential for control of thrips has been demonstrated with EPNs (Helyer *et al.*, 1995; Ebssa *et al.*, 2001a,b) and *Thripinema nicklewoodi* (Lim *et al.*, 2001; Arthurs and Heinz, 2003; see Chapter 22, this volume). Similarly, other thrip pests such as *Taeniothrips inconsequens* (a serious pest of pear and plum) may be susceptible to nematodes.

The plum curculio, *Conotrachelus nenuphar*, is a key pest of apple and stone fruits in North America. Belair *et al.* (1998) applied *S. carpocapsae* for control of *C. nenuphar* in apples and observed highly variable results ranging from 75% damage reduction to no significant reduction. In laboratory studies comparing six nematode species, Shapiro-Ilan *et al.* (2002a) reported *S. feltiae* and *S. riobrave* to be most virulent to *C. nenuphar* larvae, whereas *S. carpocapsae* and *S. riobrave* were the most virulent to *C. nenuphar* adults. In field trials in peach orchards, Shapiro-Ilan *et al.* (2004) observed, on average, greater than 90% suppression of *C. nenuphar* larvae with *S. riobrave*.

11.3. Nut Crops

11.3.1. Navel orangeworm

The navel orangeworm, *Amyelois transitella*, is a serious pest of almonds, walnuts and pistachio (Rice, 1978a,b) and the most important pest of almonds in the USA. The larval stage invades nuts during hull split and feeds on the nutmeats. The larvae infest mature nuts on the tree and nut mummies on the tree and ground. Conventional control of *A. transitella* during the growing season is through the application of organophosphate, carbamate and other insecticides. Orchard sanitation is also an important aspect of navel orangeworm control. Nut mummies are removed from the trees by shaking, polling, pruning, etc. and blown into furrows for disking (in pistachios) or flail mowing (in almonds), rendering the majority of the nuts unsuitable for development of larvae. However, some larvae survive this treatment and pose a significant threat to nuts in the following season. In addition to the need for insecticides to protect nuts from moths that have survived sanitation measures, there are air quality problems (dust) generated by disking, blowing and flail mowing. The use of EPNs offers an alternative means of control that will help reduce the use of pesticides and improve air quality. However, initial investigations on the potential of EPNs for control of the moth were not especially promising. Summer-time field application of the nematode *S. carpocapsae* to open hulled almonds resulted in over 65% mortality in baited *A. transitella* (Lindegren *et al.*, 1987), whereas dormant season (winter) application of EPNs to trees resulted in substantially lower control (Agudelo-Silva *et al.*, 1995). Siegel *et al.* (2004) studied the efficacy of *S. carpocapsae* and *S. feltiae* applied to almond and pistachio nut mummies on the ground for control of *A. transitella* larvae. Larvae were almost completely controlled with *S. carpocapsae* at 10^5 IJs/m^2 and to a lesser extent by *S. feltiae* at the same dosage. The low rate of applications used to achieve these high levels of control indicates that ground application of EPNs as a sanitation tool for *A. transitella* is a highly promising tactic and should be pursued further. EPNs persist well in this environment, offering the potential of recycling within the *A. transitella* population (Agudelo-Silva *et al.*, 1987; Siegel *et al.*, 2004).

11.3.2. Pecan weevil

The pecan weevil, *Curculio caryae*, is a key pest of pecans throughout southeastern USA as well as portions of Kansas, Oklahoma and Texas (Shapiro-Ilan, 2003). Adults emerge from soil in late July–August and feed on and oviposit in developing nuts. Larvae develop in the nuts, drop to the soil, burrowing to a depth of 8–25 cm, and form a soil-cell where they spend 1 year (and sometimes 2) before pupating and moulting to adulthood; adults spend approximately 9 additional months in the soil before emerging (Harris, 1985). Control recommendations for the pecan weevil currently consist solely of above-ground applications of chemical insecticides (mainly carbaryl) to suppress adults (Hudson *et al.*, 2003).

11.3.2.1. Potential to control larvae with entomopathogenic nematodes (EPNs)

EPNs have been reported to occur naturally in *C. caryae* larvae (Harp and Van Cleave, 1976; Nyczepir *et al.*, 1992). Yet field applications to suppress larvae (with *H. bacteriophora*, *S. carpocapsae* or *S. feltiae*) resulted in less than 35% control unless exceedingly high rates were used (Tedders *et al.*, 1973; Nyczepir *et al.*, 1992; Smith *et al.*, 1993). In order to determine if other nematode species might have greater virulence to *C. caryae* larvae than those tested previously, Shapiro-Ilan (2001a) conducted a laboratory study including nine nematode species and 13 strains. The level of *C. caryae* mortality observed was low to moderate (not more than 60%) for all nematodes tested, and no significant differences in virulence were detected among the species (Table 11.1). Additionally, Shapiro-Ilan, (2001a) demonstrated that nematode virulence to *C. caryae* larvae is substantially less compared with virulence to the Diaprepes root weevil, *D. abbreviatus*, a weevil that is currently controlled commercially by EPNs in some citrus orchards (see Section 11.4.). Susceptibility of *C. caryae* larvae to nematodes was shown to decrease further with larval age (Shapiro-Ilan, 2001a). Thus, Shapiro-Ilan (2001a) concluded that suppression of *C. caryae* larvae with EPNs is unlikely to be cost effective unless virulence can be substantially improved.

11.3.2.2. Potential to control adults

Adult pecan weevils may be more amenable to control with EPNs than larval-stage weevils (Shapiro-Ilan, 2001b, 2003). Laboratory studies conducted under parallel conditions

Table 11.1. Pecan weevil, *Curculio caryae*, control following exposure to entomopathogenic nematodes (EPNs) under laboratory conditions.[a]

Nematode (strain)	*C. caryae* stage	*C. caryae* control[b]
Heterorhabditis bacteriophora (Baine)	Larval	21.3a
H. bacteriophora (HP88)	Larval	41.0a
H. bacteriophora (NJ1)	Larval	42.7a
H. indica (Hom1)	Larval	40.9a
H. indica (original)	Larval	47.5a
H. marelatus (IN)	Larval	42.7a
H. marelatus (Point Reyes)	Larval	45.9a
H. megidis (UK211)	Larval	36.1a
H. zealandica (NZH3)	Larval	23.0a
Steinernema carpocapsae (All)	Larval	30.4a
S. feltiae (SN)	Larval	23.0a
S. glaseri (NJ43)	Larval	32.8a
S. riobrave (355)	Larval	37.7a
H. bacteriophora (Hb)	Adult	67.0b
H. bacteriophora (Oswego)	Adult	48.0bc
S. carpocapsae (All)	Adult	99.0a
S. feltiae (SN)	Adult	40.0c
S. riobrave (355)	Adult	67.0b

[a]Mortality was determined after 13-day (larvae) or 4-day (adults) exposure to 500 infective juveniles (IJs).
[b]Following correction for control mortality using Abbott's (1925) formula.
Note: Different letters following each number indicate statistical significance within each *C. caryae* stage. Data on larval control is presented with permission of the Entomological Society of America, *Journal of Economic Entomology* 94, 7–13; data on adult control is presented with the permission of the *Journal of Entomological Science* 36, 325–328.

used for the larvae (Shapiro-Ilan, 2001a) indicated high virulence of several nematodes to pecan weevil adults (Table 11.1) (Shapiro-Ilan, 2001b, 2003). *S. carpocapsae* was particularly virulent, killing close to 100% of the weevils; *S. riobrave* and *H. bacteriophora* also showed some potential (Shapiro-Ilan, 2001b, 2003). One economical approach for adult control may be to apply EPNs in a narrow (perhaps 1–2 m) band around each pecan tree to infect the weevils that crawl to the tree. If the banding method does not infect a satisfactory proportion of weevils, the application area would have to be expanded to cover the entire area of weevil emergence (i.e. under the tree's canopy).

Recent field trials (using the banding method) indicate *S. carpocapsae* (All) can provide 60–80% control of emerging *C. caryae* adults (Shapiro-Ilan, 2003; unpublished data), but this level of control is short-lived (not exceeding 1 week). The efficacy of this nematode might be improved by selecting a superior strain. Towards this end, Shapiro-Ilan *et al.* (2003) compared eight *S. carpocapsae* strains for various beneficial traits (virulence to adult weevils, environmental tolerance and reproductive capacity). Based on a novel beneficial trait ranking system, Breton, DD136, Italian, and Kapow strains were graded inferior to other strains, and Agriotos, All and Sal strains superior. Other important traits will need to be assessed (e.g. longevity) before a choice is made as to which strain(s) might be most suitable for *C. caryae* control. If none of the naturally occurring strains provide superior *C. caryae* suppression, then traits might be improved further through artificial selection (Gaugler *et al.*, 1989) or targeted hybridization (Shapiro *et al.*, 1997). In addition to research towards strain improvement, other parameters that must be investigated further to optimize control include irrigation requirements, and rate, method and area of application.

11.4. Citrus

11.4.1. Root weevils

Citrus is host to a complex of curculionid species that feed on the leaves and roots of trees. Most weevil species are of little economic importance; however, *D. abbreviatus* is a major pest responsible for annual losses of US$75–100 million by citrus growers in Florida and the Caribbean Basin (McCoy, 1999). The weevil infests more than 20,000 ha of citrus throughout Florida, and its range is expanding (Hall, 1995). The blue-green weevils, *Pachnaeus* spp., are also pests of citrus in Florida, although resulting damage is less severe than from *D. abbreviatus*. Like *D. abbreviatus*, *Pachnaeus* spp. are polyphagous. On citrus, they feed on young leaves, and eggs are oviposited on leaf surfaces (Fig. 11.1). The neonate larvae fall to the soil where they develop to adults over the next several months. Egg laying occurs from early summer until winter, and teneral adults emerge from the soil throughout the year (Nigg *et al.*, 2003). While in the soil, the larvae feed on the fibrous and major roots of citrus trees. Feeding by late instar larvae of *D. abbreviatus* causes severe damage to roots in the crown of the tree. Wounding of the root cortex also creates infection sites for *Phytophthora* spp., resulting in a pest–disease complex that severely debilitates and even kills trees (Graham *et al.*, 2003). There are no chemical pesticides registered in Florida for management of the soil-borne stages of weevil. Various commercially formulated EPNs have been used for this purpose since 1990.

11.4.2. Nematode efficacy

The earliest attempts to manage *D. abbreviatus* with EPNs involved laboratory studies and field trials to evaluate *S. glaseri*, *S. carpocapsae* and *H. bacteriophora*

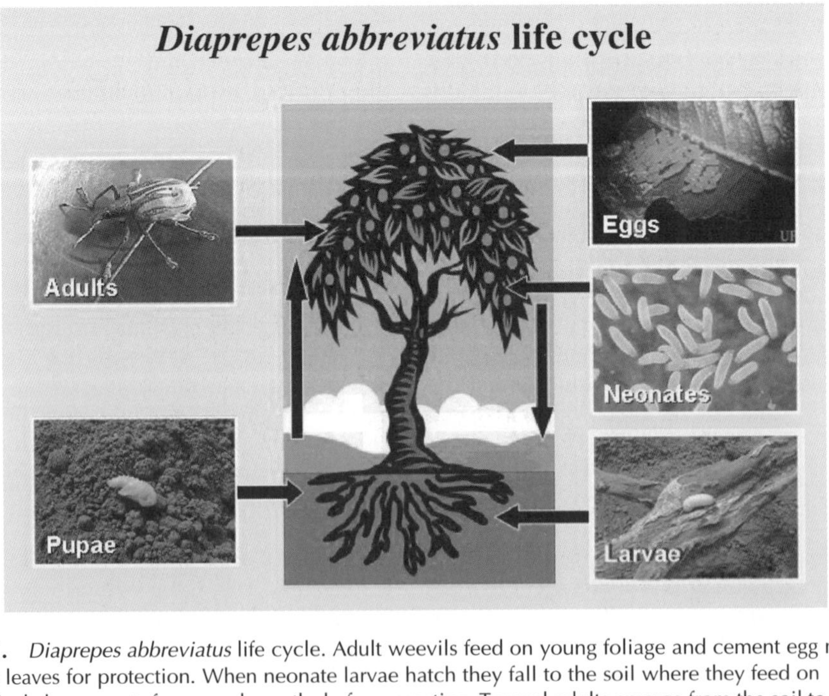

Fig. 11.1. *Diaprepes abbreviatus* life cycle. Adult weevils feed on young foliage and cement egg masses between leaves for protection. When neonate larvae hatch they fall to the soil where they feed on progressively larger roots for several months before pupating. Teneral adults emerge from the soil to reinitiate the cycle. (Note the extensive feeding channels on structural roots that promote infection by *Phytophthora* spp. Figure courtesy of Robin Stuart, University of Florida.)

(Schroeder, 1987, 1990, 1992; Downing *et al.*, 1991; Bullock and Miller, 1994). When the latter two species were used in field trials at rates ranging from 100 to 600 IJs/cm^2, the emergence of adult weevils was suppressed by as much as 60–80% for up to 1 year following treatment. Those trials resulted in widespread use of commercially formulated *S. carpocapsae* and *H. bacteriophora* by citrus growers in Florida (Fig. 11.2). Grower acceptance of EPNs resulted from the lack of effective pesticides to manage an economically important pest, and the reasonably low cost of nematode products. Despite their widespread use, the efficacy of products containing *S. carpocapsae* and *H. bacteriophora* was less apparent in subsequent field trials (Adair, 1994; Duncan and McCoy, 1996; Duncan *et al.*, 1996). In contrast, commercially formulated *S. riobrave* at rates of 100 IJs/cm^2 was found to reduce numbers of adults and weevil larvae in the rhizosphere of young trees by 80–95% within 15–30 days post-treatment (Duncan and McCoy, 1996; Duncan *et al.*, 1996; Bullock *et al.*, 1999). Laboratory trials using these and six additional EPN species revealed that *S. riobrave* and a Florida isolate of *H. indica* were significantly more effective against *D. abbreviatus* than other species evaluated, and that *H. indica* reproduces at exceptionally high levels in the insect (Schroeder, 1994; Shapiro *et al.*, 1999; Shapiro and McCoy, 2000a,b). *S. riobrave* and *H. indica* are currently the only two EPN species that are marketed in the Florida citrus industry. In 1999, approximately 20% of the hectarage infested with *D. abbreviatus* was treated with EPNs (Shapiro-Ilan *et al.*, 2002b). Populations of *Pachnaeus* spp. and *Phytophthora nicotianae* are also reduced by application of EPNs (Bullock *et al.*, 1999; Duncan *et al.*, 2002).

11.4.2.1. Factors affecting nematode efficacy

Major issues that have emerged during evaluation of commercial nematodes include product quality, application dosage, lack of persistence and regional variation in efficacy. Quality control of some nematode products has occurred periodically. Quality issues have proven correctible, but they are a serious concern because they result in reduced acceptance by growers and advisors for use of nematodes as a viable management tactic.

Generally, high levels of *D. abbreviatus* suppression have been achieved with appli-

Fig. 11.2. Methods for application of entomopathogenic nematodes (EPNs) in citrus orchards. Nematodes can be suspended in water in clean chemical-mixing tanks (A) and injected under pressure into the main irrigation line (B) for delivery via microjet-irrigation sprinklers (C). Nematodes can also be delivered via clean herbicide application equipment (D) and incorporated with irrigation. (Photographs by Gretchen Baut.)

cation rates of at least 100 IJs/cm^2. Rates recommended for *D. abbreviatus* control by commercial suppliers have tended to be considerably lower (the actual rate per unit area can depend on the size of the under-canopy being treated). Use of lower application rates can result in reduced (or absence of) efficacy and profitability (McCoy *et al.*, 2000, 2002), but the extent to which growers can increase rates in mature orchard application is constrained by cost. Suppression of larvae in soil by application of nematodes is ephemeral, in the order of 1 or 2 weeks (McCoy *et al.*, 2000, 2002). Neonate larvae soon replace many of those that were killed by nematodes. Similarly, suppression of the numbers of adult insects with non-persistent insecticides is quickly negated by recruitment of teneral adults that emerge from the soil throughout the year. Thus, there is a critical need for management tactics with greater residual activity, and the role of EPNs in future integrated pest management (IPM) programmes is unclear. The development of insect-resistant rootstocks or physical soil barriers could reduce or obviate the need for inundatively applied nematodes, as has occurred in other systems (Shapiro-Ilan *et al.*, 2002b).

Regional variation in efficacy of EPNs has become apparent (Table 11.2). Measurable efficacy has been demonstrable in most field trials conducted on Florida's central ridge (Duncan and McCoy, 1996; Duncan *et al.*, 1996, 1999, 2002, 2003), whereas efficacy has been variable, and generally poor, in trials conducted in the flatwoods regions (Adair, 1994; Stansly *et al.*, 1997; Bullock *et al.*, 1999; McCoy *et al.*, 2000, 2002). Soil texture has been implicated as a potential factor responsible for variation in efficacy (Duncan *et al.*, 2001; McCoy *et al.*, 2002). The central ridge is characterized by deep, well-drained sandy soil, whereas soils in the coastal and central flatwood areas vary in texture and factors such as salinity and drainage. Additional work is needed to characterize the edaphic factors that modulate the effectiveness of EPNs.

11.4.3. Towards regional IPM of *Diaprepes abbreviatus*

A broad continuum of damage is exhibited by orchards infested with *D. abbreviatus*. Population densities of the weevil are typically lower on the central ridge than in some regions of the flatwoods (Futch, 2002) and tree damage varies accordingly. Natural enemies may cause some of the variation in weevil population density. Endemic EPNs attack *D. abbreviatus* throughout Florida (Beavers *et al.*, 1983; Nguyen and Duncan, 2002) and were found to infect weevil larvae in soil at an average rate of 55% per week in an orchard on the central ridge compared with only 8% in an orchard on fine-textured soil in the flatwoods (Duncan *et al.*, 2003). Fine-textured, poorly drained soils are also conducive to root infection by *Phytophthora* spp. (Graham *et al.*, 2003). Flooded soil predisposes trees to greater stress from *D. abbreviatus* herbivory (Li *et al.*, 2003). Thus, a combination of edaphic factors and natural enemies may modulate the damage caused by this pest–disease complex by stressing trees and regulating the population densities of the causal agents. Additional study of regional factors that regulate damage by *D. abbreviatus* is warranted for several reasons. First, the central ridge may represent an important niche in which EPNs can be used profitably to reduce these pests below an economic threshold. Trees in some flatwood orchards may respond less favourably to nematode treatments, either because edaphic conditions are less conducive to these nematodes, tree stress is excessive, or because the pest pressure is too high to be reduced by nematodes to a non-damaging level. Second, in regions conducive to nematode activity, infection of weevil larvae by endemic nematodes can occur at a higher rate over time than that exhibited by exotic nematodes applied for insect control (Duncan *et al.*, 2003). This suggests a need to conserve the level of endemic nematodes, either by selection of application times that reduce competition with exotic nematodes or by augmentation with endemic species adapted to local conditions. Finally, a better under-

Table 11.2. Field efficacy of *Steinernema* and *Heterorhabditis* nematodes against *Diaprepes* root weevil.

Nematode	Application rate (cm^2)	% mortality[a]	References
Heterorhabditis bacteriophora	127	78	Downing *et al.*, 1991
H. bacteriophora	255	63	Downing *et al.*, 1991
H. bacteriophora	637	63	Downing *et al.*, 1991
H. bacteriophora	100	62	Schroeder, 1992
H. bacteriophora	250	0	Duncan and McCoy, 1996
H. bacteriophora	175	54	Duncan *et al.*, 1996
H. bacteriophora	255	57	Duncan *et al.*, 1996
H. bacteriophora	11	8	McCoy *et al.*, 2000
H. bacteriophora	22	8	McCoy *et al.*, 2000
H. indica	11	14	McCoy *et al.*, 2000
H. indica	22	19–21	McCoy *et al.*, 2000
H. indica	54	28	McCoy *et al.*, 2000
H. indica	11	0	McCoy *et al.*, 2002
H. indica	54	0	McCoy *et al.*, 2002
H. indica	108	27	McCoy *et al.*, 2002
Steinernema carpocapsae	250	65	Schroeder, 1987
S. carpocapsae	25	42	Schroeder, 1990
S. carpocapsae	100	50	Schroeder, 1992
S. carpocapsae	637	48	Downing *et al.*, 1991
S. carpocapsae	1666	57–82	Bullock and Miller, 1994
S. carpocapsae	153	0	Duncan *et al.*, 1996
S. carpocapsae	306	0	Duncan *et al.*, 1996
S. glaseri	250	35	Schroeder, 1987
S. riobrave	250	77–90	Duncan and McCoy, 1996
S. riobrave	120	93	Duncan *et al.*, 1996
S. riobrave	110	0–98	Bullock *et al.*, 1999
S. riobrave	ND	48–100	Bullock *et al.*, 1999
S. riobrave	11	0	McCoy *et al.*, 2002
S. riobrave	54	0–8	McCoy *et al.*, 2002
S. riobrave	108	0–36	McCoy *et al.*, 2002
S. riobrave	22	5–22	McCoy *et al.*, 2000
S. riobrave	54	30–49	McCoy *et al.*, 2000
S. riobrave	108	32–34	McCoy *et al.*, 2000
S. riobrave	216	63	McCoy *et al.*, 2000
S. riobrave	20	0–66	Duncan *et al.*, 2003

[a]Percentage mortality in treated plots, corrected for mortality in control plots. Statistical significance of treatment responses is not indicated in the table.

standing of regional factors that regulate numbers of *D. abbreviatus* may result in new insights for managing this pest in conditions that are poorly suited to the use of EPNs.

11.5. Banana

Bananas, which are grown in the tropical and subtropical areas, are a widely available fruit throughout the world. The banana stem borer, *Odoiporus longicollis*, and the banana weevil borer, *Cosmopolites sordidus*, are the most important pests. Nematodes have been used to control these pests in Australia and China, with encouraging results. These two species of insect usually occur throughout the year. The larvae and some adults feed on the base stem of the plant and bore into the stem, weakening or killing the plant.

In southern China, *O. longicollis* has six generations per year with two population peaks in March–June and November–December. Usually, the corms of the trees are cut at the base in winter after harvest. Approximately 90% of the overwintering populations in the residual stems of the banana plants are larvae that attack the banana stems the following year. EPNs are capable of migrating through living stem tissue to kill the borers; field results indicated that 76–90% of the overwintering larvae, 68–92% of the pupae and 25–80% of *O. longicollis* adults were controlled by spraying 3–6 $\times$ 10^6 IJs of *S. carpocapsae* (A24) into each residual stem base (Xu *et al.*, 1991).

C. sordidus is a major pest of bananas and plantains. Larvae burrow into corms producing severe damage, which can be exacerbated by subsequent fungal or bacterial attack. Laumond *et al.* (1979) demonstrated pathogenicity of *S. carpocapsae* to adult *C. sordidus* in laboratory trials. Figueroa (1990) demonstrated pathogenicity of several nematode species (*S. carpocapsae*, *S. feltiae* and *S. glaseri*) to *C. sordidus* larvae, and observed 100% mortality in greenhouse tests using 4000 IJs per plant. Field applications of *S. carpocapsae* in a water-thickening gel (used to keep the nematodes near the target site) to cuts or holes made in the residual banana rhizomes has provided control of larvae and adults that were attracted to cut surfaces (Treverrow *et al.*, 1991). In similar research (R. Han, 2002, unpublished data), slashing of corms at the base followed by nematode application in a polyacrylamide gel spread over the cut surface provided control of larvae as well as adults attracted to the cut corm. Kermarrec and Mauléon (1989) reported that the effects of *S. carpocapsae* on *C. sordidus* can be enhanced through synergistic interactions with chemical insecticides (e.g. chlordecone).

11.6. Litchi

Litchi is an important and high-value crop in several Asian countries, such as China, Thailand and Vietnam. In Guangdong, China, which is climatically very well suited to production of this fruit, there are over 150,000 ha of litchi orchards, comprising an estimated 30 million litchi trees. Most litchi trees are productive for 20–100 years. The value of this crop in Guangdong is over US$190 million per year, for domestic and export markets combined. The key litchi pests are the litchi stem borer, *Arbela dea*, and the litchi longhorn beetle, *Aristobia testudo*.

11.6.1. Litchi stem borer

A. dea has one generation per year, the larval stage is the damaging stage and lasts up to 9 months, beginning in June. As first instars, *A. dea* damage the lower bark of the litchi tree and then bore into the trunk as they mature. Resulting damage can weaken the trees, or cause death, depending on the litchi strain, age and location.

A. dea is susceptible to *S. carpocapsae* (A24). The nematodes are applied by conventional sprayers around the borer holes. The *A. dea* larvae are usually active just outside the borer holes at night, providing an ideal place for contact between the nematodes and the insects. Thus, the nematodes do not need to be applied directly into the borer holes (e.g. by injection), which would increase labour. Over 86% *A. dea* mortality was obtained by spraying 1000 IJs around each borer hole (Xu and Yang, 1992).

11.6.2. Litchi longhorn beetle

The most important litchi pest is *A. testudo*, which causes great economic losses to the crop. Similar to *A. dea, A. testudo* has one generation per year. The adults of this beetle lay eggs between the crotches of litchi trees. The hatched larvae, whose distribution is aggregative, bore into the stem and develop in the holes for up to 9 months (Han *et al.*, 1994). Without control the infested branches wither and die, resulting in no fruit yield.

Most tactics to control *A. testudo* are ineffective due to the inaccessibility of the larvae in tunnels. Mobile nematodes, on the other hand, actively search for the larvae of the beetle in the deepest recesses, and have been shown to produce over 73% mortality following injection of 3000 IJs *S. carpocapsae* (A24) into the freshly bored holes of the beetle (Xu *et al.*, 1995; Han *et al.*, 1996).

Successful pest control was achieved in 1700 ha of litchi orchards in Guangdong. As a result, farmers' interest in utilization of EPNs as a safe and effective control of these pests has been generated. Producers of EPNs have been less interested, however, due to competition from chemical insecticides and the relatively limited hectarage occupied by litchi pests. None the less, successful field demonstrations indicate the potential of EPNs to control these pests. Further research will focus on strain selection and formulation development to enhance migration ability and desiccation tolerance.

11.7. Summary and Conclusions

EPNs are being applied commercially for control of some important insect pests of orchards (e.g. *D. abbreviatus* and *Pachnaeus* spp. in citrus), and there are a number of cases where commercial application may be within reach: *C. pomonella* in apples, *R. indifferens* in cherries, *A. transitella* in almonds and pistachio, *C. caryae* in pecans, *C. nenuphar* in apple and stone fruits, *O. longicollis* and *C. sordidus* in banana and *A. dea* and *A. testudo* in litchi. To improve and expand the use of EPNs as inundative biocontrol agents for orchard pests, advances in research are required, particularly in reducing costs of production and application, and methodology to improve persistence of nematodes in soil or in the canopy. Additionally, inoculative or conservation approaches to biocontrol with nematodes must be explored. Various conditions associated with orchards as agroecosystems may facilitate these approaches, e.g. plant species and structural diversity, soil conditions and stability (Kaya, 1990; Barbosa, 1998; Lewis *et al.*, 1998). Various characteristics and management practices such as soil types, fertilization, irrigation, crop covers, etc. should be investigated within each specific orchard system to determine their effects on EPN ecology and potential to improve long-term efficacy.

References

Abbott, W.S. (1925) A method for computing the effectiveness of an insecticide. *Journal of Economic Entomology* 18, 265–267.

Adair, R.C. (1994) A four-year field trial of entomopathogenic nematodes for control of *Diaprepes abbreviatus* in a flatwoods citrus grove. *Proceedings of the Florida State Horticultural Society* 107, 63–68.

Agudelo-Silva, F., Lindegren, J.E. and Valero, K.A. (1987) Persistence of *Neoaplectana carpocapsae* (Kapow selection) infectives in almonds under field conditions. *Florida Entomologist* 70, 288–291.

Agudelo-Silva, F., Zalom, F.G., Hom, A. and Hendricks, L. (1995) Dormant season application of *Steinernema carpocapsae* (Rhabditida: Steinernematidae) and *Heterorhabditis* sp. (Rhabditida: Heterorhabditidae) on almond for control of overwintering *Amyelois transitella* and *Anarsia lineatella* (Lepidoptera: Gelechiidae). *Florida Entomologist* 78, 516–523.

Arthurs, S. and Heinz, K.M. (2003) Thrips parasitic nematode *Thripinema nicklewoodi* (Tylenchida: Allantonematidae) reduces feeding, reproductive fitness and tospovirus transmission by its host, *Frankliniella occidentalis* (Thysanoptera: Thripidae). *Environmental Entomology* 32, 853–858.

Barbosa, P. (1998) Agroecosystems and conservation biological control. In: Barbosa, P. (ed.) *Conservation Biological Control.* Academic Press, San Diego, California, pp. 39–54.

Beavers, J.B. and Calkins, C.O. (1984) Susceptibility of *Anastrepha suspensa* (Diptera: Tephritidae) to steinernematid and heterorhabditid nematodes in laboratory studies. *Environmental Entomology* 13, 137–139.

Beavers, J.B., McCoy, C.W. and Kaplan, D.T. (1983) Natural enemies of subterranean *Diaprepes abbreviatus* (Coleoptera: Curculionidae) larvae in Florida. *Environmental Entomology* 12, 840–843.

Bedding, R. (1990) Logistics and strategies for introducing entomopathogenic nematode technology into developing countries. In: Gaugler, R.

and Kaya, H.K. (eds) *Entomopathogenic Nematodes in Biological Control.* CRC Press, Boca Raton, Florida, pp. 233–246.

Belair, G., Vincent, C. and Chouinard, G. (1998) Foliar sprays with *Steinernema carpocapsae* against early season apple pests. *Journal of Nematology* 30, 599–606.

Belair, G., Vincent, C., Lemire, S. and Coderre, D. (1999) Laboratory and field assays of entomopathogenic nematodes for the management of the obliquebanded leafroller, *Choristoneura rosaceana* (Harris) (Tortricidae). *Journal of Nematology* 31, 684–689.

Bullock, R.C. and Miller, R.W. (1994) Suppression of *Pachnaeus litus* and *Diaprepes abbreviatus* (Coleoptera: Curculionidae) adult emergence with *Steinernema carpocapsae* (Rhabditida: Steinernematidae) soil drenches in field evaluations. *Proceedings of the Florida State Horticultural Society* 107, 90–92.

Bullock, R.C., Pelosi, R.R. and Killer, E.E. (1999) Management of citrus root weevils (Coleoptera: Curculionidae) on Florida citrus with soil-applied entomopathogenic nematodes (Nematoda: Rhabditida). *Florida Entomologist* 82, 1–7.

Cossentine, J.E., Banham, F.L. and Jensen, L.B. (1990) Efficacy of the nematode, *Heterorhabditis heliothidis* (Rhabditida: Heterorhabditidae) against the peachtree borer, *Synanthedon exitiosa* (Lepidoptera: Sesiidae) in peach trees. *Journal of the Entomological Society of British Columbia* 87, 82–84.

Cossentine, J.E., Jensen, L.B. and Moyls, L. (2002) Fruit bins washed with *Steinernema carpocapsae* (Rhabditida: Steinernematidae) to control *Cydia pomonella* (Lepidoptera: Tortricidae). *Biocontrol Science and Technology* 12, 251–258.

Deseö, K.V. and Miller, L.A. (1985) Efficacy of entomogenous nematodes, *Steinernema* spp., against clearwing moths, *Synanthedon* spp., in north Italian apple orchards. *Nematologica* 31, 100–108.

Downing, A.S., Erickson, S.G. and Kraus, M.J. (1991) Field evaluation of entomopathogenic nematodes against citrus root weevils (Coleoptera: Curculionidae) in Florida citrus. *Florida Entomologist* 74, 584–586.

Duncan, L.W. and McCoy, C.W. (1996) Vertical distribution in soil, persistence, and efficacy against citrus root weevil (Coleoptera: Curculionidae) of two species of entomogenous nematodes (Rhabditida: Steinernematidae; Heterorhabditidae). *Environmental Entomology* 25, 174–178.

Duncan, L.W., McCoy, C.W. and Terranova, A.C. (1996) Estimating sample size and persistence of entomogenous nematodes in sandy soils and their efficacy against the larvae of *Diaprepes abbreviatus* in Florida. *Journal of Nematology* 28, 56–67.

Duncan, L.W., Shapiro, D.I., McCoy, C.W. and Graham, J.H. (1999) Entomopathogenic nematodes as a component of citrus root weevil IPM. In: Polavarapu, S. (ed.) *Proceedings of Workshop on Optimal Use of Insecticidal Nematodes in Pest Management.* Rutgers University, New Brunswick, New Jersey, pp. 69–78.

Duncan, L.W., Genta, J.G., Zellers, J., Fares, A. and Stansly, P.A. (2001) Efficacy of *Steinernema riobrave* against larvae of *Diaprepes abbreviatus* in Florida soils of different texture. *Nematropica* 31, 130.

Duncan, L.W., Graham, J.H. and Zellers, J. (2002) Profitability of applications of *Steinernema riobrave*, metalaxyl and supplemental fertilisation for management of *Diaprepes abbreviatus* and *Phytophthora nicotianae* in a Florida citrus orchard. *Fourth International Congress of Nematology* 4, 192 (abstr.).

Duncan, L.W., Graham, J.H., Dunn, D.C., Zellers, J., McCoy, C.W. and Nguyen, K. (2003) Incidence of endemic entomopathogenic nematodes following application of *Steinernema riobrave* for control of *Diaprepes abbreviatus. Journal of Nematology* 35, 178–186.

Ebssa, L., Borgemeister, C., Berndt, O. and Poehling, H.M. (2001a) Efficacy of entomopathogenic nematodes against soil-dwelling life stages of western flower thrips, *Frankliniella occidentalis* (Thysanoptera: Thripidae). *Journal of Invertebrate Pathology* 78, 119–127.

Ebssa, L., Borgemeister, C., Berndt, O. and Poehling, H.M. (2001b) Impact of entomopathogenic nematodes on different soil-dwelling stages of western flower thrips, *Frankliniella occidentalis* (Thysanoptera: Thripidae), in the laboratory and under semi-field conditions. *Biocontrol Science and Technology* 11, 515–525.

Figueroa, W. (1990) Biocontrol of the banana root borer weevil, *Cosmopolites sordidus* (Gemar), with steinernematid nematodes. *Journal of Agriculture of the University of Puerto Rico* 74, 15–19.

Futch, S. (2002) Where the root weevils are? *Citrus Industry Magazine* November, 13–15.

Gaugler, R., Campbell, J.F. and McGuire, T.R. (1989) Selection for host-finding in *Steinernema feltiae. Journal of Invertebrate Pathology* 54, 363–372.

Gazit, Y., Rossler, Y. and Glazer, I. (2000) Evaluation of entomopathogenic nematodes for the control of Mediterranean fruit fly (Diptera: Tephritidae). *Biocontrol Science and Technology* 10, 157–164.

Graham, J.H., McCoy, C.W. and Bright, D.B. (2003) *Phytophthora–Diaprepes* weevil complex: *Phytophthora* spp. Relationships with citrus rootstocks. *Plant Disease* 87, 85–90.

Hall, D.G. (1995) A revision of the bibliography of the sugarcane rootstalk borer weevil, *Diaprepes abbreviatus. Florida Entomologist* 78, 364–377.

Han, R., Xu, J., Cao, L. and Liu, X. (1994) Spatial distribution pattern and sampling technique of *Aristobia testudo* (Coleoptera: Cerambycidae) in litchi orchards. *Natural Enemies of Insects* 16, 175–182.

Han, R., Liu, X., Xie, R. and Xu, J. (1996) Effects of entomopathogenic nematodes on the spatial distribution pattern of the larvae of litchi beetle *Aristobia testudo* (Coleoptera: Cerambycidae) in litchi orchards and the economic threshold. *Natural Enemies of Insects* 18, 99–105.

Harp, S.J. and Van Cleave, H.W. (1976) New records of natural enemies of the pecan weevil. *Southwestern Entomologist* 1, 38.

Harris, M.K. (1985) Pecan phenology and pecan weevil biology and management. In: Neel, W.W. (ed.) *Pecan Weevil: Research Perspective.* Quail Ridge Press, Brandon, Mississippi, pp. 51–58.

Helyer, N.L., Brobyn, P.J., Richardson, P.N. and Edmondson, R.N. (1995) Control of western flower thrips (*Frankliniella occidentalis* Pergande) pupae in compost. *Annals of Applied Biology* 127, 405–412.

Hudson, W., Bertrand, P. and Culpepper, S. (2003) *Georgia Pecan Pest Management Guide, Bulletin No. 841.* University of Georgia Cooperative Extension Service, Athens, Georgia.

Kahounova, L. and Mracek, Z. (1991) Larval mortality of *Synanthedon myopaeformis* (Lepidoptera, Sesiidae) in apple trees sprayed with *Steinernema* sp. strain *Hylobius* (*Steinernema,* Steinernematidae). *Acta Entomologica Bohemoslovaca* 88, 205–210.

Kain, D.F. and Agnello, A.M. (1999) Pest status of American plum borer (Lepidoptera: Pyralidae) and fruit tree borer control with synthetic insecticides and entomopathogenic nematodes in New York State. *Journal of Economic Entomology* 92, 193–200.

Kaya, H.K. (1990) Soil ecology. In: Gaugler, R. and Kaya, H.K. (eds) *Entomopathogenic Nematodes in Biological Control.* CRC Press, Boca Raton, Florida, pp. 93–116.

Kaya, H.K., Joos, J.L., Falcon, L.A. and Berlowitz, A. (1984) Suppression of the codling moth (Lepidoptera: Olethreutidae) with the entomogenous nematode, *Steinernema feltiae* (Rhabditida: Steinernematidae). *Journal of Economic Entomology* 77, 1240–1244.

Kermarrec, A. and Mauléon, H. (1989) Synergy between chlordecone and *Neoaplectana carpocapsae* Weiser (Nematoda; Steinernematidae) in the control of *Cosmopolites sordidus* (Coleoptera: Curculionidae). *Revue de Nematologie* 12, 324–325.

Lacey, L.A. and Chauvin, R.L. (1999) Entomopathogenic nematodes for control of codling moth in fruit bins. *Journal of Economic Entomology* 92, 104–109.

Lacey, L.A. and Unruh, T.R. (1998) Entomopathogenic nematodes for control of codling moth: effect of nematode species, dosage, temperature and humidity under laboratory and simulated field conditions. *Biological Control* 13, 190–197.

Lacey, L.A., Knight, A. and Huber, J. (2000) Microbial control of lepidopterous pests of apple orchards. In: Lacey, L.A. and Kaya, H.K. (eds) *Field Manual of Techniques in Invertebrate Pathology.* Kluwer Academic Publishers, Dordrecht, The Netherlands, pp. 557–576.

Lacey, L.A., Unruh, T.R. and Headrick, H.L. (2003) Interactions of two idiobiont parasitoids (Hymenoptera: Ichneumonidae) of codling moth (Lepidoptera: Tortricidae) with the entomopathogenic nematode *Steinernema carpocapsae* (Rhabditida: Steinernematidae). *Journal of Invertebrate Pathology* 83, 230–239.

Laumond, C., Mauléon, H. and Kermarrec, A. (1979) Données sur le spectre d'hôtes et le parasitisme du nematode entomophage *Neoaplectana carpocapsae. Entomophaga* 24, 13–27.

Lewis, E.E., Campbell, J.F. and Gaugler, R. (1998) A conservation approach to using entomopathogenic nematodes in turf and landscapes. In: Barbosa, P. (ed.) *Conservation Biological Control.* Academic Press, San Diego, California, pp. 235–254.

Li, H., Syvertsen, J.P., McCoy, C.W., Schumann, A.W. and Dunlop, J. (2003) Soil redox potential and leaf stomatal conductance of two citrus rootstocks subjected to flooding and root weevil feeding. *Proceedings of the Florida State Horticulture Society* 116, 357–368.

Lim, U.T., van Driesche, R.G. and Heinz, K.M. (2001) Biological attributes of the nematode, *Thripinema nicklewoodi,* a potential biological control agent of western flower thrips. *Biological Control* 22, 300–306.

Lindegren, J.E. and Vail, P.V. (1986) Susceptibility of Mediterranean fruit fly, melon fly, and oriental fruit fly (Diptera: Tephritidae) to the entomogenous nematode *Steinernema feltiae* in laboratory tests. *Environmental Entomology* 15, 465–468.

Lindegren, J.E., Agudelo-Silva, F., Valero, K.A. and Curtis, C.E. (1987) Comparative small-scale field application of *Steinernema feltiae* for navel orangeworm control. *Journal of Nematology* 19, 503–504.

Lindegren, J.E., Wong, T.T. and McInnis, D.O. (1990) Response of Mediterranean fruit fly (Diptera: Tephritidae) to the entomogenous nematode *Steinernema feltiae* in field tests in Hawaii. *Environmental Entomology* 19, 383–386.

McCoy, C.W. (1999) Arthropod pests of citrus roots. In: Timmer, L.W. and Duncan, L.W. (eds) *Citrus Health Management.* APS Press, St. Paul, Minnesota, pp. 149–156.

McCoy, C.W., Shapiro, D.I., Duncan, L.W. and Nguyen, K. (2000) Entomopathogenic nematodes and other natural enemies as mortality factors for larvae of *Diaprepes abbreviatus. Biological Control* 19, 182–190.

McCoy, C.W., Stuart, R.J., Duncan, L.W. and Nguyen, K. (2002) Field efficacy of two commercial preparations of entomopathogenic nematodes against larvae of *Diaprepes abbreviatus* (Coleoptera: Curculionidae) in alfisol type soil. *Florida Entomologist* 85, 537–544.

Nachtigall, G. and Dickler, E. (1992) Experiences with field applications of entomoparasitic nematodes for biological control of cryptic living insects in orchards. *Acta Phytopathologica et Entomologica Hungarica* 27, 485–490.

Nguyen, K.B. and Duncan, L.W. (2002) *Steinernema diaprepesi* n. sp. (Rhabditida: Steinernematidae), a parasite of the citrus root weevil, *Diaprepes abbreviatus* (L). *Journal of Nematology* 34, 159–170.

Nigg, H.N., Simpson, S.E., Stuart, R.J., Duncan, L.W., McCoy, C.W. and Gmitter, F.G. (2003) Abundance of *Diaprepes abbreviatus* (L.) (Coleoptera: Curculionidae) neonates falling to the soil under tree canopies in Florida citrus groves. *Journal of Economic Entomology* 96, 835–843.

Nyczepir, A.P., Payne, J.A. and Hamm, J.J. (1992) *Heterorhabditis bacteriophora*: a new parasite of pecan weevil *Curculio caryae. Journal of Invertebrate Pathology* 60, 104–106.

Patterson Stark, J.E. and Lacey, L.A. (1999) Susceptibility of western cherry fruit fly (Diptera: Tephritidae) to steinernematid and heterorhabditid nematodes in laboratory studies. *Journal of Invertebrate Pathology* 74, 206–208.

Poinar, G. O. Jr (1991) Nematode parasites. In: van der Geest, L.P.S. and Evenhius, H.H. (eds) *Tortricid Pests, Their Biology, Natural Enemies and Control.* Elsevier Science, Amsterdam, The Netherlands, pp. 273–281.

Rice, R.E. (1978a) Insect and mite pests of almonds. In: Micke, W. and Keaster, D. (eds) *Almond Orchard Management Division of Agriculture and Natural Resources Publication No. 4092.* University of California, Berkeley, California, pp. 117–124.

Rice, R.E. (1978b) Navel orangeworm, a pest of pistachio nuts in California. *Journal of Economic Entomology* 71, 822–824.

Schroeder, W.J. (1987) Laboratory bioassays and field trials of entomogenous nematodes for control of *Diaprepes abbreviatus* (Coleoptera: Curculionidae) in citrus. *Environmental Entomology* 16, 987–989.

Schroeder, W.J. (1990) Suppression of *Diaprepes abbreviatus* (Coleoptera: Curculionidae) adult emergence with soil application of entomopathogenic nematode (Nematoda: Rhabditida). *Florida Entomologist* 73, 680–683.

Schroeder, W.J. (1992) Entomopathogenic nematodes for control of root weevils of citrus. *Florida Entomologist* 75, 563–567.

Schroeder, W.J. (1994) Comparison of two steinernematid species for control of the root weevil *Diaprepes abbreviatus. Journal of Nematology* 26, 360–362.

Shapiro, D.I. and McCoy, C.W. (2000a) Susceptibility of *Diaprepes abbreviatus* (Coleoptera: Curculionidae) larvae to different rates of entomopathogenic nematodes in the greenhouse. *Florida Entomologist* 83, 1–9.

Shapiro, D.I. and McCoy, C.W. (2000b) Virulence of entomopathogenic nematodes to *Diaprepes abbreviatus* (Coleoptera: Curculionidae) in the laboratory. *Journal of Economical Entomology* 93, 1090–1095.

Shapiro, D.I., Glazer, I. and Segal, D. (1997) Genetic improvement of heat tolerance in *Heterorhabditis bacteriophora* through hybridization. *Biological Control* 8, 153–159.

Shapiro, D.I., Cate, J.R., Pena, J., Hunsberger, A. and McCoy, C.W. (1999) Effects of temperature and host age on suppression of *Diaprepes abbreviatus* (Coleoptera: Curculionidae) by entomopathogenic nematodes. *Journal of Economic Entomology* 92, 1086–1092.

Shapiro-Ilan, D.I. (2001a) Virulence of entomopathogenic nematodes to pecan weevil larvae *Curculio caryae* (Coleoptera: Curculionidae) in the laboratory. *Journal of Economic Entomology* 94, 7–13.

Shapiro-Ilan, D.I. (2001b) Virulence of entomopathogenic nematodes to pecan weevil adults (Coleoptera: Curculionidae). *Journal of Entomological Science* 36, 325–328.

Shapiro-Ilan, D.I. (2003) Microbial control of the pecan weevil, *Curculio caryae* (Coleoptera: Curculionidae). In: Dutcher, J.D., Harris, M.K. and Dean, D.A. (eds) *Integration of Chemical and Biological Insect Control in Native, Seed-*

ling, and Improved Pecan Production. Southwestern Entomologist 27(Suppl.), 101–114.

Shapiro-Ilan, D.I., Mizell, R.F. and Campbell, J.F. (2002a) Susceptibility of plum curculio, *Conotrachelus nenuphar,* to entomopathogenic nematodes. *Journal of Nematology* 24, 246–249.

Shapiro-Ilan, D.I., Gouge, D.H. and Koppenhöfer, A.M. (2002b) Factors affecting commercial success: case studies in cotton, turf and citrus. In: Gaugler, R. (ed.) *Entomopathogenic Nematology.* CAB International, Wallingford, UK, pp. 333–355.

Shapiro-Ilan, D.I., Stuart, R. and McCoy, C.W. (2003) Comparison of beneficial traits among strains of the entomopathogenic nematode, *Steinernema carpocapsae,* for control of *Curculio caryae* (Coleoptera: Curculionidae). *Biological Control* 28, 129–136.

Shapiro-Ilan, D.I, Mizell, R.F., Cottrell, T.E., and Horton, D.L. (2004) Measuring field efficacy of *Steinernema feltiae* and *Steinernema riobrave* for suppression of plum curculio, *Conotrachelus nenuphar,* larvae. *Biological Control* 30, 496–503.

Siegel, J., Lacey, L.A., Fritts, R. Jr, Higbee, B.S. and Noble, P. (2004) Use of steinernematid nematodes for postharvest control of navel orangeworm (Lepidoptera: Pyralidae, *Amyelois transitella*) in fallen pistachios. *Biological Control* 30, 410–417.

Sledzevskaya, E.R. (1987) Study of the factors determining the activity of the nematode *Neoaplectana carpocapsae* and its efficacy against orchard insect pests. In: Sonin, M.D. (ed.) *Helminths of Insects.* Amerind Publishing, New Delhi, India, pp. 152–155.

Smith, M.T., Georgis, R., Nyczepir, A.P. and Miller, R.W. (1993) Biological control of the pecan weevil, *Curculio caryae* (Coleoptera: Curculionidae), with entomopathogenic nematodes. *Journal of Nematology* 25, 78–82.

Stansly, P.A., Mizell, R.F. and McCoy, C.W. (1997) Monitoring *Diaprepes abbreviatus* with Tedder's traps in Southwest Florida citrus. *Proceedings of the Florida State Horticultural Society* 110, 22–26.

Tedders, W.L, Weaver, D.J. and Wehunt, E.J. (1973) Pecan weevil: suppression of larvae with the fungi *Metarhizium anisopliae* and *Beauveria bassiana* and the nematode *Neoaplectana dutkyi. Journal of Economic Entomology* 66, 723–725.

Treverrow, N., Bedding, R.A., Dettmann, E.B. and Maddox, C. (1991) Evaluation of entomopathogenic nematodes for control of *Cosmopolites sordidus* Gemar (Coleoptera: Curculionidae), a pest of bananas in Australia. *Annals of Applied Biology* 119, 139–145.

Unruh, T.R. and Lacey, L.A. (2001) Control of codling moth, *Cydia pomonella* (Lepidoptera: Tortricidae) with *Steinernema carpocapsae*: effects of supplemental wetting and pupation site on infection rate. *Biological Control* 20, 48–56.

Vincent, C. and Belair, G. (1992) Biocontrol of the apple sawfly, *Hoplocampa testudinea,* with entomogenous nematodes. *Entomophaga* 37, 575–582.

Weiser, J. (1955) *Neoaplectana carpocapsae,* n. sp. (Anguillaluata: Steinernematinae), novy cizopasník housenek obalece jablecného, *Carpocapsa pomonnella* L. *Bestniik Ceskoslovenske Zoologické Spolecnosti* 19, 44–51.

Xu, J. and Yang, P. (1992) The application of *Steinernema carpocapsae* nematodes for the control of the litchi stem borer, *Arbela dea. Acta Phytophylacica Sinica* 19, 217–222.

Xu, J., Yang, P. and Wu, S. (1991) The control of banana borer with *Steinernema carpocapsae* nematodes. *Chinese Journal of Tropical Crops* 12, 77–82.

Xu, J., Han, R., Liu, X., Cao, L. and Yang, P. (1995) The application of *Steinernema carpocapsae* nematodes for the control of the litchi longhorn beetle, *Aristobia testudo. Acta Phytophylacica Sinica* 22, 12–16.

Yee, W.L. and Lacey, L.A. (2003) Stage-specific mortality of *Rhagoletis indifferens* (Diptera: Tephritidae) exposed to three species of *Steinernema* nematodes. *Biological Control* 27, 349–356.

12 Soft Fruit Applications

R.S. Cowles,[1] S. Polavarapu,[2] R.N. Williams,[3] A. Thies[4] and R.-U. Ehlers[5]

[1]Connecticut Agricultural Experiment Station, Valley Laboratory, Windsor, CT 06095, USA; [2]Rutgers University, Blueberry and Cranberry Research Centre, Chatsworth, NJ 08019, USA; [3]Department of Entomology, Ohio State University, OARDC, Wooster, OH 44691, USA; [4]e-nema France, Le Columbie, 46350 Payrac, France; [5]Institute for Phytopathology, Department for Biotechnology and Biological Control, Christian-Albrechts-University Kiel, 24223 Raisdorf, Germany

12.1. Introduction

Small fruit crops comprise a diverse number of woody and herbaceous perennials belonging to several plant families (Rosaceae, Saxifragaceae, Vitaceae and Ericaceae) and are suitable for cultivation in different soil, moisture and climatic conditions. Some crops are grown worldwide, such as strawberry and grape, while cranberry production is more restricted due to its specific cultural requirements. In general, however, these crops are all affected by certain pests (root weevils and white grubs), against which entomopathogenic nematodes (EPNs) have an increasingly important role in management. Other insect pests appropriate for management with nematodes are more specific, and will be described under headings for the specific crops: blueberry, cane fruits, cranberries, currants and gooseberries, grapes and strawberries. Among these specialist pests, clearwing moths of various species appear to be particularly susceptible to infection with EPNs.

12.2. Root Weevils

Black vine weevil (*Otiorhynchus sulcatus*), strawberry root weevil (*O. ovatus*) and the rough strawberry root weevil (*O. rugosostriatus*) are the principal species of root-feeding weevils (Coleoptera: Curculionidae) injurious to small fruits (see Section 12.5.1.). Black vine weevil is the largest of these species, with the adults attaining a length of 11 mm (Fig. 12.1A). The adults of the other two species are approximately 5 mm and 7 mm long, respectively. The biology of these species is similar, with parthenogenetic females feeding extensively on the edges of leaves before laying eggs in the soil. Eggs hatch into legless larvae (Fig. 12.1A), which feed on roots through the late summer and into the autumn. Larvae complete feeding in the spring, then pupate and emerge as adults in late spring or early summer (Fig. 12.1C). Feeding of larvae on roots of plants can induce nutrient deficiencies, and can cause wilting, stunting and plant death. Internal feeding of root weevil larvae in strawberry crowns is especially devastating (Fig. 12.1B) and usually leads to plant mortality (Fig. 12.2). Root weevil larval feeding can cause girdling on plants with woody root systems, killing the plants. Feeding of adult weevils on the foliage of most small fruits is not considered likely to reduce yields for most berry crops (Garth and Shanks, 1978). The exception is when black vine weevil feeding is synchronous

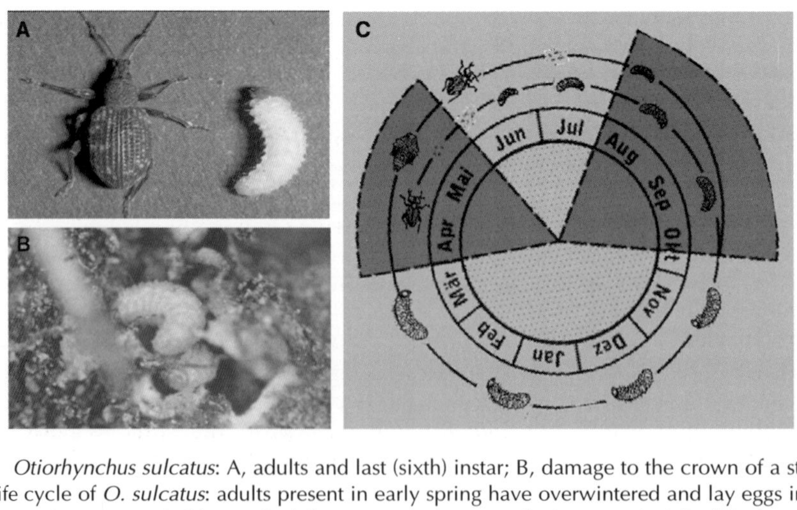

Fig. 12.1. *Otiorhynchus sulcatus*: A, adults and last (sixth) instar; B, damage to the crown of a strawberry plant; C, life cycle of *O. sulcatus*: adults present in early spring have overwintered and lay eggs in May; the new generation pupates in May and adults emerge in June; egg laying starts in July. (The best time for application of nematodes is shaded.)

Fig. 12.2. Damage caused by root weevils to strawberry, healthy (left) and wilting plant (right).

with bud break in grapes, typically in varieties that break dormancy latest in the spring. This feeding results in the loss of primary buds and new shoots, with concomitant yield reduction (Cone, 1963; Phillips, 1989). Adult weevils can become important contaminants of harvested raspberries and strawberries, so managing weevils to avoid fruit contamination may be an important management objective (Menzies, 1999).

12.2.1. Field efficacy of nematodes for weevil control

Several workers have investigated management of root weevils in field-grown small fruits, including blackcurrants (Sampson, 1994), cranberries (Shanks and Agudelo-Silva, 1990; Booth *et al.*, 2002), raspberries (Booth *et al.*, 2002) and strawberries (Simons, 1981; Scherer, 1987; Curran, 1992; Backhaus, 1994; Sampson, 1994; Berry *et al.*, 1997; Cowles, 1997; Wilson *et al.*, 1999; Booth *et al.*, 2002). Simons (1981) applied *Heterorhabditis* sp. at a rate of 5 or 10×10^9/ha in strawberries and observed 90–100% mortality in *O. sulcatus*. Scherer (1987) applied 5×10^9/ha *Heterorhabditis bacteriophora* in strawberries on sandy soils with 250 ml water/m row at 18°C and achieved 47% root weevil mortality after 25 days, 70% after 45 days and 100% after 87 days. In the second year, application was at 14°C; 97% control was achieved 24 days after application. Shanks and Agudelo-Silva (1990) evaluated *H. bacteriophora* (NC and HP88 strains) and *Steinernema carpocapsae* (All strain) against black vine weevil in cranberry bogs. Plots treated with *H. bacteriophora* on 8 April at 16×10^9/ha, with soil temperatures at the time of application of 9–12°C, did not cause reductions in black vine weevil larval populations until 10 months later, when a 70% reduction was observed. *Galleria* baiting revealed that nematodes were still present in the soil for at least the 10 months following nematode application. Using 5×10^9/ha, nematodes did not lead to black vine weevil population reductions in the same experiment. In another trial, both NC and HP88 strains of *H. bacteriophora* and *S. carpocapsae* (All strain) applied at 8.2×10^9/ha on 13 May significantly suppressed black vine weevil populations by 56%, 100% and 76%, respectively, as measured only 2 weeks after application (Shanks and Agudelo-Silva, 1990). The improved efficacy of the later application (when compared with the first experiment) was attributed to warmer soil temperatures (13–16°C) at the time of nematode application. Backhaus (1994) reported efficacy between 82% and 98% after application of

2.4×10^9/ha *H. megidis* and *H. zealandica* with 100 ml water per strawberry plant at 13°C (plant rows covered with polyethylene mulch) and also found pupae and young adults killed by the nematodes. A more recent test in cranberries comparing *S. carpocapsae* and *S. glaseri* at 3×10^9/ha resulted in 62–100% reduction in black vine weevil populations, with *S. glaseri* giving complete control (Booth *et al.*, 2002). The soil temperature during and after nematode application was 14°C or higher.

The susceptibility of strawberry root weevil to EPNs has not been studied as thoroughly as that of the black vine weevil. *H. bacteriophora* (NC-1 strain) and *S. carpocapsae* (All strain) were tested in the laboratory against strawberry root weevil, and resulted in 51% and 62% mortality, respectively (Simser and Roberts, 1994). Under field conditions, the same treatments, however, showed only 32–38% mortality. In a second field trial, *H. bacteriophora* (HP88 strain) and *S. carpocapsae* (All strain) reduced larval populations by more than 90%. Field trials with *S. carpocapsae*, *H. bacteriophora* and *H. marelatus* in strawberries resulted in consistently greater reductions in black vine weevil (36–86%) than in strawberry root weevil (0–68%) populations (Booth *et al.*, 2002).

EPNs applied to strawberry fields are likely to selectively kill species most sensitive to infection (e.g. black vine weevil) and leave many other root feeders that are less susceptible. In various tests in Connecticut, black vine weevil populations have been virtually eliminated 5 months following a May application, but smaller root weevil species and white grubs have survived into the winter. Many of these strawberry root weevils succumb in the pupal stage to nematode infection (*Heterorhabditis* spp.) during the spring – nearly 1 year after nematode application (R.S. Cowles, personal observation). The differential sensitivity of root weevil species observed in the field is supported by laboratory bioassays. Berry *et al.* (1997) found reduced susceptibility of *O. ovatus* to nematode infection compared with *O. sulcatus* to *H. marelatus* (OH10 strain) and *H. bacteriophora* (Nebraska strain), at both 14C and 18°C. The per cent mortality of the two weevil species, exposed to 5×10^9 nematodes/ha, was 20% and 40%, respectively, at 14°C, and 60% and 100% at 18°C. A field experiment directed against a mixed population of root weevil species demonstrated good efficacy of these nematodes against the root weevil larvae and pupae (50–77% mortality, 20 days after application). Another root weevil, *Phyllobius urticae*, that infests strawberries in Germany appears to be less susceptible to nematodes. Laboratory evaluation with 10 nematodes/larva of *S. carpocapsae* and *H. bacteriophora* resulted in less than 20% mortality at 20°C. An indigenous population of *S. feltiae*, found in 16.2% of the weevil larvae in the field, caused only 25% mortality in the laboratory (Pollit *et al.*, 1994). However, additional nematode species and strains need be evaluated.

12.2.2. The effects of soil temperature

Several investigations have confirmed the principle that EPNs are most effective when soil temperatures at the time of application permit them to be mobile and infect hosts. In colder regions the temperature threshold for black vine weevil larval development is lower than that required for nematode infection and/or propagation of their bacterial symbionts. This may be an important adaptation of black vine weevil for escaping infection by nematodes. In all climates, control must be achieved before the temperatures drop below 11°C, which limits the potential of most strains of *H. megidis* or *H. bacteriophora*. In a good illustration of temperature-dependent effects, Sampson (1994) related the mortality of black vine weevil from seven field experiments in strawberries to soil temperature at the time of nematode application. Control of black vine weevil with 5×10^9/ha *S. carpocapsae* improved from 36–40% at 10°C to 82–85% when soil temperatures ranged from 15°C to 19°C. Nematodes applied at intermediate soil temperatures (12°C and 13°C) resulted in intermediate (60% and 67%) larval mortality.

Optimum application rates and treatment guidelines may not yet be completely understood for management of black vine weevils with nematodes. Considerable emphasis has been placed on finding strains of nematodes that can infect and kill black vine weevil larvae at lower soil temperatures, which could permit successful application of nematodes earlier in the spring or later in the autumn (e.g. Westerman, 1998, 1999; Kakouli-Duarte and Hague, 1999; Long *et al.*, 2000; Curto *et al.*, 2001). Furthermore, finding strains of nematodes that have increased virulence (requiring fewer nematodes to kill individual larvae) could also decrease the numbers of nematodes needed to be applied to the field, which would hopefully improve their cost-effectiveness. Under laboratory conditions, *H. marelatus*, a cold temperature-adapted nematode species, is more virulent than *H. bacteriophora* against both black vine and strawberry root weevil at 14°C (Berry *et al.*, 1997). Field experiments conducted in strawberries also indicated that *H. marelata* provided 75% population reduction with an application rate of 5.2×10^9/ha, equally effective as *H. bacteriophora* applied at 38×10^9/ha. In a field trial conducted in cranberries, Berry and Liu (1999) showed that *H. marelata* and *H. megidis* applied at 5×10^9/ha provided 87% and 80% control, respectively, of black vine weevil populations. The two isolates collected from Bandon and Oregon (BPN-8 and BPS-6 strains), belonging in the *H. marelata* species group, were not effective in suppressing black vine weevil larvae.

The interaction of soil temperatures with the availability of susceptible stages of root weevils (larvae and pupae) limits the time for successful nematode application to the spring, after soil temperatures have warmed sufficiently and before eclosion of pupae to the adult stage (Shanks and Agudelo-Silva, 1990; Sampson, 1994) and the late summer and early autumn (Fig. 12.1C), when larvae are available and before soil temperatures have cooled below critical temperatures for nematode infection (Wilson *et al.*, 1999). The spring timing for nematode application may have a very brief window of opportunity, as black vine weevil development to the pupal stage is triggered by the same increase in soil temperature that is conducive for nematode infections (Smith, 1932; Garth and Shanks, 1978). Black vine weevil pupae could be easier to target for control with nematodes, based on their susceptibility to infection (Long *et al.*, 2000) and presence when soil temperatures permit infection. This strategy is only likely to result in economic benefit if the nematodes survive sufficiently in the soil (following propagation in the host) to effect control of early instar larvae of the next generation of weevils. After all, it is suppression of root injury caused by feeding of weevil larvae that is the principal goal in managing root weevils. Emphasis on preventing injury to roots has prompted efforts to find nematode strains and species that can cause infection in early instar larvae under cool soil conditions. Long *et al.* (2000) could show that *S. kraussei* was consistently more virulent than *H. megidis* at 6°C and 10°C. Willmott *et al.* (2002) compared *S. carpocapsae* with *S. kraussei* in outdoor experiments applying nematodes in early December in England. Whereas *S. carpocapsae* caused no significant mortality, *S. kraussei* killed up to 81% of the *O. sulcatus* population and was also able to persist in the soil. Field experiments in Ireland with *H. megidis* by Fitters *et al.* (2001) resulted in 76% control after autumn application. Their results indicated that application in October and November instead of September delayed weevil mortality until spring.

The reported results of several years of field trials have shown that autumn application is superior to spring application, when damage during the colder season is expected (in climatic zones where winter temperature rarely drops below 5°C). Sustainable effects can be anticipated for those species that can survive in the local soil conditions. The use of cold-active strains and species (e.g. *H. marelatus*, *S. feltiae* or *S. kraussei*) are promising approaches to overcome reduced activity at low temperature.

12.2.3. The effects of soil environment

The soil environment can influence nematode behaviour, which can translate into success or failure in biocontrol. Important abiotic factors include moisture and clay content, while biotic factors include the presence or abundance of plant roots. Increased clay content in soil can severely inhibit movement of nematodes and compromise their efficacy (Kaya, 1990). The presence of plant roots in the soil usually attracts nematodes (Choo *et al.*, 1989; Wang and Gaugler, 1998). Boff *et al.* (2001, 2002) studied the response of *H. megidis* in the presence and absence of strawberry roots and vine weevil larvae in choice experiments. In these tests, nematodes were only attracted when larvae had damaged the roots. In similar experiments using *H. bacteriophora*, the nematodes' response to roots and weevils was poor, but *S. kraussei* and *S. feltiae* were attracted to larvae. Therefore, simple generalizations about host-finding behaviour of nematodes may not adequately describe their responses, and different species may use different strategies or cues to locate hosts.

Root morphology may also influence efficacy. Sampson (1994) compared the control of black vine weevil with *S. carpocapsae* in blackcurrant and in strawberry. The poorer control in blackcurrant (34–66% larval mortality) was attributed to its larger root system, permitting black vine weevil to escape infection. In spite of poorer control, populations in blackcurrant were reduced below damage thresholds and plant vigour improved. To compensate for poorer efficacy, split applications of nematodes in the autumn and spring were suggested for blackcurrants, and were demonstrated to further improve plant growth (Sampson, 1994). Similar poor control (11–37% population reduction) of black vine weevil in red raspberry with *S. carpocapsae* (All strain) and *H. bacteriophora* (HP88 strain) could also be attributed to the size of the root system of these plants compared with cranberries or strawberries (Booth *et al.*, 2002). Root weevil larvae feeding deeper in the soil may escape infection due to two effects, lack of nematode dispersal to these greater depths and cooler soil temperatures, which could preclude infection.

The available results stress the relatively limited dispersal and host-finding potential of nematodes, and the importance of adapting application techniques to place the nematodes as close as possible to the target to achieve immediate control. Efficient placement of nematodes close to susceptible black vine weevil larvae may be of greatest importance where there is limited time for infection and recycling, or adverse soil conditions (high clay content) might limit the dispersal of the nematodes.

12.2.4. Application techniques

Spray application is suitable when nematodes are applied to soils with low clay content. The use of high volumes of water, post-application irrigation and the dissemination of straw mulch immediately after nematode application can increase nematode efficacy. Cropping systems that use polyethylene mulch pose special challenges for nematode application. Kakouli-Duarte *et al.* (1997) investigated injection of *S. carpocapsae* (Biosys strain 252) and *H. megidis* (Nemasys-H) into an irrigation system with delivery through one or two T-tape® lines per planting bed of strawberries. They obtained 88–95% mortality of black vine weevil larvae 4 weeks after applying 5000–20,000 *S. carpocapsae* or *H. megidis* per plant ($5–20 \times 10^9$/ha). Soil samples taken in a cross section across the planting bed were then baited with *Galleria* larvae to determine the distribution pattern of nematodes over time. A single irrigation line provided very poor lateral distribution of infective juvenile (IJ) nematodes, and a tenfold difference in application dosage between the first emitters and last emitters along the 80-m length of the drip tape. The double irrigation lines placed within the planting bed distributed at least some nematodes throughout the bed, to a distance of 250 m from the supply line, as measured 14 and 56 days after application. There was a dramatic reduction in nematode delivery as the distance from the supply line increased, however. Other studies recorded relatively even distributions of nematodes through drip irrigation (Kramer and Grunder, 2001; Wennemann *et al.*, 2003), particularly when the pressure in the systems was increased. At commercial plantations the situation usually changes dramatically. Testing three different drip irrigation systems at a pressure of 4 bars, the results indicated that more than 90% of nematodes were lost: most nematodes were recorded at the beginning of the tubing and none at the end (A. Thies and R.-U. Ehlers, unpublished data). These results were best explained through nematodes having settled in the tubing. Loss through settling can be reduced by adding 0.2% of the thickener carboxymethylcellulose (CMC) to the suspension (A. Peters, personal communication).

Subsurface injection could improve the effectiveness of nematodes by placing them throughout a root system using a lance (Fig. 12.3A, B). This is particularly important for effective use in soil with high clay content and/or when less active nematodes, e.g. *S. carpocapsae*, are used (Lewis *et al.*, 1992). Curran (1992) tested subsurface injection in field-grown strawberries naturally infested with black vine weevil larvae. *Heterorhabditis* sp. (isolate

Fig. 12.3. Application of nematodes: A, application with a lance connected to the tank of a commercial sprayer with nematode suspension. Bypass recirculation in the tank prevents sedimentation of the nematodes. Each plant received approximately 30,000 nematodes; B, tip of lance, the hole at the side of the tip allows exit of nematode suspension; C, application to plants by submerging roots into nematode suspension. (In this case an aquarium pump is bubbling air into the tank to prevent sedimentation. The addition of 0.4% carboxymethylcellulose (wallpaper glue) will also prevent sedimentation. Each strawberry plant requires approximately 2 ml of suspension, which should contain 40,000 nematodes.)

T390) applied at a 10-cm depth in six injections per plant was compared with a surface spray and surface application through drip emitters. In one test, the surface spray provided 86% control compared with 65% control with delivery through drip irrigation. In another test, multiple injections were more effective (79% weevil mortality) compared with application through drip irrigation or a single soil injection per plant (61% and 63% mortality, respectively). Mortality of weevils improved as the initial distribution of nematodes covered more of the root system, signifying that the nematodes did not disperse more than a few centimetres to find hosts.

A. Peters, A. Susurluk and R.-U. Ehlers (unpublished data) tested a preventive approach by applying nematodes at the time of strawberry planting by dipping roots into nematode suspensions (Fig. 12.3C). As planting is usually done at least 2 months before infestation with young weevil larvae, nematodes need to survive in the root environment in order to cause reliable control. Roots of 'Frigo' plants were submerged in a suspension of 5000 nematodes/ml of *H. bacteriophora* supplemented with 0.5% CMC to prevent nematodes from settling in the tank. Each plant received approximately 2 g of this suspension containing between 10,000 and 20,000 IJs. Planting was on 16 June. No weevil larvae were detected in the whole field at any time during the experiment. Plants were dug from the field and transferred to pots with 40 *Tenebrio molitor* larvae to assess nematode persistence. Nematodes persisted to cause approximately 60% mortality of *T. molitor* larvae 56 days after application. Even in spring of the following year over 40% larval mortality was recorded (Fig. 12.4).

A. Thies and R.-U. Ehlers (unpublished data) tested root-dipping technique in commercial strawberry fields in France. One year after planting, the number of third instar *O. ovatus* and *O. sulcatus* was not significantly different between treated and untreated rows. However, treated strawberries were in excellent condition and

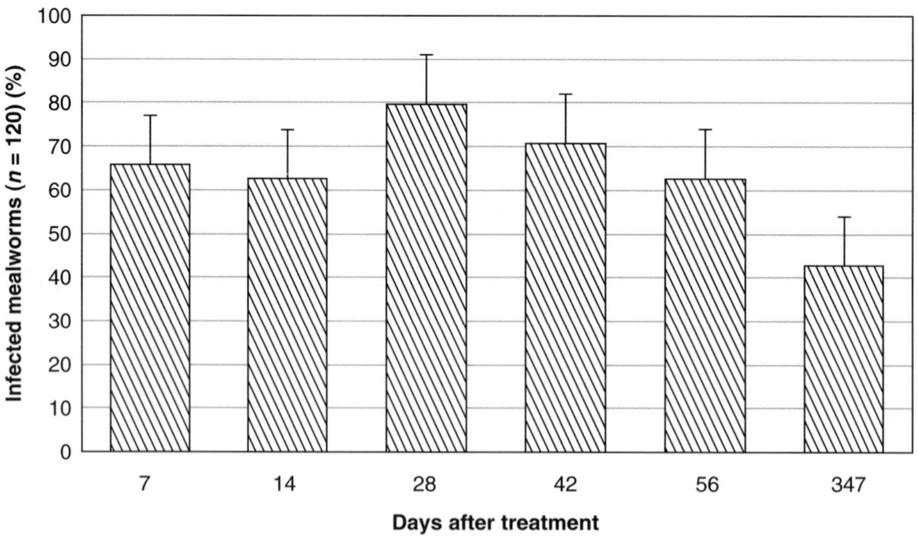

Fig. 12.4. Field persistence of *Heterorhabditis bacteriophora* assessed by *Tenebrio molitor* baiting method. Data are percent *T. molitor* larvae infected by the nematodes in pots with strawberry plants, which had received a *H. bacteriophora* treatment at 20,000 infective juveniles (IJs) per plant through submerging plant roots in nematode suspension supplemented with 0.4% carboxymethylcellulose.

yields satisfied the grower, whereas in the untreated control the plants were small or had desiccated. Weevil larvae were found in the cores of the plants, whereas in the treated rows, the larvae were found in the periphery of the root system and did not cause discernible loss of plant vigour. These effects need further investigation.

Application of nematodes, whether using an inundative or an inoculative approach, needs to produce suitable conditions for successful nematode establishment and survival (reproduction in target and alternative hosts). Factors such as soil temperature, soil type, nematode virulence towards the target insect pest, including the potential to reproduce in the host, must be considered when choosing the best application technique. In cases when weevil infestations occur routinely or when plants are grown on soils with high clay content, the preventive application by treatment of the roots with *H. bacteriophora* before planting may be the most suitable application method. Distribution through drip irrigation systems is less labour-intensive; however, the irrigation systems must be in good condition to allow for an increase in the pressure and the use of CMC to prevent sedimentation in the tubing. As more farmers use cost-intensive plastic mulching and tunnels, soil injection is the preferred method for soils that allow nematode migration. However, different behavioural strategies of nematodes must be taken into consideration (Lewis *et al.*, 1992). Spraying is most appropriate only if strawberries are in the open field without polyethylene mulch, at a temperature above 10°C and on soils with low clay content. Post-application irrigation and mulching will improve results.

12.2.5. Establishment and sustainable effects

Measuring the performance of EPNs in suppressing root weevil populations in small fruits is not always feasible, due to contamination of EPNs in untreated check plots. In strawberry field trials between 1994 and 1998, only one out of three studies conducted in Connecticut resulted in data that demonstrated control of black vine weevils following nematode application (Cowles, 1997; R.S. Cowles, unpublished data). In the successful trial, *S. carpocapsae* (All strain, applied at 7.2×10^9/ha), *S. feltiae* (Umeå strain, applied at 2.4×10^9/ha) or *H. bacteriophora* (Cruiser and Oswego strains, applied at 1.2×10^9/ha) were applied on three dates to small plots within one growing season. The 2 May 1996 application of the four treatments resulted in 33%, 20%, 20% and 36% infection of overwintered larvae, respectively, when sampled at 36 days after treatment. Evaluation of black vine weevil larval populations in late September to early October 1996 revealed populations averaging 5.5 larvae per plant in the untreated check, and reductions of black vine weevil populations in treated plots of 100%, 82%, 75% and 100%, respectively. Nematodes applied on 13 June, when black vine weevil larvae were not expected to be present, yielded complete control with the *H. bacteriophora* Cruiser strain, moderate population reductions with *S. carpocapsae* and *H. bacteriophora* Oswego strains (42% and 32%, respectively), and no significant reductions with *S. feltiae* strain. The 16 August application resulted in significant suppression of black vine weevil populations with only the *S. carpocapsae* application (100% reduction). While complete control of overwintering larvae does not appear to be a realistic objective, the propagation of nematodes in these hosts may have set up unfavourable conditions for survival of the next generation of black vine weevil larvae. Overall, a high degree of black vine weevil larval suppression appeared likely from any of the species applied, as long as there was opportunity for propagation of the nematodes in hosts within the field.

In the other two field studies in Connecticut, contamination of field plots, including untreated checks, with EPNs resulted in negligible and zero recovery of black vine weevil larvae. The surprise from these 'unsuccessful' trials was the discovery that the contaminating nematodes included species that had not been applied, including an undescribed species (*Heterorhabditis* sp., morphologically similar to *H. zealandica*, (P. Stock, 2000, Arizona, personal communication)). In some respects, these trials were exceptionally successful in demonstrating the ability of EPNs to suppress black vine weevil populations to insignificance. The discovery that EPNs were already present in growers' fields led to a survey of 55 commercial fields in September 2000, using the *Galleria* baiting technique (Bedding and Akhurst, 1975). Of the fields sampled, eight had previously been treated with EPNs. Fourteen fields were found to have nematodes, 75% (six fields) of the fields had been previously treated with nematodes and 21% (eight fields) of the fields had never been treated with nematodes. Nematode species recovered from untreated fields included *H. bacteriophora*, *S. feltiae* (both of these species were common), and single fields where *H. megidis*, *Heterorhabditis* spp. and *S. carpocapsae* occurred. Some of the strawberry fields containing native nematodes had remained productive for unusually long times (5–8 years since planting) and had moderate populations of black vine weevils during 2000. These observations, and others from field-grown nursery crops in Connecticut, suggest that the presence of moderate to large populations of black vine weevils in plantings for multiple years can permit native populations of nematodes to increase to easily detectable levels. In several situations, a collapse in the black vine weevil population coincided with the presence of EPNs in every soil sample, implicating nematodes as being partly responsible for the 'bust' in the 'boom and bust' weevil population dynamics. These observations also suggest that the long productivity of certain strawberry plantings may have partly resulted from continuous biocontrol of root weevil populations provided by the naturally occurring populations of EPNs.

12.2.6. Future perspective of weevil control with nematodes

New methods for growing strawberries, such as greenhouse or plastic tunnel production using strawberries grown in bags (Lieten and Baets, 1991), present excellent opportunities for the use of EPNs to control black vine weevil. Bag culture entails growing plants in artificial media, placing the plants in cold storage to satisfy dormancy requirements and moving the plants to a greenhouse environment to force flowering and fruiting. Black vine weevil can become a major pest under these conditions (R. Gwynn, UK, 2000, personal communication), but the warm soil temperatures and controlled growing conditions are favourable to EPNs, which permit a high level of weevil population suppression. In summary, black vine weevil is a pest especially well suited for management with EPNs. Their larvae and pupae are susceptible to infection by *S. carpocapsae*, *S. feltiae*, *S. glaseri*, *S. scarabaei*, *H. bacteriophora*, *H. marelatus* and *H. megidis* and other as yet unnamed nematode species (Klein, 1990; Cowles, 1997; R.S. Cowles, unpublished data). Along with great virulence, these nematodes also reproduce readily in black vine weevil larvae, increasing the degree of control through additional cycles of infection. Propagation of nematodes in the root weevil hosts improves the likelihood that inoculated nematodes will become naturalized in fields where they have been applied, and may be an important consideration when studying the economics of nematode use. When the objective of nematode application is to have rapid mortality of weevil larvae and an immediate high infection rate, inundative application of the most virulent species or strains, adapted to the soil temperature conditions, would be an appropriate strategy. However, even under the best soil temperature (15–20°C), moisture and application conditions, control of root weevils can be compromised by the aggregation of IJs at the most readily available hosts (Curran, 1992). This factor may make use of EPNs a poor stand-alone strategy for management of root weevils where the objective is to prevent adults from contaminating mechanically harvested fruits. Better results may be obtained if nematodes are applied preventively or if sufficient time is permitted for reproduction of nematodes in hosts, in which case less virulent species or strains of nematodes can provide excellent reductions in black vine weevil populations (Cowles, 1997). The use of recycling of nematodes in hosts to assist in long-term reductions in root weevil populations needs to be studied further, and could result in less costly inoculative rate releases of nematodes to achieve control.

12.3. Management of White Grubs

Larvae of many scarab (Coleoptera: Scarabaeidae) species are generalist root feeders. Several species of white grubs feed on the root systems of strawberries in New England, including the Japanese beetle (*Popillia japonica*), oriental beetle (*Anomala orientalis*), Asiatic garden beetle (*Maladera castanea*) and the European chafer (*Rhizotrogus majalis*) (LaMondia *et al.*, 2002). The oriental beetle is a major pest of blueberries in New Jersey and of cranberries in New England. Japanese beetle has been a focus for control efforts in Michigan due to problems with adults being mechanically harvested with blueberry fruit (R. Isaacs, Michigan, 2000, personal communication). In Europe, the major grub pests are the Garden chafer (*Phyllopertha horticola*), the June beetle (*Amphimallon* (*Rhizotrogus*) *solstitiale*) and the May beetle (*Melolontha melolontha*) (Löckener, 1994). The injury caused by white grubs is similar to that caused by root weevils: feeding of larvae on roots of plants can induce nutrient deficiencies, and can cause wilting, stunting and plant death.

The scarab pests listed above have similar life histories. For univoltine species, second or third (last) instar larvae overwinter deeply in soil, returning to feed closer to the soil surface in the spring. Some species pupate without taking up feeding again in spring. Others complete development and pupate in the soil, emerging during May or June in the northern hemisphere. The adults of different species vary in whether and to what degree they feed on foliage or flowers before mating and laying eggs. For the four species listed above as pests in New England, only Japanese beetle and Asiatic garden beetle feed as adults to any significant degree. Japanese beetles feed during the day while Asiatic garden beetle adults are nocturnal feeders. Mated females burrow into the soil to lay eggs. Eggs require adequate soil moisture to complete development and hatch. Larvae feed on roots and organic matter in the soil. Most of the species mentioned above are univoltine, but some species (e.g. some *Phyllophaga* spp., *M. melolontha* and *A. solstitiale*) may require 2–3 years to complete development and spend most of their time as larvae.

Finding a species or strain of EPNs suitable for managing a wide range of white grub species can be considered the 'Holy Grail' of many insect pathologists. Much of this focus comes from the economic importance of white grubs in turf (see Chapter 7, this volume). The deeper root systems of small fruit crops, compared with turf, might affect the distribution of white grubs in soil and thereby influence the efficacy of EPNs. However, we hope that the information developed for managing white grubs in turf can be applied towards managing these same species in small fruits. A great challenge for any practitioner wishing to manage white grubs with EPNs is their relative lack of susceptibility to currently available commercial nematode products, and the variation in susceptibility of white grub species, which can be present in mixed populations in small fruit crops (Koppenhöfer *et al.*, 2002; LaMondia *et al.*, 2002).

Japanese beetle larvae are relatively more susceptible to EPNs compared with oriental beetle, Asiatic garden beetle and European chafer (Grewal *et al.*, 2002, 2004; Koppenhöfer *et al.*, 2002). However, the principle emphasis in small fruits has been to suppress oriental beetle populations, as this species is most damaging to blueberries and strawberries in northeastern USA. There is some evidence for the feasibility of using nematodes for managing oriental beetle, based on tests in turf (Yeh and Alm, 1995; also see Chapter 7, this volume). Applications of *H. bacteriophora* (Cruiser strain) to blueberry plants at 4.9×10^9/ha in July or August gave no significant mortality (Polavarapu et al., 1998). Laboratory sand-dish (3.5 × 1.5 cm) assays demonstrated the importance of choosing nematodes with the greatest virulence – strains of *H. bacteriophora*, *H. zealandica*, *S. feltiae* and *S. glaseri* killed between 10% and 74% of test larvae (R. Stuart and S. Polavarapu, unpublished data). The most impressive results were achieved in a greenhouse test with applications of *S. scarabaei* to blueberry plants in 12 l pots. Application rates of 1.85, 3.7 and 7.4×10^9/ha nematodes resulted in a 66–86% population reduction 21 days after application. Another trial resulted in 50–56% control. The poorer results in the second trial were attributed to the larvae being much closer to pupation (S. Polavarapu and A.M. Koppenhöfer, unpublished data; Koppenhöfer and Fuzy, 2004). It remains to be seen whether the reliability of white grub biocontrol with EPNs can be improved to the point where commercially available products will be useful for managing white grubs in small fruit crops.

Results of grub control in small fruits have not been reported from Europe and we can only extrapolate from results in turf. Of the three grub species found in small fruit, *Phyllopertha horticola* can be controlled with *H. bacteriophora* (Sulistyanto and Ehlers, 1996), whereas the other two species (*A. solstitiale* and *M. melolontha*) are much less susceptible (Ehlers and Peters, 1998), leading to the available nematode products not being recommended for their control.

12.4. Blueberry Insect Management

White grubs, especially oriental beetle and Japanese beetle, cause root injury to blueberries. Tests of nematodes to control oriental beetle larvae are described in Section 12.3. The mealybug *Dysmicoccus vaccinii* Miller and Polavarapu (Homoptera: Pseudococcidae) feeds on the roots of blueberries and is typically found in association with the ants *Acanthomyops claviger* (Roger) and *Lasius neoniger* Emery (Stuart and Polavarapu, 2002). Besides stunting young plants, this insect may be a vector of red ringspot, one of the most important viral diseases of blueberries in New Jersey (Miller and Polavarapu, 1997). A large screening trial tested seven species of nematodes against this mealybug on excised roots in the laboratory (Stuart et al., 1997). Pieces of roots were placed in moist sand within Petri dishes, onto which a single mealybug was placed. Nematodes were applied in doses of 10, 50, 100 or 500 nematodes per dish, and the mortality assessed 2 and 5 days later. None of the *Steinernema* spp. or *H. bacteriophora* caused significant mortality after 2 days, but *Heterorhabditis* spp. caused infections and mortality within 5 days. In additional assessments, great variability was found between *Heterorhabditis* spp. *H. bacteriophora* (HP88 strain and two New Jersey isolates) and *H. indica* (MG-13 and EMS-13 strains) caused 65–90% mortality, but *H. zealandica* (V16 strain) and four other *H. bacteriophora* strains were ineffective. It is especially noteworthy that in this study those *Heterorhabditis* spp. that successfully infected mealybugs also reproduced. However, the production of IJs was meagre (averaging 2.8–347 nematodes per host), and would not be likely to sustain additional cycles of infection. EPNs may not be tolerant of the very acid soils (pH 3.6–5.5) where blueberries are grown (Fischer and Führer, 1990), so the best hope for using nematodes to control root mealybugs on blueberries will be through inundative releases (Stuart et al., 1997).

12.5. Cane Fruit Insect Management

Raspberries, blackberries and other *Rubus* spp. or hybrids are susceptible to a wide array of root weevil pests (see Section 12.5.1.), and two clearwing borer pests, the raspberry crown borer and the currant borer (described in Section 12.7.).

12.5.1. Additional root weevils

The clay-coloured weevil, *O. singularis* (L.), and the red-legged weevil, *O. clavipes*

(Bonsdorff), are important pests of cane fruits. *O. singularis* is damaging in the Pacific Northwest and in Europe, while *O. clavipes* is a pest in Europe (Gordon *et al.*, 1997; Menzies, 1999). The obscure root weevil, *Sciopithes obscurus* (L.), can be damaging to cane fruit and strawberries in the Pacific Northwest (Booth *et al.*, 2002). No reports are yet available on the management of these species with EPNs.

12.5.2. Raspberry crown borer

The raspberry crown borer, *Pennisetia marginata* (Harris) (Lepidoptera: Sesiidae), is native to North America, ranging across Canada and most of the USA with only a few scattered records in the Rocky Mountains and Great Plains (Beutenmüller, 1896; Duckworth and Eicklin, 1978). The wide geographic range and ubiquitous nature of this pest is probably due to wild brambles, which are omnipresent throughout its range. Birds and other animals eat berries and in turn help to transport seeds that establish new bramble patches. The sheer number of wild brambles available as a host for this pest ensures that it will continue to thrive and be a continuous threat to commercial plantings. Commercially important hosts include blackberries, loganberries, raspberries, boysenberries, thimbleberries and salmonberries (Raine, 1962). The adult moths, resembling yellowjackets, begin emerging in late summer and continue to be present through September. After mating, the adult female moths begin laying individual eggs on the underside of bramble leaves, usually along the margin. One female is capable of laying up to 200 eggs and average slightly over 100 (Slingerland and Crosby, 1915; Raine, 1962). Upon hatching, the young larvae crawl down the cane to just under the soil surface where they form a hibernaculum at the base of the stem in which they overwinter. In the spring the larvae burrow further into the cane and down into the crown. The larvae complete their second winter within the plant. In their second spring, the larvae often tunnel into the canes, causing the canes to break just above ground level. Pupation takes place in mid- to late summer, usually in old excavated galleries.

Larval feeding occurs every year, causing the vigour and yield of the planting to slowly decline as the population within the planting increases. The raspberry crown borer often goes undetected, due to the cryptic nature of its larval stage, until the plants are devastated. Reports of damage to blackberry and red raspberry plantings range from 60% to 100% of those crowns inspected (Headlee and Ilg, 1926; Clark, 1934; Hanford, 1952; R. Williams, unpublished data).

Capinera *et al.* (1986) evaluated *S. feltiae* for control of the raspberry crown borer. Individual clumps of plants were inoculated with 60,000 IJs in 150 ml water on 24 June or on 7 July. An additional 1500 ml of water was used to moisten the soil after application, and the investigators covered the drenched area with soil and leaf debris to protect the freshly applied nematodes from harmful sunlight. Subsets of ten plants were sampled at 3 and 5 days after treatment to evaluate borer mortality. There were no differences in evaluating mortality at 3 days and 5 days post-treatment, but the nematode application caused between 33% and 67% (average of 50%) mortality of raspberry crown borer larvae. The percentage of infected larvae over this short period of time, and the observation in other sesiids where recycling occurs (Miller and Bedding, 1982), suggests that *S. feltiae* may be well suited for commercial control of raspberry crown borer larvae. Further investigations will be needed to determine optimum nematode use rates, and whether repeated nematode applications can improve the degree of raspberry crown borer control.

12.6. Cranberry Insect Management

Several major pests of cranberries are potential targets for management with insecticidal nematodes. Several species of soil-dwelling immature insects mainly belonging to Lepidoptera and Coleoptera cause significant damage to roots and stems of cranberries throughout the cranberry production areas of North America. Several grub species that infest cranberries (including oriental beetle, black vine weevil and strawberry root weevil, described above) cause similar types of injury, mainly devouring fine roots, often so extensively that the plants may be easily pulled up along with the surface soil and rolled back like a carpet. This feeding can cause severe stunting and spindling of plants and, in the most severe cases, plants may die, leaving bare patches of bog. Weeds may then exploit these bare patches, making re-establishment of plants difficult and expensive. Among the coleopteran pests, only root weevils, cranberry rootworm and striped colaspis cause damage in the adult stage. In most cases, damage caused by adults is relatively unimportant.

In the Pacific Northwest, cranberry girdler and the root weevil complex are the target pests for insecticidal nematodes, whereas cranberry girdler is the target in Wisconsin. In Massachusetts, nematodes are applied for managing cranberry girdler, the root weevil complex and scarab grubs. Cranberry rootworm is the most significant root-infesting pest in New Jersey (Polavarapu and Stuart, 1997).

The resurgence of grub infestations in cranberries since the mid-1980s has been attributed to the ban on organochlorine insecticides, which were in use until the mid-1970s (Averill and Sylvia, 1998). Currently, with the exception of a granular formulation of diazinon registered in some regions for the management of cranberry girdler, there are no effective chemical insecticides registered for managing the majority of the soil insect pests on cranberries. Maintaining a flooded bog from mid-May to mid-July has been shown to be effective in managing several scarab grubs. However, this treatment will result in a total loss of crop for that year and reduced yields the following year (Averill and Sylvia, 1998).

Over the past decade, nematodes have been found to provide acceptable control of several cranberry pests. EPNs are especially suitable for use in cranberries because of some unique environmental conditions in which cranberries are grown. The cranberry root zone has high soil moisture levels and relative humidity, is protected from direct sunlight (and from ultraviolet (UV) radiation), and temperatures rarely reach levels harmful to nematodes. Insecticidal nematode use in cranberries has increased over 10 years, from 193 ha in 1989 to 831 ha in 1998. In recent years, nematode use has been mainly concentrated in Wisconsin, British Columbia and Massachusetts for managing cranberry girdler and the root weevil complex (Polavarapu, 1999a).

12.6.1. Cranberry girdler

Cranberry girdler, *Chrysoteuchia topiaria* (Zeller) (Lepidoptera: Pyralidae), larvae chew on the stems, runners and to a lesser extent on roots, during July through mid-September. Injury to stems is characterized by girdling in close association with light brown frass. Cranberry girdler injury is different from the damage caused by root-feeding coleopteran grubs. Girdled stems result in the death of individual uprights rather than the more generalized decline seen over a larger area with root-feeding grubs. The root weevil and scarab grubs feed on both roots and the bark of stems. Often, the grub damage to the bark appears similar to cranberry girdler injury, but cranberry girdler seldom causes damage as deep in the soil as that caused by the various grubs. The severity of damage to plantings is not fully apparent until the following spring. Cranberry fruitworms are univoltine, and adult moths fly from mid-June to mid-August (Dittl and Kummer, 1997).

Initial work on cranberry girdler was mainly conducted with *S. carpocapsae*. Dapsis (1993) reported that in laboratory assays, IJs of *S. carpocapsae* infected approximately 60% of newly hatched girdler larvae in 15-cm-diameter plastic arenas. In field trials, mortality of cranberry girdler larvae enclosed in $5 \times 5 \times 0.6$ cm stainless steel (40 mesh) cages placed in the field treated with formulated *S. carpocapsae* at 4.9×10^9/ha ranged between 44% and 88%. In another field trial conducted in Oregon, Smith *et al.* (1993) reported that application of formulated *S. carpocapsae* at 4.9×10^9/ha reduced the third and fourth instar larval population by 92%.

More recently, several species of heterorhabditids and steinernematids were evaluated against cranberry girdler (Berry and Liu, 1998; Henderson and Singhai, 1999). Several strains of *H. marelatus* and *H. bacteriophora* significantly reduced cranberry girdler populations in microplots (5670 cm^3) treated at $1.2–2.5 \times 10^9$/ha (Berry and Liu, 1998). In square-metre plots, *H. marelatus* significantly reduced the cranberry girdler population at both application rates, and also was found to persist for at least 6 weeks. Over the 6-week period soil samples were brought back to the laboratory and tested for presence of EPNs with wax moth larvae. A significant increase in wax moth mortality suggested that nematodes had been propagating in girdler larvae. In a laboratory study, Simard *et al.* (2002) found that *H. megidis* and *S. glaseri* have significantly lower LC_{50} values compared with *S. carpocapsae* and *S. feltiae*.

Henderson and Singhai (1999) compared the efficacy of *H. bacteriophora, S. carpocapsae* and *S. kraussei* applied at 7.4×10^9/ha. Larval populations were significantly lower in plots treated with *H. bacteriophora* and *S. carpocapsae*, but *S. kraussei* had no effect on larval populations. Both these studies have convincingly shown that *Heterorhabditis* spp. have potential in managing cranberry girdler larvae. However, considering the difficulties involved in formulating *Heterorhabditis* sp. and shorter shelf-life compared with *S. carpocapsae* formulations, there may not be any significant advantage in using *Heterorhabditis* sp. for managing this pest. None the less, attributes such as cold tolerance and efficacy at lower rates exhibited by *H. marelatus* may provide the additional incentive for its further development for cranberry girdler management.

12.6.2. Cranberry rootworm

The cranberry rootworm, *Rhabdopterus picipes* (Olivier) (Coleoptera: Chrysomelidae), adults are dark brown, shiny and about 5 mm long. The adults feed for a 2-week period in late spring or early summer on foliage of many ericaceous plants and other broadleaved evergreens, chewing curved cuts or holes in leaves and damaging new growth (Olivier and Chapin, 1980). Larval feeding is most important in cranberries – their feeding is similar to root weevils.

Several species of steinernematids and heterorhabditids were evaluated against field-collected last instar cranberry rootworms in the laboratory using Petri dishes (3.5×1.5 cm) filled with moist sand (Stuart and Polavarapu, 1997). Results indicated that various species and strains of *H. bacteriophora* and *S. glaseri* at 50–500 IJs per dish were capable of infecting and killing cranberry rootworm larvae. However, in some assays, the infection process for this insect appears to proceed relatively slowly with maximum mortality often not being achieved until about 15 days after the beginning of exposure. Heterorhabditids were generally more effective than steinernematids under laboratory conditions (Polavarapu, 1999b). More recently, a scarab-specific nematode species, *S. scarabaei*, and *H. bacteriophora* (TF strain) were evaluated against field-collected last instar cranberry rootworms using the Petri dish assay (S. Polavarapu and A.M. Koppenhöfer, unpublished data). Fifty or 100 IJs were released per dish, and mortality of larvae was recorded at 4, 7, 10, 14 and 21 days after inoculation. The rate of kill with *S. scarabaei* was much slower than with *H. bacteriophora*. With *S. scarabaei*, there was ~20% mortality at 7 days, and a maximum of 40% mortality after 21 days. *H. bacteriophora* caused 35% mortality after 4 days, and 100% mortality after 14 days. The excellent results with *H. bacteriophora* suggest that this pest may be a suitable target for management with nematodes in field situations.

In a separate field trial, *H. bacteriophora* supplied by Bio Integrated Technologies (BIT, Italy) and nematodes produced *in vivo* by Integrated BioControl Systems (IBCS, Indiana) both at 5×10^9/ha were compared with imidacloprid applied at 283–566 g a.i./ha (Polavarapu *et al.*, 2000). The cranberry rootworm grub populations were 54% and 79% lower in IBCS and BIT nematode treatments compared with untreated control plots. The cranberry rootworm mortality was equivalent to plots treated with BIT nematodes or with either rate of imidacloprid.

12.6.3. Other scarabs in cranberries

Besides Japanese beetle and oriental beetle (see Section 12.3.), several other species of scarabs feed on the root systems of cranberries. These include the cranberry root grub (*Lichnanthe vulpina* Hentz), the cranberry white grubs (*Phyllophaga georgiana* (Horn) and *P. anxia* (LeConte)), the grape anomala (*Anomala lucicola* Fab.), the striped colaspis (*Colaspis costipennis* Crotch) and two *Hoplia* spp. (*Hoplia modesta* (Haldeman) and *Hoplia equina* (LeConte)). The *Phyllophaga* spp. require 3 years to complete development (Vittum *et al.*, 1999). Two studies of nematodes have been reported for control of cranberry root grubs. Dapsis (1991) reported evaluation of *S. feltiae* (strains 27 and 980) under field conditions at 2.5 and 4.9×10^9/ha. At both rates *S. feltiae* was ineffective in suppressing the root grub populations. Weber and Henderson (1998) reported about 20% mortality of cranberry root grub with *H. bacteriophora* applied at 4.9×10^9/ha, although as many as 60% of the recovered grubs had nematode infections. The high rate of infection and low rate of mortality suggests that cranberry root grub may have a strong immune response against nematode infections.

Cranberry white grub, *P. anxia*, is the largest among the scarab grubs infesting cranberries and possibly the most difficult grub species to manage with insecticidal nematodes. Dapsis (1991) reported evaluation of *S. carpocapsae*, *S. feltiae* and *H. bacteriophora* applied in July and August against cranberry white grub in Massachusetts. None of the applications in July provided significant suppression of cranberry white grub. August applications were inconclusive, with *S. feltiae* (strain 980) providing significant suppression of grub populations in one of the two bogs. *S. glaseri* (Biosys) and *H. bacteriophora* (Ecogen) were also evaluated at 2.5×10^9/ha in Wisconsin during 1992 and 1995 (Dapsis, 1993; Dittl, 1996). In both years, these nematode species failed to suppress cranberry white grub. More recently, Weber and Henderson (1998) reported similar poor results with *H. bacteriophora* (Ecogen source).

Recent laboratory experiments have compared the virulence of *H. bacteriophora*, *S. glaseri* and *S. scarabaei* against field-collected third instar *P. georgiana* (Koppenhöfer *et al.*, 2003). Tests were conducted in a manner similar to standard turf white grub bioassays, with one grub in each 30-ml plastic container, filled with soil and provided with germinating perennial ryegrass seed for food. Treatments consisted of inoculation with 400 IJs, and an additional treatment dosage of 50 IJs of *S. scarabaei*. Mortality assessments at 7 and 14 days after inoculation revealed that *S. glaseri* was ineffective, *H. bacteriophora* was mediocre (<50% mortality after 14 days) and *S. scarabaei* gave 37% and 91% mortality after 14 days for the 50 and 400 nematode dosages, respectively. A follow-up controlled greenhouse test (S. Polavarapu and A.M. Koppenhöfer, unpublished data) investigated mortality of early instar (first and early second instar) versus late third instar larvae in cranberry plants transplanted to 2.5-l pots and dosed with 4.9×10^9/ha *S. scarabaei*. When evaluated at 21 days after inoculation, mortality was 83% and 10%, respectively, for the early and late instars (the 10% mortality was not statistically different from the untreated check). These data suggest that *S. scarabaei* may have potential for management of *Phyllophaga* spp. in cranberry, especially if directed against early instar larvae.

Hoplia grub management has been investigated by Weber and Henderson (1998), in a small plot field trial in Massachusetts. *H. bacteriophora* (Cruiser strain) was applied at a rate of 2.5 and 4.9×10^9/ha. The mortality of *Hoplia* grubs collected from treated plots was 30% and 60% for the low and high rates, respectively.

More work is needed to evaluate whether *S. scarabaei*, and other promising species or strains of nematodes, can control the full complex of white grub species feeding on cranberries under field conditions. It is likely, even if one species works well against all the white grubs, that cranberry insect management may require application of more than one species of nematode to suppress additional species of pests, such as cranberry girdler and black vine weevil larvae.

12.7. Currant and Gooseberry Insect Management

Currants and gooseberries are subject to severe stunting resulting from root feeding by black vine weevil. Experiments on the use of nematodes to control black vine weevil on currants are reported in Section 12.5.1.

The currant borer, *Synanthedon tipuliformis* (Clerck) (Lepidoptera: Sesiidae), is a pest of currants, gooseberries and raspberries in North America, Asia, Europe, Australia and New Zealand. Larvae tunnel in the stems of host plants, completing development in early spring. Adult moths in the northern hemisphere emerge in May and June, mate, and lay eggs in crevices of the bark of hosts (Taft *et al.*, 1991). Feeding within the canes of hosts reduces plant vigour, with accompanying losses in yield and quality. Up to 56% yield loss has been reported in blackcurrants, and 90% yield losses in red currants (Bedding and Miller, 1981).

EPNs were investigated for controlling the currant borer in Tasmania (Bedding and Miller, 1981; Miller and Bedding, 1982). In the first study, three nematode species were tested in controlled environmental conditions to determine whether it would be possible to disinfest canes being used as propagation material. *H. bacteriophora* (T310 strain), *S. carpocapsae* (Agriotos strain) and *S. feltiae* (T335 strain) were reared on an artificial medium, then applied to infested blackcurrant canes in doses ranging from 250 to 8000 nematodes/ml. The mortality of larvae in canes was evaluated 4 days after nematode application. Strikingly different LD_{50} values of 810, 360 and 22 nematodes, respectively, were obtained for the three nematode species. The most effective nematode, *S. feltiae*, was then included in additional tests to investigate application methods to achieve labour and nematode use efficiency. Spraying nematode suspensions was found to be more effective than dipping canes into vats. The resulting mortality of currant borer larvae was dependent on nematode concentration, with nematode concentrations exceeding 25,000/ml giving greater than 99% larval mortality. The greater efficacy of *S. feltiae*, in comparison to the other two nematode species, was attributed to their enhanced attraction to entry holes within a cane and an ability to move within larval tunnels without a film of water, via repeated coiling and uncoiling behaviour.

In the second study (Miller and Bedding, 1982), *S. feltiae* was tested for its ability to control currant borers infesting a commercial planting. Nematodes were prepared in a suspension of 1×10^8 nematodes in 10 l, applied to 50 bushes in early dormancy, just after leaf fall. This translates to a field rate of 1.7×10^{10}/ha. Bushes were sampled at 14, 26, 50 and 140 days after spraying to assess the survival of currant borer larvae. Borer infection rate increased over time, from 47% at 14 days to 72% at 140 days after inoculation. The increasing per cent infestation and the presence of recently infected larvae at 140 days post-application demonstrated that secondary cycles of nematode infections were occurring in the field. Nematodes also were applied in late dormancy or just before bloom. With all three applications, the authors were careful to choose conditions with precipitation immediately before or following the spray, which would allow nematodes time to travel over the plant surface to find openings to the currant borer galleries. Overall, the later timing (just before bloom) resulted in the highest degree of larval infection with nematodes, ranging from 68% to 90%. The authors also demonstrated that spray suspension not impinging on canes could be recovered for reuse, a concept that appears suited for adoption with modern recirculating sprayers.

12.8. Grape Insect Management

Grapes are subject to root and bud injury from black vine weevil feeding (described in Section 12.5.1.) and various white grub species. Other subterranean pests that could be targeted with EPNs include the grape root borer and grape phylloxera.

12.8.1. Grape root borer

The grape root borer, *Vitacea polistiformis* (Harris) (Lepidoptera: Sesiidae), is a major pest of grapes in eastern USA. Distribution of this species ranges from as far south as central Florida to as far north as southwestern Michigan (Snow *et al.*, 1989; Taft *et al.*, 1991). Its range includes all of the Atlantic coastal states south of Connecticut and west to Kansas. Extensive damage from this pest occurs in the southern states, gradually decreasing further north (Jubb, 1982; Alm *et al.*, 1989).

This clearwing moth attacks the roots of wild and cultivated grapes. The life cycle in northern states requires 2 years, approximately 23 months of which is spent in larval stages feeding within root tissue. In southern states the life cycle is often completed in a single season. Pupation begins in early summer of the second season, and depending on the geographic location, adults fly from mid-June through October. Damage caused by the grape root borer has resulted in enormous losses to the commercial grape industry. It has been blamed for the destruction of entire vineyards in Florida (Adlerz and Hopkins, 1981), and in South Carolina it is responsible for the total cessation of grape production (Pollet, 1975). A single grape root borer larva can destroy one of the main roots supplying the vine with nutrients, impairing winter hardiness, fruit and juice quality, and can reduce yield by 50% (All *et al.*, 1982). Two or three larvae within the root system are capable of killing the entire vine.

Early tests of EPNs against grape root borer were stimulated by the observation that naturally occurring populations of *S. carpocapsae* (All strain) were negatively correlated with population densities of grape root borer larvae (Saunders and All, 1985). Despite the demonstration of ~80% mortality in the laboratory, field applications only resulted in transient control (<7 days) of newly hatched larvae (All *et al.*, 1980; Saunders and All, 1985). Williams *et al.* (2002) screened 17 strains/species of nematodes for control of the grape root borer. Nematodes (500 IJs) were added to 30-ml sand-filled arenas in which was embedded a grape root borer feeding within an excised grape root. In these laboratory trials, *H. bacteriophora* (GPS11 strain) and *H. zealandica* (X1 strain) caused 90% and 85% mortality, respectively. Seven other strains of *H. bacteriophora* and four other *Heterorhabditis* spp. were tested, but the degree of infection varied greatly for these species or strains. Among steinernematids, *S. carpocapsae* was the most effective, causing 70% mortality of the grape root borer larvae. Greenhouse trials used potted grapes artificially infested with two grape root borer larvae per plant. In the first test, 15,000 *H. bacteriophora* or *H. zealandica* per pot caused 16% and 53% mortality, respectively. In the second test, dosages of 15,000, 30,000 and 60,000 nematodes per pot yielded mortalities of 55%, 63% and 95%. The highest rate represents a field use of 9.8×10^9/ha nematodes, but only the area around the vines needs to be treated. These results suggest that EPNs may be useful for grape root borer biocontrol, and preliminary field results are encouraging.

12.8.2. Grape phylloxera

Grape phylloxera, *Daktulosphaira vitifoliae* (Fitch) (Homoptera: Phylloxeridae), is a pest of worldwide importance in the culture of grapes. The small root form of the phylloxera sucks the sap from the roots of various grape species, leading to poor yields and plant mortality. Nearly all vinifera grapes are grafted to phylloxera-resistant rootstocks to avoid injury from this pest. New biotypes of phylloxera have overcome the resistance in some rootstocks, so alternative methods for controlling this pest are needed (English-Loeb *et al.*, 1999).

Laboratory research trials (English-Loeb *et al.*, 1999) have investigated the possibility of using EPNs to control grape phylloxera. Colonies of phylloxera on pieces of grape roots were exposed to *S. glaseri* (strain 326) or *H. bacteriophora* (Oswego strain). Phylloxera were not susceptible to infection by *S. glaseri*, but were infected with *H. bacteriophora*. The mortality due to nematode infection improved as soil moisture increased from 11% to 17%, and a dose–response relationship was obtained for phylloxera mortality relative to nematode inoculation density. Unfortunately, although *H. bacteriophora* infection caused characteristic reddish discoloration in infected individuals, there was no indication that the nematodes could reproduce in phylloxera. The experiments also may have used impractical numbers of nematodes for field application. For example, the dosage equivalent to 4.9×10^9/ha resulted in 35% mortality, and the dosage required to achieve 75% mortality in a 10-day exposure period was equivalent to 4×10^{10}/ha. Though not immediately practical, the authors point out that only two species of nematode were tested and that other species could have more favourable characteristics for targeting phylloxera for biocontrol.

12.9. Strawberry Insect Management

Strawberries are prone to injury from several root weevil and white grub species, discussed in the earlier sections. Additional pests that could eventually be targeted with EPNs include the strawberry crown borer, *Tyloderma fragariae* (Riley) (Coleoptera: Curculionidae), and the strawberry crown moth, *Synanthedon bibionipennis* (Boisduval) (Lepidoptera: Sesiidae). The strawberry crown borer is found throughout North America except in the Rocky Mountains and other high-altitude areas. The adult beetle appears in June or July and deposits eggs near the crown of the plant. After hatching, the larva burrows down into the plant where it continues to feed until fully grown. New adults begin to emerge during the month of August. Almost all infested plants die. The strawberry crown moth is a serious clearwing borer pest of strawberries and also occurs on raspberries, blackberries and loganberries in the Pacific Northwest (Slingerland and Crosby, 1915). Adult moths are active in May in the southern part of their range and are still active in July in their northern range. During this time they deposit eggs on the strawberry crown at the base of the leaves. The caterpillar burrows into the crown, at first feeding near the surface, but later eats out the whole interior of the main root, thus killing the plant (Mass, 1998). At present, nematodes have not been tested for the management of either the strawberry crown borer or crown moth.

12.10. Conclusions

Although small fruits constitute a diverse array of crops grown in many climatic and soil conditions, many pests in common are found damaging these crops. Among them, root weevils, white grubs and clearwing moths are the most important targets for EPNs. Benefits are clearly demonstrable for suppressing black vine weevil populations with nematodes in multiple crops, and nematodes have become an increasingly important tool for managing multiple pests in cranberries. It is especially promising in these crop systems that EPNs survive long after their application, suggesting that they may become permanently established in soil and will continue to provide some degree of sustained biocontrol. Along with these relatively successful uses of EPNs, there are many examples of seemingly appropriate target pests (e.g. white grubs and most species of clearwing borers) with inadequate data to demonstrate reliable and practical biocontrol. For further progress, future research will need to match the most effective nematode species with specific pest–climate–soil texture combinations. Application technology may also have to be developed to place nematodes close to the target pest. Root dips with nematode suspensions appear to be a new application method with good potential for field use. Root dips and soil injections are especially important for allowing nematodes to reach potential hosts deeper in soil or in clay soil, thereby increasing the likelihood of infection. With further research and experience, we hope that EPNs will become increasingly reliable and useful to growers for managing more species of pests.

References

Adlerz, W.C. and Hopkins, D.L. (1981) Grape insects and diseases in Florida. *Proceedings of the Florida State Horticultural Society* 94, 331–336.

All, J.N., Saunders, M.C., Dutcher, J.D. and Javid, A.M. (1980) Susceptibility of grape root borer larvae, *Vitacea polistiformis* (Lepidoptera: Sesiidae) to *Neoaplectana carpocapsae* (Nematoda: Rhabditida): potential of host kairomones for enhancement of nematode activities in grape vineyards. *Miscellaneous Publications of the Entomological Society of America* 12, 9–14.

All, J.N., Saunders, M.C. and Dutcher, J.D. (1982) Control of the grape root borer. *Eastern Grape Grower and Winery News* 8, 42–43.

Alm, S.R., Williams, R.N., Pavuk, D.M. and Heinlein, M. (1989) Distribution and seasonal flight activity of male grape root borers (Lepidoptera: Sesiidae) in Ohio. *Journal of Economic Entomology* 82, 1604–1608.

Averill, A.L. and Sylvia, M.M. (1998) *Cranberry Insects of the Northeast. A Guide to Identification, Biology and Management.* Cranberry Experiment Station, University of Massachusetts, East Wareham, Massachusetts.

Backhaus, G.F. (1994) Biological control of *Otiorhynchus sulcatus* F. by use of entomopathogenic nematodes of the genus *Heterorhabditis. Acta Horticulturae* 364, 131–142.

Bedding, R.A. and Akhurst, R.J. (1975) A simple technique for the detection of insect parasitic rhabditid nematodes in soil. *Nematologica* 21, 109–110.

Bedding, R.A. and Miller, L.A. (1981) Disinfesting blackcurrant cuttings of *Synanthedon tipuliformis*, using the insect parasitic nematode, *Neoaplectana bibionis. Environmental Entomology* 10, 449–453.

Berry, R.E. and Liu, J. (1998) *Insect Control in Cranberry with Indigenous Entomopathogenic Nematodes.* Cranberry Institute, Wareham, Massachusetts.

Berry, R.E. and Liu, J. (1999) Biological control of cranberry insects with beneficial nematodes. In: Deziel, G. and Hogan, M. (eds) *Cranberry Research Compilation. Progress and Final Reports on Cranberry Research Conducted in* 1998. Cranberry Institute, Wareham, Massachusetts, pp. 67–70.

Berry, R.E., Liu, J. and Groth, E. (1997) Efficacy and persistence of *Heterorhabditis marelatus* (Rhabditida: Heterorhabditidae) against root weevils (Coleoptera: Curculionidae) in strawberry. *Environmental Entomology* 26, 465–470.

Beutenmüller, W. (1896) Critical review of the *Sesiidae* found in America, north of Mexico. *Bulletin of the American Museum of Natural History* 8, 111–148.

Boff, M.I.C., Zoon, F.C. and Smits, P.H. (2001) Orientation of *Heterorhabditis megidis* to insect hosts and plant roots in a Y-tube sand olfactometer. *Entomologia Experimentalis et Applicata* 98, 329–337.

Boff, M.I.C., van Tol, R.H.W.M. and Smits, P.H. (2002) Behavioural response of *Heterorhabditis megidis* towards plant roots and insect larvae. *Biocontrol* 47, 67–83.

Booth, S.R., Tanigoshi, L.K. and Shanks, C.H. Jr (2002) Evaluation of entomopathogenic nematodes to manage root weevil larvae in Washington State cranberry, strawberry, and red raspberry. *Environmental Entomology* 31, 895–902.

Capinera, J.L., Cranshaw, W.S. and Hughes, H.G. (1986) Suppression of raspberry crown borer *Pennisetia marginata* (Harris) (Lepidoptera: Sesiidae) with soil applications of *Steinernema feltiae* (Rhabditida: Steinernematidae). *Journal of Invertebrate Pathology* 48, 257–258.

Choo, H.Y., Kaya, H.K., Burlando, T.M. and Gaugler, R. (1989) Entomopathogenic nematodes host-finding ability in the presence of plant roots. *Environmental Entomology* 18, 1136–1140.

Clark, H.J. (1934) A control for the raspberry crown borer. *New Jersey Agricultural Experiment Station Circular 304.*

Cone, W.W. (1963) The black vine weevil, *Brachyrhinus sulcatus*, as a pest of grapes in south central Washington. *Journal of Economic Entomology* 56, 677–680.

Cowles, R.S. (1997) Several methods reduce insecticide use in control of black vine weevil. *Frontiers of Plant Science* 49, 2–4.

Curran, J. (1992) Influence of application method and pest population size on the field efficacy of entomopathogenic nematodes. *Journal of Nematology* 24, 631–636.

Curto, G., Boselli, M. and Ricci, M. (2001) Winter field trial against black vine weevils (*Otiorhynchus sulcatus*) in potted *Euonymus japonica* using the cold active nematode *Steinernema kraussei* N0093 strain. In: Griffin, C.T., Burnell, A.M., Downes, M.J. and Mulder, R. (eds) *COST Action 819 – Developments in Entomopathogenic Nematode/Bacterial Research*. European Commission, Luxembourg, EUR 19696, 285–290.

Dapsis, L. (1991) *White Grub and Cranberry Root Grub Trials*. Cranberry Agricultural Research, Ocean Spray Cranberries, Lakeville-Middleboro, Massachusetts.

Dapsis, L. (1993) *Management of Cranberry Girdler and Cranberry White Grub with Entomopathogenic Nematodes*. Cranberry Agricultural Research, Ocean Spray Cranberries, Lakeville-Middleboro, Massachusetts.

Dittl, T. (1996) *Management of Cranberry White Grub in Wisconsin with Entomopathogenic Nematodes*. Cranberry Agricultural Research, Ocean Spray Cranberries, Lakeville-Middleboro, Massachusetts.

Dittl, T.G. and Kummer, L.D. (1997) *Major Cranberry Insect Pests of Wisconsin*. Ocean Spray Cranberries, Babcock, Wisconsin.

Duckworth, W.D. and Eicklin, T.D. (1978) The clearwing moths of California (Lepidoptera: Sesiidae). *Occasional Papers Bureau of Entomology, California Department of Agriculture* 27, 1–80.

Ehlers, R.-U. and Peters, A. (1998) Bekämpfung von Engerlingen auf Sportrasen. *Rasen/Turf/Gazon* 29, 60–67.

English-Loeb, G., Villani, M., Martinson, T., Forsline, A. and Consolie, N. (1999) Use of entomopathogenic nematodes for control of grape phylloxera (Homoptera: Phylloxeridae): a laboratory evaluation. *Environmental Entomology* 28, 890–894.

Fischer, P. and Führer, E. (1990) Effect of soil acidity on the entomophilic nematode *Steinernema kraussei* Steiner. *Biology and Fertility of Soils* 9, 174–177.

Fitters, P.F.L., Dunne, R. and Griffin, C.T. (2001) Improved control of *Otiorhynchus sulcatus* at 9°C by cold-stored *Heterorhabditis megidis* UK211. *Biocontrol Science and Technology* 11, 483–492.

Garth, G.S. and Shanks, C.H. Jr (1978) Some factors affecting infestation of strawberry fields by the black vine weevil in western Washington. *Journal of Economic Entomology* 71, 443–448.

Gordon, S.C., Woodford, J.A.T. and Birch, A.N.E. (1997) Arthropod pests of *Rubus* in Europe: pest status, current and future control strategies. *Journal of Horticultural Science* 72, 831–862.

Grewal, P.S., Grewal, S.K., Malik, V.S. and Klein, M.G. (2002) Differences in the susceptibility of introduced and native white grub species to entomopathogenic nematodes from various geographic localities. *Biological Control* 24, 230–237.

Grewal, P.S., Power, K.T., Grewal, S.K., Suggars, A. and Haupricht, S. (2004) Enhanced consistency in biological control of white grubs (Coleoptera: Scarabaeidae) with new strains of entomopathogenic nematodes. *Biological Control* 30, 73–82.

Hanford, R.H. (1952) A regional review of insects of the season 1951. *The Canadian Insect Pest Review* 30, 1–120.

Headlee, T.J. and Ilg, C. (1926) Some facts relative to the raspberry crown borer. *Journal of Economic Entomology* 19, 471–477.

Henderson, D. and Singhai, R. (1999) Efficacy study of three species of parasitic nematodes for control of cranberry girdler, *Chrysoteuchia topiaria*, in British Columbia, Canada. In: Deziel, G. and Hogan, M. (eds) *Cranberry Research Compilation. Progress and Final Reports on Cranberry Research Conducted in 1998*. Cranberry Institute, Wareham, Massachusetts, pp. 167–176.

Jubb, G.L. Jr (1982) Occurrence of the grape root borer, *Vitacea polistiformis* in Pennsylvania. *Melsheimer Entomology Series* 32, 20–24.

Kakouli-Duarte, T. and Hague, N.G.M. (1999) Infection, development, and reproduction of the entomopathogenic nematode *Steinernema arenarium* (Nematoda: Steinernematidae) in the black vine weevil *Otiorhynchus sulcatus* (Coleoptera: Curculionidae). *Nematology* 1, 149–156.

Kakouli-Duarte, T., Labuschagne, L. and Hague, N.G.M. (1997) Biological control of the black vine weevil, *Otiorhynchus sulcatus* (Coleoptera: Curculionidae) with entomopathogenic nematodes (Nematoda: Rhabditida). *Annals of Applied Biology* 131, 11–27.

Kaya, H.K. (1990) Soil ecology. In: Gaugler, R. and Kaya, H.K. (eds) *Entomopathogenic Nematodes in Biological Control*. CRC Press, Boca Raton, Florida, pp. 93–116.

Klein, M.G. (1990) Efficacy against soil-inhabiting insect pests. In: Gaugler, R. and Kaya, H.K. (eds) *Entomopathogenic Nematodes in Biological Control.* CRC Press, Boca Raton, Florida, pp. 195–214.

Koppenhöfer, A.M. and Fuzy, E.M. (2004) Effect of white grub developmental stage on susceptibility to entomopathogenic nematodes. *Journal of Economic Entomology* 97, 1842–1849.

Koppenhöfer, A.M., Cowles, R.S., Cowles, E.A., Fuzy, E.M. and Baumgartner, L. (2002) Comparison of neonicotinoid insecticides as synergists for entomopathogenic nematodes. *Biological Control* 24, 90–97.

Koppenhöfer, A.M., Fuzy, E.M., Crocker, R.L., Gelernter, W.D. and Polavarapu, S. (2003) Pathogenicity of *Heterorhabditis bacteriophora, Steinernema glaseri,* and *S. scarabaei* (Rhabditida: Heterorhabditidae, Steinernematidae) against 12 white grub species (Coleoptera: Scarabaeidae). *Biocontrol Science and Technology* 14, 87–92.

Kramer, I. and Grunder, J.M. (2001) Efficiency of irrigation systems for application of entomopathogenic nematodes. In: Griffin, C.T., Burnell, A.M., Downes, M.J. and Mulder, R. (eds) *COST Action 819 – Developments in Entomopathogenic Nematode/Bacterial Research.* European Commission, Luxembourg, EUR 19696, pp. 291–294.

LaMondia, J.A., Elmer, W.H., Mervosh, T.L. and Cowles, R.S. (2002) Integrated management of strawberry pests by rotation and intercropping. *Crop Protection* 21, 837–846.

Lewis, E.E., Gaugler, R. and Harrison, R. (1992) Entomopathogenic nematode host finding: response to host contact cues by cruise and ambush foragers. *Parasitology* 105, 309–315.

Lieten, F. and Baets, W. (1991) Greenhouse strawberry culture in peat bags. *Advances in Strawberry Production* 10, 56–57.

Löckener, A. (1994) Bodenschädlinge im Erdbeeranbau. *Obstbau* 2/94, 75–76.

Long, S.J., Richardson, P.N. and Fenlon, J.S. (2000) Influence of temperature on the infectivity of entomopathogenic nematodes (*Steinernema* and *Heterorhabditis* spp.) to larvae and pupae of the vine weevil *Otiorhynchus sulcatus* (Coleoptera: Curculionidae). *Nematology* 2, 309–317.

Mass, J.L. (1998) *Compendium of Strawberry Diseases,* 2nd edn, American Phytopathological Society, St. Paul, Minnesota, 96 pp.

Menzies, G.W. (1999) *Crop Profile for Red Raspberries in Washington.* Washington State University, 24 pp.

Miller, D.R. and Polavarapu, S. (1997) A new species of mealybug in the genus *Dysmicoccus* (Homoptera: Coccoidea: Pseudococcidae) of importance in highbush blueberries (*Vaccinium corymbosum* Ericaceae) in the eastern United States. *Proceedings of the Entomological Society of Washington* 99, 440–460.

Miller, L.A. and Bedding, R.A. (1982) Field testing of the insect parasitic nematode, *Neoaplectana bibionis* (Nematoda: Steinernematidae) against currant borer moth, *Synanthedon tipuliformis* (Lep.: Sesiidae) in black currants. *Entomophaga* 27, 109–114.

Olivier, A.D. and Chapin, J.B. (1980) The cranberry rootworm: adult seasonal history, and factors affecting its status as a pest of woody ornamentals in Louisiana. *Journal of Economic Entomology* 73, 96–100.

Phillips, P.A. (1989) Simple monitoring of black vine weevil in vineyards. *California Agriculture* 43, 12–13.

Polavarapu, S. (1999a) Insecticidal nematodes for cranberry pest management. In: Polavarapu, S. (ed.) *Optimal Use of Insecticidal Nematodes in Pest Management.* Rutgers University, New Brunswick, New Jersey, pp. 79–90.

Polavarapu, S. (1999b) *Insect Parasitic Nematodes for Cranberry Pest Management.* Rutgers Cooperative Extension, New Jersey Agricultural Experiment Station, Publication FS029, 2 pp.

Polavarapu, S. and Stuart, R. (1997) Life-history and management of root infesting insects in cranberries. *Horticultural News* 77, 7–9.

Polavarapu, S., Peng, H. and Stuart, R.J. (1998) Evaluation of Admire 2F and an entomopathogenic nematode to manage oriental beetle populations in blueberries, 1997. *Arthropod Management Tests* 23, 51.

Polavarapu, S., Stuart, R.J. and Schiffhauer, D.F. (2000) Evaluation of insecticides against cranberry rootworm. *Arthropod Management Tests* 25, 66–67.

Pollet, D.K. (1975) The grape root borer in South Carolina. *Clemson University Extension Service Circular* 550, 7.

Pollit, S., Peters, A. and Ehlers, R.-U. (1994) Control potential of a naturally occurring *Steinernema feltiae* population. In: Burnell, A.M., Ehlers, R.-U. and Masson, J.P. (eds) *COST 812 Biotechnology – Genetics of Entomopathogenic Nematode–Bacterium Complexes.* European Commission, Luxembourg, EUR 15681 EN, pp. 226.

Raine, J. (1962) Life history and behavior of the raspberry crown borer *Bembecia marginata* (Harr.) (Lepidoptera: Aegeriidae). *Canadian Entomologist* 94, 1216–1222.

Sampson, A.C. (1994) Control of *Otiorhynchus sulcatus* in soft fruit using drench treatments of *Steinernema carpocapsae. Brighton Crop Protection Conference – Pests and Diseases* 6A, 601–608.

Saunders, M.C. and All, J.N. (1985) Association of entomophilic rhabditoid nematode populations with natural control of first-instar larvae of grape root borer, *Vitacea polistiformis* in Concord grape vineyards. *Journal of Invertebrate Pathology* 45, 147–151.

Scherer, W. (1987) Bekämpfung des Gefurchten Dickmaulrüßalers in Freiland-Erdbeeren. *Obstbau* 12, 225–227.

Shanks, C.H. Jr and Agudelo-Silva, F. (1990) Field pathogenicity and persistence of heterorhabditid and steinernematid nematodes (Nematoda) infecting black vine weevil larvae (Coleoptera: Curculionidae) in cranberry bogs. *Journal of Economic Entomology* 83, 107–110.

Simard, L., Bélair, G. and Brodeur, J. (2002) Susceptibility of the European chafer (Coleoptera: Scarabaeidae) to entomopathogenic nematodes (Rhabditida: Steinernematidae, Heterorhabditidae). *Journal of Nematology* 33, 297–301.

Simons, W.R. (1981) Biological control of *Otiorhynchus sulcatus* with heterorhabditid nematodes in the glasshouse. *Netherlands Journal of Plant Pathology* 87, 149–158.

Simser, D. and Roberts, S. (1994) Suppression of strawberry root weevil, *Otiorhynchus ovatus,* in cranberries by entomopathogenic nematodes (Nematoda: Steinernematidae and Heterorhabditidae). *Nematologica* 40, 456–462.

Slingerland, M.V. and Crosby, C.R. (1915) *Manual of Fruit Insects.* Macmillan, New York, 493 pp.

Smith, F.F. (1932) Biology and control of the black vine weevil. *United States Department of Agriculture Technical Bulletin* 325, 45 pp.

Smith, K., Fisher, G.C., Poole, A. and Jackson, D. (1993) *Effectiveness of Biosafe-N, Steinernema carpocapsae for Control of Cranberry Girdler, 1992.* Cranberry Agricultural Research, Ocean Spray Cranberries, Lakeville-Middleboro, Massachusetts.

Snow, J.W., Schwarz, M. and Eichlin, T.D. (1989) Captures of clearwing moths (Lepidoptera: Sesiidae) with various octadecadienyl acetates and alcohols in central Georgia in 1993–1985. *Environmental Entomology* 18, 216–222.

Stuart, R.J. and Polavarapu, S. (1997) Susceptibility of the cranberry rootworm to entomopathogenic nematodes. *Horticultural News* 78, 15.

Stuart, R.J. and Polavarapu, S. (2002) On the relationship between the ant, *Acanthomyops claviger* and the blueberry mealybug, *Dysmicoccus vaccinii. Journal of Insect Behavior* 15, 299–304.

Stuart, R.J., Polavarapu, S., Lewis, E.E. and Gaugler, R. (1997) Differential susceptibility of *Dysmicoccus vaccinii* (Homoptera: Pseudococcidae) to entomopathogenic nematodes (Rhabditida: Heterorhabditidae and Steinernematidae). *Journal of Economic Entomology* 90, 925–932.

Sulistyanto, D. and Ehlers, R.-U. (1996) Efficacy of the entomopathogenic nematodes *Heterorhabditis megidis* and *Heterorhabditis bacteriophora* for the control of grubs (*Phyllopertha horticola* and *Aphodius contaminatus*) in golf course turf. *Biocontrol Science and Technology* 6, 247–250.

Taft, W., Smitley, D. and Snow, J.W. (1991) A guide to the clearwing borers (Sesiidae) of the north central United States. *North Central Regional Publication* 394, 30 pp.

Vittum, P.J., Villani, M.G. and Tashiro, H. (1999) *Turfgrass Insects of the United States and Canada,* 2nd edn, Cornell University Press, Ithaca, New York, 422 pp.

Wang, Y. and Gaugler, R. (1998) Host and penetration site location by entomopathogenic nematodes against Japanese beetle larvae. *Journal of Invertebrate Pathology* 72, 313–318.

Weber, D. and Henderson, D. (1998) Beneficial nematodes in cranberry pest management. Paper presented at the integrated pest management in cranberry production workshop, April 19–20, Montreal, Quebec.

Wennemann, L., Cone, W.W., Wright, L.C., Perez, J. and Conant, M.M. (2003) Distribution patterns of entomopathogenic nematodes applied through drip irrigation systems. *Journal of Economic Entomology* 96, 287–291.

Westerman, P.R. (1998) Penetration of the entomopathogenic nematode *Heterorhabditis* spp. into host insects at 9 and 20°C. *Journal of Invertebrate Pathology* 72, 197–205.

Westerman, P.R. (1999) Aggregation of entomopathogenic nematodes, *Heterorhabditis* spp. and *Steinernema* spp., among host insects at 9 and 20°C and effects on efficacy. *Journal of Invertebrate Pathology* 73, 206–213.

Williams, R.N., Fickle, D.S., Grewal, P.S. and Meyer, J.R. (2002) Assessing the potential of entomopathogenic nematodes to control the grape root borer *Vitacea polistiformis* (Lepidoptera: Sesiidae) through laboratory bioassays. *Biocontrol Science and Technology* 12, 35–42.

Willmott, D.M., Hart, A.J. and Long, S.J. (2002) Use of cold-active entomopathogenic nematode *Steinernema kraussei* to control overwintering larvae of the black vine weevil *Otiorhynchus sulcatus* (Coleoptera: Curculionidae) in outdoor strawberry plants. *Nematology* 4, 925–932.

Wilson, M., Nitzsche, P. and Shearer, P.W. (1999) Entomopathogenic nematodes to control black vine weevil (Coleoptera: Curculionidae) on strawberry. *Journal of Economic Entomology* 92, 651–657.

Yeh, T. and Alm, S.R. (1995) Evaluation of *Steinernema glaseri* (Nematoda: Steinernematidae) for biological control of Japanese and oriental beetles (Coleoptera: Scarabaeidae). *Journal of Economic Entomology* 88, 1251–1255.

13 Vegetable and Tuber Crop Applications

G. Bélair,[1] D.J. Wright[2] and G. Curto[3]
[1]*Horticultural Research and Development Centre, Agriculture and Agri-Food, Canada, St-Jean-sur-Richelieu, Quebec, Canada J3B 3E6;* [2]*Department of Biological Sciences, Imperial College London, Silwood Park Campus, Ascot, Berkshire SL5 7PY, UK;* [3]*Servizio Fitosanitario Regione emilia-Romagna, Bologna, Italy*

13.1. Introduction

Vegetables are annual productions that provide a wide variety of agricultural products recovered from various parts of the plant, including root, leaf and fruit. Numerous insect pests of economic importance are encountered on these crops. Being high-value crops, the introduction of biological pest control agents such as entomopathogenic nematodes (EPNs) has stimulated great interest worldwide for both above- and below-ground pests. Because nematodes are well adapted to soil conditions, research works have focused on many root-feeding insects, such as root weevils. Yet, success with EPN has not been achieved in field application despite promising laboratory or field trials. The objectives of the present chapter are to review significant research on EPNs against vegetable pests and provide some direction for the future use of EPNs on these crops against root and leaf feeding insects. A summary of EPN field efficacy for control of vegetable pests is presented in Table 13.1.

Table 13.1. Field efficacy of *Steinernema* and *Heterorhabditis* nematodes against major insect pests in vegetable crops.

Pests	Nematode species	Application rate (/cm^2)	% insect mortality	% damage control	References
Agrostis segetum	*S. carpocapsae*	100	67–80	ND	Yokomizo and Kashio, 1996
Delia radicum	*S. carpocapsae*	1500–5000	ND	35–64	Welch and Briand, 1961a
	S. carpocapsae	250–500	ND	6–8	Simser, 1992
	S. feltiae	55	ND	26	Hommes, 1988
	S. feltiae	300–2000	ND	16–48	Schroeder *et al.*, 1996
	H. bacteriophora	250–500	ND	0–13	Simser, 1992
Leptinotarsa decemlineata	*S. carpocapsae*	386	67	ND	Verumchuk and Danilov, 1976
	S. carpocapsae	155–310	59–71	ND	Toba *et al.*, 1983
	S. carpocapsae	93–155	79–65	ND	Wright *et al.*, 1987
	S. carpocapsae	25–76	38–69	ND	Stewart *et al.*, 1998
	H. bacteriophora	93–155	40–67	ND	Wright *et al.*, 1987
Listronotus oregonensis	*S. carpocapsae*	133–266	250	16–25	Bélair and Boivin, 1995
	S. carpocapsae	75	14	ND	Miklasiewicz *et al.*, 2002
	H. bacteriophora	75	38–80	ND	Miklasiewicz *et al.*, 2002
Pieris rapae	*S. carpocapsae*	40	ND	25–35	Bélair *et al.*, 2003
Plutella xylostella	*S. carpocapsae*	50	ND	41	Baur *et al.*, 1998
Temnorhinus mendicus	*S. carpocapsae*	50	73	85	Boselli *et al.*, 1991
	S. carpocapsae	25	93	56	Curto *et al.*, 1992
	S. carpocapsae	25–50	40–63	49–62	Boselli *et al.*, 1994
	Heterorhabditis sp.	25–50	67–75	49–57	Boselli *et al.*, 1994
	Heterorhabditis sp.	25	91	64	Boselli *et al.*, 1997
Cylas formicarius	*S. carpocapsae*	11–49	65–73	25	Jansson *et al.*, 1990
	S. carpocapsae	38	85	32	Jansson *et al.*, 1993
	S. feltiae	10–31	80	50	Jansson *et al.*, 1990
	S. feltiae	73[a]	25	42	Jansson *et al.*, 1993
	H. bacteriophora	73[a]	82	21	Jansson *et al.*, 1993

[a]Application rate in number of infected *Galleria mellonella* cadavers/m^2.
ND = not determined.

13.2. Roots and Bulbs

13.2.1. Carrot root weevil

The carrot weevil, *Listronotus oregonensis* (Coleoptera: Curculionidae), is an important pest of carrot, celery and parsley in north-eastern North America. Adults overwinter on or near the soil surface associated with plant material and debris. In the spring, they crawl over the soil surface to locate the host plant upon which they feed, females oviposit mainly on plant petioles and the young larvae bore into plant crowns and roots, or feed at the surface of larger roots. The spring migration of the adults from their overwintering sites in carrot fields provides the opportunity to infect them either through sprays or baits.

13.2.1.1. Nematodes for carrot root weevil control and factors affecting efficacy

Selection of the best EPN for a particular pest is one of the primary factors for achieving success in nematode application. In the laboratory, *Heterorhabditis bacteriophora* and *Steinernema carpocapsae* were shown to be good candidates for control of this pest (Bélair and Boivin, 1985, 1995; Miklasiewicz *et al.*, 2002). In Canada, a strain of *S. carpocapsae* isolated from carrot weevil adults has been compared with the strain DD136 (Boivin and Bélair, 1989). Both strains decreased longevity and oviposition by adults and their LT_{50} increased with decreasing temperatures. However, DD136 performed better than the local strain at 10°C.

The timing of field applications can have a marked effect on the efficacy of EPNs. For example, early season application of *H. bacteriophora* provides greater plant protection for carrot and parsley (Miklasiewicz *et al.*, 2002). Laboratory studies showed that the efficacy of EPNs against *L. oregonensis* was affected by the insect developmental stage and the age of adult weevils (Boivin and Bélair, 1989; Bélair and Boivin, 1995; Miklasiewicz *et al.*, 2002). Larvae were more susceptible than adults. Overwintered adults were substantially less susceptible than newly emerged and 2-month-old adults. Infected females still alive after 2 days stopped ovipositing (Boivin and Bélair, 1989). This last effect was especially interesting as most control approaches aim to prevent oviposition by females in the spring.

Soil type has been shown to have some influence on efficacy (Miklasiewicz *et al.*, 2002). *S. carpocapsae* caused significantly greater adult mortality in sand compared with *H. bacteriophora*, while the latter caused greater mortality in muck soil and had greater persistence. In Quebec, field application of *S. carpocapsae* as a drench or as a bait in muck-grown carrots at the rate of 4.4 billion/ha reduced carrot weevil damage by 59% (Bélair and Boivin, 1995). In Ohio, soil spray application of *S. carpocapsae* and *H. bacteriophora* in muck-grown carrot and parsley at the rate of 3.3 billion/ha had no effect on yield but *H. bacteriophora* treatments persisted longer and resulted in greater insect mortality and plant survival (Miklasiewicz *et al.*, 2002).

13.2.1.2. Current status and analysis

Although EPNs show some promise for controlling carrot weevil, they cannot compete against current management tactics using conventional pesticides. In carrot production, the economic threshold is very low at 2% of affected plants. This is mainly related to the labour costs of removing damaged roots. EPN could be used as an alternative to chemical control only under light insect pressure since the cost of EPN is still considerably higher than the cost of chemical insecticides.

13.2.2. Cabbage maggot

The cabbage maggot, *Delia radicum* (Diptera: Anthomyiidae), is a cosmopolitan pest of radish, rutabaga and other cole crops. Eggs of the economically important first generation are deposited around and on the stems of early-season (April–May) field plants. The larvae hatch in several days and tunnel into root tissue, where feeding occurs (Eckenrode and Chapman, 1971). Larvae feed by tunnelling into the roots. Plants may be killed, weakened or stunted, and yields reduced.

13.2.2.1. Nematodes for cabbage maggot control

The cabbage maggot is one the most extensively studied targets for EPN. Despite this, the level of control has remained variable and very unreliable from a commercial viewpoint. More work has been conducted on leafy crucifer crops, such as cabbage, cauliflower, broccoli or collard, than on root brassicas. *S. carpocapsae* and *S. feltiae* have been the most commonly used species in field evaluations. The level of control achieved was in most cases lower than the corresponding insecticide treatment (Welch and Briand, 1961a; Simser, 1992; Schroeder *et al.*, 1996; Vänninen *et al.*, 1999) but was sometimes comparable (Hommes, 1988; Bracken, 1990). *S. feltiae* has been reported to be slightly more effective than *S. carpocapsae* (Hommes, 1988; Schroeder *et al.*, 1996). Soil surface applications of *S. feltiae* were more effective than subsurface applications in preventing damage (Schroeder *et al.*, 1996).

13.2.2.2. Current status and analysis

Because *D. radicum* larvae are only in the soil for a brief period, the infection process for the nematode needs to be as optimum as possible. To achieve more widespread use of EPNs on brassicas more active strains will be required.

13.3. Tuber Roots and Industrial Crops

13.3.1 Sugarbeet weevil (SBW)

The sugarbeet weevil (SBW), *Temnorhinus* (= *Conorrhynchus*) *mendicus* (Coleoptera: Curculionidae), is the major insect pest of sugarbeet in all the western Mediterranean countries, especially in southern France, Italy, Spain and northern Africa. This species completes one generation in a year and overwinters as adults in the soil. Chemical insecticides are effective only against the adults. Early work on the efficacy of EPNs against SBW was conducted by Deseö (1987), Boselli *et al.* (1991) and Curto *et al.* (1992). More recent field studies have investigated the lowest effective dosage of EPNs and optimization of distribution techniques (Boselli *et al.*, 1994, 1997; Curto *et al.*, 1999). Boselli *et al.* (1997) compared *S. carpocapsae* (All), *Heterorhabditis* sp. (NL-HL81) and *H. bacteriophora* (HP88) at 7.5, 12.5, 25 and 50 infective juveniles (IJs)/cm^2, with insecticide treatments. All larval instars, pupae and newly emerged adults of *T. mendicus* were susceptible to EPNs. Greatest efficacy was achieved at first larval hatch by a direct spray on the crop following irrigation or rainfall. Nematodes applied at 25 IJs/cm^2 provided 90–95% weevil mortality, which was significantly better than insecticide treatments. In the same plots, nematode-infected weevils were observed 1 year later; persistence of EPNs being greater in clay and loamy soils compared with peat soils (G. Curto, 1994, unpublished data). EPNs could represent the best way to control SBW in organic farming or where resistance to all available insecticides has been found. Effective application with existing farm equipment and the availability of large amounts of nematodes at a low price are required.

13.3.2. Colorado potato beetle (CPB)

Colorado potato beetle (CPB), *Leptinotarsa decemlineata* (Coleoptera: Chrysomelidae), is a key pest of potatoes; both larvae and adults are phytophagous. Originally from the USA, it is now widespread. There are four instars, the last of which drops from the plant and burrows into the soil for pupation. CPB completes 1–3 generations per year, depending on the latitude. Early work by Welch (1958) demonstrated the efficacy of EPNs against CPB in soil applications. Most studies have been carried out against the fourth instars by soil treatments and *S. carpocapsae* (Agriotos, All, Breton, DD136, Mexican), *S. glaseri, S. riobrave* (TX), *S. oregonense* (OS21), *S. feltiae* (27, 980), *H. marelatus* (OH10), *H. bacteriophora* (HP88, Brecon), *H. indica* (FL2122) and *Heterorhabditis* sp. (OH23, OH95) have been tested. Welch and Briand (1961b) found that foliar application led to rapid desiccation, although antidesiccants have been shown to increase the effectiveness of *S. carpocapsae* (MacVean *et al.*, 1982).

The field use of EPNs has been simulated in cages filled with soil against spring and summer generations of CPB. Nematodes were sprayed on the soil surface a day before adding fourth instar larvae (Veremchuk and Danilov, 1976; Toba *et al.*, 1983; Wright *et al.*, 1987; Steward *et al.*, 1998). In these trials, larval mortality was generally lower than in laboratory tests (79% with *S. carpocapsae* Mexican strain at 93 IJ/cm^2, and 67% with *H. bacteriophora* at 155 IJ/cm^2). Increasing the EPN concentration did not cause a proportional increase in larval mortality.

In a greenhouse trial, *S. carpocapsae* (All) emerged successfully from a pellet formulation or 'Pesta' and killed 94% of the prepupae at 82/cm^2 against CPB prepupae (Nickle *et al.*, 1994). EPN persistence in the soil following application against spring generation of CPB larvae was low and provided no major impact on the summer generation (Toba *et al.*, 1983; Wright *et al.*, 1987; Berry *et al.*, 1997). The effectiveness of EPN in potato fields appeared to be reduced by various factors, such as the depth of beetle pupation (*c.* 1–15 cm), the migration of CPB from neighbouring plants and fields (MacVean *et al.*, 1982), and the insensitivity of CPB adults to EPN (Toba *et al.*, 1983).

13.3.3. Sweet potato weevil (SPW)

The sweet potato weevil (SPW), *Cylas formicarius* (Coleoptera: Apionidae), is the most important insect pest in sweet potatoes. It can cause damage both in the field and in storage because its whole life cycle takes place within the plant and every instar is present at the same time. Larval feeding induces terpenoid production in plants, so even slightly damaged roots become unpalatable and are not marketable. The geographical distribution of SPW is closely related to sweet potato crop areas throughout tropical and subtropical regions. SPW completes 5–8 generations in a year. There are three instars, which tunnel in both stems and tubers. Adults emerge from the pupal chamber or remain in the tuber. Since the late 1980s, a number of research projects have evaluated the pathogenicity, virulence, effectiveness and persistence of *S. carpocapsae* (Agriotos, All, Breton, G-13, Italian, Mexican, S17, S20), *S. glaseri*, *S. feltiae* (N27), *S. intermedia*, *H. bacteriophora* (HP88, NC), *H. megidis* and *Heterorhabditis* sp. (Bacardis, FL2122) against SPW. Some studies have used storage roots buried in soil in plastic boxes (Jansson *et al.*, 1990; Mannion and Jansson, 1992, 1993), and there have been a number of field trials (Jansson *et al.*, 1990, 1993), including studies on different cultivars of sweet potato (Jansson and Lecrone, 1997) and on different application methods (Jansson and Lecrone, 1994). It has been demonstrated that EPNs are able to seek out and kill SPW larvae and pupae and to reproduce in their cadavers, and a well-timed single application of EPNs provides better control than multiple applications (Jansson *et al.*, 1991). EPNs are more effective than chemicals at reducing weevil densities and heterorhabditids appear to be more effective and more persistent than steinernematids against both larvae and pupae. *H. bacteriophora* (HP88), *Heterorhabditis* sp. (Bacardis) (Jansson *et al.*, 1993) and *H. megidis* are particularly effective (Ekanayake *et al.*, 2001). Research has demonstrated that EPNs have the potential for managing SPW in the field and on stored roots. They could be a more reliable alternative to conventional insecticides against this cryptic pest but the high cost limits their use.

13.4. Leafy and Other Above-ground Vegetables

In this section, the use of nematodes to control foliar stages of some of the most important vegetable pests is discussed. More detailed information on the foliar application of nematodes is given in Chapter 5.

13.4.1. Diamondback moth (DBM)

Three million hectares of cabbages are grown worldwide (FAO, 2003). The most important cabbage pest, and the one for which resistance problems are most serious, is the diamondback moth (DBM), *Plutella xylostella* (Lepidoptera: Plutellidae) (Talekar and Shelton, 1993). Other foliar pests include cutworms (Section 13.4.3) and leafworms (e.g. *Agrotis* and *Spodoptera* spp.), cabbage looper (*Trichoplusia ni*), cabbage moths (*Mamestra brassica* and *Crocidolomia binotalis*), cabbage budworms (*Hellula* spp.) and cabbage butterflies (*Pieris* spp.).

In laboratory leaf disc assays, *S. carpocapsae* was particularly effective against DBM larvae (Baur *et al.*, 1995), but was less effective against larvae on plants unless the relative humidity was very high (Baur *et al.*, 1997a). Nematodes have been suggested as possible components of integrated pest management (IPM) programmes for DBM (Baur *et al.*, 1998). Their survival and efficacy on foliage can be enhanced by spray adjuvants (Baur *et al.*, 1997b; Mason *et al.*, 1998a) and by improvements in their placement on foliage through optimization of spray equipment (Lello *et al.*, 1996; Mason *et al.*, 1998b, 1999; Piggott *et al.*, 2003). Field studies on cabbage in the Malaysian highlands confirmed that nematodes have potential for the control of DBM within IPM programmes (Mason *et al.*, 1999).

13.4.2. Dipteran leafminers

The use of nematodes to control the cabbage maggot or cabbage root fly (*D. radicum*) was considered in Section 13.2.2. Other important dipteran pests include the agromyzid leafminers (e.g. *Liriomyza* spp., *Chromatomyia* spp.) (Diptera: Agromyzidae), polyphagous species that are increasingly important foliar pests of vegetables worldwide. Glasshouse trials in the UK have shown that *S. feltiae* can give effective control of *Liriomyza huidobrensis*, *L. bryoniae* and *Chromatomyia syngenesiae* on other vegetables (lettuce, tomato) and ornamentals under glass, most notably at high humidity (Williams and MacDonald, 1995; Williams and Walters, 2000). In leafminer control, once the IJ enters a mine in search of a host larva it is effectively protected from the environment. The aim is therefore to maximize the density and distribution of nematodes on leaf surfaces to enable as many nematodes as possible to locate a mine entrance.

13.4.3. Cutworms

Cutworms (Lepidoptera: Noctuidae) (e.g. *Agrotis* spp.) are polyphagous insects, which attack numerous vegetable crops. Soil-dwelling larvae feed at night on the leafstalk or petiole of plants and cut them at or below the soil surface. One larva can destroy many plants in a single night. Damage is often highly concentrated in the field. A large number of field studies have established the potential of EPNs for cutworm control (Lössbroek and Theunissen, 1985; Capinera *et al.*, 1988; Levine and Oloumi-Sadeghi, 1992; Yokomizo and Kashio, 1996; Shapiro *et al.*, 1999). Aqueous suspension of either *S. carpocapsae* or *S. feltiae* at rates ranging from 1 to 10 billion/ha provides a level of control similar to or better than chemical insecticides. In a carrot field test, a single ground spray of *S. carpocapsae* at 1 billion/ha or two applications of 0.5 billion/ha with an 8-day interval between sprays caused 80% and 67% mortality of *Agrotis segetum* larvae, respectively (Yokomizo and Kashio, 1996). Cutworm problems in vegetable crops tend to be very specific, and mainly occur in the second year following a return from pasture. The field borders, along ditches and the areas infested with tall grasses will suffer from early season damage by cutworms. When based on good scouting, only limited areas may need to be sprayed with EPNs. The rapid loss of nematode efficiency suggests that improved formulations, with enhanced longevity, are necessary to acquire this niche market for cutworm control.

13.4.4. Cucumber beetles/rootworms

Cucumber beetle/rootworms (*Diabrotica* spp., *Acalymma vittatum*) attack a variety of crops in the Cucurbitaceae such as squash, gourd, pumpkin and cucumber. Laboratory assays have shown that *S. carpocapsae* and *H. bacteriophora* are potential control agents for *Diabrotica undecimpunctata* (Coleoptera: Chrysomelidae) larvae, and were particularly effective when host larvae were reared on squash compared with groundnut and maize, respectively (Barberchek, 1993; Barbercheck *et al.*, 1995). Laboratory and field studies have also demonstrated the potential of *Steinernema* spp. for use within IPM systems against *A. vittatum* in commercial cucumber production (Ellers-Kirk *et al.*, 2000).

13.4.5. Current status and analysis

The use of nematodes by growers to control foliar pests on vegetables will require optimization of formulations, application technology and spray regimes. This is most likely to be attainable in humid conditions for protected crops and in the humid tropics and subtropics on high-value crops, such as Chinese cabbage, where the high relative cost of nematodes compared with chemical insecticides is a less significant factor. Nematodes could therefore be particularly useful components of IPM programmes for DBM and other lepidopteran pests on brassicas. The potential of nematode use is likely to be greater in niche organic markets, e.g. for cutworm control. Nematodes can also be effective substitutes for some chemical treatments for the control of leafminer and other cryptic species, and are already used successfully by some growers on ornamentals to control leafminers and thrips. The withdrawal of approvals for agrochemicals on many horticultural food crops within Europe, North America and elsewhere is likely to represent an increasing market opportunity for biopesticide products, including nematodes.

13.5. Summary and Conclusions

For many vegetable insect pests, organic production is seen as the most favourable niche for the implementation of EPNs. Demand for organic vegetables has increased manifold since the mid-1990s, and it is likely that the potential for using EPNs in this market sector will increase. High-value horticultural crops in general, such as brassicas, where in many areas of the tropics and subtropics excessive use of chemical insecticides has led to major resistance, residue and pest resurgence problems, represent another potential area for nematodes. To achieve such goals more laboratory and, especially, field studies need to be conducted. The examination of new application methods, including cadavers (Shapiro-Ilan *et al.*, 2003) and nemabags (Menzler-Hokkanen and Hokkanen, 2003), need to be conducted, together with the optimization of methods such as band application, baiting, irrigation, soil spray and foliar spray technology. One feature of vegetable crops is the high number of plants per hectare that have to be protected. New species or isolates with higher levels of virulence are needed. The best matches tend to be for nematodes that have high virulence towards hosts in a protected environment. Improvements in production technology, distribution and application will be a key to reducing nematode costs and ensuring quality, thereby increasing the competitiveness of EPNs and opening up new markets. The integration of EPNs with other biopesticides, such as *Bacillus thuringiensis* and *Beauveria* spp., should also be actively pursued for the development of sustainable strategies for the management of pest complexes.

References

Barbercheck, M.E. (1993) Tritrophic level effects on entomopathogenic nematodes. *Environmental Entomology* 22, 1166–1171.

Barbercheck, M.E., Wang, J. and Hirsh, I.S. (1995) Host-plant effects on entomopathogenic nematodes. *Journal of Invertebrate Pathology* 66, 169–177.

Baur, M.E., Kaya, H.K. and Thurston, G.S. (1995) Factors affecting entomopathogenic nematode infection of *Plutella xylostella* on a leaf surface. *Entomologia Experimentalis et Applicata* 77, 239–250.

Baur, M.E., Kaya, H.K., Gaugler, R. and Tabashnik, B.E. (1997a) Effects of adjuvants on entomopathogenic nematode persistence and efficacy against *Plutella xylostella. Biocontrol Science and Technology* 7, 513–525.

Baur, M.E., Kaya, H.K. and Tabashnik, B.E. (1997b) Efficacy of a dehydrated steinernematid nematode against black cutworm (Lepidoptera: Noctuidae) and diamondback moth (Lepidoptera:

Plutellidae). *Journal of Economic Entomology* 90, 1200–1206.

Baur, M.E., Kaya, H.K., Tabashnik, B.E. and Chilcutt, C.F. (1998) Suppression of diamondback moth (Lepidoptera: Plutellidae) with an entomopathogenic nematode (Rhabditidae: Steinernematidae) and *Bacillus thuringiensis* Berliner. *Journal of Economic Entomology* 91, 1089–1095.

Bélair, G. and Boivin, G. (1985) Susceptibility of the carrot weevil (Coleoptera: Curculionidae) to *Steinernema feltiae, S. bibionis,* and *Heterorhabditis heliothidis. Journal of Nematology* 17, 363–366.

Bélair, G. and Boivin, G. (1995) Evaluation of *Steinernema carpocapsae* Weiser for control of carrot weevil adults, *Listronotus oregonensis* (LeConte) (Coleoptera: Curculionidae), in organically grown carrots. *Biocontrol Science and Technology* 5, 225–231.

Bélair, G., Fournier, Y. and Dauphinais, N. (2003) Efficacy of steinernematid nematodes against three pests of crucifers in Quebec. *Journal of Nematology* 35, 259–265.

Berry, R.E., Liu, J. and Reed, G. (1997) Comparison of endemic and exotic entomopathogenic nematode species for control of Colorado potato beetle (Coleoptera: Chrysomelidae). *Journal of Economic Entomology* 90, 1528–1533.

Boivin, G. and Bélair, G. (1989) Infectivity of two strains of *Steinernema feltiae* (Rhabditida: Steinernematidae) in relation to temperature, age, and sex of carrot weevil (Coleoptera: Curculionidae) adults. *Journal of Economic Entomology* 82, 762–765.

Boselli, M., Curto, G.M. and Lazzari, G.L. (1991) Monitoraggio del cleono della barbabietola da zucchero (*Temnorrhinus mendicus* Gyll.) (Coleoptera Curculionidae) e prove preliminari di lotta con nematodi entomopatogeni. *Informatore fitopatologico* 7/8, 55–59.

Boselli, M., Curto, G.M., Tacconi, R., Pollini, A. and Santi, R. (1994) Field trials with *Steinernema carpocapsae* and *Heterorhabditis* sp. juveniles against the sugarbeet weevil, *Conorrhynchus mendicus* Gyll., in Italy. *Nematologica* 40, 463–465.

Boselli, M., Curto, G.M. and Tacconi, R. (1997) Field efficacy of entomopathogenic nematodes against the sugar-beet weevil *Temnorhinus* (= *Conorrhynchus*) *mendicus* Gyll. (Coleoptera: Curculionidae). *Biocontrol Science and Technology* 7, 231–238.

Bracken, G.K. (1990) Susceptibility of first-instar cabbage maggot, *Delia radicum* (L.) (Anthomyiidae: Diptera) to strains of the entomogenous nematodes *Steinernema feltiae* Filipjev, *S. bibionis* (Bovien), *Heterorhabditis bacteriophora* Poinar, and *H. heliothidis* (Khan, Brooks, and Hirschmann). *Canadian Entomologist* 122, 633–639.

Capinera, J.L., Pelissier, D., Menout, G.S. and Epsky, N.D. (1988) Control of black cutworm, *Agrostis ipsilon* (Lepidoptera: Noctuidae) with entomogenous nematodes (Nematoda: Steinernematidae, Heterorhabditidae). *Journal of Invertebrate Pathology* 52, 427–435.

Curto, G.M., Boselli, M., Paganini, U. and Ghedini, R. (1992) Prove di lotta contro il cleono della barbabietola da zucchero (*Conorrhynchus mendicus* Gyll.) (Coleoptera Curculionidae) con nematodi entomopatogeni. *Informatore Fitopatologico* 12, 45–49.

Curto, G.M., Boselli, M. and Tacconi, R. (1999) Field effectiveness of entomopathogenic nematodes against the sugar-beet weevil *Temnorhinus* (= *Conorrhynchus*) *mendicus* Gyll. (Coleoptera: Curculionidae) in Italy. In: *Proceedings of COST 819 Workshop* 16 to 20 May 1995 *'Application and Persistence of Entomopathogenic Nematodes'* Todi, Perugia, Italy, pp. 107–115.

Deseö, K.V. (1987) Entomoparasitic nematodes in the control of curculionids. Microbial control of weevils and environmental persistence of pathogens and nematodes. *WPRS Bulletin* XII, 4, Versailles, 2–4 September.

Eckenrode, C.J. and Chapman, R.K. (1971) Observations on cabbage maggot activity under field conditions. *Annals of the Entomological Society of America* 64, 1226–1230.

Ekanayake, H.M.R.K., Abeysinghe, A.M.C.P. and Toida, Y. (2001) Potential of entomopathogenic nematodes as bio-control agents of sweet potato weevil, *Cylas formicarius* (Fabricius) (Coleoptera: Brenthidae). *Japanese Journal of Nematology* 31, 19–25.

Ellers-Kirk, C.D., Fleischer, S.J., Snyder, R.H. and Lynch, J.P. (2000) Potential of entomopathogenic nematodes for biological control of *Acalymma vittatum* (Coleoptera: Chrysomelidae) in cucumbers grown in conventional and organic soil management systems. *Journal of Economic Entomology* 93, 605–612.

FAO (2003) FAOSTAT agriculture data. Last updated 10 June 2003. Available at: http://apps.fao.org/cgi-bin/nph-db.pl?subset=agriculture

Hommes, M. (1988) Alternative possibilities of cabbage root fly control after the efficiency of chlorfenvinphos and carbofuran has begun to fail in some growing areas of the Federal Republic of Germany. *Acta Horticulturae* 219, 47–51.

Jansson, R.K. and Lecrone, S.H. (1994) Application methods for entomopathogenic nematodes

(Rhabditida: Heterorhabditidae): aqueous suspensions versus infected cadavers. *Florida Entomologist* 77, 281–284.

Jansson, R.K. and Lecrone, S.H. (1997) Effect of sweet potato cultivar on efficacy and persistence of entomopathogenic nematodes for control of *Cylas formicarius*. *Nematropica* 27(1), 41–52.

Jansson, R.K., Lecrone, S.H., Gaugler, R. and Smart, G.C. Jr (1990). Potential of entomopathogenic nematodes as biological control agents of sweet potato weevil (Coleoptera: Curculionidae). *Journal of Economic Entomology* 83, 1818–1826.

Jansson, R.K., Lecrone, S.H. and Gaugler, R. (1991) Comparison of single and multiple releases of *Heterorhabditis bacteriophora* Poinar (Nematoda: Heterorhabditidae) for control of *Cylas formicarius* (Fabricius) (Coleoptera: Apionidae). *Biological Control* 1, 320–328.

Jansson, R.K., Lecrone, S.H. and Gaugler, R. (1993) Field efficacy and persistence of entomopathogenic nematodes (Rhabditida: Steinernematidae, Heterorhabditidae) for control of sweet potato weevil (Coleoptera: Apionidae) in Southern Florida. *Journal of Economic Entomology* 86, 1055–1063.

Lello, E.R., Patel, M.N., Matthews, G.A. and Wright, D.J. (1996) Application technology for entomopathogenic nematodes against foliar pests. *Crop Protection* 15, 567–574.

Levine, E. and Oloumi-Sadeghi, H. (1992) Field evaluation of *Steinernema carpocapsae* (Rhabditida: Steinernematidae) against the black cutworm (Lepidoptera: Noctuidae) larvae in field corn. *Journal of Entomological Science* 27, 427–435.

Lössbroek, T.G. and Theunissen, J. (1985) The entomogenous nematode *Neoaplectana bibionis* as a biological control agent of *Agrostis segetum* in lettuce. *Entomologia Experimentalis et Applicata* 39, 261–264.

MacVean, C.M., Brewer, J.W. and Capinera, J.L. (1982) Field test of antidesiccants to extend the infection period of an entomogenous nematode, *Neoaplectana carpocapsae*, against the Colorado potato beetle. *Journal of Economic Entomology* 75, 97–101.

Mannion, C.M. and Jansson, R.K. (1992) Movement and postinfection emergence of entomopathogenic nematodes from sweet potato weevil, *Cylas formicarius* (F.) (Coleoptera: Apionidae). *Biological Control* 2, 297–305.

Mannion, C.M. and Jansson, R.K. (1993) Within-root mortality of *Cylas formicarius* (Coleoptera: Apionidae) by entomopathogenic nematodes. *Journal of Economic Entomology* 86, 722–729.

Mason, J.M., Matthews, G.A. and Wright, D.J. (1998a) Appraisal of spinning disc technology for the application of entomopathogenic nematodes. *Crop Protection* 17, 453–461.

Mason, J.M., Matthews, G.A. and Wright, D.J. (1998b) Screening and selection of adjuvants for the spray application of entomopathogenic nematodes against a foliar pest. *Crop Protection* 17, 463–467.

Mason, J.M., Matthews, G.A. and Wright, D.J. (1999) Evaluation of spinning disc technology for the application of entomopathogenic nematodes against a foliar pest. *Journal of Invertebrate Pathology* 73, 282–288.

Menzler-Hokkanen, I. and Hokkanen, H.M.T. (2003) Entomopathogenic nematode delivery systems for biological control of pests on major outdoor crops: the case of oilseed rape. *Program and Abstracts for the 36th Annual Meeting of Society for Invertebrate Pathology*, Burlington, Vermont, pp. 60.

Miklasiewicz, T.J., Grewal, P.S., Hoy, C.W. and Malik, V.S. (2002) Evaluation of entomopathogenic nematodes for suppression of carrot weevil. *Biocontrol* 47, 545–561.

Nickle, W.R., Connick, W.J. and Cantelo, W.W. (1994) Effects of Pesta-pelletized *Steinernema carpocapsae* (All) on western corn rootworms and Colorado potato beetles. *Journal of Nematology* 26, 249–250.

Piggott, S.J., Liu, Q.Z., Glazer, I. and Wright, D.J. (2003) Does osmoregulatory behaviour in entomopathogenic nematodes predispose desiccation tolerance. *Nematology* 4, 483–487.

Schroeder, P.C., Ferguson, C.S., Shelton, A.M., Wilsey, W.T., Hoffmann, M.P. and Petzoldt, C. (1996) Greenhouse and field evaluations of entomopathogenic nematodes (Nematoda: Heterorhabditidae and Steinernematidae) for control of cabbage maggot (Diptera: Anthomyiidae) on cabbage. *Journal of Economic Entomology* 89, 1109–1115.

Shapiro, D.I., Lewis, L.C., Obrycki, J.J. and Abbas, M. (1999) Effects of fertilizers on suppression of black cutworm (*Agrostis ipsilon*) damage with *Steinernema carpocapsae*. *Journal of Nematology* 31, 690–693.

Shapiro-Ilan, D.I., Lewis, E.E., Tedders, W.L. and Son, Y. (2003) Superior efficacy observed in entomopathogenic nematodes applied in infected-host cadavers compared with application in aqueous suspension. *Journal of Invertebrate Pathology* 83, 270–272.

Simser, D. (1992) Field application of entomopathogenic nematodes for control of *Delia radicum* in collards. *Journal of Nematology* 24, 374–378.

Stewart, J.G., Boiteau, G. and Kimpinski, J. (1998) Management of late-season adults of the Colorado potato beetle (Coleoptera: Chrysomelidae) with entomopathogenic nematodes. *The Canadian Entomologist* 130, 509–514.

Talekar, N.S. and Shelton, A.M. (1993) Biology, ecology, and management of the diamondback moth. *Annual Review of Entomology* 38, 275–301.

Toba, H.H., Lindegren, J.E., Turner, J.E. and Vail, P.V. (1983) Susceptibility of the Colorado potato beetle and the sugarbeet wireworm to *Steinernema feltiae* and *S. glaseri*. *Journal of Nematology* 15, 597–601.

Vänninen, I., Hokkanen, H. and Tyni-Juslin, J. (1999) Screening of field performance of entomopathogenic fungi and nematodes against cabbage root flies (*Delia radicum* L. and *D. floralis* (Fall); Diptera, Anthomyiidae). *Acta Agriculturae Scandinavica* 49, 167–183.

Veremchuk, G.V. and Danilov, L.G. (1976) Entomopathogenic nematodes. *Zashchita Rastenii* 8, 22.

Welch, H.E. (1958) Test of a nematode and its associated bacterium for control of the Colorado potato beetle, *Leptinotarsa decemlineata* (Say). *Annual Report of Entomological Society of Ontario* 88, 53–54.

Welch, H.E. and Briand, L.J. (1961a) Field experiment on the use of a nematode for the control of vegetable crop insects. *Annual Report of Entomological Society of Ontario* 91, 197–202.

Welch, H.E. and Briand, L.J. (1961b) Tests of the nematode DD136 and an associated bacterium for control of the Colorado potato beetle, *Leptinotarsa decemlineata* (Say). *The Canadian Entomologist* 93, 759–763.

Williams, E.C. and MacDonald, O.C. (1995) Critical factors required by the nematode *Steinernema feltiae* for the control of the leafminers *Liriomyza huidobrensis, Liriomyza bryoniae* and *Chromatomyia syngenesiae. Annals of Applied Biology* 127, 329–341.

Williams, E.C. and Walters, K.F.A. (2000) Foliar application of the entomopathogenic nematode *Steinernema feltiae* against leafminers on vegetables. *Biocontrol Science and Technology* 10, 61–70.

Wright, R.J., Agudelo-Silva, F. and Georgis, R. (1987) Soil applications of steinernematid and heterorhabditid nematodes for control of Colorado potato beetle, *Leptinotarsa decemlineata* (Say). *Journal of Nematology* 19, 201–206.

Yokomizo, K. and Kashio, T. (1996) Application of an entomogenous nematode, *Steinernema carpocapsae*, for control of the cutworm, *Agrostis segetum*, in carrot fields. *Proceedings of the Association for Plant Protection of Kyushu* 42, 89–92 (in Japanese).

14 Cereal, Fibre, Oilseed and Medicinal Crop Applications

H.E. Cabanillas,[1] R.J. Wright[2] and R.V. Vyas[3]
[1]*USDA, ARS, Kika de la Garza SARC, Beneficial Insects Research Unit, 2413 E.Hwy 83, Weslaco, TX 78596, USA;* [2]*Department of Entomology, 202 Plant Industry Building, University of Nebraska-Lincoln, Lincoln, NE 68583-0816, USA;* [3]*Department of Nematology, Gujarat Agricultural University, Anand 388 110, India*

14.1. Introduction

Entomopathogenic nematodes (EPNs) can be effective tools to manage insect pests attacking cereal, fibre, oilseed and medicinal crops. However, their adoption in agroecosystems in general has been slow. Recent trends towards precision agriculture (Grisso *et al.*, 2002; Whelan *et al.*, 2003), conservation tillage (Bull and Sanderetto, 1996), organic farming (Yussefi and Willer, 2003) and a growing interest in medicinal crops limit the use of chemical pesticides and thus have provided fresh impetus for the development of effective alternative measures to control insect pests. This chapter reviews the research on EPNs with respect to their efficacy and use for pest control in cereal, fibre, oilseed and medicinal crops.

14.2. Cereal Crops (Maize, Barley, Oats, Wheat)

14.2.1. Maize earworm

The maize earworm (*Helicoverpa zea*) is a serious pest of many crops. Larvae damage maize by feeding on whorls, silks, tassels or kernels, eventually dropping into the soil to pupate. Emerging adults attack maize, cotton, sorghum, tobacco and numerous vegetable crops. In earlier studies, poor to moderate levels of suppression were achieved when nematodes were applied to foliage or silks of the maize plants to control maize earworm in maize (Tanada and Reiner, 1962; Bong and Sikorowski, 1983; Bong, 1986; Richter and Fuxa, 1990; Purcell *et al.*, 1992). More recent studies indicate that the control strategy should be focused on the prepupal and pupal stages of maize earworm in the soil to prevent adult emergence and demonstrate the potential of controlling the maize earworm with soil applications of *Steinernema riobrave* (= *S. riobravis*), a natural control agent of the maize earworm in Texas (Raulston *et al.*, 1992; Cabanillas and Raulston, 1994a,b; Cabanillas *et al.*, 1994). The following factors play an important role in the successful use of *S. riobrave* against maize earworm. First, nematode applications should be matched with the most susceptible stage of the maize earworm. Cabanillas and Raulston (1995) obtained insect mortalities of 100% and 95% in maize fields by applying *S. riobrave* to the soil when 50% of the larvae were late instars (still in the maize ears) and when 10% of the larvae had left the ears to pupate, respectively (Fig. 14.1). Second, irrigation method and timing and nematode concentration should be optimum. *S. riobrave*, at the most effective nematode concentration of 200,000 nematodes/m^2, caused higher insect mortalities when it was applied via in-furrow irrigation (95%) than when it was applied after

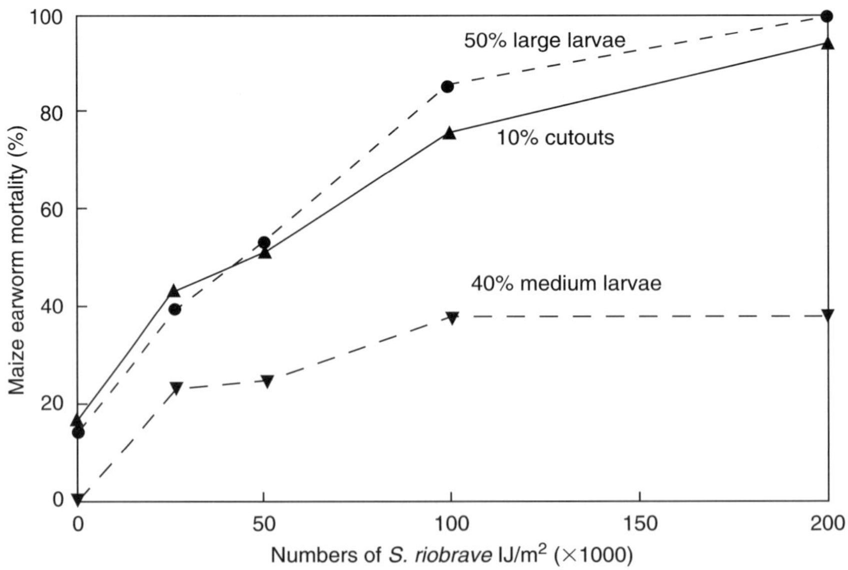

Fig. 14.1. Effect of *Steinernema riobrave* concentration and timing of soil application on parasitism of maize earworm *Heliothis zea* prepupae and pupae in maize.

irrigation (84%) or before irrigation (56%) (Fig. 14.2; Cabanillas and Raulston, 1996a). Nematode application through in-furrow irrigation may provide better moisture conditions for the nematode distribution, dispersal and survival than nematode application to dry soil before irrigation or after irrigation (Kaya, 1990). Third, the nematode species and the application method should be matched with the target ecosystem. *S. riobrave* (TX strain), at the most effective concentration of 200,000 nematodes/m^2, caused 95% maize earworm prepupae and pupae mortality while *S. carpocapsae* Weiser (All strain) did not cause any insect mortality in maize fields (Fig. 14.3; Cabanillas and Raulston, 1996b). This superiority of *S. riobrave* was attributed to its greater tolerance of warm soil temperatures (> 38°C) compared with *S. carpocapsae* (Gray and Johnson, 1983; Grewal *et al.*, 1994). Cabanillas and Raulston (1996a) found that subsurface nematode incorporation produced higher insect infections than soil surface applications in the fields that received nematodes before or after irrigation. Subsurface application probably provides greater nematode protection against desiccation and sunlight than soil surface application (Gaugler, 1988). Similarly, Feaster and Steinkraus (1996) achieved excellent results, by applying *S. riobrave* to the soil in Arkansas maize to control maize earworm. They demonstrated that this nematode has potential as an inundative biocontrol agent for this pest (Table 14.1). Mean mortalities from *S. riobrave* infections were 79.2% and 91.3% at nematode levels of 3.7×10^6 and 1.2×10^7 nematodes/m^2 of soil, respectively. Although similar results were obtained in irrigated and non-irrigated plots, higher infection occurred in the plots receiving flood irrigation (Table 14.1).

14.2.2. Maize rootworm

The maize rootworm complex (northern *Diabrotica barberi* and western D. *virgifera*) are important pests of maize in North America. Eggs are laid in the soil around the base of maize plants and the developing grubs feed on the roots. Variable results have been obtained on nematode efficacy against this insect pest. Early studies showed poor

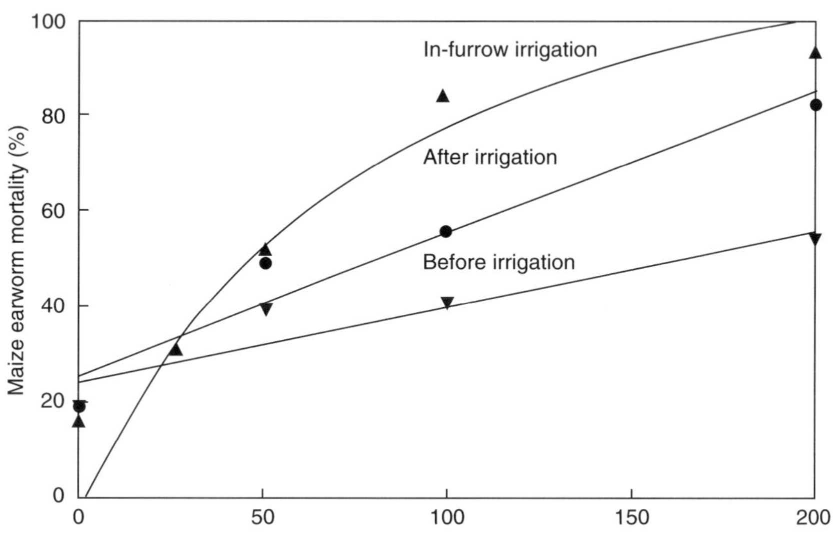

Fig. 14.2. Effects of irrigation timing and concentration of *Steinernema riobrave* on mortality of maize earworm *Heliothis zea* prepupae buried within 6 days after nematode application in soil in a maize field.

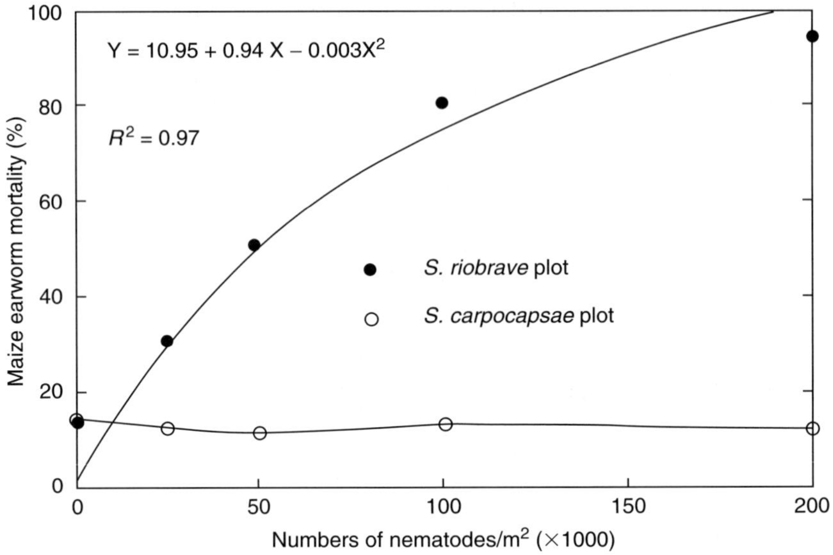

Fig. 14.3. Parasitism of maize earworm *Heliothis zea* prepupae and pupae by *Steinernema riobrave* in field plots receiving applications of either *S. riobrave* or *S. carpocapsae*.

Table 14.1. Mean percentage (SE) of maize earworm larvae or pupae infected with *Steinernema riobrave* and percentage survival to adults after collection from soil.[a]

Number of nematodes/m^2 of soil	After-treatment irrigation[b]	% infected	% maize earworm survival to adult (± SEM)
Application after larval release			
0	–	0 (0) a	59.4 (1.3) a
	+	0 (0) a	66.9 (4.9) a
3.7×10^6	–	72.6 (2.6) b	17.1 (5.9) b
	+	79.2 (7.6) bc	8.5 (4.5) bc
1.2×10^7	–	85.9 (2.7) bc	9.4 (3.0) bc
	+	91.3 (5.0) c	1.1 (1.1) c
Application before larval release			
0	–	0 (0) a	66.2 (4.9) a
	+	0 (0) a	60.6 (8.2) a
5.2×10^5	–	69.7 (12.6) b	11.1 (4.9) b
	+	78.5 (5.5) b	7.6 (4.6) bc
5.3×10^6	–+	89.4 (4.7) b	0 (0) c
		94.8 (3.1) b	0 (0) c

[a]Column values followed by the same letter are not significantly different (Fisher's LSD, $P < 0.05$. Data were transformed to arcsine square root (%/100) before analysis).
[b]+ = flood irrigated; – = non-irrigated

control of western maize rootworm when *S. carpocapsae* (DD-136 strain) at 20,000 nematodes/plant was applied at planting and after plant emergence (Rohrbach, 1969; Munson and Helms, 1970). Rohrbach (1969) attributed this poor efficacy to low soil moisture and high temperature when nematodes were applied. Later studies indicated that nematodes could provide protection against this pest when application timing is targeted to the susceptible larval stage (Thurston and Yule, 1990; Georgis *et al.*, 1991). Field trials showed that soil application of *S. carpocapsae* (Mexican and All strains) and *Heterorhabditis bacteriophora* (Lewiston strain) was more effective against the western maize rootworm larvae when applied after planting (second instar phenology) than when applied at planting time (Jackson, 1996). Nematode placement and application rates are important efficacy factors to be considered in controlling this insect pest. Jackson and Hesler (1995) observed that application rates of 100,000 or 200,000 nematodes/plant (about 8.5 billion and 17 billion nematodes/ha) were more effective (reduction of root injury and adult emergence) than the control, when nematodes were applied after planting (against second instars). Based on root injury and adult rootworm emergence, nematodes applied within a 15-cm-diameter circle centred on the plant base were more effective than two other placement patterns covering a wider area, or an area further away from the plant base (Jackson and Hesler, 1995). Wright *et al.* (1993) demonstrated that nematode protection from maize rootworm feeding was equal to the insecticide chlorpyrifos (Lorsban 4E) when both the insecticide and *S. carpocapsae* (All strain) were applied through a centre-pivot irrigation system at a rate of 1.2–2.5 × 10^9 nematodes/ha during the appearance of second instar maize rootworm larvae. Similar results were obtained when *S. carpocapsae* (All and Mexican strains) was applied through a lateral-move irrigation system (Ellsbury *et al.*, 1996). Journey and Ostlie (2000) obtained good efficacy by timing the effective rates of *S. carpocapsae*, (All strain) with the vulnerable insect stage in Minnesota dryland maize to control western maize rootworm. *S. carpocapsae*, at the most effective rates (1 million and 10 million nematodes/30.5 row-cm), resulted in greater reduction of both root injury (Fig. 14.4) and insect adult emergence (Fig. 14.5) when it was applied to late second and early third instars than when it was applied to younger second instars.

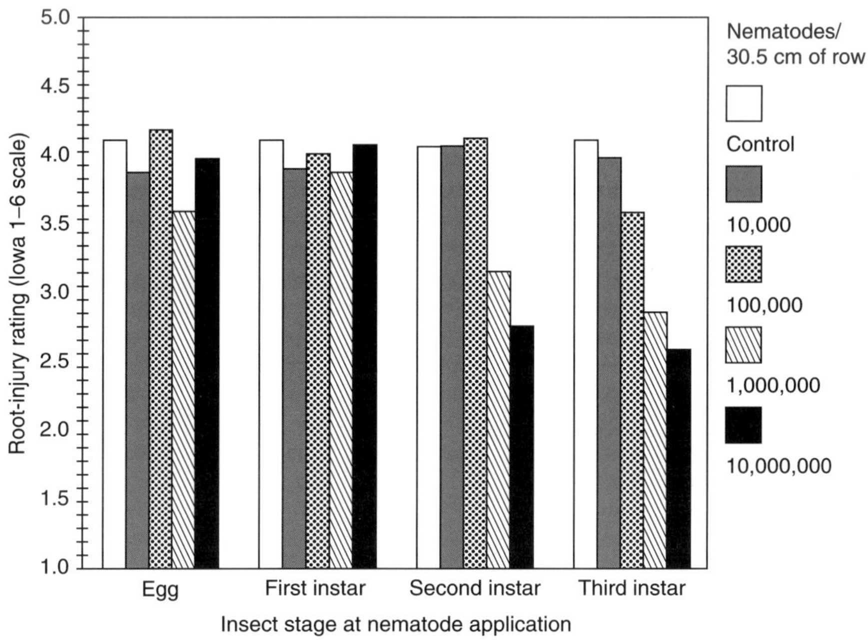

Fig. 14.4. Effects of *Steinernema carpocapsae* application rate and timing on maize root injury by western maize rootworm *Diabrotica virgifera*, 1992.

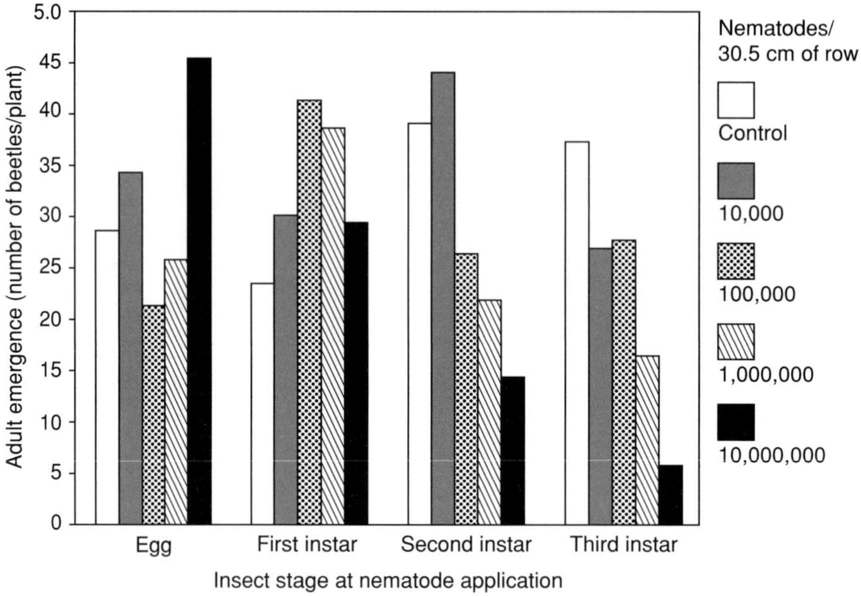

Fig. 14.5. Effects of *Steinernema carpocapsae* application rate and timing on adult emergence of western maize rootworm *Diabrotica virgifera*, 1992.

14.2.3. Black cutworm

The black cutworm (*Agrotis ipsilon*) is a periodic pest of seedling maize in the US maize belt. Adults migrate north in the spring and lay eggs in maize fields before planting time. Larvae feed on the leaves and at the base of seedlings. Small plants may be killed when larvae feed below the plant growing point. Insecticides applied at or after planting may be used to control this insect pest. Field application of *S. carpocapsae* (Mexican strain) in different formulations at 5.35×10^5 nematodes/m^2 against black cutworm larvae in seedling maize plots reduced insect damage by 50% (Capinera *et al.*, 1988). Levine and Oloumi-Sadeghi (1992) found that nematode application at 1.25×10^9 nematodes/ha reduced the number of cut plants by 76–83% during 1–10 days after treatment. Nematode control was equal to the best planting-time insecticides (chlorpyrifos, tefluthrin, fonophos) and the liquid insecticide (permethrin) applied after plant emergence. The use of manure and other fertilizers can affect nematode efficacy against black cutworm. Shapiro *et al.* (1999) found that composted manure and urea application at two rates (280 kg N/ha and 560 kg N/ha) did not affect the efficacy of *S. carpocapsae* (All strain) when applied in water at 1.25×10^5 nematodes/m^2 before larval introduction; however, the high rates of fresh manure reduced nematode efficacy against black cutworms.

14.3. Fibre Crops (Cotton, Kenaf, Flax, Hemp)

14.3.1. Boll weevil

The boll weevil (*Anthonomus grandis*) is one of the most destructive pests of cotton in the USA and other parts of the world (Ridgway and Lloyd, 1983). It can survive and reproduce only on cotton and a few related plant species. The adults feed on young leaf buds and squares (floral buds). The female weevils oviposit by puncturing squares and young bolls with their ovipositor, and lay the eggs inside. One or two larvae may complete development in each square or boll. Although infested bolls do not typically abscise, infested squares commonly do, and thus weevil development frequently occurs at the soil surface. Recently, laboratory studies showed that *S. riobrave* and other EPNs were pathogenic against the third instar weevils (Cabanillas, 2003). Further tests on nematode concentration and moisture levels demonstrated the ability of *S. riobrave* to kill the boll weevil inside abscised squares and bolls (Fig. 14.6; Cabanillas, 2003). However, the nematode's ability to kill weevils depends on soil moisture levels. Applications of 200,000 and 400,000 nematodes/m^2 in buried bolls or squares produced higher insect mortalities in pots with 20% soil moisture either in bolls (94% and 97% infectivity) or squares (92% and 100% infectivity) than those with 10% soil moisture in bolls (44% and 58% infectivity) or squares (0% and 13% infectivity).

14.3.2. Pink bollworm

The pink bollworm (*Pectinophora gossypiella*) is another destructive pest of cotton

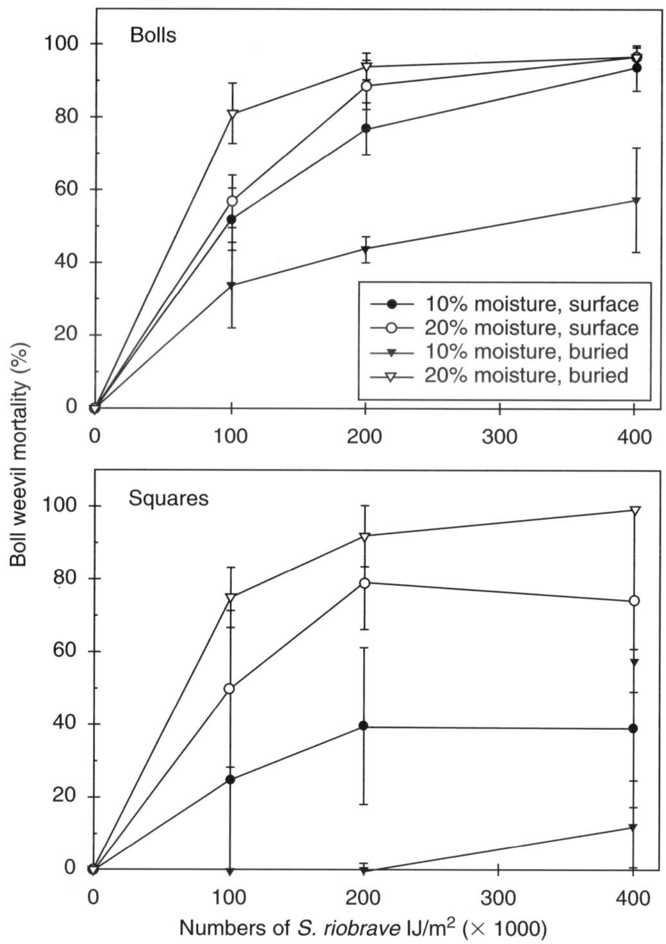

Fig. 14.6. Effects of *Steinernema riobrave* applied to soil on the control of the boll weevil inside abscised squares and bolls of cotton located on the soil surface or buried as a response to nematode concentration and soil moisture. Means are statistically different if their standard error confidence intervals do not overlap.

in large areas of southwestern USA. It damages squares and bolls and prefers cotton but will feed on okra, kenaf and hibiscus. The larvae bore into the cotton bolls and feed from 10 to 14 days on the seed. One larva eats a whole seed or parts of several seeds. When larvae finish feeding, they either drop to the ground or remain in the seed to pupate and, later, pupae emerge as adults. Laboratory studies indicated that uninjured pupae of pink bollworm were not susceptible to *S. carpocapsae* or *S. riobrave* because of the small size of the pupa spiracular orifices (Henneberry *et al.*, 1995). However, larvae diapausing in soil were susceptible to nematodes (Gouge *et al.*, 1999). For example, *Heterorhabditis bacteriophora* infected insects during autumn, but *S. riobrave* failed to locate this insect within the cotton bolls during autumn or winter. This difference is attributed to positive geotropism or more suitable moisture levels (Gouge *et al.*, 1999). It appears that *H. bacteriophora* has greater cold tolerance than *S. riobrave* to control pink bollworm during autumn or winter (Gouge *et al.*, 1999). Irrigating a cotton field immediately after application of *S. riobrave* (1 billion nematodes/acre) through a spray rig with dropped nozzles produced 19% yield increase and infested cotton bolls were reduced by about 30% (Gouge *et al.*, 1997).

14.3.3. Tobacco budworm

The tobacco budworm (*Heliothis virescens*) and the cotton bollworm (*Helicoverpa zea*) are serious insect pests of several economically important row crops grown in most warm regions of the world, including cotton, maize, tobacco and soybean. Bell (1995) reported that the emergence of adult *H. virescens* was reduced by 66% and 57% by *S. riobrave* when late instars were buried in the soil of seedling cotton plants, and when second and late instars were placed on terminals and bolls of late-season cotton plants, respectively.

14.3.4. Foliar pests

Although many crop plants can tolerate some threshold level of leaf injury from foliar pests before the crop is affected, in practice foliar pests can cause damage that can destroy the entire crop. For example, cotton pests such as the beet armyworm, *Spodoptera exigua*, larvae can destroy seedlings and skeletonize leaves. Other important foliar pests of cotton are *Heliothis armigera* and *Earias insulana*. The gypsy moth, *Lymantria dispar*, is one of North America's most devastating forest pests. It feeds on the foliage of hundreds of species, including cotton, but its most common hosts are oaks and aspen. During heavy infestations, trees may be completely defoliated. Males are strong fliers, but females do not fly. This pest continues to spread and it is a threat to cotton in western North America.

Effective control of the foliage pests *Heliothis* spp. and *Spodoptera* spp. on cotton was achieved under greenhouse conditions (i.e. relative humidity >90% and moderate temperatures) by applying nematodes on the plant (Samsook and Sikora, 1981; Ishibashi, 1987). It was shown that nematode survival will depend on the conditions provided on the plant surface. Shapiro *et al.* (1985) demonstrated that antidesiccants improved nematode persistence. For example, *S. carpocapsae* (DD-136 strain) when mixed with glycerol (2.5% wt per wt) or Folicote (6% wt per wt) resulted in 25% and 74% control of gypsy moth *L. dispar* larvae, respectively, after a 19-h drying at 55% RH. Glazer and Navon (1990) found that nematode activity was enhanced by selecting nematode strains with greater desiccation tolerance in combination with an antidessicant. Higher levels of control resulted with *S. feltiae* (Pye strain) (75% with glycerol and 95% with Folicote) than with *S. carpocapsae* (= *S. feltiae*) 'All' strain (10% with glycerol and 40% with Folicote). Nematode efficacy can be improved in foliar applications by selecting strains that tolerate drying conditions. It appears that the 'Pye' strain has better ability to withstand drying conditions than the 'All' strain. Glazer *et al.* (1992) reported that foliar applications of nematodes at 500 and 1000 nematodes/ml combined with antidesiccants resulted in higher than 85% control of the cotton pests *E. insulana* and *Spodoptera littoralis*, respectively, on bean plants.

To achieve satisfying control, EPNs' persistence and infectivity on the foliage must be enhanced too. Recent studies indicated that penetration of *S. carpocapsae* into the diamondback moth (DBM) larvae was greater through active invasion than by insect ingestion on cabbage leaves (Schroer and Ehlers, 2003). This shows the importance of enhancing nematode persistence against foliar pests. Schroer and Ehlers (2003) found that formulating EPNs with 0.3% surfactant based on castor oil and 0.5% xanthan gum improved the efficacy compared to water. The infective dose (ID_{50}) is lowered from 20.3 to 6.7 *S. carpocapsae* per larva.

14.4. Oil Crops

Oil crops including castor, soybean, sunflower, safflower, groundnut, olive and jojoba are important in Asia, Africa, Australia and the Americas, and are the primary source of cash or barter, especially for millions of the rural poor. Groundnut oil comprises 55% of India's vegetable oil production where cooking oil is the second largest import item. Leaf-eating caterpillars and white grubs are significant pests in many agricultural systems, including groundnut and other oil crops.

14.4.1. Leaf-eating caterpillars

The leaf-eating caterpillar, *Spodoptera litura*, causes serious yield losses in castor in India. Patel (1999) conducted laboratory trials and found that *Steinernema* sp. (Vatrak isolate nr. *carpocapsae*) caused similar larval mortality (59%) as nuclear polyhedrosis virus and azadirachtin (1500 ppm) and was the second best treatment to endosulfan. No field testing has yet occurred.

The beet armyworm, *S. exigua*, has a wide host range including soybean and sunflower among other crops. It is mainly a pest of late-planted seedling soybeans. Small larvae skeletonize the lower leaves and large larvae feed over the whole plant. Skeletonization and often profuse silk webbing, which gives the plants a shiny appearance, are characteristic signs of infestation by this species. The beet armyworm has few effective parasites, diseases or predators to lower its population and is resistant to several insecticides. Gothama *et al.* (1996) demonstrated the potential use of combining two pathogens in beet armyworm management on soybean. The combination of *S. carpocapsae* and nuclear polyhedrosis virus produced higher larval mortality (62%) than either the nematodes (25–35%) or the virus alone (27–34%). The poor efficacy of using nematodes alone was attributed in part to the foliar environment, which exposes them to adverse moisture conditions that result in their rapid desiccation and death. The foliage persistence of *S. carpocapsae* was 12–24 h. Similarly, Sezhian *et al.* (1996) found additive effects when *S. carpocapsae* was combined with an insect phagostimulant against *S. litura* on sunflower. The combination of nematodes and the phagostimulant produced higher larval mortality (22%) than nematodes alone (7%).

14.4.2. Scarab pests

Scarab pests include about 30,000 species distributed throughout the world. Many species of scarab beetles attack oil crops such as groundnut. For example, the scarab beetle, *Maladera matrida*, first detected in Israel in 1983, attacks groundnuts and other crops. During its life cycle, the adults emerge from the soil at sunset and aggregate to feed and mate. This behaviour is important to consider in biocontrol to suppress its populations. White grubs such as the root grub, *Holotrichia consanguinea*, are significant pests in many agricultural systems, including groundnut, throughout the semiarid tropics and subtropics. The larvae feed on roots, killing seedlings and sometimes older plants, thus reducing crop yields. Selection of the nematode species and timing of application can be important efficacy factors in suppressing grubs of *M. matrida* on groundnuts. In one trial, applications of *H. bacteriophora* (10 million nematodes/m^2) at 5 weeks after planting reduced the grub population by 60% as compared with heptachlor (90%). Nematodes applied 5 weeks after planting were more effective than after 8 weeks. In another trial, application of *H. bacteriophora* NC strain at 300,000 nematodes/m^2 at 6 weeks after planting caused greater insect mortality (90%) than *S. carpocapsae* 'All' strain (40%) applied at 750,000 nematodes/m^2 (Glazer and Gol'berg, 1993). Plot size and confined conditions may enhance nematode activity against these insect pests. Vyas and Yadav (1993) found that soil applications of 10,000 to 1 million *S. glaseri*/m^2 against root grub, *H. consanguinea*, on groundnut, resulted in 40% and 100% mortality at 10 and 20 days after treatment, respectively, at the highest dose.

14.5. Medicinal Crops

The medicinal plant industry is growing worldwide. Like any other agricultural crop, medicinal plants such as cress, chrysanthemum, echinacea, valerian, milk thistle and feverfew are subject to attack by insect pests. The following are selected cases of the use of nematodes to control insect pests that attack medicinal plants.

14.5.1. Diamondback moth (DBM)

Among the few medicinal crops, cress (*Lepidium sativum*) is an important commercial crop grown in Southeast Asia and Africa. The medicinal values attributed to this crop include diuretic properties found in leaves, nourishing, laxative and poultice properties found in seeds and antibacterial properties found in root and stalk extracts (Patel, 1998). The DBM, *Plutella xylostella*, is an important pest of cress because it reduces seed yield drastically. The younger larvae attack the foliage and the older larvae (third and fourth instars) bore into the fruit capsules and damage the developing seeds. Successful results on the use of nematodes against this pest have been obtained in India. Vyas *et al.* (2000) demonstrated that applications of *S. glaseri* (100,000 nematodes/m^2) with adjuvants reduced *P. xylostella* larvae on cress by 59%, which resulted in 43.3% yield increase (Table 14.2). This efficacy was similar to those obtained with the chemical insecticide monocrotophos and the bacterium *Bacillus thuringiensis* var. *kurstaki* (Vyas *et al.*, 2000). Also, foliar spray of *S. thermophilum*

Table 14.2. Comparative efficacy of *Steinernema glaseri* with *Bacillus thuringiensis* var. *kurstaki* and an insecticide for the control of the diamondback moth (DBM) larvae, *Plutella xylostella*, on cress crop in India.

Treatments	Larval population/plant					Yield kg/ha	Yield increase (%)
	Days after treatment						
	0	1	2	3	4		
S. glaseri	14a[a]	9b	7b	7b	8ab	972c	25
Btk	15a	8b	3b	2b	2b	1414a	82
Monocrotophos	15a	7b	1b	4b	3b	1217ab	61
Control	14a	15a	15a	15a	15a	776d	–

[a]Column values followed by the same letter are not significantly different ($P < 0.05$, Duncan's new multiple range test).
Btk, *Bacillus thuringiensis* var. *kurstaki*.

has been found to be very effective, causing 37–45% mortality of DBM on cabbage under field conditions during winter when the minimum temperature was 5°C (Ganguly and Somvanshi, 2003).

14.5.2. Mint root borer

The essential oils obtained from some mint species (*Mentha pulegium, M. spicata*) exhibit antimicrobial properties against some strains of Gram-positive and Gram-negative bacteria. Insects cause severe damage on spearmint and peppermint plants and in most cases must be controlled to obtain maximum yields. The mint root borer (*Fumibotrys fumalis*) is a serious pest of peppermint in Idaho and other growing areas worldwide. Larvae hatch from eggs deposited on the undersides of leaves and then feed for a short time before moving to the soil to feed inside the rhizomes. Once inside, the larvae hollow out and eventually kill the plants. This insect overwinters as a prepupa within an earthen cell 1.2–3.8 cm below the soil surface and pupation occurs within the cell. Nematodes provide effective control against the mint root borer. Grewal and Georgis (1999) reported that application timing is a key factor to obtain effective control of larvae because of the limited persistence of nematodes, prolonged emergence of adults and the formation of resistant hibernacula. Nematodes should be applied before the formation of hibernacula. However, care must be taken to not apply the nematodes prematurely. Application of *S. carpocapsae* can effectively control the mint root borer by using two applications of 1 billion/acre before and after harvest, rather than one preharvest application of 2 billion nematodes (J. Takeyasu, 1992, unpublished data).

14.5.3. Mint flea beetle

Mint flea beetles (*Longitarsus ferrugineus* and *L. waterhousei*) attack spearmint fields and cause serious damage. It is a tiny, oval, brown beetle with large hind legs for jumping. Eggs overwinter in the soil and hatch in the spring. The slender white larvae bore into mint roots, causing the most severe damage. This larval feeding is first noted in the spring, when the mint stand does not 'green-up'. Close examination reveals 0.8 mm holes or tunnels in the roots. Heavy populations can cause large areas of dead or weak mint. The adult flea beetle emerges in early July and feeds on the leaf surface causing a shot-hole appearance as the mint matures. Insecticidal control of the larval stage in the soil is ineffective; however, applying the most effective nematode species to the soil could reduce the insect population before adult emergence. Grewal and Georgis (1999) indicate that in a field test, *H. bacteriophora* and *S. carpocapsae* provided 94% and 67% control, respectively, of *L. waterhousei*.

14.5.4. Leafminer

Chrysanthemum is an important medicinal crop. Dry flowers of *Chrysanthemum* are used to induce menstrual flow, cause abortion, and cure intestinal worms and indigestion. Its leaves when chewed are a remedy for colds, indigestion and diarrhoea (Bhattacharyee, 2001). The leafminer (*Liriomyza trifolii*) causes significant damage to chrysanthemums. Foliar applications of *S. carpocapsae* (500 million nematodes/ha) suppressed this pest to levels comparable with the chemical insecticide abamectin (Harris *et al.*, 1990).

14.5.5. Weevils

The cultivated strawberry, and particularly the wild strawberry (*Fragaria vesca* L.), have nutritional and medicinal values because of their vitamins and laxative, diuretic and astringent properties. Their leaves and fruit contain malic and citric acids. Similarly, cranberry (*Vaccinium macrocarpon*) is used not only as food but also to prevent recurrent urinary tract infections and other human illnesses. Harmful insects such as the root weevil, the black vine weevil and other insect pests attack these crops. For example, the strawberry root weevil (*Otiorhynchus ovatus*), and black vine weevil (*O. sulcatus*), are important pests of mint, strawberries and cranberries. Adults feed on leaves, and larvae chew on tunnel roots. Nematodes are effective control agents against these pests and may replace chemical insecticides as a better approach to control these weevils. *H. bacteriophora* seems more effective than *S. carpocapsae* against both weevils (Klinger, 1988; Shanks and Agudelo-Silva, 1990; Miduturi *et al.*, 1994). *H. bacteriophora* NC and HP88 strains reduced larvae and pupae of the black vine weevil by 70% and 100%, respectively; and *S. carpocapsae* (All strain) reduced pest populations by 75% (Shanks and Agudelo-Silva, 1990). Additional discussion of EPNs use for suppression of *Otiorhynchus* spp. may be found in Chapter 12.

14.6. Conclusion and Future Needs

Understanding the factors affecting nematode efficacy to control pests that attack cereal, fibre, oilseed and medicinal crops is an important step to improving pest control. Since many nematode species are prime candidates for biocontrol of a number of soil pests, the selection of the appropriate species to match the environmental conditions of the target agroecosystem is of paramount importance. For example, both nematodes, *S. carpocapsae* and *S. riobrave*, are excellent control agents against the maize earworm under laboratory conditions; however, under high soil temperatures in the field, *S. riobrave* is more effective than *S. carpocapsae* to control this pest. Knowledge of the temperature limits and optima of each nematode species are important for effective field application. Although the greatest effect of this nematode on the control of this insect pest can be attributed to the features of *S. riobrave* (high mobility, tolerance to low soil moisture), other factors also contribute to its success. Using an effective nematode concentration applied via in-furrow irrigation at the critical time in relation to the target insect life cycle are the other key efficacy factors in suppressing the maize earworm populations. Application via irrigation enhances nematode effectiveness compared with when nematodes are applied to dry soils and then irrigated. The same principle applies in using this nematode to control the boll weevil in cotton and other pests. The nematode effectiveness against other pests such as the maize rootworm in maize depends on soil moisture, nematode concentration, nematode placement and application timing. Applications of *S. carpocapsae* and *H. bacteriophora* are more effective against the maize root worm larvae when applied to the soil after planting (second instar) within 15 cm around the plant base than when applied at planting time and further away from the plant base. Despite the success obtained with certain nematodes, their use has not been implemented in the field for several reasons:

1. Control of certain pests against crops can be considered uneconomical. This may be the case for the use of *S. riobrave* against the maize earworm in maize and against the boll weevil in cotton.
2. Lack of public awareness of the importance of nematodes for control of these insect pests.
3. Limited study on nematodes to control insect pests on new emerging economical crops such as oil crops, including flax, linseed (*Linum usitatissimum* L.) and medicinal crops (echinacea, valerian, milk thistle, feverfew etc.). However, cropping systems used for these crops are sometimes different from those used for conventional crops and may be more conducive for nematode application. The high economic value of these crops may also be more favourable to nematode use.
4. Limited nematode research on the control of insect pests that attack crops grown under conservation tillage systems compared with conventional systems.
5. Availability of reliable chemical products at low cost.

6. Difficulty in integrating EPNs into crop pest management programmes currently used by growers.

The recent emphasis on conservation tillage and organic farming that limits the use of chemical pesticides serves as a strong motivation to use EPNs. In the 21st century, farmers and agricultural managers are adopting precision agriculture, also referred to as precision farming practices, or variable rate technology (Grisso *et al.*, 2002; Whelan *et al.*, 2003). Thus, the effective use of nematodes as biocontrol agents will require the knowledge of precision agriculture tools. Although the above review indicates the potential of EPNs for the control of pests in the agroecosystems, more research and development effort is needed to exploit the full potential of nematodes, especially in the new production systems.

References

Bell, M.R. (1995) Effects of an entomopathogenic nematode and a nuclear polyhedrosis virus on emergence of *Heliothis virescens* (Lepidoptera: Noctuidae). *Journal of Entomological Science* 30, 243–250.

Bhattacharyee, S.K. (2001) *Handbook of Medicinal Plants.* Pointer Publishers, Jaipur, India, 478 pp.

Bong, C.F.J. (1986) Field control of *Heliothis zea* (Boddie) (Lepidoptera: Noctuidae) using a parasitic nematode. *Insect Science and Its Application* 7, 23–25.

Bong, C.F.J. and Sikorowski, P.P. (1983) Use of the DD 136 strain of *Neoaplectana carpocapsae* Weiser (Rhabditidae: Steinernematidae) for control of corn earworm (Lepidoptera: Noctuidae). *Journal of Economic Entomology* 76, 590–593.

Bull, L. and Sanderetto, C. (1996) Crop residue management and tillage system trends. *Statistical Bulletin No. 930,* US Department of Agriculture, Washington, DC.

Cabanillas, H.E. (2003) Susceptibility of the boll weevil to *Steinernema riobrave* and other entomopathogenic nematodes. *Journal of Invertebrate Pathology* 82, 188–197.

Cabanillas, H.E. and Raulston, J.R. (1994a) Pathogenicity of *Steinernema riobravis* against corn earworm, *Helicoverpa zea* (Boddie). *Fundamentals and Applied Nematology* 17, 219–223.

Cabanillas, H.E. and Raulston, J.R. (1994b) Evaluation of the spatial pattern of *Steinernema riobravis* in corn plots. *Journal of Nematology* 26, 25–31.

Cabanillas, H.E. and Raulston, J.R. (1995) Impact of *Steinernema riobravis* (Rhabditida: Steinernematidae) on the control of *Helicoverpa zea* (Lepidoptera: Noctuidae) in corn. *Journal of Economic Entomology* 88, 58–64.

Cabanillas, H.E. and Raulston, J.R. (1996a) Effects of furrow irrigation on the distribution and infectivity of *Steinernema riobravis* against corn earworm in corn. *Fundamentals and Applied Nematology* 19, 273–281.

Cabanillas, H.E. and Raulston, J.R. (1996b) Evaluation of *Steinernema riobravis, S. carpocapsae,* and irrigation timing for the control of corn earworm, *Helicoverpa zea. Journal of Nematology* 28, 75–82.

Cabanillas, H.E., Poinar, G.O. Jr and Raulston, J.R. (1994) *Steinernema riobravis* n. sp. (Rhabditida: Steinernematidae) from Texas. *Fundamentals and Applied Nematology* 17, 123–131.

Capinera, J.L., Pelissier, L.D., Menout, G.S. and Epsky, N.D. (1988) Control of black cutworm, *Agrotis ipsilon,* with entomogenous nematodes (Nematoda: Steinernematidae, Heterorhabditidae). *Journal of Invertebrate Pathology* 52, 427–435.

Ellsbury, M.M., Jackson, J.J., Woodson, W.D., Beck, D.L. and Stange, K.A. (1996) Efficacy, application, distribution, and concentration by stemflow of *Steinernema carpocapsae* (Rhabditida: Steinernematidae) suspensions applied with a lateral-move irrigation system for corn rootworm (Coleoptera: Chrysomelidae) control in maize. *Journal of Economic Entomology* 89, 74–81.

Feaster, M.A. and Steinkraus, D.C. (1996) Inundative biological control of *Helicoverpa zea* (Lepidoptera: Noctuidae) with the entomopathogenic nematode *Steinernema riobravis* (Rhabditida: Steinernematidae). *Biological Control* 7, 38–43.

Ganguly, S. and Somvanshi, V.S. (2003) Implementation of entomopathogenic nematodes in India. *Third International Symposium on Entomopathogenic Nematodes and Symbiotic Bacteria,* 4–7 September 2003. Ohio State University, Wooster, Ohio (Abstr.).

Gaugler, R. (1988) Ecological considerations in the biological control of soil-inhabiting insects with entomopathogenic nematodes. *Agriculture Ecosystems and Environment* 24, 351–360.

Georgis, R., Kaya, H.K. and Gaugler, R. (1991) Effect of steinernematid and heterorhabditid nematodes (Rhabditida: Steinernematidae and Heterorhabditidae) on nontarget arthropods. *Environmental Entomology* 20, 815–822.

Glazer, I. and Gol'berg, A. (1993) Field efficacy of entomopathogenic nematodes against the beetle, *Maladera matrida. Biocontrol Science and Technology* 3, 367–376.

Glazer, I. and Navon, A. (1990) Activity and persistence of entomoparasitic nematodes tested against *Heliothis armigera* (Lepidoptera: Noctuidae). *Journal of Economic Entomology* 83, 1795–1800.

Glazer, I., Klein, M., Navon, A. and Nakache, Y. (1992) Comparison of efficacy of entomopathogenic nematodes combined with antidesiccants applied by canopy against three cotton pests (Lepidoptera: Noctuidae). *Journal of Economic Entomology* 85, 1636–1641.

Gothama, A.A.A., Lawrence, G.W. and Sikorowski, P.P. (1996) Activity and persistence of *Steinernema carpocapsae* and *Spodoptera exigua* nuclear polyhedrosis virus against *S. exigua* larvae on soybean. *Journal of Nematology* 28, 68–74.

Gouge, D.H., Smith, K.A., Payne, C., Lee, L.L., Van Berkum, J.R., Ortega, D. and Henneberry, T.J. (1997) Control of pink bollworm, *Pectinophora gossypiella* Saunders (Lepidoptera: Gelechiidae) with biocontrol and biorational agents. In: *Proceedings of the Beltwide Cotton Conference*, Vol. 2, National Cotton Council, Memphis, Tennessee, pp. 1066–1072.

Gouge, D.H., Lee, L.L. and Henneberry, T.J. (1999) Parasitism of diapausing pink bollworm *Pectinophora gossypiella* (Lepidoptera: Gelechiidae) larvae by entomopathogenic nematodes (Nematoda: Steinernematidae, Heterorhabditidae). *Crop Protection* 18, 531–537.

Gray, P.A. and Johnson, D.T. (1983) Survival of the nematode *Neoaplectana carpocapsae* in relation to soil temperature, moisture and time. *Journal of Georgia Entomological Society* 18, 454–460.

Grewal, P.S. and Georgis, R. (1999) Entomopathogenic nematodes. In: Hall, F.R. and Menn, J.J. (eds) *Biopesticides: Use and Delivery*. Humana Press, Totowa, New Jersey, pp. 271–299.

Grewal, P.S., Selvan, S., and Gaugler, R. (1994) Thermal adaptation of entomopathogenic nematodes: niche breadth for infection, establishment, and reproduction. *Journal of Thermal Biology* 19, 245–253.

Grisso, R., Alley, M., McClellan, P., Brann, D. and Donohue, S. (2002) *Precision Farming: A Comprehensive Approach*. Virginia Cooperative Extension Publication Number 442–500, 7 pp.

Harris, M.A., Begley, J.W. and Warkentin, D.L. (1990) *Liriomiza trifolii* (Diptera: Agromyzidae) suppression with foliar applications of *S. carpocapsae* and abamectin. *Journal of Economic Entomology* 83, 2380–2384.

Henneberry, T.J., Lindegren, J.E., Forlow, J.L. and Burke, R.A. (1995) Pink bollworm (Lepidoptera: Gelechiidae), cabbage looper, and beet armyworm (Lepidoptera: Noctuidae) pupal susceptibility to steinernematid nematodes (Rhabditida: Steinernematidae). *Journal of Economic Entomology* 88, 835–839.

Ishibashi, N. (1987) Desiccation survival of *Steinernema feltiae* str. DD-136 and possible desiccation protectants for foliar application of nematodes. In: Ishibashi, N. (ed.) *Recent Advances in Biological Control of Insect Pests by Entomogenous Nematodes in Japan*. Saga University, Japan, pp. 139–144.

Jackson, J.J. (1996) Field performance of entomopathogenic nematodes for suppression of western corn rootworm. *Journal of Economic Entomology* 89, 366–372.

Jackson, J.J. and Hesler, L.S. (1995) Placement and application rate of the nematode *Steinernema carpocapsae* for the suppression of the western corn rootworm. *Journal of Kansas Entomological Society* 68, 461–467.

Journey, A.M. and Ostlie, K.R. (2000) Biological control of the western corn rootworm (Coleoptera: Chrysomelidae) using the entomopathogenic nematode, *Steinernema carpocapsae. Environmental Entomology* 29, 822–831.

Kaya, H.K. (1990) Soil ecology. In: Gaugler, R. and Kaya, H.K. (eds) *Entomopathogenic Nematodes in Biological Control*. CRC Press, Boca Raton, Florida, pp. 93–115.

Klinger, J. (1988) Investigation on the parasitism of *Otiorhynchus salicicola* and *O. sulcatus* (Coleoptera: Curculionidae) by *Heterorhabditis* sp. (Nematoda). *Entomophaga* 33, 325–331.

Levine, E. and Oloumi-Sadeghi, H. (1992) Field evaluation of *Steinernema carpocapsae* (Rhabditida: Steinernematidae) against the black cutworm larvae in field corn. *Journal of Entomological Science* 27, 427–435.

Miduturi, J.S., De Clerq, R. and De Grisse, A. (1994) Greenhouse and field control of black vine weevil *Otiorhynchus sulcatus* F. with *Steinernema carpocapsae* and *Heterorhabditis* sp. In: *Bulletin of the International Organization for Biological and Integrated Control of Noxious Animals and Plants* 17, 148–154.

Munson, J.D. and Helms, T.J. (1970) Field evaluation of a nematode (DD-136) for control of corn rootworm larvae. *Proceedings of the North Central Branch Entomological Society of America* 25, 97–99.

Patel, K.V. (1998) Chemical composition of *Asalia* (*Lepidium sativum*) seed. In: Kothari, I.L., Parabia, M.H., Jadeja, G.C. and Rao, T.V. (eds) *Medicinal Plants Indian Herbal Heritage*. KTC Corporation, Ahmedabad, India, pp. 138–141.

Patel, N.S. (1999) Evaluation of native entomopathogenic nematodes (*Steinernema* spp.) against key pests of Gujarat. MSc thesis, Gujarat Agricultural University, Gujarat, India.

Purcell, M., Johnson, M.W., LeBeck, L.M. and Hara, A.H. (1992) Biological control of *Helicoverpa zea* (Lepidoptera: Noctuidae) with *Steinernema carpocapsae* (Rhabditida: Steinernematidae) in corn used as a trap crop. *Environmental Entomology* 21, 1441–1447.

Raulston, J.R., Pair, S.D., Loera, J. and Cabanillas, H.E. (1992) Prepupal and pupal parasitism of *Helicoverpa zea* and *Spodoptera frugiperda* (Lepidoptera: Noctuidae) by *Steinernema* sp. in cornfields in the Lower Rio Grande Valley. *Journal of Economic Entomology* 85, 1666–1670.

Richter, A.R. and Fuxa, J.R. (1990) Effect of *Steinernema feltiae* on *Spodoptera frugiperda* and *Heliothis zea* (Lepidoptera: Noctuidae) in corn. *Journal of Economic Entomology* 83, 1286–1291.

Ridgway, R.L. and Lloyd, E.P. (1983) Evolution of cotton insect management in the United States. *USDA Agriculture Handbook No.* 589, pp. 3–27.

Rohrbach, M.D. (1969) Use of the nematode *Neoaplectana carpocapsae* DD-136 in controlling northern and western corn rootworms. MSc thesis, University of Missouri, Columbia.

Samsook, V. and Sikora, R.A. (1981) Influence of parasite density, host developmental stage and leaf-wetness duration on *Neoaplectana carpocapsae* parasitism of *Heliothis virescens. Med Fac Landbouww Universitat de Gent* 46, 685–693.

Schroer, S. and Ehlers, R.-U. (2003). Improvement of persistence and infectivity of entomopathogenic nematodes using formulation adjuvants for foliar application against *Plutella xylostella. Proceedings of the Third International Symposium on Entomopathogenic Nematodes and Symbiotic Bacteria.* 4–7 September 2003. The Ohio State University, Wooster, Ohio (Abstr. Poster Code: C015).

Sezhian, N., Sivakumar, C.V. and Venugopal, M.S. (1996) Alteration of effectiveness of *Steinernema carpocapsae* Weiser against *Spodoptera litura* F. larvae on sunflower by addition of an insect phagostimulant. *Indian Journal of Nematology* 26, 77–81.

Shanks, C.H. and Agudelo-Silva, F. (1990) Field pathogenicity and persistence of heterorhabditid and steinernematid nematodes (Nematoda) infesting black vine weevil larvae (Coleoptera: Curculionidae) in cranberry bogs. *Journal of Economic Entomology* 83, 107–110.

Shapiro, D.I., Lewis, L.C., Obrycki, J.J. and Abbas, M. (1999) Effects of fertilizers on suppression of black cutworm (*Agrotis ipsilon*) damage with *Steinernema carpocapsae. Journal of Nematology* 31 (Suppl.), 690–693.

Shapiro, M., Melane, W. and Bell, R. (1985) Laboratory evaluation of selected chemicals and antidesiccants for protection of the entomogenous nematode, *Steinernema feltiae* (Rhabditidae: Steinernematidae) against *Lymantria dispar* (Lepidoptera: Lymantriidae). *Journal of Economic Entomology* 78, 1437–1441.

Tanada, Y. and Reiner, C. (1962) The use of pathogens in the control of the corn earworm, *Heliothis zea* (Boddie). *Journal of Insect Pathology* 4, 139–154.

Thurston, G.S. and Yule, W.N. (1990) Control of larval northern corn rootworm (*Diabrotica barberi*) with two steinernematid species. *Journal of Nematology* 22, 127–131.

Vyas, R.V. and Yadav, D.N. (1993) *Steinernema glaseri* (Steiner) Travassor, a biological control agent of root-grub *Holotrichia consanguinea* Blan. *Pakistan Journal of Nematology* 11, 41–44.

Vyas, R.V., Patel, N.S., Joshi, N.R., Patel, N.B. and Patel, D.J. (2000) Bioefficacy of *S. glaseri* against DBM, *P. xylostella* on cress, recommendation for scientific community. In: *Proceedings of 35th Plant Protection Sub Committee,* AGRESCO, Gujarat Agricultural University, Anand, 3–4 April 2000.

Whelan, B., McBratney, A. and Boydell, B. (2003) *The Impact of Precision Agriculture.* Australian Center for Precision Agriculture, University of Sydney, Australia, 7 pp.

Wright, R.J., Witkowski, J.F., Echtenkamp, G. and Georgis, R. (1993) Efficacy and persistence of *Steinernema carpocapsae* applied through a center-pivot irrigation system against larval corn rootworms. *Journal of Economic Entomology* 86, 1348–1354.

Yussefi, M. and Willer, H. (2003) *The World of Organic Agriculture – Statistics and Future Prospects,* 11th edn, Tholey–Theley: International Federation of Organic Agriculture Movements (IFOAM®; Available at: http://www.ifoam.org), 129 pp.

15 Forestry Applications

P. Torr,[1] M.J. Wilson[1] and S. Heritage[2]
[1]*School of Biological Sciences, University of Aberdeen, Cruickshank Building, St Machar Drive, Aberdeenshire, AB24 3UU, UK;* [2]*Entomology Branch, Forest Research, Northern Research Station, Roslin, Midlothian, EH25 9SY, UK*

15.1. Introduction

The use of entomopathogenic nematodes (EPNs) in plantation forestry has grown over recent years, with research into use of nematodes for control of species such as the larch sawfly, *Cephalcia lariciphila* (Georgis and Hague, 1981, 1988), the spruce budmoth, *Zeiraphera canadensis* (Turgeon and Finney-Crawley, 1991) and the pine processionary caterpillar, *Thaumetopoea pityocampa* (Triggiani and Tarasco, 2002). The most concerted effort in this area, certainly in Europe, has been regarding the control of the large pine weevil, *Hylobius abietis*. *H. abietis* is a widely distributed pest of plantation forestry occurring throughout Europe and Asia (Scott and King, 1974) and is often regarded as the most serious pest in conifer plantation (e.g. Bratt *et al.*, 2001). A similar pest status is occupied by the generic species *H. congener* in North America, where limited work using EPNs as control agents has been undertaken (Eidt *et al.*, 1995a,b).

H. abietis develops in the stumps and roots of dying and dead conifers, and emerges to feed as adults on the bark and cambium of any woody plant, showing a preference for conifers (Munro, 1928; Scott and King, 1974). It is the feeding on young conifer transplants in plantations that gives the large pine weevil its pest status, weakening the trees by removing patches of bark and underlying soft tissue, often killing them if the stems are girdled (Fig. 15.1). In

 Nematodes as Biocontrol Agents (eds P.S. Grewal, R.-U. Ehlers and D.I. Shapiro-Ilan)

Fig. 15.1. An adult large pine weevil on a lodgepole pine needle (left). Girdling of stem on a conifer transplant due to *Hylobius abietis* feeding (right).

the absence of protection, losses of plants used for restocking can be up to 100%, with an estimated national average of 50% of untreated plants being killed over the first few years of establishment (Heritage and Moore, 2001). These losses would cost the forest industry around £12 million per year. It is estimated that *H. abietis* costs the Forestry Commission, the UK's largest forest operator, approximately £2 million per year on control measures and replanting when protection fails. The area of conifers in Britain being felled and restocked is forecast to increase from 10,000 to 15,000 ha/year by 2010 (Anonymous, 2002) and there is no reason to believe that proportionate losses will not be at least as great as present levels.

Species often become pests simply because populations are no longer limited by the biotic or abiotic factors of their environment, and this is the case with the large pine weevil. *H. abietis* will not breed in live plants in natural forests and must exploit the relatively sparse resource of damaged or fallen conifer trees. Thus, their populations are usually limited to the carrying capacity of the available resources and this implies that the weevils are well adapted to exploiting such resources quickly and effectively. The predominant way in which coniferous forests in northern Europe are harvested and regenerated involves the clear cutting of a site before replanting (Örlander *et al.*, 2000). This practice produces a large supply of dead wood in the form of stumps and roots that are perfect for the development of the insects. At the same time, material suitable as food for the adults is reduced by the removal of small twigs in the canopy of standing trees. It is therefore obvious that the potential for damage from *H. abietis* is intensified by the current management of plantations.

15.2. Biology of *Hylobius abietis*

Spring sees large numbers of weevils migrate to new sites, attracted to conifer volatiles emanating from freshly cut areas. The immigrants arrive by walking (Eidmann, 1968; Mráček, 1988) or by flight (Solbrek and Gyldberg, 1979; Solbrek, 1980; Mráček, 1988; Örlander *et al.*, 1997). Suitable roots for oviposition are located by olfactory orientation to conifer volatiles (Nordlander *et al.*, 1986). Females oviposit throughout spring and summer. Eggs are placed either in the soil surrounding roots of freshly killed host trees, or in small notches on the bark of roots excavated by adult weevils (Scott and King, 1974; Nordenhem and Nordlander, 1994) when the surrounding material is dry or likely to dry out (Örlander *et al.*, 1997). Throughout the oviposition period adults feed voraciously, and as oviposition becomes less intensive fat reserves are built up in the body, ready for hibernation (Guslits, 1969). As the temperature decreases, around October time, adult weevils move below ground and overwinter in the soil (Nordenhem, 1989). Large proportions of weevils arriving at a fresh clear cutting overwinter and remain there throughout the following season.

The length of the developmental period from egg to emergent adult shows wide variation across Europe, depending largely on the predominant climatic conditions. When established in the root, the larva makes a long tunnel that increases in diameter as the weevils pass through five or six larval moults before pupation (Bejer-Petersen *et al.*, 1962). Feeding during this time takes place in the cambial region, scarring the bark (Scott and King, 1974), although larvae have been observed feeding on the heartwood of twigs with thin bark under experimental conditions (Salisbury, 1998). The larva packs waste material (frass) densely behind itself in the tunnel and just before pupation hollows out a pupal chamber, which it thatches tightly with wood fibres (Scott and King, 1974). Generally, after overwintering as a full-grown larva, *H. abietis* pupates throughout the summer when the soil temperatures are high (Bejer--Petersen *et al.*, 1962; Scott and King, 1974). Therefore, adult weevils of the new generation emerge to a large extent 18 months after oviposition (Bejer-Petersen *et al.*, 1962; Scott and King, 1974; Nordenhem, 1989). Other weevils of the new generation remain in their pupal chambers until the second year, when there is synchronous emergence of weevils of both categories on the clear-cutting. In northern latitudes, a developmental time of 3 years is not uncommon (Bejer-Petersen *et al.*, 1962).

The adult weevils may live for up to 4 years (Eidmann, 1979; Örlander *et al.*, 1997) and this longevity of the injurious stage makes *H. abietis* a serious pest. Add to this the difficulty in controlling the weevils due to several features of their biology and behaviour, and the scale of the *H. abietis* problem becomes clear. First, the adults can be present on the site in large numbers unnoticed because of their nocturnal habits; unless specialist sampling systems are used (e.g. Moore, 2001) their presence is indicated only by the death of plants. Second, there are two peaks of adult-feeding activity each year (Fig. 15.2). The first coincides with adults coming out of hibernation in spring, and the second concurs with the period of maturation feeding by newly emerged adults combined with feeding by older adults in preparation for winter hibernation in late summer/early autumn. The precise timing and magnitude vary considerably between forests and from year to year, and it is thus difficult to predict the need for protective measures and their timing. Third, because the eggs, larvae and pupae develop under the bark of roots and stumps, they are difficult to monitor and control using chemicals. For all these reasons, prophylactic treatment of plants with insecticide has been adopted as a necessary precaution to protect them from damage.

15.3. *H. abietis* Control Methods

Currently, the most effective control methods applied involve the use of insecticides (Leather *et al.*, 1999; Heritage and Moore, 2001; Wainhouse *et al.*, 2001). In fact, the large pine weevil is so damaging to young trees that it is the only UK forest pest for which prophylactic applications of insecticides are used (Heritage and Johnson, 1997; Wainhouse *et al.*, 2001). The main insecticides used in forestry to control the large pine weevil are broad-spectrum, contact and ingested pyrethroids. Permethrin is currently the recommended insecticide for these operations in the UK, but its use in forestry will no longer be allowed in any EC country after the year 2003. The

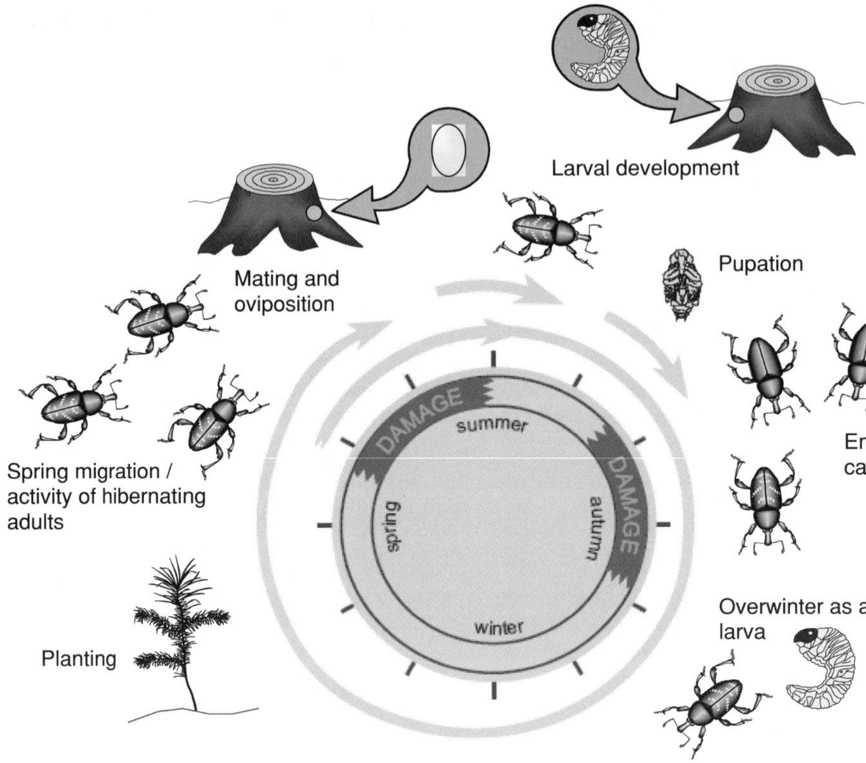

Fig. 15.2. Generalized life cycle of *Hylobius abietis* showing oviposition periods, development time from egg to adult and main damage periods. Circle showing development time from egg to adult, and broken arrows showing pupation period.

insecticide that is most likely to be adopted as a replacement is alpha-cypermethrin, which will be considerably more expensive than its predecessor. A typical regime consisting of a preplanting treatment followed by two postplanting applications in the second year is likely to cost around £340/ha. In addition, many forests are now being managed within the parameters set by the Forestry Stewardship Council (FSC). The FSC directive states that 'Management systems shall promote the development and adoption of environmentally friendly non-chemical methods of pest management and strive to avoid the use of chemical pesticides'. The strategy adopted is to protect the plants directly with the insecticides, a method that provides only short-term pest management. The insecticides are repellent to the adult weevils and there is no evidence to suggest that current control measures have any significant effect on overall insect populations (Leather *et al*., 1999). It is therefore apparent that a suitable alternative pest control strategy must be adopted if the *H. abietis* problem is to be reduced.

15.3.1. Alternative control methods

Various attempts have been made to suppress weevil populations by reduction of larval or adult numbers. *H. abietis* is susceptible to a range of natural enemies and some of these have been studied with regard to use as biocontrol agents. The wasp *Bracon hylobii* parasitizes *H. abietis* larvae, and under natural conditions has shown favourable reductions in larval numbers (Munro, 1928; Gerdin, 1977). The fungi *Beauveria bassiana* and *Metarhizium anisopliae* have both been found to naturally infect and kill adults (Gerdin, 1977). However, a disadvantage of using fungi for *H. abietis* control is that the fungal spores develop in contact with a suitable host. The immature stages of *H. abietis* are protected by the bark whilst the adult weevils are very mobile, spend much of their time under the soil surface and are therefore difficult targets to reach.

Preliminary studies using a range of wood-rotting fungi inoculated into freshly felled logs showed that *H. abietis* larvae actively avoid established fungal lesions (Armendáriz *et al.*, 2002), suggesting that competitive exclusion may be a promising line of research. Again, disadvantages with this have been shown. The growth rates of the fungi are very slow and *H. abietis* has the capability to migrate between roots if the breeding material becomes unsuitable. A larva can enter the soil and orient to host odours in order to find a more suitable substrate in which it can complete its development (Nordenhem and Nordlander, 1994). This ability to migrate between roots implies that the larvae have great potential for avoiding competition for food and means that control via competitive exclusion may not be that good an option.

15.4. Entomopathogenic Nematodes (EPNs) as Biocontrol Agents of *H. abietis*

Arguably the most promising line of research has come from the use of EPNs. They have shown much potential for controlling the large pine weevil, both in the laboratory (Pye and Burman, 1978; Armendáriz *et al.*, 2002) and, more importantly, in field trials (Pye and Pye, 1985; Collins, 1993; Brixey, 2000). These parasites are very attractive control agents for the forest industry as indigenous, unmodified types are exempt from registration requirements in many countries (Richardson, 1996).

15.4.1. Susceptibility of *H. abietis* at different life stages

Until more recently, the majority of research into biocontrol of *H. abietis* has focused on direct plant protection that is targeting the adult weevils (Pye and Pye, 1985; Collins, 1993). Understandably, the thinking was that the most economic method of reducing *H. abietis* feeding damage would be to substitute the chemical insecticide application with EPNs. Adult weevils spend relatively long periods within the soil (Munro, 1928), the natural environment for nematodes, yet this strategy proved ineffective due to the length of time required to establish infection (Brixey, 2000). For this approach to be effective, rapid killing of adults would be necessary to prevent feeding damage and oviposition leading to a subsequent generation. Moreover, the different *H. abietis* life stages vary to the degree of susceptibility to nematode infection, with the larval and pupal stages of *H. abietis* being more susceptible to control by EPNs (Pye and Burman, 1978; Brixey, 2000). It has been ascertained that at least 25 days should be allowed to achieve the maximum level of infection of *H. abietis* larvae (Brixey, 2000). Comparing this with *Galleria mellonella* larvae, which usually die within 48 h of exposure to nematodes, *H. abietis* seems to have a relatively high resistance to nematode infection. Understanding the key to this resistance might enable nematode efficacy to be improved in the future.

By reducing the larval population, the potential for damage is removed before the insects reach the adult stage and the population of weevils should diminish. The larval development period is lengthy: at least 1 year of their life cycle occurs in the moist environment of pine stumps located below soil level. This habitat is unreachable for chemical insecticides and this is where nematodes have a significant advantage. The insecticides only act by contact and when applied to the surface of the stumps would fail to affect most of the population of *H. abietis* larvae. However, cryptic habitats such as this generally harbour conditions more favourable for nematode survival and infectivity. The most consistent, efficacious results with nematodes have been obtained in cryptic habitats, especially against insects that bore into plants (Begley, 1990). The potential of infective juveniles (IJs) to search for and infect larvae within insect galleries under the bark of a stump (e.g. Moore, 1970) is key to why EPNs have shown so much promise in controlling *H. abietis*.

15.4.2. Window of application

Considering the life cycle of *H. abietis*, and that the larval and pupal stages of development are the most susceptible to nematode infection, it is clear that there are potentially two times in the year when nematode applications would be most effective. The first opportunity for nematode application coincides with early instar larvae developing in newly cut stumps. Early instar larvae are targeted by nematode treatment to stumps throughout summer of the season after felling. At this stage the integrity of the bark and limited insect activity may present the nematodes with difficulty in finding the host. The second opportunity for nematode application in the field coincides with the pupal stage. Pupae are targeted via applications during summer 12 months after felling. A possible problem with applications at this time is that *H. abietis* pupae present a more difficult target for nematodes to locate, as they develop within chambers in the sapwood, the entrance to which is packed with wood fibres and frass (Scott and King, 1974). However, Pye and Burman (1978) demonstrated that nematodes could penetrate packed sawdust to infect *H. abietis* larvae, suggesting that this may not be a problem. Pupation by *H. abietis* is fairly synchronous throughout the UK regardless of the rates of larval development. Although the pupal stage is relatively brief, lasting only a few weeks, field trials have demonstrated that the window for effective application of nematodes is between mid-May and early July.

15.4.3. Field trials

Three nematode types commercially available in the UK have been shown to infect, kill and reproduce in *H. abietis* larvae under laboratory conditions. Results shown in Fig. 15.3 demonstrate that *S. carpocapsae* and *S. feltiae* gave similar levels of infection in field populations (53–56%) but the efficacy of *H. megidis* was substantially lower (Brixey, 2000). Further trials comparing *S. carpocapsae* with *S. feltiae* applied using a high-pressure spray demonstrated that *S. carpocapsae* gave the most

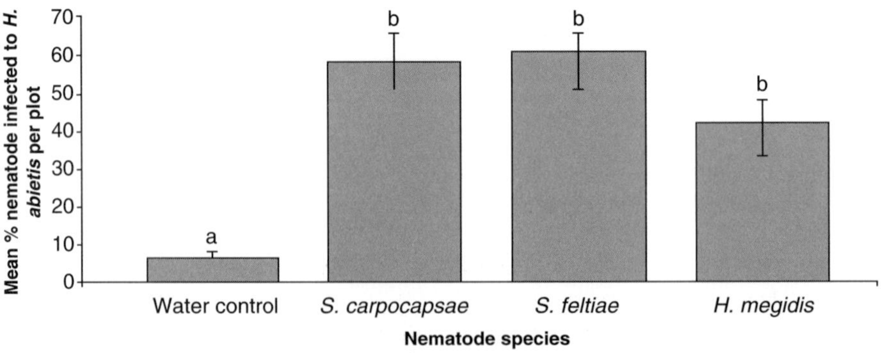

Fig. 15.3. Comparative field efficacy of three entomopathogenic nematode (EPN) species applied at a dose of 3.5×10^6 IJs in 500 ml of water around the base of every stump, at the time of final instar larvae and pupae. Same letter means no significant difference between treatments. (Adapted from Brixey, 2000.)

consistent level of control of the large pine weevil, although the difference in efficacy was not statistically significant (S. Heritage, unpublished data). This nematode is also the easiest and cheapest to produce of the two and has therefore been chosen as the principal control agent for further trials. This level of recorded infection has translated to around 70% reduction in emergence (Fig. 15.4).

15.4.4. Current practice

Nematodes are targeted at the insect as it develops in the stumps. Because they reduce the number of new insects emerging from the stumps, nematodes have two effects:

1. Most plant damage is caused by the emergent population and therefore damage within the treated area will be reduced.
2. After feeding, the newly emerged insects migrate to colonize more recently felled areas. By reducing the numbers of these insects, the population within the forest unit will be gradually suppressed.

H. abietis adults may live several years and move considerable distances in this time, and as a result may re-invade treated sites from adjacent untreated areas. Thus, nematodes may be slightly less effective when used in compartments on the edge of the treated area. Where possible, entire forest blocks should be managed using nematodes to minimize this edge effect. To optimize the impact of nematodes, they must be applied close to the time that the insects are pupating and are most vulnerable. For this reason they cannot be applied as part of the felling operation. Only a single application of nematodes to each restocking site is necessary to control *H. abietis*. The specification has been derived from a number of field experiments and has consistently provided a good level of insect control. *S. carpocapsae* is applied at a dose of 3.5×10^6 nematodes in 500 ml of water around the base of every stump (Fig. 15.5). The spot treatment is to reduce nematode application to non-target areas and minimize the environmental impact of the operation. Reducing the number of nematodes or the volume of water applied to each stump would lead to considerable cost savings and this is the subject of further research.

Access to forest restocking sites can be problematic. Sites are often on soft ground with substantial debris remaining after the felling, ditches and very high stumps. The nematode–water mixture is carried across the site by a forwarder-mounted spray rig and delivered to the target through hand-held lances (Fig. 15.6). Forwarders

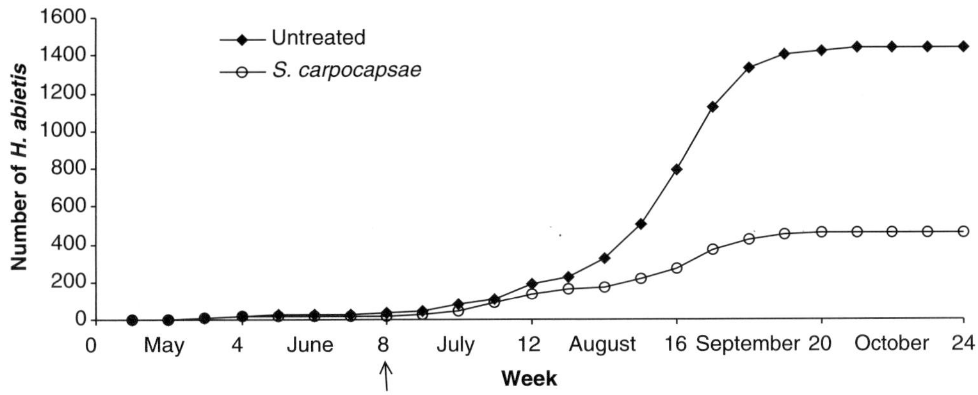

Fig. 15.4. Cumulative numbers of adult *Hylobius abietis* caught in emergence traps from a clear-felled site of lodgepole pine in Moray, Scotland. Arrow indicates timing of nematode application. (S. Heritage, 2001, unpublished data.)

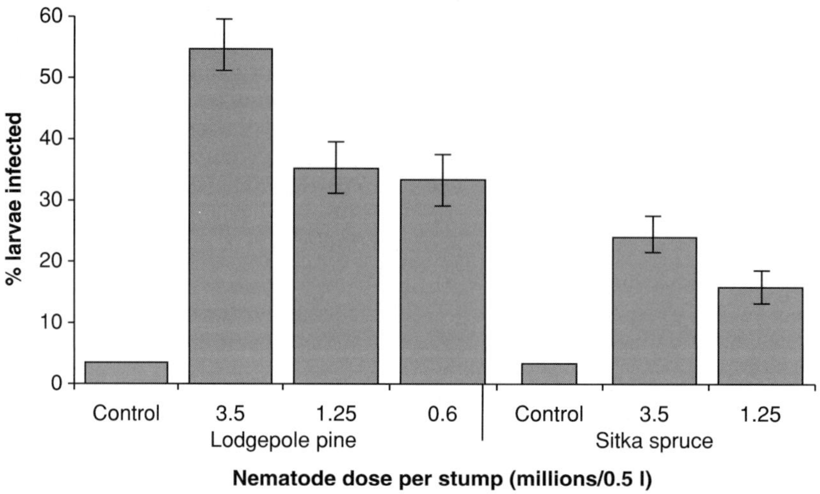

Fig. 15.5. Field efficacy of different doses of *Steinernema carpocapsae* against *Hylobius abietis* developing in lodgepole pine and Sitka spruce stumps after 4 weeks' exposure per plot (± SE), where $n = 7$ replicate blocks (pine) and $n = 5$ replicate blocks (spruce). (From Brixey, 2000.)

are used to extract the timber from the site after felling and have good ground clearance, often eight-wheel drive and balloon tyres to reduce bogging. The forwarder travels along the original brash mat (extraction tracks formed from tree branches during the felling operation), which must be left undisturbed until after the site treatment. Any site felled by a harvester should be suitable for management using the current application system (equals approximately 70% of the felled coniferous plantations in the UK). Each forwarder unit is fitted with a 500-l nematode spray

Fig. 15.6. Workers from Forest Research applying *Steinernema carpocapsae* at a clear-felled Sitka spruce plantation in Moray Firth, Scotland, using a forwarder-mounted spray rig fitted with bulk water tanks and four spray lances.

tank, which is enough to treat approximately 0.5 ha of a clear-cut site before refilling. It also carries 2000 l of clean water and an insulated box containing additional nematodes. The unit can therefore treat around 2.5 ha before returning to base. On sites with firm dry ground it will be possible to mount the rig on a large tractor unit that has adequate ground clearance, thus reducing the application costs significantly.

The spray rig is equipped with four hand-held lances at the end of flexible hoses around 10 m long. Each can deliver the required dose within 1 s, allowing treatment to take place at a reasonable walking pace given the terrain. The unit is accompanied by a supervisor who marks the extent of stumps treated using spray paint and keeps in contact with the driver by radio for safety. Using this method around 5 ha can be treated each day with a resultant reduction of 60–75% in adult emergence. During 2003, around 200 ha of UK restocking was treated using this system, and the annual area treated is likely to increase rapidly once a reduction in damage to plants has been demonstrated.

15.4.5. The management of large-scale treatment programmes

Effective use of nematodes requires careful management to ensure that the correct quantities of good quality nematodes are applied to the stumps at the correct biological time. In addition, the FSC directive states that 'Use of biocontrol agents shall be documented, minimized, monitored and strictly controlled in accordance with national laws and internationally accepted scientific protocols.' For these reasons, nematodes are used on sites within the framework of an integrated management system. The use of the technical skill base of local field stations maintains the quality necessary for effective site management. Participating field stations provide forest managers with a complete package of services. This includes site monitoring, the management of nematode application, quality control and full documentation. The unit organizes the use of nematodes as part of the management of *H. abietis* populations in coniferous restocking areas and provides forest managers with all information necessary to confirm that the system is effective and safe.

15.4.6. Factors affecting success/failure of nematodes

Nematode survival and infectivity in this environment is of paramount importance in determining their potential for *H. abietis* control. Conifer forest soils are generally very high in organic matter and are acidic in nature. Very little information is available on nematode movement and efficacy in such soils, although Brixey (2000) commented that efficacy was compromised. Factors such as soil texture and moisture (Kaya, 1990), soil pH (Kung *et al.*, 1990), temperature (Grewal *et al.*, 1994), ultraviolet (UV) radiation (Gaugler and Boush, 1978; Gaugler *et al.*, 1992), natural antagonists (Kaya, 2002) and resistance to desiccation (Patel *et al.*, 1997; Grewal, 2000) have all been demonstrated to be important in the persistence of nematodes in the soil. Greater persistence of a nematode would constitute a more prolonged dose, and as *H. abietis* control appears to be dependent on dose (Brixey, 2000), it follows that a greater capacity for field survival will result in increased efficacy. It is generally regarded that heterorhabditids tend to be less tolerant of environmental stress than steinernematids (Grewal, 2000, 2002), and this could be an important factor when considering reasons why *S. carpocapsae* and *S. feltiae* seem to outperform *H. megidis* in the field. Further research into this area is required.

Nematode species are adapted to exploit a much narrower range of hosts than laboratory tests have suggested (Peters, 1996; Simões and Rosa, 1996). *Steinernema* spp. and *Heterorhabditis* spp. have generally been considered to have a broad host range, a belief based mainly on bioassays. For example, *S. carpocapsae* has been known to infect 250 insects in 75 families and 11 orders (Poinar, 1979). However, killing an insect in a laboratory bioassay does not necessarily mean that a nematode will be effective under field conditions, as in the laboratory host contact is ensured, environmental conditions are optimum and no ecological or behavioural barriers to infection exist (Gaugler, 1981; Gaugler *et al.*, 1997). The occurrence of a nematode in a particular habitat is likely to be comparable with that of its natural host. With ever increasing accuracy in identification, it is becoming clear that habitat preferences for EPNs are apparent (e.g. Hominick *et al.*, 1996) and this is a valid point when considering field efficacy. Under field conditions, the host range of a nematode species is restricted to insect species with a similar temporal and spatial occurrence and distribution. *S. carpocapsae* seem to be found in woodlands, *S. feltiae* prefer fields and grassland but are also found in woodlands and *H. megidis* have been isolated mainly in sandy coastal soils (Hominick, 2002). The occurrence of steinernematids in habitats more similar to that of *H. abietis* could go some way to explaining why they give better control than *H. megidis*.

15.5. Conclusions

With careful consideration given to the timing of nematode treatment and application technique, an average 70% infection in *H. abietis* using *S. carpocapsae* has been achieved (Brixey, 2000). These promising results suggest that the use of EPNs could provide an important opportunity to reduce the forest industry's dependency on chemicals. However, as in any insect management scheme, the silvicultural possibilities should be given great attention, and the use of nematodes for *H. abietis* control should be part of an integrated management system including improving silvicultural techniques currently employed (Heritage and Moore, 2001).

An effective biocontrol strategy against *H. abietis* will require the monitoring of felling and *H. abietis* development to predict accurately the time of pupation. Nematode applications should occur at least 4 weeks before *H. abietis* emergence from the stumps and is predicted to optimize efficacy. A dose of 3.5×10^6 nematodes in 500 ml of water per stump (equivalent to 7.5×10^9/ha) has been found to be effective. It is probable that higher doses would increase the level of control, but this is not an economically viable option. The rate at which the system is adopted will depend on a number of factors. Initially, the availability of suitable nematodes at an acceptable cost may be the main restriction. Once the success of nematodes has been demonstrated, the total cost of their use compared with alternative systems is likely to be important. *S. carpocapsae* formulated for use in horticulture will cost roughly £850/ha with an additional £120/ha for their application. The use of alpha-cypermethrin to protect plants is likely to cost around £340/ha. To reduce the overall cost of the use of nematodes in forestry, the Forestry Commission has invested in a different production system that should reduce the overall cost to less than the equivalent cost of chemicals. Pressure to use non-chemical means of forest protection to retain FSC certification may accelerate the rate of adoption. It is therefore clear that a major hurdle that has to be overcome is the cost of nematodes; more precisely, production of the nematodes at a lower cost along with increased efficacy of application.

Additional work is required to improve the predictability and efficacy of EPNs in forest situations. The correct nematode must be found for the job, in that both host and parasite must have similar ecological requirements. More biocontrol failures with nematodes can be attributed to making releases in inappropriate habitats than for any other reason (Gaugler *et al.*, 1997). Recently, *S. kraussei* has been made commercially available as a biocontrol agent in the UK. This nematode has been found to favour woodlands, especially under coniferous trees (Stock *et al.*, 1999) and is known to have a thermal niche lower than the other available nematodes (Grewal *et al.*, 1994; Mráček *et al.*, 1999). This could be very important in the northern European climate, where temperatures are often low. Such similarity in habitat preference for pest and control agent is very encouraging and could result in *S. kraussei* being a more effective nematode species for the control of *H. abietis*.

References

Anon. (2002) *Forestry Facts and Figures 2002. A Summary of Statistics about Woodland and Forestry in Great Britain.* Economics and Statistics Unit, Forestry Commission, Edinburgh, UK.

Armendáriz, I., Downes, M.J. and Griffin, C.T. (2002) Effect of timber condition on parasitization of pine weevil (*Hylobius abietis* L.) larvae by entomopathogenic nematodes under laboratory conditions. *Biocontrol Science and Technology* 12, 225–233.

Begley, J.W. (1990) Efficacy against insects in habitats other than soil. In: Gaugler, R. and Kaya, H.K. (eds) *Entomopathogenic Nematodes in Biological Control.* CRC Press, New York, pp. 215–231.

Bejer-Petersen, B., Juutinen, E., Bakke, A., Butovitsch, V., Eidman, H., Heqvist, K.J. and Lekander, B. (1962) Studies on *Hylobius abietis* L. 1. Development in the Nordic countries. *Acta Entomologica Fennica* 17, 1–107.

Bratt, K., Sunnerheim, K., Nordenhem, H., Nordlander, G. and Långström, B. (2001) Pine weevil (*Hylobius abietis*) antifeedants from lodgepole pine (*Pinus contorta*). *Journal of Chemical Ecology* 27(11), 2253–2262.

Brixey, J. (2000) The use of entomopathogenic nematodes to control the immature stages of the large pine weevil *Hylobius abietis* (L.). PhD thesis. University of Reading, Reading, UK.

Collins, S.A. (1993) The potential of entomopathogenic nematodes to reduce damage by *Hylobius abietis* (L.). PhD thesis. Imperial College of Science, Technology and Medicine, London, UK.

Eidmann, H.H. (1968) Invasion of conifer plantations by radioactively labelled *Hylobius abietis* L. *Current Topics in Forest Entomology.* USDA Forest Service General Technical Report 8, 103–109.

Eidmann, H.H. (1979) Integrated management of pine weevil (*Hylobius abietis* L.) populations in Sweden. *Current Topics in Forest Entomology.* USDA Forest Service General Technical Report WO-8, pp. 103–109.

Eidt, D.C., Zervos, S. and Finney-Crawley, J.R. (1995a) Susceptibility of adults of *Hylobius congener* Dalletorre, Shenkling and Marshall (Coleoptera: Curculionidae) to entomopathogenic nematodes. *Canadian Entomologist* 127(3), 439–441.

Eidt, D.C., Zervos, S., Pye, A.E. and Finney-Crawley, J.R. (1995b) Control of *Hylobius congener* Dalletorre, Shenkling and Marshall (Coleoptera: Curculionidae) using entomopathogenic nematodes. *Canadian Entomologist* 127(3), 431–438.

Gaugler, R. (1981) Biological control potential of Neoaplectanid nematodes. *Journal of Nematology* 13(3), 241–249.

Gaugler, R. and Boush, G.M. (1978) Effects of ultraviolet radiation on and sunlight on the entomogenous nematode *Neoaplectana carpocapsae. Journal of Invertebrate Pathology* 32, 291–296.

Gaugler, R., Bednarek, A. and Campbell, J.F. (1992) Ultraviolet inactivation of heterorhabditid and steinernematid nematodes. *Journal of Invertebrate Pathology* 59, 155–160.

Gaugler, R., Lewis, E.E. and Stuart, R.J. (1997) Ecology in the service of biological control, the case of entomopathogenic nematodes. *Oecologia* 109, 403–489.

Georgis, R. and Hague, N.G.M. (1981) A Neoaplectanid nematode in the larch sawfly *Cephalcia lariciphila* (Hymenoptera: Pamphiliidae). *Annals of Applied Biology* 99, 171–177.

Georgis, R. and Hague, N.G.M. (1988) Field evaluation of *Steinernema feltiae* against the web-spinning larch sawfly *Cephalcia lariciphila. Journal of Nematology* 20(2), 317–320.

Gerdin, S. (1977) Observations on pathogens and parasites of *Hylobius abietis* (Coleoptera, Curculionidae) in Sweden. *Journal of Invertebrate Pathology* 30, 263–264.

Grewal, P.S. (2000) Anhydrobiotic potential and long-term storage of entomopathogenic nematodes (Rhabditida, Steinernematidae). *International Journal for Parasitology* 30, 995–1000.

Grewal, P.S. (2002) Formulation and application technology. In: Gaugler, R. (ed.) *Entomopathogenic Nematology.* CAB International, Wallingford, UK, pp. 265–287.

Grewal, P.S., Selvan, S. and Gaugler, R. (1994) Thermal adaptation of entomopathogenic nematodes, niche breadth for infection, establishment and reproduction. *Journal of Thermal Biology* 19(4), 245–253.

Guslits, I.S. (1969) Morphological and physiological description of the pine weevil, *Hylobius abietis* L. (Coleoptera, Curculionidae), during the period of maturation and oviposition. *Entomologist Review* 48(1), 52–55.

Heritage, S. and Johnson, D. (1997) The use of post planting sprays to improve the protection of plants from damage by *Hylobius abietis*. *Research Information Note 272*. Forestry Commission, Edinburgh, UK.

Heritage, S. and Moore, R. (2001) The assessment of site characteristics as part of a management strategy to reduce damage by *Hylobius*. *Information Note 38*. Forestry Commission, Edinburgh, UK.

Hominick, W.M. (2002) Biogeography. In: Gaugler, R. (ed.) *Entomopathogenic Nematology*. CAB International, Wallingford, UK, pp. 115–143.

Hominick, W.M., Reid, A.P., Bohan, D.A. and Briscoe, B.R. (1996) Entomopathogenic nematodes, biodiversity, geographical distribution and the convention on biological diversity. *Biocontrol Science and Technology* 6, 317–331.

Ishibashi, N. and Kondo, E. (1990) Behaviour of infective juveniles. In: Gaugler, R. and Kaya, H.K. (eds) *Entomopathogenic Nematodes in Biological Control*. CRC Press, New York, pp. 139–150.

Kaya, H.K. (1990) Soil ecology. In: Gaugler, R. and Kaya, H.K. (eds) *Entomopathogenic Nematodes in Biological Control*. CRC Press, New York, pp. 93–115.

Kaya, H.K. (2002) Natural enemies and other antagonists. In: Gaugler, R. (ed.) *Entomopathogenic Nematology*. CAB International, Wallingford, UK, pp. 189–203.

Kung, S.P., Gaugler, R. and Kaya, H.K. (1990) Influence of soil pH and oxygen on persistence of *Steinernema* spp. *Journal of Nematology* 22, 440–445.

Leather, S.R., Day, K.R. and Salisbury, A.N. (1999) The biology of the large pine weevil *Hylobius abietis* (Coleoptera, Curculionidae), a problem of dispersal? *Bulletin of Entomological Research* 89, 3–16.

Moore, G.E. (1970) *Dendroctonus frontalis* infection by the DD-136 strain of *Neoaplectana carpocapsae* and its bacterium complex. *Journal of Nematology* 2, 341–344.

Moore, R. (2001) Emergence trap developed to capture adult large pine weevil *Hylobius abietis* (Coleoptera, Curculionidae) and its parasite *Bracon hylobii* (Hymenoptera, Braconidae). *Bulletin of Entomological Research* 91, 109–115.

Mráček, Z. (1988) Migrace klikoroha borového (*Hylobius abietis* L.). *Lesnictví* 34, 817–836.

Mráček, Z., Bečvář, S., Kindlmann, P. and Webster, J.M. (1999) Factors influencing the infectivity of a Canadian isolate of *Steinernema kraussei* (Nematoda, Steinernematidae) at low temperature. *Journal of Invertebrate Pathology* 73, 243–247.

Munro, J.W. (1928) The biology and control of *Hylobius abietis* (L.). *Forestry* 2, 31–39.

Nordenhem, H. (1989) Age, sexual development, and seasonal occurrence of the pine weevil *Hylobius abietis* (L.). *Journal of Applied Entomology* 108, 260–270.

Nordenhem, H. and Nordlander, G. (1994) Olfactory oriented migration through soil by root-living *Hylobius abietis* (L.) larvae (Col., Curculionidae). *Journal of Applied Entomology* 117, 457–462.

Nordlander, G., Eidmann, H.H., Jacobsson, U., Nordenhem, H. and Sjödin, K. (1986) Orientation of the large pine weevil *Hylobius abietis* to underground sources of host volatiles. *Entomologia Experimentalis et Applicata* 41, 91–100.

Örlander, G., Nilsson, U. and Nordlander, G. (1997) Pine weevil abundance on clear-cuttings of different ages, a 6-year study using pitfall traps. *Scandinavian Journal of Forest Research* 12, 225–240.

Örlander, G., Nordlander, G., Wallertz, K. and Nordenhem, H. (2000) Feeding in the crowns of Scots Pine by the pine weevil *Hylobius abietis*. *Scandinavian Journal of Forest Research* 15, 194–201.

Patel, M.N., Perry, R.N. and Wright, D.J. (1997) Desiccation survival and water contents of entomopathogenic nematodes, *Steinernema* spp. (Rhabditida, Steinernematidae). *International Journal for Parasitology* 27, 67–70.

Peters, A. (1996) The natural host range of *Steinernema* and *Heterorhabditis* spp. and their impact on insect populations. *Biocontrol Science and Technology* 6, 389–402.

Poinar, G.O. Jr (1979) *Nematodes for Biological Control of Insects*. CRC Press, New York, pp. 150–154.

Pye, A.E. and Burman, M. (1978) *Neoaplectana carpocapsae*, infection and reproduction in large pine weevil larvae, *Hylobius abietis*. *Experimental Parasitology* 46, 1–11.

Pye, A.E. and Pye, N.L. (1985) Different applications of the insect parasitic nematode *Neoaplectana carpocapsae* to control the large pine weevil, *Hylobius abietis*. *Nematologica* 31, 109–116.

Richardson, P.N. (1996) British and European legislation regulating rhabditid nematodes. *Biocontrol Science and Technology* 6, 449–463.

Salisbury, A. (1998) Some observations on the large pine weevil *Hylobius abietis* (L.) (Coleoptera:

Curculionidae). *Entomologist's Gazette* 49, 195–197.

Scott, T.M. and King, C.J. (1974) The large pine weevil and black pine beetles. *Forestry Commission Leaflet number* 58, London HMSO.

Simões, N. and Rosa, J.S. (1996) Pathogenicity and host specificity of entomopathogenic nematodes. *Biocontrol Science and Technology* 6, 403–411.

Solbrek, C. (1980) Dispersal distances of migrating pine weevils, *Hylobius abietis* (L.), Coleoptera: Curculionidae. *Entomologia Experimentalis et Applicata* 28, 123–131.

Solbrek, C. and Gyldberg, B. (1979) Temporal flight pattern of the large pine weevil, *Hylobius abietis* L. (Coleoptera, Curculionidae), with special reference to the influence of weather. *Zeitschrift für Angewandte Entomologie* 88, 532–536.

Stock, S.P., Pryor, B.M. and Kaya, H.K. (1999) Distribution of entomopathogenic nematodes (Steinernematidae and Heterorhabditidae) in natural habitats in California, USA. *Biodiversity and Conservation* 8, 535–549.

Triggiani, O. and Tarasco, E. (2002) Efficacy and persistence on entomopathogenic nematodes in controlling larval populations of *Thaumetopoea pityocampa* (Lepidoptera: Thaumetopoeidae). *Biocontrol Science and Technology* 12, 747–752.

Turgeon, J.J. and Finney-Crawley, J.R. (1991) Susceptibility of first and second instar larvae of the spruce budmoth, *Zeiraphera canadensis* (Lepidoptera: Tortricidae), to the entomogenous nematode *Heterorhabditis heliothidis* under controlled conditions. *Journal of Invertebrate Pathology* 57, 126–127.

Wainhouse, D., Evans, H., Moore, R., Webber, J., Thorpe, K. and Staley, J. (2001) The integrated Forest Management Programme. *Forest Research Annual Report and Accounts* 2000–2001, Edinburgh, UK.

16 Applications for the Control of Pests of Humans and Animals

I. Glazer,[1] M. Samish[2] and F.G. del Pino[3]
[1]*Department of Nematology, ARO, Volcani Center, Bet-Dagan, 50250, Israel;* [2]*Kimron Veterinary Institute, Bet-Dagan, PO Box 12, 50250, Israel;* [3]*Department of Animal Biology, Faculty of Science, Universitat Autònoma de Barcelona 08193 Bellaterra, Barcelona, Spain*

16.1. Introduction

Thousands of species of insects, mites and ticks worldwide have the ability to attack animals. Fortunately, only a few are significant within a given area as pests of pets, domestic livestock and poultry (Lancaster and Meisch, 1986). For example, in the USA only 45 species are reported to be pests of cattle and 75 of fowl. Economic damage caused by such pests is in the range of tens of billions of US dollars. The damage is related not only to the direct effect of the pests on animal mortality, reduction in productivity and reproduction, but also to the ability of certain pests to transmit diseases. During the past 100 years, the control

(eds P.S. Grewal, R.-U. Ehlers and D.I. Shapiro-Ilan)

of veterinary pests and the diseases they transmit has been largely through the application of insecticides and acaricides as sprays or dips. Short-interval applications of pesticides, in conjunction with control of animal movement, quarantine and slaughter, can prevent transmission of the parasites. However, the development of pesticide resistance has been a major problem. This has been compounded by the increasing cost of the pesticides and poor management and also by inadequate maintenance. Another complication associated with the use of pesticides in husbandry practice is their role as environmental pollutants, which may also contaminate animal products like milk and meat.

Insects and other arthropods also parasitize humans. The most important pests for humans are mosquitoes, black flies, fleas, ticks, lice, houseflies and cockroaches. As in the case of animals the devastating effect of these pests is not only their direct damage and nuisance but also transmission of diseases. Control measures for human pests include chemicals (pesticides and repellents) and physical methods (nets and elimination of incubation sites). Whereas the control of animal pests is applied mainly to the animal directly or to the site it inhabits, in the case of human pests the application is also done at the breeding sites of the pests.

The biocontrol of plant insect pests is developing rapidly, but relatively little attention is being paid as yet to its potential use against veterinary and human pests (Samish and Rehacek, 1999). The only worldwide success story of a biocontrol agent against human pests is the use of the bacterium *Bacillus thuringiensis israelensis* (Bti) for control of mosquito larvae (Margalit and Ben-Dov, 2000). Although the pathogenicity of entomopathogenic nematodes (EPNs) was tested against thousands of insect species, little attention was given by entomopathogenic nematologists to veterinary and human pests. Entomophilic mermithids were intensively studied for use against mosquitoes during the 1960s (see Chapter 23, this volume), but the initiation of use of steinernematids and heterorhabditids against veterinary and human pests has been explored only in the last decade. This chapter provides current knowledge on this particular issue and is intended to encourage further exploration of EPN use against veterinary and human pests.

16.2. Ticks

Ticks (Ixodidae, approximately 850 species) inhabit highly variable ecological niches. Ticks are economically very important pests worldwide, mainly as vectors of animal and human diseases. Approximately 80% of the world's cattle population of 1281 million is at risk from ticks and tick-borne diseases (TBD) (Sutherst *et al.*, 1982). Over a decade ago McCosker (1979) estimated the global cost of their control and of productivity losses to be US $7000 million per year (= US $7/head/year). In Africa, with 186 million head of cattle, ticks and TBD are the most serious constraints to increased production. TBDs also potentially affect wildlife (Ginsberg 1993; Sonenshine, 1993). Furthermore, in the USA, TBDs are by far the most commonly reported vector-borne illnesses affecting humans, especially Lyme disease and Rocky Mountain spotted fever (Center for Disease Control, 1996).

Ticks are the sole or major vectors of many domestic animal pathogens, i.e. *Anaplasma, Babesia, Cowdria, Ehrlichia* and *Theileria* and human diseases such as African tick-borne relapsing fever, Q fever, tick-borne encephalitis, Lyme disease and Rocky Mountain spotted fever. In addition, ticks can also provoke anaemia, toxicosis and sometimes also tick paralysis.

Ticks are obligatory blood-sucking arthropods. They have three blood-sucking stages, i.e. larvae, nymphs and adults. In some species all three stages drop off the animal to the ground when fully engorged, while in others only two stages or only fully engorged adults drop off. During most of their life cycle the ticks remain in the upper layer of the ground. Engorged female ticks require several pre-oviposition days after they drop off the host. The other tick stages that drop off their hosts to the ground also need several days before they moult, finish their prefeeding period, and become active. During these resting periods, the ticks in the upper layer of the ground can serve as ideal targets for nematode attack (Fig. 16.1). Tick control is based worldwide nearly solely on chemical acaricides. Application of acaricides often causes not only the development of resistance, but also acaricide poisoning of domestic animals. Little attention is being paid as yet to potential use of nematodes against ticks (Samish and Rehacek, 1999).

16.2.1. Infectivity

Samish and Glazer (1991) were the first to report that EPNs are capable of killing engorged females of the cattle ticks, e.g. *Boophilus annulatus*. Studies during the last decade show that EPNs are also pathogenic

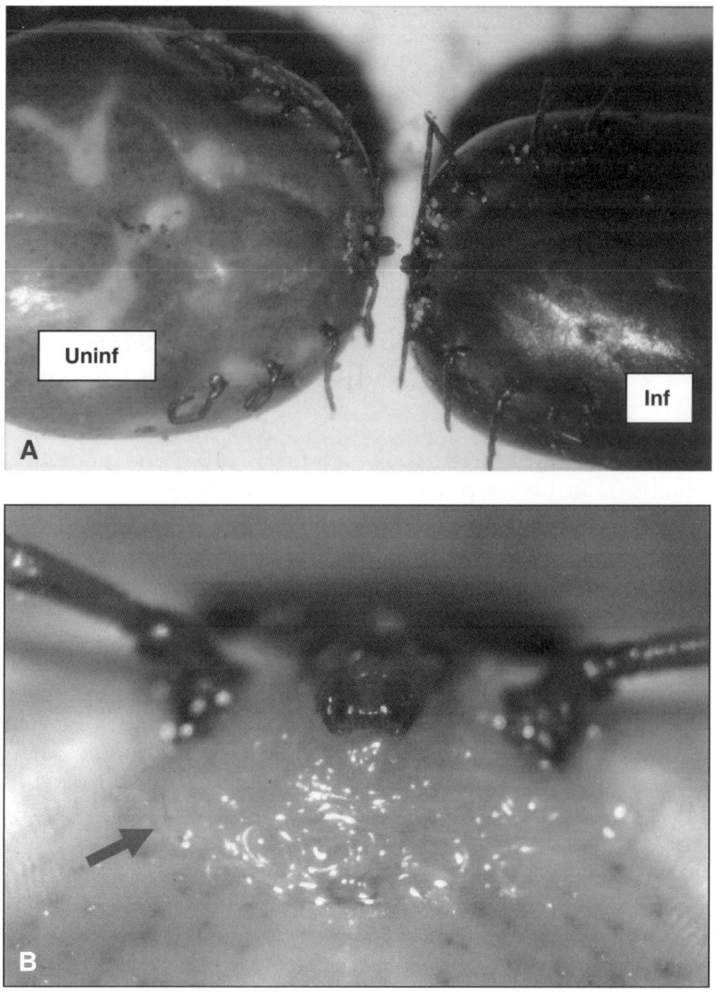

Fig. 16.1. Infectivity of entomopathogenic nematodes (EPNs) to ticks. (A) Infected (Inf) and uninfected (Uninf) engorged females of the cattle tick *Boophilus annulatus*. (B) Agregation of infective juveniles (IJs) of *Steinernema carpocapsae* near the mouth region of engorged female of *Amblyomma americanum* (courtesy of Prof. K.M. Kocan and Dr. E.F. Blouin).

to other tick species (El-Sadawy *et al.*, 1998; Samish *et al.*, 2001). Out of 16 ixodid tick species from six genera and three argasid species from two genera tested, only one species seemed not susceptible to nematodes (Table 16.1). However, engorged female ticks from several species demonstrated in Petri dish trials very large differences in susceptibility to nematodes (Table 16.1). Studies in soil-filled cups showed that engorged Ixodidae (*Hyalomma dromedarii*) females were more susceptible to nematodes than argasid ticks (Hassanian *et al.*, 1999; Kaaya *et al.*, 1999). Fully engorged argasid and ixodid female ticks were generally most sensitive to EPNs, unfed adult ticks being less sensitive and pre-imaginal stages the least sensitive, while the eggs were fully resistant (Samish and Glazer, 1991, 1992; Samish *et al.*, 1996; Samish *et al.*, 2001). Ovipositing females seem to be more sensitive than their pre-oviposition stage (Mauleon *et al.*, 1993; Samish *et al.*, 2001). During their feeding stage on a host, ticks are resistant to nematodes except on very moist feeding sites (Kocan *et al.*, 1998a; M. Samish *et al.*, 2002, unpublished data).

The 42 nematode strains tested for their antitick activity showed varying degrees of virulence (Table 16.1). In laboratory tests, heterorhabditids were generally more virulent to ticks than steinernematid nematodes (Mauleon *et al.*, 1993; Hill, 1998; Hassanian *et al.*, 1999; Kaaya *et al.*, 1999; Glazer *et al.*, 2001). Strains virulent to one tick species and one stage were found, in most cases, also to be highly virulent to other tick species and stages (Hassanian *et al.*, 1999; Kaaya *et al.*, 1999; Samish *et al.*, 2001). The quantity of nematodes and the time required to kill 50% of the ticks (LC_{50} and LT_{50}, respectively) were lowest in trials with engorged *B. annulatus* (Table 16.1). The LC_{50} for engorged females was mostly lower than for unfed adults. Also, unfed ixodid females were killed up to six times more quickly than engorged ticks. The LT_{50} for unfed *Rhipicephalus bursa* females is only 1 day compared with 6 days for engorged females (Samish *et al.*, 2000b). The rate of mortality post infection may be affected by the antibacterial activity that was found in ticks after engorgement (Samish *et al.*, 2000c). At high nematode concentrations and optimum conditions, the nematodes killed the engorged females before they had time to lay eggs (Samish *et al.*, 2000a). However, this has still to be demonstrated under field conditions with ticks that have a shorter pre-oviposition period.

16.2.2. Host–parasite interactions

Nematodes are known to enter the body of insects mainly via natural orifices. Nematodes virulent to engorged *Amblyomma americanum* ticks were attracted towards the natural apertures of the females (Zhioua *et al.*, 1995). However, their mode of invasion has not yet been proven. No obvious relationship was observed between the size of spiracles, genital openings or cuticle thickness and the relative susceptibility to two strains of nematodes (Mauleon *et al.*, 1993). Exposure of *B. annulatus* ticks to nematodes for only 1 h resulted in some mortality. The mortality increased linearly up to 100% when the exposure time was increased up to 32 h (Samish *et al.*, 1996; Kocan *et al.*, 1998b). Ticks can be killed by the injection of a single nematode (Glazer and Samish, 1993) but axenic nematodes are unable to kill ticks even though they are pathogenic to insects (M. Samish *et al.*, 2002, unpublished data). This demonstrates the crucial role of the symbiotic bacteria (*Photorhabdus* and *Xenorhabdus*) of nematodes in killing the ticks (Kocan *et al.*, 1998a,b).

The number of nematodes that were recovered from infected ticks increased in the course of the first 3–4 days of exposure, but longer exposure did not increase their quantity per cadaver (Kocan *et al.*, 1998a,b; Samish *et al.*, 1996). Between 16 and 140 nematodes were found in each engorged *B. annulatus* female infected with various nematode strains. The LD_{50} of the various nematode strains to ticks was not related to the average number of nematodes recovered (Samish and Glazer, 1991). A few days after

Table 16.1. Infectivity data – ticks (Petri dish tests).

Species	Number of strains tested	Most virulent nematodes: Species	Strain	Susceptibility of developmental stages: Engorged nymphs	Unfed adults	Engorged females[a]	References
Amblyomma americanum	5	*Steinernema riobrave*	TX	—	—	< 35	Kocan *et al.*, 1998b
A. cajennense	1	*S. riobrave*	TX	—	—	> 180	Kocan *et al.*, 1998a
A. gemma	4	*S. carpocapsae*	DT	++	—	250	Kaaya *et al.*, 1999
A. maculatum	1	*S. riobrave*	TX	—	—	> 180	Kocan *et al.*, 1998a
A. variegatum	21	*S. carpocapsae*	Mex	+	+	20	Samish and Glazer 1992; Mauleon *et al.*, 1993; Samish *et al.*, 2000c
Argas persicus	3	*Heterorhabditis bacteriophora*	HP88	+	+	70	Hassanain *et al.*, 1999
Boophilus annulatus	7	*Heterorhabditis* sp.	IS-5	++	++	< 2.6	Samish and Glazer, 1992; Mauleon *et al.*, 1993; Samish *et al.*, 2000
B. annulatus	9	*S. carpocapsae*	Mex	—	—	25	Mauleon *et al.*, 1993
B. decoloratus	4	*S. riobrave*	TX	—	—	< 50	Kaaya *et al.*, 2000
B. microplus	14	None	—	NS	NS	NS[a]	Mauleon *et al.*, 1993
Dermacentor variabilis	2	*S. riobrave*	TX	—	—	> 180	Kocan *et al.*, 1998a
Hyalomma dromedarii	5	*Steinernema* sp.	S1	+	—	50	El-Sadawy, 1998
Heterorhabditis excavatum	5	*Heterorhabditis* sp.	IS-5	+	+	—	Samish *et al.*, 1996, 2000b
Ixodes scapularis	15	*H. megidis*	M145	—	—	—	Hill, 1998; Zhioua *et al.*, 1995
Rhipicephalus appendiculatus	5	*S. riobrave*	TX	+	+	< 50	Kaaya *et al.*, 2000
R. bursa	5	*Heterorhabditis* sp.	IS-3	—	—		Samish *et al.*, 1999, 2000
R. evertsi	5	*S. carpocapsae*	DT	+	+	< 50	Kaaya *et al.*, 2000
R. sanguineus	7	*S. riobrave*	TX	+	+	> 180	Samish *et al.*, 1999; Samish and Glazer, 1992; Kocan *et al.*, 1998a
Omithodoros moubata	3	*S. carpocapsae*	DT	+	—	—	M. Samish *et al.*, unpublished data
O. tholozani	4	*S. carpocapsae*	DT	+	—	—	M. Samish *et al.*, unpublished data

[a]LC_{50} – IJs/cm^2.

juvenile nematodes penetrated or were injected into ticks, all or most of them died inside their tick host. However, in rare cases, they survived as infective juveniles (IJs) or even started to develop within the tick, but did not complete their life cycle (Mauleon *et al.*, 1993; El-Sadawy *et al.*, 1998; Hill, 1998; Kaaya *et al.*, 1999). When the cuticle of ticks was slit artificially before their infection, the nematodes were able to complete their life cycle (Zhioua *et al.*, 1995; Samish *et al.*, 1998).

16.2.3. Application

Nematodes have been used successfully to control ticks also in simulated field conditions, but have not yet been tried in large-scale field trials. For example, ticks were killed in 10-l buckets with soil inoculated with any of the five nematode strains (Mauleon *et al.*, 1993; Hassanain *et al.*, 1999; Kaaya *et al.*, 1999).

There is a tenuous connection between virulence assays in Petri dishes containing filter paper and those conducted in soil environments such as buckets (Samish *et al.*, 1997, 2000b). For example, the steinernematids *Steinernema carpocapsae* DT and *S. carpocapsae* Mexican killed *B. annulatus* ticks on soil more rapidly than in Petri dishes. Heterorhabditid strains (HP88, IS-3 and IS-5), however, were more efficient in Petri dishes than on soil (Kocan *et al.*, 1998a). These differences could be due to different search strategies of the nematodes (Lewis, 2002) or different abilities to stay virulent under various soil environment stress factors (Glazer, 2002). All nematode strains tested were most efficient against ticks at about 26°C. Some strains were far less efficient at 22°C or 30°C, while others had a wider range of efficiency (between 18°C and 34°C) (Samish *et al.*, 1996). Placing ticks on sandy soil 3 days after it was sprayed with nematodes resulted in 100% mortality 10 days post-infection. However, the mortality of ticks on sandy soil with 25% v/v cattle manure or soil containing 40–50% silt was only 45% or 25%, respectively (Samish *et al.*, 1995). Some ticks prefer humid environments such as the upper soil layer, under stones or in leaf litter conditions, which also favour EPNs. Thus nematodes could be used as part of an integrated tick-control strategy in target sites such as human and animal rest areas, animal paths, water sources, etc.

16.3. Flies

The housefly, *Musca domestica*, is a well-known cosmopolitan pest of animals. This is the most common species found on pig and poultry farms, horse stables and ranches. Houseflies are always found in association with humans or activities of humans. Not only are they a nuisance, but they also can transport disease-causing organisms. Excessive fly populations are obnoxious to farm workers, and a public health hazard to nearby human habitations. Flies commonly develop in large numbers as a serious problem in poultry manure of caged hens. The control of *M. domestica* is vital to human health and comfort in many areas of the world. The most important damage related to this insect is the annoyance and indirect damage produced by the potential transmission of more than 100 pathogens associated with this fly, which may cause disease in humans and animals, including typhoid, cholera, bacillary dysentery, tuberculosis, anthrax, ophthalmia and infantile diarrhoea, as well as parasitic worms. Pathogenic organisms are picked up by flies from garbage, sewage and other sources of filth, and then transferred on their mouthparts and other body parts, through their vomitus and faeces, to human and animal food. The housefly has a complete metamorphosis with distinct egg, larva or maggot, pupal and adult stages. The housefly overwinters either in the maggot or pupal stage under manure piles or in other protected locations. Warm summer conditions are generally optimum for the development of the housefly; it can complete its life cycle in as little as 7–10 days, and have as many as 10–12 generations in one summer.

16.3.1. Infectivity

Geden *et al.* (1986) demonstrated the infectivity of EPNs to different developmental stages of *M. domestica*. Second and third instar larvae and adults of the muscid were highly susceptible to *S. carpocapsae* and *Heterorhabditis heliothidis* (= *H. bacteriophora*) when hosts were confined in Petri dishes containing nematode-treated filter paper. The maggots were not susceptible to *S. glaseri*, and the fly pupae were refractory to infection by all three species. When second and third instar larvae were exposed to 5000 nematodes/host in rearing medium, *S. carpocapsae* caused higher mortality (55–61%) than *H. heliothidis* (11–26%). Both *S. carpocapsae* and *H. heliothidis* were more infective for third instar larvae (21–29%) than for second instar larvae (2–6%) at this dosage in poultry droppings. When adult flies were offered *S. carpocapsae* suspensions in 5% sucrose bait in cotton balls, mortality ranged from 53% to 67% at dose rates ranging from 1000 to 100,000 nematodes/ml bait (Table 16.1b).

Taylor *et al.* (1998) screened 40 strains representing eight species of *Heterorhabditis* and five species of *Steinernema* for virulence towards third instar larvae of *M. domestica* in a filter paper assay. None of the 22 strains of *Heterorhabditis* infecting maggots caused significant levels of mortality in a filter paper assay. Ten strains of *Steinernema* infected maggots, of which seven strains (four *S. carpocapsae*, two *S. feltiae* and one *S. scapterisci*) caused significant mortality. Ten *Heterorhabditis* strains and ten *Steinernema* strains successfully reproduced for two generations in maggots. Taylor *et al.* (1998) also selected six strains of *Steinernema* for ten generations on maggots and then compared them with the unselected parent strains. No difference in pathogenicity between selected and unselected strains was observed (Table 16.2).

16.3.2. Host–parasite interactions

Renn (1998a) determined the routes of penetration of the *S. feltiae* IJs into larval and adult houseflies. IJs aggregate on the proboscis and anal aperture of male and female houseflies after 1 h. The nematodes penetrate female flies after 2 h by moving through the cloaca, then along the oviduct, and through the ovaries. Male houseflies are penetrated via the cloaca, and then the nematodes enter the haemocoel by penetrating the wall of the ejaculatory sac. All larval stages are penetrated via the anal aperture. Nematodes then move through the hindgut and penetrate the wall of the ileum, immediately posterior to the pylorus.

Histopathological analysis of the effect of different dosages (50, 100, 200, 500 and 1000 IJs/larvae) of *S. feltiae* on the larval tissues of *M. domestica* was conducted by Ghally *et al.* (1991). *S. feltiae* nematodes invade the fat tissue, gut, cuticle and muscle tissue of the host. All of these tissues, along with the gut epithelium, show signs of disintegration before death of the host. The tissue of the gut and the fat body are the most severely damaged by the presence of *S. feltiae*. The damage described in this investigation depended mainly upon the time and the intensity of infection.

16.3.3. Application

EPNs were evaluated as a control measure for housefly populations in various animal farming environments.

16.3.3.1. Cattle

Two strains of *S. feltiae* (SN and UNK-36) and two of the best *Heterorhabditis* strains (*H. bacteriophora* OSWEGO and *H. megidis* HF-85) were tested in a fresh bovine manure substrate (Taylor *et al.*, 1998). All the four strains produced significant fly mortality in the manure substrate; the *S. feltiae* strains had significantly lower LC_{50} values than did the *Heterorhabditis* spp. The most promising strain, *S. feltiae* SN, gave LC_{50} and LC_{99} values of 4 and 82 IJs/maggot, respectively. These doses were equivalent to 2.7 and 55 IJs/g of manure and 5.1 and 104 IJs/cm^2 of surface area (Taylor *et al.*, 1998). Shapiro *et al.* (1996, 1999) reported

Table 16.2. Infectivity data – insects.

Pests						Most virulent nematodes		% mortality of developmental stages				
Common name	Species	Bioassay	Infectivity measure	Nematode concentration	Number of strains tested	Species	Strain	Young larvae	Old larvae	Pupae	Adult	References
Housefly	*Musca domestica*	PDA	Insect mortality	5000 IJ/host	3	*Steinernema feltiae*[a]	DD-136	100	92	—	—	Geden *et al.*, 1986
		PDA	"	100	40	*S. feltiae*	SN	—	50.4	—	—	Taylor *et al.*, 1998
		SFC – Poultry manure	"	5000 IJ/host	3	*S. feltiae*[a]	DD-136	2.4	28.2	—	—	Geden *et al.*, 1986
		SFC – Poultry manure	"	1000 IJ/host	3	*S. feltiae*[a]	All	—	36.9	—	—	Georgis *et al.*, 1987
		SF – Poultry manure	Emerging adults	4×10^6 IJ/m^2	1	*Heterorhabditis heliothidis*	NC1	—	—	—	86	Belton *et al.*, 1987
Louse	*Pediculus humanus humanus*	PDA	Insect mortality	1500 IJ/dish	4	*S. glaseri*	—	75–90	80–95	—	92–97	Weiss *et al.*, 1993
	P. humanus capitis	PDA	"	100 IJ/host	4	*H. bacteriophora*	RN	100	100	—	84	Doucet *et al.*, 1998
Cat flea	*Ctenocephalides felis*	PDA + soil	Insect mortality	50 IJ/cm^2	1	*S. carpocapsae*	All	87–100	87–100	95–100		Manweiler, 1994
		SFC	Emerging adults	50 IJ/cm^2	1	*S. carpocapsae*	All	—	—	—	67	Manweiler, 1994

[a]*Steinernema feltiae* = *Steinernema carpocapsae*.
PDA = Petri-dish assay; SFC = Simulated field conditions

fresh cow manure to be detrimental to *S. carpocapsae* virulence, whereas composted cow manure had no effect.

16.3.3.2. Pigs

The efficacy of *S. feltiae* and *H. megidis* after formulation into a housefly bait was compared with a commercial bait formulation of methomyl for the control of houseflies in a UK pig farm (Renn, 1998b). The housefly infestation was confined to the farrowing unit, which consisted of ten farrowing houses, where two adjacent houses were sequentially restocked with pregnant sows at weekly intervals. Shortly after restocking, one house was baited with one of the nematode species and the other with methomyl. Significantly fewer flies were counted in the houses baited with either *S. feltiae* or *H. megidis* than those baited with methomyl. The efficacy of *S. feltiae* sprayed on to the manure was also compared with methomyl bait. Counts of houseflies carried out in the farrowing cycle before this treatment were not significantly different; however, significantly fewer flies occurred after *S. feltiae* was sprayed.

16.3.3.3. Poultry

Belton *et al.* (1987) evaluated the use of *H. heliothidis* (= *H. bacteriophora*) (North Carolina strain) for control of *M. domestica* maggots in chicken manure. Laboratory tests showed that *H. heliothidis* applied at rates of $2.0 \times 10^6/m^2$ and $4.0 \times 10^6/m^2$ significantly decreased fly emergence over 24 days compared with the control (310 and 227 versus 1570, respectively). Treatment of chicken manure in barns with the nematode after a 10-week period showed fly populations of about 10×10^6 in the control barns and 2×10^6 in the treated barn. No fly pupae were found parasitized by *H. heliothidis* at any time. Georgis *et al.* (1987) demonstrated that the nematodes are severely hampered by poultry manure. Exposure of IJs of *S. carpocapsae, S. feltiae* and *H. bacteriophora* to manure resulted in high nematode mortality (70–100%) within 18 h. Nematodes exposed to manure slurry for more than 6 h had drastically reduced efficacy against maggots of *M. domestica*. It was concluded that 'poor survival and limited movement of nematodes in the poultry manure appear to make them unlikely candidates for biocontrol of filth flies in this habitat' (Georgis *et al.*, 1987). To overcome this problem, Renn (1995) tested the use of the nematodes *S. feltiae* and *H. megidis* encapsulated in calcium alginate against the larvae of the housefly in chicken manure. Aliquots of capsules (15 ml) containing either 1,000,000, 500,000, 250,000 or 125,000 nematodes were added to 70-ml portions of grassmeal diet containing either eggs of first, second or third instar larvae. Complete mortality of the larvae was achieved within 6 days. Another experiment was carried out where immature houseflies were placed in chicken manure. The emergence of houseflies as adults was used to measure the effect of the encapsulated nematodes. The treatment with encapsulated *S. feltiae* resulted in 55–96% reduction in adult housefly emergence, whereas treatment with encapsulated *H. megidis* resulted in 35–98% reduction in emergence. Finally, when encapsulated nematodes were presented as bait to adult houseflies, little infectivity was observed. Renn and Wright (2000) further evaluated the effect of artificial substrates on the pathogenicity of *S. feltiae* to adult houseflies, demonstrating that certain substrates with high porosity are most effective, giving LD_{50} values of 2812–8912 IJs. In contrast, poor substrates with low porosity required 531, 450–899, 930 *S. feltiae* IJs to achieve LD_{50}. The use of advanced formulation and baiting approaches may allow the use of EPNs against this important pest.

16.4. Fleas

Fleas are annoying and medically important pests of human and animals. In addition to their bites and induced dermatitis, fleas are important vectors of dog tapeworm, murine typhus and plague (Strand, 1977). Fleas can be a significant problem outdoors near homes as well as in indoor areas. Persistence of outdoor populations of fleas can be a continuous source of flea re-infestations for companion animals (Henderson *et al.*, 1995).

16.4.1. Infectivity

Silverman *et al.* (1982) were the first to demonstrate susceptibility of the cat flea, *Ctenocephalides felis*, to *S. carpocapsae*. Henderson *et al.* (1995) evaluated the effect of *S. carpocapsae* on cat flea larvae and pupae in different substrates. Application of the nematodes to potting soil, sand, or gravel substrates containing different developmental stages of cat flea reduced adult flea emergence by 70–100%. It was demonstrated in this study that the silk cover of the cocoon did not prevent infection by nematodes (Henderson *et al.*, 1995).

16.4.2. Application

In the mid-1990s, 'Biosys' developed and launched a *S. carpocapsae*–based product for outdoor control of fleas (Manweiler, 1994; see also Table 16.1). It was suggested that nematode application would be a part of a flea control programme including the use of chemicals (IGRs) for indoor treatments against adults, shampoos or animal dips, and EPNs to the area where fleas were developing, to eliminate outdoor sources of re-infestations (Manweiler, 1994). The nematode-containing product that was sold in substantial quantities in southern USA was withdrawn from the market due to competition by new and effective chemicals, and widespread fungal contamination of the newly developed water-dispersible granular formulation (Grewal and Georgis, 1999). Flea control using EPNs is continuing in the USA currently on a smaller scale (Shapiro-Ilan, 2003, personal communication).

16.5. Lice

Other than malaria-carrying mosquitoes, no other insect has caused more death to mankind than lice, particularly the body louse, *Pediculus humanus humanus*. This ectoparasite that thrives in conditions of overcrowding, filth and famine may have a devastating effect on humans due to its association with endemic typhus (*Rickettsia prowazeki*), trench fever (*Rochalimaea Quintana*) and louse-borne relapsing fever (*Borrelia recurrentis*), diseases that probably have changed the course of history of mankind.

16.5.1. Infectivity

The susceptibility of the body louse *P. humanus humanus* to infection by EPNs was first demonstrated by Weiss *et al.* (1993). Exposure of female lice to IJs of *S. carpocapsae* and *S. glaseri* in a Petri dish assay resulted in > 85% mortality within 24 h. *H. bacteriophora* HP88 strain was found to be less effective, causing only 45% mortality after 42-h exposure. Mortality of lice females was directly related to increased dosage of *S. glaseri* IJs from 100 to 800 IJs/dish (Table 16.2). Complete mortality was achieved at a concentration of 400 IJs/dish. Exposure of lice to 800 IJs/dish of *H. bacteriophora* HP88 caused only 27% mortality (Table 16.1). Nematodes could be observed in the transparent body of the adult louse (Fig. 16.2).

Doucet *et al.* (1999) showed that the head louse *P. humanus capitis* is also susceptible to EPN infection. The authors showed substantial differences in infectivity of various nematode strains. The least infectious was *S. feltiae*, while *H. bacteriophora* strains readily killed adults and nymphs of the insect. The above studies show that the small size of the insect hinders nematode penetration into the insect (Doucet *et al.*, 1998) and causes abnormal development of the nematode inside the body (Weiss *et al.*, 1993).

16.5.2. Application

No further development of EPNs for control of lice has been reported. It seems that the

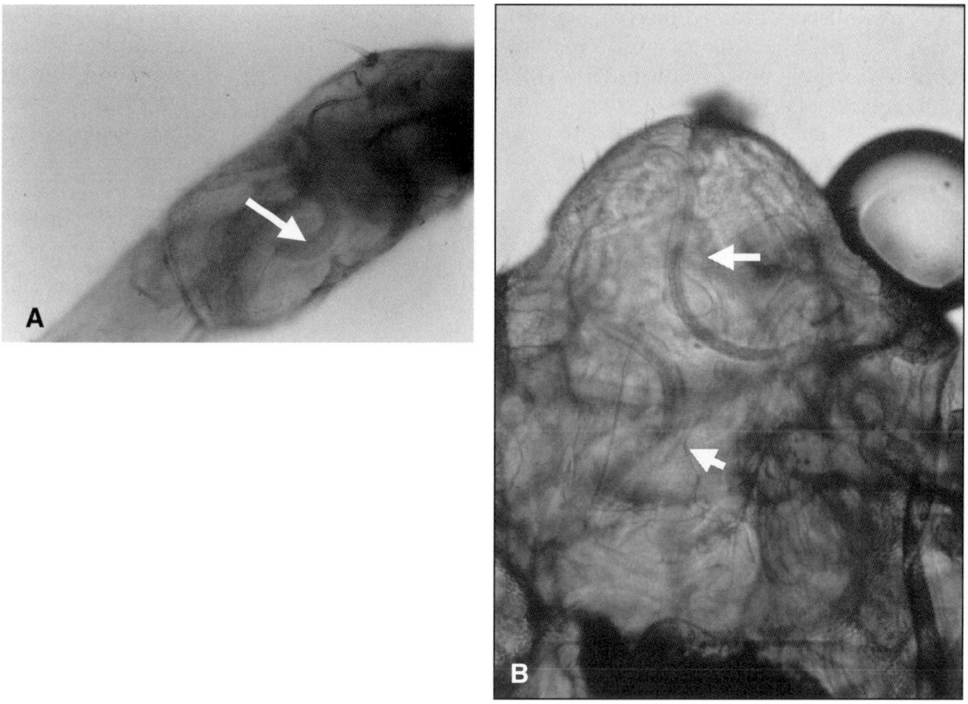

Fig. 16.2. Infective juveniles (IJs) (arrows) of *Steinernema carpocapsae* in the legs (A) and abdomen (B) of adult of the body louse *Pediculus humanus humanus*.

use of nematodes on human head or body will be impractical unless an appropriate formulation is developed. Doucet *et al.* (1998) suggested using the toxins produced by the nematodes. It may also be possible to use toxins from the symbiotic bacteria as antilouse agents.

16.6. Cockroaches

Cockroaches are one of the oldest and more important household and commercial establishment pests associated with humans around the world. Of the approximately 4000 living species of cockroaches in the world, only a few species inhabit human dwellings. The most common are the German, American and oriental cockroaches. The German cockroach, *Blattella germanica*, is the most important domiciliary pest worldwide, especially in regions that provide a favourable environment during cold or dry months. German cockroach infestations are associated with warm, humid environments such as commercial food preparation and storage facilities, or private kitchens (Cornwell, 1976). The American cockroach, *Periplaneta americana*, is one of the largest of the house-infesting cockroaches with worldwide distribution. It generally prefers warm and humid conditions, and is active from 21°C to 33°C. American cockroaches are found in residences, but are more common in larger commercial buildings such as restaurants, bakeries, food-processing plants and grocery stores. During the summer, American cockroaches can be found outdoors in yards and alleys. They are the most common species found in city sewer systems (Smith and Whitman, 1992). The oriental cockroach, *B. orientalis*, is adapted to areas that combine summer heat and moderately cold winter temperatures. It can successfully infest outdoor habitats year-round where it finds harbourage and perhaps food in plant material or crevices in sidewalks and buildings, but will move indoors or to protected harbourages during cold weather (Robinson, 1996). Indoors, they are typically found in crawl spaces, cellars, basements and on the first floor but at times on higher floors, especially around water pipes on which they climb (Smith and Whitman, 1992).

The importance of cockroaches as a pest is traditionally linked to their medical importance. They are known to be capable of carrying many common disease pathogens such as fungi, viruses, protozoa and about 40 species of bacteria that are pathogenic to vertebrates. The more common of such diseases include bacteria such as *Salmonella* and *Escherichia coli* (food poisoning), *Mycobacterium leprae* (leprosy), *Yersinia pestis* (bubonic plague), *Shigella* spp. (dysentery and diarrhoea in children), *Pseudomonas* spp. (urinary tract infections), *Staphylococcus*, *Streptococcus*, *Bacillus* and *Clostridium*. Cockroaches also cause diverse allergic responses in sensitive individuals. Fragments of dead domestic cockroaches and exuvia, cockroach faecal material and saliva have been recognized as strong indoor allergenic agents (Pope *et al.*, 1993) and are equally as important as dust mites in the etiology of house dust allergies (Hullet and Dockhorn, 1979).

Currently available pesticides reduce cockroach populations by 85–100%, but resistance to organophosphates, carbamates and pyrethroids is already widespread within cockroach populations (Scharf *et al.*, 1995). Also, the continuous use of pesticides in human living spaces can harm humans upon exposure. Biocontrol with nematodes could be a good tool if incorporated into an integrated pest management (IPM) approach for cockroaches.

16.6.1. Infectivity

The susceptibility of cockroaches to infection by EPNs was first observed by Skierska *et al.* (1976). Exposure of adult *B. germanica* to 200–800 IJs of *S. carpocapsae* (DD 136) in a Petri dish assay resulted in 95% mortality. However, Laumond *et al.* (1979), in an analogous experiment using 10,000 IJs/dish of the same nematode strain, reported that *B. germanica* was resistant to these nematodes. Subsequent studies have shown that EPNs are pathogenic to several cockroaches. *S. carpocapsae* is the most widely tested species in infectivity studies with household cockroaches (Table 16.3). Other species of steinernematids have been tested against cockroaches in laboratory studies. *S. feltiae* evaluation in Petri dish assays resulted in 50% mortality for adults and 30% mortality for nymphs of *B. germanica* (García del Pino and Morton, 2001), and between 92% and 100% mortality against *B. orientalis* (Kotlarska-Mordzinska *et al.*, 2000). In similar experiments, *S. arenarium* caused between 45% and 63% mortality of *B. germanica* (García del Pino and Morton, 2001), and *S. scapterisci* killed 66% of *B. germanica* but is ineffective against *P. americana* (Grewal *et al.*, 1993).

In general heterorhabditids have been less used in infectivity studies with cockroaches. *H. bacteriophora* caused 50–70% mortality of the German cockroach (García del Pino and Morton, 2001) and 73–95% mortality of the oriental cockroach (Koehler *et al.*, 1992). *H. heliothidis* (= *H. bacteriophora*) has been tested against American cockroaches, with 88% mortality of nymphs but only 9% mortality of adults (Zervos and Webster, 1989). The German cockroach is more susceptible to EPN infection. Appel *et al.* (1993) observed a high susceptibility (LT_{50}: 2.1 days) of adults of this cockroach to 0.5×10^6 IJs of *S. carpocapsae* in a Petri dish bioassay. Locatelli and Parleaz (1987) showed that 500 IJs/dish of *S. feltiae* (= *S. carpocapsae*), *S. bibionis* (= *S. feltiae*) and *Heterorhabitis* sp. caused 100% mortality of the German cockroach. Koehler *et al.* (1992) determined the susceptibility of *P. americana*; smokybrown cockroach, *P. fuliginosa*; oriental cockroach, *B. orientalis*; *B. germanica*; and brownbanded cockroach, *Supella longipalpa* to *S. carpocapsae* (All strain) (Fig. 16.3). Two different experiments were conducted: (i) a forced-exposure bioassay in

Table 16.3. Infectivity data – cockroaches.

Species	Bioassay	Number of strains tested	Nematodes: Species	Nematodes: Strain	Nematode concentration	Infectivity measure	% mortality of developmental stages: Nymphs	Adults females	Adults males	References
Blattella germanica	PDA	7	*Steinernema feltiae*[a]	All	500/dish	% mortality	—	100	100	Locatelli and Parleaz, 1987
	"	1	*S. bibionis*	Bib	500/dish	"	—	100	100	Locatelli and Parleaz, 1987
	"	1	*Heterorhabditis heliothidis*	HH	1,000/dish	"	—	40	80	Locatelli and Parleaz, 1987
	"	3	*Heterorhabditis* sp.	HW	500/dish	≪	—	100	100	Locatelli and Parleaz, 1987
	PDA	1	*S. carpocapsae*	All	500,000/dish	LT_{50} (days)	< 1	< 1	< 1	Koehler *et al.*, 1992
	Bait	1	*S. carpocapsae*	All	500,000/bait	LT_{50} (days)	4.9	3.5	3.5	Koehler *et al.*, 1992
	PDA	1	*S. scapterisci*	Uruguay	8,000/dish	% mortality	—	66		Grewal *et al.*, 1993
	PDA	1	*S. carpocapsae*	All	0.5×10^6/dish	LT_{50} (days)	—	—	2.1	Appel *et al.*, 1993
	SFC (Baits)	1	*S. carpocapsae*	All	2×10^6/bait	% reduction	—	67.1		Appel *et al.*, 1993
	PDA	4	*S. feltiae*	e-nema	40,000/dish	% mortality	30	50		Garcia del Pino and Morton, 2001
	"	1	*H. bacteriophora*	e-nema	40,000/dish	"	50	70		Garcia del Pino and Morton, 2001
	"	1	*S. arenarium*	S2	40,000/dish	"	45	62.5		Garcia del Pino and Morton, 2001
	Terraria (Bait)	1	*S. carpocapsae*	—	0.27×10^6	% mortality	—		80	Pye *et al.*, 2001
Periplaneta americana	PDA	1	*H. heliothidis*	T327	1,000/dish	% mortality	88	9		Zervos and Webster, 1989
	PDA	1	*S. carpocapsae*	All	500,000/dish	LT_{50} (days)	1	5	2.6	Koehler *et al.*, 1992

continued

Table 16.3. *Continued.* Infectivity data – cockroaches.

Species	Bioassay	Number of strains tested	Nematodes: Species	Nematodes: Strain	Nematode concentration	Infectivity measure	% mortality of developmental stages: Nymphs	Adults females	Adults males	References
	SFC (Bait)	1	*S. carpocapsae*	All	500,000/bait	LT_{50} (days)	0	0	0	Koehler *et al.*, 1992
	PDA	1	*S. scapterisci*	Uruguay	8,000/dish	% mortality	—	0		Grewal *et al.*, 1993
P. fuliginosa	PDA	1	*S. carpocapsae*	All	500,000/dish	LT_{50} (days)	< 1	< 1	1.2	Koehler *et al.*, 1992
	Bait	1	*S. carpocapsae*	All	500,000/bait	LT_{50} (days)	0	6.7	9.8	Koehler *et al.*, 1992
	SFC (spray)	1	*S. carpocapsae*	—	500,000/l	% reduction	—	0	—	Smith *et al.*, 1997
P. brunnea	Glass jars (3,75 liter)	1	*S. carpocapsae*	All	300,000/jar	% mortality	—	—	92	Corpus and Sikorowski, 1992
Supella longipalpa	PDA	1	*S. carpocapsae*	All	500,000/dish	LT_{50} (days)	< 1	< 1	< 1	Koehler *et al.*, 1992
	SFC (Bait)	1	*S. carpocapsae*	All	500,000/bait	LT_{50} (days)	4	2.6	3.7	Koehler *et al.*, 1992
Blatta orientalis	PDA	1	*S. carpocapsae*	All	500,000/dish	LT_{50} (days)	1.4	1	< 1	Koehler *et al.*, 1992
	SFC (Bait)	1	*S. carpocapsae*	All	500,000/bait	LT_{50} (days)	0	12	7	Koehler *et al.*, 1992
	PDA	1	*S. feltiae*	Low87	1,000/dish	% mortality	96.7	91.7	100	Kotlarska-Mordzinska *et al.*, 2000
	"	1	*H. bacteriophora*	PL	1,000/dish	"	73.3	78.3	95	Kotlarska-Mordzinska *et al.*, 2000

[a] *Steinernema feltiae* = *Steinernema carpocapsae.*

PDA = Petri dish assay; SFC = simulated field conditions.

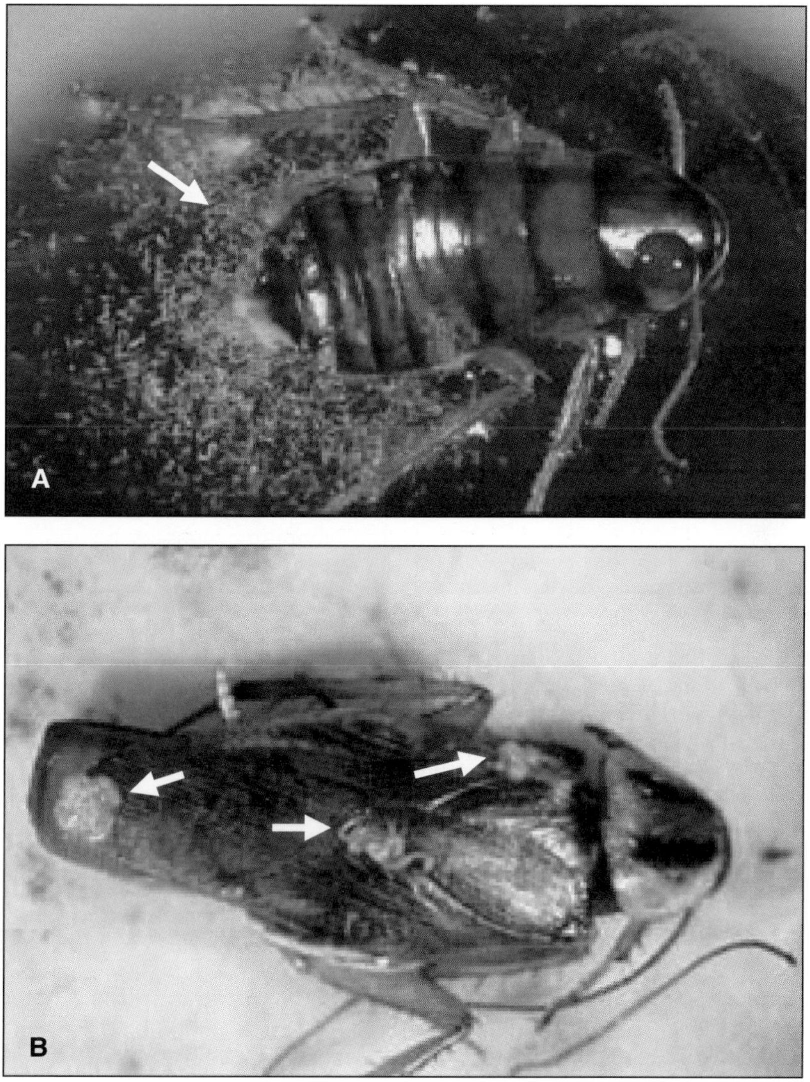

Fig. 16.3. Infection of nymphs of the American cockroach *Periplaneta americana* (A) and the German cockroach *Blattella germanica* (B) by entomopathogenic nematodes (EPNs) (arrows).

Petri dishes with 500,000 IJs and (ii) a non-forced-exposure bioassay with bait stations. In forced-exposure bioassay the relative order of susceptibility was: German = brownbanded > smokybrown > oriental > American. In bait bioassay the susceptibility was: brownbanded > German > oriental > smokybrown > American. In both bioassays, the American cockroach was the least susceptible to *S. carpocapsae* and in bait stations the nematodes were ineffective in killing late stage nymphs, males or females. Behavioural, morphological, physiological or immunological responses may be the reason for the lower susceptibility of *P. americana* to infection. Zervos and Webster (1989) observed encapsulation and melanization as immunological responses in adult American cockroaches to *H. bacteriophora*. These immunological responses caused nematode mortality but not before bacteria were released. Koehler *et al.* (1992) observed that in a forced bioassay, late stage nymphs, males and females of the German and brownbanded cockroaches have equal susceptibility to *S. carpocapsae*. Nevertheless, different instars of cockroaches usually have different susceptibility to EPNs. In general, EPNs are less effective in killing nymphs than adults, and in adults, males are more susceptible than females. In smokybrown cockroach the relative order of susceptibility was: females > males > nymphs; in oriental cockroach: males > females > nymphs; and in American cockroaches: nymphs > males > females (Koehler *et al.*, 1992). Appel and Benson (1994) showed that mature ootheca of the German cockroach are not affected by *S. carpocapsae* and there is no transoothecal transmission of this nematode in infected females.

16.6.2. Application

Although the pathogenicity of *Steinernema* and *Heterorhabditis* spp. to cockroaches has been proved in many laboratory bioassays with Petri dish or bait stations, only a few experiments have been carried out in field conditions. Smith *et al.* (1997) compared the efficacy of *S. carpocapsae* and other chemical insecticides against *P. fuliginosa* in outdoor applications. They applied the nematodes as a spray solution (500,000 IJs/l) in the perimeter area with vegetation or with soil and permanent shade of single-family detached houses. Nematodes did not affect the abundance of smokybrown cockroaches, and no cockroaches were recovered that exhibited nematode infection. The use of EPNs against German cockroaches in indoor applications was tested by Appel *et al.* (1993). They treated apartments with baits containing 2×10^6 nematodes presented in moisture-retaining stations patented by Weber *et al.* (1992). This nematode concentration had low repellence and good formulation and survival characteristics. Nematode treatment of the apartments was as good as the standard insecticide bait treatment, with 67% of reduction at 8 weeks after treatment. They concluded that the optimization of the exposure station, the number of stations needed per apartment and the station location will likely increase field performance. Pye *et al.* (2001) evaluated a commercial bait station with *S. carpocapsae* against *B. germanica* for use in structures. The bait station contained 0.27×10^6 IJs in polysaccharide gel with food and pheromone attractant. It killed > 80% of small populations ($n = 25$) in 4 days, and the killing activity continued for about 1 month.

16.7. Conclusions

Important veterinary and human pests are highly susceptible to EPNs. In some cases the laboratory bioassay results led to commercial utilization (i.e. fleas and houseflies). In general, most of the off-host stages of these pests hide for days and some even for weeks in the dark, humid upper layer of the ground, and this particular environment is also the natural habitat of the IJs (Kaya, 1990). Most pests can be found in confined areas, where nematodes can be applied in high concentrations. Particularly, the off-host stages of most pests drop off mainly at the resting areas of the mammalian host. EPNs are especially well suited for an integrated control regime since they are resistant to many insecticides (Glazer *et al.*, 1997; see Chapter 20, this volume). In addition, since certain compounds are applied mainly on the host and the nematodes are commonly applied on the ground, the two control methods could serve as complementary treatments.

The wide genetic diversity of nematode strains suggests that screening, selection and/or genetic manipulation could further improve their virulence. This, and the frequent similarity of the preferred ecological habitats of the nematodes and the off-host stages, makes it desirable to devote further efforts towards the development of a commercial nematode agent.

A comparison of the susceptibility level of various developmental stages in the course of the pest life cycle with their natural habitat (Scheme 16.1) indicates that in

Housefly (*Musca domestica*)

Life cycle 10–12[a] days	Eggs	First larvae	Second larvae	Third larvae	Pupae	Adult
Location						
Susceptibility to EPNs	–	+ + +	+ + +	+ + +	+ + +	+ + (In baits)

Cat flea (*Ctenocephalides felis*)

Life cycle 14–25 days	Eggs	First larvae	Second larvae	Third larvae	Pupae	Adult
Location						
Susceptibility to EPNs	–	+ + +	+ + +	+ + +	+ + +	+ + +

Body louse (*Pediculus humanus humanus*)

Life cycle 21–27[a] days	Eggs	First nymph	Second nymph	Third nymph	Adult
Location					
Susceptibility to EPNs	–	+ +	+ + +	+ + +	+ + +

German cockroach (*Blattella germanica*)

Life cycle 80–92[a] days	Eggs	First nymph	Second nymph	Third nymph	Fourth nymph	Fifth nymph	Sixth nymph	Adult
Location	Encountering EPN is possible only in bait							
Susceptibility to EPNs	–	+ + +	+ + +	+ + +	+ + +	+ + +	+ + +	+ + +

Ticks

One host (*Boophilus annulatus*)

Life cycle 60–120 days	Eggs	Larvae			Nymphs			Adults		
		Unfed	Fed	Engorged	Unfed	Fed	Engorged	Unfed	Fed	Engorged
Location										
Susceptibility to EPNs	–	+ + +	–	–	–	–	–	–	+	+ + +

Two hosts (*Rhipicephalus bursa*)

Life cycle 141–223 days	Eggs	Larvae			Nymph			Adults		
		Unfed	Fed	Engorged	Unfed	Fed	Engorged	Unfed	Fed	Engorged
Location										
Susceptibility to EPNs	–	+ + +	–	–	–	–	+	+ +	–	+ +

Three hosts (*Rhipicephalus sanguineus*)

Life cycle 71–128 days	Eggs	Larvae			Nymph			Adults		
		Unfed	Fed	Engorged	Unfed	Fed	Engorged	Unfed	Fed	Engorged
Location										
Susceptibility to EPNs	–	+ + +	–	+ +	+ + +	–	+	+ + +	–	+ + +

[a]Under laboratory conditions.

Susceptibility to EPNs: – = not susceptible; + = low; + + = moderate; + + + = high

In/on ground | On/off host | On host

Scheme 16.1

Table 16.4. Comparative considerations for the control of animal and human pests with those for control of plant pests.

Consideration	Plant pests	Veterinary and human pests
Inhabiting site of the pest	• In the soil • Plant foliage	• On host • On soil surface • In the soil
Hazardous and antagonistic factors affecting EPNs efficacy	• Desiccation • UV radiation • Soil fauna	• Body heat of host • Animal manure
Application of EPN	• Large surface in the field or greenhouse • Formulations for storage and/or UV radiation and desiccation protection	• Small surface on hosts and also where hosts are confined • Formulation for survival on and off host body
Economic	• Relatively lower value • Direct damage of pests to the crop – higher economic threshold	• Higher value and ability to pay more per treatment • Indirect damage as disease vectors – lower economic threshold
End-user	• Open for testing new technologies and pest management approaches	• Conservative and resistant for new technologies and pest management approaches

many cases the most susceptible stages are inhabiting an environment in which EPNs can be applied; mainly the soil. For certain pests, such as lice and cockroaches, particularly those inhabiting substrates like manure, special formulations are required to ensure the effectiveness of the nematodes. When development of EPNs for the control of veterinary pests is considered, there is a natural tendency to use the knowledge and approaches used in plant protection. However, there are substantial differences between the considerations for plants versus veterinary and human pest management considerations (see Table 16.4).

References

Appel, A.G. and Benson, E.P. (1994) Pathogenicity and limited transoothecal transmission of *Steinernema carpocapsae* (Nematoda: Steinernematidae) in adult female German cockroaches (Dictyoptera: Blattellidae). *Journal of Medical Entomology* 31, 127–131.

Appel, A.G., Benson, E.P., Ellenberger, J.M. and Manweiler, S.A. (1993) Laboratory and field evaluations of entomogenous nematode (Nematoda: Steinernematidae) for German cockroach (Dictyoptera: Blattellidae) control. *Journal of Economic Entomology* 86, 777–784.

Belton, P., Rutherford, T.A., Trotter, D.B. and Webster, J.M. (1987) *Heterorhabditis heliothidis*: a potential biological control agent of house flies in caged-layer poultry barns. *Journal of Nematology* 19, 263–266.

Centers for disease control (CDC). Anonymous summary of notifiable diseases, US (1996) *Morbidity and Mortality Report* 45, 1–88.

Cornwell, P.B. (1976) *The Cockroach*, Vol. 2. Associated Business Programmes, London.

Doucet, M.M.A., Miranda, M.B. and Bertolotti, M.A. (1998) Infectivity of entomogenous nematodes (Steinernematidae and Heterorhabditidae) to *Pediculus humanus capitis* De Geer (Anoplura: Pediculidae). *Fundamental and Applied Nematology* 21, 13–16.

Doucet, M.M.A., Bertolotti, M.A., Giayetto, A.L. and Miranda, M.B. (1999) Host range, specificity, and virulence of *Steinernema feltiae, Steinernema rarum,* and *Heterorhabditis bacteriophora* (Steinernematidae and Heterorhabditidae) from Argentina. *Journal of Invertebrate Pathology* 73, 237–242.

El-Sadawy, H.A.E. (1998) Testing some entomopathogenic nematodes for the biocontrol of *Hyalomma dromedrii* Koch (Acarina: Ixodidae). *Journal of Union Arab Biology*, Cairo 10A, 1–11.

García del Pino, F. and Morton, A. (2001) Los nematodos entomopatógenos para el control biológico de la cucaracha alemana, *Blattella germanica* (Dictyoptera: Blattellidae). *Proceedings of III Congreso Nacional de Entomología Aplicada Pamplona Spain* 2001, pp. 122.

Geden, C.J., Axtell, R.C. and Brooks, W.M. (1986) Susceptibility of the house fly, *Musca domestica* (Diptera: Muscidae), to the entomogenous nematodes *Steinernema feltiae, S. glaseri* (Steinernematidae), and *Heterorhabditis heliothidis* (Heterorhabditidae). *Journal of Medical Entomology* 23, 326–332.

Georgis, R., Mullens, B.A. and Meyer, J.A. (1987) Survival and movement of insect parasitic nematodes in poultry manure and their infectivity against *Musca domestica. Journal of Nematology* 19, 292–295.

Ghally, S.E., Kamel, E.G. and Nasr, N.M. (1991) *Steinernema* parasitism of larvae of the house fly *Musca domestica* Linnaeus. *Journal of Egyptian Society of Parasitology* 21, 633–640.

Glazer, I. (2002) Survival biology. In: Gaugler, R. (ed.) *Entomopathogenic Nematology.* CAB International, Wallingford, UK, pp. 169–188.

Glazer, I. and Samish, M. (1993) Suitability of *Boophilus annulatus* repleted female ticks as hosts of the nematode *Steinernema carpocapsae. Journal of Invertebrate Pathology* 61, 220–222.

Glazer, I., Salame, L. and Segal, D. (1997) Genetic enhancement of nematicidal resistance of entomopathogenic nematodes. *Biocontrol Science and Technology* 7, 499–512.

Glazer, I., Alekseev, E. and Samish, M. (2001) Factors affecting the virulence of entomopathogenic nematodes to engorged female *Boophilus annulatus* ticks. *Journal of Parasitology* 87, 808–812.

Ginsberg, H.S. (1993) Transmission risk of Lyme disease and implications for tick management. *American Journal of Epidemiology* 138, 65–73.

Grewal, P.S. and Georgis, R. (1999) Entomopathogenic nematodes. In: Hall, F.R. and Menn, J. (eds) *Methods in Biotechnology: Biopesticides: Use and Delivery.* Humana Press, Totawa, New Jersey, pp. 271–299.

Grewal, P.S., Gaugler, R., Kaya, H.K. and Wusaty, M. (1993) Infectivity of the entomopathogenic nematode *Steinernema scapterisci* (Nematoda: Steinernematidae). *Journal of Invertebrate Pathology* 62, 22–28.

Hassanain, M.A., Derbala, A.A., Abdel-Barry, N.A., El-Sherif, M.A. and El-Sadawe, H.A. (1999) Biological control of ticks (Argasidae) by entomopathogenic nematodes. *Egyptian Journal of Biological Pest Control* 7, 41–46.

Henderson, G., Manweiler, S.A., Lawrence, W.J., Tempelman, R.J. and Foil, L.D. (1995) The effect of *Steinernema carpocapsae* (Weiser) application to different life stages on adult emergence of the cat flea *Ctenocephalides felis* (Bouche). *Veterinary Dermatology* 6, 159–163.

Hill, D.E. (1998) Entompathogenic nematodes as control agents of developmental stages of the black-legged tick, *Ixodes scapularis. Journal of Parasitology* 84, 1124–1127.

Hullet, A.C. and Dockhorn, R.J. (1979) House dust mite (*D. Farinae*) and cockroach allergy in a midwestern population. *Annals of Allergy* 42, 160–165.

Kaaya, G.P., Glazer, I. and Samish, M. (1999) Laboratory evaluation of pathogenicity of entomopathogenic nematodes to African tick species. *Annals of the New York Academy of Science* 916, 303–308.

Kaya, H.K. (1990) Soil ecology. In: Gaugler, R. and Kaya, H.K. (eds) *Entomopathogenic Nematodes in Biological Control.* CRC Press, Boca Raton, Florida, pp. 93–115.

Kocan, K.M., Pidherney, M.S., Bluin, E.F. and Claipool, P.L. (1998a) Interaction of entomopathogenic nematodes (Steinernematidae) with selected species of ixodid ticks (Acari: Ixodidae). *Journal of Medical Entomology* 35, 514–520.

Kocan, K.M., Bluin, E.F., Pidenherney, M.S., Claipool, P.L., Samish, M. and Glazer, I. (1998b) EPNs as a potential biological control method for ticks. *Annals of the New York Academy of Science* 849, 355–356.

Koehler, P.G., Patterson, R.S. and Martin, W.R. (1992) Susceptibility of cockroaches (Dictyoptera: Blattellidae, Blattidae) to infection by *Steinernema carpocapsae. Journal of Economic Entomology* 85, 1184–1187.

Kotlarska-Mordzinska, J., Goszczynski, W. and Bednarek, A. (2000) Susceptibility of the cockroach, *Blatta orientalis* L. to entomopathogenic nematodes. *IOBC WPRS Bulletin* 23, 39–40.

Lancaster, J.L. Jr and Meisch, M.V. (1986) *Arthropods in Livestock and Poultry Production.* Ellis Horwood, London, p. 402.

Laumond, C., Mauléon, H. and Kermarrec, A. (1979) Données nouvelles sur le spectre d'hôtes et le parasitisme du nématode entomophage *Neoaplectana carpocapsae. Entomophaga* 24, 13–27.

Lewis, E.E. (2002) Behavioral ecology. In: Gaugler, R. (ed.) *Entomopathogenic Nematology.* CAB International, Wallingford, UK, pp. 205–224.

Locatelli, D.P. and Parleaz, E. (1987) Valutazione in laboratorio de ll'attivitá di *Steinernema* spp.

e *Heterorhabditis* spp. su *Blattella germanica* (L.). *La difensa delle piante* 10, 339–348.

Manweiler, S.A. (1994) Development of the first cat flea biological control product employing the entomopathogenic nematode *Steinernema carpocapsae. Proceedings of the Brighton Crop Protection Conference – Pests and Diseases* 3, 1005–1012.

Margalit, Y. and Ben-Dov, E. (2000) Biological control by *Bacillus thuringiensis* subs. *israelensis.* In: Rechcigel, J.E. and Rechcigel, N.A. (eds) *Insect Pest Management – Techniques for Environmental Protection.* CRC Press, Boca Raton, Florida, pp. 243–301.

Mauleon, H., Barre, N. and Panoma, S. (1993) Pathogenicity of 17 isolates of entomophagous nematodes (Steinernematidae and Heterorhabditidae) for the ticks *Amblyomma variegatum* (Fabricius), *Boophilus microplus* (Canestrini) and *Boophilus annulatus* (Say). *Experimental and Applied Acarology* 17, 831–838.

McCosker, P.J. (1979) Global aspects of the management and control of ticks of veterinary importance. *Recent Advances in Acarology II.* Academic Press, New York, pp. 45–53.

Pope, A.M., Patterson, R. and Burge, H. (1993) *Indoor Allergens: Assessing and Controlling Adverse Health Affects.* National Academy Press, Washington, DC, 308 pp.

Pye, N., Holbrook, G. and Pye, A. (2001) New bait station with *Steinernema carpocapsae* nematodes kills cockroaches. *Proceedings of 34th Annual Meeting of the Society of Invertebrate Pathology,* Nordwijkerhout, The Netherlands, p. 62.

Renn, N. (1995) Mortality of immature houseflies (*Musca domestica* L.) in artificial diet and chicken manure after exposure to encapsulated entomopathogenic nematodes (Rhabditida: Steinernematidae, Heterorhabditidae). *Biocontrol Science and Technology* 5, 349–359.

Renn, N. (1998a) Routes of penetration of the entomopathogenic nematode *Steinernema feltiae* attacking larval and adult houseflies (*Musca domestica*). *Journal of Invertebrate Pathology* 72, 281–287.

Renn, N. (1998b) The efficacy of EPNs for controlling housefly infestations of intensive pig units. *Medical Veterinary Entomology* 12, 46–51.

Renn, N. and Wright, E. (2000) The effect of artificial substrates on the pathogenicity of *Steinernema feltiae* (Rhabditida: Steinernematidae) to adult *Musca domestica* (Diptera: Muscidae). *Nematology* 2, 217–222.

Robinson, W.H. (1996) *Urban entomology: Insect and mite pests in the human environment.* Chapman and Hall, London, 430 pp.

Samish, M. and Glazer, I. (1991a) Killing ticks with parasitic nematodes of insects. *Journal of Invertebrate Pathology* 58, 281–282.

Samish, M. and Glazer, I. (1991b) Pathogenicity of parasitic nematodes to ticks. In: Dusbabek, R. and Bukva, V. (eds) *Tick-Borne Diseases and their Vectors.* SPB Academic Publishers, The Hague, The Netherlands, pp. 629—632.

Samish, M. and Glazer, I. (1992) Infectivity of EPNs (Steinernematidae and Heterorhabditidae) to female ticks of *Boophilus annulatus* (Arachnida: Ixodidae). *Journal of Medical Entomology* 29, 614–618.

Samish, M. and Rehacek, J. (1999) Pathogens and predators of ticks and their potential in biological control. *Annual Review Entomology* 44, 159–182.

Samish, M., Alekseev, E. and Glazer, I. (1995). The development of entomopathogenic nematodes in the tick *Boophilus annulatus.* In: Coons, L. and Rothschild, M. (eds) *Tick Borne Pathogens at the Host Vector Interface: A Global Perspective.* Krugel, National Park, South Africa, pp. 2–4.

Samish, M., Glazer, I. and Alekseev, E.A. (1996) The susceptibility of the development stages of ticks (Ixodidae) to entomopathogenic nematodes. In: Rodger, M., Horn, D.J., Needham, G.R. and Welbourn, W.C. (eds) *Acarology IX.* Ohio Biological Survey, Columbus, Ohio, pp. 121–123.

Samish, M., Alekseev, E. and Glazer, I. (1997) The effect of substrate composition on the success of entomopathogenic nematodes to kill ticks. *Israel Journal of Medical Science* 33, 705.

Samish, M., Alekseev, E. and Glazer, I. (1998) The effect of the soil environment upon anti-tick activity of entomopathogenic nematodes. *Annual New York Academy of Science* 849, 402–404.

Samish, M., Alekseev, E. and Glazer, I. (2000a) Efficacy of entomopathogenic nematode strains against engorged *Boophilus annulatus* (Say) female ticks (Arachnida: Ixodida) under seminatural conditions. *Journal of Medical Entomology* 36, 727–732.

Samish, M., Alekseev, E. and Glazer, I. (2000b) Interaction between ticks and pathogenic nematodes. Susceptibility of tick species at various developmental stages. *Journal of Medical Entomology* 36, 733–740.

Samish, M., Alekseev, E. and Glazer, I. (2000c) Mortality rate of adult ticks due to infection by entomopathogenic nematodes. *Journal of Parasitology* 86, 679–684.

Samish, M., Alekseev, E. and Glazer, I. (2001) EPNs for the biocontrol of ticks. *Trends in Parasitology* 17, 368–371.

Scharf, M.E., Bennett, G.W., Reid, B.L. and Qui, C. (1995) Comparison of three insecticide resistance detection methods for the German cockroach (Dictyoptera: Blattellidae). *Journal of Economic Entomology* 88, 536–542.

Shapiro, D.I., Tylka, G.L. and Lewis, L.C. (1996) Effects of fertilizers on virulence of *Steinernema carpocapsae. Applied Soil Ecology* 3, 27–34.

Shapiro, D.I., Tylka, G.L. and Lewis, L.C. (1999). The effects of fertilizers on black cutworm, *Agrotis ipsilon* (Lepidoptera: Noctuidae) suppression by *Steinernema carpocapsae. Journal of Nematology* 31(Suppl.), 690–693.

Silverman, J., Platzer, E.G. and Rust, M.K. (1982) Infection of cat flea, *Ctenocephalides felis* (Bouche) by *Neoaplectana carpocapsae* (Weiser). *Journal of Nematology* 14, 394–497.

Skierska, B., Szadziewska, M. and Stanuszek, S. (1976) Laboratory test for the usability of the entomophilic nematodes Steinernematidae Chitwood and Chitwood 1937 in biological control of some noxious arthropods. *Bulletin of the Institute of Maritime Tropical Medicine, Gdynia* 27, 207–227.

Smith, E.H. and Whitman, R.C. (1992) *Field Guide to Structural Pests.* National Pest Control Association, Dunn Loring, Virginia.

Smith, L.M., Appel, A.G., Mack, T.P., Keever, G.J. and Benson, E.P. (1997) Evaluation of methods of insecticide application for control of smokybrown cockroaches (Dictyoptera: Blattidae). *Journal of Economic Entomology* 90, 1232–1242.

Sonenshine, D.E. (1993) *Biology of Ticks,* Vol. 2. Oxford University Press, New York, 465 pp.

Strand, M.A. (1977) Pathogens of Siphonaptera. In: Roberts, D.W. and Strand, M.A. (eds) *Pathogens of Medically Important Arthropods.* Bulletin of World Health Organization, pp. 55.

Sutherst, R.W., Jones, R.J. and Schnitzerling, H.J. (1982) Tropical legumes of the genus *Stylosanthes* immobilize and kill cattle ticks. *Nature* 295, 320–321.

Taylor, D.B., Szalanski, A.L., Adams, B.J. and Peterson, R.D. (1998) Susceptibility of house fly (Diptera: Muscidae) larvae to EPNs (Rhabditida: Heterorhabditidae, Steinernematidae). *Environmental Entomology* 7, 1514–1519.

Weber, T., Georgis, R., Pruitt, P. and Wren, J. (1992) *Insect Trap.* US Patent: 5,172,514.

Weiss, M., Glazer, I., Mumcuoglu, K.Y., Elking, Y. and Galun, R. (1993) Infectivity of steinernematid and heterorhabditid nematodes for the human body louse *Pediculus humanus humanus* (Anoplura: Pediculidae). *Fundamental and Applied Nematology* 16, 489–493.

Zervos, S. and Webster, J.M. (1989) Susceptibility of the cockroach *Periplaneta americana* to *Heterorhabditis heliothidis* (Nematoda: Rhabditoidea) in the laboratory. *Canadian Journal of Zoology* 67, 1609–1611.

Zhioua, E., Lebrum, R.A., Ginsberg, H.S. and Aeschlimann, A. (1995) Pathogenicity of *Steinernema carpocapsae* and *S. glaseri* (Nematoda: Steinernematidae) to *Ixodes scapularis* (Acari: Ixodidae). *Journal of Medical Entomology* 32, 900–905.

17 Applications for Social Insect Control

D.H. Gouge
University of Arizona, MAC, Maricopa, AZ 85239, USA

17.1. Introduction

> For so work the honeybees, creatures that by a rule in nature teach the act of order to a peopled kingdom. (William Shakespeare, Henry V 1599)

Social insects are undoubtedly one of the most fascinating and mysterious insect groups. The concept of a superorganism constructed of a colony described as a multicellular animal with individual members of the colony being analogous to individual cells conjures up almost supernatural expectations. Indeed, the degree of communication and cooperation within extensive communities has generated some of the most complex ecological interactions emerging in science.

Social behaviour in the Class Insecta covers a broad range of activities and interactions. Subsocial behaviour is widespread in many insect orders including cockroaches, earwigs, webspinners, true bugs, aphids and various beetles. However, this chapter focuses on insect groups classed as truly social or eusocial in nature. Eusocial systems are characterized by parental care of young (young are dependent on adults), overlapping generations (essential for care of young) and the reproductive division of labour (caste formation). The objective of this chapter is to examine and review literature pertaining to the impact, or potential impact, of entomopathogenic nematodes (EPNs) on eusocial insects that harm humans, structures and plants. Various diverse groups of nematodes are associated with species in these groups. Bedding (1984) offers a comprehensive account of the life cycle and pathogenicity of mermithids, sphaerularids, steinernematids and other nematodes parasitic on ants, bees, wasps, sandflies, woodwasps and insect parasitoids. Only four groups of insects are considered here: bees (Apidae), wasps (Vespidae), ants (Formicidae) and termites (Rhinotermitidae). Bees, wasps and ants of the Aculeate Hymenoptera have the ability to sting and as such can be hazardous to people. Termites and certain ants can be devastating crop and structural pests. Honeybees, *Apis mellifera*, are essentially beneficial to humans and are generally considered a valuable non-target insect when screening the biocidal activity of chemicals and biocontrol agents.

As the majority of individuals in social colonies do not contribute genetically to the formation of offspring, they commonly exhibit altruistic behaviour motivated towards the preservation of the colony reproductives. An example of this would be defensive honeybee workers attacking a perceived threat. So we are left to consider whether the social character itself affords a species added protection against potential pathogens such as EPNs. Polyethism (behavioural differences among castes) may also serve to protect colonies against parasites as groups within the colony become highly specialized and efficient at specific jobs such as cleaning. The special difficulties associated with using biocontrol agents against social insects have been documented. Pereira and Stimac (1997) studied ant species and noted special difficulties in assessing the achieved level of control. Relatively low genetic variability among individuals in a social insect colony could mean that the efficacy of an EPN is compromised or even enhanced (Sherman *et al.*, 1988). Military-style colony (self- and allogrooming) and nest hygiene could eliminate or facilitate the spread of pathogens (Oi and Pereira, 1993). Additionally, the highly volatile chemical nature of social insect communication could also influence microorganisms (Schildknecht and Koob, 1971; Story *et al.*, 1991). Once within a social nest however, a pathogen is in a highly regulated and protected environment, usually experiencing a stable temperature and humidity, high host density and little to no ultraviolet (UV) light. Perhaps as a reflection of the challenges, Peters (1996), in a comprehensive study on the natural host range of *Steinernema* spp. and *Heterorhabditis* spp., found only two references to the natural infection of social insects.

17.2. Apidae – Honeybees and Bumble Bees

The original reports by Dutky and Hough (1955) and Cantwell *et al.* (1972) indicated that adult bees were resistant to *Steinernema carpocapsae*. However, Hackett and Poinar (1973) showed that adult worker honeybees could be parasitized and killed by *S. carpocapsae*, when infective juvenile (IJ) nematodes were offered to the bees in sugar, honey or fruit solutions. Kaya *et al.* (1982) attributed the low susceptibility of the honeybee brood stages to high temperatures in the hives between 33.3°C and 35.2°C and low humidity (40–78%). During the days after nematode application, there was a notable increase in adult worker mortality in the colonies that had received the highest nematode concentration, but colony activity was not affected. Direct spraying of nematodes on to caged workers resulted in 4–10% mortality. It was concluded that *S. carpocapsae* can safely be used against insect pests in areas where bees occur by following normal spray precautions. Close-proximity spraying might kill some bees but would not be detrimental to the colony. Similarly, Stock (1996) tested the pathogenicity of *Heterorhabditis argentinensis* on insects associated with lucerne crops. Stock concluded that *H. argentinensis* could be used against damaging insects of lucerne crops without causing serious damage to the beneficial insects surveyed, including *A. mellifera*. To reduce exposure of bees, nematodes may be applied while bees are not actively foraging.

Baur *et al.* (1995) established honeybees as resistant by exposing workers and brood to four EPN species under conditions normally encountered in the hive by spraying nematodes directly on to combs. The mortality of adult bees exposed to any of the nematode species was less than 10%, and there was no evidence of nematode infection when dead adults were dissected. The authors decided to assess the impact of nematodes on brood, by looking at smaller-size honeycombs placed in the second storey (super) of a hive and large brood combs placed in the main section of the hive. The results were inconsistent between these two experiments and did not support the hypothesis that high-temperature-tolerant species of nematodes (*S. riobrave*) are more infective to honeybees. All things considered, EPNs are not a threat to natural or commercial bee colonies and would not be a good control option for treating problem colonies.

17.3. Vespidae – Yellowjackets

Vespula spp. in particular can pose a serious hazard to humans and domesticated animals due to their attraction to food items and their ability to sting. In reality, human deaths are rare but a significant proportion of society is hypersensitive to the venom, as in the case of bee stings. Akhurst (1980) was the first to report naturally infected *Vespula* sp. in Tasmania. Later, Bedding (1984) discovered a naturally infected queen German yellowjacket, *Vespula germanica*. He found the wasp hibernating beneath eucalyptus bark and identified *S. carpocapsae* from the insect.

During the 1970s and 1980s large numbers of insect species were screened for susceptibility to EPNs. Laumond *et al.* (1979), Wassink and Poinar (1984) and Wojcik and Georgis (1987) included vespid species, finding some susceptibility to the nematodes. Poinar and Ennik (1972) reported that the workers of the Western yellowjacket *V. pennsylvanica*, the entirely insectivorous *V. rufa*, *V. atropilosa* from northern California and *Vespula* sp. from The Netherlands were all susceptible to infection by the Agriotos and Leningrad strains of *S. carpocapsae*. Infection occurred after the adult yellowjackets ingested the IJs placed in a fruit concentrate or on sugar cubes. The authors indicate that most of the yellowjackets placed with nematode-infested food died after the juveniles of *S. carpocapsae* penetrated into their thorax and liberated cells of *Xenorhabdus nematophilus* into the haemocoel. Similarly, Wojcik and Georgis (1988) mixed IJs of *S. feltiae* in a sucrose solution and allowed adult *V. pensylvanica* to drink the solution. Nematodes were recovered from the gut of all treated insects following just 1-h exposure. By 18 h all the insects were dead.

Gambino (1984) tested the susceptibility of *V. pensylvanica* to three species of EPNs in the laboratory. Full-grown larvae were removed from their combs and placed in Petri dishes, and exposed to approximately 2000 IJs/10 larvae. Within 2 days, yellowjacket mortality was 100% in those exposed to *S. carpocapsae* (Italian strain) and *Steinernema* sp. (Hopland strain) and 98% in those exposed to *H. bacteriophora*. All three nematode species were able to reproduce inside the wasp hosts, and preliminary tests indicated that the same three nematode species could kill and reproduce in the common yellowjacket *V. vulgaris*.

Guzman (1984) evaluated the biocontrol potential of the nematode *S. feltiae* against *V. germanica*. Wasp larvae treated in the laboratory with 20 or 200 IJs/larva began to die within 24 h, and all were dead in 48 h. When adult workers captured in the field were kept in wire cages and fed with about 200 IJs/adult in a sucrose bait, mortality was 76–100%. The characteristic septicaemia resulting from nematode infection was present in several treated wasps that were dissected, and adults and second generation juvenile nematodes were found in the wasps' haemocoel, indicating that reproduction had occurred.

In the 1990s research focused less on nematode host ranges under laboratory conditions and to a greater extent investigated natural host range and considered the impact of augmentation of nematode populations on natural insect populations (Peters, 1996). Field tests using *S. feltiae* drenches against yellowjacket colonies were undertaken by Gambino *et al.* (1992). They found that in all treated colonies, worker activity was reduced by at least 50% after 1 week. Some treated colonies partially recovered, while some were destroyed. *Vespula* workers were observed removing nematode-infected wasps from the colony. The results suggest that under certain conditions, nematodes could be used as biocontrol agents of yellowjackets; however, prompt removal of infected individuals by workers would eliminate recycling of the nematodes and reduce ongoing infection. Therefore, efficacy was obviously limited due to the social cooperation of the workers.

Despite the limited control effect, both Drlik (1994) and Lind (1998) are proponents of pouring *S. feltiae* into the holes of ground-nesting yellowjackets as a least toxic method of management. Rose *et al.* (1999) agree and conclude from their pathogen-screening tests that EPNs may be useful as inundative control agents but do not have the potential to establish and provide long-term or permanent suppression. As yellowjackets in many circumstances may be considered beneficial (predators of plant feeding pests), the impermanent effect of EPNs can be considered an advantage from an integrated pest management (IPM) point of view.

Thus, two application strategies have been employed to manage yellowjacket pests using EPNs. One is the application of IJs directly to the nest and the other is to attract workers to nematode-treated baits. Wojcik and Georgis (1987) used desiccated nematodes successfully in their studies with *V. pennsylvanica*. Although the use of baits to deliver IJs may reduce worker numbers, it is unlikely to affect the colony as a whole. The use of baits to reduce populations is a sound approach and repeated drench applications to nests is a useful management option for areas not requiring immediate control.

17.4. Formicidae – Ants

In the USA, ants are the most frequently occurring home pest that clients report to pest management professionals. Globally, the ant species that are considered pests vary greatly between different regions. Many can be considered simply as nuisance infestations, but effects that are more serious do occur. Pharaoh ants, *Monomorium pharaonis*, are considered one of the top forensic pests as they can phoretically transmit many food-borne pathogens. Carpenter ants, *Camponotus* spp., can be a damaging pest of buildings and several stinging ant species can cause injury to humans, especially by eliciting hypersensitivity reactions. In the USA, almost as many people die each year because of both wasp and ant stings, as because of honeybee stings. Fire ants (*Solenopsis* spp.), especially the red imported fire ant (*Solenopsis invicta*), injure people, domesticated animals and livestock to a significant extent.

The individual ant itself appears as a veritable fortress against invading pathogens. Furthermore, adult ant anatomy must limit the ability of IJs to infect via the oral route; debris in the food, such as dirt and other particular matter, is filtered before it enters the prepharynx and is collected in a tiny trap known as the infra-buccal pocket. When this pocket becomes full the ant empties it into an area within or outside the nest that is designated as a waste product area. Many ant species have an infra-buccal filter, but usually in the adult stage only. Spinules (spines) in the spiracular chambers would limit access as would pilosity (setae) around the anus. Many of the adaptations are considered to be primarily water conservation structures but certainly double as defensive structures against relatively large invading parasites. Also, most ant species have an efficient colony protection system, in that the loss of many foraging ants has a minimal effect on the colony in general. Management is achieved when foraging ants collect the toxin (usually in bait form) and return it to the colony for dissemination. Chemical bait stations are commonly used and consist of small structures with small openings. A species-targeted attractive bait is contained within the bait station, which contains a slow-acting pesticide. Many entomopathogens are naturally slow-acting and ideal for bait station delivery. However, most research in this area has concentrated on the delivery of fungal pathogens as opposed to EPNs. Georgis (1987) did show that desiccated IJs of *S. carpocapsae* All strain in a sugar solution bait were infective to carpenter and harvester (*Pogonomyrmex* sp.) ants. However, few others have pursued the idea of baited IJ formulations and no further publications are apparent.

Early studies discovered that *S. carpocapsae* was able to infect adult tropical fire ants *Solenopsis geminata*; larvae and adults of *Camponotus* spp.; larvae, pupae and adults of parasol ants, *Acromyrmex octospinosus* (leaf-cutter); and workers of *Myrmica* sp. (Kermarrec, 1975; Laumond *et al.*, 1979; Bedding, 1984). Poole (1976) conducted studies using *S. carpocapsae* on the black imported fire ant *So. richteri* and the red imported fire ant *So. invicta*. Both reproductives and brood were susceptible under laboratory conditions. Poole found that workers were less susceptible. In the autumn, the nematodes were capable of causing about 35% mortality of the colony, including approximately 12% worker infection after 90 days. The same tests repeated in the spring achieved much greater mortality (80% colony, including 22% worker). Considering *Solenopsis richteri* data alone, the nematodes destroyed 80% of spring mounds and 36% of autumn mounds. Poole suggested that the worker ants have a high grooming activity level and also observed workers regurgitate nematodes to the alates and larvae or the colony. Poole found that a single alate ant could produce up to 3000 IJs. Poole concluded that *S. carpocapsae* could be an effective biocontrol agent if timing of application, mound size (small mounds were more easily controlled) and climatic conditions were favourable.

Quattlebaum (1980) evaluated the known fungal and nematode pathogens of the red imported fire ant *So. invicta*. He applied the DD-136 strain of *S. carpocapsae* and the NC strain of *H. bacteriophora* (= *heliothidis*) to *So. invicta* and *So. richteri* fire ant mounds. The IJs were applied using a compressed air sprayer 12–24 inches deep into the centre of mounds. Quattlebaum (1980) concluded that both nematode species were able to reduce field populations of *So. invicta*. *S. carpocapsae* had the greatest efficacy, which was shown to vary with different soil textures. He further established synergistic effects when *S. carpocapsae* was applied along with various carbamate and organophosphate insecticides.

Drenching outdoor ant nests with EPNs has produced mixed results. Drees *et al.* (1992) applied IJs of *Steinernema* spp. and *Heterorhabditis* spp. to reproductive stages of *So. invicta* under laboratory conditions. The mortality of reproductive larvae, pupae and alates ranged from 28% to 100% after 96 h at 23–25°C. *S. carpocapsae* All was the most consistent species tested; this nematode caused 82–94%, 64–96% and 38–99% mortality of fire ant larvae, pupae and alates, respectively. Drees *et al.* (1992) also found the workers to be unsusceptible to nematode infection and vigorously preened nematodes from brood, alates and themselves. In a field study, *S. carpocapsae* was applied to active fire ant mounds in water suspensions. Six weeks after treatment, activity in mounds treated with nematodes ranged from 52% to 80%. Satellite mound activity accounted for 0–24% of overall activity. They concluded that soil drench applications of *S. carpocapsae* show potential as a control method for the red imported fire ant, but colony relocation after nematode treatment could limit overall efficacy unless application techniques are developed to overcome or take advantage of colony movement. Jouvenaz *et al.* (1990) and Jouvenaz and Martin (1992) evaluated commercially available *S. carpocapsae* and *S. feltiae* for the control of *So. invicta*. The earlier study indicated moderate (40–58%) queen mortalities and in separate tests they observed that colonies of ants vacated soil areas treated with high concentrations of nematodes. The latter test involved the application of nematodes to *So. invicta* nesting in nursery pots containing *Pittosporum* sp. shrubs. However, the fire ants could not be eliminated from 1-gallon pots treated with 30,000 or 300,000 nematodes (1000 or 10,000 nematodes per square inch of soil surface). A bioassay using larvae of *Galleria mellonella* as host traps indicated that the distribution of nematodes within the pots was not uniform. Morris *et al.* (1990) showed that treatment of *So. invicta* with *S. carpocapsae* gave control comparable with that obtained with amidinohydrazone over a 6-week period. Castellanos *et al.* (1997) tested the susceptibility of *So. germinata* to *H. bacteriophora* and found that the IJs were only able to infect adult males, but not the eggs, workers and young ants.

Kermarrec *et al.* (1986) discussed apparent host resistance of leaf-cutter ants (*A. octospinosus*) to *S. carpocapsae*. They observed a decline in IJ penetration into mature pupae and adult ants; also noting a decrease in nematode development. Infection and development of the nematodes occurred in 95–100% of the third and fourth stage larvae, penetration in older pupae occurred in approximately 5% of hosts and very few supported the development of the nematodes to the adult stage. When millions of IJs were introduced into the fungus gardens of the ants, intense social grooming, together with nest cleaning and building activities followed, resulting in the successful elimination of the nematodes after 10 days.

Baur *et al.* (1998) observed that ants were the most apparent invertebrate scavengers observed foraging on EPN-killed insects. Workers of the Argentine ant, *Linepithema humile*, scavenged nematode-killed insects on the surface and those buried 2 cm below the soil surface. They noted that ant workers scavenged significantly more steinernematid-killed (60–85%) than heterorhabditid-killed (10–20%) insects, and more 4-day-post-infected cadavers (hosts died within 48 h after exposure to nematodes) were scavenged than 10-day-post-infected cadavers. The 10-day-post-infected hosts contained live IJ nematodes and the authors hypothesized that the ants may serve as phoretic agents. Other ant species, including seed harvester ants *Messor andrei* (formerly *Veromessor*), *Pheidole vistana*, *Formica pacifica* and *Monomorium ergatogyna*, also scavenged nematode-killed insects. These ant species removed or destroyed about 45% of the steinernematid-killed insects. The authors concluded that survival of steinernematid nematodes may be more significantly impacted by invertebrate scavengers, especially ants, than that of heterorhabditid nematodes, and placement of steinernematid-killed insects in the field for biocontrol may be an ineffective release strategy. None of the insects killed by the symbiotic bacterium, *Photorhabdus luminescens* from *H. bacteriophora*, were scavenged, whereas 70% of the insects killed by the symbiotic bacterium, *X. nematophilus* from *S. carpocapsae*, and 90% of the insects killed by *Bacillus thuringiensis* were scavenged by the Argentine ant. Later studies, however, indicated that *X. nematophila* also deters ants (Zhou *et al.*, 2002).

Zhou *et al.* (2002) studied the response of ants to deterrent factors produced by *X. nematophila* and *P. luminescens*, by the symbiotic bacteria of the EPNs. They showed that activity is present in the supernatants of bacterial cultures, and is filterable, heat-stable, acid-sensitive and passes through a 10-kDa-pore-size membrane. Thus, the factor appears to comprise small, extracellular and possibly non-proteinaceous compound(s). They also established that the amount of repellency depends on the ant species, the sucrose concentration (*in vitro* assays), and the strain, form and age of the bacteria. They conclude that the symbiotic bacteria of some species of EPNs produce compounds that deter ants and thus protect nematodes from being eaten during reproduction within insect cadavers. Dudney (1997), later updated by Gouge *et al.* (2001), filed for use patents based on the insecticidal effect of *X. nematophila* on *S. invicta*. Further research is needed to isolate, synthesize and evaluate the potential of ant-deterrent factors for the management of ants.

As the ecology and biology of different ant species is so diverse, it is difficult to predict whether EPNs would or would not be useful in a management programme. Repeated drench applications of large doses would perhaps be more effective on species less mobile as a colony. Nematode-bait stations would, however, be a fascinating consideration, although ant species that are repelled by specific nematode species may rapidly develop bait station aversion.

17.5. Rhinotermitidae – Termites

Subterranean termites (Rhinotermitidae) are of ecological interest to scientists because they play an essential role as decomposers of dead plant material (Ebeling, 1978). However, a number of termite species are of economic importance as pests in agriculture, forestry and urban situations. Prophylactic as well as mitigation measures against termites are not always effective due to a fragmentary understanding of termite biology and ecology. A partial review by Myles (2002) lists five nematodes as pathogens of termites. Diseased termite colonies are rarely encountered in the field, although at any time even a healthy, vigorous termite colony will harbour some pathogenic organisms. However, sanitary measures within a colony, such as allogrooming, removing, entombing or feeding on cadavers and the production of antibiotics ensure that disease outbreaks are kept in check. Generally, only when colony vigour is weakened by age or chemical/environmental stress do epizootics readily develop and colonies perish from diseases. Some of the modern termiticides are even known to act synergistically with soil microorganisms to cause a more rapid decline in termite populations (Lenz, 2004).

17.5.1. *Coptotermes* termites

Some of the earliest work on termites was conducted by Reese (1971), who infected Formosan termites, *Coptotermes formosanus*, with nematodes and returned them to the colonies. Individuals in some colonies were seen to collect and isolate the parasitized termites and wall them behind earthen barriers. The overall lack of success was attributed to the nest cleaning and very large number of termites in a single colony. Fujii (1975) also describes *S. carpocapsae*-infected *C. formosanus* individuals being walled off and the strategy proved to limit colony infection. Relatively few investigations have been undertaken to establish if this is an effective strategy against the spread of steinernematid or heterorhabditid nematodes within termite colonies of different species.

In Australia, *Heterorhabditis* spp. have also been used to eliminate residual populations of subterranean *Coptotermes* sp. trapped in buildings after a perimeter chemical has been applied. IJs kill the trapped termites and can move from the site of application inside the building to the nest of the colony. The reported temperatures of above 30°C in the centre of nests of *Coptotermes* where reproductives and brood are housed prove lethal for the nematodes. Hence the impact with currently used isolates of the nematode may never go beyond killing termites in the outer parts of the nest or within the tunnel system in the soil, although in some cases apparent colony elimination has occurred (R.A. Bedding, 2004, personal communication). Nematode isolates or species that are tolerant to higher temperatures need to be evaluated for control of subterranean termites that form central compact nests such as the species of *Coptotermes*.

The susceptibility of constrained *C. formosanus* and *Reticulitermes speratus* to *S. feltiae* was investigated by Wu *et al.* (1991). They found that nematode invasion was via the mouth and all termite castes were very susceptible to the nematodes with the exception of the eggs. The susceptibility of the termites to *S. feltiae* was evaluated under an imitative artificial nest environment. Hosts were exposed to concentrations of *S. feltiae* ranging from 5000 to 1,500,000/nest. Mean nematode-associated host mortality 7 and 20 days after exposure was dose-related, ranging from 4.8% at the lowest to 97.9% at the highest dosage level. The termites could detect the nematodes and sought ways to avoid nematode attacks by extending their earthen tunnels to the outside of the nest; however, a considerable number of the termites had been parasitized before escaping.

17.5.2. *Reticulitermes* termites

Most pest species of subterranean termites in North America belong to the endemic Holoarctic genus *Reticulitermes*. *Reticulitermes* spp. are found in every state in continental USA except Alaska, but are most common in the warm and humid south-eastern region. Samarasinghe (1996) demonstrated susceptibility of *R. santonensis* and *Zootermopsis* to *Metarhizium anisopliae, Heterorhabditis* sp. and *Steinernema* sp. Mix and Beal (1985) covered areas of a heavily *Reticulitermes*-infested pine forest floor with pine boards attractive to the termites. Relatively low levels of nematodes were applied and it was noted that certain nematode species and strains were more effective against termites than others; overall however, EPNs did not look very promising. Differential susceptibility to nematode species was also noted by Poinar and Georgis (1989), who detailed the comparison of *S. carpocapsae* (All) and *H. bacteriophora* (= *heliothidis*) (NC) efficacy against *R. hesperus* in Petri dish experiments. They concluded that both nematode species caused the same level of mortality but *S. carpocapsae* killed the termites twice as quickly.

A study testing the efficacy of *S. carpocapsae* against foraging workers of *R. tibialis* in pastureland was published by Epsky and Capinera (1988). They established LD_{50} values for specific nematode–termite combinations, showing that large numbers of *S. carpocapsae* (Breton) were required ($LD_{50} \times 1.5 \times 10^4$ nematodes per *R. tibialis*) in filter paper assays. When IJs were applied directly beneath baited traps ($1 \times 10^7/m^2$) a significant decrease in the number of termites per trap resulted. The pattern was apparent for 2–3 weeks. Termites attacking traps on treated sites entered traps at a corner or from the top, suggesting that termites may be avoiding contact with the nematodes. The reinvasion potential of termites shown in this study suggests either that nematode applications should be made frequently or that the entire colony rather than only the feeding site must be treated. The researchers did report avoidance of nematodes and the exploitation of gaps in EPN barriers to access a food source.

Mauldin and Beal (1989) determined the efficacy of nematodes in preventing or eliminating eastern subterranean termites *R. flavipes* in the laboratory and infestations of *Reticulitermes* spp. in the field. Nematodes tested in a laboratory study were two strains (Breton and All) of *S. carpocapsae*, *S. feltiae* and *H. bacteriophora* (= *heliothidis*). The same nematodes were tested in field studies except that the Mexican strain of *S. carpocapsae* was used instead of the Breton strain. In the laboratory study, termites quickly moved from a nest container through a tube containing a mixture of sand, vermiculite and water to reach a chamber in which nematodes had been released. After 9.5 weeks, termite survival rates in the nematode treatments and in the untreated control did not differ significantly. In field studies, nematodes did not eliminate or control termites in a simulation of soil treatments under concrete slabs or in logs naturally infested with termites. Conversely, Poinar and Georgis (1989) found that the *Reticulitermes* spp. were susceptible (80–87% mortality) when applying high doses of nematodes and carefully maintained moisture levels in the laboratory.

Nguyen and Smart (1994) isolated a new EPN *Neosteinernema longicurvicauda* (Rhabditida: Steinernematidae) from naturally infected subterranean termites, *R. flavipes*. Adult nematodes were found outside the termite cadaver. Diagnosis of the Family Steinernematidae was emended to accommodate the new species (Nguyen and Smart, 1994), but no studies on the effectiveness of this species are available.

The infectivity of *S. carpocapsae* (Breton), *S. riobrave* (TX), *H. bacteriophora* (HP88) and *H. indica* (Coimbatore) in *R. flavipes* and *C. formosanus* was investigated by Wang *et al.* (2002). In Petri dish tests, they were all effective against *C. formosanus* at a dose of 400 nematodes/ termite. *S. riobrave* had no detectable effect against *R. flavipes* even at a rate of 2000 nematodes/termite. The highest levels of infection were recorded for *H. indica* and *H. bacteriophora* in both termite species. The investigators observed that *H. indica* repelled termites at high concentrations in sand and vermiculite medium. The length of repellency varied with the nematode concentration. Nematodes were able to reproduce in *R. flavipes* and *C. formosanus*.

17.5.3. Other termites

Poinar (1979) established the susceptibility of *Coptotermes*, *Nasutitermes* and *Termes* to *S. carpocapsae*. Bedding and Stanfield (1981) reported that large colonies of the Australian genus *Mastotermes* could be killed using *Heterorhabditis* sp. but noted that termites tended to be repelled from treated areas. Nematode-infected termites were recovered from neighbouring untreated trees within a week of treatment. The cadavers yielded approximately 10,000 IJs, and the dead termites were not walled off sufficiently to prevent exit and spread of the nematode progeny. After injections of larvae of a *Heterorhabditis* isolate from tropical Australia into eucalyptus trunks in which *Mastotermes darwiniensis* foragers were active, masses of dead termites were found. However, due to the complex biology of *M. darwiniensis*, including its diffuse nest system, the presence of multiple sets of reproductives, large territory size and simultaneous use of many feeding sites, it remained uncertain what the impact of the treatment on the colonies as a whole was (R.A. Bedding, 2004, personal communication). Georgis *et al.* (1982) reported high mortality of *Zootermopsis* and *Reticulitermes* exposed to *S. carpocapsae* and *H. bacteriophora* (= *heliothidis*) in Petri dish assays.

Using a *Heterorhabditis* isolate from Darwin, Australia, Danthanarayana (1983), and later Danthanarayana and Vitarana (1987), successfully managed live-wood tea termites. *Glyptotermes dilatatus* is the main pest of lowland-grown tea in Sri Lanka and chemical control in the 1980s was largely ineffective. The nematodes were mass-produced by CSIRO. The Australian nematode infected the termites and was able to kill and breed in the cadaver both in the laboratory (at 22°C and 100% RH) and under extreme climatic conditions in the field (mean temperature 28.3°C; temperature range 19.5–38.4°C; only 7 days low rainfall between 20 December 1981 and 28 February 1982). Within each cadaver, up to about 3500 juveniles were produced, which were able to infect healthy termites in the laboratory and in the field, creating a chain of infection leading to the eradication of the colony. The cost of control was $2.39 (Sri Lankan Rs 50.20)/1000 infected bushes (using 1987 values). Field success was attributed to the application of large numbers of nematodes (approximately 120,000–240,000 IJs/bush). Entire colonies were killed within 3 months.

Likewise, for species of the dampwood termite in the genus *Neotermes* on islands of the South Pacific, nematodes showed potential in eliminating infestations in the unbranched trunks of coconut palms, but their effectiveness was less guaranteed in branched trees of citrus, cocoa or American mahogany (*Swietenia macrophylla*) (Lenz and Runko, 1992; Lenz *et al.*, 2000). These branches allowed parts of the population to occasionally retreat and block off the connection to the main trunk, which had received injections of infective nematode larvae, thus preventing the spread of the nematodes to all areas occupied by a colony.

Amarasinghe and Hominick (1993a) found that the native Sri Lankan *Heterorhabditis* spp. and *S. carpocapsae* isolates were more effective against the live-wood termite *Postelectrotermes militaris* compared with the non-native nematodes. Amarasinghe and Hominick (1993b) demonstrated that the potential use of nematodes as biocontrol agents of termites was limited by termite behaviour, which may be overcome with the use of high nematode rates.

Two species of fungus-growing termites that destroy tropical crops were used to test the efficiency of three strains of EPNs. Rouland *et al.* (1996) studied the sensitivity of *Macrotermitinae* termite reproductives to *Heterorhabditis* and *Steinernema* spp. The alates of the two termite species showed a very high sensitivity to the different strains of nematodes tested. Later, Benmoussa-Haichour *et al.* (1998) measured the sensitivity of the different castes of higher termites to EPNs. Workers, soldiers, nymphs and alates of *Macrotermitinae* were infected with *Heterorhabditis* and *Steinernema*. The researchers found that the biochemical composition of different castes influenced the development of the nematodes. Due to the different biochemical compositions of the castes, the development of EPNs only occurs in alates.

In the USA, various pest management professionals have confirmed the efficacy of EPNs for termite control. Olkowski *et al.* (1991) reported positive data generated by pest management professionals who applied EPNs at very high rates into trenches around the foundations of termite-infested homes. Other similar reports exist based on the experiences of pest management professionals who also demonstrate good success treating termite-infested homes. Therefore, commercial companies sell nematode products for subterranean termite management.

17.6. Summary and Conclusions

Social insects in general present many challenges in the development of management strategies. They demonstrate the most spectacular evolutionary transitions: competition to cooperation and individualism to altruism between adults. Certainly, the greatest barrier to social pest population manipulation (management) is our own limited level of understanding of social insect behavioural ecology. The pest species may have complex behaviour linked with their social structure; they may have cryptic or diffuse nest sites. Relatively large doses of IJs appear to be an overall requirement for the management of social insects. Alternatively, IJs must be delivered in a bait formulation. While honeybees and bumble bees may be resistant or may effectively evade infection by EPNs, yellowjackets appear to be amenable to management by the use of nematodes. Nematodes appear to be quite effective for the management of species of ants that are less mobile as a colony. Several studies also document the effectiveness of nematodes for the control of fire ants. Successful protection of trees or crops from termites using nematodes has also been documented. However, the protection of homes and buildings against termites with nematodes is not very practical. Low economic threshold for termite damage to homes, the large number of nematodes required to kill termites, the presence of secondary reproductives and the behaviour of socially oriented termites (utilizing barrier gaps and removing infected individuals) preclude nematodes as a viable termite management option (Klein, 1990). However, further research in this area is warranted.

References

Akhurst, R.J. (1980) Morphological and functional dimorphism in *Xenorhabdus* spp., bacteria symbiotically associated with insect pathogenic nematodes *Neoaplectana* and *Heterorhabditis*. *Journal of General Microbiology* 121, 303–309.

Amarasinghe, L.D. and Hominick, W.M. (1993a) Efficacy of entomopathogenic nematodes to control up-country live-wood termite. *Sri Lanka Journal of Tea Science* 62, 16–24.

Amarasinghe, L.D. and Hominick, W.M. (1993b) Potential of using entomopathogenic nematodes to control up-country live-wood termite (*Postelectrotermes militaris*). *Sri Lanka Journal of Tea Science* 62, 66–78.

Baur, M.E., Kaya, H.K., Peng, Y.S. and Jiang, J. (1995) Nonsusceptibility of the honeybee, *Apis mellifera* (Hymenoptera: Apidae) to steinernematid and heterorhabditid nematodes. *Journal of Nematology* 27, 378–381.

Baur, M.E., Kaya, H.K. and Strong, D.R. (1998) Foraging ants as scavengers on entomopathogenic nematode-killed insects. *Biological Control* 12, 231–236.

Bedding, R.A. (1984) Nematode parasites of Hymenoptera. In: Nickle, W.R. (ed.) *Plant and Insect Nematodes.* Marcel Dekker, New York, pp. 755–795.

Bedding, R.A. and Stanfield, M. (1981) Nematodes for insect control. *Mastotermes. Annual Report.* CSIRO Division of Entomology, pp. 79–80.

Benmoussa-Haichour, D., Reversat, G. and Rouland, C. (1998) Sensitivity of the different castes of higher termites to an infestation with entomopathogenic nematodes – role of the lipids composition. *Actes des Colloques Insectes Sociaux* 11, 45–52.

Cantwell, G.E., Lehnert, T. and Fowler, J. (1972) Are biological insecticides harmful to the honey bee? *American Bee Journal* 112, 255–258.

Castellanos, L., Rodriguez, I. and Martinez, M. (1997) Susceptibility of the ant *Solenopsis germinata* (Hymenoptera: Formicidae) to the parasitism by *Heterorhabditis bacteriophora* (Rhabditida: Heterorhabditidae). *Revista de Proteccion Vegetal* 12, 159–162.

Danthanarayana, W. (1983) Control of live-wood tea termite, *Glyptotermes dilatatus* (Bugnion & Popoli) with the entomogenous nematode, *Heterorhabditis* sp. *XVth Congress of the Pacific Science Association*, Dunedin, New Zealand.

Danthanarayana, W. and Vitarana, S.I. (1987) Control of the live-wood tea termite *Glyptotermes dilatatus* using *Heterorhabditis* sp. (Nemat.). *Agricultural Ecosystems and Environment* 19, 333–342.

Drees, B.M., Miller, R.W., Vinson, S.B. and Georgis, R. (1992) Susceptibility and behavioral response of red imported fire ant (Hymenoptera: Formicidae) to selected entomogenous nematodes (Rhabditida: Steinernematidae & Heterorhabditidae). *Journal of Economic Entomology* 85, 365–370.

Drlik, T. (1994) Solving yellowjacket problems. *IPM Practitioner* 16, 19.

Dudney, R.A. (1997) Use of *Xenorhabdus nematophilus* Im/1 and 19061/1 for fire ant control. US Patent US 5616318 970401. (Application Information US 95-488820 950609.)

Dutky, S.R. and Hough, W.S. (1955) Note on a parasitic nematode from codling moth larvae, *Carpocapsa pomonella*. *Proceedings of the Entomological Society of Washington* 57, 244.

Ebeling, W. (1978) *Urban Entomology*. Division of Agricultural Sciences, University of California No. 4057.

Epsky, N. and Capinera, J.L. (1988) Efficacy of the entomogenous nematode *Steinernema feltiae* against a subterranean termite, *Reticulitermes tibialis* (Isoptera: Rhinotermitidae). *Journal of Economic Entomology* 81, 1313–1317.

Fujii, J.K. (1975) Effects of an entomogenous nematode, *Neoaplectana carpocapsae* Weiser, on the Formosan subterranean termite, *Coptotermes formosanus* Shiraki, with ecological and biological studies on *C. formosanus*. PhD thesis, University of Hawaii, Honolulu.

Gambino, P. (1984) Susceptibility of the western yellowjacket *Vespula pensylvanica* to three species of entomogenous nematodes. *International Research Communications System Medical Science: Microbiology, Parasitology and Infectious Diseases* 12, 264.

Gambino, P., Pierluisi, G.J. and Poinar, G.O. Jr (1992) Field test of the nematode *Steinernema feltiae* (Nematoda: Steinernematidae) against yellowjacket colonies (Hym.: Vespidae). *Entomophaga* 37, 107–114.

Georgis, R. (1987) Nematodes for biological control of urban insects. *Proceedings of the American Chemical Society, Division of Environmental Chemistry* 27, 816–821.

Georgis, R., Poinar, G.O. Jr and Wilson, A.P. (1982) Susceptibility of damp-wood termites and soil and wood-dwelling termites to the entomogenous nematode *Neoaplectana carpocapsae. International Research Communications System Medical Science: Microbiology, Parasitology and Infectious Diseases* 10, 563.

Gouge, D.H., Dudney, R.A. and Smith, K.A. (2001) Bacteria for insect control. US Patent. Patent Expired.

Guzman, R.F. (1984) Preliminary evaluation of the potential of *Steinernema feltiae* for controlling *Vespula germanica. New Zealand Journal of Zoology* 11, 100.

Hackett, K.J. and Poinar, G.O. Jr (1973) The ability of *Neoaplectana carpocapsae* Weiser (Steinernematidae: Rhabditida) to infect adult honeybees (*Apis mellifera*, Apidae: Hymenoptera). *American Bee Journal* 113, 100.

Jouvenaz, D.P. and Martin, W.R. (1992) Evaluation of the nematode *Steinernema carpocapsae* to control fire ants in nursery stock. *Florida Entomologist* 75, 148–151.

Jouvenaz, D.P., Lofgren, C.S. and Miller, R.W. (1990) Steinernematid nematode drenches for control of fire ants, *Solenopsis invicta*, in Florida. *Florida Entomologist* 73, 190–193.

Kaya, H.K., Marston, J.M., Lindegren, J.E. and Peng, Y.S. (1982) Low susceptibility of the honey bee, *Apis mellifera* L. (Hymenoptera: Apidae), to the entomogenous nematode, *Neoaplectana carpocapsae* Weiser. *Environmental Entomology* 11, 920–924.

Kermarrec, A. (1975) Etude des relations synécologiques entres les nématodes et la fourmi manioc: *Acromyrmex octospinosus* Reich. *Annales de Zoologie Ecologie Animale* 7, 27–44.

Kermarrec, A., Febvay, G. and Decharme, M. (1986) Protection of leaf-cutting ants from biohazards: is there a future for microbiological control? In: Lofgren, C.S. and Vander, R.K. (eds) *Fire Ants and Leaf-Cutting Ants: Biology and Management.* Westview Press, Boulder, Colorado, 247 pp.

Klein, M.G. (1990). Efficacy against soil-inhabiting insect pests. In: Gaugler, R. and Kaya, H.K. (eds) *Entomopathogenic Nematodes in Biological Control.* CRC Press, Boca Raton, Florida, 385 pp.

Laumond, C., Mauléon, H. and Kermarrec, A. (1979) New data on the host spectrum and parasitism of the entomophagous nematode, *Neoaplectana carpocapsae*. *Entomophaga* 24, 13–27.

Lenz, M. (2004) Biological control in termite management – the potential of nematodes and fungal pathogens. *Proceedings of the 35th Annual Meeting, The International Research Group on Wood Preservation.* Ljubljana, Slovenia IRG/WP 04 10521.

Lenz, M. and Runko, S. (1992) Use of microorganisms to control colonies of the Coconut Termite *Neotermes rainbowi* (Hill) on Vaitupu, Tuvalu. *Commonwealth Scientific and Industrial research Organisation, Division of Entomology, Termite Group Report* 92/16, 47 pp.

Lenz, M., Kamath, M.K., Lal, S. and Senivasa, E. (2000) Status of the tree-damaging *Neotermes* sp. In: *Fiji's American Mahogany Plantation and Preliminary Evaluation of the Use of Entomopathogens for their Control.* ACIAR Small Project No. FST/96/205, Project Report, 110 pp.

Lind, P. (1998) Solving yellowjacket problems. *Journal of Pesticide Reform* 18, 22–23.

Mauldin, J.K. and Beal, R.H. (1989) Entomogenous nematodes for control of subterranean termites, *Reticulitermes* spp. (Isoptera: Rhinotermitidae). *Journal of Economic Entomology* 82, 1638–1642.

Mix, J. and Beal, R.H. (1985) Can nematodes halt subterranean termites? *Pest Control* 53, 22–23.

Morris, J.R., Stewart, K.W. and Hassage, R.L. (1990) Use of the nematode *Steinernema carpocapsae* for control of the red imported fire ant (Hymenoptera: Formicidae). *Florida Entomologist* 73, 675–677.

Myles, T.G. (2002) Isolation of *Metarhizium anisopliae* (Deuteromycotina: Hyphomycetes) from *Reticulitermes flavipes* (Isoptera: Rhinotermitidae) with convenient methods for its culture and collection of conidia. *Sociobiology* 40, 257–262.

Nguyen, K.B. and Smart, G.C. Jr (1994) *Neosteinernema longicurvicauda* n. gen., n. sp. (Rhabditida: Steinernematidae), a parasite of the termite *Reticulitermes flavipes* (Koller). *Journal of Nematology* 26, 162–174.

Oi, D.H. and Pereira, R.M. (1993) Ant behavior and microbial pathogens. *Florida Entomologist* 76, 63–74.

Olkowski, W., Daar, S. and Olkowski, H. (1991) *Common-Sense Pest Control, Least-Toxic Solutions for Your Home, Garden, Pets and Community.* The Taunton Press, Newtown, Connecticut.

Pereira, R.M. and Stimac, J.L. (1997) Biocontrol options for urban pest ants. *Journal of Agricultural Entomology* 14, 231–248.

Peters, A. (1996) The natural host range of *Steinernema* and *Heterorhabditis* spp. and their impact on insect populations. *Biocontrol Science and Technology* 6, 389–402.

Poinar, G.O. Jr (1979) *Nematodes for Biological Control of Insects.* CRC Press, Boca Raton, Florida.

Poinar, G.O. Jr and Ennik, F. (1972) The use of *Neoaplectana carpocapsae* (Steinernematidae: Rhabditida) against adult yellowjackets (*Vespula* spp., Vespidae: Hymenoptera). *Journal of Invertebrate Pathology* 19, 331–334.

Poinar, G.O. Jr and Georgis, R. (1989) Biological control of social insects with nematodes. In: Leslie, A.R. and Metcalf, R.L. (eds) *Integrated Pest Management for Turfgrass and Ornamentals.* US Environmental Protection Agency, Washington, DC.

Poole, M.A. (1976) Survey and control efficacy of endoparasites of *Solenopsis richteri* Forel and *Solenopsis invicta* Buren in Mississippi. PhD thesis, Mississippi State University, Starkville, Mississippi.

Quattlebaum, E.D. (1980) Evaluation of fungal and nematode pathogens to control the red imported fire ant, *Solenopsis invicta* Buren. PhD thesis, Clemson University, Clemson, South Carolina.

Reese, K.M. (1971) Navy fights Formosan termite in Hawaii. *Chemical and Engineering News* 52.

Rose, E.A.F., Harris, R.J. and Glare, T.R. (1999) Possible pathogens of social wasps (Hymenoptera: Vespidae) and their potential as biological control agents. *New Zealand Journal of Zoology* 26, 179–190.

Rouland, C., Benmoussa, D., Reversat, G. and Laumond, C. (1996) Study of the sensitivity of sexual termites Macrotermitinae to an infestation with entomopathogenic nematodes of genera *Heterorhabditis* and *Steinernema. Comptes Rendus de l Academie des Sciences Serie Iii-Sciences de la Vie-Life Sciences* 319, 997–1001.

Samarasinghe, S.M. (1996) Pathogenicity of entomopathogenic agents against termites employing insecticidal bioassay with borax-toxicity method. *Sri Lanka Journal of Tea Science* 63, 78–93.

Schildknecht, H. and Koob, K. (1971) Myrmicacin, the first insect herbicide. *Angewandte Chemie* 10 (International Edition), 124–125.

Sherman, P.W., Seeley, T.D. and Reeve, H.K. (1988) Parasites, pathogens, and polyandry in social Hymenoptera. *American Naturalist* 131, 602–610.

Stock, S.P. (1996) Pathogenicity of the entomopathogenic nematode *Heterorhabditis argentinensis* (Nematoda: Heterorhabditidae) on insects associated with alfalfa crops. *Revista de la Sociedad Entomologica Argentina* 55, 175–177.

Story, G.K., Van der Meer, R.K., Boucias, D.G. and McCoy, C.W. (1991) Effect of fire ant (*Solenopsis invicta*) venom alkaloids on the *in vitro* germination and development of selected entomopathogenic fungi. *Journal of Invertebrate Pathology* 58, 88–95.

Wang, C., Powell, J.E. and Nguyen, K. (2002) Laboratory evaluations of four entomopathogenic nematodes for control of subterranean termites (Isoptera: Rhinotermitidae). *Environmental Entomology* 31, 381–387.

Wassink, H. and Poinar, G.O. Jr (1984) Use of the entomogenous nematode, *Neoaplectana carpocapsae* Weiser (Steinernematidae: Rhabditida), in Latin America. *Nematropica* 14, 97–109.

Wojcik, W.F. and Georgis, R. (1987) Infection of selected insect species with desiccated *Steinernema feltiae* (Nematoda). *Proceedings of the Annual Meeting of the Society of Invertebrate Pathology*, Gainesville, Florida, 80.

Wojcik, W.F. and Georgis, R. (1988) Infection of adult western yellowjackets with desiccated *Steinernema feltiae* (Nematoda). *Journal of Invertebrate Pathology* 52, 183–184.

Wu, H.J., Wang, Z.N., Ou, C.F., Tsai, R.S. and Chow, Y.S. (1991) Susceptibility of two Formosan termites to the entomogenous nematode *Steinernema feltiae* Filipjev. *Bulletin of the Institute of Zoology, Academia Sinica* 30, 31–39.

Zhou, X., Kaya, H.K., Heungens, K. and Goodrich-Blair, H. (2002) Response of ants to a deterrent factor(s) produced by the symbiotic bacteria of entomopathogenic nematodes. *Applied and Environmental Microbiology* 68, 6202–6209.

18 A Systems Approach to Conservation of Nematodes

M.E. Barbercheck[1] and C.W. Hoy[2]
[1]*Department of Entomology, Pennsylvania State University, University Park, PA 16802,USA;* [2]*Department of Entomology, Ohio State University, Wooster, OH 44691, USA*

18.1. Introduction

Among the major types of biocontrol practices (classical, augmentative, inoculative, conservation), conservation biocontrol, especially for insect pathogens, has been relatively understudied by researchers and underutilized by growers. Conservation biocontrol theory suggests that the efficacy of natural enemies, as measured by the control of target pests, can be enhanced through direct manipulation of the environment. If habitat is managed so that natural enemy success, e.g. survival, reproduction and foraging behaviour, is enhanced in the vicinity of the target pest, then natural enemies that colonize should be able to persist in a system and increase the likelihood of maintaining pest populations below economic thresholds (Barbosa, 1998; Letourneau and Altieri, 1999). Conservation biocontrol has the potential to be long term and cost effective (Gurr and Wratten, 1999).

The total extent of nematode–insect interactions, including non-parasitic and parasitic relationships, is not known. There are more than 30 nematode families associated with terrestrial and aquatic insects and these associations probably occur in most insect orders (Nickle, 1984). To understand how habitat can be manipulated to favour particular insect–entomopathogenic nematode (EPN) interactions, considerable knowledge about how the environment affects these relationships is needed. Although many aspects of the biology of several EPN species have been examined, relatively little is known about the ecology of this broad group of organisms. Much of our recent knowledge about nematode ecology has been produced from studies

 Nematodes as Biocontrol Agents (eds P.S. Grewal, R.-U. Ehlers and D.I. Shapiro-Ilan)

on nematodes that have been considered for, or are currently used as, biocontrol agents produced commercially for inundative application. EPNs in the families Steinernematidae and Heterorhabditidae fall within this group. Reports and studies of natural occurrence and ecology of EPNs are relatively uncommon. Information from manipulative experiments, however, can help assess the effects of environmental factors, point out gaps in our knowledge and help us understand how this important group of natural enemies of insects can be conserved (Lewis *et al.*, 1998).

18.2. Natural Recycling of Nematodes and Conservation Biocontrol

EPNs have been found in many types of natural and managed habitats (Poinar, 1979; Hominick, 2002). All EPNs possess a survival or infective stage that is not associated with an insect but rather dwells in or on soil. Soil and sediments are the home of diverse assemblages of organisms that interact in complex trophic webs (Moore *et al.*, 1988; Hawksworth, 1991; Freckman *et al.*, 1997). In surveys and field studies where soil characteristics have been measured, soil texture does not appear to be a major limiting factor for most nematodes. However, there is not an extensive field-based literature that explicitly examines texture and few that address structure. Among EPNs, heterorhabditids appear to be more restricted by soil type than do steinernematids (Hominick, 2002). In Florida citrus groves, soil type was not correlated with infection of root weevils by *Steinernema carpocapsae* (Beavers *et al.*, 1983), but suppression of root weevils by *S. riobrave* is greatest in coarse, sandy soils compared with fine-textured soils (Shapiro *et al.*, 2000; Duncan *et al.*, 2001). In no-till and conventional-till maize fields in North Carolina, no significant relationship was detected between occurrence of endemic *S. carpocapsae* and *Heterorhabditis bacteriophora* and soil organic matter, pH or soil texture (Millar and Barbercheck, 2002). In a survey of Spanish soils for EPNs, Garcia del Pino and Palomo (1996) concluded that soil moisture and temperature regimes were more important than other factors in determining the prevalence of EPNs, which were associated with soils with udic moisture regimes and cryic temperature regimes.

The widespread occurrence of entomogenous nematodes indicates that they can be significant factors in the natural regulation of some insect populations. The natural 'value' of nematodes, the service they provide in the regulation of insect pests, has not frequently been estimated but may be considerable (references in Nickle, 1984; Peters, 1996). Reported naturally occurring insect mortality from EPNs ranges from 8% to 71% (Georgis and Hague, 1981; Mrácek, 1986; Klein, 1990; Akhurst *et al.*, 1992; Raulston *et al.*, 1992; Cabanillas and Raulston, 1994; Campbell *et al.*, 1998). In a natural system in coastal California, endemic EPNs in the family Heterorhabditidae were dynamically linked with populations of a root-feeding insect and its host plant (Strong *et al.*, 1995, 1996, 1999). The EPN *H. marelatus* indirectly protects bush lupine, *Lupinus arboreus*, by killing root-feeding hepialid larvae. Lupines suffer heavy root damage and subsequent mortality from these larvae. *H. marelatus* causes high mortality of the hepialid larvae, and the spatial distribution of *H. marelatus* was correlated with the long-term fluctuation in coverage of bush lupine (Strong *et al.*, 1995).

Additional studies provide evidence for both reproduction in a pest insect and population increase over time. Richardson and Grewal (1991) demonstrated that *S. feltiae* reproduced in the sciarid fly, *Lycoriella auripila*, in mushroom houses and provided more effective control of the second than first generation larvae. Klein and Georgis (1992) observed that *H. bacteriophora* applications made in the autumn provided 60–65% control of the Japanese beetle grubs in turfgrass in Ohio, and the control increased to 95% in the following spring. Successful inoculative releases of *S. scapterisci* against the mole crickets in pastures in Florida have also been documented (Parkman and Smart, 1996).

These direct observations of EPN recycling and insect pest suppression suggest that conservation biocontrol by EPNs should be feasible. Despite the widespread occurrence of EPNs at a global scale, however, they are typically very patchy and difficult to predict both in space and time at local scales. For example, Campbell *et al.* (1995), sampling *H. bacteriophora* and *S. carpocapsae* in 1.5-m^2 turfgrass plots at random from a 1000-cell grid, found low percentages of positive samples at any given sample date, and three orders of magnitude differences in numbers of infective juveniles (IJs) in positive samples. But if population density within all samples is averaged for each sample date, and sample dates are compared over time, no seasonal pattern of abundance emerges. EPNs apparently exist as metapopulations of many dynamic patches within a site (Lewis *et al.*, 1998), but possibly with little seasonal variation for the site as a whole. Other studies (Garcia del Pino and Palomo, 1997; Efron *et al.*, 2001) have documented seasonal variation in nematode population density, however, and suggested that it was related to seasonal abundance of insect hosts. Despite this positive correlation with insect hosts over a season, correlation between nematodes and hosts at a particular time during the season at sampled sites can be negative. Campbell *et al.* (1998), for example, found a negative correlation within a sampling transect between *H. bacteriophora* and *Popillia japonica* density, and suggested that this pattern resulted from suppression of the scarab larvae. There is little information available in the literature that directly links nematodes with host population dynamics (but see Strong *et al.*, 1995, 1996, 1999), and this linkage may need to be modelled explicitly given the complex field observations of nematode population dynamics.

18.3. Systems Analysis of Population Management

EPN population densities in unmanaged systems are observed consistently to vary in both space and time. For effective biocontrol, however, population density would need to be managed to provide sufficient control when and where insect pests occur. Habitat and population management can be designed, and the level of effort required and expected benefit in terms of pest suppression can be predicted only if EPN population dynamics can be quantitatively described and predicted, with the impact of habitat manipulation and management actions included in that description. We focus first on previous mathematical modelling studies of EPN populations and suggest a general model of spatial and temporal dynamics. We then place available ecological data into the context of mathematical models of nematode population dynamics. Finally, we suggest how such models could be developed and refined to provide the quantitative understanding of EPN population ecology needed to design a conservation biocontrol approach.

18.3.1. Previous modelling studies

Some of the first mathematical modelling of EPN population dynamics explicitly included spatial effects to explore ways to improve inundative applications for control of black vine weevil, *Otiorhynchus sulcatus* F. (Van der Werf *et al.*, 1995; Westerman and Van der Werf, 1998). Vertical movement of nematodes over time applied to a sand-column surface was simulated. The model included rates of movement and aggregation near hosts and penetration, and was calibrated with these rates measured for different heterorhabditid strains at different temperatures and for different insect hosts. Simulation results suggested that rates of vertical movement had little effect on control whereas rates of aggregation near hosts and penetration had large and moderate effects on control, respectively. Because genetic variation in aggregation was judged to be lacking but proportion of infectious nematodes could be improved through storage and handling techniques, the latter was suggested as a viable means of enhancing black vine weevil control. Field observations of vertical distributions of *S. carpocapsae* and *H. bacteriophora* (Ferguson *et al.*, 1995; Campbell *et al.*, 1996) have matched well with the observed foraging strategies of these species and might be predicted with such process-oriented models. However, Garcia del Pino and Palomo (1997) found significant seasonal differences in the vertical distribution of both *H. bacteriophora* and *S. feltiae*, and attributed these to seasonal variation in soil temperature and moisture conditions. Therefore, further refinement of models of vertical distribution in response to the soil environment may be required to account for the impact of changing soil moisture in field situations.

Horizontal distributions of EPNs have also been observed in field studies and have been mathematically described with statistical distributions of abundance in random samples. Within a site, nematodes often have a patchy or aggregated distribution (Cabanillas and Raulston, 1994; Stuart and Gaugler, 1994; Spiridonov and Voronov, 1995; Campbell *et al.*, 1996; Campbell *et al.*, 1998; Lewis *et al.*, 1998). This could be a function of either the spatial or the temporal availability of hosts, the number of progeny produced, nematode dispersal behaviour, establishment and persistence of EPNs under different local conditions, local extinctions and re-introductions, natural enemies or competition from other nematodes or insect pathogens (Mrácek, 1980; Timper *et al.*, 1988, 1991; Barbercheck and Kaya, 1990, 1991a,b; Ishibashi and Kondo, 1990; Timper and Kaya, 1992; Stuart and Gaugler, 1994; Rosenheim *et al.*, 1995; Strong *et al.*, 1995, 1996; Baur *et al.*, 1998; Campbell *et al.*, 1998). Naturally occurring heterorhabditid and steinernematid nematodes are aggregated at scales of less than a metre (Cabanillas and Raulston, 1994; Spiridonov and Voronov, 1995; Efron *et al.*, 2001), representing the results of a recent insect infection and a low-level background population density of juveniles maintained over longer periods in a non-infectious state with some future potential of entering an infectious state. The statistical models describing aggregation, however, provide only spatial information at a single point or averaged over a discrete period of time. Models of temporal dynamics require a rather different formulation.

The first detailed models that described and predicted changes in EPN populations over multiple generations were developed to describe the interaction between an EPN population and a single insect species in a spatially and temporally uniform environment. Following these general analytical models describing host–parasite interactions (Anderson and May, 1981; Bowers *et al.*, 1993), Fenton *et al.* (2000) developed a model to explore the use of EPNs as biocontrol agents for sciarid flies in mushroom crops. The environment in this model was assumed to be spatially and temporally homogeneous, and this is reasonable for the controlled environment monoculture it describes, particularly when compared with other cropping systems. The model focused on three populations: nematodes in susceptible hosts; in infected cadavers; and in the soil as free-living juveniles. Using parameter estimates from the literature for rates of host infection, nematode and host reproduction, and nematode mortality, conditions were sought under which steinernematid and heterorhabditid nematodes could regulate a host population at a stable equilibrium. Long-term persistence of the host and nematode population, i.e. over many years in the model output, was predicted to be unlikely. Using realistic parameter values, the model predicts cyclic behaviour in both nematode and host populations, with a high probability of extinction for one or the other. This initial generalized model, however, indicated that further research is needed on the influence of alternate hosts and spatial heterogeneity on stability. Subsequent studies (Fenton *et al.*, 2001; Fenton *et al.*, 2002) extended the model by adding larval and adult stages for the insect host and focusing the analysis on the shorter-term dynamics typical of greenhouse crops. Optimum strategies for inundative application of nematodes for control of sciarid flies in mushroom production were explored with the model and then compared with data from commercial mushroom houses to predict, validate and explain the best application strategy for *S. feltiae*.

18.3.2. A generic model

In most field situations, the environment is both spatially and temporally heterogeneous, and nematode population density is correspondingly patchy both spatially and temporally, depending on the scale considered. Mathematical models for this typical field situation will require both biologically detailed models of EPN temporal dynamics over multiple generations, as in the work of Fenton *et al.* (2000, 2001, 2002), and a way of handling spatial variation, as in the work of Westerman and Van der Werf (1998). A spatially explicit population model that can simultaneously describe changes in population size over time and space would require 'replicate' population models that describe population change for each of a number of different locations over time.

Spatially explicit models are required to examine such impacts as the influence of patterns of environmental heterogeneity and dispersal on population dynamics. A simplified diagrammatic form of such a model for EPN populations is given in Fig. 18.1. State variables, those representing

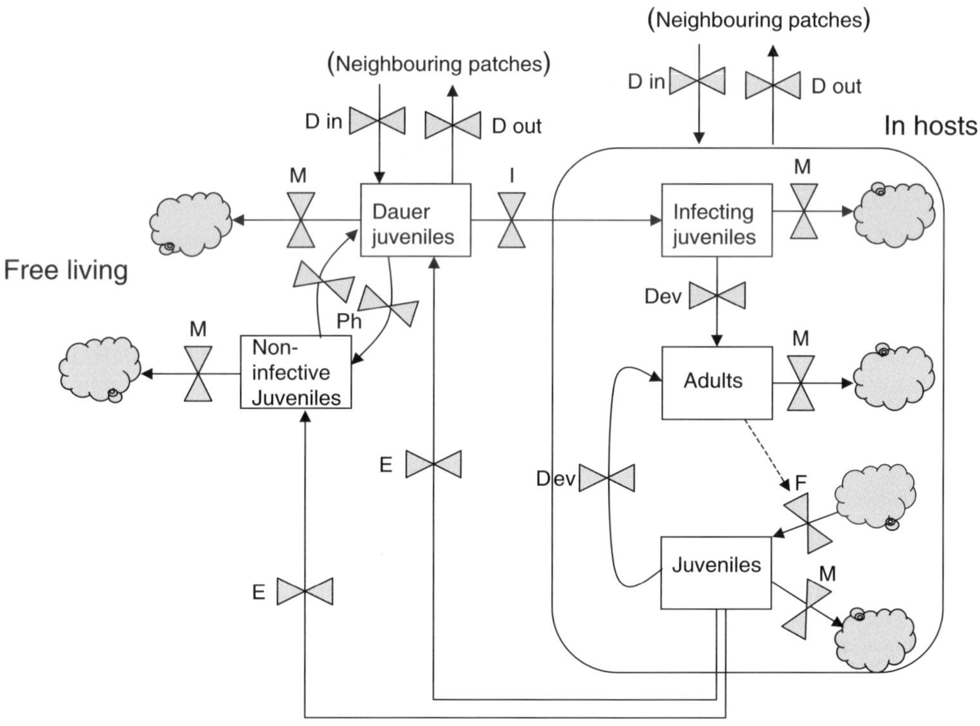

Fig. 18.1. A generalized model of entomopathogenic nemotode (EPN) population dynamics. Rates of change in the numbers of nematodes in various life stages include dispersal (D in, D out), development (Dev), reproduction or adult fecundity (F), emergence (E), infection (I), phasing from infective to non-infective juveniles (Ph) and mortality (M). The rates of change are influenced by both abiotic and biotic factors as described in this chapter.

the state of the system and that interact and change over time, are the various life stages of the nematode population both inside hosts and in the soil. Their changes over time are described by a combination of rates including infection, development, reproduction, emergence, mortality and dispersal. These changes would need to be modelled for each location being described, typically a two-dimensional (horizontal distribution) or three-dimensional (horizontal and vertical distributions) grid, with grid cell size and number chosen strategically according to the objectives of the analysis and computing capabilities. As discussed above, grid cell size of less than a metre may be required to describe the spatial component of EPN population dynamics, placing constraints on the extent of the area described. Furthermore, the spatial pattern of conditions relevant to pathogenic nematodes may have to be represented accurately to result in accurate representation of spatial dynamics (Söndergrath and Schröder, 2002). An accurate accounting of conditions associated with each cell of a grid representing the nematode's habitat would be required so that the spatial pattern of these conditions reflects the field environment.

18.3.3. Rate equations and parameter values from the literature

Mathematical expressions for the rate equations and their parameter values, as well as initial values for the state variables, are required to progress from the diagrammatic model in Fig. 18.1 to a mathematical model that could describe or predict dynamics of a population. Much of the biological detail needed to suggest the form and even parameter values for these rate equations is available from past research. Although this literature has been briefly reviewed with reference to particular rate functions (Fenton *et al.*, 2000), we highlight some of the literature that most directly relates to the rates needed for a spatially explicit mathematical model of EPN population dynamics.

18.3.3.1. Infection rates

Infection rates for at least some EPNs are complicated by a dynamic division of the population into infectious and non-infectious subpopulations (Fan and Hominick, 1991), as depicted in Fig. 18.1. Only a fraction of the IJ population appears to be infectious at any given time, and this proportion is dynamic over time. Because IJs exist as aggregated groups, phased infectivity may reduce intraspecific competition and deleterious density-dependent and genetic effects (Fan and Hominick, 1991; Selvan *et al.*, 1993; Stuart and Gaugler, 1994; Bohan and Hominick, 1995). The proportion likely to infect changes over time is due to a combination of intrinsic factors and cues from infected hosts (Fairbairn *et al.*, 2000) and the changes over time can be described well by an infection model (Bohan and Hominick, 1996, 1997). If the proportion of IJs depends on host density, however, using such functions in a model may require incorporating detailed models of the insect host population in addition to the nematode population. Probably the most important biotic factor influencing the occurrence and persistence of a nematode at a location is the presence of suitable insect hosts (Bednarek and Mrácek, 1986; Mrácek and Webster, 1993; Peters, 1996; Mrácek *et al.*, 1999). When EPN populations interact with entire host communities, the variation in infection rates for different host insects in the community would need to be considered (Gouge *et al.*, 1999).

Abiotic conditions can also influence infection rates. For example, virulence increased with soil moisture content for *H. bacteriophora*, *S. glaseri*, *S. feltiae* and *S. carpocapsae* in laboratory experiments (Grant and Villani, 2003). Hudson and Nguyen (1989) tested the infectivity of *S. scapterisci* to the mole crickets, *Scapteriscus vicinus* and *Scapteriscus acletus*, under a variety of conditions in the laboratory and found that soil moisture in the range of 5–15% had no effect on infection of the mole crickets. Infection of *Galleria mellonella* by *H. bacteriophora* and *S. glaseri* showed no significant variation in relation to bulk density, but infection rates for *S. carpocapsae* increased (Portillo-Aguilar *et al.*, 1999). Rates of movement and infection by the nematodes were strongly correlated with the amount of soil pore space having dimensions similar to or greater than the diameters of the nematodes. Temperature influences infection rates as well with an optimum that varies by nematode species or strain (Grewal *et al.*, 1994a,b; Mason and Hominick, 1995; Henneberry *et al.*, 1996; Gouge *et al.*, 1999).

18.3.3.2. Development and reproductive rates

The process of development within *G. mellonella* has been described in some detail (Mason and Hominick, 1995; Wang and Bedding, 1996) and density-dependent effects on this process have also been described (Selvan *et al.*, 1993). Development rates in the much more varied host communities typical in the field are not well described and limit our ability to predict population dynamics. Likewise, reproductive rates for EPNs have been described as a function of temperature in *G. mellonella* (Mason and Hominick, 1995; Gouge *et al.*, 1999), but not in the wide range of insect species that could be infected in the field, a deficit in the available literature on naturally cycling EPN populations.

18.3.3.3. Emergence rates

EPNs may survive inhospitable humidity conditions in the soil within an infected host (Brown and Gaugler, 1997; Koppenhöfer *et al.*, 1997; Brown *et al.*, 2002). Brown and Gaugler (1997) found that EPN IJs survived adverse environmental conditions by remaining in the host cadaver for up to 50 days. Survival varied among the species and was dependent on the environmental conditions to which the cadaver was exposed. In *G. mellonella* infected at 25°C and then exposed to RH of 75%, 85%, 96% or 100%, significantly fewer *H. bacteriophora* and *S. feltiae* emerged at low RHs, but RH did not affect emergence of *S. glaseri* or *S. carpocapsae* (Brown and Gaugler, 1997). At 25°C, all IJs of *S. carpocapsae* died by day 28 in insects infected with *S. carpocapsae* at 75% RH. The authors concluded that IJs of EPNs can survive adverse environmental conditions for limited periods in the host cadaver, but low temperatures and RHs prevent emergence and the IJs eventually die (Brown and Gaugler, 1997). Nematodes may also survive in living insects with a latent infection. Low temperature induced infection latency in *G. mellonella* exposed to *S. carpocapsae* and *S. riobrave*. Host death was delayed until optimum temperature conditions resumed (Brown *et al.*, 2002). Again, similar studies on a range of insect hosts that may be infected in the field would be needed to quantify survival and emergence rates for a naturally occurring population.

18.3.3.4. Mortality rates

Given the importance in evaluating constraints on inundative applications of EPNs, mortality rates of EPNs have been relatively well studied. We focus here on biotic interactions affecting the life stage spent outside the host, although there is evidence that predation on nematode-infected insects can contribute considerably to mortality of EPNs (Rosenheim *et al.*, 1995; Baur *et al.*, 1998). Estimates of mortality rates in a variety of laboratory and field settings have been reviewed recently, presented in terms of half-lives for a nematode population (Baur and Kaya, 2001; Strong, 2002) or exponential rates of decline in populations without access to hosts (Fenton *et al.*, 2000), which simply assume that a constant proportion of the population dies at any given time due to the combination of all factors influencing mortality. Half-lives can be used to calculate exponential rates of decline, and vice versa. Mortality rates may have to be simulated as a function of changing biotic and abiotic environmental conditions to accurately reflect a given field situation, or permit thorough analysis of how these conditions can best be manipulated to enhance biocontrol.

Predation and parasitism may play an important role in regulating populations of

EPNs in soil. The widespread distribution of nematophagous fungi, bacteria, nematodes, mites, collembolans and other microarthropods (e.g. symphylans, diplurans, centipedes) in soil, their abundance and the high rates of predation observed for some species in the laboratory suggests that these organisms may have a considerable impact on nematodes in the natural environment (Stirling, 1991). Even specialist nematophagous invertebrates will attack a variety of nematode prey (Muraoke and Ishibashi, 1976; Walter *et al.*, 1987; Chapter 26, this volume). Under laboratory conditions, omnivorous and nematophagous predators can be voracious feeders (Gilmore and Raffensperger, 1970). The capacity of a predator to exert a regulatory effect on a population of nematodes is determined partly by their ability to increase their population level and/or predation rate as prey density increases. Many nematophagous organisms have rapid development times and high reproductive capacities, and many species, e.g. predatory mesostigmatid mites, exhibit at least some degree of specificity towards nematodes, and many are capable of reproducing rapidly by parthenogenesis (Walter *et al.*, 1987; Walter, 1988a,b). The potential impact of natural enemies of nematodes has generally been assessed in observation chambers or in pots of sterilized soil, but there is little documented evidence that activity observed in these simple systems is highly correlated with effects in the field (Gilmore and Raffensperger, 1970; Epsky *et al.*, 1988; Gilmore and Potter, 1993).

Temperature-dependent mortality rates (Kung *et al.*, 1991) have been described in sufficient detail to estimate parameters for these rate functions at least for some nematode species and strains and model insect hosts. These rates would need to be adjusted, however, for species and strain, as well as the rates of change in conditions like humidity and soil moisture (see reviews by Gaugler and Kaya, 1990; Kaya and Gaugler, 1993; Baur and Kaya, 2001; Glazer, 2002; Shapiro-Ilan *et al.*, 2002), pH (Kung *et al.*, 1990b; Ghally, 1995), oxygen conditions, e.g. in waterlogged soils (Kung *et al.*, 1990b; Qiu and Bedding, 1999) and salinity of the soil solution (Thurston *et al.*, 1994).

18.3.3.5. Movement rates

Considerable progress has been made in quantifying dispersal of entomogenous nematodes. All soil nematodes can actively disperse (Moyle and Kaya, 1981; Georgis and Poinar 1983a,b,c; Mannion and Jansson, 1992; Cabanillas and Raulston, 1994; Parkman and Smart, 1996; Hsiao and All, 1998), be moved passively on or in an organism (Epsky *et al.*, 1988; Timper *et al.*, 1988; Shapiro *et al.*, 1995) or with soil or water. Active dispersal may be affected by the search behaviour of the infective or dispersal stages (Lewis *et al.*, 1992, 1993; Grewal *et al.*, 1994a). Different steinernematid and heterorhabditid nematode species find and infect insects at a variety of horizontal and vertical distances from the point of application to soil in laboratory assays (Moyle and Kaya, 1981; Schroeder and Beavers, 1987; Alatorre-Rosas and Kaya, 1990; Barbercheck and Kaya, 1991a,b; Schroeder *et al.*, 1993; Koppenhöfer *et al.*, 1995). These laboratory estimates of dispersal may reflect rates of dispersal in the field in similar soil and temperature conditions. Quantifying rates of dispersal through phoresy or movement of infected hosts would be context-dependent, a function of the host community and its behavioural characteristics. Rates of passive dispersal in surface water flow or runoff currently are not well described, but if these rates could be manipulated by irrigation then some control over nematode distributions in the field may be possible. Movement within the soil profile in response to moisture conditions (Westerman, 1998; Gouge *et al.*, 2000) may also need to be considered if vertical distribution is a key to successful biocontrol and, accordingly, part of the model. In laboratory studies, nematodes are differentially affected by soil texture and structure (Georgis and Poinar, 1983a,b,c; Kung *et al.*, 1990a; Barbercheck and Kaya, 1991a; Barbercheck, 1993). In general, movement is more restricted in soils with restrictive pore space (heavy or poorly structured soils) than in soils with a more porous structure (Portillo-Aguilar *et al.*, 1999).

18.3.4. Incorporating and analysing management options

Biotic interactions and their mediation by physical factors are critical to conservation biocontrol. The choice of practices that favour EPNs and soil biodiversity in general may also in turn favour their natural enemies (Sayre and Walter, 1991; Stirling, 1991; Bellows, 1999). In fact, in laboratory and glasshouse experiments, nematodes that appear to give effective control of soil-dwelling pests in non-soil media or soil depauperate in biota often show lower efficacy in native soil (Ishibashi and Kondo, 1986; Timper *et al.*, 1991; Timper and Kaya, 1992). A goal of conservation biocontrol is to identify the type of biodiversity that is needed to maintain and/or enhance biocontrol. Therefore, the choice of production practices that can improve the contribution of endemic nematodes as insect control agents needs to be selected carefully and based on a careful analysis of the complex interactions described above. In the following sections we review what is known about how various management practices could influence the rates describing nematode population dynamics, a necessary step for using models of nematode population dynamics to choose a combination of practices that conserves EPN populations and improves biocontrol.

18.3.4.1. Physical disturbance and ground cover in managed systems

Lack of physical disturbance (stability) and favourable soil conditions (adequate moisture, aeration, structure) have favoured successful use of some inundatively applied EPNs (Shapiro-Ilan *et al.*, 2002). Under a conventional tillage regime the soil surface tends to have greater fluctuations in temperature and moisture than under no-till or reduced tillage management, and EPNs are often more frequently detected in reduced tillage regimes (Brust, 1991; Hsiao and All, 1998; Shapiro *et al.*, 1999b; Hummel *et al.*, 2002; Millar and Barbercheck, 2002). Brust (1991) suggested that no-till may create a diverse environment that allows alternate hosts and soil conditions conducive to nematode survival and recycling. Surface residues (Shapiro *et al.*, 1999b) could benefit nematode persistence through protection from desiccation or ultraviolet (UV) radiation, increasing insect pest suppression by EPNs (Shapiro *et al.*, 1999a), or enhance nematode movement (Hsiao and All, 1998).

18.3.4.2. Fertilizers – synthetic and organic

The application of fertilizers to soil represents a nutrient disturbance that can have profound direct and indirect effects on the abundance and community composition of soil biota (reviewed in Verhoef and Brussard, 1990; Neher and Barbercheck, 1999). Prolonged (10- to 20-day) exposure to high inorganic fertilizer concentrations inhibited EPN infectivity and reproduction, whereas short (1-day) exposures increased infectivity (Bednarek and Gaugler, 1997). Application of lime or magnesium fertilizers that raise the soil pH may also increase the activity of EPNs (Jaworska, 1993). Additions of organic matter effectively change such soil factors as bulk density, porosity and moisture-holding capacity, and can increase the diversity of organisms in the soil. Even though organic amendments have been used successfully to create phytopathogen-suppressive soils, almost no documentation exists on the effects of these amendments on populations of EPNs. In field experiments, organic manure used as fertilizer has both increased (Bednarek and Gaugler, 1997) and decreased (Hsiao and All, 1997; Shapiro *et al.*, 1999a) nematode establishment and recycling.

18.3.4.3. Pesticides

Compatibility of EPNs with agricultural chemicals has been reviewed (see Chapter 20, this volume) and tolerance to specific insecticides is variable. Several commonly used insecticides, fungicides, herbicides, miticides and synthetic fertilizers are not detrimental to EPNs and can be applied as a tank mix, but nematicides, e.g. fenamiphos, are generally not compatible with the application of EPNs. These results are from laboratory studies on laboratory-reared nematodes, however, and effects of pesticides applied to endemic EPNs have generally not been tested.

18.3.4.4. Crop species and variety

Crop varieties directly affect the soil abiotic environment, e.g. soil temperature and moisture through shading, and the biotic environment through the provision of particular insect hosts associated with the crop. Root density in a system can affect the ability of a nematode to find an insect host (Choo and Kaya, 1991). The efficacy of natural enemies of herbivorous insects can often be related to plant secondary chemistry, and this phenomenon has been demonstrated for several insect pathogen groups, including EPNs (Barbercheck, 1993; Epsky and Capinera, 1994; Barbercheck *et al.*, 1995; Grewal *et al.*, 1995).

18.4. Conclusion and Future Direction

In most habitats, we expect to find well-adapted native nematodes. Field surveys and multivariate analysis can help with postulating habitat characteristics that are associated with naturally occurring endemic EPNs. Empirical work will be needed, however, to explore ways to manipulate habitats for improved biocontrol. Field studies that document and evaluate naturally occurring biocontrol are critical because the knowledge gained also can be relevant to applied biocontrol and pest management. Our ability to develop successful biocontrol programmes will be enhanced by field studies that address the complexity of trophic interactions in agricultural systems (Cohen *et al.*, 1993). For habitat manipulation to be a viable control option, basic research on several key areas needs to be addressed in a systems framework.

Studies that are specifically designed to determine rate functions and estimate their parameters will be required to explain and predict EPN population dynamics in any given ecosystem. Although the conditions that influence the rates are generally known from existing literature, the response of EPN species and strains typically has not been measured over a sufficient range of conditions to estimate rate function parameters. Because these functions depend on the species or strain and habitat they tend to be context-dependent and need to be estimated for the specific system being studied.

The dynamics of propensity towards infection by dauer juveniles (DJ) and its relationship to longevity in the soil are not well understood and could play a very substantial role in observable population dynamics, particularly when observation of nematode populations requires infection of bait insects. From the available literature, there appears to be a trade-off between the two traits, both of which could vary over time for an individual as well as for a population. The gene by environment interaction that influences this trade-off and the resulting observable infectious population needs to be better researched and understood to accurately model population dynamics. New methodology may also be required to study highly resilient but non-infectious nematodes in their natural habitat.

The host range and food web relationships for naturally occurring EPNs are very poorly understood. Even naturally occurring epizootics in well-known host species are observed very infrequently, because observation is quite difficult. The food web is likely to be variable and highly dependent on habitat, which would influence the insect community. Again, this research can be exceedingly difficult and labour-intensive, because of low and variable population densities of both nematodes and insect hosts, as well as the difficulty of sampling both in the soil and over time, and sufficiently context-dependent to be of relatively modest scientific value. But such information is absolutely critical for describing and predicting the population dynamics of naturally occurring EPN populations. If overall biological diversity of a system is increased through reduction of the frequency or intensity of disturbance, what are the outcomes for infection and reproductive rates of EPN populations? If communities of non-pest alternate insect hosts are needed to sustain nematode populations over time, how can such communities be supported without also supporting more pests? If different nematode species occupy the same habitat, what are the effects of competition on pest management? These questions cannot be satisfactorily addressed without understanding EPN hosts at the community level.

Studies of nematode movement have largely been restricted to the laboratory (although see Wilson *et al.*, 2003) and have not permitted a landscape point of view. Numerous studies have been conducted to determine the effects of sown and natural edge habitats on survival of natural enemies and their movement into adjacent crops, a key feature of conservation biocontrol for foliar pests. Can 'refugia' in uncultivated borders serve as a donor of either nematodes or hosts for neighbouring cultivated areas? If nematodes can diffuse out of a refuge, then what are the dynamics of this movement and how can it be manipulated? Is there movement from more distant non-crop areas, as with aerial biocontrol agents? If so, what is the extent and dynamics of immigration? What factors limit immigration of EPNs into a field? Are some species competitive dominants while others are disturbance-tolerant or -adapted? These questions have not been addressed largely because they are not relevant to inundative applications, but they are important questions with regard to conservation biocontrol.

Systems analysis can be used to elucidate patterns, understand complex relationships and better describe the spatial and temporal variation in EPN populations in a complex environment. Mathematical models of a system can be used as complex hypotheses that focus on particular aspects of nematode ecology in a field setting. This chapter has highlighted the complexity of the soil food web and the many factors influencing the role of EPNs in the soil ecosystem. Mathematical models need not incorporate all of the biological detail described so far about EPNs into mathematical form to conduct a useful analysis of the system. The level of biological detail captured in each of the rate equations in a model is dictated by the hypotheses of the modeller and the objective of the analysis. Focus on a particular aspect of nematode ecology could result in a very mechanistic, detailed and process-oriented treatment of some rate equations, and a more descriptive and less mechanistically detailed treatment of others. The point is to generate a prediction based on a hypothesis for some functional aspect of population dynamics that can be compared with observations of nematode biology and ecology in field data. This comparison between model prediction and observation leads to the explanation of EPN population dynamics needed to design conservation biocontrol approaches. The systems approach has only recently been applied with regard to EPNs, and provides an excellent opportunity for progress in understanding EPN ecology. Biological detail can be added as it becomes available. The only real constraint to the systems approach in nematode ecology now is the number of scientists using it.

References

Akhurst, R.J., Bedding, R.A., Bull, R.M. and Smith, D.R.J. (1992) An epizootic of *Heterorhabditis* spp. (Heterorhabditidae: Nematoda) in sugar cane Scarabaeidae (Coleoptera). *Fundamental and Applied Nematology* 15, 71–73.

Alatorre-Rosas, R. and Kaya, H.K. (1990) Interspecific competition between entomopathogenic nematodes in the genera *Heterorhabditis* and *Steinernema* for an insect host in sand. *Journal of Invertebrate Pathology* 55, 179–188.

Anderson, R.M. and May, R.M. (1981) The population dynamics of microparasites and their invertebrate hosts. *Philosophical Transactions of the Royal Society of London, Series B*, 291, 451–524.

Barbercheck, M.E. (1993) Tritrophic level effects on entomopathogenic nematodes. *Environmental Entomology* 22, 1166–1171.

Barbercheck, M.E. and Kaya, H.K. (1990) Interactions between *Beauveria bassiana* and the entomogenous nematodes, *Steinernema feltiae* and *Heterorhabditis heliothidis*. *Journal of Invertebrate Pathology* 55, 225–234.

Barbercheck, M.E. and Kaya, H.K. (1991a) Effect of host condition and soil texture on host finding by the entomogenous nematodes *Heterorhabditis bacteriophora* (Rhabditida: Heterorhabditidae) and *Steinernema carpocapsae* (Rhabditida: Steinernematidae). *Environmental Entomology* 20, 582–589.

Barbercheck, M.E. and Kaya, H.K. (1991b) Competitive interactions between entomopathogenic nematodes and *Beauveria bassiana* (Deuteromycotina: Hyphomycetes) in soilborne larvae of *Spodoptera exigua* (Lepidoptera: Noctuidae). *Environmental Entomology* 20, 707–712.

Barbercheck, M.E., Wang, J. and Hirsh, I.S. (1995) Host plant effects on entomopathogenic nematodes. *Journal of Invertebrate Pathology* 66, 169–177.

Barbosa, P. (1998) *Conservation Biological Control.* Academic Press, San Diego, California, 396 pp.

Baur, M.E. and Kaya, H.K. (2001) Persistence of entomopathogenic nematodes. In: Baur, M.E., Kaya, H.K. and Fuxa, J. (eds) *Environmental Persistence of Entomopathogens and Nematodes.* Southern Cooperative Series Bulletin 398, Oklahoma Agricultural Experiment Station, Stillwater, Oklahoma, pp. 47.

Baur, M.E., Kaya, H.K. and Strong, D.R. (1998) Foraging ants as scavengers on entomopathogenic nematode-killed insects. *Biological Control* 12, 231–236.

Beavers, J.B., McCoy, C.W. and Kaplan, D.T. (1983) Natural enemies of subterranean *Diaprepes abbreviatus* (Coleoptera: Curculionidae) larvae in Florida. *Environmental Entomology* 12, 840–843.

Bednarek, A. and Gaugler, R. (1997) Compatibility of soil amendments with entomopathogenic nematodes. *Journal of Nematology* 29, 220–227.

Bednarek, A. and Mrácek, Z. (1986) The incidence of nematodes of the family Steinernematidae in *Cephalcia falleni* Dalm. (Hymenoptera: Pamphilidae) habitat after an outbreak of the pest. *Journal of Applied Entomology-Zeitschrift fur Angewandte Entomologie* 102, 527–530.

Bellows, T.S. (1999) Controlling soil-borne plant pathogens. In: Bellows, T.S. and Fisher, T.W. (eds) *Handbook of Biological Control.* Academic Press, San Diego, California, pp. 699–711.

Bohan, D.A. and Hominick, W.M. (1995) Intrapopulation infectious structure and temporal variation in *Steinernema feltiae*. In: Griffin, C.T., Gwynn, R.L. and Masson, J.P. (eds) *Ecology and Transmission Strategies of Entomopathogenic Nematodes.* European Commission, Luxembourg, pp. 83–94.

Bohan, D.A. and Hominick, W.M. (1996) Investigations on the presence of an infectious proportion amongst populations of *Steinernema feltiae* (site 76 strain) infective stages. *Parasitology* 112, 113–118.

Bohan, D.A. and Hominick, W.M. (1997) Long-term dynamics of infectiousness within the infective-state pool of the entomopathogenic nematode *Steinernema feltiae* (site 76 strain) Filipjev. *Parasitology* 114, 301–308.

Bowers, R.G., Begon, M. and Hodgkinson, D.E. (1993) Host-pathogen population cycles in forest insects – lessons from simple models reconsidered. *Oikos* 67, 529–538.

Brown, I.M. and Gaugler, R. (1997) Temperature and humidity influence, emergence and survival of entomopathogenic nematodes. *Nematologica* 43, 363–375.

Brown, I.M., Lovett, B.J., Grewal, P.S. and Gaugler, R. (2002) Latent infection: a low temperature survival strategy in steinernematid nematodes. *Journal of Thermal Biology* 27, 531–539.

Brust, G.E. (1991) Augmentation of an endemic entomogenous nematode by agroecosystem manipulation for the control of a soil pest. *Agriculture, Ecosystems and Environment* 36, 175–184.

Cabanillas, H.E. and Raulston, J.R. (1994) Evaluation of the spatial pattern of *Steinernema riobravis* in corn plots. *Journal of Nematology* 26, 25–31.

Campbell, J.F., Lewis, E.E., Yoder, F. and Gaugler, R. (1995) Entomopathogenic nematode (Heterorhabditidae and Steinernematidae) seasonal population dynamics and impact on insect populations in turfgrass. *Biological Control* 5, 598–606.

Campbell, J.F., Lewis, E.E., Yoder, F. and Gaugler, R. (1996) Entomopathogenic nematode (Heterorhabditidae and Steinernematidae) spatial distribution in turfgrass. *Parasitology* 113, 473–482.

Campbell, J.F., Orza, G., Yoder, F., Lewis, E.E. and Gaugler, R. (1998) Spatial and temporal distribution of endemic and released entomopathogenic nematode populations in turfgrass. *Entomologia Experimentalis et Applicata* 86, 1–11.

Choo, H.Y. and Kaya, H.K. (1991) Influence of soil texture and presence of roots on host finding by *Heterorhabditis bacteriophora. Journal of Invertebrate Pathology* 58, 279–280.

Cohen, J.E., Beaver, R.A., Cousins, S.H., De Angelis, D.L., Goldwasser, L., Heong, K.L., Holt, R.D., Kohn, A.J., Lawton, J.H., Martinez, N., O'Malley, R., Page, L.M., Patten, B.C., Pimm, S.L., Polis, G.A., Rejmanek, M., Schoener, T.W., Schoenly, K., Sprules, W.G., Teal, J.M., Ulanowicz, R.E., Warren, P.H., Wilbur, H.M. and Yodzis, P. (1993) Improving food webs. *Ecology* 74, 252–258.

Duncan, L.W., Genta, J.G., Zellers, J., Fares, A. and Stansly, P.A. (2001) Efficacy of *Steinernema riobrave* against larvae of *Diaprepes abbreviatus* in Florida soils of different texture. *Nematropica* 31, 130.

Efron, D., Nestel, D. and Glazer, I. (2001) Spatial analysis of entomopathogenic nematodes and insect hosts in a citrus grove in a semi-arid region in Israel. *Environmental Entomology* 30, 254–261.

Epsky, N.D. and Capinera, J.L. (1994) Influence of herbivore diet on the pathogenesis of *Steinernema carpocapsae* (Nematoda: Steinernematidae). *Environmental Entomology* 23, 487–491.

Epsky, N.D., Walter, D.E. and Capinera, J.L. (1988) Potential role of nematophagous microarthropods as biotic mortality factors of entomopathogenic nematodes (Rhabditida: Steinernematidae and Heterorhabditidae). *Journal of Economic Entomology* 81, 821–825.

Fairbairn, J.P., Fenton, A., Norman, R. and Hudson, P.J. (2000) Re-assessing the infection strategies of the entomopathogenic nematode *Steinernema feltiae* (Rhabditidae; Steinernematidae). *Parasitology* 121, 211–216.

Fan, X. and Hominick, W.M. (1991) Efficiency of the Galleria (wax moth) baiting technique for recovering infective stages of entomopathogenic rhabditids (Steinernematidae and Heterorhabditidae) from sand and soil. *Revue de Nématologie* 14, 381–387.

Fenton, A., Norman, R., Fairbairn, J.P. and Hudson, P.J. (2000) Modelling the efficacy of entomopathogenic nematodes in the regulation of invertebrate pests in glasshouse crops. *Journal of Applied Ecology* 37, 309–320.

Fenton, A., Norman, R., Fairbairn, J.P. and Hudson, P.J. (2001) Evaluating the efficacy of entomopathogenic nematodes for the biological control of crop pests: a nonequilibrium approach. *American Naturalist* 58, 408–425.

Fenton, A., Gwynn, R.L., Gupta, A., Norman, R., Fairbairn, J.P. and Hudson, P.J. (2002) Optimal application strategies for entomopathogenic nematodes: integrating theoretical and empirical approaches. *Journal of Applied Ecology* 39, 481–492.

Ferguson, C.S., Schroeder, P.C. and Shields, E.J. (1995) Vertical distribution, persistence, and activity of entomopathogenic nematodes (Nematoda: Heterorhabditidae and Steinernematidae) in alfalfa snout beetle (Coleoptera: Curculionidae) infested fields. *Environmental Entomology* 24, 149–158.

Freckman, D.W., Blackburn, T.H., Brussaard, L., Hutchings, P., Palmer, M.A. and Snellgrove, P. (1997) Linking biodiversity and ecosystem functioning of soils and sediments. *Ambio* 26, 556–562.

Garcia del Pino, F. and Palomo, A. (1996) Natural occurrence of entomopathogenic nematodes (Rhabditida: Steinernematidae and Heterorhabditidae) in Spanish soils. *Journal of Invertebrate Pathology* 68, 84–90.

Garcia del Pino, F. and Palomo, A. (1997) Temporal study of natural populations of Heterorhabditid and Steinernematid nematodes in horticultural crop soils. *Fundamental and Applied Nematology* 5, 473–480.

Gaugler, R. and Kaya, H.K. (1990) *Entomopathogenic Nematodes in Biological Control.* CRC Press, Boca Raton, Florida, 365 pp.

Georgis, R. and Hague, N.G.M. (1981) A neoaplectanid nematode in the larch sawfly, *Cephalcia lariciphila* (Hymenoptera: Pamphilidae). *Annals of Applied Biol*ogy 99, 171–177.

Georgis, R. and Poinar, G.O. Jr (1983a) Effect of soil texture on the distribution and infectivity of *Neoaplectana carpocapsae* (Nematoda: Steinernematidae). *Journal of Nematology* 15, 308–311.

Georgis, R. and Poinar, G.O. Jr (1983b) Effect of soil texture on the distribution and infectivity of *Neoaplectana glaseri* (Nematoda: Steinernematidae). *Journal of Nematology* 15, 329–332.

Georgis, R. and Poinar, G.O. Jr (1983c) Effect of soil texture on the distribution and infectivity of *Heterorhabditis heliothidis* (Nematoda: Steinernematidae). *Journal of Nematology* 15, 652–654.

Ghally, S.E. (1995) Some factors affecting the activity and pathogenicity of *Heterorhabditis heliothidis* and *Steinernema carpocapsae* nematodes. *Journal of the Egyptian Society of Parasitology* 25, 125–135.

Gilmore, S.K. and Potter, D.A. (1993) Potential role of collembola as biotic mortality agents for entomopathogenic nematodes. *Pedobiologica* 37, 30–38.

Gilmore, S.K. and Raffensperger, E.M. (1970) Foods ingested by *Tomocerus* spp. (Collembola, Entomobryidae), in relation to habitat. *Pedobiologica* 10, 135–140.

Glazer, I. (2002) Survival biology. In: Gaugler, R. (ed.) *Entomopathogenic Nematology.* CAB International, Wallingford, UK, pp. 169–187.

Gouge, D.H., Lee, L.L. and Henneberry, T.J. (1999) Effect of temperature and lepidopteran host species on entomopathogenic nematode (Nematoda: Steinernematidae, Heterorhabditidae) infection. *Environmental Entomology* 28, 876–883.

Gouge, D.H., Smith, K.A., Lee, L.L. and Henneberry, T.J. (2000) Effect of soil depth and moisture on the vertical distribution of *Steinernema riobrave* (Nematoda: Steinernematidae). *Journal of Nematology* 32, 223–228.

Grant, J.A. and Villani, M.G. (2003) Soil moisture effects on entomopathogenic nematodes. *Environmental Entomology* 32, 80–87.

Grewal, P.S., Lewis, E.E., Gaugler, R. and Campbell, J.F. (1994a) Host finding behaviour as a predictor of foraging strategy in entomopathogenic nematodes. *Parasitology* 108, 207–215.

Grewal, P.S., Selvan, S. and Gaugler, R. (1994b) Thermal adaptation of entomopathogenic nematodes: niche breadth for infection, establishment, and reproduction. *Journal of Thermal Biology* 19, 245–253.

Grewal, S.K., Grewal, P.S. and Gaugler, R. (1995) Endophytes of fescue grasses enhance susceptibility of *Popillia japonica* larvae to an entomopathogenic nematode. *Entomologia Experimentalis et Applicata* 74, 219–224.

Gurr, G.M. and Wratten, S.D. (1999) Integrated biological control: a proposal for enhancing success in biological control. *International Journal of Pest Management* 45, 81–84.

Hawksworth, D.L. (1991) *The Biodiversity of Microorganisms and Invertebrates: Its Role in Sustainable Agriculture.* CAB International, Wallingford, UK, 302 pp.

Henneberry, T.J., Forlow, J., Burke, R.A. and Lindegren, J.E. (1996) Temperature effects on infection and mortality of *Pectinophora gossypiella* (Lepidoptera: Gelechiidae) larvae by two entomopathogenic nematode species. *Environmental Entomology* 25, 179–183.

Hominick, W.M. (2002) Biogeography. In: Gaugler, R. (ed.) *Entomopathogenic Nematology.* CAB International, Wallingford, UK, pp. 115—143.

Hsiao, W.F. and All, J.N. (1997) Effect of animal manure on the survival and pathogenicity of the entomopathogenic nematode, *Steinernema carpocapsae. Zhonghua Kunchong* 17, 53–65.

Hsiao, W.F. and All, J.N. (1998) Survey of the entomopathogenic nematode, *Steinernema carpocapsae*: (Rhabditida: Steinernematidae) natural populations and its dispersal in the field. *Chinese Journal of Entomology* 18, 39–49.

Hudson, W.G. and Nguyen, K.B. (1989) Effects of soil moisture exposure time nematode age and nematode density on laboratory infection of *Scapteriscus vicinus* and *Scapteriscus acletus* (Orthoptera: Gryllotalpidae) by *Neoaplectana* sp. (Rhabditida: Steinernematidae). *Environmental Entomology* 18, 719–722.

Hummel, R., Walgenbach, J.F., Barbercheck, M.E., Kennedy, G.G., Hoyt, G.D. and Arellano, C. (2002) Effects of production practices on soil-borne entomopathogens in western North Carolina vegetable systems. *Environmental Entomology* 31, 84–91.

Ishibashi, N. and Kondo, E. (1986) *Steinernema feltiae* (DD-136) and *S. glaseri*: persistence in soil and bark compost and their influence on native nematodes. *Journal of Nematology* 18, 310–316.

Ishibashi, N. and Kondo, E. (1990) Dynamics of the entomogenous nematode *Steinernema feltiae* applied to soil with and without nematicide treatment. *Journal of Nematology* 19, 404–412.

Jaworska, M. (1993) Investigations on the possibility of using entomophilic nematodes in reduction of *Cephalcia abietis* (L.) (Hym., Pamphilidae) population. *Polskie Pismo Entomologiczne* 62, 201–213.

Kaya, H.K. and Gaugler, R. (1993) Entomopathogenic nematodes. *Annual Review of Entomology* 38, 181–206.

Klein, M.G. (1990) Efficacy against soil-inhabiting insect pests. In: Gaugler, R. and Kaya, H.K. (eds) *Entomopathogenic Nematodes in Biological Control.* CRC Press, Boca Raton, Florida, pp. 195–214.

Klein, M.G. and Georgis, R. (1992). Persistence of control of Japanese beetle (Coleoptera: Scarabaeidae) with steinernematid and heterorhabditid nematodes. *Journal of Economic Entomology* 85, 727–730.

Koppenhöfer, A.M., Kaya, H.K., Shanmugam, S. and Wood, G.L. (1995) Interspecific competition between steinernematid nematodes within an insect host. *Journal of Invertebrate Pathology* 66, 99–103.

Koppenhöfer, A.M., Baur, M.E., Stock, S.P., Choo, H.Y., Chinnasri, B. and Kaya, H.K. (1997) Survival of entomopathogenic nematodes within host cadavers in dry soil. *Applied Soil Ecology* 6, 231–240.

Kung, S.P., Gaugler, R. and Kaya, H.K. (1990a) Soil type and entomopathogenic nematode persistence. *Journal of Invertebrate Pathology* 55, 401–406.

Kung, S.P., Gaugler, R. and Kaya, H.K. (1990b) Influence of soil pH and oxygen on persistence of *Steinernema* spp. *Journal of Nematology* 22, 440–445.

Kung, S.P., Gaugler, R. and Kaya, H.K. (1991) Effects of soil temperature, moisture, and relative humidity on entomopathogenic nematode persistence. *Journal of Invertebrate Pathology* 57, 242–249.

Letourneau, D.K. and Altieri, M.A. (1999) Environmental management to enhance biological control in agroecosystems. In: Bellows, T.S. and Fisher, T.W. (eds) *Handbook of Biological Control.* Academic Press, San Diego, California, pp. 319–354.

Lewis, E.E., Gaugler, R. and Harrison, R. (1992) Entomopathogenic nematode host finding: response to contact cues by cruise and ambush foragers. *Parasitology* 105, 309–315.

Lewis, E.E., Gaugler, R. and Harrison, R. (1993) Response of cruiser and ambusher entomopathogenic nematodes (Steinernematidae) to host volatile cues. *Canadian Journal of Zoology* 71, 765–769.

Lewis, E.E., Campbell, J.F. and Gaugler, R. (1998) A conservation approach to using entomopathogenic nematodes in turf and landscapes. In: Barbosa, P. (ed.) *Conservation Biological Control.* Academic Press, San Diego, California, pp. 235–254.

Mannion, C.M. and Jansson, R.K. (1992) Movement and postinfection emergence of entomopathogenic nematodes from sweetpotato weevil, *Cylas formicarius* (F.) (Coleoptera: Apionidae). *Biological Control* 2, 297–305.

Mason, J.M. and Hominick, W.M. (1995) The effect of temperature on infection, development and reproduction of heterorhabditids. *Journal of Helminthology* 69, 337–345.

Millar, L.C. and Barbercheck, M.E. (2002) Effects of tillage practices on entomopathogenic nematodes in a corn agroecosystem. *Biological Control* 25, 1–11.

Moore, J.C., Walter, D.E. and Hunt, H.W. (1988) Arthropod regulation of micro- and mesobiota in below-ground detrital food webs. *Annual Review of Entomology* 33, 419–439.

Moyle, P.L. and Kaya, H.K. (1981) Dispersal and infectivity of the entomogenous nematode, *Neoaplectana carpocapsae* Weiser (Rhabditida: Steinernematidae), in sand. *Journal of Nematology* 13, 295–300.

Mrácek, Z. (1980) The use of '*Galleria* traps' for obtaining nematode parasites of insects in Czechoslovakia (Lepidoptera: Nematoda, Steinernematidae). *Acta entomologica bohemoslavaca* 77, 378–382.

Mrácek, Z. (1986) Nematodes and other factors controlling the sawfly, *Cephalcia abietis* (Pamphilidae: Hymenoptera), in Czechoslovakia. *Forest Ecology and Management* 15, 75–79.

Mrácek, Z. and Webster, J.M. (1993) Survey of Heterorhabditidae and Steinernematidae (Rhabditida: Nematoda) in western Canada. *Journal of Nematology* 25, 710–717.

Mrácek, Z., Becvár, S. and Kindlmann, P. (1999) Survey of entomopathogenic nematodes from the families Steinernematidae and Heterorhabditidae (Nematoda: Rhabditida) in the Czech Republic. *Folia Parasitologica* 46, 145–148.

Muraoke, M. and Ishibashi, N. (1976) Nematode feeding mites and their feeding behavior. *Applied Entomology and Zoology* 11, 1–7.

Neher, D. and Barbercheck, M.E. (1999) Diversity and function of soil mesofauna. In: Collins, W. and Qualset, C.O. (eds) *The Biodiversity of Agroecosystems.* CRC Press, Boca Raton, Florida, pp. 27–47.

Nickle, W.R. (1984) *Plant and Insect Nematodes.* Marcel Decker, New York, 925 pp.

Parkman, J.P. and Smart, G.C. (1996) Entomopathogenic nematodes, a case study: introduction of *Steinernema scapterisci* in Florida. *Biocontrol Science and Technology* 6, 413–419.

Peters, A. (1996) The natural host range of *Steinernema* and *Heterorhabditis* spp. and their impact on insect populations. *Biocontrol Science and Technology* 6, 389–402.

Poinar, G.O. Jr (1979) *Nematodes for Biological Control of Insects.* CRC Press, Boca Raton, Florida, pp. 277.

Portillo-Aguilar, C., Villani, M.G., Tauber, M.J., Tauber, C.A. and Nyrop, J.P. (1999) Entomopathogenic nematode (Rhabditida: Heterorhabditidae and Steinernematidae) response to soil texture and bulk density. *Environmental Entomology* 28, 1021–1035.

Qiu, L. and Bedding, R. (1999) The relationship between energy metabolism and survival of the infective juveniles of *Steinernema carpocapsae* under unstressed-aerobic and anaerobic conditions. In: Glazer, I., Richardson, P., Boemare, N. and Coudert, F. (eds) *Survival Strategies of Entomopathogenic Nematodes.* EUR 18855 EN Report, pp. 149–156.

Raulston, J.R., Pair, S.D., Loera, J. and Cabanillas, H.E. (1992) Prepupal and pupal parasitism of *Helicoverpa zea* and *Spodoptera frugiperda* (Lepidoptera: Noctuidae) by *Steinernema* sp. in cornfields in the lower Rio Grande valley. *Journal of Economic Entomology* 85, 1666–1670.

Richardson, P.N. and Grewal, P.S. (1991) Comparative assessment of biological (Nematoda: *Steinernema feltiae*) and chemical methods of control of the mushroom fly *Lycoriella auripila* (Diptera: Sciaridae). *Biocontrol Science and Technology* 1, 217–228.

Rosenheim, J.A., Kaya, H.K., Ehler, L.E., Marois, J.J. and Jaffee, B.A. (1995) Intraguild predation among biological control agents: theory and evidence. *Biological Control* 5, 303–335.

Sayre, R.M. and Walter, D.E. (1991) Factors affecting the efficacy of natural enemies of nematodes. *Annual Review of Phytopathology* 29, 149–166.

Schroeder, W.J. and Beavers, J.B. (1987) Movement of the entomogenous nematodes of the families Steinernematidae and Heterorhabditidae in soil. *Journal of Nematology* 19, 257–259.

Schroeder, P.C., Villani, M.G., Ferguson, C.S., Nyrop, J.P. and Shields, E.J. (1993) Behavioral interactions between Japanese beetle (Coleoptera: Scarabaeidae) grubs and an entomopathogenic nematode (Nematoda: Heterorhabditidae) within turf microcosms. *Environmental Entomology* 22, 595–600.

Selvan, S., Campbell, J.F. and Gaugler, R. (1993) Density-dependent effects on entomopathogenic nematodes (Heterorhabditidae and Steinernematidae) within an insect host. *Journal of Invertebrate Pathology* 62, 278–284.

Shapiro, D.I., Tylka, G.L., Berry, E.C. and Lewis, L.C. (1995) Effects of earthworms on the dispersal of Steinernema spp. *Journal of Nematology* 27, 21–28.

Shapiro, D.I., Lewis, L.C., Obrycki, J.J. and Abbas, M. (1999a) Effects of fertilizers on suppression of black cutworm (*Agrotis ipsilon*) damage with *Steinernema carpocapsae. Journal of Nematology* 31, 690–693.

Shapiro, D.I., Obrycki, J.J., Lewis, L.C. and Jackson, J.J. (1999b) Effects of crop residue on the persistence of *Steinernema carpocapsae. Journal of Nematology* 31, 517–519.

Shapiro, D.I., McCoy, C.W., Fares, A., Obreza, T. and Dou, H. (2000) Effects of soil type on virulence and persistence of entomopathogenic nematodes in relation to control of *Diaprepes abbreviatus. Environmental Entomology* 29, 1083–1087.

Shapiro-Ilan, D.I., Gouge, D.H. and Koppenhöfer, A.M. (2002) Factors affecting commercial success: case studies in cotton, turf and citrus. In: Gaugler, R. (ed.) *Entomopathogenic Nematology*. CAB International, Wallingford, UK, pp. 333–355.

Söndergrath, D. and Schröder, B. (2002) Population dynamics and habitat connectivity affecting the spatial spread of populations – a simulation study. *Landscape Ecology* 17, 1–14.

Spiridonov, S.E. and Voronov, D.A. (1995) Small scale distribution of *Steinernema feltiae* juveniles in cultivated soil. In: Griffin, C.T., Gwynn, R.L. and Masson, J.P. (eds) *Ecology and Transmission Strategies of Entomopathogenic Nematodes*. European Commission, Luxembourg, pp. 36–41.

Stirling, G.R. (1991) *Biological Control of Plant Parasitic Nematodes*. CAB International, Wallingford, UK, 282 pp.

Strong, D.R. (2002) Populations of entomopathogenic nematodes in foodwebs. In: Gaugler, R. (ed.) *Entomopathogenic Nematology*. CAB International, Wallingford, UK, pp. 225–240.

Strong, D.R., Maron, J.L., Connors, P.G., Whipple, A., Harrison, S. and Jeffries, R.L. (1995) High mortality, fluctuation in numbers, and heavy subterranean insect herbivory in bush lupine, *Lupinus arboreus. Oecologia* 104, 85–92.

Strong, D.R., Kaya, H.K., Whipple, A.V., Child, A.L., Kraig, S., Bondonno, M., Dyer, K. and Maron, J.L. (1996) Entomopathogenic nematodes: natural enemies of root-feeding caterpillars on bush lupine. *Oecologia* 108, 167–173.

Strong, D.R., Whipple, A.V., Child, A.L. and Dennis, B. (1999) Model selection for a subterranean trophic cascade: root-feeding caterpillars and entomopathogenic nematodes. *Ecology* 80, 2750–2761.

Stuart, R.J. and Gaugler, R. (1994) Patchiness in populations of entomopathogenic nematodes. *Journal of Invertebrate Pathology* 64, 39–45.

Thurston, G.S., Ni, Y. and Kaya, H.K. (1994) Influence of salinity on survival and infectivity of entomopathogenic nematodes. *Journal of Nematology* 26, 345–351.

Timper, P. and Kaya, H.K. (1992) Impact of a nematode-parasitic fungus on the effectiveness of entomogenous nematodes. *Journal of Nematology* 24, 1–8.

Timper, P., Kaya, H.K. and Gaugler, R. (1988) Dispersal of the entomogenous nematode *Steinernema feltiae* (Rhabditida: Steinernematidae) by infected adult insects. *Environmental Entomology* 17, 546–550.

Timper, P., Kaya, H.K. and Jaffee, B.A. (1991) Survival of entomogenous nematodes in soil infested with the nematode-parasitic fungus *Hirsutella rhossiliensis* (Deuteromycotina: Hyphomycetes). *Biological Control* 1, 42–50.

Van der Werf, W., Smits, P.H., Downes, M. and Griffin, C. (1995) Concepts and prospects for modelling the efficacy and ecology of entomopathogenic nematodes. In: Griffin, C.T., Gwynn, R.L. and Masson, J.P. (eds) *Ecology and Transmission Strategies of Entomopathogenic Nematodes*. European Commission, Luxembourg, pp. 42–51.

Verhoef, H.A. and Brussard, L. (1990) Decomposition and nitrogen mineralization in natural and agro-ecosystems: the contribution of soil animals. *Biogeochemistry* 11, 175–211.

Walter, D.E. (1988a) Predation and mycophagy by endeostigmatid mites (Acariformes: Prostigmata). *Experimental and Applied Acarology* 4, 159–166.

Walter, D.E. (1988b) Nematophagy by soil arthropods from the shortgrass steppe, Chihauahuan desert and rocky mountains of the central United States. *Agriculture Ecosystems and Environment* 24, 307–316.

Walter, D.E., Hunt, H.W. and Elliot, E.T. (1987) The influence of prey type on the development and reproduction of some predatory soil mites. *Pedobiologica* 30, 419–424.

Wang, J. and Bedding, R. (1996) Population development of *Heterorhabditis bacteriophora* and *Steinernema carpocapsae* in the larvae of *Galleria mellonella*. *Fundamental and Applied Nematology* 19, 363–367.

Westerman, P.R. (1998) Penetration of the entomopathogenic nematode *Heterorhabditis* spp. into host insects at 9 and 20°C. *Journal of Invertebrate Pathology* 72, 197–205.

Westerman, P.R. and Van der Werf, W. (1998) Exploring the scope for improving biocontrol of black vine weevil, *Otiorhynchus sulcatus*, with *Heterorhabditis* spp. at low temperatures: a simulation study. *Agricultural Systems* 57, 43–63.

Wilson, M.J., Lewis, E.E., Yoder, F. and Gaugler, R. (2003) Application pattern and persistence of the entomopathogenic nematode *Heterorhabditis bacteriophora*. *Biological Control* 26, 180–188.

19 Interactions with Plant-parasitic Nematodes

E.E. Lewis[1] and P.S. Grewal[2]
[1]*Department of Nematology and Department of Entomology, University of California Davis, Davis, CA 95616, USA;* [2]*Department of Entomology, Ohio State University, Wooster, OH 44691, USA*

19.1. Introduction

The interaction between entomopathogenic nematodes (EPN) and plant-parasitic nematodes has received a great deal of research in field, greenhouse and laboratory studies. The topic of whether or not EPNs represent an alternative to currently used plant-parasitic nematode management methods remains controversial. At one extreme of the array of viewpoints is the argument that there is no predictable and defined interaction at all; that despite anecdotal instances of plant-parasitic nematode reduction following EPN release, there is no potential for control. At the other extreme are advocates of developing EPNs specifically to control plant-parasitic nematodes in the field. Not surprisingly, each of these arguments has strengths and weaknesses. Our goal in this chapter is to try to provide an unbiased review of field and greenhouse efficacy studies. This goal is particularly challenging because of the paucity of 'published' accounts of trials where the results of the tests did not support the hypothesis that there is an interaction.

Control of plant-parasitic nematodes is vital to agriculture. Crop damage caused by plant-parasitic nematodes accounts for approximately 12% of the annual loss of food and fibre production in the world (Barker *et al.*, 1994). Annually, this translates to US$8 billion of damage in the USA alone. The great economic losses to agriculture due to plant-parasitic nematodes have resulted in significant chemical nematicide use. Current dissatisfaction with chemical nematicides, due to safety issues, environmental concerns and limited use of many products (e.g. methyl bromide) and the time it takes for the development of nematode- or disease-resistant plant cultivars, has stimulated interest in control strategies that are ecologically compatible with current production systems. In fact, developing alternatives to hazardous chemical nematicides is one of the top priorities for the future of nematology (Barker *et al.*, 1994).

In addition to determining whether or not EPNs can reduce plant-parasitic nematode populations to acceptable densities in agricultural situations, there are other compelling reasons to determine what actually occurs in the soil after EPN release. For example, when the interactions between EPNs and plant-parasitic nematodes occur they represent an unexpected off-target impact of EPN applications. Off-target impacts of biocontrol agents are usually thought of in terms of whether or not a natural enemy might switch hosts or prey, thereby causing harm to desirable species. Indeed, guidelines have been proposed to standardize and measure this type of risk (Van Lenteren *et al.*, 2003). However, the interaction between EPNs and plant-parasitic nematodes is farther off-target than what is generally considered by these guidelines, and has nothing to do with EPNs exploiting unexpected resources. Another question asked might address what other members of the soil fauna and flora are impacted by EPN release (see Chapter 6 in this volume for more in depth treatment of this subject).

19.2. Background and History

Antagonism between plant-parasitic nematodes and EPNs was first documented in 1986 by two unrelated studies. In one of the studies, applications of *Steinernema glaseri* to greenhouse tomatoes were shown to suppress *Meloidogyne javanica* populations (Bird and Bird, 1986). In the other, Ishibashi and Kondo (1986) observed that application of *S. feltiae* and *S. glaseri* to non-sterilized field soil or bark compost reduced numbers of plant-parasitic nematodes but increased rhabditoid nematodes. Plant-parasitic nematode suppression has been repeated in greenhouse and field studies since this first report.

One of the most difficult challenges of this chapter is to present the data in a way that allows patterns to be seen and trends to be recognized. There are several reasons for this situation:

- The application methods of EPNs have been wildly variable among studies.
- Some studies apply EPNs in a single application, whereas others apply EPNs multiple times.
- The dose rates differ among studies.
- There are several different EPN species tested.
- The plant-parasitic nematode species tested are many, and their life histories differ significantly.
- Cropping systems and soil types differ.
- The methods of measuring the outcomes of the interactions differ.

For example, various studies measure the plant-parasitic nematode population in the soil, the rate of root penetration, the level of plant-parasitic nematode reproduction or some aspect of plant growth. Obviously, relating one study to another poses significant difficulty.

19.3. Nematode Life Histories

When EPNs were first found to affect plant-parasitic nematode populations, the data were surprising because there are no obvious and direct ecological links between these two nematode groups. Steinernematid and heterorhabditid nematodes are lethal parasites of insects. The infective stage juvenile searches for and infects a suitable insect host (Gaugler, 1981). The nematodes penetrate into the host haemocoel and release their symbiotic bacteria (usually *Xenorhabdus* spp. for Steinernematidae and *Photorhabdus* spp. for Heterorhabditidae) and kill the insect host within about 48 h. The nematodes develop into adults by feeding on the symbiont bacteria and degraded host tissue, mate and produce two to three generations inside the same host.

The next group of infective juveniles (IJs) develops 7–14 days later, emerges and searches for a new insect host.

Plant-parasitic nematodes, as a group, have a tremendous array of life histories. Plant-parasitic nematodes are often categorized into groups based on feeding strategies. Endoparasitic nematodes undergo at least one stage of development inside their plant host. They can be either migratory or sedentary within the root tissue. Semi-endoparasitic nematodes feed at a single site on the root, with the anterior embedded inside the root and the posterior protruding from the root. Ectoparasitic nematodes feed on roots without entering the root tissue.

The diversity of life history strategies of plant-parasitic nematodes combined with interspecific variation among EPNs and their bacteria may underlie some of the variability seen among plant-parasitic nematode species in their level of suppression by EPNs. However, multiple trials against the same plant-parasitic nematode species often yield variable results as well. The results sometimes vary even among tests against the same nematode species in the same location, and we remain largely in the dark as to why this occurs. Until we understand more of the sources of variation in these results, using EPNs against plant-parasitic nematodes will not be widely accepted by growers. Hopefully, suppliers will not oversell the results that we have as a panacea for plant-parasitic nematode management.

19.4. Proposed Mechanisms

Three different, but not mutually exclusive, mechanisms have been proposed to explain the interaction between EPNs and plant-parasitic nematodes. Each of these hypotheses has proponents and detractors. In fact, it is likely that more than one of these proposed mechanisms (or mechanisms not considered here) can act in concert. Bird and Bird (1986) suggested that EPNs attracted to plant roots by carbon dioxide may accumulate along the roots and force plant-parasitic nematodes away. However, Grewal *et al.* (1999) demonstrated that when dead EPNs were applied to sand containing a tomato seedling and *M. incognita* root penetration was reduced. When live EPNs were applied, there was no reduction of *M. incognita* infection. These studies are difficult to compare, however, because of the different timescales involved. Grewal *et al.* (1999) measured root penetration for up to 7 days after application, whereas Bird and Bird (1986) counted egg masses on tomato roots 8 weeks after application only.

A second hypothesis is that the massive doses of EPNs applied to soil for insect control (2.5 billion/ha is a standard dose) may lead to a build-up of general nematode antagonists that result in nematode-suppressive soils (Ishibashi and Kondo, 1986; Ishibashi and Choi, 1991). Contrary evidence comes from experiments carried out in sterile sand (Grewal *et al.*, 1999; Lewis *et al.*, 2001; Perez and Lewis, 2002). In all of these studies, significant reduction of *M. incognita* infection to tomato roots was measured when EPNs were applied. No natural enemy populations were present at the beginning of the studies. Further, the reductions in the laboratory, greenhouse and field tests in several studies have occurred within days of application, making the activity of natural enemies an unlikely mechanism. However, some of the longer-term reductions measured (Grewal *et al.*, 1997, for example) may be due in part to natural enemies. No serious test of this hypothesis has been conducted.

Finally, allelochemicals produced by the EPNs or their symbiotic bacteria have been shown to both repel and intoxicate plant-parasitic nematodes in laboratory studies, and have been hypothesized to be one reason for plant-parasitic nematode decline after EPN applications (Grewal *et al.*, 1999; Hu *et al.*, 1999). Dead (crushed or autoclaved) IJs of *S. carpocapsae* can cause significant mortality of foliar nematodes (*Aphelenchoides fragariae*) when held together in water suspension (Fig. 19.1; G.B. Jagdale and P.S. Grewal, unpublished data). Allelochemicals may also be released from nematode-infected insect cadavers in the field, killing plant-parasitic nematodes in

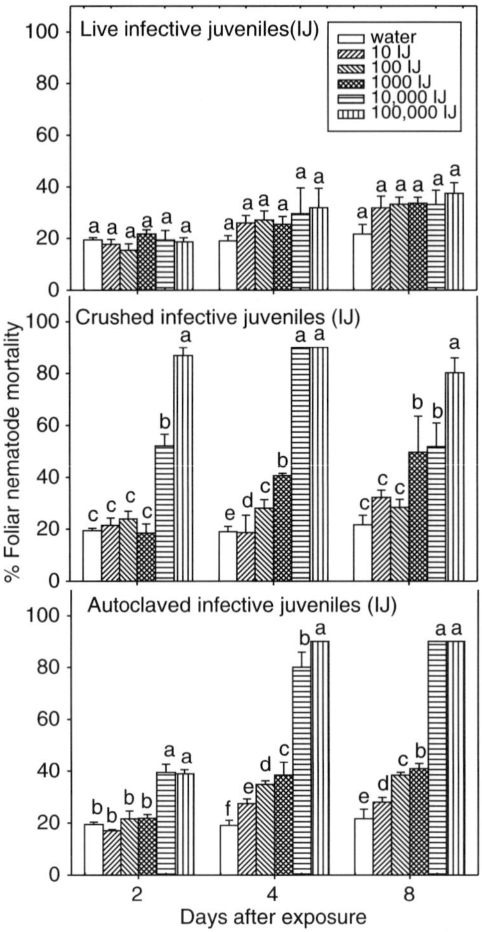

Fig. 19.1. Effect of live, crushed and autoclaved infective juveniles (IJs) of *Steinernema carpocapsae* on mixed stages of foliar nematode *Aphelenchoides fragariae*. The extract was prepared by crushing 25 mg cadaver/ml of water. Treatments included either live, crushed or autoclaved IJs: (i) 10 IJs; (ii) 100 IJs; (iii) 1000 IJs; (iv) 10,000 IJs; (v) 100,000 IJs; and (vi) water as control. Experiment was conducted in 24-well plates containing 500 foliar nematodes; treatments were arranged in a randomized block design with four replicates. Percentage mortality of foliar nematodes was recorded at 2, 4 and 8 days after treatment.

the vicinity (McInerny *et al.*, 1991a,b). G.B. Jagdale and P.S. Grewal (unpublished data) have observed that *S. carpocapsae*-infected wax moth cadavers and their extracts are toxic to foliar nematodes, and the activity of the intact cadaver increases as the time of exposure increases from 2 to 8 days (Fig. 19.2). Although neither *Xenorhabdus* nor *Photorhabdus* has been isolated from soil in the absence of either an infected insect or an IJ EPN, there are anecdotal reports of *Photorhabdus* being associated with wounds in human patients (Farmer *et al.*, 1989; Peel *et al.*, 1999). Perhaps by using PCR-based diagnostic techniques, isolating these bacteria from soil where they may be associated with alternative substrates will be possible.

19.5. Evidence

The studies that have been conducted in the greenhouse or the field are summarized in Table 19.1. This table does not list all the laboratory studies conducted since the focus of this book is on applications of EPNs for biocontrol. Providing an exhaustive list of all studies conducted is impossible at present due to several studies that were conducted but not published. There is great variation in the degree of impact on plant-parasitic nematode populations in the studies listed. For the most part, determining the cause of these differences is not the focus of any of the studies listed in Table 19.1, although there is usually some mention of local conditions and limited comparison with other studies in most papers published in this area.

Of the 29 combinations of EPNs/symbiotic bacteria and plant-parasitic nematodes listed in Table 19.1, 16 show at least some level of interaction between EPNs and plant-parasitic nematodes that result in a decline in some aspect of the plant-parasitic nematode population. However, the presence of some level of interaction does not necessarily suggest that the interaction results in a decrease in plant-parasitic nematode populations that would be acceptable from an agricultural production point of view. In some studies, there may be a measurable, even statistically significant, impact on plant-parasitic nematode populations, but the effect might not lead the authors to consider EPNs a viable plant-parasitic nematode management strategy. We try to

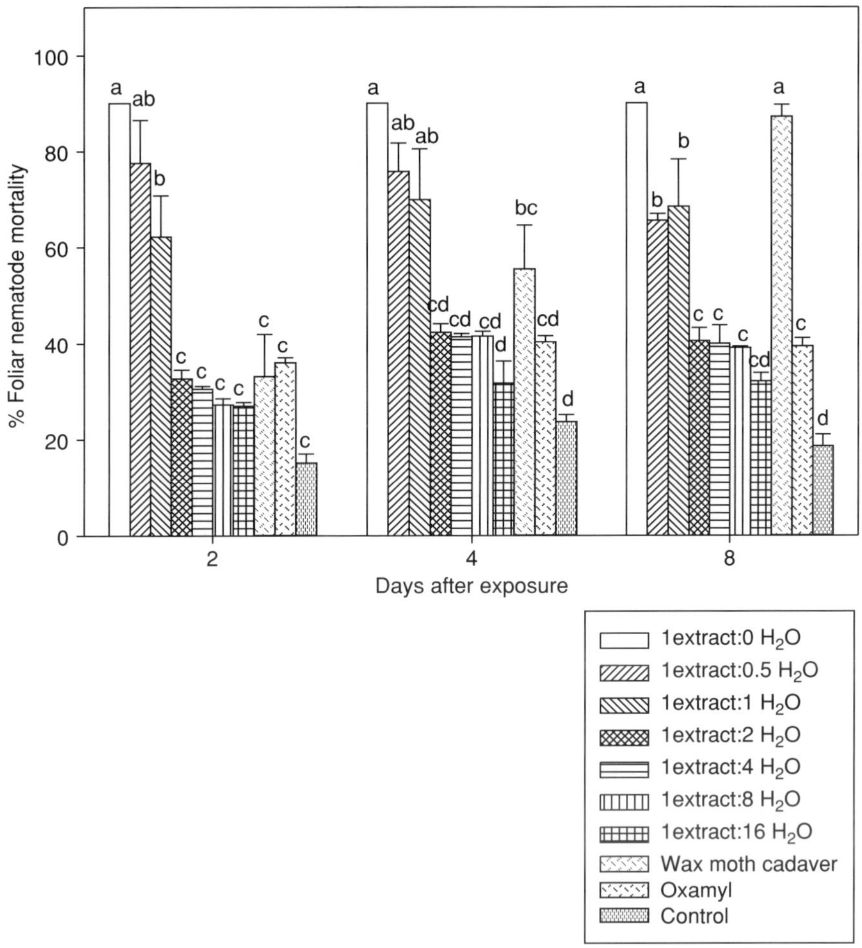

Fig. 19.2. Effect of *Steinernema carpocapsae* infected wax moth (*Galleria mellonella*) cadaver on mixed stages of foliar nematode *Aphelenchoides fragariae*. The cadaver extract was prepared by crushing 25 mg cadaver/ml of water and mixed to the following concentrations. Treatments included: (i) 100% extract; (ii) 66.67% extract; (iii) 50% extract; (iv) 33.33% extract; (v) 20% extract; (vi) 11.11% extract; (vii) 5.88% extract; (viii) 1 wax moth cadaver; (ix) oxamyl; and (x) water (control). The experiment was conducted in 24-well plates containing 500 foliar nematodes; treatments were arranged in a randomized block design with four replicates. Percent mortality of foliar nematodes was recorded at 2, 4 and 8 days after treatment.

catalogue the conditions that may have led to the discrepancies that produced these different categories where possible.

19.6. Field Trials

Three studies have shown that field applications of EPNs had a suppressive effect on several plant-parasitic nematode species in turfgrass, although the suppression recorded was not for all combinations of entomopathogenic and plant-parasitic nematode species tested, and varied among trials (Smitley *et al.*, 1992; Grewal *et al.*, 1997; Somasekhar *et al.*, 2002). All of these studies measured plant-parasitic nematode density in soil cores as the assessment of the effect of EPN application, and

Table 19.1. Summary of field or greenhouse trials conducted to test the impact of entomopathogenic nematode (EPN) releases on plant-parasitic nematode populations.

EPN or bacteria species	Plant-parasitic nematode species	Crop (field (F)/greenhouse (GH))	Impact	References
Steinernema carpocapsae	*Radopholus similis*	Banana	Y	Aalten, 1996
S. carpocapsae All	*Belonolaimus* sp. *Meloidogyne* sp. *Criconemella* sp.	Turf (F)	Y, N	Grewal *et al.*, 1997
S. carpocapsae All	*Globodera rostochiensis*	Potato (GH/F)	Y	Perry *et al.*, 1998
S. carpocapsae All	*Aphelenchoides* *Criconemella* *Helicotylenchus* *Hoplolaimus* *Pratylenchus* *Rotylenchus* *Trichodorus* *Tylenchorynchus* *Tylenchus* *Xiphinema*	Boxwood (F)	?	Jagdale *et al.*, 2002
S. carpocapsae	*Pratylenchus penetrans*	Strawberry (F)	N	LaMondia and Cowles, 2002
S. carpocapsae + *Heterorhabditis bacteriophora* (Nematodes added together as one treatment)	*Tylenchorynchus* spp. *Pratylenchus projectus* *Criconemella rustica*	Turf (F)	N	Smitley *et al.*, 1992
S. carpocapsae + *H. bacteriophora*	*P. penetrans*	Turf (F)	Y	Smitley *et al.*, 1992
S. feltiae	*G. rostochiensis*	Potato (GH/F)	Y	Perry *et al.*, 1998
S. feltiae SN	*Meloidogyne javanica*	Tomato and soybean (GH)	Y	Fallon *et al.*, 2002
S. feltiae MG-14	*M. javanica*	Tomato and soybean (GH)	Y	Fallon *et al.*, 2002
S. feltiae	*Pratylenchus penetrans*	Strawberry (F)	N	LaMondia and Cowles, 2002
S. glaseri	*M. javanica*	Tomato (GH)	Y	Bird and Bird, 1986
S. glaseri NC	*Aphelenchoides* sp. *Mesocriconema* sp. *Helicotylenchus* sp. *Hoplolaimus* sp. *Longidorus* sp. *Pratylenchus* sp. *Rotylenchus* sp. *Trichodorus* sp. *Tylenchorynchus* sp. *Xiphinema* sp.	Turf (F)	?	G.B. Jagdale and P.S. Grewal, unpublished data
S. riobrave TX	*Belonolaimus* sp. *Meloidogyne* sp. *Criconemella* sp.	Turf (F)	Y	Grewal *et al.*, 1997
S. riobrave	*M. incognita*	Cotton (F)	Y	Gouge *et al.*, 1997
S. riobrave	*M. incognita*	Tomato (GH)	Y	Perez and Lewis, 2002

S. riobrave TX	*M. javanica*	Tomato and soybean (GH)	N	Fallon *et al*., 2002
S. riobrave	*Mesocriconema xenoplax*	Peach (GH)	N	Nyczepir *et al*., 2004
		Pecan (F)		
S. riobrave	*M. incognita*	Squash (F)	N	Riegel *et al*., 1998
H. bacteriophora	*M. xenoplax*	Peach (GH)	N	Nyczepir *et al*., 2004
		Pecan (F)		
H. bacteriophora	*M. incognita*	Squash (F)	N	Riegel *et al*., 1998
H. bacteriophora	*M. incognita*	Tomato (GH)	Y	Perez and Lewis, 2002
H. bacteriophora	*Mesocriconema* sp.	Turf (F)	Y	G.B. Jagdale and P.S. Grewal, unpublished data
GPS 11	*Helicotylenchus* sp. *Trichodorus* sp. *Tylenchorynchus* sp.			
H. indica MG-13	*M. javanica*	Tomato and soybean (GH)	Y	Fallon *et al*., 2002
H. zealandica X1	*Aphelenchoides* sp.	Turf (F)	?	G.B. Jagdale and P.S. Grewal, unpublished data
	Mesocriconema sp. *Helicotylenchus* sp. *Hoplolaimus* sp. *Longidorus* sp. *Pratylenchus* sp. *Rotylenchus* sp. *Trichodorus* sp. *Tylenchorynchus* sp. *Xiphinema* sp.			
Photorhabdus luminescens (from *H. bacteriophora*)	*Mesocriconema* sp.	Turf (F)	Y	G.B. Jagdale and P.S. Grewal, unpublished data
	Helicotylenchus sp. *Trichodorus* sp.			
GPS 11	*Tylenchorynchus* sp.			
Pseudomonas oryzihabitans (from *S. abbasi*)	*M. javanica*	Tomato seedling (GH)	Y	Samaliev *et al*., 2000
Xenorhabdus nematophilus (from *S. carpocapsae*)	*M. javanica*	Tomato seedling (GH)	N	Samaliev *et al*., 2000
X. nematophilus (from *S. carpocapsae* All)	*Mesocriconema* sp.	Turf (F)	Y	G.B. Jagdale and P.S. Grewal, unpublished data
	Helicotylenchus sp. *Trichodorus* sp. *Tylenchorynchus* sp.			

Y = there was a measurable decline of plant-parasitic nematode populations after an EPN application. We do not mean to indicate that the authors of the original papers suggest that EPNs are viable as plant-parasitic nematode control agents, just that there was a measurable effect; N = there was no measurable decline; ? = when species were grouped together there was a decrease of plant-parasitic nematodes. However, not every species was suppressed.

so are comparable. Smitley *et al.* (1992) were the first to test the impact of EPNs on plant-parasitic nematodes in a field trial using EPN densities that were similar to the dose recommended for insect control. They conducted their trials in Michigan, USA, where the target nematode species, soil characteristics and climate differ from the Grewal *et al.* (1997) study, which was conducted in south-eastern USA. Smitley *et al.* (1992) had irrigation regime as a central variable to the test, and had two separate field trials included in the paper. The first trial tested the effect of irrigation on the impact of a combined release of 2.5 billion IJs/ha of *Heterorhabditis bacteriophora* and 2.9 billion IJs/ha of *S. carpocapsae* against stunt nematode, *Tylenchorhynchus* spp., and lesion nematode, *Pratylenchus penetrans*. Four weeks after treatment in irrigated plots, stunt nematode populations were significantly decreased by approximately half after EPN applications, but lesion nematode populations were not. The second field trial tested the effects of 2.5 billion *H. bacteriophora* IJs/ha only, with and without irrigation, on lesion, stunt and ring nematode populations. In this trial, *H. bacteriophora* reduced numbers of lesion nematode in irrigated plots but did not reduce either stunt or ring nematode populations. Despite the inconsistent results, the authors did show that EPNs impacted plant-parasitic nematodes in the field, and that the impact was limited to areas that were irrigated, or at least moist enough for the EPNs to survive.

Grewal *et al.* (1997) tested the effects of *S. riobrave* and *S. carpocapsae* on three turfgrass nematode pests, ring nematode (*Criconemella* spp.), sting nematode (*Belonolaimus longicaudatus*) and root-knot nematode (*Meloidogyne* sp.) in two locations in south-eastern USA. In the first test an application rate of 6 billion IJs/acre (six times the standard application rate for insect control) of each nematode species was compared with fenamiphos (Nemacure), a chemical nematicide commonly used in turfgrass for plant-parasitic nematode control. *S. riobrave* applications were equally or more effective than fenamiphos against all plant-parasitic nematode species tested for up to 8 weeks, whereas *S. carpocapsae* had no effect on plant-parasitic nematode populations at any time tested. The authors suggested that *S. carpocapsae*, because of their habit of foraging near the soil surface, did not occur in a location where they would likely interact with the root-feeding plant-parasitic nematodes. The second trial in this study compared two doses of these two EPN species: 1 billion versus 6 billion IJs/acre. Dose had no effect on plant-parasitic nematode suppression, but species did. Though *S. carpocapsae* did suppress all plant-parasitic nematode species to some degree, *S. riobrave* consistently outperformed it. The results of this field trial suggest that EPNs might be useful as a nematode management product for plant-parasitic nematodes in turfgrass. The upcoming restrictions on using chemical nematicides in turfgrass may spur more interest in EPNs as plant-parasitic nematode management products.

Somasekhar *et al.* (2002) took a different approach to studying the interaction. They hypothesized that the soil nematode community was altered by the inundative application of EPNs, and aimed to determine which members of the community were impacted most significantly and how the disruption caused by EPNs compared with the disruption caused by an insecticide (trichlorfon) that is commonly used in turfgrass. They took soil cores before treatment and 30 and 60 days after treatment (DAT) and extracted the nematodes. The data collected from the soil cores were subjected to a number of community-level analyses. The nematode communities of the turfgrass ecosystem were indeed altered, according to the measures preformed. However, the alteration was the result of decreases in plant-parasitic nematode numbers, but not in the numbers of free-living nematodes extracted from soil cores. Trichlorfon reduced all groups of soil nematodes. This was the first indication that EPNs reduced plant-parasitic nematodes, but may not have the same effect on all nematodes.

G.B. Jagdale and P.S. Grewal (unpublished data) tested the effects of *S. carpo-*

capsae All strain and *H. bacteriophora* GPS11 strain and their symbiotic bacteria *Xenorhabdus nematophilus* and *Photorhabdus luminescens*, respectively, against the plant-parasitic nematodes associated with turfgrass in Ohio during 2002 and 2003. The site was a golf course and turf consisted of a 50:50 Kentucky bluegrass and perennial ryegrass blend and the soil was sandy. Total plant-parasitic nematode numbers did not differ significantly among any of the treatments in 2003, but there were some differences in 2002. IJs of *H. bacteriophora* and both the species of bacteria caused 32–57% reduction in total populations of plant-parasitic nematodes relative to the control 30 DAT in 2002. Only four plant-parasitic nematode genera, *Mesocriconema*, *Helicotylenchus*, *Trichodorus* and *Tylenchorynchus*, were significantly influenced by the application of the EPNs and bacteria. The populations of *Mesocriconema* were significantly reduced by *S. carpocapsae* (65%), *H. bacteriophora* (80%) and *Ph. luminescens* (76%) treatments relative to the control 60 DAT in 2002 but not in 2003. The populations of *Helicotylenchus* were significantly reduced (57–65%) in the plots treated with both bacteria 60 DAT only in 2002. The population of *Trichodorus* was significantly reduced (64%) in *S. carpocapsae* treatment relative to the control 30 DAT in 2003 but not in 2002. The population of *Tylenchorynchus* was significantly reduced in *H. bacteriophora* (40%) and *Ph. luminescens* (40%) treatments relative to the control 15 DAT and in *Ph. luminescens* (67%) 30 DAT only in 2002. Populations of *Tylenchus* spp., a nematode associated with many different host plants but not causing economic damage, also were significantly reduced by the *H. bacteriophora* treatment 15 DAT relative to the control in 2002 and in all the treatments 30 DAT in 2003. The non-stylet bearing nematodes were not affected by the application of either EPNs or bacteria in either year.

Jagdale *et al.* (2002) compared the impact of live and dead *S. carpocapsae* and another chemical nematicide, ethoprop (Mocap), on the plant-parasitic nematode community in soil associated with boxwood (*Buxus* spp.) plantings during two field seasons. Soil cores were collected before treatment and 15 and 30 DAT, and subjected to Baerman funnel extraction. The results were variable. Some of the genera of plant-parasitic nematodes, or suspected plant-parasitic nematodes, were decreased after treatment (*Criconemella*, *Helicotylenchus*, *Hoplolaimus*, *Longidorus*, *Rotylenchus*, *Trichodorus*, *Tylenchorynchus* and *Xiphinema*), whereas others were not reduced (*Aphelenchoides*, *Hemicycliophora*, *Pratylenchus* and *Tylenchus*; see Table 19.1 for summary). However, when the nematodes were divided into ‘stylet-bearing plant-parasitic nematodes’, ‘stylet-bearing non-plant-parasitic nematodes’ and ‘non-stylet-bearing nematodes’, the plant-parasitic nematodes as a group were reduced in all trials and the non-stylet-bearing nematodes were not reduced in any trials. This reinforces the contention of Somasekhar *et al.* (2002) that free-living nematodes do not suffer ill effects due to EPN applications.

In strawberries *P. penetrans* feeds on the plant roots, reducing vigour and increasing the severity of strawberry black root rot disease. LaMondia and Cowles (2002) tested the effects of two inundative releases of *S. feltiae* or *S. carpocapsae* per year on *P. penetrans* in strawberries in the same plots over a period of 3 years. *P. penetrans* populations were assessed by extraction of the nematodes from the strawberry roots at the end of each season. No impact of either EPN species at any dose tested on *P. penetrans* density was recorded. It is interesting to note that Jagdale *et al.* (2002) also found that *P. penetrans* was among the nematode species not reduced by applications of EPNs in boxwood plantings. However, Smitley *et al.* (1992) found *H. bacteriophora* applications to reduce lesion nematode in turf.

Gouge *et al.* (1997) applied a mid-season treatment of oxamyl or *S. riobrave* at a rate of 2.5 or 5 billion IJs/ha to control *M. incognita* in cotton in Arizona. The EPNs were applied via drip irrigation lines in the field. *M. incognita* and *Tylenchorhynchus* spp. soil stages were extracted from soil samples weekly for 6 weeks after treatment to assess the effects of the treatments. One week after application of 2.5 billion *S. riobrave*/ha *M. incognita* was reduced by 83% and *Tylenchorhynchus* spp. were reduced by 85%. Interestingly, the higher rate of *S. riobrave* did not work as well as the lower one. Once again, these data agree with other studies that have shown that EPNs reduce populations of *Tylenchorhynchus* spp. (Jagdale *et al.*, 2002) and *Meloidogyne* spp. (Bird and Bird, 1986; Lewis *et al.*, 2001; Perez and Lewis, 2002).

Yet other studies with *Meloidogyne* spp. have shown no interactions (Riegel *et al.*, 1998). The study by Riegel *et al.* (1998), published as an abstract only, was conducted in microplots. There are no specific numbers available; however, the authors' conclusion was that neither *S. riobrave* nor *H. bacteriophora* provided adequate control of *M. incognita* on squash in Florida. The authors report that no difference in shoot length was measured between *S. riobrave*–treated microplots and controls. However, shoot length was longer and root-galling indexes were lower on plants treated with 1,3-dichloropropene.

Nycziper *et al.* (2004) tested the effect of *H. bacteriophora* or *S. riobrave* on *Mesocriconema xenoplax* populations on peach (in the greenhouse) and pecan (in the field) and concluded that no significant reduction could be measured in either of these systems. Two applications of 50 EPNs/cm^2 were conducted 3 months apart in the greenhouse study. After 180 days, *Me. xenoplax* populations were measured and no differences were found. In the field microplot study on pecan, *S. riobrave* was applied at a rate of 200 IJs/cm^2 three times and again, after 150 days, no differences were found in the *Me. xenoplax* populations. In contrast, in turf, ring nematode populations were decreased by *S. carpocapsae* and *S. riobrave* applications (Grewal *et al.*, 1997).

A field trial testing *S. carpocapsae* against *Radopholus similis* on banana resulted in reduced penetration rates into banana roots (Aalten, 1996). When the application rate of *S. carpocapsae* was increased, no resulting increase in the level of suppression was recorded.

19.7. Greenhouse Studies

Most greenhouse studies explored interactions between EPNs and *M. javanica* or *M. incognita* or both. Bird and Bird (1986) were the first to show suppression of *M. javanica* by *S. glaseri*. The application rates in this study were very high. In one test, *S. glaseri* was applied as a daily dose of 200,000 for 5 days, which led to a 2.5-fold decrease in the number of *M. javanica* egg masses/plant. In a second experiment, when *S. glaseri* was applied at a rate of 5,000,000/plant in single or multiple doses, *M. javanica* egg masses (expressed as egg masses/g of root) were reduced by a factor of six. Further, fresh weights of whole plants and roots were also significantly increased.

Other studies have tested other species of EPNs, different application methods or have evaluated the effects on plant-parasitic nematode populations in different ways. Perez and Lewis (2002) compared the impact on *M. incognita* of three species of EPNs applied at two rates, either before or after *M. incognita* was introduced to the system. Application of *H. bacteriophora, S. feltiae or S. riobrave* at either 25 IJs/cm^2 (recommended rate for insect control) or 125 IJs/cm^2 reduced both penetration into tomato roots and egg production by *M. incognita* on a per-plant basis. The greenhouse results corresponded with results from a scaled-down laboratory-based companion study described in the same paper. Perez and Lewis (2002) showed that applying EPNs either before or at the same time as *M. incognita* provided better control than if the entomopathogenic nematodes were applied after *M. incognita* had already infected the plants. Fallon *et al.* (2002) showed that applications of *S. riobrave* or *H. indica* reduced root penetration by *M. javanica* 3 days after inoculation, but when the experiment was conducted on tomato, a corresponding reduction in egg production by *M. javanica* was not seen. In addition, no increases in characters associated with plant vigour were recorded in EPN-treated pots. An interesting observation of this study was that significant numbers of EPNs penetrated the root tissue in the presence of *M. javanica*.

Perry *et al.* (1998) conducted a pot study in both the greenhouse and outdoors to test the impact of *S. feltiae* and *S. carpocapsae* on the potato cyst nematode, *Globodera rostochiensis*. In the greenhouse, potatoes were planted in soil that was infested with cysts of *G. rostochiensis* and EPNs were applied the week of planting, 1 week or 2 weeks after planting. A fourth treatment, in the greenhouse trial only, was an application of EPNs on all of these 3 weeks. There were minor reductions in some aspects of *G. rostochiensis* populations. In the greenhouse trial, application of *S. feltiae* 2 weeks after planting and application of *S. carpocapsae* 0 and 1 week after planting reduced the *G. rostochiensis* invasion rate, which was measured 4 weeks after planting. However, none of the treatments had an impact on either cyst production or any measure of plant health 16 weeks after planting. In the outdoor trial two rates of EPNs were applied: one was the recommended rate for insect control and the other was three times the recommended rate. This trial produced similar results. Of the 12 treatments, eight resulted in a reduction of invasion, but only one reduced cyst production. Again, there was no measurable effect on any plant characters. The short-term reduction in nematode penetration rate, the lack of long-term suppression of reproduction and the lack of positive effects on the plants mirrors the results from Fallon *et al.* (2002). The authors' conclusion was that EPNs would not be a viable biocontrol agent for cyst nematodes, based on their data.

Two greenhouse trials have tested the symbiotic bacteria, without the nematode partner, against plant-parasitic nematodes. Grewal *et al.* (1999) conducted a pot trial against *M. incognita* on tomato plants, applying two formulations of cell-free culture broth from *Xenorhabdus* sp. 'R' (the symbiont of *S. riobrave*). One treatment was 10 ml of the broth and for the second treatment 10 ml of broth was oven-dried to a paste and formulated on to corncob grit. Two weeks after treatment, root penetration assays showed that the culture filtrate formulated on to corncob grit had significantly reduced root penetration. Four weeks after treatment, no differences in root penetration among the bacteria-based treatments were observed. A treatment of Nemacure reduced penetration at both time periods. Samaliev *et al.* (2000) tested the nematicidal effects of *Pseudomonas oryzihabitans* (the bacterial symbiont of *S. abbasi*) and *X. nematophilus* in laboratory and greenhouse assays. In their pot trial, *Ps. oryzihabitans* applied at a rate of 20 ml culture broth containing 10^6 cells/ml decreased root galling whereas the same rate of *X. nematophilus* broth did not.

19.8. Summary and Conclusions

Of the six species of EPNs that have been tested against various species of plant-parasitic nematodes, all have caused some reduction in plant-parasitic nematode numbers or reproduction in at least one trial. Members of both genera of EPNs have yielded positive and negative results. Only *S. feltiae* has caused some suppression in every case where it was tested. The foraging strategy of *S. carpocapsae* may be the cause of this species' relative poor performance in these assays, due to the fact that they are usually located in the upper 2 cm of soil (Campbell *et al.*, 1996).

Of the seven papers that have a root-knot nematode as one of the targets, suppression was recorded in all but a single case. These studies varied significantly in the application methods used, and the results from studies that have extremely high doses at multiple times may be of limited value when trying to use EPNs as biocontrol agents. The cost of the material may be prohibitively expensive unless the dose can be reduced. It remains to be seen whether the level of reduction caused will satisfy growers for management of root-knot nematode.

On the other hand, *P. penetrans* has been a target in three studies and was not reduced in two of them. Perry *et al.* (1998) argued that since cyst nematode penetration was decreased, but cyst production was not, EPNs did not represent a reasonable control method for cyst nematodes.

The lack of any impact on free-living nematodes, which has been documented twice, is interesting. Though plant-parasitic nematodes are often decreased after field applications of EPNs, free-living nematodes are not. This emphasizes how little is really known of how EPN applications impact soil ecology in general.

The future commercial potential of using EPNs to manage plant-parasitic nematodes is not clear. Based on the research to date, most growers would be reluctant to rely on this method of nematode control. However, there may be certain situations where there are few alternatives. One example is management of plant-parasitic nematodes in turfgrass in the USA. Current management practices are based on chemicals that may not be available for use in turfgrass due to government regulation in the near future. The lack of available chemical nematicides, combined with the relatively consistent suppression in turfgrass that has been documented, may make this a viable market. Other markets are more questionable at this time.

The final question concerns what a nematode control product based on EPNs would look like. One possibility is to market EPN products in the same way as they are currently marketed for insect control. However, since the application of dead EPNs seems to work nearly as well as applying live ones does, the shelf-life problems that plague insect control products may not be as important. There is also the possibility of developing a nematode management product based on the bacteria (or metabolites) alone. This would be especially attractive since the material could be treated in ways similar to chemical pesticides. However, the broad-spectrum activity of *Xenorhabdus* and *Photorhabdus* bacteria spp. would warrant studies of how these applications would impact non-target species.

References

Aalten, M.P. (1996) Effect of the entomopathogenic nematode *Steinernema carpocapsae* on *Radopholus similis* invasion into banana roots. *Brighton Crop Protection Conference – Pests and Disease* 1–3, 439–440.

Barker, K.R., Hussey, R.S., Krusberg, L.R., Bird, G.W., Dunn, R.A., Ferris, H., Ferris, V.R., Freckman, D.W., Gabriel, C.J., Grewal, P., MacGuidwin, A.E., Riddle, D.L., Roberts, P.A. and Schmitt, D.P. (1994) Plant and soil nematodes: societal impact and focus for the future. *Journal of Nematology* 26, 127–137.

Bird, A. and Bird, J. (1986) Observations on the use of insect parasitic nematodes as a means of biological control of root-knot nematodes. *International Journal of Parasitology* 10, 511–516.

Campbell, J.F., Lewis, E.E., Yoder, F. and Gaugler, R. (1996) Spatial and temporal distribution of entomopathogenic nematodes in turf. *Parasitology* 113, 473–482.

Fallon, D.J., Kaya, H.K., Gaugler, R. and Sipes, B.S. (2002) Effects of entomopathogenic nematodes on *Meloidogyne javanica* on tomatoes and soybeans. *Journal of Nematology* 34, 239–245.

Farmer, J.J. III, Jorgensen, J.H., Grimont, P.A.D., Akhurst, R.J., Poinar, G.O. Jr, Ageron, E., Pierce, G.V., Smith, J.A., Carter, G.P., Wilson, K.L. and Hickman-Brenner, F.W. (1989) *Xenorhabdus luminescens* (DNA hybridization group 5) from human clinical specimens. *Journal of Clinical Microbiology* 27, 1594–1600.

Gaugler, R. (1981) Ecological considerations in the biological control of soil-inhabiting insects with entomopathogenic nematodes. *Journal of Nematology* 13, 241–249.

Gouge, D.H., Smith, K.A., Lee, L.L., Van Berkum, J.R. and Henneberry, T.J. (1997) Suppression of plant parasitic nematodes in cotton using the entomopathogenic nematode *Steinernema riobravis* (Cabanillas, Poinar, and Raulston) (Rhabditida: Steinernematidae). *Proceedings of the Beltwide Cotton Conference* 2, National Cotton Council, Memphis, Tennessee, 1324–1326.

Grewal, P.S., Martin, W.R., Miller, R.W. and Lewis, E.E. (1997) Suppression of plant-parasitic nematode populations in turfgrass by application of entomopathogenic nematodes. *Biocontrol Science and Technology* 7, 393–399.

Grewal, P.S., Lewis, E.E. and Venkatachari, S. (1999) Allelopathy: a possible mechanism of suppression of plant-parasitic nematodes by entomopathogenic nematodes. *Nematology* 1, 735–743.

Hu, K.J., Li, J.X. and Webster, J.M. (1999) Nematicidal metabolites produced by *Photorhabdus luminescens* (Enterobacteriaceae), bacterial symbiont of entomopathogenic nematodes. *Nematology* 1, 457–469.

Ishibashi, N. and Choi, D.-R. (1991) Biological control of soil pests by mixed application of entomopathogenic and fungivorous nematodes. *Journal of Nematology* 23, 175–181.

Ishibashi, N. and Kondo, E. (1986) *Steinernema feltiae* (DD-136) and *S. glaseri*: persistence in soil and bark compost and their influence on native nematodes. *Journal of Nematology* 18, 310–316.

Jagdale, G.B., Somasekhar, N., Grewal, P.S. and Klein, M.G. (2002) Suppression of plant-parasitic nematodes by application of live and dead infective juveniles of an entomopathogenic nematode, *Steinernema carpocapsae*, on boxwood (*Buxus* spp.). *Biological Control* 24, 42–49.

LaMondia, J.A. and Cowles, R.S. (2002) Effect of entomopathogenic nematodes and *Trichoderma harzianum* on the strawberry black root rot pathogens *Pratylenchus penetrans* and *Rhizoctonia fragariae*. *Journal of Nematology* 34, 351–357.

Lewis, E.E., Grewal, P.S. and Sardanelli, S. (2001) Interactions between *Steinernema feltiae-Xenorhabdus bovienii* insect pathogen complex and root-knot nematode *Meloidogyne incognita*. *Biological Control* 21, 55–62.

McInerney, B.V., Gregson, R.P., Lacey, W.C., Akhurst, R.J., Lyons, G.R., Rhodes, S.H. and Smith, D.R.J. (1991a) Biologically active metabolites from *Xenorhabdus* spp. Part 1. Dithiolopyrrolone metabolites with antibiotic activity. *Journal of Natural Products* 54, 748–774.

McInerney, B.V., Taylor, W.C., Lacey, M.J., Akhurst, R.J. and Gregson, R.P. (1991b) Biologically active metabolites from *Xenorhabdus* spp. Part 2. Benzopyran-1-one derivatives with gastroprotective activity. *Journal of Natural Products* 54, 785–795.

Nyczepir, A.P., Shapiro-Ilan, D.I., Lewis, E.E. and Handoo, Z.A. (2004) Effect of entomopathogenic nematodes on *Mesocriconema xenoplax* populations in peach and pecan. *Journal of Nematology* 36, 181–185.

Peel, M.M., Alfredson, D.A., Gerrard, J.G., Davis, J.M., Robson, J.M., McDougall, R.J., Scullie, B.L. and Akhurst, R.J. (1999) Isolation, identification, and molecular characterization of strains of *Photorhabdus luminescens* from infected humans in Australia. *Journal of Clinical Microbiology* 37, 3647– 3653.

Perez, E.E. and Lewis, E.E. (2002) Effects of entomopathogenic nematode application to *Meloidogyne incognita* penetration and egg production in laboratory and greenhouse experiments. *Journal of Nematology* 34, 171–174.

Perry, R.N., Hominick, W.M., Beane, J. and Briscose, B. (1998) Effects of the entomopathogenic nematodes, *Steinernema feltiae* and *S. carpocapsae*, on the potato cyst nematode, *Globodera rostochiensis*, in pot trials. *Biocontrol Science and Technology* 8, 175–180.

Riegel, C., Dickson, D.W., Nguyen, K.B. and Smart, G.C. (1998) Management of root-knot nematodes with entomopathogenic nematodes. *Journal of Nematology* 30, 512–513.

Samaliev, H.Y., Andreoglou, F.I., Elawad, S.A., Hague, N.G.M. and Gowen, S.R. (2000) The nematicidal effects of the bacteria *Pseudomonas oryzihabitans* and *Xenorhabdus nematophilus* on the root-knot nematode *Meloidogyne javanica*. *Nematology* 2, 507–514.

Smitley, D.R., Warner, W.R. and Bird, G.W. (1992) Influence of irrigation and *Heterorhabditis bacteriophora* on plant-parasitic nematodes in turf. *Journal of Nematology* 24 (Suppl.), 637–641.

Somasekhar, N., Grewal, P.S., De Nardo, E.A.B. and Stinner, B.R. (2002) Non-target effects of entomopathogenic nematodes on the soil nematode community. *Journal of Applied Ecology* 39, 735–744.

Van Lenteren, J.C., Babendreier, D., Bigler, F., Burgio, G., Hokkanen, H.M.T., Kuske, S., Loomans, A.J.M., Menzler-Hokkanen, I., Van Rijn, P.C.J., Thomas, M.B., Tommasini, M.G. and Zeng, Q.Q. (2003) Environmental risk assessment of exotic natural enemies used in inundative biological control. *Biocontrol* 48, 3–38.

20 Compatibility and Interactions with Agrochemicals and Other Biocontrol Agents

A.M. Koppenhöfer[1] and P.S. Grewal[2]
[1]Department of Entomology, Rutgers University, New Brunswick, NJ 08901, USA; [2]Department of Entomology, Ohio State University, Wooster, OH 44691, USA

20.1. Introduction

Entomopathogenic nematodes (EPNs) may be combined with other agricultural chemicals and control agents for various purposes. First, nematodes and other control agents may be applied simultaneously, or within a short time interval of each other, to control different pest species or stages of a pest. For convenience, nematodes may also be tank-mixed with other control agents, i.e. combined in the tank of application equipment, thus increasing the chances of interactions due to the higher concentration of both agents. Finally, nematodes may be combined with other control agents to achieve better control of a single pest through additive or, preferably, synergistic effects on pest mortality. The types of interactions between nematodes and other agents may range from effects on control agent viability and/or virulence, control agent recycling in the target pest, to effects on the pests' susceptibility to the control agents. Section 20.2 discusses effects of combinations on the viability of EPNs and Section 20.3 the effects of combinations on pest mortality and control agent recycling.

20.2. Compatibility

EPN infective juveniles (IJs) can tolerate short-term exposure (2–24 h) to many chemical and biological insecticides, fungicides, herbicides, fertilizers and growth regulators, and can thus be tank-mixed and applied together (see Tables 20.1–20.4). This offers a cost-effective alternative to pest control, and facilitates the use of nematodes in integrated pest management (IPM) systems. However, the actual concentration of the chemical to which the nematodes will be exposed will vary depending upon the application volume and system used (Alumai and Grewal, 2004). Incompatibility between agrochemicals and EPNs can be managed by choosing an appropriate time interval between the applications (Grewal *et al.*, 1998), the length of which may depend on the persistence of chemicals in the target substrate. Although specific information on appropriate application intervals is limited, it is usually recommended to wait for 1 and 2 weeks after the application of chemical insecticides and chemical nematicides, respectively, before nematode application. Tables 20.1–20.4 show that nematode species differ in compatibility with specific chemical or biological materials. Because some chemicals used as inert ingredients or adjuvants in formulations can be toxic to nematodes (Krishnayya and Grewal, 2002), compatibility of each formulation with the specific nematode species should be evaluated. The effect of pesticides on the natural populations of EPNs is discussed in Chapter 18 of this volume.

20.3. Interaction with Other Control Agents

The combination of two control agents against a given pest can result in antagonistic, additive or synergistic effects on speed of kill and mortality of the pest. Additive (also termed complementary) effects occur when the agents act independently of each other, i.e. there is no interaction. Synergistic or antagonistic effects occur when the interaction between agents renders the combination more or less effective in control than in the case of an additive effect. Synergism (or potentiation) is defined as the joint (or supplemental) action of two agents resulting in a greater effect than the sum of the activities of the agents acting alone. Unless the mechanism of the interaction is known, the determination of the kind of interaction is based upon statistical tests that determine whether the observed effect is significantly different from an additive effect. It has to be noted that to determine interactions, particularly synergism, the mortality caused by the individual agents should be low enough to allow for statistically significant improvements.

Observations on interaction between EPNs and chemicals or other pathogens are summarized in Tables 20.5 and 20.6, respectively. In Sections 20.3.1–20.3.4 we generalize the observations and add any additional observations made on the mechanisms responsible for the observed interaction and the effect of the chemicals or pathogens on nematode reproduction in the host cadavers resulting from combined exposure.

20.3.1. Chemicals

Most insecticides listed in Table 20.5 did not interact with the respective nematode species (additive effect). The organophosphate oxamyl increased *Steinernema carpocapsae* efficacy against *Agrotis segetum* synergistically, but only in fumigated soil, probably by enhancing the nematodes' nictation behaviour (Ishibashi, 1993). The pyrethroid tefluthrin had a weak synergistic effect on efficacy of *S. carpocapsae* and *Heterorhabditis bacteriophora* against *Diabrotica virgifera virgifera* larvae, probably because sublethal tefluthrin doses caused a paralytic and convulsive response in the insects that may have increased their nematode susceptibility (Nishimatsu and Jackson, 1998). The synergism between *Heterorhabditis* spp. and *S. glaseri* and the neonicotinoid imidacloprid in scarab larvae is well documented. Imidacloprid reduces the grubs' defensive behaviours resulting in increased nematode attachment and penetration (Koppenhöfer *et al.*, 2000b). However, *S. kushidai* and *S. scarabaei*, two rather scarab-specific species, generally do not interact with imidacloprid. Imidacloprid does not compromise nematode recycling in grubs (Koppenhöfer *et al.*, 2003).

Table 20.1. Compatibility of entomopathogenic nematodes (EPNs) with acaricides, insecticides and nematicides.

Active ingredient	Trade name/formulation	Nematode[a]	Compatibility	References
Abamectin	Dynamec	Sf	–	Head *et al.*, 2000
Acephate	Orthene 97 PE	Hb	–	Rovesti *et al.*, 1988
		Sf	–	De Nardo and Grewal, 2003
Aldicarb	Temik	Hb	–	Rovesti *et al.*, 1988
		Sc, Sf	–	Rovesti and Deseö, 1990
Amitraz	Edrizar 20.5 EC	Hb	+	Rovesti *et al.*, 1988
		Sc, Sf	–	Rovesti and Deseö, 1990
Azadirachtin	Azatin 3 EC	Sf	+	Grewal *et al.*, 1998
Azadirachtin	Margosan-O/0.25%	Sf	+	Grewal *et al.*, 1998
Azadirachtin	Nimbecidine	Sf	–	Krishnayya and Grewal, 2002
Bacillus thuringiensis (Bt) *israelensis*	ABG-6193	Hb, Sc	+	Poinar *et al.*, 1990
Bt *kurstaki*	Delfin WG	Sc	+	Barbarossa *et al.*, 1996
Bt *kurstaki*	Dipel	Hb, Sc	+	Poinar *et al.*, 1990
Bt *san diego*	M-One	Hb, Sc	+	Poinar *et al.*, 1990
Bt	Gnatrol	Hb, Sc	+	Poinar *et al.*, 1990
		Sf	+/–	De Nardo and Grewal, 2003
Bendiocarb	Turcam 76W	Sf, Sc, Hb	–	Zimmerman and Cranshaw, 1990
Binapacryl	Morocode	Sc	–	Prakasa Rao *et al.*, 1975
Carbaryl	Sevin SL	Hb	–	Zimmerman and Cranshaw, 1990
		Sc	+	Alumai and Grewal, 2004; Das and Divakumar, 1987; Zimmerman and Cranshaw, 1990
		Sf	+	Zimmerman and Cranshaw, 1990
Carbofuran	Curater 5 G	Hb	–	Rovesti *et al.*, 1988; Rovesti and Deseö, 1991
		Sc, Sf	+	Rovesti and Deseö, 1990
Carbosulfan	Marshal 5 G	Hb	–	Rovesti *et al.*, 1988; Rovesti and Deseö, 1991
		Sc, Sf	+	Rovesti and Deseö, 1990
Chlorfenvinphos	Birlane	Sc	–	Prakasa Rao *et al.*, 1975
Chloropyriphos-ethyl	Terial 40 L/40.8 EC	Hb	+	Rovesti *et al.*, 1988
		Sc	+	Rovesti and Deseö, 1990
		Sf	–	Rovesti and Deseö, 1990
Chlorpyrifos	Dursban 4 E	Hb, Sc, Sf	–	Zimmerman and Cranshaw, 1990
Chlorpyrifos	Dursban Pro 23.5%	Hb, Sc	+	Alumai and Grewal, 2004
Cyfluthrin	Baythroid 50	Sc	–	Barbarossa *et al.*, 1996
Cyhexatin	Plictran L 51.8 Flo	Hb	+	Rovesti *et al.*, 1988
		Sc, Sf	+	Rovesti and Deseö, 1990
Cyproconazole	Alto 100 SL	Sc	+	Barbarossa *et al.*, 1996
Deltamethrin	Decis 2.8 EC	Hb	+	Rovesti *et al.*, 1988
		Sc	+	Rovesti and Deseö, 1990
		Sf	+/–	Rovesti and Deseö, 1990; Head *et al.*, 2000
Diazinon	Basudin 20 E	Hb	+/–	Rovesti *et al.*, 1988; Rovesti and Deseö, 1991

continued

Table 20.1. *Continued.* Compatibility of entomopathogenic nematodes (EPNs) with acaricides, insecticides and nematicides.

Active ingredient	Trade name/formulation	Nematode[a]	Compatibility	References
Diazinon	Diazinon AG 500	Hb	–	Rovesti and Deseö, 1991; Zimmerman and Cranshaw, 1990
		Sc, Sf	–/+	Rovesti and Deseö, 1990; Zimmerman and Cranshaw, 1990
Dicofol	Kelthane MF 42 E	Hb	+	Rovesti *et al.*, 1988
		Sc, Sf	+	Rovesti and Deseö, 1990
Difenoconzole	Score 250 EC	Sc	–	Barbarossa *et al.*, 1996
Diflubenzuron	Adept IGR	Sc	+	Rovesti and Deseö, 1990
		Sf	+	Rovesti and Deseö, 1990; De Nardo and Grewal, 2003
Diflubenzuron	Dimilin 5 WP	Hb	+	Rovesti *et al.*, 1988
		Sf	+	Grewal and Richardson, 1993
Dimethoate	Danadim	Sf	+	Head *et al.*, 2000
DNOC	Sclinon 50 EC	Hb	+	Rovesti *et al.*, 1988
		Sc, Sf	+	Rovesti and Deseö, 1990
Endosulfan	Thiodan 32.9 WP	Hb	+	Das and Divakumar, 1987
		Sc	+	Das and Divakumar, 1987; Rovesti and Deseö, 1990
		Sf	+	Rovesti and Deseö, 1990
Exitiazox	Matacar 10 WP	Hb	+	Rovesti *et al.*, 1988
		Sc, Sf	+	Rovesti and Deseö, 1990
Fenamiphos	Nemacur 4.8 G	Hb	–	Rovesti *et al.*, 1988
		Sc, Sf	–	Rovesti and Deseö, 1990
Fenitrothion	Sumithion	Sc	–	Prakasa Rao *et al.*, 1975
Fenoxycarb	Precision 25 WP	Sf	–	De Nardo and Grewal, 2003
Fenthion	Lebaycid	Sc	+	Prakasa Rao *et al.*, 1975
Flubenzimine	Cropotex 50 WP	Hb	–	Rovesti *et al.*, 1988; Rovesti and Deseö, 1991
		Sc, Sf	–	Rovesti and Deseö, 1990
Fluroxypyr	Starane 180	Sc	+	Barbarossa *et al.*, 1996
Fluvalinate	Klartan	Sc	+	Barbarossa *et al.*, 1996
Fonofos	Dyfonate 5 G	Hb	+	Rovesti *et al.*, 1988
		Sc, Sf	+	Rovesti and Deseö, 1990
Formothion	Anthio	Sc	–	Prakasa Rao *et al.*, 1975
Halofenozide	Mach2 1.5 G	Hb	–	Alumai and Grewal, 2004
		Sc	+	Alumai and Grewal, 2004
Haloxyfop	Gallant super	Sc	+	Barbarossa *et al.*, 1996
Heptenophos	Hostacquick	Sc	–	Barbarossa *et al.*, 1996
		Sf	–	Head *et al.*, 2000
Imidacloprid	Merit 75 WP	Hb	+	Koppenhöfer and Kaya, 1998; Koppenhöfer *et al.*, 2003; Alumai and Grewal, 2004
		Sc	+	Alumai and Grewal, 2004; Koppenhöfer *et al.*, 2003
		Hm, Sf, Sg	+	Koppenhöfer *et al.*, 2003
Isofenphos + phoxim	Oftanol combination	Hb	+/–	Poinar *et al.*, 1990
L-Cyhalothrin	Karate	Sc	–	Barbarossa *et al.*, 1996
Lecithin	Bio blatt	Sc	–	Barbarossa *et al.*, 1996

Lindane (gamma HCH)	Lintox 3G	Hb	+	Rovesti *et al.*, 1988
		Sc, Sf	+	Rovesti and Deseö, 1990
Malathion	Malathion 50 EC	Sc	–	Das and Divakumar, 1987
Methamsodium	Geort 32.7 WSL	Hb	–	Rovesti *et al.*, 1988
		Sc, Sf	–	Rovesti and Deseö, 1990
Methomyl	Lannate 25 WP	Hb	–	Rovesti *et al.*, 1988; Rovesti and Deseö, 1991
		Sc, Sf	–	Rovesti and Deseö, 1990
Methoprene	Apex 5E	Sf	+	Grewal *et al.*, 1998
Metramitron	Goltix 70 WG	Sc	–	Barbarossa *et al.*, 1996
Monocrotophos	Azodrin	Sc	–	Prakasa Rao *et al.*, 1975
Neem	Neem kernel	Hb	–	Prakasa Rao *et al.*, 1975
		Sc, Sf, Sg	+	Rovesti *et al.*, 1988
Neem oil	Neem oil	Sf	+	Krishnayya and Grewal, 2002
Neem oil	Neem oil + soap	Sf	–	Krishnayya and Grewal, 2002
Parathion	E 605 FT	Hb	–	Rovesti *et al.*, 1988
		Sc, Sf	–	Rovesti and Deseö, 1990
Penconazole	Omnex WP 10	Sc	–	Barbarossa *et al.*, 1996
Phenthioate	Elsan	Sc	+	Prakasa Rao *et al.*, 1975
Phorate	Thimet 5G	Hb	–	Rovesti *et al.*, 1988
		Sc	–	Rovesti and Deseö, 1990; Gupta and Siddiqi, 1999
		Sf	+	Rovesti and Deseö, 1990
Phosalone	Zolone Flo	Sc	–/+	Prakasa Rao *et al.*, 1975; Barbarossa *et al.*, 1996
Phosmet	Imidan	Sc	–	Barbarossa *et al.*, 1996
Phosphamidon	Dimecron 20 EC	Hb	+/–	Rovesti *et al.*, 1988; Rovesti and Deseö, 1991
		Sc	+	Rovesti and Deseö, 1990
		Sf	–	Rovesti and Deseö, 1990
Propargite	Kelaran 57 EC	Hb	+/–	Rovesti *et al.*, 1988; Rovesti and Deseö, 1991
		Sc	–	Rovesti and Deseö, 1990
		Sf	+	Rovesti and Deseö, 1990
Pyridimine phosphate	PP-511	Sc	–	Prakasa Rao *et al.*, 1975
Spinosad	Conserve SC	Hb, Sc	+	Alumai and Grewal, 2004
		Sf	+	De Nardo and Grewal, 2003
Tebuconazole	Folicur 250 EC	Sc	–	Barbarossa *et al.*, 1996
Teflubenzuron	Nomolt	Sc	–	Barbarossa *et al.*, 1996
Terbufos	Gyanater 2G	Hb	+/–	Rovesti *et al.*, 1988; Rovesti and Deseö, 1991
		Sc, Sf	+	Rovesti and Deseö, 1990
Thiamethoxam	Meridian 25%	Hb	–/+	Alumai and Grewal, 2004; Koppenhöfer *et al.*, 2003
		Hm, Sf, Sg	+	Koppenhöfer *et al.*, 2003
		Sc	+	Alumai and Grewal, 2004; Koppenhöfer *et al.*, 2003
Trichlorfon	Dipterex 80	Sf	+	Head *et al.*, 2000
Trichlorfon	Dylox 80	Hb, Sc	–	Alumai and Grewal, 2004
Vamidothion	Kilval	Sc	+	Prakasa Rao *et al.*, 1975
Verticillium lecanii	Micro Germin	Sc	–	Barbarossa *et al.*, 1996
White oil	Biancolio 80 E	Hb	+	Rovesti *et al.*, 1988
		Sc, Sf	+	Rovesti and Deseö, 1990

[a]Sf = *Steinernema feltiae*; Hb = *Heterorhabditis bacteriophora*; Sc = *S. carpocapsae*; Hm = *H. megidis*; Sg = *S. glaseri*. – = non-compatible interaction; + = compatible interaction; / = both interaction types were observed.

Table 20.2. Compatibility of entomopathogenic nematodes (EPN) with fungicides.

Active ingredient	Trade name/ formulation	Nematode[a]	Compa- tibility	References
Abamectin	Vertimec	Sc	–	Barbarossa *et al.*, 1996
Aluminum tris	Aliette 80 WDG	Hb, Sc	–	Alumai and Grewal, 2004
Azoxystrobin	Abound 22.9%	Sf	+	Krishnayya and Grewal, 2002
Azoxystrobin	Heritage 50%	Sf	+	De Nardo and Grewal, 2003
Benomyl	Benlate 76 W	Hb, Sc, Sf	+	Zimmerman and Cranshaw, 1990
Bupirimate	Nimrod 25.6 EC	Hb	+	Rovesti *et al.*, 1988
		Sc, Sf	+	Rovesti and Deseö, 1990
Calcium hypochlorite	Chlorine	Sf	–	Grewal *et al.*, 1998
Carbendazin	Bavistin 50 WP	Hb	–	Rovesti *et al.*, 1988
		Sc	+/–	Rovesti and Deseö, 1990; Barbarossa *et al.*, 1996; Gupta and Siddiqi, 1999
		Sf	+	Rovesti and Deseö, 1990
Carboxin + thiram	Blekritt 31+30 WP	Hb	+	Rovesti *et al.*, 1988
		Sc, Sf	+	Rovesti and Deseö, 1990
Chloridazon	Pyramin	Sc	+	Barbarossa *et al.*, 1996
Chlorothalonil	Bravo 500 40.4%	Sf	+	Grewal *et al.*, 1998
Chlorothalonil	Clortosip 40 L Flo	Hb, Sc, Sf	+	Rovesti *et al.*, 1988; Zimmerman and Cranshaw, 1990
Cinameldehyde	Cinamate 30%	Sf	–	Krishnayya and Grewal, 2002
Cinameldehyde	Vertigo 50%	Sf	–	Grewal *et al.*, 1998
Copper sulphate	Poltiglia bord	Hb	+	Rovesti *et al.*, 1988
Cycloxydim	Focus	Sc	+	Barbarossa *et al.*, 1996
Cymoxanil + folpet	Folcarb combination 8.4 + 64 WP	Hb	–	Rovesti *et al.*, 1988
		Sc, Sf	+	Rovesti and Deseö, 1990
Diafenthiuron	Polo	Sc	+	Barbarossa *et al.*, 1996
Dinocap	Karathane PB 19.5 WP	Hb	+	Rovesti *et al.*, 1988
		Sc, Sf	+	Rovesti and Deseö, 1990
Dithianon	Delan 75 WP	Hb	+	Rovesti *et al.*, 1988
		Sc	+	Rovesti and Deseö, 1990; Barbarossa *et al.*, 1996
		Sf	+	Rovesti and Deseö, 1990
Dodine	Melprex 65 WP	Hb	–	Rovesti *et al.*, 1988
		Sc, Sf	–	Rovesti and Deseö, 1990
Etridiazole	Terrazole 35 WP	Sf	+	De Nardo and Grewal, 2003
Fenarimol	Rubigan 6 PB 6 WP	Hb	+	Rovesti *et al.*, 1988
		Sc, Sf	+	Rovesti and Deseö, 1990
Fenpropimorph	Corbel 79.5 EC	Hb	+	Rovesti *et al.*, 1988
		Sc, Sf	+	Rovesti and Deseö, 1990
Fentin-acetate	Brestan	Hb	–	Rovesti *et al.*, 1988
		Sc	+	Rovesti and Deseö, 1990
		Sf	–	Rovesti and Deseö, 1990
Flutdioxonil	Medallion 50%	Sf	+	De Nardo and Grewal, 2003
Flutolanil	Prostar 70 WP	Sf	+	De Nardo and Grewal, 2003
Folpet	Foltan 47.5 WP	Hb	+	Rovesti *et al.*, 1988
		Sc, Sf	+	Rovesti and Deseö, 1990
Formaldehyde	Formalin 37%	Sf	–	Grewal *et al.*, 1998

Fosethyl	Aliette	Sc	+	Barbarossa *et al.*, 1996
Hydrogen peroxide and peroxyacetic acid mixture	ZeroTol	Sf	–	Krishnayya and Grewal, 2002
Iprodione	Chipco 26GT	Sf	+	De Nardo and Grewal, 2003
Mannose	Dithane M 45 80 WP	Hb	+	Rovesti *et al.*, 1988
		Sc, Sf	+	Rovesti and Deseö, 1990
Mefenoxam	Subdue 2X 45 WSP	Hb, Sc	+	Alumai and Grewal, 2004
Mefenoxam	Subdue Maxx 21.3 ME	Sf	+	De Nardo and Grewal, 2003
Mepronil	?[b]	Hb, Sg	+	Lee *et al.*, 1999
Mercoprop-p	Duplosan	Sc	+	Barbarossa *et al.*, 1996
Mercurious chloride (60%) + mercuric chloride (30%)	Calco Clor	Hb, Sc, Sf	–	Zimmerman and Cranshaw, 1990
Metalaxyl + folpet	Eucritt F 10 + 40 WP	Hb	–	Rovesti *et al.*, 1988
		Sc, Sf	+	Rovesti and Deseö, 1990
Metazachlor	Butisan	Sc	+	Barbarossa *et al.*, 1996
Methoprene	Apex 5E 65.7%	Sf	+	Grewal *et al.*, 1998
Pentachloronitrobenzene	Terraclor 75%	Hb, Sc, Sf	+	Zimmerman and Cranshaw, 1990
Prochloraz	Sportak 40 EC	Hb	–	Rovesti *et al.*, 1988
		Sc, Sf	+	Rovesti and Deseö, 1990
Prochloraz-Manganese	Sporgon 50%	Sf	+	Grewal *et al.*, 1998
Procymidone	Sialex 50 WP	Hb	–	Rovesti *et al.*, 1988
		Sc, Sf	+	Rovesti and Deseö, 1990
Propamocarb	Previcur 66.5 WSL	Hb	–	Rovesti *et al.*, 1988
		Sc, Sf	+	Rovesti and Deseö, 1990
Propiconazole	Tilt 10 EC	Hb	+	Rovesti *et al.*, 1988
		Sc, Sf	+	Zimmerman and Cranshaw, 1990
Pyrazophos	Afugan 15 EC	Hb	+	Rovesti *et al.*, 1988
		Sc, Sf	+	Rovesti and Deseö, 1990
Rapeseed oil	Telmion	Sc	+	Barbarossa *et al.*, 1996
Streptomyces griseoviridis strain K61	Mycostop 30%	Sf	+	De Nardo and Grewal, 2003
Sulphur	Kumulus	Sc	+	Barbarossa *et al.*, 1996
Sulphur	Tiolvit 79.9 WP	Hb	+	Rovesti *et al.*, 1988
		Sc, Sf	+	Rovesti and Deseö, 1990
Thiobendazole	Mertect 340-F 42.28%	Sf	–	Grewal *et al.*, 1998
Thiophanate-methyl	Fungo Flo 4.5 F	Sf	+	De Nardo and Grewal, 2003
Thiophanate-methyl	Topsin M	Sf	+	Barbarossa *et al.*, 1996
Thiram	Pomarsol F	Sc	+	Barbarossa *et al.*, 1996
Thiram	TMTD 49 WP	Hb	+	Rovesti *et al.*, 1988
		Sc, Sf	+	Rovesti and Deseö, 1990
Tolclofos-methyl	Rizolex 50 WP	Hb	+	Rovesti *et al.*, 1988; Lee *et al.*, 1999
		Sc, Sf	+	Rovesti *et al.*, 1988
		Sg	+	Lee *et al.*, 1999
Tolylfluanide	Euparen M	Sc	+	Barbarossa *et al.*, 1996
Trichoderma harzianum	Rootshield 1.15%	Sf	+	De Nardo and Grewal, 2003
Triflumizole	Terraguard 50 W	Sf	+	De Nardo and Grewal, 2003
Triforine	Saprol 18 EC	Hb	+	Rovesti *et al.*, 1988
		Sc, Sf	+	Rovesti and Deseö, 1990

[a]Sc = *Steinernema carpocapsae*; Hb = *Heterorhabditis bacteriophora*; Sf = *S. feltiae*; Sg = *S. glaseri*.

[b] = not mentioned in paper.

– = non-compatible interaction; + = compatible interaction; / = both interaction types were observed.

Table 20.3. Compatibility of entomopathogenic nematodes (EPNs) *Steinernema* and *Heterorhabditis* with herbicides.

Active ingredient	Trade name/ formulation	Nematode[a]	Compatibility	References
2-4-D sodium salt	2-4-D	Sc	–	Gupta and Siddiqi, 1999
Alachlor	Lasso 43.1 EC	Hb	–	Rovesti *et al.*, 1988
		Sc, Sf	–	Rovesti and Deseö, 1990
Atrazine	Gesaprim 44.4 Flo	Hb	+	Rovesti *et al.*, 1988
		Sc, Sf	+	Rovesti and Deseö, 1990
Chloridazon	Pyramin 39 Flo	Hb	–	Rovesti *et al.*, 1988
		Sc, Sf	+	Rovesti and Deseö, 1990
Chlorsulfuron	Glean 75 Df	Hb	+	Rovesti *et al.*, 1988
		Sc, Sf	+	Rovesti and Deseö, 1990
Chlorthal-dimethyl	Dacthal 75 WP	Hb	+	Rovesti *et al.*, 1988
		Sc, Sf	+	Rovesti and Deseö, 1990
Chlortoluron	Dicuran L 45 Flo	Hb	+	Rovesti *et al.*, 1988
		Sc, Sf	+	Rovesti and Deseö, 1990
Clethodim	Envoy 12.6%	Sf	+	De Nardo and Grewal, 2003
Clopyralid	Cirloxin 10 WSL	Hb	+	Rovesti *et al.*, 1988
		Sc, Sf	+	Rovesti and Deseö, 1990
Dalapon	Dalascam 85 WP	Hb	+	Rovesti *et al.*, 1988
		Sc, Sf	+	Rovesti and Deseö, 1990
Dicamba	Banvel	Hb, Sc, Sf	+	Zimmerman and Cranshaw, 1990
Diphenamid	Enide 50 WP	Hb	+	Rovesti *et al.*, 1988
		Sc, Sf	+	Rovesti and Deseö, 1990
Fluazifop butyl	Fusilade 26 EC	Hb	+	Rovesti *et al.*, 1988
		Sc, Sf	+	Rovesti and Deseö, 1990
Fomesafen	Flex 29 WSL	Hb	+	Rovesti *et al.*, 1988
		Sc, Sf	+	Rovesti and Deseö, 1990
Glyphosate	Roundup 30.4 WSL	Hb	+	Rovesti *et al.*, 1988; Rovesti and Deseö, 1991
		Sc, Sf	+	Rovesti and Deseö, 1990
Ioxynil	Cipotril 24.8 E	Hb	+	Rovesti *et al.*, 1988
		Sc, Sf	+	Rovesti and Deseö, 1990
L-flamprop-isopropyl	Effix 20.9 EC	Hb	+	Rovesti *et al.*, 1988
		Sc, Sf	+	Rovesti and Deseö, 1990
Lenacil	Venazar 29 WSL	Hb	+	Rovesti *et al.*, 1988
		Sc, Sf	+	Rovesti and Deseö, 1990
Linuron	Afalon DS 37.6 Flo	Hb	+/–	Rovesti *et al.*, 1988; Rovesti and Deseö, 1991
MCPP (mecoprop)	U 46 KV 51 WSL	Hb	+	Rovesti *et al.*, 1988
		Sc, Sf	+	Rovesti and Deseö, 1990
Metrobuzin	Sencor 35 WP	Hb	+	Rovesti *et al.*, 1988
		Sc, Sf	+	Rovesti and Deseö, 1990
Oxyfluorfen	Goal 23.6 EC	Hb	+	Rovesti *et al.*, 1988
		Sc, Sf	–	Rovesti and Deseö, 1990
Paraquat	Gramoxone 17.9 WSL	Hb	–	Rovesti *et al.*, 1988; Rovesti and Deseö, 1991
		Sc, Sf	–	Rovesti and Deseö, 1990
Pendimethalin	Stomp 31.7 E	Hb	–	Rovesti *et al.*, 1988
		Sc, Sf	+	Rovesti and Deseö, 1990
Phenmedipham	Betanal 15.9 EC	Hb	+	Rovesti *et al.*, 1988
Propyzamide	Kerb 50 WP	Hb	–	Rovesti *et al.*, 1988
		Sc	+	Rovesti and Deseö, 1990
		Sf	–	Rovesti and Deseö, 1990
Trifluralin	Treflan 45.8 EC	Hb	–	Rovesti *et al.*, 1988; Rovesti and Deseö, 1991
		Sc	–	Rovesti and Deseö, 1990

[a]Sc = *Steinernema carpocapsae*; Hb = *Heterorhabditis bacteriophora*; Sf = *S. feltiae*.
– = non-compatible interaction; + = compatible interaction; / = both interaction types were observed.

Table 20.4. Compatibility of entomopathogenic nematodes (EPNs) *Steinernema* and *Heterorhabditis* with fertilizers, growth regulators and surfactants.

Active ingredient	Trade name/ formulation	Nematode[a]	Compa- tibility	References
Ammonium sulphate	Ammonium sulphate	Sc	+	Prakasa Rao *et al.*, 1975
Ancymidol	A-Rest 0.4%	Sf	+	De Nardo and Grewal, 2003
Antibacterial liquid soap	Ajax	Sf	–	Krishnayya and Grewal, 2002
Nitrogen and potassium oxide mixture	Gromore	Sc	+	Prakasa Rao *et al.*, 1975
Nitrogen and potassium oxide mixture	Sulphala	Sc	+	Prakasa Rao *et al.*, 1975
Paclobutrazol	Bonzi 0.4%	Sf	+	De Nardo and Grewal, 2003
Potassium oxide	Muriate of Potash	Sc	+	Prakasa Rao *et al.*, 1975
Potassium oxide	Superphosphate	Sc	+	Prakasa Rao *et al.*, 1975
Uniconazole-P	Sumagic 0.055%	Sf	+	De Nardo and Grewal, 2003
Urea	Urea	Sc	+	Prakasa Rao *et al.*, 1975

[a]Sc = *Steinernema carpocapsae*; Sf = *S. feltiae*.
+ = compatible interaction; – = non-compatible interaction.

20.3.2. Pathogens

20.3.2.1. Nematode–nematode interactions

Combination of two nematode species against one target pest generally results in additive target mortality (Table 20.6). Under laboratory conditions, two nematode species may infect the same host individual (Kondo, 1989; Alatorre-Rosas and Kaya, 1991; Koppenhöfer *et al.*, 1995), but under natural conditions, dual infections are unlikely because steinernematid nematodes are repelled from hosts infected with other nematodes within 6–9 h of establishment of the first species in the host's haemocoel (Glazer, 1997; Grewal *et al.*, 1997). While this avoidance reduces interference, combination of two nematode species against one target is likely to lead to competitive exclusion of one species in the long run (Koppenhöfer *et al.*, 1996). However, combination of two nematode species may provide effective control of two pests (Kaya *et al.*, 1993) and the nematode species may coexist (Koppenhöfer and Kaya, 1996) if the targets differ in susceptibility to the two nematodes due to differences in the nematodes' search strategies and/or pathogenicity.

20.3.2.2. Nematode–bacterium interactions

Combination of EPNs with *Bacillus thuringiensis* (Bt) against several lepidopteran pests resulted in additive or antagonistic interactions (Table 20.6). In the European crane fly, *Tipula paludosa*, combination of *S. carpocapsae* and the β-exotoxin of Bt serovar *thuringiensis* caused a strong synergistic effect on larval mortality (Lam and Webster, 1972), but the nematode rates required were not economically feasible. Synergistic combinations of *S. glaseri* or *H. bacteriophora* with Bt serovar *japonensis* (Btj) may be feasible when applied against a Btj-sensitive scarab species (Koppenhöfer *et al.*, 1999). Nematode reproduction was reduced (Bednarek, 1986) or inhibited (Kaya and Burlando, 1989; Poinar *et al.*, 1990) by combination with various Bt strains unless the nematodes were inoculated before the bacterium.

Milky disease bacterium, *Paenibacillus* (= *Bacillus*) *popilliae*, infection of third instar masked chafer, *Cyclocephala hirta*, enhanced the efficacy of *H. bacteriophora* and *S. glaseri* by facilitating penetration of the nematode into the midgut (Thurston *et al.*, 1993, 1994). Neither *H. bacteriophora* nor *P. popilliae* reproduction in the host

Table 20.5. Effect of combining entomopathogenic nematodes (EPNs) with insecticides on insect control.

Active ingredient	Trade name/ formulation[a]	Rate (a.i./ha)	Nematode[b]	Target pest	Arena[c]	Interaction[d]	References
Acephate	?/?	250 ppm[e]	Sc	*Pieris rapae crucivora*	F	+	Ishibashi, 1993
				Spodoptera litura	F	+	Ishibashi, 1993
Chlorpyrifos-methyl	Dursban 2 EC	9.4 kg	HG, Sl	*Exomala orientalis*	F	*	Lee *et al*., 2002
Cryomazine	Trigard 75WP	0.8–50 g/kg seed[f]	Hb	*Delia antiqua*	L	+	Yildirin and Hoy, 2003
Diazinon	Diazinon 4 EC	0.7 kg	Sk	*Cyclocephala hirta*	G	+	Koppenhöfer *et al*., 2000c
				E. orientalis	G	+	Koppenhöfer *et al*., 2000c
Dichlorvos	?/?	250 ppm[e]	Sc	*P. rapae crucivora*	F	+	Ishibashi, 1993
				S. litura	F	+	Ishibashi, 1993
Fonofos	Techn. grade	0.06–0.25 ppm[g]	Sc	*Diabrotica virgifera virgifera*	L	−/+	Nishimatsu and Jackson, 1998
Halofenozide	Mach2/60 WP	1.1–2.3 kg	Hm	*Popillia japonica*	L	+	Mannion *et al*., 2000
Imidacloprid	Merit 75 WP	0.2 kg	Hb	*C. borealis*	G	*	Koppenhöfer *et al*., 2000a
				C. hirta	G, F	*/+	Koppenhöfer and Kaya, 1998; Koppenhöfer *et al*., 2000a
				C. pasadenae	G	*	Koppenhöfer and Kaya, 1998; Koppenhöfer *et al*., 2000a
				E. orientalis	G, F	*	Koppenhöfer *et al*., 2000a, 2002
				P. japonica	G, F	*/+	Koppenhöfer *et al*., 2000a, 2002
			Sg	*C. borealis*	G, F	*	Koppenhöfer and Kaya, 1998; Koppenhöfer *et al*., 2000a
				C. hirta	G, F	*	Koppenhöfer and Kaya, 1998; Koppenhöfer *et al*., 2000a
				C. pasadenae	G, F	*	Koppenhöfer and Kaya, 1998; Koppenhöfer *et al*., 2000a

				E. orientalis	G	*	Koppenhöfer *et al.*, 2000a, 2002
				P. japonica	G, F	*	Koppenhöfer *et al.*, 2000a
			Hm, He	*E. orientalis*	G	*	Koppenhöfer *et al.*, 2002
			Sk	*C. borealis*	G	+	Koppenhöfer *et al.*, 2000a
				C. hirta	G	+	Koppenhöfer *et al.*, 2000a
				C. pasadenae	G	+	Koppenhöfer *et al.*, 2000a
				P. japonica	G	+	Koppenhöfer *et al.*, 2000a
			Ss	*C. borealis*	G	+	Koppenhöfer and Fuzy, 2003a
				E. orientalis	G, F	*/+	Koppenhöfer and Fuzy, 2003a
				P. japonica	G	+	Koppenhöfer and Fuzy, 2003a
				Maladera castanea	G	+	Koppenhöfer and Fuzy, 2003b
Oxamyl	?/1% G	3 kg	Sc	*Agrotis segetum*	F	+/*	Ishibashi, 1993
Tefluthrin	Techn. grade	0.06–0.25 ppm[g]	Sc, Hb	*D. virgifera virgifera*	L	+/*	Nishimatsu and Jackson, 1998
Terbufos	Techn. grade	0.025–0.1 ppm[g]	Sc	*D. virgifera virgifera*	L	–/+	Nishimatsu and Jackson, 1998
Thiamethoxam	Meridian 25 WG	0.2 kg	Sg	*E. orientalis*	G	*	Koppenhöfer *et al.*, 2002
			Hb	*E. orientalis*	G, F	+	Koppenhöfer *et al.*, 2002
			Hm, He	*E. orientalis*	G	+	Koppenhöfer *et al.*, 2002

[a]? = not mentioned in paper.
[b]Sc = *S. carpocapsae*; HG = *Heterorhabditis* sp. Gyeongsan; Sl = *S. longicaudum*; Hb = *H. bacteriophora*; Sk = *S. kushidai*; Hm = *H. marelatus*; Sg = *S. glaseri*; He = *H. megidis*; Ss = *S. scarabaei*.
[c]F = field; L = laboratory; G = greenhouse/pot experiment.
[d]+ = additive; * = synergistic; – = antagonistic; / = both interaction types were observed.
[e]Foliar application.
[f]a.i. was film-coated on onion seed.
[g]Mixed in soil.

Table 20.6. Effect of combining entomopathogenic nematodes (EPNs) with other pathogens on insect control.

Other pathogen	Nematode[a]	Target pest	Arena[b]	Interaction[c]	References
Nematodes					
Heterorhabditis bacteriophora	Sk, Sg	*Cyclocephala hirta*	G, F	+	Lam and Webster, 1972
Steinernema glaseri	Sk	*Exomala orientalis*	G	+	Lam and Webster, 1972
H. bacteriophora	Sc, Sr, SH	*Diabrotica undecimpunctata undecimpunctata*	L, G	+	Choo *et al.*, 1996
Steinernema sp. Hanrim	Hb, Sr	*D. undecimpunctata undecimpunctata*	L, G	+	Choo *et al.*, 1996
Bacteria					
Bacillus thuringiensis (Bt) *kurstaki*	Sc	*Plutella xylostella*	L	+/–[d]	Baur *et al.*, 1998
Bt *aizawai*	Sc	*P. xylostella*	F	+	Baur *et al.*, 1998
Bt *kurstaki*	Sc	*Platyptilia carduidactyla*	F	–	Bari and Kaya, 1984
Bt *kurstaki*	Sc	*Lymantria dispar*	L	–	Bednarek, 1986
Bt *thuringiensis* (β-exotoxin)	Sc	*Tipula paludosa*	L	*	Lam and Webster, 1972
Bt *japonensis*	Hb	*C. hirta*	L, G, F	+/*	Koppenhöfer and Kaya, 1997; Koppenhöfer *et al.*, 1999
		C. pasadenae	L, G, F	+/*	Koppenhöfer and Kaya, 1997; Koppenhöfer *et al.*, 1999
		Popillia japonica	F	+	Koppenhöfer *et al.*, 2000c
	Sg	*C. hirta*	L	+/*	Koppenhöfer and Kaya, 1997
		E. orientalis	L, F	*/+	Koppenhöfer *et al.*, 1999
	Sk	*C. hirta*	L	+	Koppenhöfer and Kaya, 1997
Paenibacillus popilliae[e]	Hb	*C. hirta*	G	+	Thurston *et al.*, 1994
	Sg	*C. hirta*	G	*	Thurston *et al.*, 1994

Serratia marcescens	Hi, Sc	*Curculio caryae*	L	–	Shapiro-Ilan *et al.*, 2004.
Fungi					
Beauveria bassiana	Sc	*Galleria mellonella*	L	accel.	Kamionek *et al.*, 1974a
		Tribolium castaneum	L	decel.	Kamionek *et al.*, 1974a
Paecilomyces farinosus	Sc	*G. mellonella*	L	accel.	Kamionek *et al.*, 1974b
		T. castaneum	L	+	Kamionek *et al.*, 1974b
		Trogoderma granarium	L	+	Kamionek *et al.*, 1974b
Paecilomyces fumosoroseus	Hi, Sc	*Cu. caryae*	L	–	Shapiro-Ilan *et al.*, 2004.
B. bassiana	Sc, Hb	*Spodoptera exigua*	L	+	Barbercheck and Kaya, 1991a
B. bassiana	Hi	*Cu. caryae*	L	–	Shapiro-Ilan *et al.*, 2004.
B. bassiana	Sc	*Cu. caryae*	L	+/–	Shapiro-Ilan *et al.*, 2004.
Metarhizium anisopliae	Hi	*Cu. caryae*	L	+	Shapiro-Ilan *et al.*, 2004.
M. anisopliae	Sc	*Cu. caryae*	L	+/–	Shapiro-Ilan *et al.*, 2004.
Viruses					
L. dispar NPV[f]	Sc	*L. dispar*	L	–	Bednarek, 1986
S. exigua NPV	Sc	*S. exigua*	L, F	+	Gothama *et al.*, 1995, 1996

[a]Sk = *S. kushidai*; Sg = *S. glaseri*; Sc = *S. carpocapsae*; Sr = *S. riobrave*; SH = *Steinernema* sp. Hanrim; Hb = *H. bacteriophora*; Hi = *H. indica*.
[b]G = greenhouse/pot experiment; F = field; L = laboratory.
[c]+ = additive; / = both interaction types were observed; – = antagonistic; * = synergistic; accel. = speed of kill accelerated; decel. = speed of kill decelerated.
[d]Additive effect in Bt-susceptible *P. xylostella* strain, antagonistic effect in Bt-resistant strain.
[e]Third instar larvae were exposed for 15 days to *Paenibacillus popilliae* before exposure to nematodes.
[f]Nucleopolyhedrovirus.

was affected (Thurston *et al.*, 1993). However, the slow establishment of milky disease in white grub populations and the lack of *in vitro* production methods for the bacterium limit the feasibility of this combination.

Simultaneous application of *Serratia marcescens* with *H. indica* or *S. carpocapsae* against pecan weevil, *Curculio caryae*, resulted mostly in antagonism (Shapiro-Ilan *et al.*, 2004). Reduced mortality in *C. caryae* may have been due to negative interactions between the *Se. marcescens* and the nematode's symbiont. Indeed, Martin (2002) reported that *Photorhabdus luminescens* (a symbiont of *Heterorhabditis* spp.) and *Se. marcescens* inhibited each other *in vitro*, and exhibited antagonistic toxicity to the Colorado potato beetle, *Leptinotarsa decemlineata*.

20.3.2.3. Nematode–fungus interactions

Nematode–fungus combinations generally result in additive effects on target mortality (Table 20.6), but in most studies mortality caused by the individual agents was too high to allow for significant improvement, and often only effects on speed of kill could be determined. Nematodes and fungi rarely coproduced progeny in infected hosts (Barbercheck and Kaya, 1991a) because the nematodes' symbiotic bacteria and the fungi excluded each other, the agent infecting first, often excluding the other. Avoidance by *H. bacteriophora* and *S. carpocapsae* of *Beauveria bassiana*-infected hosts may reduce antagonistic interactions between fungus and nematodes in the field (Barbercheck and Kaya, 1991b). Shapiro-Ilan *et al.* (2004) reported that combinations of *Be. bassiana* with *S. carpocapsae* or *H. indica* tended to be antagonistic for suppression of *Cu. caryae*, whereas combinations with *Metarhizium anisopliae* tended to be additive (particularly with *H. indica*); the authors suggested that further studies on combinations of *M. anisopliae* with EPNs are warranted. Additionally, Shapiro-Ilan *et al.* (2004) observed antagonism between *Paecilomyces fumosoroseus* combined with *H. indica or S. carpocapsae.*

20.3.2.4. Nematode–virus interactions

Combination of *S. carpocapsae* with the nucleopolyhedrovirus (NPV) of the beet armyworm, *Spodoptera exigua*, caused additive mortality of *Sp. exigua* larvae without affecting *S. carpocapsae* reproduction (Gothama *et al.*, 1995, 1996). *S. carpocapsae* infected and developed in NPV-infected larvae of *Lymantria dispar* (Bednarek, 1986) and *Sp. exigua* (Kaya and Burlando, 1989). However, if nematode infection occurred only a few days before the insect was killed by the NPV, the developing nematodes were exposed to desiccation when the insect's cuticle disintegrated (Bednarek, 1986; Kaya and Burlando, 1989). In sixth instars of the common armyworm, *Pseudaletia unipuncta*, infected with a granulovirus (Oregonian strain), *S. carpocapsae* reproduced normally (Kaya and Brayton, 1978).

20.3.3. Parasitoids

Larval survival of the braconid *Apanteles militaris* (Kaya, 1978) and the ichneumonid *Hyposoter exiguae* (Kaya and Hotchkin, 1981) inside *Ps. unipuncta* larvae, and of the braconid *Meteorus rubens* inside black cutworm, *Agrotis ipsilon*, larvae (Zaki *et al.*, 1997) was reduced by host infection with *S. carpocapsae*, although parasitoid survival increased with age of the larvae at infection. *S. carpocapsae* reproduced in *Ps. unipuncta* larvae that were parasitized or from which the parasitoid had already emerged (Kaya and Hotchkin, 1981). After cocoon formation, pupae of *Ap. militaris* (Kaya, 1978) and *Hy. exiguae*, *Cotesia medicaginis* and *Chelonus* sp. (Kaya and Hotchkin, 1981) were resistant to nematode infection. *S. carpocapsae* also reduced survival of *Cardiochiles diaphaniae* in *Diaphania* spp. hosts, but the parasitoid's pupae were immune to nematode infection (Shannag and Capinera, 2000). Larvae of the ichneumonids *Mastrus ridibundus* and *Liotryphon caudatus*, ectoparasitic on cocooned codling moth larvae, were susceptible to *S. carpocapsae*, but were not affected once their own tightly woven cocoons were formed (Lacey *et al.*, 2003). In addition, adult female *Ma. ridibundus* and *Li. caudatus* avoided ovipositing on *S. carpocapsae*-infected codling moth larvae.

Ganaspidium utilis, a larval–pupal parasitoid of the leafminer *Liriomyza trifolii*, was susceptible to *S. carpocapsae*, but combination of *G. utilis* with *S. carpocapsae* caused higher leafminer mortality than either agent alone (Hara *et al.*, 1989). *Diglyphus begini* fed on, but did not oviposit on, *L. trifolii* larvae infected with *S. carpocapsae*, thus avoiding unsuitable hosts (Sher *et al.*, 2000). *Steinernema* spp. applied to the soil to control spruce web-spinning sawfly, *Cephalcia arvensis*, reduced emergence of the sawfly parasitoid, *Xenoschesis fulvipes*, by 66.6% (Battisti, 1994).

In adult tawny mole cricket, *Scapteriscus vicinus*, combination of *S. scapterisci* with larvae of the tachinid *Ormia depleta* resulted in a higher percentage of hosts producing nematode progeny but a lower percentage of tachinid progeny compared with exposure to either agent alone. About 10–31% of hosts produced both nematode and tachinid progeny (Parkman and Frank, 2002). *S. carpocapsae* infected > 3-day-old larvae of the tachinid armyworm parasitoid, *Compsilura concinnata*, inside the armyworm host and adult flies as they eclosed from the host, but did not infect pupae (Kaya, 1984). When armyworms contained ≤ 3-day-old tachinid larvae, both nematode and parasitoid could develop.

20.3.4. Predators

Although larvae and pupae of several species of carabids, cicindelids, staphylinids and hover fly larvae entering the ground for pupation were found to be moderately susceptible to several species of nematodes (Georgis *et al.*, 1991; Pölking and Heimbach, 1992; Bathon, 1996), the effect of nematode application on natural populations of carabids, staphylinids, histerids and formicids was generally negligible, or were temporally, as well as spatially, restricted (Georgis *et al.*, 1991; Bathon, 1996; Wang *et al.*, 2001).

20.4. Synthesis and Recommendations

While nematodes are compatible with many agrochemicals even to the extent of being tank-mixed, many other agrochemicals negatively affect nematode survival. Since there appears to be no clear trend as to the sensitivity of different nematode species to different classes of agrochemicals, any new nematode species–agrochemical combination needs to be evaluated for compatibility. Otherwise, the two agents should be applied with 1–2 weeks time interval between them.

Combined application of two agents is only useful if target mortality is synergistically increased. Where synergistic interactions between insecticides and nematodes have been observed, the insecticide either affected nematode behaviour or, more commonly, the behaviour of the target pest. This is not surprising as the insecticides are mostly nerve poisons. In studying such interactions, it is important to use experimental arenas that allow for the normal expression of both nematode and insect behaviours. Results observed in oversimplified laboratory arenas may not be reproducible in the field.

Combinations of nematodes with other nematode species, fungi and viruses generally result in additive effects on pest mortality, whereas nematode–bacteria combination interactions range from antagonistic to synergistic. However, as only a few studies have been conducted for these combinations, different trends may be observed with different insect or pathogen species, and/or environmental conditions. Where two organisms are coapplied against the same insect target, there will always be competition for resources even though under field conditions coinfections are less likely and avoidance is more likely.

Only the interactions of nematodes with Btj or imidacloprid in white grubs have been sufficiently studied in greenhouse and field trials to allow for recommendations to be made at this time. Unfortunately, Btj is presently not commercially available. The nematode–imidacloprid combination is an effective tool for curative white grub control. However, because both control agents are expensive, the economic feasibility of the combination will depend on how much nematode and/or imidacloprid rate can be reduced compared with their recommended application rates. This will depend on the nematode and imidacloprid susceptibility of the target white grub species and will require further studies.

References

Alatorre-Rosas, R. and Kaya, H.K. (1991) Interaction between two entomopathogenic nematode species in the same host. *Journal of Invertebrate Pathology* 57, 1–6.

Alumai, A. and Grewal, P.S. (2004) Tankmix-compatibility of the entomopathogenic nematodes, *Heterorhabditis bacteriophora* and *Steinernema carpocapsae*, with selected chemical pesticides used in turfgrass. *Biocontrol Science and Technology* 14, 725–730.

Barbarossa, B., Frulla, R., Grazzi, G. and Rovesti, L. (1996) Compatibility of *Steinernema carpocapsae* with pesticides. *International Organization for Biological and Integrated Control Bulletin* 19, 136–139.

Barbercheck, M.E. and Kaya, H.K. (1991a) Competitive interactions between entomopathogenic nematodes and *Beauveria bassiana* (Deuteromycotina: Hyphomycetes) in soilborne larvae of *Spodoptera exigua* (Lepidoptera: Noctuidae). *Environmental Entomology* 20, 707–712.

Barbercheck, M.E. and Kaya, H.K. (1991b) Effect of host condition and soil texture on host finding by the entomogenous nematodes *Heterorhabditis bacteriophora* (Rhabditida: Heterorhabditidae) and *Steinernema carpocapsae* (Rhabditida: Steinernematidae). *Environmental Entomology* 20, 582–589.

Bari, M.A. and Kaya, H.K. (1984) Evaluation of the entomogenous nematode *Neoaplectana carpocapsae* (=*Steinernema feltiae*) Weiser (Rhabditida: Steinernematidae) and the bacterium *Bacillus thuringiensis* Berliner var. *kurstaki* for suppression of the artichoke plume moth (Lepidoptera: Pterophoridae). *Journal of Economic Entomology* 77, 225–229.

Bathon, H. (1996) Impact of entomopathogenic nematodes on non-target hosts. *Biocontrol Science and Technology* 6, 421–434.

Battisti, A. (1994) Effects of entomopathogenic nematodes on the spruce web-spinning sawfly *Cephalcia arvensis* Panzer and its parasitoids in the field. *Biocontrol Science and Technology* 4, 95–102.

Baur, M.E., Kaya, H.K., Tabashnik, B.E. and Chilcutt, C.F. (1998) Suppression of diamondback moth (Lepidoptera: Plutellidae) with an entomopathogenic nematode (Rhabditida: Steinernematidae) and *Bacillus thuringiensis* Berliner. *Journal of Economic Entomology* 91, 1089–1095.

Bednarek, A. (1986) Development of the *Steinernema feltiae* (Filipjev) entomogenous nematode (Steinernematidae) in the conditions of occurrence in the insect's body cavity of other pathogens. *Annals of Warsaw Agricultural University – Animal Science* 20, 69–74.

Choo, H.Y., Koppenhöfer, A.M. and Kaya, H.K. (1996) Combination of two entomopathogenic nematode species for suppression of an insect pest. *Journal of Economic Entomology* 89, 97–103.

Das, J.N. and Divakumar, B.J. (1987) Compatibility of certain pesticides with DD-136 nematode. *Plant Protection Bulletin* 39, 20–22.

De Nardo, E.A.B. and Grewal, P.S. (2003) Compatibility of *Steinernema feltiae* (Nematoda: Steinernematidae) with microbial and chemical pesticides and plant growth regulators used in glasshouse plant production. *Biocontrol Science and Technology* 13, 441–448.

Georgis, R., Kaya, H.K. and Gaugler, R. (1991) Effect of steinernematid and heterorhabditid nematodes (Rhabditida: Steinernematidae and Heterorhabditidae) on nontarget arthropods. *Environmental Entomology* 20, 815–822.

Glazer, I. (1997) Effects of infected insect on secondary invasion of steinernematid entomopathogenic nematodes. *Parasitology* 114, 597–604.

Gothama, A.A.A., Sikorowski, P.P. and Lawrence, G.W. (1995) Interactive effects of *Steinernema carpocapsae* and *Spodoptera exigua* nuclear polyhedrosis virus on *Spodoptera exigua* larvae. *Journal of Invertebrate Pathology* 66, 270–276.

Gothama, A.A.A., Lawrence, G.W. and Sikorowski, P.P. (1996) Activity and persistence of *Steinernema carpocapsae* and *Spodoptera exigua* nuclear polyhedrosis virus against *S. exigua* larvae on soybean. *Journal of Nematology* 28, 68–74.

Grewal, P.S. and Richardson, P.N. (1993) Effects of application rates of *Steinernema feltiae* (Nematoda: Steinernematidae) on biological control of the mushroom fly *Lycoriella auripila* (Diptera:

Sciaridae). *Biocontrol Science and Technology* 3, 29–40.

Grewal, P.S., Lewis, E.E. and Gaugler, R. (1997) Response of infective stage parasites (Nematoda: Steinernematidae) to volatile cues from infected hosts. *Journal of Chemical Ecology* 23, 503–515.

Grewal, P.S., Weber, T.A. and Betterley, D.A. (1998) Compatibility of *Steinernema feltiae* with chemicals used in mushroom production. *Mushroom News* 46, 6–10.

Gupta, P. and Siddiqi, M.R. (1999) Compatibility studies on *Steinernema carpocapsae* with some pesticidal chemicals. *Indian Journal of Entomology* 61, 220–225.

Hara, A.H., Mello, C.L., Tsutsumi, L.H.A. and Kaya, H.K. (1989) Laboratory susceptibility of some tropical pest species and a non-target organism to the entomopathogenic nematode, *Steinernema carpocapsae. Journal for Hawaiian and Pacific Agriculture* 2, 6–9.

Head, J., Walters, K.F.A. and Langton, S. (2000) The compatibility of the entomopathogenic nematode, *Steinernema feltiae,* and chemical insecticides for the control of the South American leafminer, *Liriomyza huidobrensis. Biocontrol* 45, 345–353.

Ishibashi, N. (1993) Integrated control of insect pests by *Steinernema carpocapsae.* In: Bedding, R.A., Akhurst, R.J. and Kaya, H.K. (eds) *Nematodes and the Biological Control of Insects.* CSIRO Publications, East Melbourne, Australia, pp. 105–113.

Kamionek, M., Sandner, H. and Seryczynska, H. (1974a) The combined action of *Beauveria bassiana* (Bals/Vuill) (Fungi Imperfecti: Moniliales) and *Neoaplectana carpocapsae* Weiser (Nematoda: Steinernematidae). *Bulletin de l'Academie Polonaise des Sciences* 22, 253–257.

Kamionek, M., Sandner, H. and Seryczynska, H. (1974b) Combined action of *Paecilomyces farinosus* Dicks (Brown et Smith) (Fungi Imp.: Moniliales) and *Neoaplectana carpocapsae* Weiser, 1955 (Nematoda: Steinernematidae). *Acta Parasitologica* 22, 357–363.

Kaya, H.K. (1978) Interaction between *Neoaplectana carpocapsae* (Nematoda: Steinernematidae) and *Apanteles militaris* (Hymenoptera: Braconidae), a parasitoid of the armyworm *Pseudaletia unipuncta. Journal of Invertebrate Pathology* 31, 358–364.

Kaya, H.K. (1984) Effect of the entomogenous nematode *Neoaplectana carpocapsae* on the Tachinid parasite *Compsilura concinnata. Journal of Nematology* 16, 9–13.

Kaya, H.K. and Brayton, M.A. (1978) Interaction between *Neoaplectana carpocapsae* and a granulosis virus of the armyworm *Pseudaletia unipuncta. Journal of Nematology* 10, 350–354.

Kaya, H.K. and Burlando, T.M. (1989) Development of *Steinernema feltiae* (Rhabditida: Steinernematidae) in diseased insect hosts. *Journal of Invertebrate Pathology* 53, 164–168.

Kaya, H.K. and Hotchkin, P.G. (1981) The nematode *Neoaplectana carpocapsae* Weiser and its effect on selected ichneumonid and braconid parasites. *Environmental Entomology* 10, 474–478.

Kaya, H.K., Burlando, T.M. and Thurston, G.S. (1993) Two entomopathogenic nematode species with different search strategies for insect suppression. *Environmental Entomology* 22, 859–864.

Kondo, E. (1989) Studies on the infectivity and propagation of entomogenous nematodes, *Steinernema* spp., (Rhabditida: Steinernematidae), in the common cutworm, *Spodoptera litura* (Lepidoptera: Noctuidae). *Bulletin of the Faculty of Agriculture, Saga University* 67, 1–88.

Koppenhöfer, A.M. and Fuzy, E.M. (2003a) *Steinernema scarabaei* for the control of white grubs. *Biological Control* 28, 47–59.

Koppenhöfer, A.M. and Fuzy, E.M. (2003b) Biological and chemical control of the Asiatic garden beetle, *Maladera castanea* (Coleoptera: Scarabaeidae). *Journal of Economic Entomology* 96, 1076–1082.

Koppenhöfer, A.M. and Kaya, H.K. (1996) Coexistence of two steinernematid nematode species (Rhabditida: Steinernematidae) in the presence of two host species. *Applied Soil Ecology* 4, 221–230.

Koppenhöfer, A.M. and Kaya, H.K. (1997) Additive and synergistic interaction between entomopathogenic nematodes and *Bacillus thuringiensis* for scarab grub control. *Biological Control* 8, 131–137.

Koppenhöfer, A.M. and Kaya, H.K. (1998) Synergism of imidacloprid and an entomopathogenic nematode: a novel approach to white grub (Coleoptera: Scarabaeidae) control in turfgrass. *Journal of Economic Entomology* 91, 618–623.

Koppenhöfer, A.M., Kaya, H.K., Shanmugam, S. and Wood, G.L. (1995) Interspecific competition between steinernematid nematodes within an insect host. *Journal of Invertebrate Pathology* 66, 99–103.

Koppenhöfer, A.M., Baur, M.E. and Kaya, H.K. (1996) Competition between two steinernematid nematode species for an insect host at different soil depths. *Journal of Parasitology* 82, 34–40.

Koppenhöfer, A.M., Choo, H.Y., Kaya, H.K., Lee, D.W. and Gelernter, W.D. (1999) Increased field and greenhouse efficacy against scarab grubs with a combination of an entomopathogenic nematode and *Bacillus thuringiensis. Biological Control* 14, 37–44.

Koppenhöfer, A.M., Brown, I.M., Gaugler, R., Grewal, P.S., Kaya, H.K. and Klein, M.G. (2000a) Synergism of entomopathogenic nematodes and imidacloprid against white grubs: greenhouse and field evaluation. *Biological Control* 19, 245–251.

Koppenhöfer, A.M., Grewal, P.S. and Kaya, H.K. (2000b) Synergism of imidacloprid and entomopathogenic nematodes against white grubs: the mechanism. *Entomologia Experimentalis Applicata* 94, 283–293.

Koppenhöfer, A.M., Wilson, M.G., Brown, I., Kaya, H.K. and Gaugler, R. (2000c) Biological control agents for white grubs (Coleoptera: Scarabaeidae) in anticipation of the establishment of the Japanese beetle in California. *Journal of Economic Entomology* 93, 71–80.

Koppenhöfer, A.M., Cowles, R.S., Cowles, E.A., Fuzy, E.M. and Baumgartner, L. (2002) Comparison of neonicotinoid insecticides as synergists for entomopathogenic nematodes. *Biological Control* 24, 90–97.

Koppenhöfer, A.M., Cowles, R.S., Cowles, E.A., Fuzy, E.M. and Kaya, H.K. (2003) Effect of neonicotinoid insecticide synergists on entomopathogenic nematode fitness. *Entomologia Experimentalis et Applicata* 106, 7–18.

Krishnayya, P.V. and Grewal, P.S. (2002) Effect of neem and selected fungicides on viability and virulence of the entomopathogenic nematode *Steinernema feltiae. Biocontrol Science and Technology* 12, 259–266.

Lacey, L.A., Unruh, T.R. and Headrick, H.L. (2003) Interactions of two idiobiont parasitoids (Hymenoptera: Ichneumonidae) of codling moth (Lepidoptera: Tortricidae) with the entomopathogenic nematode *Steinernema carpocapsae* (Rhabditida: Steinernematidae). *Journal of Invertebrate Pathology* 83, 230–239.

Lam, A.B.Q. and Webster, J.M. (1972) Effect of the DD-136 nematode and of a β-exotoxin preparation of *Bacillus thuringiensis* var. *thuringiensis* on leatherjackets, *Tipula paludosa* larvae. *Journal of Invertebrate Pathology* 20, 141–149.

Lee, D.W., Choo, H.Y. and Choi, E.J. (1999) Effect of aqueous solution of pesticides on survival of entomopathogenic nematodes. *Korean Journal of Turfgrass Science* 13, 213–222.

Lee, D.W., Choo, H.Y., Kaya, H.K., Lee, S.M., Smitley, D.R., Shin, H.K. and Park, C.G. (2002) Laboratory and field evaluation of Korean entomopathogenic nematode isolates against the oriental beetle *Exomala orientalis* (Coleoptera: Scarabaeidae). *Journal of Economic Entomology* 95, 918–926.

Mannion, C.M., Winkler, H.E., Shapiro, D.I. and Gibb, T. (2000) Interaction between halofenozide and entomopathogenic nematode *Heterorhabditis marelatus* for control of Japanese beetle (Coleoptera: Scarabaeidae) larvae. *Journal of Economic Entomology* 93, 48–53.

Martin, P.A.W. (2002) An *in vitro* inhibition test that predicts toxicity of bacterial pathogen combinations in the Colorado potato beetle. *Biocontrol Science and Technology* 12, 643–647.

Nishimatsu, T. and Jackson, J.J. (1998) Interaction of insecticides, entomopathogenic nematodes, and larvae of the western corn rootworm (Coleoptera: Chrysomelidae). *Journal of Economic Entomology* 91, 410–418.

Parkman, J.P. and Frank, J.H. (2002) Interaction between *Ormia depleta* (Diptera: Tachinidae) and *Steinernema scapterisci* (Nematoda: Steinernematidae), natural enemies of pest mole crickets (Orthoptera: Gryllotalpidae). *Environmental Entomology* 31, 1226–1230.

Poinar, G.O. Jr, Thomas, G.M. and Lighthart, B. (1990) Bioassay to determine the effect of commercial preparation of *Bacillus thuringiensis* on entomogenous rhabditoid nematodes. *Agriculture Ecosystems and Environment* 30, 195–202.

Pölking, A. and Heimbach, U. (1992) Wirkung einiger biologischer Pflanzenschutzmittel auf zwei räuberische Käferarten in Laborversuchen. *Mitteilungen aus der Biologischen Bundesanstalt für Land- and Forstwirtschaft* 283, 378.

Prakasa Rao, P.S., Das, P.K. and Pandhi, G. (1975) Note of compatibility of DD-136 (*Neoaplectana dutkyi*), an insect parasitic nematode with some insecticides and fertilizers. *Indian Journal of Agricultural Sciences* (publ. 1977) 45, 275–277.

Rovesti, L. and Deseö, K.K. (1990) Compatibility of chemical pesticides with entomopathogenic nematodes, *Steinernema carpocapsae* Weiser and *S. feltiae* Filipjev (Nematoda: Steinernematidae). *Nematologica* 36, 237–245.

Rovesti, L. and Deseö, K.K. (1991) Compatibility of pesticides with entomopathogenic nematodes, *Heterorhabditis heliothidis. Nematologica* 37, 113–122.

Rovesti, L., Heinzpeter, E.W., Tagliente, F. and Deseö, K.V. (1988) Compatibility of pesticides with the entomopathogenic nematode *Heterorhabditis bacteriophora* Poinar (Nematoda: Heterorhabditidae). *Nematology* 34, 462–476.

Shannag, H.K. and Capinera, J.L. (2000) Interference of *Steinernema carpocapsae* (Nematoda: Steinernematidae) with *Cardiochiles diaphaniae* (Hymenoptera: Braconidae), a parasitoid of melonworm and pickleworm (Lepidoptera: Pyralidae). *Environmental Entomology* 29, 612–617.

Shapiro-Ilan, D.I., Jackson, M., Reilly, C.C. and Hotchkiss, M.W. (2004) Effects of combining an entomopathogenic fungi or bacterium with entomopathogenic nematodes on mortality of *Curculio caryae* (Coleoptera: Curculionidae). *Biological Control* 30, 119–126.

Sher, R.B., Parrella, M.P. and Kaya, H.K. (2000) Biological control of the leafminer *Liriomyza trifolii* (Burgess): implications for intraguild predation between *Diglyphus begini* Ashmead and *Steinernema carpocapsae* (Weiser). *Biological Control* 17, 155–163.

Thurston, G.S., Kaya, H.K., Burlando, T.M. and Harrison, R.E. (1993) Milky disease bacterium as a stressor to increase susceptibility of scarabaeid larvae to an entomopathogenic nematode. *Journal of Invertebrate Pathology* 61, 167–172.

Thurston, G.S., Kaya, H.K. and Gaugler, R. (1994) Characterizing the enhanced susceptibility of milky disease-infected scarabaeid grubs to entomopathogenic nematodes. *Biological Control* 4, 67–73.

Wang, Y., Crocker, R.L., Wilson, L.T., Smart, G., Wei, X., Nailon, W.T. Jr and Cobb, P.P. (2001) Effect of nematodes and fungal treatments on nontarget turfgrass-inhabiting arthropod and nematode populations. *Environmental Entomology* 30, 196–203.

Yildirin, E. and Hoy, C.W. (2003) Interaction between cryomazine and the entomopathogenic nematode *Heterorhabditis bacteriophora* Poinar ‘GPS11’ for control of onion maggot, *Delia antiqua* (Meigen). *Crop Protection* 22, 923–927.

Zaki, F.N., Awadallah, K.T. and Gesraha, M.A. (1997) Competitive interaction between the braconid parasitoid *Meteorus rubens* Nees and the entomogenous nematode, *Steinernema carpocapsae* (Weiser) on larvae of *Agrotis ipsilon* Hufn. (Lepidoptera, Noctuidae). *Journal of Applied Entomology* 121, 151–153.

Zimmerman, R.J. and Cranshaw, W.S. (1990) Compatibility of three entomogenous nematodes (Rhabditida) in aqueous solutions of pesticides used in turfgrass maintenance. *Journal of Economic Entomology* 83, 97–100.

Part III

Entomophilic Nematodes

21 Application of *Beddingia siricidicola* for Sirex Woodwasp Control

R.A. Bedding[1] and E.T. Iede[2]
[1]*Division of Entomology, CSIRO, GPO Box 1700, Canberra, ACT 2601, Australia;* [2]*EMBRAPA Florestas, Brazil*

21.1. Introduction

The use of the nematode *Beddingia (Deladenus*[1]*) siricidicola* is now recognized as the most important means of controlling *Sirex noctilio*, a serious pest threatening nearly 8 million ha of pine plantations in the southern hemisphere (Iede *et al.*, 2000; Carnegie *et al.*, 2003).

S. noctilio (Hymenoptera: Siricidae), originally from Europe, is the only one of some

[1] Those species of *Deladenus* having both free-living and parasitic life cycles were assigned to a new genus, *Beddingia*, by Blinova and Korenchenko (1986) and this nomenclature was adopted by Remillet and Laumond (1991).

40 species of woodwasp found throughout the world that can kill relatively healthy pine trees. The tree species most susceptible to sirex, *Pinus radiata*, *P. taeda*, *P. elliottii* and *P. patula*, all of which originated from North America, were long ago adopted for major plantations in Australia, now with 1 million ha; New Zealand, 1.8 million ha; Brazil, 2.2 million ha; Chile, 1.5 million ha; Argentina, 0.3 million ha; Uruguay, 0.1 million ha; and South Africa, 0.7 million ha (Iede *et al.*, 2000; Wood *et al.*, 2001; M. Wingfield, 2003, personal communication).

Later, *S. noctilio* was accidentally introduced into each of these regions (New Zealand during the early 1900s, Tasmania during the early 1950s, the mainland of Australia at the beginning of 1960s, South America during the 1980s and South Africa during the 1990s) so that there is now an unfortunate combination of the most virulent woodwasp, highly susceptible tree species, sometimes high density of plantings with inadequate forest management, originally an absence of natural enemies and, at least in Australia, a climate disposed to make trees periodically even more susceptible to attack.

Initially, during the early 1960s, it was hoped to eradicate sirex on the mainland of Australia by having mandatory reporting and by seeking out and destroying all sirex-infested trees; none the less, sirex spread at a rate of about 20–30 km/year, reaching borders of the state of Victoria after about 20 years. Sirex has now spread to the major part of Australia's 1 million ha of pine plantations, reaching most plantations in Tasmania, Victoria, New South Wales and South Australia and has just (Carnegie *et al.*, 2003) been found near the border of Queensland, but not yet in Western Australia. In South America, sirex is now well established in Uruguay, Brazil and Argentina (Iede *et al.*, 2000) and was reported from Chile in 2001, while in South Africa it has recently migrated from the Cape to near Durban (M. Wingfield, 2003, personal communication).

B. siricidicola was first introduced into the Australian state of Victoria during the early 1970s and from then on there were relatively few serious outbreaks of sirex. This led to complacency so that even though sirex arrived in the 113,000-ha 'Green Triangle' forests of southwest Victoria/southeast South Australia during 1979, no serious attempt was made to introduce nematodes for the next 8 years (Fig. 21.1). By then it was almost too late; in 1987 1.8 million trees were killed by sirex and during the next 2 years a further 3 million were killed. This area was thus rather like a huge control plot showing that in the absence of natural enemies, sirex could kill up to 80% of trees in some areas. Fortunately, as a result of an AUS$1.3 million operation to introduce nematodes during 1987 (Haugen and Underdown, 1990a), levels of nematode parasitism of up to 100% were reached within 2 years and the sirex population crashed, but not before millions of dollars worth of timber was lost and the quality of many of the remaining trees impaired (M.G. Underdown, 1992, personal communication). On evidence from this outbreak, it has been calculated (M.G. Underdown, 1992, personal communication) that in the absence of control agents, sirex had the potential to cause an average loss of timber from the total pine plantations in Australia valued at between US$16 and US$60 million per year. In Brazil, where 350,000 ha of pine are currently infested by sirex, it is estimated that, on an average, US$6.6 million would have been lost each year had companies not adopted an integrated pest management (IPM) programme based mainly on nematode release.

21.2. Biology of Sirex

Sirex usually has a 1-year life cycle, with adults emerging from late December to March in Australia, but as early as October with peak emergence from November to December in Brazil (Iede *et al.*, 1998), and living for about 1 week. Female sirex drill 10–20 mm into the living pine tree and insert toxic mucus and spores of a pathogenic fungus, *Amylostereum areolatum* (Coutts, 1969a,b; Gaut, 1969). If the tree is suitable,

Fig. 21.1. An area of the 50,000-ha 'Green Triangle' in southern Australia, where up to 80% of trees were killed over 2 years by sirex.

one or more eggs are also laid nearby. Depending upon the size of the female, sirex oviposit from 30 to 450 eggs (Madden, 1974). The toxic mucus prevents sugar from being passed down from the leaves. Normally sugar is converted by the tree into fungal poisons (polyphenols) at the site of fungal infection, thus stopping the fungus spreading, but this cannot happen after a successful sirex attack.

Fungal spores germinate and can now grow into the wood; within a few weeks/months a successfully attacked tree will die as a result of the mucus and fungus combined. The fungus then grows throughout the dead tree while the eggs laid by sirex hatch and the resulting sirex grubs bore into the wood, feeding on the growing fungus.

21.3. Biology of the Nematode

Only *B. siricidicola*, of seven species of *Beddingia* (Bedding, 1968, 1975; Akhurst 1975) parasitizing 31 siricid and parasitoid hosts from 31 tree species and 29 countries (Bedding and Akhurst, 1978), was found to be suitable for the control of sirex (Bedding, 1984). This nematode can achieve levels of parasitism approaching 100% when the density of sirex-infested trees is high. It has an unusual and complicated biology (Bedding, 1967, 1972, 1993) that has been exploited for the biocontrol of sirex (Bedding and Akhurst, 1974; Bedding, 1979, 1984, 1993; Iede *et al.*, 1998).

B. siricidicola is extraordinary in having two separate life cycles associated with two morphologically very different adult female types (Bedding, 1967, 1968; Fig. 21.2). There is a parasitic cycle in which from 1 to 100, 0.5- up to 2.5-cm-long, cylindrical, often green-coloured females release thousands of juvenile nematodes into the body cavity of adult sirex wasps, and a free-living cycle where 1- to 2-mm-long females feed on the symbiotic fungus as it grows in the tree and lay eggs in the wood fibres (tracheids) (Fig. 21.3). These two types of females are so morphologically different

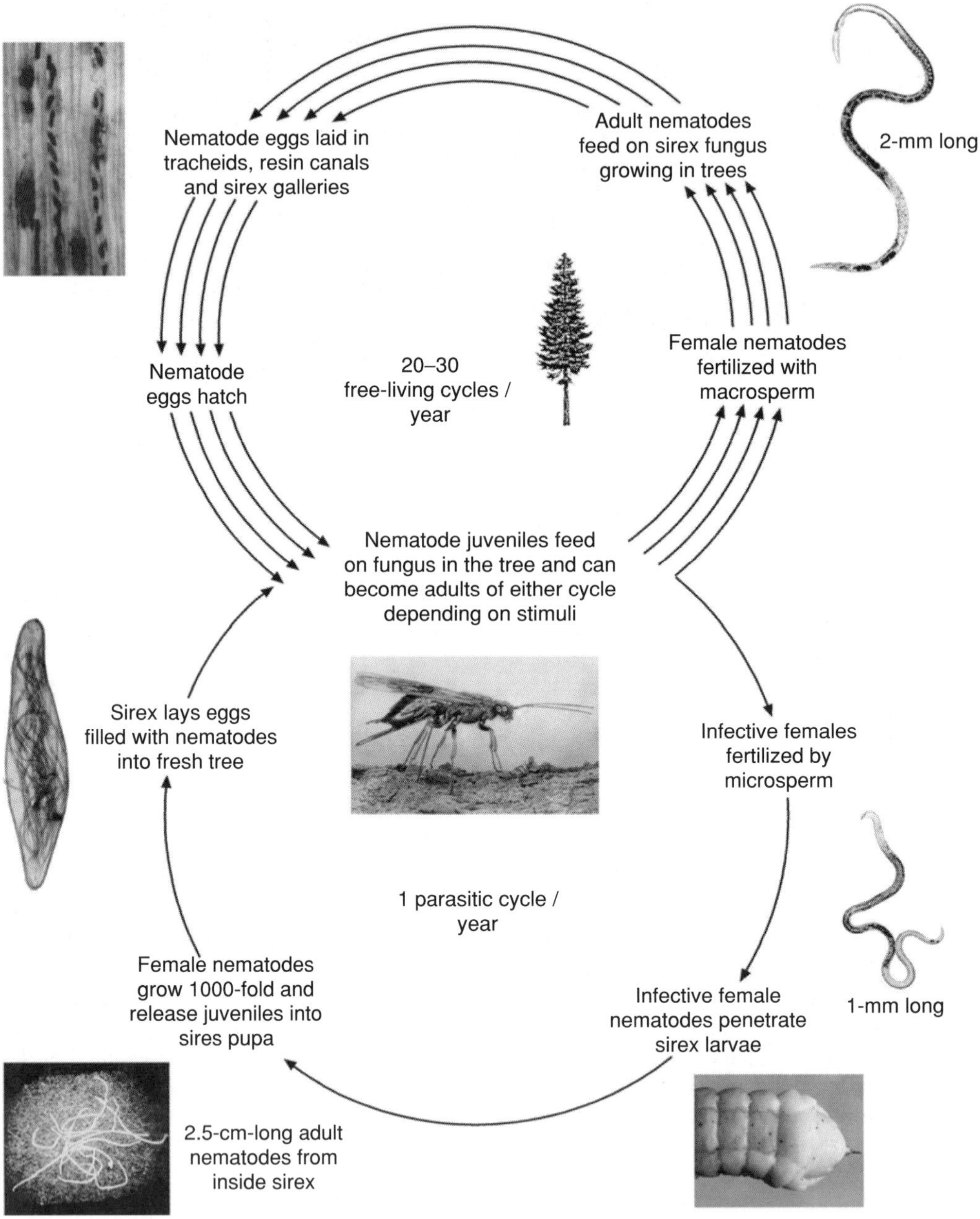

Fig. 21.2. Biology of the nematode parasite of sirex, *Beddingia siricidicola*.

that each on its own would have been placed in a separate family of nematodes.

At about the time adult parasitized sirex emerge from infested trees, adult nematodes have usually released most of the juveniles that are within them into the insect's blood cavity and the juveniles have migrated to the insect's reproductive organs. In the male sirex the testes become greatly enlarged, often fused and filled with thousands of juveniles. However, this is a dead end for the nematodes because by this time the testes

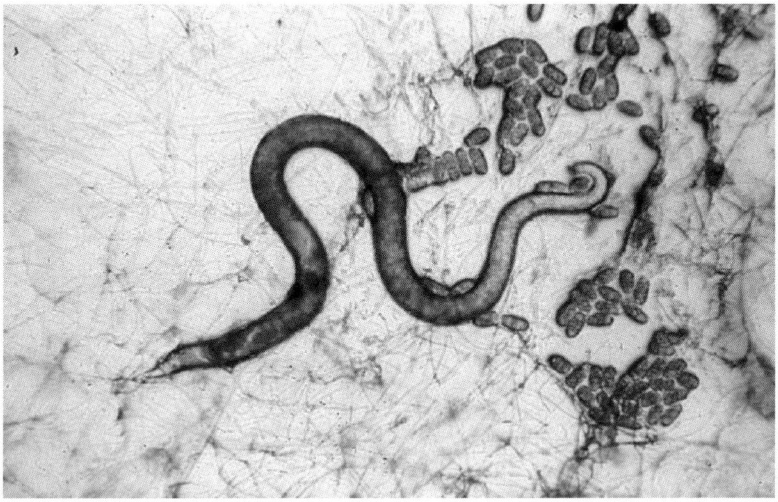

Fig. 21.3. Fungal-feeding female of *Beddingia siricidicola* with eggs on potato dextrose agar (PDA) plate.

have already emptied sperm, but not nematodes, into the insect's seminal vesicles so that parasitized males can still fertilize females but not transmit nematodes. On the other hand, female sirex are effectively sterilized because, apart from ovarial development being suppressed to various degrees, each egg that is produced is filled with up to 200 juvenile nematodes. Nevertheless, the parasitized female sirex still oviposits readily, and often in several different trees, but lays packets of nematodes instead of viable eggs. Since many sirex often attack the same trees, larval progeny of unparasitized sirex can eventually become infected with nematodes, but this is only made possible by the intervening free-living cycle.

A large tree may contain many hundreds of millions of tracheids (hollow fibres) weighing from 1000 kg to 5000 kg but rarely more than a few hundred sirex larvae. Nematodes have to migrate up and down the tracheids and through the infrequent holes in these (degenerate bordered pits) to move from tracheid to tracheid. Each egg from a nematode-infected female sirex contains not more than 200, 0.5-mm-long juvenile nematodes and the chances of many of these nematodes reaching and penetrating sirex larvae would be slim indeed, except that *B. siricidicola* breeds in vast numbers (possibly hundreds of millions) during many generations and spreads to all parts of the tree while feeding on the sirex symbiotic fungus as it grows in the tree.

It is only when larval nematodes reach the immediate microenvironment around sirex larvae that they are stimulated by high CO_2 and low pH (Bedding, 1993) to develop into the pre-parasitic kind of female rather than the fungal-feeding form. This infective female mates with special males, formed under the same conditions, that produce microspermatozoa instead of the amoeboid spermatozoa produced by normal males. After mating the infective female uses its large anterior spear-like stylet to bore into sirex larvae, after which it migrates in the insect's blood cavity for a few days before shedding its cuticle to leave its entire body surface covered with microvilli (Riding, 1970), which rapidly absorb food. Within a few weeks of penetrating a sirex larva the nematode grows up to 1000-fold in volume, depending on the size of its host. It now remains, often for several months, relatively unchanged, until the sirex larva pupates. The nematode's reproductive system then develops rapidly from a few cells to produce many thousands of juvenile nematodes in little more than a week, and these are released into the insect's blood cavity at about the time the insect emerges from the tree. It is of interest to note that in some other species of host and with some other strains and species of the nematode, timing of juvenile nematode release can be later so that the host's eggs' shells have already formed and nematodes cannot enter the eggs. In these cases nematodes are deposited on the outside of eggs and they can then later infect larvae derived from their own host, which is essential where the siricid is a solitary species.

21.4. Application

21.4.1. General

Unlike most of the nematodes featured in this book, *B. siricidicola* is used as a classical biocontrol agent. Once it has been introduced, most of its dispersal is by females of the pest insect, sirex. However, human intervention has been and is still required to monitor and assist its spread, particularly to new sirex-infested plantations and also within them. In order to introduce the nematodes into new areas it has been necessary to isolate the best strain and develop the methods for rearing, storing, formulation, inoculation and distribution, as described below. In Australia, there is a National Strategy document (Haugen *et al.*, 1990), detailed standard operating procedures for rearing, storing, formulation and quality control (Calder and Bedding, 2002) and operations worksheets (National Sirex Coordination Committee, 2000) covering various aspects of sirex control including inoculation and distribution of nematodes. In South America, particularly in Brazil, there is a similar provision of technical and other publications (Iede *et al.*, 1998).

21.4.2. Choice of species and strain

Several hundred isolates of seven species of *Beddingia* were screened for potential to control sirex during the early 1970s. Most siricids are associated with *Amylostereum chaillettii* and this is also the only fungus on which five species of *Beddingia* can feed (Bedding and Akhurst, 1978). *B. wilsoni* fed on *A. chaillettii* and *A. areolatum* but frequently parasitized the insect parasitoids *Rhyssa* spp. and so this left only strains of *B. siricidicola* for further consideration. Many strains of this species parasitized but did not fully sterilize Australian *S. noctilio* (Bedding, 1972) and were eliminated.

Then, hundreds of randomly selected sirex-infested logs were inoculated with the various remaining strains of *B. siricidicola*. Four of these nematode strains from Corsica, Thasos, Sopron and New Zealand parasitized nearly 100% of the emerging sirex (R.A. Bedding and R.J. Akhurst, 1971, unpublished data), but it was found that sirex parasitized by the 198 strain from Sopron, Hungary, were significantly larger than those parasitized by other species. Using flight mills it was found that while parasitism itself had no significant effect on the flying abilities of sirex, insect size, which was often affected by parasitism, had a major effect. Whereas very large sirex females could fly up to 200 km on the flight mills, very small females could fly only about 2 km. Not only could large infested sirex fly further and probably infest many more trees, they also produce more eggs and more nematodes. As a result of these findings most of the releases in Australia have been of the 198 strain, and this is the only strain that has been supplied to South America and South Africa.

21.4.3. Even the best strain deteriorates

Bedding and Akhurst (1974) showed that using their methods (described below), sirex emerging from correctly inoculated sirex-infested trees were over 98% parasitized by nematodes. However, during the late 1980s, after the Victorian Forest Commission had been inoculating sirex-infested trees for over 15 years, it was found (as a result of the Green Triangle outbreak) that sirex emerging from these trees were only about 25% parasitized (M.G. Underdown, 1992, personal communication). In Brazil the defective strain was used for inoculations from 1990 to 1994. This obviously had implications not only for the current release programme but even more importantly for the effectiveness of sirex control as a whole. The possibilities were either that incorrect inoculation procedures had been gradually adopted or that there had been genetic change in the nematode used. It was soon determined that declining parasitism in inoculated logs over several years was certainly a result of genetic change in the nematode used. While use of the fungal-feeding cycle to maintain cultures and mass-produced *B. siricidicola* is an essential part in the use of this nematode for biocontrol, it also resulted in this major problem (Bedding, 1992).

Because *Beddingia* had been cultured in the free-living form for over 20 years without intervention of the parasitic life cycle, this led to the selection of a strain that rarely formed the pre-parasitic infective stage. Even at high concentrations of CO_2 and low pH such cultures will rarely produce infective females. While there is little or no selection against a predisposition to develop into infective females in the field, the opposite is true when *B. siricidicola* is cultured artificially. The nematodes pass through repeated generations without intervention of a parasitic cycle (stock cultures had been through hundreds of generations). Infective female nematodes produced in these cultures could not reproduce because there was no insect host and so there was a strong selection pressure against their production.

The production of low levels of parasitism in inoculated logs was unfortunate and costly (four times as many trees needed to be inoculated), but of far greater significance was what this meant in terms of the ability of this 'defective' strain of nematodes to control sirex populations in plantations. There was reason to believe that nematode control with the defective strain would not occur until sirex infestations were severe (perhaps $>$ 10% tree death), whereas the original strain produced high levels of parasitism at very much lower tree death (probably $<$ 1% tree death). Results from New South Wales and South Australia tended to confirm this (R.H. Eldridge, New South Wales, 2000; M.G. Underdown, 1996, personal communication). Although the defective strain was used very effectively in the Green Triangle it was almost certainly only effective there because of the very high density of sirex infestation (up to 80% tree death) and intense nematode inoculation (20% of all infested trees).

21.4.4. Re-isolation of the nematode from area of original release

It was obviously important to obtain non-defective nematodes and this was achieved in 1991 by collecting sirex-infested timber from the Kamona forest in Scottsdale, Tasmania, where it was first liberated in 1970 but where no subsequent liberations had taken place (Bashford, 1991, personal communication). Nematodes were found and extracted from just one of the only nine sirex-infested trees found and then established in monoxenic culture on *A. areolatum* as described below. After a series of test inoculations in sirex-infested billets, it was found that this new culture, called the 'Kamona' strain, produced over 95% parasitism in the sirex that subsequently emerged compared with the defective strain where emerging sirex were only 23% parasitized. Since then the Kamona strain has been used for all liberations in Australia, South America and South Africa.

In spite of the success of the Kamona strain, there are still two major problems: (i) large areas of parts of southeast Australia were inoculated, over many years, with the 198 strain as it became progressively defective; and (ii) even the Kamona strain had several years of only fungal culturing before liberation, and we have recently found (R.A. Bedding and J. Calder, 2001, unpublished data) that it starts to become defective when subcultured in the laboratory for as little as 6 months. The first issue, to be discussed later, is being addressed by endeavouring to swamp the defective strain already in the field with the Kamona strain. In the second case, Bedding (1993) developed methods for storing *B. siricidicola* in liquid nitrogen and stored hundreds of ampules of the Kamona strain in Dewars so that at the beginning of each season, it has been possible to begin fresh fungal cultures of the nematodes and then mass-rear them before any decline in infectivity. In Brazil, the original strain introduced in 1989 from Australia became defective and the Kamona strain was introduced in 1995; nematodes have since been re-isolated from the field every year.

21.4.5. Storage in liquid nitrogen

Although entomopathogenic nematodes (EPNs) can be readily stored in liquid nitrogen using methods developed by Popiel *et al.* (1988), Popiel and Vasquez (1991) and Curran *et al.* (1992), these methods resulted in 100% mortality of *B. siricidicola*. However, when various larval stages of *B. siricidicola* are sterilely suspended in 5% glycerol solution and water is slowly evaporated in a laminar flow cabinet, to achieve 50% glycerol after several days, the nematodes can be successfully frozen in liquid nitrogen. Vials, each containing 300 μl of the suspension, are plunged directly into liquid nitrogen, and even after 10 years there has been over 75% survival with every prospect that the remaining nematodes will survive indefinitely. The nematodes are revived by exposing a vial to running lukewarm water to ensure rapid thawing, adding and mixing 500 μl of sterile water and then placing 100-μl aliquots along the fungal front of 5-day-old cultures of *A. areolatum* growing on one-third strength potato dextrose agar (PDA) plates.

21.5. Nematode Culture

Culture of *B. siricidicola* relies upon the fungal-feeding cycle of this nematode. Because it is released inoculatively, unlike EPNs, only relatively moderate numbers of *B. siricidicola* are required even for millions of hectares of pine forest, so large-scale rearing methods have not been required. Cultures are originally established and subcultured monoxenically on PDA plates, which are then used to inoculate 500-ml flasks containing autoclaved grain. All procedures in culturing are conducted under fully sterile conditions (using sterilized equipment and media in a laminar flow cabinet with the operator wearing sterilized gloves and spraying inside the cabinet and all objects put into it with 70% ethanol).

A complicating factor when culturing *Beddingia* sp. is that the nematodes feed well only on the advancing front of the fungus; if the fungus grows too rapidly it smothers the nematode and the culture is lost; if there are too many nematodes the fungus may be unable to grow adequately.

21.5.1. Establishment of cultures of the fungus

Cultures of the fungus *A. areolatum* alone are required when nematode cultures are to be established either from liquid nitrogen or from parasitized sirex and are usually made by subculturing from already established cultures. Initially, cultures are made from live sirex females by plunging them into 100% ethanol within a laminar flow cabinet, igniting the insect and then plunging it into a Petri dish of sterile ringers in which it is dissected using sterile instruments and a dissecting microscope sprayed with 70% alcohol. The two ooidial glands found inside the insect at the base of the ovipositor are removed and streaked on PDA plates, and 4 or 5 days later areas of uncontaminated fungus are removed and placed on fresh plates.

21.5.2. Establishment of the nematode cultures from sirex

Most culturing, after initial removal from liquid nitrogen, is directly from monoxenic plates to fresh plates or, for mass rearing, to flasks. However, the initial cultures had to

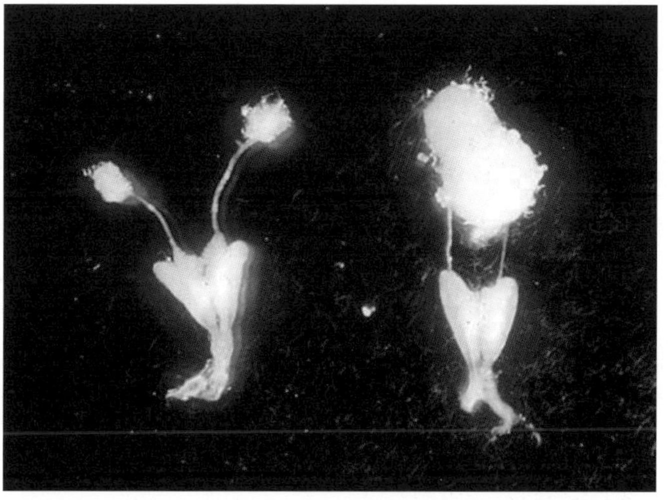

Fig. 21.4. Male reproductive organs from parasitized sirex (right) and unparasitized sirex (left). It is convenient to use the infected testes to establish monoxenic cultures of *Beddingia siricidicola* on the symbiotic fungus *Amylostereum areolatum.*

be made from live parasitized sirex adults; this is still necessary when nematodes are removed from field insects to test for the infectivity and/or strain of nematodes involved using molecular biology (Fig. 21.4).

In the same way as ooidial glands were removed sterilely from the female, testes filled with juvenile nematodes are removed from male sirex and placed centrally on 5-day-old fungal cultures growing on one-third strength PDA plates.

21.5.3. Mass culture and dispatch

Stoppered, 500-ml conical flasks, each containing 90 g of an equal mixture of brown rice and wheat and 150 ml of tap water, are autoclaved for 30 min at 121°C, cooled rapidly and each inoculated by sterile spatula with about one-half the growing front from a mature (about 2-week-old) culture plate of nematodes (Fig. 21.5). Flasks are then kept at 23°C until mature (5–8 weeks) when they are harvested with water, washed, sieved and dispatched in batches of 1 or 5 million in breathable plastic bags together with

Fig. 21.5. Culture flasks just after inoculation with nematode/fungus culture (left) and just before harvest (right).

packets of finely ground (< 600 μm) polyacrylamide gel (5 g for each million nematodes).

21.6. Nematode Application

21.6.1. Plantation inoculation

In Australia, inoculation is conducted as part of a national strategy of sirex awareness, quarantine, detection, monitoring and silvicultural control (Haugen *et al.*, 1990) and this is similar in South America (Iede *et al.*, 2000).

When sirex populations are moderate (1–3%) or severe (> 3% infested) it might be necessary to inoculate 20% (every fifth row) of infested trees to obtain rapid control. However, if pre-emptive action is taken it is usually only necessary to inoculate a small number of easily accessible trap trees for each 20–30 ha of forest to be protected. Generally speaking, pine trees do not become susceptible to sirex until they are about 12 years old, and this is when action should be taken to protect such forest compartments using nematodes when sirex infestations are in the vicinity (as determined by aerial and ground surveys and/or results from trap tree plots).

To introduce nematodes, trap tree plots are established near the roadside of each 20- to 30-ha compartment where, by injecting at the base of five trees just enough weedicide (e.g. DiCamba) to almost kill them, they become highly susceptible to sirex (Madden and Irvine, 1971; Neumann *et al.*, 1982, 1989). This is conducted 2–3 months before the expected flight season and trees that become infested with sirex can then be inoculated with nematodes during April to August. It is particularly important that nematode inoculations are as well dispersed as possible throughout plantations since most nematode dispersal occurs only to the infested trees nearby; Bedding and Akhurst (1973) liberated 50 parasitized sirex at the corner of a 1000-ha plantation during 1970 and found that while nematode parasitism in the 30-ha compartment of liberation rose from 31% in 1970, 47% in 1971 to 92% in 1972, in the whole forest it had reached only 37% during those 3 years, with 6% in the first year and 14% in the second.

21.6.2. Tree inoculation

How sirex-infested trees are inoculated is of the utmost importance if most of the emerging sirex are to be parasitized without size being adversely affected (Bedding and Akhurst, 1974). During initial experiments, holes were drilled into sirex-infested billets and a suspension of nematodes was added to each hole; this resulted in negligible parasitism in the emerging sirex. The drilling resulted in the cut ends of the tracheids (wood tubes) being twisted and water was rapidly absorbed into the wood, leaving the nematodes 'high and dry'. However, when tracheids are cut cleanly, nematodes are added in a gel suspension, the wood is moist enough and the spacing of inoculation is optimum, nearly 100% of emerging sirex are parasitized.

Inoculation is currently achieved using a specifically designed, frequently sharpened, rebound, hammer punch (Fig. 21.6) that cuts the tracheids cleanly and nematodes suspended in a 1% finely ground (< 600 μm) polyacrylamide gel (Australia), or as per Bedding and Akhurst (1978) in foamed, 10% gelatine solution (Brazil). Sirex-infested trees are felled and the branches trimmed off. Inoculation holes approximately 10 mm deep are made every 30 cm with one row where tree diameter is less than 15 cm and two rows of staggered holes where tree diameter is greater than 15 cm. One million nematodes are mixed in 500 ml of 1% gel (enough for 10–20 trees) and dispensed from a sealant gun, syringe or sauce bottle so that 2000 nematodes are added per inoculation hole in about 1 ml gel. The gel is further pressed into the hole using a finger. Two operators are usually involved with one making holes and the other dispensing the gel containing the nematodes.

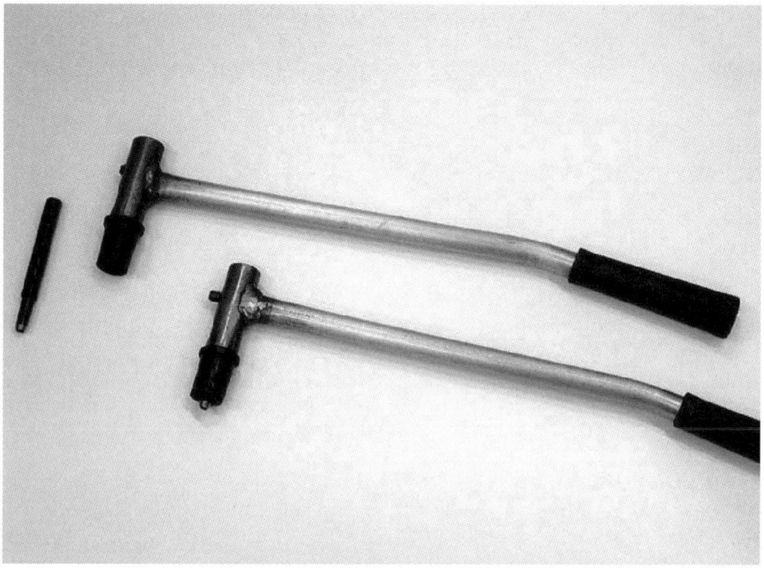

Fig. 21.6. Wad punch mounted in a hammer enables clean cutting of the tree's tracheids so that nematodes can enter.

21.6.3. Monitoring

Even after nematodes have been liberated in an area it is important to ensure that they are established. The main method of assessing nematode levels is to dissect sirex emerging from caged logs. The presence of nematodes can also be detected soon after trees have died by cutting chips from sirex-infested trees and standing these in shallow water for 24 h. Where there is moderate to severe sirex infestation and there is no nematode parasitism it is recommended that 20% of sirex-infested trees be inoculated; where 1–5% of sirex are parasitized, 10% of trees should be inoculated; where 5–10%, 5% should be inoculated while those with greater than 10% parasitism, no further inoculation is worthwhile.

21.6.4. Replacing defective strain

As described above, some areas of Australia and Brazil may have the defective strain of *B. siricidicola* present in sirex populations. R.A. Bedding and J. Calder (2000, unpublished data) have found that unfortunately it requires several back-crosses between Kamona and defective strains before crosses become fully infective and this is reflected in randomly amplified polymorphic DNAs (RAPDs) of the back-crosses (see Fig. 21.7). This has meant that in such areas, the Kamona strain has had to be repeatedly reintroduced until it dominates. This is achieved much more readily when levels of sirex infestation are low. Use of RAPDs to distinguish between Kamona and defective strains has proved invaluable to determine what strains dominate in the field. J. Calder (1998, unpublished data) found that out of 100 PCR primers tested, the two strains (and also various strains of EPNs) could be readily separated using the primers OP-AO4, OP-X11 and OP-FO3. In addition, the ability to form infective females can be tested by sterilely harvesting eggs from the mycetophagous cultures of *B. siricidicola* and placing these on *A. areolatum* growing on 0.2% lactic acid/PDA plates inside desiccators containing 10% CO_2.

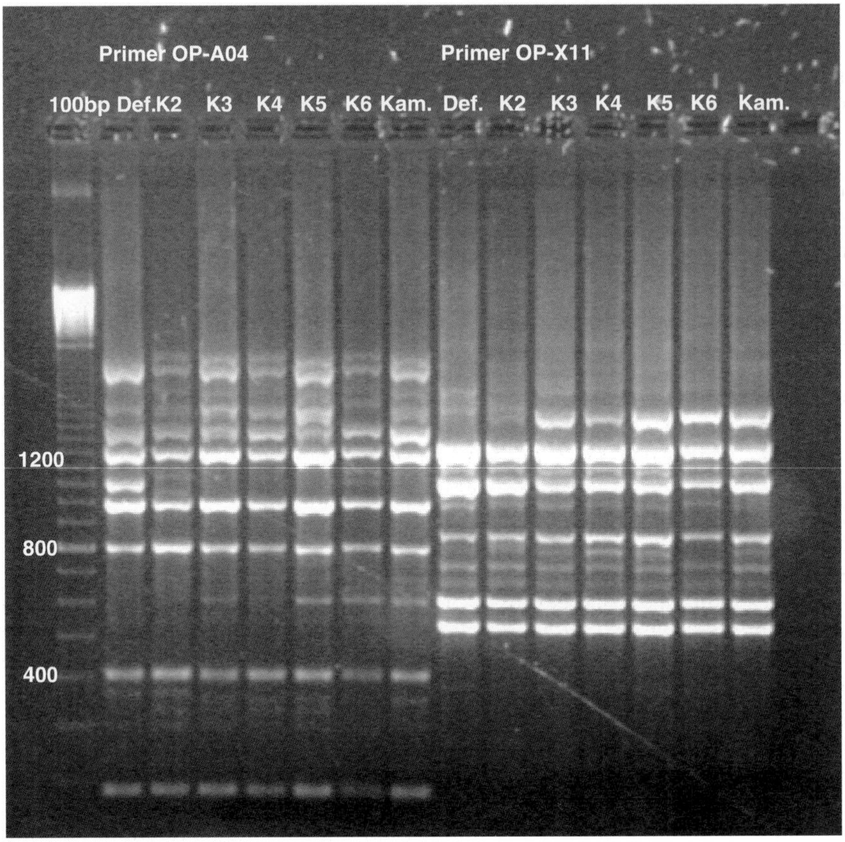

100bp = 100bp ladder
Def. = Defective strain
K2 = Kamona males x defective females
K3 = Kamona males x K2 females
K4 = Kamona males x K3 females
K5 = Kamona males x K4 females
K6 = Kamona males x K5 females
Kam. = Kamona strain (K98-B)

Fig. 21.7. Randomly amplified polymorphic DNAs (RAPDs) using two primers to distinguish between the 'defective' and Kamona strains derived from the original *Beddlingia siricidicola* isolate from Sopron, Hungary, and also a series of back-crosses between these two strains.

21.7. Evidence of Control

It is believed that nematode parasitism is density dependent and that provided nematodes are adequately distributed and established, control occurs before there are high numbers of tree deaths provided non-defective nematodes are in place (Bedding, 1993). During 1972, in a 400-ha forest of *P. radiata* in northern Tasmania where about 5–10% of trees had been killed annually by sirex over several years, all infested trees from every 10th row were inoculated with nematodes. In 1973, based on sampling 64 widely dispersed infested trees, parasitism was 86%, while in the following year no sirex-killed trees could be located by ground and aerial surveys. Similarly, in the Green Triangle, from 1988 to 1989 nearly 100% parasitism resulted after inoculating sirex-infested trees in one row out of every five, and ever since it has been difficult to find any sirex-infested trees. In a

12,000-ha plantation in Encruzilhado do Sul in Brazil, where sirex infested about 30% of trees in some compartments in 1991, nematodes were released from 1990 to 1993, resulting in levels of parasitism of 45% in 1991, 75% in 1992 and more than 90% in 1994. In 1995 it was difficult to find any sirex-infested trees in this area. Generally, in Brazil, where parasitism is evaluated annually in seven localities, parasitism ranged from 17% in one locality, 39%, 57% and 65% in three others to over 92% in another three localities. This sort of variation also occurs annually in the numerous sites examined in Australia, with higher levels of parasitism apparently related to higher density of sirex-infested trees (R.H. Eldridge, 2000; Underdown, 1996, personal communication). A major factor contributing to the successful control of sirex is that, although at the beginning of an infestation of a plantation there are usually plenty of suppressed or otherwise susceptible trees that are readily killed by one or a few sirex females, as these trees are utilized by sirex, and as only the more resistant trees remain, it takes more and more sirex to kill each tree.

21.8. Discussion and Conclusions

Nematode control of sirex commenced in the early 1970s and was the first commercial use of nematodes to control any pest. In terms of the value of crop saved by nematode control, it is currently undoubtedly worth more than all other uses of nematodes. However, because it is a classical biocontrol agent usually requiring only an initial inoculation into plantations, sales of the nematode amount to less than about US$40,000 per year. While this is an example of one of the most successful biocontrol projects of its kind, there is no room for complacency, even though this may be understandable since major infestations of sirex may only occur at intervals of many years, even in the absence of nematodes. There is always the possibility of nematode strain deterioration, inadequate distribution of nematodes throughout each plantation and major calamities (fire, wind and hail damage or inadequate or untimely thinning) within forests, leading to massive build-up of sirex populations before nematodes can exert control. It is also possible that strains of sirex could develop where nematodes are released too late into the pupal haemolymph to penetrate eggs because shells have already formed (as already occurs in some other host species and even other strains of *S. noctilio*). This was suspected of an isolated New Zealand population of sirex, but it turned out to be a change in the nematode strain that was responsible (R.A. Bedding, 1990, unpublished data).

Change in the nematode strain so that it becomes defective in infectivity is one of the most important problems. Annual use of liquid nitrogen-stored material should adequately deal with this problem, but whether continual re-isolation from the field followed by many fungal-feeding generations for mass production in the laboratory is satisfactory is a matter for conjecture. Currently, it is still necessary in Australia and perhaps Brazil to swamp defective strain nematodes that are already in the field, and it is now possible to use genetic probes to test for defectiveness in field-collected isolates.

Hopefully, if the situation is continually monitored, *B. siricidicola* can be used for hundreds of years to control this very serious pest that may spread to many other areas of the world.

References

Akhurst, R.J. (1975) Cross-breedings to facilitate the identification of *Deladenus* spp., nematode parasites of woodwasps. *Nematological* 21, 267–272.

Bedding, R.A. (1967) Parasitic and free-living cycles in entomogenous nematodes of the genus *Deladenus*. *Nature* 214, 174–175.

Bedding, R.A. (1968) *Deladenus wilsoni* n.sp. and *D. siricidicola* n.sp. (Neotylenchidae), entomophagous nematodes parasitic in siricid woodwasps. *Nematologica* 14, 515–525.

Bedding, R.A. (1972) Biology of *Deladenus siricidicola* (Neotylenchidae) an entomophagous

nematode parasitic in siricid woodwasps. *Nematologica* 18, 482–493.
Bedding, R.A. (1975) Five new species of *Deladenus* (Neotylenchidae), entomophagous-mycetophagous nematodes parasitic on siricid woodwasps. *Nematologica* 20, 204–225.
Bedding, R.A. (1979) Manipulating the entomophagous-mycetophagous nematode, *Deladenus siricidicola* for the biological control of the woodwasp *Sirex noctilio* in Australia. In: Waters, W.E. (ed.) *Current Topics in Forest Entomology.* USDA Forest Service, General Technical Paper WO-8, 144–147.
Bedding, R.A. (1984) Nematode parasites of Hymenoptera. In: Nickle, W.R. (ed.) *Plant and Insect Parasitic Nematodes.* Marcel Dekker, New York, pp. 755–795.
Bedding, R.A. (1992) Strategy to overcome the crisis in control of sirex by nematodes. *Australian Forest Grower* Summer 1991/92, 15–16.
Bedding, R.A. (1993) Biological control of *Sirex noctilio* using the nematode *Deladenus siricidicola.* In: Bedding, R.A., Akhurst, R.J. and Kaya, H.K. (eds) *Nematodes and the Biological Control of Insect Pests.* CSIRO Publications, East Melbourne, Australia, pp. 11–20.
Bedding, R.A. and Akhurst, R.J (1973) Nematode parasites of *Sirex. CSIRO Division of Entomology Annual Report.* 1972–73, 52.
Bedding, R.A. and Akhurst, R.J. (1974a) Biological Control of Sirex Wasp. *CSIRO Annual Report,* p. 88.
Bedding, R.A. and Akhurst, R.J. (1974b) Use of the nematode *Deladenus siricidicola* in the biological control of *Sirex noctilio* in Australia. *Journal of the Australian Entomological Society* 13, 129–135.
Bedding, R.A. and Akhurst, R.J. (1978) Geographical distribution and host preferences of *Deladenus* species (Nematoda: Neotylenchidae) parasitic in siricid woodwasps and associated hymenopterous parasitoids. *Nematologica* 24, 286–294.
Blinova, S.L. and Korenchenko, E.A. (1986) *Phaenopsitylenchus lacicis* g.n. and sp.n. (Nematoda: Phaenopsitylenchidae fam. N.) parasite of *Phenops guttulata* and remarks on taxonomy of nematodes of the superfamily Shaerularioidea. *Acad. Nauk. SSSR trudy Gelmint. Lab.* 34, 14–23.
Calder, J. and Bedding, R.A. (2002) Standard operating procedures for production, experimental analysis and quality assurance testing of *Beddingia siricidicola. National Sirex Co-ordination Committee,* Canberra, Australia, 102 pp.
Carnegie, A.J., Waterson, D.G. and Eldridge, R.H. (2005) The history and management of sirex wood wasp, *Sirex noctilio* (Hymenoptera: Siricidae), in pine plantations in New South Wales, Australia. *New Zealand Journal of Forestry Science* 35(1), 3–24.
Coutts, M.P. (1969a) The mechanism of pathogenicity of *Sirex noctilio* on *Pinus radiata.* 1. Effects of the symbiotic fungus *Amylostereum* sp. (Thelophoraceae). *Australian Journal of Biological Sciences* 22, 915–924.
Coutts, M.P. (1969b) The mechanisms of pathogenicity of *Sirex noctillo* on *Pinus radiata.* 2. Effects of *S. noctilio* mucus. *Australian Journal of Biological Sciences* 22, 1153–1161.
Curran, J., Gilbert, C. and Butler, K. (1992) Routine cryopreservation of isolates of *Steinernema* and *Heterorhabditis* spp. *Journal of Nematology* 24, 269–270.
Gaut, I.P.C. (1969) Identity of the fungal symbiont of *Sirex noctilio. Australian Journal of Biological Sciences* 22, 905–914.
Haugen, D.A. (1990) Control procedures for *Sirex noctilio* in the Green Triangle: review from detection to severe outbreak (1977–1987). *Australian Forestry* 53, 24–32.
Haugen, D.A. and Underdown, M.G. (1990a) *Sirex noctilio* control program in response to the 1987 Green Triangle outbreak. *Australian Forestry* 53, 33–40.
Haugen, D.A. and Underdown, M.G. (1990b) Woodchip sampling for the nematode *Deladenus siricidicola* and the relationship with the percentage of *Sirex noctilio* infected. *Australian Forestry* 54, 3–8.
Haugen, D.A., Bedding, R.A., Underdown, M.G. and Neumann, F.G. (1990) National strategy for control of *Sirex noctilio* in Australia. *Australian Forest Grower* 13, No. 2.
Iede, E.T., Penteado, S., Do, R.C. and Bisol, J.C. (1988) Primeiro registro de ataque de *Sirex noctilio* em *Pinus taeda* no Brazil. *Circular Tecniuca N0* 20. EMBRAPA.
Iede, E.T., Penteado, S., Do, R.C. and Schaitza, E.G. (1998) *Sirex noctilio* in Brazil: detection, evaluation, and control. In: *Proceedings of a Conference: Training in the Control of Sirex noctilio by the Use of Natural Enemies,* Colombo, Brazil, pp. 45–52.
Iede, E.T., Klasmer, P., Penteado, S. and Do, R.C. (2000) *Sirex noctilio* in South America: distribution, monitoring and control. In: *XXI International Congress of Entomology.* Foz do Iguaçu, Pr. agosto 2000. Embrapa Soja Anais. Londrina-PR. 2000 1, 474.
Madden, J.L. (1968) Physiological aspects of host tree favourability for the woodwasp, *Sirex noctilio* F. *Proceedings of the Ecological Society of Australia* 3, 147–149.

Madden, J.L. (1974) Oviposition behaviour of the woodwasp, *Sirex noctilio* F. *Australian Journal of Zoology* 22, 341–351.

Madden, J.L. (1977) Physiological reactions of *Pinus radiata* to attack by the woodwasp, *Sirex noctilio F.* (Hymenoptera: Siricidae). *Bulletin of Entomological Research* 67, 405–426.

Madden, J.L. and Irvine, C.J. (1971) The use of lure trees for the detection of *Sirex noctilio* in the field. *Australian Forestry* 35, 164–166.

National Sirex Co-ordination Committee (2000) *National Sirex Control Strategy, Operations Worksheet No.1*. Melbourne, Australia, 20pp.

Neumann, F.G., Harris, J.A., Kassaby, F.Y. and Minko, G. (1982) An improved technique for early detection and control of the sirex woodwasp in radiata pine plantations. *Australian Forestry* 47, 218–224.

Neumann, F.G., Andrew, P., Shepherd, R. and Stewart, H. (1989) Guidelines for the control of sirex wasp in Victoria (report to regional management). *Lands and Forests Technical Report* 1, 1–17.

Popiel, I. and Vasquez, E.M. (1991) Cryopreservation of *Steinernema carpocapsae* and *Heterorhabditis bacteriophora. Journal of Nematology* 23, 432–437.

Popiel, I., Holtemann, K.D., Glazer, I. and Womersley, C. (1988) Commercial storage and shipment of entomogenous nematodes. *International Patent* No. WO 88/011344.

Remillet, M. and Laumond, C. (1991) Sphaerularioid nematodes of importance in agriculture. In: Nickle, W.R. (ed.) *Manual of Agricultural Nematology*. Marcel Dekker, New York, pp. 967–1024.

Riding, I.L. (1970) Microvilli on the outside of a nematode. *Nature* 226, 179–180.

Wood, M.S., Stephens, N.C., Allison, B.K. and Howell, C.I. (2001) *Plantations of Australia – a report from the National Plantation Inventory and the National Farm Forest Inventory (abridged version).* National Forest Inventory, Bureau of Rural Sciences, Canberra.

22 The Entomophilic *Thripinema*

J. Funderburk and K.S. Latsha
North Florida Research and Education Center, University of Florida, Quincy, FL 32351, USA

22.1. Introduction

There are over 5000 described species of thrips (Moritz *et al.*, 2001). These insects comprise the Order Thysanoptera. The life history strategy presumably is pre-adapted from the mycetophagous ancestral group, and the need to succeed in a habitat in which optimum conditions are brief (Mound, 1997). Species of thrips generally are *r*-selected with population attributes that include a short generation time, a moderately broad food tolerance, a tendency towards parthenogenesis, vagility and a competitive breeding structure. Agricultural crops provide new opportunities for quick colonization and establishment of opportunistic thrips, and some species are important pests in the field or greenhouse. Since the classic studies on *Thrips imaginis* in Australia (Davidson and Andrewartha, 1948a,b), researchers have looked for biotic and abiotic factors that account for the large-scale spatial and temporal patterns displayed by populations of thrips (Reitz *et al.*, 2002). In the book edited by Lewis (1997), there were no clear examples showing that natural enemies are able to suppress populations of thrips. Mound (1997) speculated that the population attributes of thrips outstripped the capacities of natural enemies to suppress populations. Recent research, however, indicates that *Orius* spp. (Heteroptera: Anthocoridae) are important natural enemies that suppress populations and cause local extinctions of thrips (Funderburk, 2002). Other research is showing that the entomophilous *Thripinema* (Tylenchida: Allantonematidae) are important natural enemies. In this chapter we describe the host–parasite biology of *Thripinema* and thrips, and evaluate the potential of *Thripinema* sp. as biocontrol agents for flower thrips in greenhouse and field crops.

The tylenchoid ancestors of the Family Allantonematidae progressed towards mycetophagy and eventually parasitism of the insect haemocoel (Siddiqi, 2000). The generalized life cycle of a *Thripinema* sp. parasitizing a flower thrips is shown in Fig. 22.1. The development of a large robust stylet and the correlated transformation of the broad corpus into a muscular organ were the keys to allow the entomoparasitic female to penetrate the exoskeleton of the insect and develop in the host haemocoel. Once inside, this spermaticized, slender female goes through a radical transformation, where its development is synchronized with that of the host. The parasitic female increases in size and the genital organs gradually fill most of the body cavity. The eggs are set free in the host's haemocoel. The eggs hatch and the juveniles moult to the fourth stage and exit through the anus or oviduct of the host. Supposedly, the male undergoes a final moult in the free life and mates with the female. In a single heterosexual cycle, both free-living and parasitic forms are found, the former partially free-living, as they do not feed or multiply outside the host. Sterility of the thrips host is the outcome of parasitism (Loomans *et al.*, 1997).

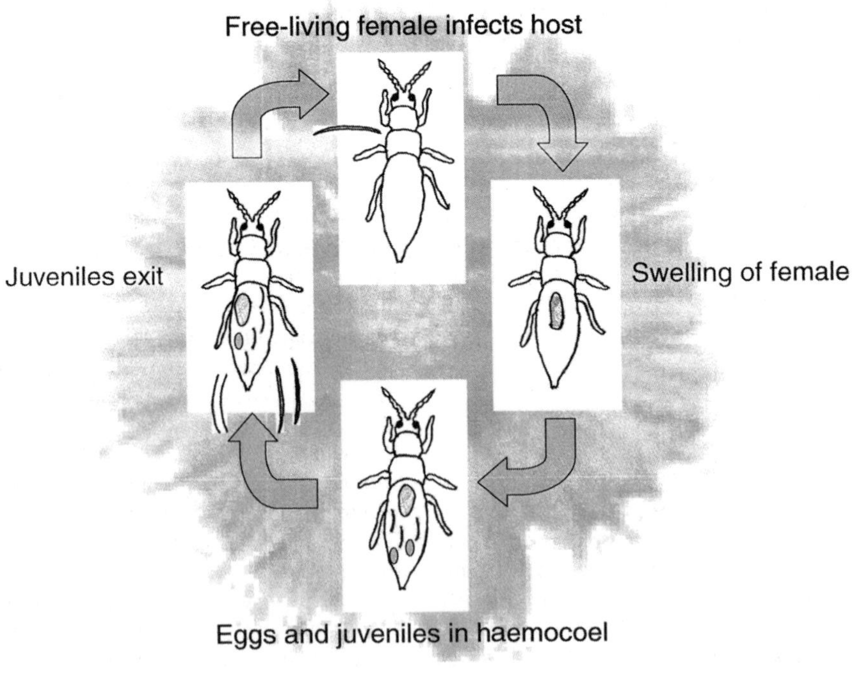

Fig. 22.1. The generalized life cycle of a *Thripinema* species parasitizing a flower thrips.

22.2. Thrips

Species of thrips emerged as crop pests in the 1980s with the widespread movement of live plants. More than 20 species of thrips are cosmopolitan. Mound (1997) lists 35 species that are significant crop pests. Nearly all are in the Subfamily Thripinae, particularly species of flower thrips in the genera *Thrips* and *Frankliniella*. Pollen provides nutrients for greater egg production of flower thrips, and the adults feed on the flowers of a wide range of available plant species, including those that are not suitable reproductive hosts (Funderburk, 2002). Leaves are exploited by the adults when flowers are scarce, and leaves may be preferred as a more stable source of food for the developing larvae. Two pests of great economic importance are *Frankliniella occidentalis* and *F. fusca*. At least seven species of thrips, including *F. fusca* and *F. occidentalis*, are competent vectors of tomato spotted wilt virus (Ullman *et al.*, 1997), which is the type species in the genus *Tospovirus* in the Family Bunyaviridae (German *et al.*, 1992). The disease is spread in a field by adult thrips that develop and acquire the virus as larvae from infected plants either outside the field (hereafter primary spread) or inside the field (hereafter secondary spread). The severity and timing of epidemics in a particular crop system are the result of interactions between the thrips vector, the pathogen host plant and the pathogen.

22.3. *Thripinema*

22.3.1. Host–parasite biology

Thripinema species are obligate parasites of thrips and parasitize their host in a moist microhabitat such as a leaf gall, flower perianth or foliage terminal (Loomans *et al.*, 1997). Unlike nematodes in the families Steinernematidae and Heterorhabditadae, *Thripinema* species need to keep their host alive for their own survival and transmission. All known species within the genus render female thrips sterile, although the physiological mechanisms are not understood. Lysaght (1937) presumed that the parasites cause sterility by either depriving the thrips of protein required for normal egg development or by secreting a toxin that damages the reproductive organs. Green and Parrella (1995) hypothesized that the stretch receptors in the thrips abdomen that regulate oogenesis may respond to the increasing number of developing nematodes by signalling the ovaries to halt oogenesis as if maximum egg production had been attained. Recent research refutes theories that the developing juveniles are responsible for stopping oogenesis, and it suggests that host oogenesis is actively regulated by the female *Thripinema* (Sims *et al.*, 2005).

The parasitic females of *Thripinema* sp. actively locate thrips by rotating the anterior part of their body, while holding their posterior part to a plant structure until making contact with the host. The nematode penetrates the insect cuticle through an intersegmental membrane or through a coxal cavity (Tipping *et al.*, 1998; Lim *et al.*, 2001), and it swells to the characteristic oval shape of the infective female. Fourth stage juveniles exit the host through the ovipositor or anus (Reddy *et al.*, 1982; Mason and Heinz, 2002). The sex ratio is female-biased (Sharga, 1932; Mason and Heinz, 2002), which suggests that species of *Thripinema* have a tendency towards parthenogenesis. Whether mating occurs inside the host or outside on plant structures is unknown. Females are oviparous and complete one generation per host (Mason and Heinz, 2002). Development is synchronized with the host, as indicated by differences in generation time when different stages of the host are parasitized (Sims *et al.*, 2005).

Previous reports indicated that male thrips are not parasitized because none were recovered from field populations (Sharga, 1932; Loomans *et al.*, 1997). However, larval thrips and adult males and females are parasitized, with the adult males the least preferred and the adult females the most preferred (Tipping *et al.*, 1998; Lim *et al.*, 2001; Funderburk *et al.*, 2002a,b; Mason and Heinz, 2002). The number of nematodes emerging from a thrips is influenced by the stage and sex of the host (Sims *et al.*, 2005). As many as 320 juveniles were observed in a single thrips host (Reddy *et al.*, 1982).

There are no physical or behavioural signs of parasitism from *Thripinema*, and parasitism has no effect on mortality or longevity of a female thrips host (Lim *et al.*, 2001). Male longevity is reduced, possibly due to their infrequent feeding, which allows the nematodes to deplete the body fluids (Lim *et al.*, 2001). Parasitism results in nearly complete sterility of female thrips (Sims *et al.*, 2005). Even females parasitized as adults do not lay eggs (or at most only a few).

22.3.2. Species of *Thripinema*

The first record of a described entomogenous nematode occurring in thrips was in Germany in 1895 (Uzel, 1895), and the first detailed ecological study was Sharga (1932) on *Howardula aptini*. Morphological differences and host range between *Howardula* sp. prompted a taxonomic revision and the creation of the genus *Thripinema* (Siddiqi, 2000). There are five described species that are known to parasitize 11 species of thrips in five genera (Table 22.1). The geographical range of *Thripinema* sp. includes England (Sharga, 1932), New Brunswick in Canada (Nickle and Wood, 1964), India (Reddy *et al.*, 1982), Moscow in Russia (Chizhov *et al.*, 1995), New Zealand (Teulon *et al.*, 1997), Chile (Funderburk *et al.*, 2002a) and the USA, including California (Wilson and Cooley, 1972) and Florida (Tipping *et al.*, 1998). There are undoubtedly other undescribed species in this little-studied genus.

The host specificity of *Thripinema* remains unclear. Some species are known to parasitize more than one species of thrips (Table 22.1). *Thripinema khrustalevi* isolate Chile parasitizes and develops in *F. australis*, but not *F. occidentalis* (Funderburk *et al.*, 2002a). Under laboratory conditions, *T. fuscum* parasitizes but is unable to develop in the larvae and adults of *F. bispinosa* and *F. occidentalis* (K.S. Latsha, J. Funderburk and D. Boucias, unpublished data). Parasitism of *F. tritici* and *F. occidentalis* under field conditions is rare (Funderburk *et al.*, 2002b).

Table 22.1. Species of *Thripinema* and thrips host species.

Thripinema species	Thrips host(s)	References
T. nicklewoodi	*Frankliniella vaccinii*	Nickle and Wood, 1964
	Taeniothrips vaccinophilus	
	F. occidentalis	Wilson and Cooley, 1972
T. aptini	*Aptinothrips rufus*	Sharga, 1932
		Lysaght, 1936
T. reniraoi	*Megaluriothrips* sp.	Reddy *et al.*, 1982
T. krustalevi	*Thrips trehernei*	Chizhov *et al.*, 1995
	Thrips physapus	
T. khrustalevi isolate Chile	*F. australis*	Funderburk *et al.*, 2002a
T. fuscum	*F. fusca*	Tipping *et al.*, 1998
	F. tritici	Funderburk *et al.*, 2002b
	F. occidentalis	
Thripinema sp. unidentified	*Thrips obscuratus*	Teulon *et al.*, 1997

22.4. *Thripinema* as a Biocontrol in Greenhouse Crops

22.4.1. *Frankliniella occidentalis,* the western flower thrips (WFT)

Before the 1980s, WFT was a damaging pest of a wide range of crops in its native geographic regions of western North America (Kirk, 2002). The species thereafter was spread by the international plant trade, and it is a serious worldwide pest of over 200 vegetable and ornamental crops grown in the greenhouse and field. Feeding reduces the aesthetic quality or the yield of a crop. The spread of WFT may have been facilitated by the appearance of a new insecticide-resistant biotype (Kirk, 2002). Further, its spread appears to be responsible for the emergence of tomato spotted wilt virus as a threat to global agriculture.

22.4.2. *Thripinema nicklewoodi,* a natural enemy of western flower thrips (WFT)

Parasitism of WFT by *T. nicklewoodi* occurs naturally on many crops in western North America. It is the primary natural enemy of WFT populations in greenhouse floricultural crops (Heinz *et al.*, 1996). High levels of parasitism occur in field populations, but effects of parasitism on population dynamics of WFT under field conditions have not been adequately studied.

22.4.3. *Thripinema nicklewoodi* as a component of integrated pest management (IPM)

Arthurs and Heinz (2002) describe a procedure for *in vivo* rearing of *T. nicklewoodi.* The procedure produces a doubling of infected WFT from one generation to the next. No species of *Thripinema* has been successfully reared *in vitro.* The lack of an *in vitro* procedure for commercial mass production likely precludes the use of inundative releases of *Thripinema* as a practical tactic in integrated pest management (IPM) programmes.

The natural establishment of *T. nicklewoodi* in greenhouse-grown carnations, chrysanthemum, and roses in California (K.M. Heinz, unpublished data, as reported in Arthurs and Heinz, 2002) suggests the feasibility of innoculative releases for biocontrol of WFT. Arthurs *et al.* (2003) reported that temperature conditions in commercial greenhouses in Texas are favourable for the establishment of *T. nicklewoodi.* Patterns of aggregation of WFT in the flowers, buds and foliar terminals in floricultural crops favour the survival of the free-living females of *T. nicklewoodi* and their ability to rapidly locate an uninfected host.

Arthurs and Heinz (2003) found that parasitism reduced feeding of WFT adults on chrysanthemums and beans by about 90% and, as a result, transmission of tomato spotted wilt virus by about 50%. This is an especially intriguing find in that insecticides are not effective in preventing primary spread of the disease. The lack of effective controls for tomato spotted wilt greatly enhances the commercial potential of *Thripinema* sp. Because time lags in measurable thrips suppression will occur when using innoculative release procedures, other tactics such as resistant cultivars, predatory mites, and biorational insecticides may need to be integrated with *T. nicklewoodi* in a WFT/tomato spotted wilt IPM programme.

22.5. *Thripinema* as a Biocontrol in Field Crops

22.5.1. *Frankliniella fusca,* the tobacco thrips

The tobacco thrips of eastern North America feeds on a wide range of plants. The species is widespread in the USA northward into Canada and south to Puerto Rico and Mexico (Moritz *et al.*, 2001). Natural hosts are many wild grasses and other flowering plants (Moritz *et al.*, 2001). The tobacco thrips reproduces in small grains, tobacco, snap beans and cowpeas. It is the most abundant species on seedlings of groundnut in southern USA (Funderburk *et al.*, 2002b). Injury to the leaves can be severe when large numbers are present. Even severe injury is not usually economically important, although certain groundnut cultivars are less tolerant, especially when injury is combined with other stress factors such as drought and herbicide injury (Brecke *et al.*, 1996). Depending on the crop cultivar and the number of days in the growing season, at least two additional generations of tobacco thrips develop in a groundnut field.

In south-eastern USA, the tobacco thrips is the key vector of tomato spotted wilt in groundnut (Funderburk *et al.*, 2002b). Severe epidemics are common in groundnut during mid- or late season, and it is the key pest of groundnut in the region. Viruliferous adults immigrating into a groundnut field during the vegetative stages of crop growth are responsible for the first cycle of disease. Additional cycles of disease are mostly secondary spread (Funderburk *et al.*, 2002b).

22.5.2. *Thripinema fuscum*, a natural enemy of tobacco thrips

Parasitism of tobacco thrips by *T. fuscum* occurs naturally in groundnut in south-eastern USA. The intrinsic capacity of increase of *T. fuscum* is several times greater than that of the host (Sims *et al.*, 2005). The flowers of groundnut are the primary site for aggregation of the adults of *F. fusca* and for the free-living females of *T. fuscum* to locate new hosts.

Figure 22.2 shows the relationship in groundnut between parasitism and population dynamics of the host (Funderburk *et al.*, 2002b). Parasitism is low during the vegetative crop-growth stages, and populations of tobacco thrips increase rapidly. Populations are suppressed to near-extinction levels after the groundnuts begin flowering. Parasitism remains high for the remainder of the growing season, and populations of thrips are unable to rebuild.

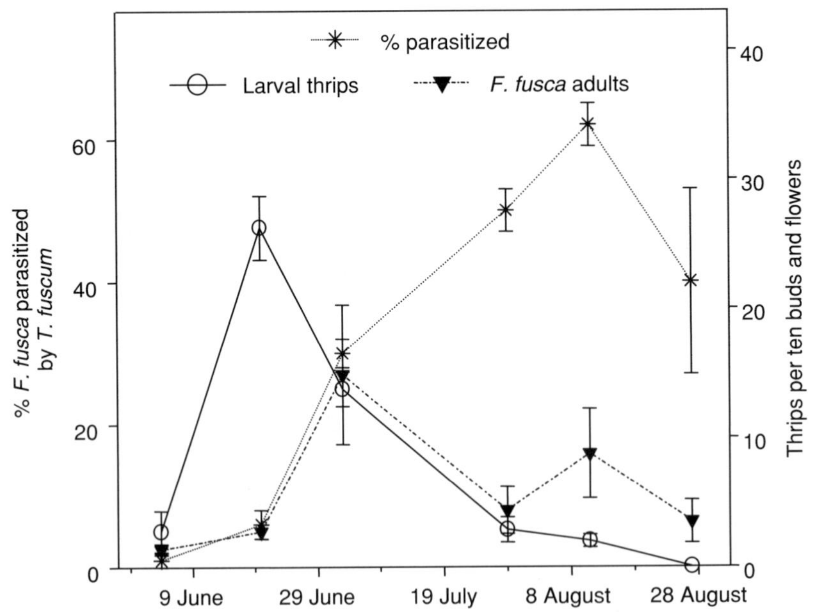

Fig. 22.2. Means (± standard errors) of % parasitism by *Thripinema fuscum* and population densities of *Frankliniella fusca* in groundnut (adapted from Funderburk *et al.*, 2002b).

22.5.3. *Thripinema fuscum* as a component of integrated pest management (IPM)

Biocontrol by *T. fuscum* is integrated with insecticidal, cultural and plant resistance tactics to manage tobacco thrips, and, more importantly, tomato spotted wilt disease. Parasitism of *T. fuscum* during mid- and late season reduces cycles of secondary spread under normal weather conditions (Funderburk *et al.*, 2002b). Additional control of larvae with insecticides is useful in further reducing cycles of secondary spread.

Figure 22.3 illustrates the relationship between parasitism and the population dynamics of *F. fusca* in groundnut planted on three dates (Stavisky *et al.*, 2002). The overall abundance of tobacco thrips in the landscape decreases as parasitism increases during late spring, and the planting date can be delayed as a cultural tactic to maximize the benefits of biocontrol. The later the planting of groundnuts, the lower the number of immigrating adults into the field. Further, increases in parasitism result in fewer larvae in delayed plantings. Little or no build-up in populations occurs in late plantings.

Figure 22.4 illustrates the impact of planting date on the incidence of tomato spotted wilt disease (Stavisky *et al.*, 2002). The shapes of the disease progress curves are similar for each groundnut planting. Rapid increases in disease due to primary spread

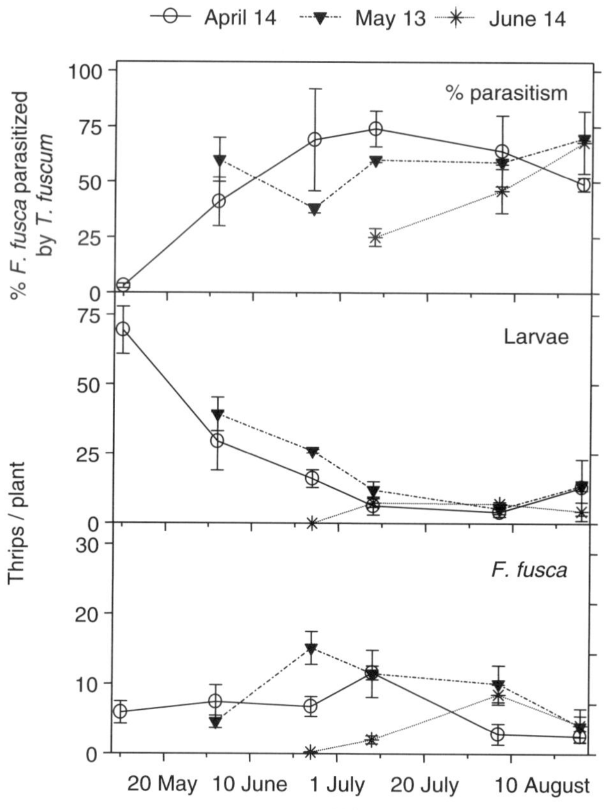

Fig. 22.3. Means (± standard errors) of % parasitism of the adult female *Frankliniella fusca* by *Thripinema fuscum* (top) and the population densities of *F. fusca* larvae (middle) and adults (bottom) in groundnuts planted on three dates (adapted from Stavisky *et al.*, 2002).

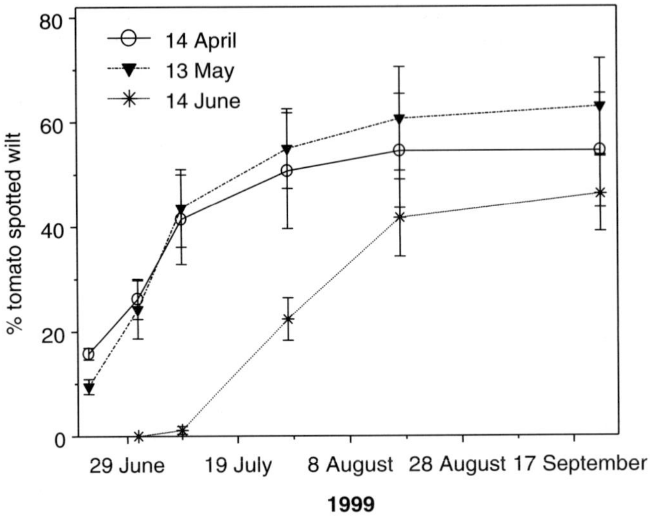

Fig. 22.4. The cycles of tomato spotted wilt virus (% incidence ± standard error) in groundnut planted on three dates (adapted from Stavisky *et al.*, 2002).

by viruliferous immigrating adults of tobacco thrips occur during the vegetative stages. Additional increases in disease incidence occur during the early reproductive crop-growth stages. Later planting results in lower disease incidence because of the effects of parasitism on the vector's populations, as described earlier.

Flowers are the sites where tobacco thrips adults become parasitized, and the groundnuts usually flower for the remainder of the growing season. If flowering is suppressed, however, due to extreme drought, parasitism of the adults declines, and a build-up in vector populations occurs (J. Funderburk, J. Stavisky and T. Momol, unpublished data). There is an additional tertiary cycle of tomato spotted wilt attributable to secondary spread. Under conditions of extreme drought, irrigation of the crop is recommended, when possible, to prevent such an additional cycle in disease incidence.

Groundnut cultivars with partial resistance to tomato spotted wilt are commercially available. The mechanisms of resistance are not related to effects on populations of the vector (J. Funderburk, J. Stavisky and T. Momol, unpublished data). The population dynamics of the vector and the shape of the disease progress curves are very similar on susceptible and resistant cultivars. Parasitism is important in reducing secondary spread of tomato spotted wilt virus in susceptible and resistant cultivars. Further, incorporating biocontrol (e.g. with *T. fuscum*) into a groundnut IPM programme reduces the potential for tomato spotted wilt virus strains to overcome resistance factors in the plant.

22.6. Summary and Conclusions

Species of *Thripinema* keep the host alive and render the female thrips sterile. They are intrinsically capable of suppressing populations of thrips. The free-living females need to survive and locate new hosts for *Thripinema* to be effective as biocontrol agents. Species are successfully produced *in vivo*. The host cues necessary for parasite development need to be understood before *in vitro* procedures for mass production are possible. Further, the host specificity of *Thripinema* sp. needs to be elucidated so that their potential against individual species of thrips pests can be evaluated. Research has shown that inoculative releases of *T. nicklewoodi* for biocontrol of *F. occidentalis* in greenhouse crops is feasible. Other research has shown that *T. fuscum* is an important natural enemy of *F. fusca* in field crops. Their usefulness as biocontrol tactics in IPM programmes is greatly enhanced by their proven ability to limit primary and secondary spread of tomato spotted wilt. Additional research is needed to evaluate the effect of *Thripinema* sp. on the population dynamics of pest thrips in more geographic regions and other production systems. We conclude that the *Thripinema* are important natural enemies of thrips, and they show promise as biocontrol agents of thrips pests under field and greenhouse conditions.

References

Arthurs, S. and Heinz, K.M. (2002) *In vivo* rearing of *Thripinema nicklewoodi* (Tylenchida: Allantonematidae) and prospects as a biological control agent of *Frankliniella occidentalis* (Thysanoptera: Thripidae). *Journal of Economic Entomology* 95, 668–674.

Arthurs, S. and Heinz, K.M. (2003) Thrips parasitic nematode *Thripinema nicklewoodi* (Tylenchida: Allantonematidae) reduces feeding, reproductive fitness, and tospovirus transmission by its host, *Frankliniella occidentalis* (Thysanoptera: Thripidae). *Environmental Entomology* 32, 853–858.

Arthurs, S., Heinz, K.M., Thompson, S. and Krauter, P.C. (2003) Effect of temperature on infection, development and reproduction of the parasitic nematode *Thripinema nicklewoodi* in *Frankliniella occidentalis*. *Biocontrol* 48, 417–429.

Brecke, B.J., Funderburk, J.E., Teare, I.D. and Gorbet, D.W. (1996) Interaction of early-season herbicide injury, tobacco thrips injury, and cultivar on peanut. *Agronomy Journal* 88, 14–18.

Chizhov, V.N., Subbotin, S.A. and Zakharenkova, N.N. (1995) *Thripinema khrustalevi* sp. n. (Tylenchida: Allantonematidae), a parasite of Thrips (Thysanoptera) in Moscow. *Russian Journal of Nematology* 3, 89–94.

Davidson, J. and Andrewartha, H.G. (1948a) Annual trends in natural population of *Thrips imaginis* (Thysanoptera). *Journal of Animal Ecology* 17, 193–199.

Davidson, J. and Andrewartha, H.G. (1948b) The influence of rainfall, evaporation and atmospheric temperature fluctuations on a natural population of *Thrips imaginis* (Thysanoptera). *Journal of Animal Ecology* 17, 200–222.

Funderburk, J.E. (2002) Ecology of thrips. In: Marullo, R. and Mound, L.A. (eds) *Thrips and Tospoviruses: Proceedings of the 7th International Symposium on Thysanoptera*. Australian National Insect Collection, Canberra, Australia, pp. 121–128.

Funderburk, J.E., Ripa, R., Espinoza, F. and Rodriguez, F. (2002a) Parasitism of *Frankliniella australis* (Thysanoptera: Thripidae) by *Thripinema khrustalevi* (Tylenchida: Allantonematidae) isolate Chile. *Florida Entomologist* 85, 645–649.

Funderburk, J.E., Stavisky, J., Tipping, C., Gorbet, D., Momol, T. and Berger, R. (2002b) Infection of *Frankliniella fusca* (Thysanoptera: Thripidae) in peanut by the parasitic nematode *Thripinema fuscum* (Tylenchidae: Allantonematidae). *Environmental Entomology* 31, 558–563.

German, T.L., Ullman, D.E. and Moyer, J.W. (1992) Tospoviruses: diagnosis, molecular biology, phylogeny, and vector relationships. *Annual Review of Phytopathalogy* 30, 315–348.

Green, I.D. and Parrella, M.P. (1995) Two new natural enemies of western flower thrips in California. In: Parker, B.L., Skinner, M. and Lewis, T. (eds) *Thrips Biology and Management*. Plenum Press, New York, pp. 277–280.

Heinz, K.M., Heinz, L.M. and Parrella, M.P. (1996) Natural enemies of western flower thrips indigenous to California ornamentals. *International Organization of Biological Control/Western Palearctic Regional Section Bulletin* 19(1), 51–54.

Kirk, W.D.J. (2002) The pest and vector from the west: *Frankliniella accidentalis*. In: Marullo, R. and Mound, L.A. (eds) *Thrips and Tospoviruses: Proceedings of the 7th International Symposium on Thysanoptera*. CSIRO Entomology, Canberra, Australia, pp. 33–42.

Lewis, T. (ed.) (1997) *Thrips as Crop Pests*. CAB International, Wallingford, UK, 740 pp.

Lim, U.T., Van Driesche, R.G. and Heinz, K.M. (2001) Biological attributes of the nematode, *Thripinema nicklewoodi*, a potential biological control agent of western flower thrips. *Biological Control* 22, 300–306.

Loomans, A.J.M., Murai, T. and Greene, I.D. (1997) Interactions with hymenopterous parasitoids and parasitic nematodes. In: Lewis, T. (ed.) *Thrips as Crop Pests*. CAB International, Wallingford, UK, pp. 355–397.

Lysaght, A.M. (1936) A note on the adult female of *Anguillulina aptini* (Sharga), a nematode parasitizing *Aptinothrips rufus* Gmelin. *Parasitology* 28, 290–292.

Lysaght, A.M. (1937) An ecological study of a thrips (*Aptinothrips rufus*) and its nematode parasite (*Anguillulina aptini*). *Journal of Animal Ecology* 6, 169–192.

Mason, K. and Heinz, K.M. (2002) Biology of *Thripinema nicklewoodi* (Tylenchida), an obligate *Frankliniella occidentalis* (Thysanoptera) parasite. *Journal of Nematology* 34, 332–339.

Moritz, G., Morris, D. and Mound, L.A. (2001) *Thrips ID Pest Thrips of the World.* (CD-rom) ©Australian Center for International Agricultural Research, Collingwood, Australia.

Mound, L.A. (1997) Biological diversity. In: Lewis, T. (ed.) *Thrips as Crop Pests.* CAB International, Wallingford, UK, pp. 197–215.

Nickle, W.R. and Wood, G.S. (1964) *Howardula aptini* (Sharga 1932) parasitic in blueberry thrips in New Brunswick. *Canadian Journal of Zoology* 42, 843–846.

Reddy, Y.N., Nickle, W.R. and Rao, P.N. (1982) Studies on *Howardula aptini* (Nematoda-Sphaerulariidae) parasitic in *Megaluriothrips* sp. in India. *Indian Journal of Nematology* 12, 1–5.

Reitz, S.R., Funderburk, J.E., Hansen, E.A., Baez, I., Waring, S. and Ramachandran, S. (2002) Interspecific variation in behavior and its role in thrips ecology. In: Marullo, R. and Mound, L.A. (eds) *Thrips and Tospoviruses: Proceedings of the 7th International Symposium on Thysanoptera.* Australian National Insect Collection, Canberra, Australia, pp. 133–140.

Sharga, U.S. (1932) A new nematode, *Tylenchus aptini* n. sp. parasite of Thysanoptera (Insecta: *Aptinothrips rufus* Gmelin). *Parasitology* 24, 268–279.

Siddiqi, M.R. (2000) *Tylenchida Parasites of Plants and Insects,* 2nd edn, CAB International, Wallingford, UK.

Sims, K., Funderburk, J. and Boucias, D. (2005) Host-parasite biology of *Thripinema fuscum* (Tylenchida: Allantonematidae) and *Frankliniella occidentalis* (Thysanoptera: Thripidae). *Journal of Nematology* 37 (in press).

Stavisky, J., Funderburk, J.E., Momol, T. and Gorbet, D. (2002) Influence of parasitism by *Thripinema fuscum* on dynamics of local populations of *Frankliniella fusca.* In: Marullo, R. and Mound, L.A. (eds) *Thrips and Tospoviruses: Proceedings of the 7th International Symposium on Thysanoptera.* Australian National Insect Collection, Canberra, Australia, pp. 141–143.

Teulon, D.A.J., Wouts, W.M. and Penman, D.R. (1997) A nematode parasite of the New Zealand flower thrips (Thysanoptera: Thripidae). *New Zealand Entomologist* 20, 67–69.

Tipping, C., Nguyen, K.B., Funderburk, J.E. and Smart, G.C. Jr (1998) *Thripinema fuscum* n. sp. (Tylenchida: Allantonematidae), a parasite of the tobacco thrips, *Frankliniella fusca* (Thysanoptera). *Journal of Nematology* 30, 232–236.

Ullman, D.E., Sherwood, J.L. and German, T.L. (1997) Thrips as vectors of plant pathogens. In: Lewis, T. (ed.) *Thrips as Crop Pests.* CAB International, Wallingford, UK, pp. 539–565.

Uzel, H. (1895) *Monographie der Ordnung Thysanoptera,* Koniggratz, 472 pp.

Wilson, T.H. and Cooley, T.A. (1972) A chalcidoid planidium and an entomophilic nematode associated with the western flower thrips. *Annals of the Entomological Society of America* 65, 414–418.

23 Mermithid Nematodes

E.G. Platzer,[1] B.A. Mullens[1] and M.M. Shamseldean[2]
[1]*Department of Nematology, University of California, Riverside, CA 92521, USA;*
[2]*Department of Agricultural Zoology and Nematology, Faculty of Agriculture, Cairo University, Giza 12311, Egypt*

23.1. Introduction

Mermithids are generally long, thin and translucent white nematodes. They vary in length from 10 mm to 100 mm and are parasitic in the Phylum Arthropoda. Mermithids produce preparasitic juveniles that enter host species by ingestion of eggs, as encysted juveniles within ingested paratenic hosts or by direct penetration of the host integument and entrance into the haemocoel. Here the parasitic juvenile stage grows and moults once. After storing sufficient sustenance for its adult phase, it emerges and kills the host. Further development to the adults and reproduction take place outside the host. Reviews of mermithids in biocontrol (Petersen, 1985; Kaiser, 1991; Popiel and Hominick, 1992; Federici, 1995; Baker and Capinera, 1997; and Kerry and Hominick, 2002) have listed attributes such as ease of application, environmental safety, host specificity, laboratory manipulation of life history traits, lethality, mass rearing *in vivo* and potential for long-term recycling in the environment. However, rearing mermithids may be costly, and the population dynamics are not compatible with the characteristics of an ideal biocontrol organism (Popiel and Hominick, 1992; Federici, 1995; Kerry and Hominick, 2002). None the less, research on mermithids continues to show that mermithids do indeed exert long-term effects on host populations (Baker and Capinera, 1997), and mass rearing *in vivo* in developing countries can be a viable and attractive alternative for control of vector insects (Santamarina and Perez, 1997).

(eds P.S. Grewal, R.-U. Ehlers and D.I. Shapiro-Ilan)

Three case studies describing the control of aquatic and terrestrial insects are presented to illustrate the scope encompassed by three mermithid species in biocontrol.

23.2. *Romanomermis* Case Study

Mermithid nematodes parasitizing mosquitoes have substantial potential for vector control (Platzer, 1981). Of the seven currently recognized genera in the Mermithidae that are parasitic in mosquitoes (Poinar, 2001), the best-known genus is *Romanomermis*. The genus contains at least 13 species. Twelve are parasites of mosquitoes and, interestingly, the type species is a parasite of an amphipod (Petersen, 1985). The current biogeographic distribution includes five species in North America, five in China and one each in India, Kazakhstan and Romania. *Romanomermis culicivorax* is the most thoroughly researched species and was first isolated at Lake Charles, Louisiana (Petersen, 1985). Basic rearing concepts refined by Petersen (1985) led to an exceptional series of publications on biocontrol of mosquitoes with mermithids and further development of basic biological concepts (e.g. sex ratio regulation) for mermithids parasitizing mosquitoes.

23.2.1. Life history

Following the initial discovery of *R. culicivorax* infecting larvae of *Anopheles crucians* and *Psorophora confinnis* in Louisiana, the nematodes were established in a laboratory culture (Petersen, 1985). The life history of *R. culicivorax* involves only larval stages of mosquitoes. Preparasitic juveniles attack the mosquito larvae and penetrate into the haemocoel via the cuticle (Fig. 23.1). Mosquito larvae die 7–10 days post-infection. At that time the postparasitic juveniles penetrate the host's body wall and emerge to complete their development in the bottom sediment of the mosquito pool. Here the mermithids moult, mate and the females oviposit in the bottom sediments. When the mermithids are maintained at 27°C, the minimum time for maturation, oviposition, embryonation, hatching of the eggs and emergence of the infectious preparasitic juveniles is 3 weeks (Platzer, 1981).

Unlike most mermithids, *R. culicivorax* has a broad host range comprising over 90 species of mosquitoes in 13 genera (Petersen, 1985; Peng *et al.*, 1992). Only a few mosquito species inhibited the development of *R. culicivorax* by encapsulation (Petersen, 1985). Susceptibility of host mosquitoes is related to larval age; the first instars are most susceptible and the fourth instars least susceptible (Petersen, 1973). In mixed mosquito populations, anophelines are more susceptible to parasitism than culicines (Petersen, 1985). This difference probably results from differences in larval behaviour because larvae of anophelines spend more time at the surface than

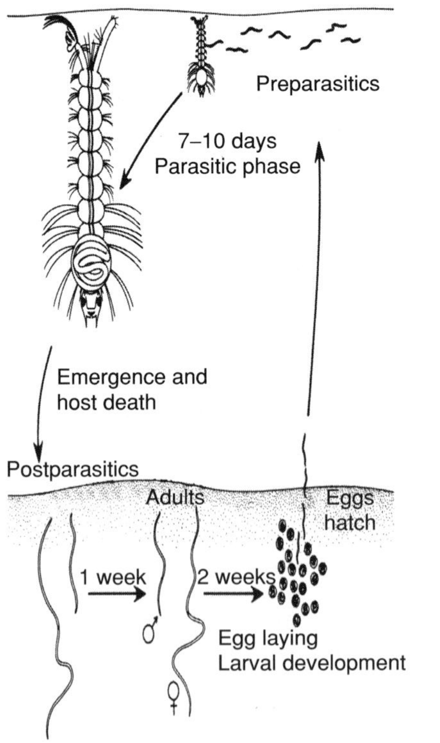

Fig. 23.1. Diagram of the life cycle of *Romanomermis culicivorax* in *Culex pipiens* under laboratory conditions at 27°C (from Platzer, 1981).

larvae of culicines, and the infectious preparasites are located in surface waters.

Temperature constitutes the major environmental influence in the life history of *R. culicivorax*. Petersen (1985) established 18°C as the lower limit for infectivity of *R. culicivorax*. A number of environmental constraints beyond temperature affect *R. culicivorax*. Preparasitic juveniles do not tolerate mild salinity (Petersen, 1973), the toxicity of which depends on the particular ion and concentration thereof (see Platzer, 1981).

23.2.2. Biocontrol

Subsequent to the development and establishment of effective mass rearing technology for *R. culicivorax,* field trials were conducted to assess the efficacy of the nematodes in mosquito control (Petersen, 1985). Petersen's group established the basic parameters for control of anopheline mosquito larvae. Preparasites (1000/m^2 of water surface) of *R. culicivorax* were applied with a standard pesticide sprayer, and 64% of all anopheline instars were infected. Subsequent annual sampling demonstrated persistent establishment of the mermithid in the mosquito population, although the prevalence was generally lower than initial infection levels. Studies on fallow rice fields in Louisiana produced infections in up to 38% and 61% of *P. confinnis* and *A. quadrimaculatus*, respectively. In Cuba, Santamarina and Perez (1997) demonstrated the efficacy of *R. culicivorax* in sewage-settling ponds and natural ponds.

In the first successful large-scale attempt to control mosquitoes with a parasite, Petersen (1985) controlled anopheline larvae with *R. culicivorax* in a 14.4-ha lake in El Salvador. The mosquito developmental sites were treated 11 times during 7 weeks with 2400–4800 preparasites/m^2 at each application. The anopheline larval population declined by 17-fold from the first preparasite application to the end of the parasite release period. Unfortunately, no subsequent studies on persistence of the parasite were possible because of political unrest.

As an alternative to application of preparasites, *R. culicivorax* can be applied as eggs or postparasites (Petersen, 1985). Kerwin and Washino (1985) inoculated 20 km of linear transects in rice fields in California with postparasites over a period of 2 years. Sentinel cages showed that the nematode established successfully, provided relatively high levels of control and overwintered.

Although this review emphasizes *R. culicivorax*, there are excellent studies on species from China, India and Argentina that show the continuing promise of mermithid nematodes for mosquito control. However, this potential will be achieved only if there is a continued sustained effort to conduct quantitative studies on the basic ecology of the mermithids and develop appropriate population models. Preliminary efforts on population modelling have been reviewed by Kerry and Hominick (2002), and although these writers project a lacklustre future for mermithids in biocontrol, it is premature to dismiss mermithids with the current ecological database. Most mermithid research has been descriptive, but with an increased effort to quantitate the nature of mermithid habitats and the interactions with their hosts and other biota, it should be possible to successfully harness the potential of this fascinating group for biocontrol.

23.3. *Heleidomermis* Case Study

The unusual mermithid genus *Heleidomermis* was described by Rubtsov in 1970. Three species are known and are all parasites of biting midges in the Family Ceratopogonidae (formerly Heleidae) (see Poinar and Mullens, 1987). *Heleidomermis vivipara* was described from Karelia in the former Soviet Union, while *H. ovipara* was found in the Tunkin Valley of the Burjat, also former Soviet Union. The third species was originally reported from Allegheny Co., New York, as *H. vivipara*, but later was recognized as *H. magnapapula*, which appears to be widespread in populations of the

Culicoides variipennis complex in North America. It has been found in *C. variipennis* and *C. sonorensis*, both of which inhabit shallow mud habitats at the edges of ponds or slowly flowing water, and which are often polluted by animal waste. The status of *C. sonorensis* as a primary vector of bluetongue virus to North American ruminants led to detailed investigations of *H. magnapapula* as a biocontrol agent.

23.3.1. Life history

The biology of the genus is quite unusual. The nematodes mature within the host, and adults mate immediately after emergence. This contrasts with the usual pattern for the family, in which free-living postparasitic juveniles complete development to adult several days to weeks after emergence from the host. While *H. ovipara* is oviparous, the other two species (*H. vivipara* and *H. magnapapula*) are ovoviviparous. Partly as a result of life history traits, the life cycle of *H. magnapapula* needs as little as 2 weeks (Mullens and Velten, 1994). This is an adaptation to a host insect which itself is a typical *r*-selected species; *C. sonorensis* develop rapidly as well, and are able to disperse and utilize widely scattered ephemeral habitats. Like its host, *H. magnapapula* is resistant to polluted or saline habitats that would be impossibly harsh for most other mermithids. Even the thermal ecology of the parasite is matched to the host, with lower developmental threshold temperatures of 9–10°C (Mullens and Luhring, 1998).

The life cycle of *H. magnapapula* (Fig. 23.2) has been described by Mullens and Velten (1994). Midges are reared in pans of water with pads of polyester batting for substrate. Nutrient broth plus a microorganism starter culture (mainly bacteria and fungi), provide food for *C. sonorensis* larvae (Hunt, 1994). When hosts are in the late second or early third instar, gravid (mated) female nematodes are added. Mature host larvae are harvested 5–9 days later and held in Petri dishes of water for parasite

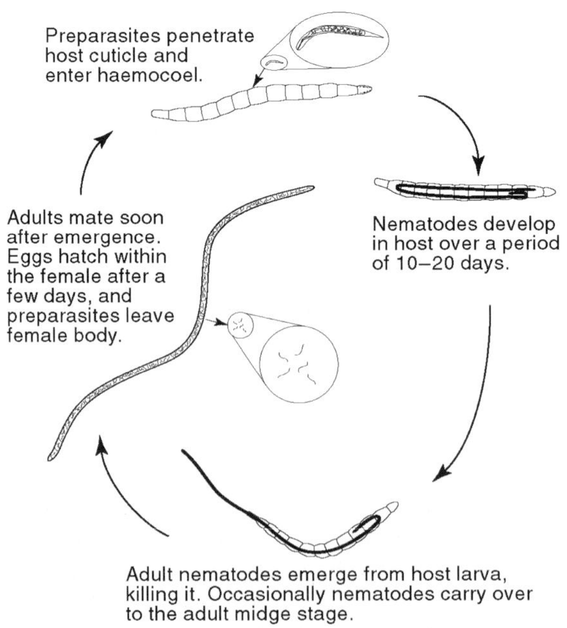

Fig. 23.2. Life cycle of *Heleidomermis magnapapula* in biting midges of the *Culicoides variipennis* complex in North America. Typical habitat is manure-polluted surface mud at the margins of ponds or slowly moving water.

emergence. Male emergence precedes females (average 12.2 versus 13.4 days after parasites were added to pans), and mating occurs immediately. Eggs hatch internally in 3–5 days, and preparasitic juveniles are expelled into the environment by the female over the next 1–2 days, after which the female dies. On moist agar, females crawl a few millimetres, back up briefly and then proceed forward again. This results in several preparasites being expelled every few millimetres as the female crawls forward and the opening of the barrel-shaped vagina flexes. The preparasites are short-lived (1–2 days), but the period of infectivity is shorter (Mullens and Luhring, unpublished data). It probably is imperative that preparasites find a host within 6–12 h.

Preparasites penetrate the cuticle to access the haemocoel, and sex determination is density-dependent. If a single preparasite enters, a female develops about 60–70% of the time (Mullens and Velten, 1994; Paine and Mullens, 1994). Hosts with two parasites usually yield males, and this is invariably the case if more than two preparasites occur in the same host. While as many as nine small males can emerge from a host, their fitness is dubious. This mechanism of sex ratio determination allows the parasite to track host availability in time; an excess abundance of parasites produces more males and parasite reproduction declines. Sex ratios from field-collected hosts are male-biased (Paine and Mullens, 1994). In the field, early stage host larvae are more abundant immediately above the waterline, while later stage hosts are widely distributed but common below the waterline. Free-living, adult *H. magnapapula* are found at the waterline (males) or just above the waterline (females). It is speculated that mating occurs near the waterline, but females then disperse above the waterline to distribute preparasitic juveniles in areas occupied by early stage host larvae, which are the preferred host stage (Mullens and Luhring, 1998).

H. magnapapula is primarily a larval parasite, invading early instars and emerging from the last (fourth) instar. However, if a later stage larva is attacked, nematode development carries over into the adult host. This is rather rare in nature (Paine and Mullens, 1994) but easily induced in the laboratory (Mullens and Velten, 1994). This is important for dispersal, and may be both adaptive and encouraged by constricting water levels in a drying, ephemeral habitat.

23.3.2. Biocontrol

H. magnapapula invariably kills the host as nematodes emerge. In the field, parasitism rates (point prevalence based on parasite emergence from mature, field-collected fourth instar larvae) range as high as 69%, although average levels of 7–20% are more common over time (Paine and Mullens, 1994; Mullens and Luhring, 1998). This estimate is conservative and underestimates true impact by up to 50%. The natural role and potential of *H. magnapapula* for biocontrol of *C. sonorensis* was discussed by Mullens and Luhring (1998). Due to differential mortality of parasitized hosts during their development, and the strong possibility of many mature (fourth instar) hosts already having been killed by parasites before collection, the most accurate estimate of parasite impact is dissection of third instar hosts. The host range of *H. magnapapula* is not strictly limited to the *C. variipennis* complex, but other natural hosts are unknown. The *C. variipennis* complex contains physically the largest members of the genus in North America. Eight different *Culicoides* spp. exposed in the laboratory were readily attacked by the parasite, although it was able to mature fully only in two (Mullens *et al.*, 1997). Preparasites ignored larvae of most other aquatic Diptera that might share the habitat. Preparasites did penetrate *Chironomus* sp., but were melanized (Poinar and Mullens, 1987; Mullens *et al.*, 1997). *H. magnapapula* tracks the host in time (through adjusting sex ratios), and thus would not be expected to overexploit host resources. Since the mermithid must be produced *in vivo*, inundative releases for control would probably be impractical. Even if a very high level of short-term control were achieved, it would likely not persist. The more logical use would be in inoculative releases. The mermithid distribution among even nearby, disjunct habitats (such as manure wastewater ponds or stock tank overflows) appears to be sporadic (Paine and Mullens, 1994). Given the low parasitism in adult midges (a likely dispersal mechanism), colonization of habitats by *H. magnapapula* may be inefficient and subject to stochastic effects. In such a scenario, intentional releases of mermithid adults could 'seed' habitats with this important natural enemy. Such releases could be a part of an integrated pest management (IPM) strategy.

23.4. *Mermis* Case Study

The terrestrial mermithid genus *Mermis* was described in 1842 but probably had been observed in 1623 (Poinar, 1975). The genus includes about ten recognized species (Kaiser, 1991; Baker and Capinera, 1997). These mermithids are medium to large, robust nematodes, 15–200 mm long, and up to 0.5 mm in diameter. Known hosts include Orthoptera, Coleoptera, Dermaptera, Hymenoptera and Lepidoptera. The genus has been reported from Africa, Asia, Australia, Pacific Islands, Papua New Guinea, Philippines, India, and one species, *Mermis nigrescens*, has a holarctic distribution.

23.4.1. Life history

The type species, *Mermis nigrescens*, conforms to the classic model for mermithids. Juveniles are parasitic in Orthoptera and Dermaptera, whereas the postparasitic stages are free-living and exist on stored nutrients. The life history is based on *M. nigrescens* in grasshoppers (Christie, 1937). During late spring and early summer, gravid female *M. nigrescens* emerge from soil, ascend low vegetation and lay eggs (Fig. 23.3). Eggs have byssi on branching appendages at the polar ends, which facilitate attachment to the vegetation. When deposited, eggs contain fully developed infectious juveniles (second stage juveniles, J2). Eggs remain viable throughout the summer and, if moist, can be infectious for a year. Grasshoppers are infected by ingesting egg-contaminated vegetation. The eggs hatch within the alimentary tract. The juvenile employs a stylet, penetrates the intestinal wall and enters the haemocoel. The average number of juveniles/grasshopper is one to five (maximum of 100/host). Crowded conditions during nematode development result in all males. If conditions are not crowded ($\leq$ 2/host), all emerging nematodes are females. Developmental time in grasshoppers at ambient temperatures is 4–10 weeks. When postparasites emerge, in summer or autumn, the host dies. Postparasites move 15–45 cm below the soil surface. They do not aggregate, and moult the following spring. Females soon develop eggs, even without males. Viable eggs are produced, but deposition does not take place until the second spring, nearly 2 years after deposition of the original eggs.

23.4.2. Biocontrol

Since infection with *M. nigrescens* and congeneric species suppresses host gonadal growth and results in host death, they are significant in biocontrol. Early observers reported 75% infection in *Melanoplus* sp. in Vermont and up to 36% infection of grasshoppers in western Canada (Baker and Capinera, 1997). Bland (1976) reported that *Melanoplus femurrubrum* in Michigan was infected at a level of 71% with *M. nigrescens* and this mermithid exerted some population control. *Hexamermis* sp. in Spain was reported as an important parasite of *Dociostaurus maroccamus* with infection levels from 10% to 45% in certain regions (Baker and Capinera, 1997). The most extensive data on the impact of mermithids on grasshopper populations are summarized in Baker and Capinera (1997). Population regulation of grasshoppers is provided by *Mermis* sp. and related species in the moist uplands of Australia. Major

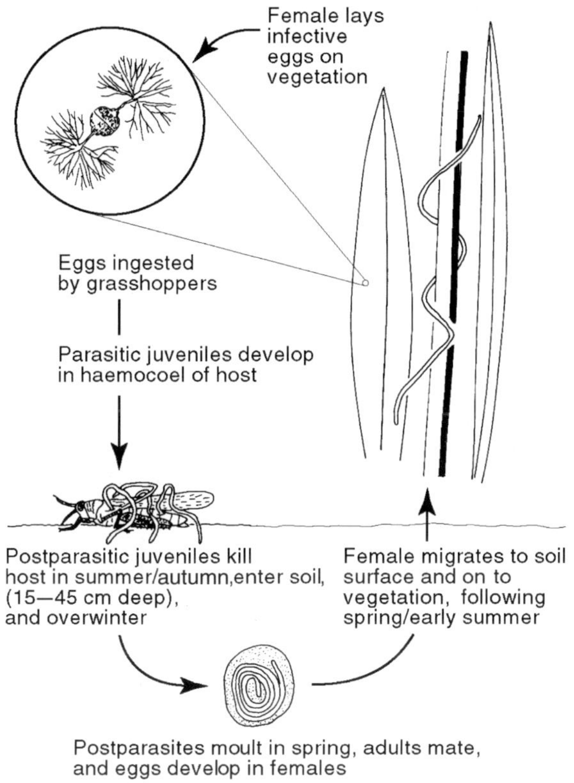

Fig. 23.3. Life cycle of *Mermis nigrescens* under typical field conditions (redrawn in part from Christie, 1937).

grasshopper outbreaks in the Central Tablelands collapsed under the pressure of mermithid infections (Baker and Capinera, 1997). The protracted life history of *Mermis* sp. and related species has precluded mass production, although augmentation of natural grasshopper infections was demonstrated in small-screened plots that acted as nurseries for biocontrol (Christie, 1937). Christie increased the number of infected grasshoppers by supplementing the natural grasshopper population in prior years with infected grasshoppers, thereby increasing the number of *M. nigrescens* in the soil. This demonstration provided early evidence for augmentative control and the possibility of establishing natural production sites or nurseries for *M. nigrescens* that could serve as supply sites for inoculation of mermithid-free or mermithid-depauperate areas.

Environment, especially rainfall, limits terrestrial mermithids (Kaiser, 1991; Baker and Capinera, 1997). Mermithids were absent in regions receiving less than 400 mm rainfall yearly; optimum conditions occur in wet areas receiving 800–1200 mm precipitation yearly. Mermithid populations in grasshoppers were highest after the monsoons in Pakistan (Kaiser, 1991). Degradation of land by erosion, lowering of the water table and deforestation has adverse effects on the abundance of mermithid populations (Baker and Capinera, 1997). Interestingly, Popiel and Hominick (1992) point out that the use of pesticides in IPM may lower host populations to the point where mermithid populations may decline precipitously, even to the point of extinction. These observations suggest that IPM programmes need to consider further development of mermithid augmentation programmes to prevent localized losses of natural control. In general, the relatively protracted life history of *Mermis* and related species has not engendered intense interest by biocontrol researchers, who often are pressed to seek quick-acting agents to resolve urgent pest problems. However, Australian research has provided a solid foundation for the necessity of long-term studies on the population dynamics of terrestrial mermithids, and the hope is that researchers in other biogeographic regions can develop further insights on the biocontrol impact through long-term population studies.

References

Baker, G.L. and Capinera, J.L. (1997) Nematodes and nematomorphs as control agents of grasshoppers and locusts. *Memoirs of the Entomological Society of Canada* 171, 157–211.

Bland, R.G. (1976) Effects of parasites and food plants on the stability of populations of *Melanoplus femurrubrum. Environmental Entomology* 5, 724–728.

Christie, J.R. (1937) *Mermis subnigrescens,* a nematode parasite of grasshoppers. *Journal of Agricultural Research* 55, 353–364.

Federici, B.A. (1995) The future of microbial insecticides as vector control agents. *Journal of the American Mosquito Control Association* 11, 260–268.

Hunt, G.J. (1994) *A Procedural Manual for the Large-Scale Rearing of the Biting Midge,* Culicoides variipennis (*Diptera: Ceratopogonidae*). ARS-121, USDA-ARS, Washington, DC.

Kaiser, H. (1991) Terrestrial and semiterrestrial Mermithidae. In: Nickle, W.R. (ed.) *Manual of Agricultural Nematology.* Marcel Dekker, New York, pp. 899–965.

Kerry, B.R. and Hominick, W.M. (2002) Biological control. In: Lee, D.L. (ed.) *The Biology of Nematodes.* Taylor & Francis, London, pp. 483–509.

Kerwin, J.L. and Washino, R.K. (1985) Recycling of *Romanomermis culicivorax* (Mermithidae: Nematoda) in rice fields in California, USA. *Journal of Medical Entomology* 22, 637–643.

Mullens, B.A. and Luhring, K.A. (1998) Age-dependent parasitism of *Culicoides variipennis sonorensis* (Diptera: Ceratopogonidae) by *Heleidomermis magnapapula* (Nematode: Mermithidae) and considerations for assessing parasite impact. *Biological Control* 11, 49–57.

Mullens, B.A. and Velten, R.K. (1994) Laboratory culture and life history of *Heleidomermis magnapapula* in its host, *Culicoides variipennis* (Diptera: Ceratopogonidae). *Journal of Nematology* 26, 1–10.

Mullens, B.A., Luhring, K.A. and Breidenbaugh, M.S. (1997) Experimental host range studies with *Heleidomermis magnapapula* (Mermithidae), a parasite of *Culicoides variipennis* (Ceratopogonidae). *Journal of American Mosquito Control Association* 13, 398–401.

Paine, E.O. and Mullens, B.A. (1994) Distribution, seasonal occurrence, and pattern of parasitism of *Heleidomermis magnapapula* (Nematoda: Mermithidae), a parasite of *Culicoides variipennis* (Diptera: Ceratopogonidae) in California. *Environmental Entomology* 23, 154–160.

Peng, Y., Platzer, E.G., Song, J. and Peloquin, J. (1992) Susceptibility of thirty-one species of mosquitoes to *Romanomermis yunanensis. Journal of West China University of Medical Sciences* 23, 412–415.

Petersen, J.J. (1973) Relationship of density, location of hosts, and water volume to parasitism of larvae of the southern house mosquito by a mermithid nematode. *Mosquito News* 33, 516–520.

Petersen, J.J. (1985) Nematodes as biological control agents: Part I. Mermithidae. *Advances in Parasitology* 24, 307–346.

Platzer, E.G. (1981) Biological control of mosquitoes with mermithids. *Journal of Nematology* 13, 257–262.

Poinar, G.O. Jr (1975) *Entomogenous Nematodes. A Manual and Host List of Insect-Nematode Associations.* E.J. Brill, Leiden, The Netherlands.

Poinar, G.O. Jr (2001) Nematoda and Nematomorpha. In: Thorp, J.H. and Covich, A.P. (eds) *Ecology and Classification of North American Freshwater Invertebrates,* 2nd edn, Academic Press, San Diego, California, pp. 255–295.

Poinar, G.O. Jr and Mullens, B.A. (1987) *Heleidomermis magnapapula* n. sp. (Mermithidae: Nematoda) parasitizing *Culicoides variipennis* (Ceratopogonidae: Diptera) in California. *Revue de Nematologie* 10, 387–391.

Popiel, I. and Hominick, W.M. (1992) Nematodes as biological control agents: Part II. *Advances in Parasitology* 31, 381–433.

Santamarina, A.M. and Perez, R.P. (1997) Reduction of mosquito larval densities in natural sites after introduction of *Romanomermis culicivorax* (Nematoda: Mermithidae) in Cuba. *Journal of Medical Entomology* 34, 1–4.

Part IV

Slug-Parasitic Nematodes

24 Biology, Production and Formulation of Slug-parasitic Nematodes

M.J. Wilson[1] and P.S. Grewal[2]

[1]*School of Biological Sciences, University of Aberdeen, Aberdeen, AB24 3UU, UK;* [2]*Department of Entomology, Ohio State University, Wooster, OH 44691-4096, USA*

24.1. Introduction

Slugs and snails are molluscs, belonging to the Class Gastropoda. Generally, animals are classed as snails when there is a large external shell and as slugs when there is no external shell, or the shell is very small in comparison with body size. Many slugs and snails are serious pests of agriculture and horticulture and comprehensive reviews of their biology and pest status can be found in Barker (2001) and Barker (2002), respectively.

Many species of nematodes are known to be associated with slugs and snails and several of these are known to be parasites (see Mengert, 1953; Grewal *et al.*, 2003a; Morand *et al.*, 2003, for reviews) but only one species, *Phasmarhabditis hermaphrodita* (Schnieder), has been developed as a biocontrol agent and its use has been targeted as a control agent for slugs. It should be noted that there has been some recent interest in the use of other rhabditid and cephalobid nematodes to control pest snails in Australia (Charwat and Davies, 1999; Charwat *et al.*, 2000), although this work is still in its very early stages.

P. hermaphrodita has been on sale as a biological molluscicide in the UK since 1994, and since then production and sales have increased and it is now sold in several European countries. *P. hermaphrodita* is quite possibly the least well researched of all commercially available biocontrol

(eds P.S. Grewal, R.-U. Ehlers and D.I. Shapiro-Ilan)

agents, with most research being funded by Becker Underwood (formerly MicroBio Ltd) and focusing on bringing a product to market rapidly. Very little is known about its ecology and distribution, and equally little is known about its associations with bacteria in nature and the host–parasite interactions. This chapter reviews the biology of *P. hermaphrodita* and aspects related to commercialization (mass production, formulation) while Chapter 25 deals with field application and efficacy.

24.2. Historical Aspects

Surprisingly, even though the amount of published studies on *P. hermaphrodita* is tiny compared with that of the entomopathogenic nematodes (EPNs), research on *P. hermaphrodita* dates back much farther. The nematode was first described as being associated with the slug *Arion ater* by Schneider in 1859, more than 60 years before the first reference to EPN (Steiner, 1923). Later, Maupas (1900), in a study of nematode development and reproduction, established cultures of *P. hermaphrodtia* (which he called *Rhabditis causenelli*) that he found as dauer larvae, also in the intestine of *A. ater*. The cultures were maintained on 'rotting flesh'. He found *P. hermaphrodita* to be a protandrous autogamous hermaphrodite with males occurring rarely (1 male/715 females). Despite much observation he never saw them mating with females. Mengert (1953) conducted a detailed study of nematodes associated with terrestrial molluscs. Mengert (1953) did not find *P. hermaphrodita* but did find the closely related species *P. neopapillosa* in another large slug *Limax cinereoniger*.

While all the above authors observed the nematodes to be associated with slugs, none of them considered *P. hermaphrodita* to be a parasite. Mengert (1953) considered that the three closely related species *P. papillosa, P. neopapillosa* and *P. hermaphrodita* could live within slugs but remained as dauer larvae until the slug died, not harming slugs and thus not living as parasites.

Apart from some passing taxonomic references (Osche, 1954; Andrássy, 1983) there were no further publications on these nematodes until the early 1990s, when Wilson *et al.* (1993c) patented the use of *Phasmarhabditis* nematodes as biological molluscicides, following a 5-year research programme funded by MicroBio Ltd (now Becker Underwood). The patent was based on *P. hermaphrodita* being capable of parasitizing and killing a range of agricultural and horticultural pest slug species that had not been investigated by Maupas (1900) or Mengert (1953). This patent and the subsequent launch of the product Nemaslug® led to a spate of publications on this nematode, with the majority concentrating on field testing.

24.3. Taxonomic Relations

Andrássy (1983) published a comprehensive review of the Suborder Rhabditina that proposed the new genus *Phasmarhabditis*. The genus comprises five species, three of which are known to be associated with slugs or snails: *P. papillosa, P. neopapillosa* and *P. hermaphrodita. P. papillosa* can be easily distinguished from the other two species by having a cupola-shaped tail with a pointed tip. The other two species have elongate conoid tails, 3–4 anal body diameters in length.

The species *P. neopapillosa* and *P. hermaphrodita* are morphologically indistinguishable, and are separated by numbers of males present in the population. In *P. neopapillosa* males and females are equally common, whereas in *P. hermaphrodita* males are extremely rare. When Andrássy (1983) erected the genus he noted that their separate species status needed clarification, as environmental sex determination is not uncommon in nematodes. More recently, Hooper *et al.* (1999) examined specimens of the two species both morphologically and by protein profile electrophoresis. As with previous examinations, there were no morphological differences between the two species, but total protein electrophoresis of the two species gave distinct banding patterns and isoenzyme analysis revealed differences in the mobility of phosphoglucose isomerase between species. Since isoenzyme electrophoresis is a well-established tool for identification and separation of species (Dalmasso and Bergé, 1983) these authors considered that the two species were valid. DNA sequence data from all species of *Phasmarhabditis* are sadly lacking.

It should be noted that all workers do not accept the validity of the genus *Phasmarhabditis*. For example, Sudhaus (1976) considers these two nematodes to be separate species, but refers to them as *Rhabditis* (*Pelliodìtis*) *hermaphrodita* and *R.* (*P.*) *neopapillosa*.

The genus *Phasmarhabditis* belongs in the same superfamily as the genera *Heterorhabditis* and *Steinernema*. However, since neither Sudhaus (1976) nor Andrássy (1983) included the EPNs in their taxonomic reviews and with the absence of DNA sequence data for *Phasmarhabditis* spp. it is not possible to draw firm conclusions on the phylogenetic relationships between the three genera. While all three *Phasmarhabditis* species named above are associated with slugs, all work on commercial development has used *P. hermaphrodita* and the remainder of this chapter focuses on this species.

24.4. Life Cycle

The life cycle of *Phasmarhabditis* nematodes in nature is poorly studied, but it would appear to be more varied than the life cycle of EPNs and seems to depend on slug species (if any) it encounters. There have been three distinct life cycles described: saprobic, necromenic and parasitic.

24.4.1. Saprobic life cycle

Maupas (1900) reared the nematodes continuously on rotting flesh for 2 years, and Tan and Grewal (2001a) were able to grow the nematode on homogenized slugs and slug faeces, suggesting that the nematode can live saprobically. Unlike the EPNs, *P. hermaphrodita* can reproduce on a wide range of bacteria (Wilson *et al.*, 1995a) and it seems likely that if dauer larvae were to encounter, for example, a dead invertebrate in the soil, they may well be able to recover and reproduce saprobically. However, there are no data available to demonstrate the occurrence of this in nature. The ability of *P. hermaphrodita* to live saprobically led Tan and Grewal (2001a) to suggest that if this nematode could persist in the environment in the absence of live hosts, it may be suitable for long-term inoculative slug control. This is an interesting possibility and one that should be further explored. This approach could be of particular benefit in wheat, maize and oilseed crops that are grown in broad-scale agriculture. These crops are severely damaged by slugs, but because of the low crop value, inundative applications of *P. hermaphrodita* are not economically feasible.

24.4.2. Necromenic life cycle

This is the life cycle that was first observed by Maupas (1900) and Menger (1953). Mengert (1953) states that when the opportunity arises dauer larvae can enter a slug and can survive there, without further development until the slug dies. The dauer larvae then recover, develop and reproduce while feeding on the slug cadaver, eventually re-forming dauer larvae when the food source becomes depleted. Dauer larvae can be found in this arrested state of development in the mantle cavity, the general body cavity or the digestive tract of slugs. It would appear that this is the life cycle adopted when the nematode encounters larger species of slugs and smaller members of the Arionidae family. Mengert (1953) was working with *L. cinereoniger* (maximum length 30 cm) and Maupas worked with *A. ater* (maximum length 15 cm), both much larger than the most widely distributed pest species *Deroceras reticulatum* (maximum length 6 cm). It may well be that many of the larger species and smaller members of the Arionidae that appear not to be susceptible to parasitic life cycle of *P. hermaphrodita* in laboratory bioassays are species that induce *P. hermaphrodita* to adopt the necromenic life cycle. Certainly, live *P. hermaphrodita* dauer larvae can be found within adult *A. ater* exposed to *P. hermaphrodita* and showing no disease symptoms if the slugs are sacrificed and dissected. The mechanisms regulating whether *P. hermaphrodita* enters the parasitic or necromenic life cycle once in a slug are not understood.

24.4.3. Parasitic life cycle

The parasitic life cycle has been studied by Wilson *et al.* (1993a) and Tan and Grewal (2001a) in the common pest slug *D. reticulatum*. The primary route of entry of dauer larvae into the slugs is through the dorsal integumental pouch, through a short canal and into the slug's shell cavity below the mantle. Once inside, they recover, develop into adults and reproduce. During this period the slug usually develops a characteristic swelling of the rear half of the mantle (Fig. 24.1).

If many nematodes invade simultaneously, the infecting nematodes may spread to other body regions at an early stage. In cases where only low numbers of nematodes enter, as appears to be common in nature, the shell cavity seems to be the main site of development for the first generation. *P. hermaphrodita* typically produce about 250–300 offspring, and once the second generation is produced these offspring spread throughout the slug's body and develop. The slug dies and a third generation is usually produced, which feed on the slug cadaver before forming dauer larvae. Slug death tends to occur between 4 and 21 days after infection depending on exposure rate and temperature. However, from a very short time after infection the slug feeding is strongly inhibited (Glen *et al.*, 2000; Grewal *et al.*, 2001, 2003b), so when used for biocontrol purposes crop protection is rapid even if host mortality is not.

24.5. Host Range

All host range studies have been done to investigate whether *P. hermaphrodita* can infect and enter the parasitic life cycle, thus killing the host. No detailed work has been done to distinguish if 'non-susceptible' slugs are invaded by nematodes that enter the necromenic life cycle, or whether the nematodes are not capable of invading.

Fig 24.1. Healthy (left) and nematode-infected (right) *Deroceras reticulatum* showing the characteristic swelling of the rear half of the mantle where nematodes reproduce.

Wilson *et al.* (1993a) showed a single high dose of nematodes applied to slugs kept in soil caused significant mortality of *D. reticulatum*, *D. panormitanum*, *A. silvaticus*, *A. distinctus*, *A. intermedius*, *A. ater* (juveniles), *Tandonia budapestensis* and *T. sowerbyi*, representing three pest families of slugs. This dose rate was extremely high (in most cases 1.9×10^4/slug) and it cannot be concluded that all the above species could be controlled economically. Grewal *et al.* (2003b) showed that *P. hermaphrodita* caused significant mortality of *D. reticulatum*, *D. leave* and *Leidyula floridana* at 300–2700 infective juveniles (IJs)/slug in filter paper and soil bioassays. Some species appear to be susceptible as juveniles but become resistant as they mature. This is the case for the snail *Helix aspersa* (Glen *et al.*, 1996), and the pest slugs *A. lusitanicus* (Speiser *et al.*, 2001; Grimm, 2002) and *A. hortensis* (Grewal *et al.*, 2003b). However, *P. hermaphrodita* seems to have a necromenic life cycle in *A. subfuscus* and *L. maximus*, as up to 10 IJs nematodes invaded but did not kill these slug species (Grewal *et al.*, 2003b).

Coupland (1995) showed that four species of snails that are pests in Australia (*Theba pisana*, *Cernuella virgata*, *Cochlicella acuta* and *Cochlicella barbara)* were all killed rapidly by 300 dauer larvae/snail when exposed in Petri dishes on filter paper. In an investigation into the susceptibility of non-target snail species, Wilson *et al.* (2000) showed soil treated with the recommended field dose of *P. hermaphrodita* killed the snail *Monacha cantiana*, but not six other species of common field-margin snails. Treatment of the soil with five times the recommended field dose also killed the snail *Cepaea hortensis*, but not the remaining five snail species.

24.6. Associations with Bacteria and Mechanisms of Pathogenicity

Little is known about the association of *P. hermaphrodita* with bacteria in nature. Because of the many similarities between *P. hermaphrodita* and EPNs, much early research concentrated on the role of bacteria in promoting nematode growth and the pathogenicity to slugs of the resulting nematode/bacterium complexes (Wilson *et al.*, 1995a,b).

Initial studies had shown that it was possible to grow *P. hermaphrodita in vitro* with a xenic (unknown number of unknown species) mix of bacteria (Wilson *et al.*, 1993b) and that nematodes produced this way were capable of killing slugs (Wilson *et al.*, 1993a). However, for reasons of commercial reproducibility and to avoid the danger of the xenic mix containing pathogens, it is preferable that a single bacterial isolate be used. Wilson (2002) collected over 100 bacterial isolates (selected by colonial morphology) from *P. hermaphrodita* dauer larvae, xenic cultures of *P. hermaphrodita* and slug cadavers that had died following infection with *P. hermaphrodita.* A subset of 13 of these bacterial isolates were identified to species and used in growth experiments (Wilson *et al.*, 1995a). *P. hermaphrodita* was capable of growth and reproduction on the vast majority of these bacteria. Highest yields were obtained with *Providencia rettgeri*, *Moraxella osloensis* (referred to as *M. phenylpyruvica* by Wilson *et al.*, 1995a), and two isolates of *Pseudomonas fluorescens*. Nematode/bacterium combinations of *P. hermaphrodita* grown with all the above bacterial species were bioassayed against *D. reticulatum*, but only nematodes grown with *Mor. osloensis* or *Ps. fluorescens* were found to be consistently pathogenic (Wilson *et al.*, 1995b). Neither of these bacteria alone were found to be pathogenic to *D. reticulatum* when 16-h nutrient broth cultures grown at 25°C were injected into the mantle cavity. However, more recently Tan and Grewal (2001b) found 48-h Petri-dish cultures of *Mor. osloensis* were pathogenic. Findings from these studies suggest that it is unlikely that *P. hermaphrodita* forms highly specific mutualistic associations with bacteria as do the EPNs. No specific bacterium has been isolated consistently from *P. hermaphrodita*. In addition, the slow progression of disease in slugs caused by *P. hermaphrodita* in comparison with the rapid mortality caused by the bacteria from EPNs makes a mutualistic bacterial association seem unlikely. Furthermore, the lack of bacterial specificity as a food source and lack of a rigid cuticle in slugs also suggest a more general association with bacteria for *P. hermaphrodita*. When EPNs kill insects in soil, the insect cuticle remains intact throughout the nematode's reproductive cycle. This allows the antibiotic-producing *Xenorhabdus*/*Photorhabdus* to thrive largely free of competition from other soil microorganisms. The majority of *P. hermaphrodita* reproduction takes place on the slug cadaver. There is no rigid cuticle on slugs and shortly after death, slug cadavers form a small pool of mucilaginous material that is easily invaded by soil microbes and free-living nematodes. In such a situation an ability to thrive on a variety of soil microorganisms would be highly beneficial, as is the case for *P. hermaphrodita*. However, studying the bacteria associated with *P. hermaphrodita* in nature from a wide range of isolates remains a priority for future research. It may be that *P. hermaphrodita* has a loose association with a specific bacterium, similar in some ways to that between *Steinernema glaseri* and *X. poinarii*.

As a result of the work of Wilson *et al.* (1995a,b), *Mor. osloensis* was chosen as the bacterium for use in commercial development of *P. hermaphrodita* and this system is now the focus of intense studies on mechanisms of pathogenicity. Recent research shows that *Mor. osloensis* plays a major role in the pathogenicity of the nematodes to the slugs (Tan and Grewal, 2001b, 2002, 2003). Axenic nematodes do not kill slugs and the number of bacteria carried by the IJs directly correlates with the nematode-induced slug mortality (Tan and Grewal, 2001b). The number of viable cells of *Mor. osloensis* varies with the age of IJs, with the older nematodes carrying fewer viable cells (Tan and Grewal, 2001b). Investigations into the molluscicidal toxins produced by *Mor. osloensis* revealed that the bacteria produce a heat-stable endotoxin consisting of a lipopolysaccharide (LPS) (Tan and Grewal, 2002). The purified LPS is lethal to slugs when administered into the haemocoel or shell cavity, with an estimated LD_{50} of 48 μg/slug (Tan and Grewal, 2002). The LPS is a rough-type LPS with an estimated molecular weight of 5300 (Tan and Grewal, 2003). Toxicity of the LPS resides in the lipid A moiety and was quantified to contain about 6×10^7 endotoxin units/mg. Coinjection of galactosamine with the LPS increased LPS toxicity to the slug by two to four times. The galactosamine-induced sensitization of the slug to the LPS was reversed completely by uridine, thus indicating that slug hepatopancreas may be the potential site of action of the LPS.

24.7. Production and Formulation

One of the reasons that allowed *P. hermaphrodita* to be brought to the marketplace rapidly is that the technology developed for mass production and formulation of EPNs can be transferred for use with *P. hermaphrodita* with only minimal modification. Wilson *et al.* (1993b) showed that *P. hermaphrodita* could be grown in xenic culture using solid foam chip culture, based on the methods of Bedding (1984), and also in deep liquid culture on a flask shaker, suitable for scale up to fermenter production. Later, in monoxenic culture, Wilson *et al.* (1995a) achieved yields approaching 100,000 IJs/ml medium. Wilson *et al.* (1994) used a variety of media including both partially and wholly soluble media. The nematode is now routinely mass-produced in air-lift fermenters of up to 20,000 l in a proprietary medium and yielding in excess of 100,000/ml. Once maximum yields of IJs are obtained the nematodes are concentrated by centrifugation before formulation (Young *et al.*, 2002). The nematode can be formulated using any of Becker-Underwood's vermiculite, polymer or clay formulation. Shelf-life tends to be lower than many *Steinernema* spp. and is more typical of *Heterorhabditis* spp. All current formulations need to be refrigerated during storage.

24.8. Effects on Non-target Organisms

Little work has been done on the effects of *P. hermaphrodita* on non-target organisms. While many snail species are pests in many parts of the world, some species in the UK are of conservation interest (Kerney and Stubbs, 1980), and there is concern about the potential spread of *P. hermaphrodita* into field margins and adjacent woodlands, with possible detrimental effects on these mollusc populations. Wilson *et al.* (2000) tested seven species of common hedgerow snails in laboratory studies for susceptibility and found only *C. hortensis* and *M. cantiana* to be susceptible. However, when *P. hermaphrodita* was applied to plots of oilseed rape adjacent to field margins, no reduction in numbers of any species of snail was found in the adjacent hedgerows, even though *C. hortensis* and *M. cantiana* were both present. It is unlikely that many snail species would come into contact with *P. hermaphrodita* as they tend to live in plants above the ground, unlike the shell-less slugs that live in the soil (Mengert, 1953). Thus, the threat to non-target snails is unlikely to be high.

There have been many investigations of nematode parasites associated with soil insect pests, but *P. hermaphrodita* has never been found associated with insects in nature. Wilson *et al.* (1994), in a study of the effect of EPNs on slugs, bioassayed *P. hermaphrodita* against larvae of the tenebrionid beetles *Zophobas morio* and *Tenebrio molitor* and found neither to be susceptible to this species. In addition, Wilson *et al.* (1993d) reported that adults of the beneficial predatory carabid beetle, *Pterostichus melanarius*, were not killed in laboratory assays using high doses of *P. hermaphrodita*.

Earthworms are important components of the soil fauna that improve the nutritional and physical status of soils. Thus, the isolation of a *Phasmarhabditis*-like nematode that appeared to be a lethal parasite of the earthworm *Lumbricus terrestris* (Zaborski *et al.*, 2001) caused great concern. These authors were able to cultivate this nematode on dog food agar and found it was able to kill several other species of earthworms in the laboratory. Zaborski *et al.* (2001) did not formally identify their nematode, but their description was closer to *P. neopapillosa* than any other species of *Phasmarhabditis. P. neopapillosa* has not been bioassayed against earthworms, but the preliminary reports of Wilson *et al.* (1993d), and the more detailed studies of Grewal and Grewal (2003) and De Nardo *et al.* (2003), show that the commercially available strain of *P. hermaphrodita* does not kill the earthworms *L. terrestris* and *Eisenia fetida*, respectively.

References

Andrássy, I. (1983) *A Taxonomic Review of the Suborder Rhabditina (Nematoda: Secernentia).* ORTSTOM, Paris, 241 pp.

Barker, G.M. (2001) *The Biology of Terrestrial Gastropods.* CAB International, Wallingford, UK, 468 pp.

Barker, G.M. (2002) *Molluscs as Crop Pests.* CAB International,Wallingford, UK, 468 pp.

Bedding, R.A. (1984) Large scale production, storage and transport of the insect parasitic nematodes *Neoaplectana* spp. and *Heterorhabditis* spp. *Annals of Applied Biology* 104, 117–120.

Charwat, S.M. and Davies, K.A. (1999) Laboratory screening of nematodes isolated from South Australia for potential as biocontrol agents of helicid snails. *Journal of Invertebrate Pathology* 74, 55–61.

Charwat, S.M., Davies, K.A. and Hunt, C.H. (2000) Impact of a rhabditid nematode on survival and fecundity of *Cernuella virgata* (Mollusca: Helicidae). *Biocontrol Science and Technology* 10, 147–155.

Coupland, J.B. (1995) Susceptibility of helicid snails to isolates of the nematode *Phasmarhabditis hermaphrodita* from southern France. *Journal of Invertebrate Pathology* 66, 207–208.

Dalmasso, A. and Bergé, J.B. (1983) Enzyme polymorphisms and the concept of parthenogenetic species, exemplified by *Meloidogyne.* In: Stone, A.R., Platt, H.M. and Khalil, L.F. (eds) *Concepts in Nematode Systematics.* Academic Press, London, pp. 187–196.

De Nardo, E.A.B., Grewal, S.K., Sinderman, A. and Grewal, P.S. (2003) Non-susceptibility of the earthworm *Eisenia fetida* to slug-parasitic nematode *Phasmarhabditis hermaphrodita. Biocontrol Science and Technology* 14, 93–98.

Glen, D.M. and Wilson, M.J. (1997) Slug-parasitic nematodes as biocontrol agents for slugs. *Agro Food Industry Hi-Tech* 8, 23–27.

Glen, D.M., Wilson, M.J. and Hughes, L. (1996) Exploring and exploiting the potential of the rhabditid nematode *Phasmarhabditis hermaphrodita* as a biocontrol agent for slugs. In: Henderson, I. (ed.) *Slug and Snail Pests in Agriculture.* Symposium Proceedings No. 66, British Crop Protection Council, Farnham, UK, pp. 271–280.

Glen, D.M., Wilson, M.J., Brain, P. and Stroud, G. (2000) Feeding activity and survival of slugs, *Deroceras reticulatum*, exposed to the rhabditid nematode, *Phasmarhabditis hermaphrodita*: a model of dose response. *Biological Control* 17, 73–81.

Grewal, S.K. and Grewal, P.S. (2003) Survival of earthworms exposed to the slug-parasitic nematode *Phasmarhabditis hermaphrodita. Journal of Invertebrate Pathology* 82, 72–74.

Grewal, P.S., Grewal, S.K., Taylor, R.A.J. and Hammond, R.B. (2001) Application of molluscicidal nematodes to slug shelters: a novel approach to economic biological control of slugs. *Biological Control* 22, 72–80.

Grewal, P.S., Grewal, S.K., Tan, L. and Adams, B. (2003a) Parasitism of mollusks by nematodes: types of associations and evolutionary trends. *Journal of Nematology* 35, 146–156.

Grewal, S.K., Grewal, P.S. and Hammond, R.B. (2003b) Susceptibility of slugs (Mollusca: Gastropoda) native and non-native to North America to *Phasmarhabditis hermaphrodita* (Nematoda: Rhabditidae). *Biocontrol Science and Technology* 13, 119–125.

Grimm, B. (2002) Effect of the nematode *Phasmarhabditis hermaphrodita* on young stages of the pest slug *Arion lusitanicus. Journal of Molluscan Studies* 68, 25–28.

Hooper, D.J., Wilson, M.J., Rowe, J.A. and Glen, D.M. (1999) Some observations on the morphology and protein profiles of the slug-parasitic nematodes *Phasmarhabditis hermaphrodita* and *P. neopapillosa* (Nematoda: Rhabditidae). *Nematology* 1, 173–182.

Kerney, M.P. and Stubbs, A. (1980) *The Conservation of Snails, Slugs and Fresh Water Mussels.* The Nature Conservancy Council, Shrewsbury, UK, 23 pp.

Maupas, E. (1900) Modes et formes de reproduction des nématodes. *Archives de Zoologie Expérimentale et Générale* 7, 563–628.

Mengert, H. (1953) Nematoden und Schnecken. *Zeitschrift für Morphologie und Ökoligie der Tiere* 4, 311–349.

Morand, S., Wilson, M.J. and Glen, D.M. (2003) Nematode parasites. In: Barker, G.M. (ed.) *Natural Enemies of Terrestrial Molluscs.* CAB International, Wallingford, UK.

Osche, G. (1954) Über die gegenwärtig ablaufende Entstehund von Zwillings und Kmomplementärten bei Rhabditiden (Nematodes). (Fotalisatin und Artbildung). *Zoologische Jahrbücher Systems* 82, 618–624.

Schneider, A.F. (1859) Über eine Nematodenlarve und gewisse Verscheidenheiten in dn Geschlechsorganen der Nematoden. *Zeitschrift für wissenschaftliche Zoologie* 10, 176–178.

Speiser, B., Zaller, J.G. and Neudecker, A. (2001) Size-specific susceptibility of the pest slugs *Deroceras reticulatum* and *Arion lusitanicus* to the nematode biocontrol agent *Phasmarhabditis hermaphrodita. Biocontrol* 46, 311–320.

Steiner, G. (1923) *Aplectana kraussei* n. sp., eine in der Blattsespe *Lyda* sp. Parasitierende Nematodenform, nebst Bomerkungen uber das Seitenorgan der parasitischen Nematoden. *Zentralblatt für Bakteriologie Parasitenkunde Infecktionskankheiten und Hygiene Abteilung I Original* 59, 14–18.

Sudhaus, W. (1976) Vergleichende Untersuchungen zur Phylogenie, Systematik, Ökologie, Biologie und Ethologie der Rhabditidae (Nematoda). *Zoologia* 43, 1–229.

Tan, L. and Grewal, P.S. (2001a) Infection behavior of the rhabditid nematode *Phasmarhabditis hermaphrodita* to the grey garden slug *Deroceras reticulatum. Journal of Parasitology* 87, 1349–1354.

Tan, L. and Grewal, P.S. (2001b) Pathogenicity of *Moraxella osloensis,* a bacterium associated with the nematode *Phasmarhabditis hermaphrodita,* to the slug *Deroceras reticulatum. Applied and Environmental Microbiology* 67, 5010–5016.

Tan, L. and Grewal, P.S. (2002) Endotoxin activity of *Moraxella osloensis* against the grey garden slug, *Deroceras reticulatum. Applied and Environmental Microbiology* 68, 3943–3947.

Tan, L. and Grewal, P.S. (2003) Characterization of the first molluscicidal lipopolysaccharide from *Moraxella osloensis. Applied and Environmental Microbiology* 69, 3646–3649.

Wilson, M.J. (2002) A nematode parasite for biological control of slugs. PhD Thesis, University of Bristol.

Wilson, M.J., Glen, D.M. and George, S.K. (1993a) The rhabditid nematode *Phasmarhabditis her-*

maphrodita as a potential biological-control agent for slugs. *Biocontrol Science and Technology* 3, 503–511.

Wilson, M.J., Glen, D.M., George, S.K. and Butler, R.C. (1993b) Mass cultivation and storage of the rhabditid nematode *Phasmarhabditis hermaphrodita*, a biocontrol agent for slugs. *Biocontrol Science and Technology* 3, 513–521.

Wilson, M.J., Glen, J.D. and Pearce, J.D. (1993c) *Biological Control of Molluscs.* World Intellectual Property Organization, Patent No. WO 93/00816, 38 pp.

Wilson, M.J., George, S.K., Glen, D.M., Pearce, J.D. and Rodgers, P.B. (1993d) Biological control of slug and snail pests with a novel parasitic nematode. In: *ANPP Third International Conference on Pests in Agriculture,* Montpellier, France, pp. 425–432.

Wilson, M.J., Glen, D.M., Hughes, L.A., Pearce, J.D. and Rodgers, P.B. (1994) Laboratory tests of the potential of entomopathogenic nematodes for the control of field slugs (*Deroceras reticulatum*). *Journal of Invertebrate Pathology* 64, 182–187.

Wilson, M.J., Glen, D.M., George, S.K. and Pearce, J.D. (1995a) Selection of a bacterium for the mass-production of *Phasmarhabditis hermaphrodita* (Nematoda, Rhabditidae) as a biocontrol agent for slugs. *Fundamental and Applied Nematology* 18, 419–425.

Wilson, M.J., Glen, D.M., Pearce, J.D. and Rodgers, P.B. (1995b) Monoxenic culture of the slug parasite *Phasmarhabditis hermaphrodita* (Nematoda, Rhabditidae) with different bacteria in liquid and solid-phase. *Fundamental and Applied Nematology* 18, 159–166.

Wilson, M.J., Hughes, L.A., Hamacher, G.M. and Glen, D.M. (2000) Effects of *Phasmarhabditis hermaphrodita* on non-target molluscs. *Pest Management Science* 56, 711–716.

Young, J.M., Dunnill, P. and Pearce, J.D. (2002) Separation characteristics of liquid nematode cultures and the design of recovery operations. *Biotechnology Progress* 18, 29–35.

Zaborski, E.R., Gittenger, L.A.S. and Roberts, S.J. (2001) A possible *Phasmarhabditis* sp. (Nematoda: Rhabditidae) isolated from *Lumbricus terrestris* (Oligochaeta: Lumbricidae). *Journal of Invertebrate Pathology* 77, 284–287.

25 Application of Slug-parasitic Nematodes

A. Ester[1] and M.J. Wilson[2]
[1]Applied Plant Research Ltd, Business Unit Arable Farming and Field Production of Vegetables, PO Box 430, 8200 AK Lelystad, The Netherlands;
[2]School of Biological Sciences, University of Aberdeen, Aberdeen, AB24 3UU, UK

25.1. Introduction

Slugs and snails are pests of agriculture and horticulture throughout the world (Barker, 2002). Slugs are pests distributed widely across the globe, with several families causing damage. However, by far the most widespread pest species is *Deroceras reticulatum*, which has been reported as a pest throughout Europe, Australasia, North and South America. This pest species is particularly susceptible to *Phasmarhabditis hermaphrodita* and most applications are made to control this species. Slugs are particularly problematic in the cool moist areas in north-west Europe, where they damage a huge range of crops. Over recent years changes in agricultural practices has led to a large increase in slug problems, particularly in arable agriculture. Such changes include reductions in cultivations, use of more varied crop rotations, incorporation of crop residues and the use of cover crops, all of which favour build-up of slug populations.

Slugs tend to be controlled by the use of baited pellets containing either metaldehyde or carbamate compounds. More recently pellets containing iron phosphate have been introduced. These pesticides are not always effective for many reasons including poor targeting when subterranean slugs are causing damage (e.g. in potatoes), breakdown of pellets and slugs failing to consume a lethal dose of active ingredient, because these compounds are repellent (Bailey, 2002).

The nematode *P. hermaphrodita* was first described as a potential biocontrol agent for slugs by Wilson *et al.* (1993). It was released as a commercial product for use by home gardeners in 1994 under the trade name 'Nemaslug®' (Glen *et al.*, 1994; 1996). The nematode is now sold in the UK, Ireland, the Netherlands, Belgium, France, Germany, Czech Republic, Italy and Switzerland.

The biology of this nematode and its methods of production and formulation are discussed in Chapter 24 of this volume. The aim of this chapter is to review field efficacy of the nematode in different crops and present some previously unpublished experimental data. In addition, we shall review the various application strategies and environmental conditions needed for success, and look to future opportunities to broaden the commercial use of this nematode.

25.2. Application Strategies

There has been little research on the environmental conditions needed to successfully apply *P. hermaphrodita*, but it is quite likely that optimum environmental conditions are similar to that for the more widely studied entomopathogenic nematodes (EPNs) (Wilson and Gaugler, 2000) and thus it is recommended that protocols for EPNs should be followed (Koppenhöfer, 2000). Briefly, EPNs are known to be sensitive to ultraviolet (UV) radiation damage and it is recommended that nematodes be applied preferably in early evening to prevent this damage. In addition, nematodes are sensitive to desiccation and should be applied to damp soils wherever possible, or irrigation should be applied immediately after application. If this is not possible, nematode efficacy can be enhanced by cultivating the soil immediately after application (Wilson *et al.*, 1996; Hass *et al.*, 1999b). This inverts the soil, removing nematodes from the surface and thus slowing desiccation. In addition this shields the nematodes from potential damage caused by UV radiation. It should be noted that not all types of cultivation benefit nematode efficacy equally (Hass *et al.*, 1999b).

In the many field experiments reported in the literature, *P. hermaphrodita* has been applied to soil with a variety of equipment ranging from simple watering cans, through knapsack sprayers, to tractor-mounted sprayers. The nematodes seem tolerant of the pressures associated with hydraulic spray systems but, as with EPNs, it is recommended that screen filters be removed from the spray equipment.

As with EPNs, the tendency is to apply nematodes evenly over the ground surface, although other application strategies have been attempted. Wilson *et al.* (1999) demonstrated the tendency of *D. reticulatum* to avoid soil treated with *P. hermaphrodita*. This suggested that it may be possible to protect crops by treating soil with either spots of nematodes around individual plants, or in narrow bands centred on the crop rows in row crops. With this approach it may be possible to reduce the area of soil treated and hence cost of application. However, in field experiments such benefits have not been realized. Hass *et al.* (1999a), in a field scale experiment with winter wheat, found no benefit associated with band treatment as opposed to uniform application. In a more detailed mini-plot experiment using *P. hermaphrodita* to control slug damage in Chinese cabbage, Hass *et al.* (1999b) found no evidence of slugs being repelled by *P. hermaphrodita*.

Another novel approach to increasing efficacy and reducing dose of *P. hermaphrodita* is the application of nematodes to slug shelters. Grewal *et al.* (2001) used a mini-plot experiment to test this application strategy to protect ornamental plants (*Impatiens* or *Hosta*), and found that application below artificial shelters at a rate of $0.6 \times 10^6/m^2$ gave similar plant protection to a uniform application on the whole plot at a rate of $0.3 \times 10^6/m^2$, representing a 63% reduction in the numbers of nematodes applied.

Since *P. hermaphrodita* takes between 1 and 3 weeks to kill infected slugs (Wilson *et al.*, 1993) it would seem logical to apply the nematodes approximately 2–3 weeks before damage is likely to occur, e.g. 2 weeks before drilling in oilseed rape. However, since it has been shown that nematodes strongly inhibit feeding by slugs almost immediately after infection (Glen *et al.*, 2000a) this does not appear to be necessary. In the vast majority of successful field experiments described, *P. hermaphrodita* was applied immediately before the crop became vulnerable, and in studies where different application timings have been tested there seems to be little benefit associated with early application (see Section 25.6.1).

The nematode can be used as part of an integrated crop management system, as it is compatible with the cultivation and techniques used to suppress slug populations. In addition, the nematode can be combined with metaldehyde bait pellets, as these pellets do not adversely affect the nematodes even at very high concentrations, unlike the carbamate-based methiocarb pellets (Wilson *et al.*, 2000).

25.3. Glasshouse Flowers

The slug *Lehmannia valentiana* is the most serious pest in *Cymbidium*, which is grown for cut flower production in greenhouses in the UK, the Netherlands and North America (Gittenberger *et al.*, 1984; South, 1992). This represents an extremely high-value crop under protected conditions and thus has much potential for implementation of *P. hermaphrodita*. *L. valentiana* originates from Spain and Portugal and has been widely distributed by human activities, mostly in association with movement of plant material. Once established in the greenhouse the slugs are difficult to control. In midsummer, plants are divided and transplanted into new pots, resulting in distribution of slugs and eggs throughout the greenhouse. *Cymbidium* plants are mostly grown in 10 l pots in a rock wool medium. Growers use a drip irrigation system, which keeps the pot at optimum moisture content. The moist artificial medium, crowded with roots, is a particularly suitable habitat for slugs, with many places to shelter. From autumn until spring the plants produce flowers. During the evening and night slugs move along stems and leaves to the flower and mainly consume the edges of the flower petals. Crops are severely damaged (see Fig. 25.1) despite intensive use of molluscicides. No alternative measures are currently available. Large amounts of slug pellets, applied frequently, keep the slug population low in many greenhouses, nevertheless many growers report variable efficacy of this approach.

Fig. 25.1. *Cymbidium* flowers showing characteristic slug damage to petals.

In a glasshouse experiment *Cymbidium* cv. Beauty Fred 60 was grown in 16-cm pots filled with 0.1 kg green grow cubes (rock wool) mixed with 1.5 kg shingle. In each pot one small tube (2-cm diameter) was placed, filled with water and one flower stem. Four healthy slugs of different sizes were added to each pot. The day–night cycle was 16:8 L:D and the temperature was 20°C and 15°C. The stems were replaced after 7 days. Treatments tested included metaldehyde 6.4% a.i. (trade name Caragoal®) slug bait pellets applied by hand on the surface of the medium. *P. hermaphrodita* was applied with a pipette with 21.4 ml water again at several different rates and numbers of applications (Table 25.1). In the combined treatment, nematodes were applied before the metaldehyde slug pellets. In the double nematode treatments the interval between the applications was 3 days. Numbers of flower petals and damaged flower petals per stem were counted after 7 and 14 days (Table 25.1).

Treatments with metaldehyde 44.8 mg or 300,000 nematodes/pot resulted in significantly fewer flower petals damaged by small slugs than the comparable untreated pots after 7 and 14 days (Table 25.1). No differences were found between the same treatments with two different slug weights. The combined treatments of metaldehyde plus nematodes significantly decreased slug damage caused by both small and large slugs. Within the combined treatments no differences were found after 2 weeks; all treatments gave less damage than the untreated (Ester *et al.*, 2003a).

25.4. Vegetable Crops

Slugs are difficult to control in a wide range of vegetable crops. Many crops are extensively damaged despite intensive use of molluscicides, but no adequate control measures are currently available. Thus, *P. hermaphrodita* has been tested as a biocontrol agent in a broad range of vegetable crops (Glen *et al.*, 2000b).

25.4.1. Butterhead lettuce (*Lactuca sativa* var. *capitata*)

Most cultivars of butterhead lettuce are highly susceptible to leaf feeding by slugs

Table 25.1. Mean number of *Cymbidium* flower petals per stem and number of slug-damaged petals after 7 and 14 days. After 7 days new stems were used.

Treatment	Slug weight (g)	Dose mg a.i./pot	After 7 days: Flower petals	After 7 days: Damaged flower petals	After 14 days: Flower petals	After 14 days: Damaged flower petals
Untreated	0.27	0	55.0	17.5a	60.0	29.0a
	0.66	0	61.2	11.3ab	61.2	34.2a
Metaldehyde	0.29	44.8 mg	63.8	0.3de	65.0	0.5b
	0.65	44.8 mg	62.5	2.3bcde	63.8	4.8b
Nematodes	0.27	300,000	65.0	0.5de	56.2	0.3b
	0.65	300,000	61.2	3.0bcde	60.0	1.3b
Metaldehyde + nematodes	0.38	44.8 mg + 300,000	62.5	0.3de	53.8	0.0b
	0.38	44.8 mg + 150,000	62.5	0.3de	52.5	0.0b
	0.36	44.8 mg + 50,000	65.0	0.0e	57.5	0.0b
	0.37	44.8 mg + 2 × 50,000	57.5	1.8bcde	58.8	0.0b
	0.36	44.8 mg + 10,000	56.2	3.3bcde	55.0	1.5b
	0.37	44.8 mg + 2 × 10,000	67.5	1.3cde	62.5	1.0b

Means followed by the same letter do not differ significantly ($P > 0.05$).

throughout their development and are particularly vulnerable during the first couple of weeks after transplanting the seedlings. Slugs can destroy the plants completely, resulting in a high percentage of plant loss. In addition, contamination of harvested product by slugs can result in rejection of the crop by buyers. Lettuce can be grown in the field, or under protected conditions, particularly in polythene tunnels. Wilson *et al.* (1995a) described two experiments in which several doses of *P. hermaphrodita* were used to control slug damage to butterhead lettuce (cv. Titania) grown in polythene tunnels. In the first experiment no reduction in slug damage was caused by any nematode application rate, but rates of 300,000/m^2 and greater significantly reduced slug numbers. In the second experiment, the rate of 300,000 nematodes/m^2 significantly reduced slug damage but the higher rate did not. In this experiment, there was a significant negative linear relationship between slug numbers and nematode rate at the end of the experiment. In addition, other successful uses of *P. hermaphrodita* in lettuce have been documented. Speiser and Andermatt (1996) described a small outdoor plot trial where a single high rate (1×10^{10}/ha) of *P. hermaphrodita* reduced damage to butterhead lettuce.

25.4.2. Iceberg lettuce (*Lactuca sativa*)

During crop growth, particularly the last 4 weeks before harvest, it is a common practice to irrigate iceberg lettuces several times per week. Slugs assume importance after formation of the heads, when closure traps water between the leaves. *D. reticulatum* uses the heads as a shelter and easily devalues the crop by feeding, which encourages decay and is cosmetically unacceptable (Port and Ester, 2002). In iceberg lettuce, presence of slugs, feeding damage and faeces must be avoided in the saleable head.

A semi-field trial was conducted in four replicates to assess control of *D. reticulatum* with *P. hermaphrodita* in the Netherlands. An iron fence with a copper barrier at the top surrounded plots of 1 m^2. Forty slugs were added per plot, except for one control without slugs. Nematodes were diluted in 2000 l water/ha and applied at a range of rates (Table 25.2) using either a watering can or sprayer. After nematode treatment 16 lettuce seedlings (cv. Anouk) with four leaves were transplanted. As a chemical reference iron (III) phosphate (a.i. 1%) and metaldehyde (a.i. 6.4%) were applied after planting. Iceberg lettuce plants were replaced after 2 weeks. The total number of

Table 25.2. Mean percentage of healthy leaves of iceberg lettuce, following treatment with *Phasmarhabditis hermaphrodita* or metaldehyde.

Treatment	Rate/m^2	Number of slugs	Treatment dates	Planting dates	% healthy leaves	
					23 July	30 July
Untreated	0	0	—	5 and 19 July	97a	97a
	0	40	—	5 and 19 July	35e	36e
Nematodes (sprayer)	300,000	40	5 July	5 and 19 July	83abc	75abcd
Nematodes (can)	300,000	40	11 July	19 July	84abc	85ab
	300,000	40	5 July	5 and 19 July	92ab	93a
	150,000	40	5 July	5 and 19 July	76bcd	77abcd
	75,000	40	5 July	5 and 19 July	64d	61d
	37,500	40	5 July	5 and 19 July	65d	65cd
Iron (III) phosphate	0.0500 g	40	5 July	5 and 19 July	87ac	82abc
Metaldehyde	0.0448 g	40	5 July	5 and 19 July	75cd	67bcd

Means followed by the same letter do not differ significantly ($P > 0.05$).

leaves and the numbers of undamaged leaves were counted 4 and 11 days after the second transplantation and calculated into percentages. Table 25.2 shows that good plant protection can be achieved with much lower numbers of nematodes than the currently recommended rate.

25.4.3. Cabbage (*Brassica oleracea* var. *capitata*)

Cabbage, like lettuce, can be grown in the field or under glass or polythene and is frequently damaged by slugs. Again, most severe slug damage occurs to the newly emerged or transplanted seedlings. The main slug species that cause damage are *Arion rufus* and *D. reticulatum*.

An experiment with cabbage seedlings in a plastic tunnel was carried out in Croatia in April 2003. This consisted of only two treatments of *P. hermaphrodita* applied at a rate of 300,000 nematodes/m^2 and methiocarb slug pellets at a standard rate. Each treatment consisted of plots of 6 m^2 with four replicates. The soil temperature was measured at a depth of 10 cm and ranged from 10°C to 21°C. At each assessment the percentage of leaf area consumed by slugs was estimated. The plants were assessed eight times in the period from 3 to 30 days after treatment.

Up to 8 days after treatment *P. hermaphrodita* and methiocarb demonstrated significantly better results than the untreated plants. But from the 10th to the 30th day *P. hermaphrodita* showed significantly better results in preventing leaf damage caused by slugs than methiocarb, i.e. 2% and 15% leaf damage, respectively, while the damage to the untreated plants increased to 40% on day 30 (transplanting) (Grubisic *et al.*, 2003). The soil temperatures ranged from 10°C to 21°C, which is suitable for *P. hermaphrodita* survival in the soil and for efficient parasitism of slugs.

25.4.4. Brussels sprouts

Slugs are the most serious pests of Brussels sprouts in western Europe. The field slug (*D. reticulatum*) is the most harmful species but *A. circumscriptus* is also responsible for crop damage in many western European countries. Brussels sprouts are mostly grown on clay soils, which contain many niches for sheltering slugs from environmental conditions. Damage can result from seedling losses but the main damage is deformation, rot and contamination with slime and faeces of the harvested product resulting in serious loss of crop value. Slugs are active just before button (sprout) formation. The slugs rest and feed upon the maturing crop (Fig. 25.2A). During the evening and night the slugs move on to the plant and attack the young buttons by eating the

A B

Fig. 25.2. (A) Slug damage to buttons of Brussels sprouts. (B) Application of *Phasmarhabditis hermaphrodita* to Brussels sprouts using helicopter.

outer leaves. Port and Ester (2002) mentioned that fields with a moderate population density of slugs may have 60–80% of buttons attacked.

We briefly present results from two large-scale experiments on Brussels sprouts in the Netherlands. The first was done in 1999 at Westmaas (clay soil with 20–30% silt) in the south-western part of the Netherlands. This location had a high density of *D. reticulatum* and used the Brussels sprout variety Romulus. The second was done in 2002 and treatments were applied on four commercial farms, with each farm being treated as a separate replicate. Plots were each 400 m^2. The fields were planted between the second week of April and up to the last week of May. The treatments in both these experiments are summarized in Table 25.3. Nematodes (Nemaslug®) were applied as 0.6-m-wide band applications between the crop rows. Metaldehyde pellets were broadcast by hand. The nematodes were applied as a suspension by knapsack sprayer in 1999 for the small-scale experiment. In 2002, the nematodes were broadcast applied with a 3-m spraymatic 10 S spray boom in 1000 l water/ha at 4 bar pressure using Teejet XR 11006 nozzles without filters.

Crop damage by slugs was assessed during the autumn on four different occasions (every 4 weeks) by cutting the stems and counting the number of damaged and undamaged sprouts from ten plants in each plot. In 2002, crop damage was assessed on three occasions by counting the numbers of damaged and undamaged sprouts from 60 plants.

In 1999, treatments with nematodes and metaldehyde decreased the plant damage by slugs in all experiments (Table 25.4). There were no significant differences in slug damage between nematode application rates; all nematode treatments resulted in greater reduction in leaf damage compared with metaldehyde.

In 2002, 50,000 nematodes/m^2 applied three or six times gave significant protection against slug damage (Table 25.5). Nematode treatments showed the same level of protection as the metaldehyde pellets. Among the treatments, nematodes applied three times at a rate of 50,000/m^2 in 2002 proved to be most effective in protecting Brussels sprouts against slug damage. These treatments resulted in levels of damaged sprouts similar to metaldehyde pellets at a rate of 448 g a.i./ha applied six times in all years (Ester *et al.*, 2003b). This represents a 50% reduction in the typical rate of nematodes used in single broadcast applications.

In addition to the above-described experiments, Glen *et al.* (2000b) described a field experiment showing a reduction in slug

Table 25.3. Nematode rates per square metre and metaldehyde in grams per hectare used, number of applications and different intervals and the period of treatment in Brussels sprouts to control *Deroceras reticulatum*.

Treatment	Rate/m^2	Number of applications	Interval (weeks)	1999	2002
Nematodes	50,000	3	4	—	1/7–10/9
		6	2	—	1/7–10/9
	150,000	7	2	15/7–7/10	—
	300,000	5	2	12/8–7/10	—
		7	2	15/7–7/10	—
	450,000	5	2	12/8–7/10	—
Metaldehyde	448 g a.i./ha	7	2	15/7–7/10	—
		5	2	12/8–9/10	—
		6	2	—	1/7–10/9
		7	2	15/7–7/10	—

Table 25.4. Mean % leaf damage to Brussels sprouts in 1999 following multiple applications of metaldehyde or *Phasmarhabditis hermaphrodita*.

Treatment	Rate/m^2	Number of applications	Week 8	Week 12	Week 16	Week 20
Nematodes	150,000	7	1.0c	5.7b	11.8c	9.6c
	300,000	7	2.3bc	5.8b	11.2c	10.9c
	450,000	7	1.9c	1.9b	7.8c	5.5c
Metaldehyde	448 g a.i./ha*	7	1.4c	5.5b	9.0c	12.9c
Nematodes	300,000	5	7.6a	7.4b	10.0c	7.3c
	450,000	5	2.6bc	3.0b	4.2c	1.8c
Metaldehyde	448 g a.i./ha*	5	7.2ab	24.7a	34.2b	35.0b
Untreated	0	0	8.9a	37.8a	63.9a	52.1a

Means followed by the same letter do not differ significantly ($P > 0.05$).

damage to Brussels sprouts grown in the north-west of Spain. Broadcast application at the recommended rate reduced slug damage when the nematodes were applied 3 days before planting.

25.4.5. Green asparagus

Green asparagus (*Asparagus officinalis*) is a high-value crop, especially in the first 4 weeks of harvesting. It is damaged extensively by slugs because it is grown for 10 years or more on the same field. Green asparagus spears are deformed by tiny feeding marks at the growing tips, resulting in an unmarketable product. Green asparagus was introduced in the Netherlands in the 1980s. Traditionally, white asparagus was grown mainly on sandy soils. Green asparagus, however, is mostly grown on clay soils and is thus more susceptible to slug damage. The surface of clay soils is completely closed by a crust in winter. Slugs have restricted capability to pass through this crust so there is little movement of slugs from the soil during this period. Early in spring, slugs start moving below the soil surface and affect newly developed spears. In response to increasing temperatures towards the end of March, the spears start to grow and push the clay clods, enabling both spears and slugs to appear above ground level. In the next 2 months temperatures in the Netherlands are often low, leading to slow development of the crop. This is the most vulnerable stage as slugs feed for a long time in the same region of the spear, causing severe deformation. These early spears have a high value. In June, temperatures increase and spears grow fast and are harvested once or twice a day, thus reducing slug damage. Current control methods based on metaldehyde are inadequate.

Field experiments were conducted in 2000, at Oudkarspel (soil with 22% silt) in the western part of the Netherlands, testing a range of nematode application rates (Table 25.6). This location had a high density of *D. reticulatum*. The asparagus variety

Table 25.5. Mean % damage to Brussels sprouts on four commercial farms in 2002 following multiple applications of *Phasmarhabditis hermaphrodita* or metaldehyde pellets.

Treatment	Rate/m^2	Number of applications	Week 35	Week 39	Week 44
Nematodes	50,000	6	5.6a	6.3a	17.0a
	50,000	3	5.3a	4.4a	11.6a
Metaldehyde	448 g/ha	6	2.9a	8.7a	25.4ab
Untreated	0	0	13.9b	23.5b	37.8b

Means followed by the same letter do not differ significantly ($P > 0.05$).

Table 25.6. Percentage slug-damaged green asparagus spears divided in four harvesting periods in 2000 following single or multiple applications of *Phasmarhabditis hermaphrodita* or metaldehyde pellets.

Treatment	Rate/m^2	Number of applications	% affected spears 26/4–5/5	6/5–15/5	16/5–25/5	26/5–6/6
Nematodes	10,000/m^2	3	79.3a	42.5ab	22.8ab	6.5bc
	50,000/m^2	3	28.6bcd	16.3de	11.3abcd	5.2bc
	100,000/m^2	3	19.5d	15.8de	8.9bcd	5.0c
	300,000/m^2	1	16.3d	14.3de	7.9cd	15.0abc
Metaldehyde	350 g a.i./ha	4	29.5bcd	6.5e	1.1d	5.0c
Untreated	0	—	76.0a	47.1a	25.9a	11.4abc

Means followed by the same letter do not differ significantly ($P > 0.05$).

Gijnlim was used. The asparagus fields used in 2000 were planted in 1997. The experimental layout was in randomized blocks with five replicates. Each plot consisted of one asparagus bed of 5-m length and 1.5-m width. The experimental row application was 0.3 m wide; this means each plot consisted of 1.5-m^2 treated area. The first treatment was carried out at the time the soil surface crust was breaking as the first asparagus spears emerged (11 April). Additional treatments were made on 18 April, 25 April and 2 May. The nematodes (Nemaslug®) and metaldehyde were applied as 0.3-m-wide band applications. The nematodes were applied at 300,000/m^2 as a suspension to the soil surface using a watering can in 15 l (6 mm) water. Metaldehyde formulated as bait pellets at standard rates were included as a control. Asparagus spears were harvested daily, by cutting the spears at 21 cm length above soil level. The harvested spears were counted and divided into slug-damaged and unaffected spears. In the first period, 26 April to 5 May, nematodes applied once at 300,000/m^2, or three times at 100,000/m^2 or 50,000/m^2 gave significant protection against slugs. Nematodes at a rate of 10,000/m^2 applied three times did not significantly control slugs in any of the periods. Metaldehyde pellets strongly decreased the percentage of slug-damaged spears, especially in the second and third period of harvesting.

P. hermaphrodita applied three times at 50,000/m^2 as a row application reduced slug damage in asparagus significantly. The row application covers 20% of the soil surface, which is a reduction of 80% when compared with the recommended broadcast application. Thus, row application three times at 50,000/m^2 results in a reduction to a total application rate of only 10% of the recommended rate, whilst protecting the asparagus spears to the same extent as the molluscicide pellets. Nematodes at lower rates were insufficient (Table 25.6). Thus, Brussels sprouts represent a large market for *P. hermaphrodita* (Ester and Rozen, 2003; Ester *et al.*, 2003c).

25.5. Arable Crops

Arable crops represent the largest cropped area damaged by slugs and the greatest overall value. Because of the low value of these crops per unit area, commercial use of *P. hermaphrodita* in arable agriculture is not currently practised. However, experimental results suggest that the nematode has potential for use in arable agriculture, particularly on organic farms where chemical molluscicides cannot be used. In addition, repeated application methods, as described above for outdoor vegetable crops, have much potential to reduce the numbers of nematodes applied.

25.5.1. Oilseed rape

Oilseed rape is a major oilseed crop in much of western Europe and is often used as a break crop in predominantly cereal rotations to aid weed and soil-borne disease control. The crop is also grown in much of North America (where it is called canola) but slug problems are rarely recorded in this part of the world. Two factors have led to the increased pest status of slugs in this crop in Europe: (i) the move from spring-sown crops to autumn-sown crops, a time when large slug populations may be present and active; and (ii) the trend to breed varieties that are low in glucosinolates, the natural sulphur-containing compounds that are toxic to many polyphagous herbivores. This step was taken to make the by-product of the crop, rapeseed meal, more palatable as an animal feed. While slugs can feed on this crop throughout its growing cycle, they only do economic damage at the early stages of crop establishment. Slugs do not feed on the seeds, but feed on the newly emerged seedlings resulting in loss of the apical meristem and death of the plant. In extreme cases this can lead to loss of the entire crop and redrilling is usually not feasible because of the short window of establishment for this crop. Thus, crops are frequently treated prophylactically with molluscicide pellets at the time of drilling. By far, the most important pest species in this crop is *D. reticulatum*, although it frequently occurs with one or more of the smaller species of arionid slugs. The pest status of slugs in this crop and their means of control have been reviewed by Moens and Glen (2002).

There is no published work on use of *P. hermaphrodita* to control slugs in oilseed rape, although the fact that *D. reticulatum* is the predominant pest suggests that the nematodes will have potential. Conversely, the low value of the crop and the fact that it tends to be drilled in August, when soils may be too dry to use nematodes, may preclude widespread use. One experiment has been done looking at timing of application of *P. hermaphrodita* to oilseed rape in which the nematodes were applied either at drilling, or at 1 or 2 weeks before drilling. The experiment used replicate plots 12 m × 6 m wide of each treatment arranged in randomized blocks. Oilseed rape (cv. Apex) was direct-drilled into the plots on three dates separated by 1 week (18 August 1993, 25 August 1993 and 1 September 1993). On all dates nematodes (300,000/m^2) and methiocarb pellets were applied to plots immediately after drilling. In addition, nematodes were applied 1 week before the second drilling and 1 and 2 weeks before the third drilling. Four weeks after the final drilling date, numbers of established oilseed rape seedlings were counted (Table 25.7). While it is clear that the nematode can control slugs in this crop, the problems of dry soils at time of application are highlighted as being far more important than time of application. Even the best results for nematodes were significantly less than the methiocarb pellets.

Another problem associated with oilseed crops is the large build-up of slugs below their dense canopy during the cropping season as food is plentiful. Thus, the following crops (usually cereals) are particularly at risk. The moist shaded conditions that promote build-up of slug populations should also encourage build-up of *P. hermaphro-*

Table 25.7. Mean square root numbers of rape seedlings/0.25 m^2 quadrat in an experiment investigating timing of nematode application to oilseed rape.

	Untreated	Nematodes applied 2 weeks before drilling	Nematodes applied 1 week before drilling	Nematode applied at drilling	Methiocarb pellets applied at drilling
First drilling	3.514a	*	*	3.646a	4.122a
Second drilling	4.062a	*	3.949a	3.663a	3.924a
Third drilling	2.393a	3.709b	3.310b	3.905b	4.731c

Numbers within each row followed by the same letters are not significantly different ($P > 0.05$). It is not possible to compare drilling dates because of differences in numbers of plants drilled on each date.

dita populations. An area of future research may be using low-rate inoculative releases of *P. hermaphrodita* into oilseed rape crops to protect the following cereal crop.

25.5.2. Wheat

In western Europe autumn-sown wheat (winter wheat) is the most extensively grown field crop and is particularly vulnerable to damage by slugs. The crop is sown in early autumn when slug populations are high, and temperatures are still suitable for slug activity. While there is still much spring-sown wheat grown in Europe this is rarely attacked by slugs. Slugs attack the seeds of autumn-sown wheat immediately after drilling and eat both the embryo and the endosperm, resulting in plant death. Slugs also attack the developing seedlings during the early stages of crop establishment, often causing loss of the apical meristem and plant death. Severe slug damage results in farmers having to redrill entire fields. As with oilseed rape, the species of slug that causes most damage is considered to be *D. reticulatum*, but this species frequently occurs in conjunction with small *Arion* sp. or more rarely species from the Milacidae. A large proportion of winter wheat in Europe is treated prophylactically with molluscidal baits. Because this crop represents the most economically important crop damaged by slugs there has been much interest in the use of *P. hermaphrodita* in this crop. Wilson *et al.* (1994) were the first to publish data on use of *P. hermaphrodita* to control slugs in wheat. They investigated the efficacy of five application rates of *P. hermaphrodita* ranging from 10^8 to 10^{10} nematodes/ha applied to the wheat field immediately after the seeds were sown. They found that the two highest rates of application (3×10^9 and 1×10^{10}/ha) gave protection to the crop equivalent to standard molluscicidal bait pellets applied at the recommended rate. The lower of these two rates was later shown to be efficacious in another trial (Wilson *et al.*, 1996) investigating the benefits of soil incorporation following application. Further experiments showed the benefits of applying nematodes to winter wheat to reduce slug damage, and particularly how postapplication incorporation could be beneficial (Hass *et al.*, 1999b). While there is no doubt that *P. hermaphrodita* is capable of controlling slugs in winter wheat, the high cost of nematode application has precluded widespread use of the nematodes in this low-value crop.

25.5.3. Sugarbeet

Slug problems are frequent in sugarbeet crops, especially when the soil is a clay or loam and the land surface undulates. To protect the soil from water erosion, sugarbeet is frequently grown in a cover crop. Beet grows in a cover of dead organic material in an undisturbed layer of topsoil, which favours survival of slugs. Covers can be Italian rye grass, rye, yellow mustard or black radish, which are either killed by herbicides or winter frost depending on the crop.

As with other arable crops, slugs are pests during crop establishment when slug grazing may kill developing plants, particularly between the 4- and 6-leaf stage. The seeds and recently germinated seedlings are not susceptible to slugs. The main species involved is *D. reticulatum*. An experiment was conducted in the Netherlands in autumn 1993. One week after drilling the sugarbeet seeds, the nematodes were applied as a furrow application. The seed furrow was 10 cm wide. The nematodes were added at a rate of 300,000/m^2 as a suspension by a watering can, during a rainy period. Methiocarb pellets (240 g a.i./ha) were added to the seed furrow during drilling. Slug numbers were recorded under shelter traps (Hommay *et al.*, 1991). These traps measured 40 × 40 cm and numbers of slugs in these traps were multiplied by 6.25 to calculate slugs per square metre. The damage to the crop by slugs was assessed in May, June and July by counting the number of sugarbeet plants of 30 m^2 from each plot. The plant development was assessed

Table 25.8. Crop stand (score 0–10) on 30 May and the average number of sugarbeet plants/ha at Wijnansrade (W) and Heerlen (H) on 14 June 1994; average number of trapped slugs/m^2, Wijnansrade, 1994.

Treatments	Rate	Crop stand (W)	Number of plants (W)	Number of plants (H)	Slugs 20 May	Slugs 1 June	Slugs 14 June
Untreated	0	5.4a	6,100a	7,000a	3.1a	7a	16a
Nematodes	300,000/m^2	7.5b	8,600b	8,900b	0.4b	2b	11ab
Methiocarb	0.24 kg a.i./ha	7.0b	8,200b	8,800b	0.5b	2b	9b

Means followed by the same letter do not differ significantly ($P > 0.05$).

by estimating the crop stand with a score (from 1 to 10) for leaf quantity and uniformity of the crop. A low score for leaf quantity indicates a low number of leaves and a non-uniform crop. In June, plots treated with nematodes or methiocarb showed significantly ($P < 0.05$) more plants in comparison with the untreated plots at both sites (Table 25.8). At Wijnansrade, plots treated with nematodes or methiocarb pellets resulted in a significantly ($P < 0.05$) lower number of slugs compared with the untreated plots in May and June 1994 (Ester and Geelen, 1996). Slug activity was only high in the drill furrow. Clearly, the nematode can control slugs in this crop.

25.6. Conclusions

It can be seen clearly that *P. hermaphrodita* has potential to control slugs in a wide range of crops, from glasshouse-grown cut flowers to field vegetable and arable crops. One factor that helps success of *P. hermaphrodita* is that its optimum conditions mimic those of slug pests in western Europe, where it has achieved high success. Both slugs and nematodes favour high moisture content soils and mild temperatures (optimum temperature approximately 17°C). Price and production capacity are likely to limit nematode use in the near future, but overall the outlook is bright, with Becker Underwood, the producer of *P. hermaphrodita* forecasting a considerable increase in sales over the next few years. Many of the experiments described above show that there is much potential to reduce the current recommended rate of 300,000/m^2, and hence reduce application costs (Ester *et al.*, 2005). This is particularly well illustrated in the case of Brussels sprouts being treated. The recommended rate to control slugs in Brussels sprout crops is 500 million nematodes/ha applied three times separated by 1-month intervals. This represents a 50% reduction in the previously recommended single application. As a result of the experiments described above, in 2003, 35 Brussels sprout growers used nematodes for slug control in Belgium and the Netherlands for the first time. This represented about 150 ha Brussels sprouts. In most of the nematode-treated fields, nematodes were applied with the standard spray equipment. A small number of growers in the south-west of the Netherlands applied the nematodes by helicopter (Fig. 25.2B).

There are many other crops damaged by slugs for which *P. hermaphrodita* has potential to be used, in addition to those described above. For crops in which slugs cause subterranean damage, the potential of nematodes has not really been explored. A good example is potatoes, where slugs damage the maturing tubers before harvest (Bus and Ester, 1996). Slugs are notoriously difficult to control under such conditions, as pellets placed on the soil surface have no effect on the slugs causing damage. Pellets incorporated into the soil at the time of sowing break down before slugs become a problem. Potatoes are frequently irrigated and this exacerbates the slug problem. We believe potatoes and other similar crops represent another future market for *P. hermaphrodita*.

References

Bailey, S.E.R. (2002) Molluscicidal baits for control of terrestrial gastropods. In: Barker, G.M. (ed.) *Molluscs as Crop Pests.* CAB International, Wallingford, UK, pp. 33–54.

Barker, G.M. (ed.) (2002) *Molluscs as Crop Pests.* CAB International, Wallingford, UK, 558 pp.

Bus, C.B. and Ester, A. (1996) Slug problems in potatoes at emergence and in storage. *Abstracts of Conference Papers of the 13th Triennial Conference of the EAPR,* Veldhoven, 67–68.

Ester, A. and Geelen, P.M.T.M. (1996) Integrated control of slugs in a sugar beet crop growing in a rye cover crop. In: Henderson, I.F. (ed.) *Slug and Snail Pests in Agriculture.* Monograph No. 41, British Crop Protection Council, Thornton Heath, UK, pp. 445–450.

Ester, A. and Rozen, van K. (2003) Effect of Nemaslug®, salt or carvone on the slug damage in green asparagus. *Asparagus Research Newsletter* 19, 10–23.

Ester, A., Rozen, van K. and Hazendonk, A. (2003a) Efficacy of pesticides to control *Lehmannia valentiana* (Férussac) in orchids (*Cymbidium*) in greenhouse experiments. In: *Slugs & Snails: Agricultural, Veterinary & Environmental Perspectives.* Monograph No. 80, British Crop Protection Council, Thornton Heath, UK, pp. 89–94.

Ester, A., Huiting, H.F., Molendijk, L.P.G. and Vlaswinkel, M.E.T. (2003b) The rhabditid nematode *Phasmarhabditis hermaphrodita* Schneider as a potential biological agent to control field slugs *Deroceras reticulatum* (Müller) in Brussels sprouts. In: *Slugs & Snails: Agricultural, Veterinary & Environmental Perspectives.* Monograph No. 80, British Crop Protection Council, Thornton Heath, UK, pp. 313–318.

Ester, A., Rozen, van K. and Molendijk, L.P.G. (2003c) Field experiments using the rhabditid nematode *Phasmarhabditis hermaphrodita* or salt as control measures against slugs in green asparagus. *Crop Protection* 22, 689–695.

Ester, A., Huiting, H.F. and Rozen, van K. (2005) Slug parasitic nematodes in vegetable crops. *IOBC/WPRS - Insect Pathogens and Entomoparasitic Nematodes Bulletin,* Kiel, Germany (in press).

Gittenberger, E., Backhuys, W. and Ripken, Th.E.J. (1984). De landslakken van Nederland. In: van Bohemen, H.D., Buizer, D.A.G. and Littel, A. (eds) KNNV No. 37, Amsterdam.

Glen, D.M., Wilson, M.J., Pearce, J.D. and Rodgers, P.B. (1994) Discovery and investigation of a novel nematode parasite for biological control of slugs. In: *Proceedings of the Brighton Crop Protection Conference, Pests and Diseases.* pp. 617–624.

Glen, D.M., Wilson, M.J., Hughes, L., Cargeeg, P. and Hajjar, A. (1996) Exploring and exploiting the potential of the rhabditid nematode *Phasmarhabditis hermaphrodita* as a biocontrol agent for slugs. In: Henderson, I.F. (ed.) *Slug and Snail Pests in Agriculture.* Monograph No. 66, British Crop Protection Council, Thornton Heath, UK, pp. 271–280.

Glen, D.M., Wilson, M.J., Brain, P. and Stroud, G. (2000a) Feeding activity and survival of slugs, *Deroceras reticulatum,* exposed to the rhabditid nematode *Phasmarhabditis hermaphrodita*: a model of dose response. *Biological Control* 17, 73–81.

Glen, D.M., Wiltshire, C.W., Hughes, L., Ester, A., Rozen, van K., Castillejo, J., Iglesias, J., Speiser, B., Coupland, J. and Gwynn, R. (2000b) The use of slug-parasitic nematodes and other techniques for control of slug and snail damage in horticultural crops. *The BCPC Conference – Pests and Diseases,* 345–350.

Grewal, P.S., Grewal, S.K., Taylor, R.A.J. and Hammond, R.B. (2001) Application of molluscicidal nematodes to slug shelters: a novel approach to economic biological control of slugs. *Biological Control* 22, 72–80.

Grubisic, D., Ostrec, L.J. and Dusak, I. (2003) Biological control of slugs in vegetable crops in Croatia. In: Dussart, G.B.J. (ed.) *Slugs & Snails: Agricultural, Veterinary & Environmental Perspectives.* Monograph No. 80, British Crop Protection Council, Thornton Heath, UK, pp. 115–120.

Hass, B., Glen, D.M., Brain, P. and Hughes, L.A. (1999a) Targeting biocontrol with the slug-parasitic nematode *Phasmarhabditis hermaphrodita* in slug feeding areas: a model study. *Biocontrol Science and Technology* 9, 587–598.

Hass, B., Hughes, L.A. and Glen, D.M. (1999b) Overall versus band application of the nematode *Phasmarhabditis hermaphrodita* with and without incorporation into soil, for biological control of slugs in winter wheat. *Biocontrol Science and Technology* 9, 579–586.

Hommay, G., Peureux, D. and Verbeke, D. (1991) Un Réseau d'observations sur les limaces dans l'est de la France. *Phytoma – La Défense des végétaux* 431, 14–19.

Koppenhöfer, A.M. (2000) Nematodes. In: Lacey, L.A. and Kaya, H.K. (eds) *Field Manual of Techniques in Invertebrate Pathology.* Kluwer Academic Publishers, Dordrecht, The Netherlands, pp. 283–301.

Moens, R. and Glen, D.M. (2002) Agriolimacidae, Arionidae and Milacidae as pests in west European oilseed rape. In: Barker, G.M. (ed.) *Molluscs as Crop Pests.* CAB International, Wallingford, UK, pp. 425–441.

Port, G. and Ester, A. (2002) Gastropods as pests in vegetable and ornamental crops in western Europe. In: Barker, G.M. (ed.) *Molluscs as Crop Pests.* CAB International, Wallingford, UK, pp. 337–352.

South, A. (1992). *Terrestrial Slugs, Biology, Ecology and Control.* Chapman and Hall, London, 428 pp.

Speiser, B. and Andermatt, M. (1996) Field trials with *Phasmarhabditis hermaphrodita* in Switzerland. In: Henderson, I.F. (ed.) *Slug and Snail Pests in Agriculture.* Monograph No. 66, British Crop Protection Council, Thornton Heath, UK, pp. 419–424.

Wilson, M.J. and Gaugler, R. (2000) Terrestrial mollusc pests. In: Lacey, L.A. and Kaya, H.K. (eds) *Field Manual of Techniques in Invertebrate Pathology.* Kluwer Academic Publishers, Dordrecht, The Netherlands, pp. 787–804.

Wilson, M.J., Glen, D.M. and George, S.K. (1993) The rhabditid nematode *Phasmarhabditis hermaphrodita* as a potential biological control agent for slugs. *Biocontrol Science and Technology* 3, 503–511.

Wilson, M.J., Glen, D.M., George, S.K., Pearce, J.D. and Wiltshire, C.W. (1994) Biological control of slugs in winter wheat using the rhabditid nematode *Phasmarhabditis hermaphrodita. Annals of Applied Biology* 125, 377–390.

Wilson, M.J., Glen, D.M., George, S.K. and Hughes, L.A. (1995a) Biocontrol of slugs in protected lettuce using the rhabditid nematode *Phasmarhabditis hermaphrodita. Biocontrol Science and Technology* 5, 233–242.

Wilson, M.J., Hughes, L.A. and Glen, D.M. (1995b) Developing strategies for the nematode, *Phasmarhabditis hermaphrodita,* as a biological control agent for slugs in integrated crop management systems. In: McKinlay, R.G. and Atkinson, D. (eds) *Integrated Crop Protection: Towards sustainability?* Monograph No. 63, British Crop Protection Council, Thornton Heath, UK, 33–40.

Wilson, M.J., Hughes, L.A., Hamacher, G.M., Barahona, L.D. and Glen, D.M. (1996) Effects of soil incorporation on the efficacy of the rhabditid nematode, *Phasmarhabditis hermaphrodita,* as a biological control agent for slugs. *Annals of Applied Biology* 128, 117–126.

Wilson, M.J., Hughes, L.A., Jefferies, D. and Glen, D.M. (1999) Slugs (*Deroceras reticulatum and Arion ater* agg.) avoid soil treated with the rhabditid nematode *Phasmarhabditis hermaphrodita. Biological Control* 16, 170–176.

Wilson, M.J., Hughes, L.A., Major, B.M. and Glen, D.M. (2000) Integrating chemical and biological control of slugs. *Proceedings of the Brighton Crop Protection Conference – Pests and Diseases.* Vol. 3, pp. 1007–1012.

Part V

Predatory Nematodes

26 Potential of Predatory Nematodes to Control Plant-parasitic Nematodes

A.L. Bilgrami and C. Brey
Department of Entomology, Rutgers University, New Brunswick, NJ 08901, USA

26.1. Introduction

> Picture these ferocious little mononchs engaged in a ruthless chase in the midst of stygian darkness. We may imagine them taking up the scent of various small animals upon which they feed, among which almost anything they can lay mouth seems not to amiss, and pursuing them with a relentless zeal that knows no limit but repletion.
>
> (Cobb, 1917)

Written 85 years ago, these elegant words were drawn upon to present the whimsical view of soil life and recreate the 'stygian' world of N.A. Cobb. The above words apply equally well to the constant struggle among all types of predatory nematodes competing for food and space with other nematodes and microorganisms living side by side in the soil environment.

Biocontrol of nematodes has been oriented almost exclusively to microbial pathogens (Kerry, 2000). New approaches need to be explored beyond the present narrow bacterial–fungal base if nematode biocontrol is to move forward and evolve into an established subdiscipline remotely comparable with insect biocontrol. While pathogens have received significant research emphasis for biocontrol of plant-parasitic nematodes, predatory nematodes have been ignored. Although a wide range of soil invertebrates including mites, insects, turbellarians and tardigrades prey on nematodes, the most important predators are the nematodes themselves (Bilgrami, 1997). Predatory nematodes have shown biocontrol potential against plant-parasitic nematodes and established themselves as an important entity of the soil food web.

26.2. Types of Predatory Nematodes

Predatory nematodes belonging to the orders Mononchida, Dorylaimida, Diplogasterida, Aphelenchida, Enoplida and Rhabditida are classified into three categories depending on their feeding apparatus, food and feeding habits (Bilgrami, 1997). Predators commonly known as mononchs (Mononchida) possess highly sclerotized feeding apparatus with a large pointed dorsal tooth, small teeth or denticles. Their feeding apparatus is a cutting and engulfing type (e.g. *Mononchus, Iotonchus, Mylonchulus*), where they engulf prey whole and intact (Bilgrami *et al.*, 1986).

The second type is referred to as the stylet-bearing predators, e.g. dorylaim, nygolaim and aphelenchid (Dorylaimida, Aphelenchida). These predators cannot engulf or swallow prey intact or cut prey into pieces due to the type of the feeding apparatus, which is a piercing and sucking type. These predators puncture the prey with their narrow needle-like feeding apparatus that sucks the prey body contents. Feeding apparatus in dorylaim predators (e.g. *Labronema*) is axial in position but in nygolaims (e.g. *Aquatides*) it is non-axial. The former has a dorsal aperture and a groove but the latter does not have any such structures. Feeding apparatus of an aphelenchid predator (e.g. *Seinura*) is narrow and pointed (Jairajpuri and Bilgrami, 1990).

The third type of feeding is the cutting and sucking type that is represented by diplogasterid predators, e.g. *Mononchoides, Butlerius* (Diplogasterida). Their feeding apparatus, commonly known as the buccal cavity, is small but well armed with a strong claw-like movable dorsal tooth. Teeth or denticles may also be present to help cut prey cuticle and grind food particles (Jairajpuri and Bilgrami, 1990). The feeding apparatus and feeding mechanisms of actinolaim (e.g. *Actinolaimus*), enoplid (e.g. *Ironus*) and pelagnematoid (e.g. *Thalassogenus*) predators are similar to those of diplogasterid predators. In actinolaim predators the vestibule is reinforced with plate-like or ribbed-basket-like structures accompanied by large onchia with or without denticles. Onchia help predators slit open the prey's cuticle. The enoplid predators (e.g. *Ironus*) have three sharply pointed teeth to tear open the prey (van der Heiden, 1974). The action of buccal armature is supported by the feeding apparatus, muscles and the oesophageal suction.

26.3. Prey Capturing and Feeding Abilities

Prey capturing and feeding mechanisms of the predatory nematodes are divided into five phases: (i) encounter with prey; (ii) attack response; (iii) attack; (iv) extra corporeal digestion; and (v) ingestion (Bilgrami and Jairajpuri, 1989b).

26.3.1. Encounter with prey

Predators encounter prey by chance contact (e.g. mononchs) (Grootaert and Maertens, 1976; Fig. 26.1A) or by wilful movements in response to prey-emitted kairomones (diplogasterid, dorylaim or nygolaim predators) (Bilgrami and Jairajpuri, 1988a; Bilgrami *et al.*, 2000, 2001). Contrary to earlier beliefs that predation is aleatory, the stylet-bearing (e.g. *Mesodorylaimus*, *Aquatides*) or cutting and sucking type of predators (e.g. *Mononchoides*, *Butlerius*) establish contacts with the prey in response to attractants (Bilgrami, 1997). For example, the predators were attracted and aggregated at feeding sites around an excised/injured prey in response to their attractants (Wyss and Grootaert, 1977; Shafqat *et al.*, 1987; Bilgrami *et al.*, 2001). Positive attraction towards prey and during and post-feeding aggregation activities of the predators at the feeding site suggest more than a casual role of chemoattractants in establishing predator prey contacts. Interestingly, the diplogasterid predators are attracted towards bacteria besides prey nematodes (Bilgrami and Jairajpuri, 1988a) in order to feed.

26.3.2. Attack response

Head probing, feeding apparatus movements and oesophageal pulsations generate an attack response in predatory nematodes. The successful attacks are made at right angles to prey (Fig. 26.1A). Glancing contacts or contacts other than right angles do not result in successful prey puncture. Attack response may be aggressive as in *Prionchulus punctatus* or *Mylonchulus dentatus*, vigorous but confined as in *Labronema vulvapapillatum* and gradual and restricted as in *Aquatides thornei* or *Dorylaimus stagnalis*. Probing in *Mononchus aquaticus* is rapid side-to-side lip rubbing for short durations. *Mononchoides fortidens* probes its prey vigorously for short duration in comparison to *Mononchoides longicaudatus*. *Butlerius* spp. show attack response by head shaking and lip rubbing against the prey's body.

26.3.3. Attack

Predators initiate side-to-side lip rubbing over the body of prey simultaneously with few feeding apparatus movements that cut or penetrate the prey cuticle (Bilgrami and Jairajpuri, 1989b). Predators search either another spot on the prey body if the attack is unsuccessful or they begin to search for another prey individual. Prey is attacked by the stylet (e.g. *Mesodorylaimus*, *Discolaimus*, *Seinura*), mural tooth (e.g. *Aquatides*), dorsal tooth (e.g. *Mylonchulus*), onchia (*Actinolaimus*), teeth (e.g. *Ironus*) or combined actions of movable dorsal tooth and high oesophageal suction (e.g. *Mononchoides*, *Butlerius*).

L. vulvapapillatum puncture prey with quick feeding apparatus movements (Wyss and Grootaert, 1977) (Fig. 26.1B) whereas *Aquatides* and *Dorylaimus* achieve perforation by gradual and intermittent thrusting of their feeding apparatus (Shafqat *et al.*, 1987). *Deplenteron* (Yeates, 1969), *Butlerius* (Grootaert *et al.*, 1977) or *Mononchoides* (Bilgrami and Jairajpuri, 1989b) use their movable dorsal tooth and oesophageal suction to slit open the prey cuticle (Fig. 26.1C). *Mononchus*, *Iotonchus* and other mononchs engulf and swallow their prey whole, or they shred their prey before feeding (Fig. 26.1D) (Bilgrami *et al.*, 1986).

L. vulvapapillatum and *A. thornei* require 5–6 feeding apparatus thrusts (Wyss and Grootaert, 1977) whereas *D. stagnalis* needs 6–8 thrusts to puncture and penetrate prey cuticle (Shafqat *et al.*, 1987). *Mesodorylaimus bastiani* needs fewer thrusts (6–9) than *Aporcelaimellus nivalis* (7–12) (Khan *et al.*, 1991) to perforate prey cuticle. *Seinura* injects toxic oesophageal secretions to paralyse prey to feed (Hechler, 1963). Other stylet-bearing predators disorganize internal body organs of prey structures to make them immobile, whereas mononchs inactivate their prey by holding them firmly with the buccal armature and high oesophageal suction.

26.3.4. Extracorporeal digestion

The feeding apparatus lumen of piercing and sucking types of predators is too narrow to ingest large food molecules. Therefore,

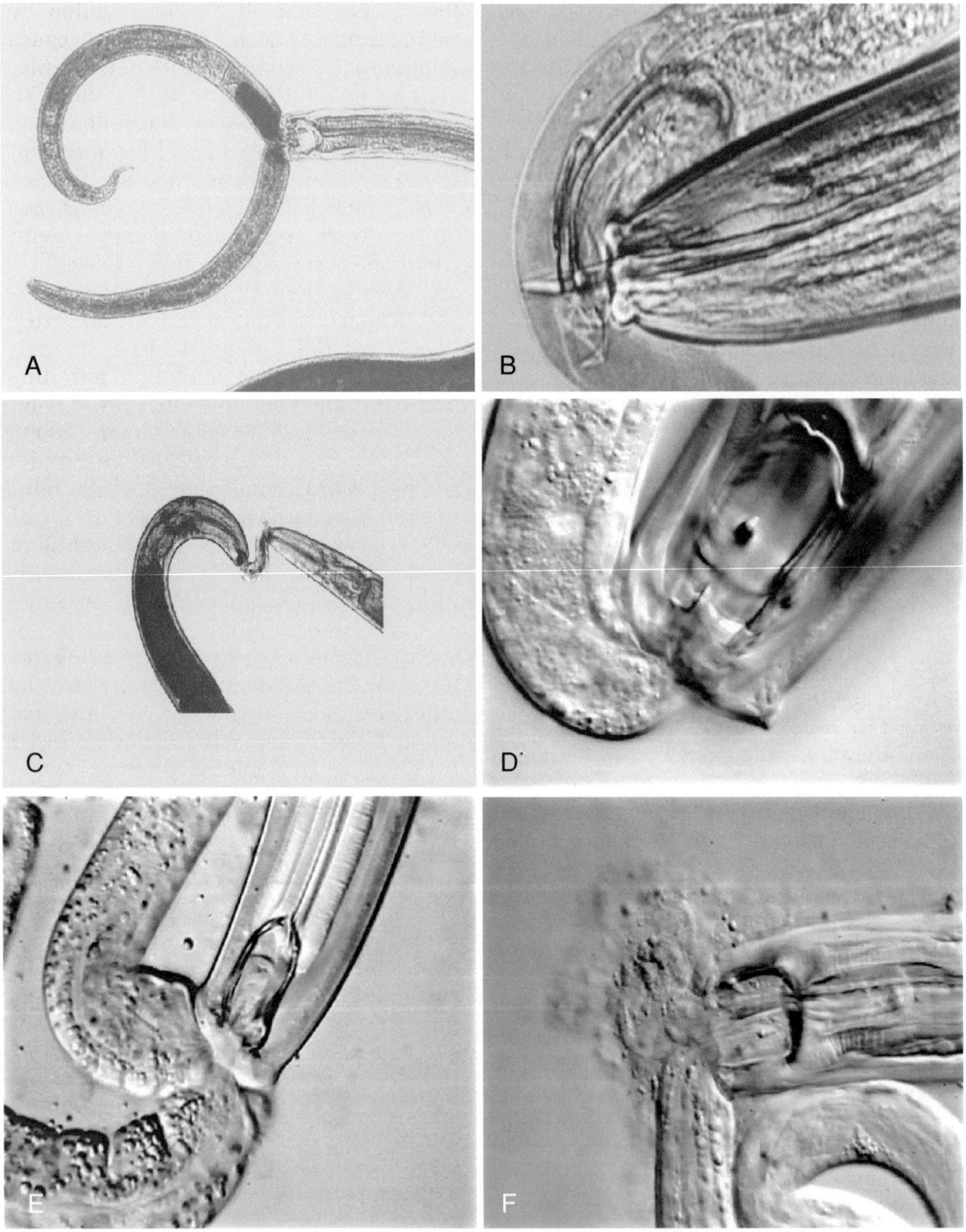

Fig. 26.1. (A) *Iotonchus* sp. catching a prey individual. (B) *Labronema vulvapapillatum* feeding at the posterior region of *Panagrellus redivivus.* (C) Two individuals of *Diplenteron colobocercus* feeding together on a single *Mesorhabditis littoralis.* (D) *Anatonchus tridentatus* attaching an individual of *Panagrellus redivivus.* (E) *A. tridentatus* feeding on *P. redivivus.* (F) *A. tridentatus* ingesting body contents of *P. redivivus.* (Fig. 26.1A: Courtesy of Eisenback and Zunke, 1997, NemaPix Vol. I, *A Journal of Nematological Images,* Mactode Publications; Fig. 26.1B,D,E and F: Courtesy of U. Wyss.)

the food is partially digested outside the oesophagus before ingestion. Such a phenomenon is known as extracorporeal digestion. It is reported to occur in plant-parasitic (Wyss, 1971) and predatory nematodes (Bilgrami and Jairajpuri, 1989b). Predatory mononchs do not predigest food since they can swallow a prey whole or ingest its pieces through the wide oral aperture. In contrast, diplogasterid predators partially digest food molecules before the ingestion by releasing oesophageal gland secretions (Bilgrami and Jairajpuri, 1989b). Complex food globules are broken down into small particles that are ingested through the feeding apparatus lumen en route to the intestine. *Diplenteron, Dorylaimus, Aquatides, Seinura, Mononchoides* show extracorporeal digestion of food molecules.

26.3.5. Ingestion

Many species of mononchs engulf entire prey or ingest it after shredding it into pieces (e.g. *Iotonchus*) (Fig. 26.1E,F), but few (e.g. *Mylonchulus*) feed by cutting and sucking (Bilgrami *et al.* 1986) their prey. Swallowing of prey is supported by the oesophageal muscle contractions that pull prey into the buccal cavity through vertically positioned plates. Some individuals show periods of inactivity after devouring an entire prey, while others initiate further attacks. Dorylaim, nygolaim and aphelenchid predators cannot engulf prey or shred it into pieces; instead they penetrate and rupture internal prey structures by making sideways movements of the feeding apparatus. Prey contents are ingested by the feeding apparatus that pass into the intestine through the oesophago–intestinal junction by simultaneous relaxation and contraction of the oesophageal bulb. Once the contents are ingested, predators detach their lips from the prey, retract the feeding apparatus and move in search of another prey. Dorylaim and nygolaim predators also feed on the eggs of other nematode species but not conspecific eggs. When in contact with conspecific eggs, these predators probe in an exploratory fashion by making side-to-side lip rubbing but cause no harm to the eggs (Esser, 1987). The diplogasterids could devour intact first stage juveniles of small prey nematodes (e.g. *Acrobeloides* or *Cephalobus*) but must cut larger prey into pieces to feed. The process of food ingestion in *Neoactinolaimus, Ironus* or *Thalassogenus* is identical with other groups of predators.

An injured prey attracts predators to aggregate at the feeding site. Predators struggle among themselves to feed if their number exceeds more than two at a feeding site. Aggregation at the feeding site is common in dorylaim (Bilgrami *et al.*, 2000), nygolaim (Bilgrami *et al.*, 2001) and diplogasterid predators (Bilgrami and Jairajpuri, 1988a). Up to eight individuals of diplogasterid predators may be seen aggregating at the feeding site. Diplogasterid predators, e.g. *Mononchoides*, are most active in showing during and post-feeding aggregation at the feeding site (Fig. 26.2A–D). Feeding in a group allows predators to quickly finish the feeding before hunting other prey. Aggregation at the feeding sites is most pronounced at low prey densities. Feeding is completed when prey contents are completely ingested.

Predatory nematodes *Seinura paynei*, which have been recovered from mushroom substrates, were found feeding on free-living nematodes, e.g. *Acrobeloides* sp. and *Bursilla labiata* (Grewal *et al.*, 1991). The widespread distribution of *S. paynei* and their feeding on nematodes in mushroom substrate suggest that these predators may also control populations of the nematode parasite of mushroom *Aphelenchoides richardsoni* (Grewal *et al.*, 1991). However, more studies are needed to understand true predatory potential of aphelenchid nematodes.

26.4. Biocontrol Potential

Mononch, dorylaim, nygolaim, diplogasterid and other groups of predators show differential predatory potential. Their

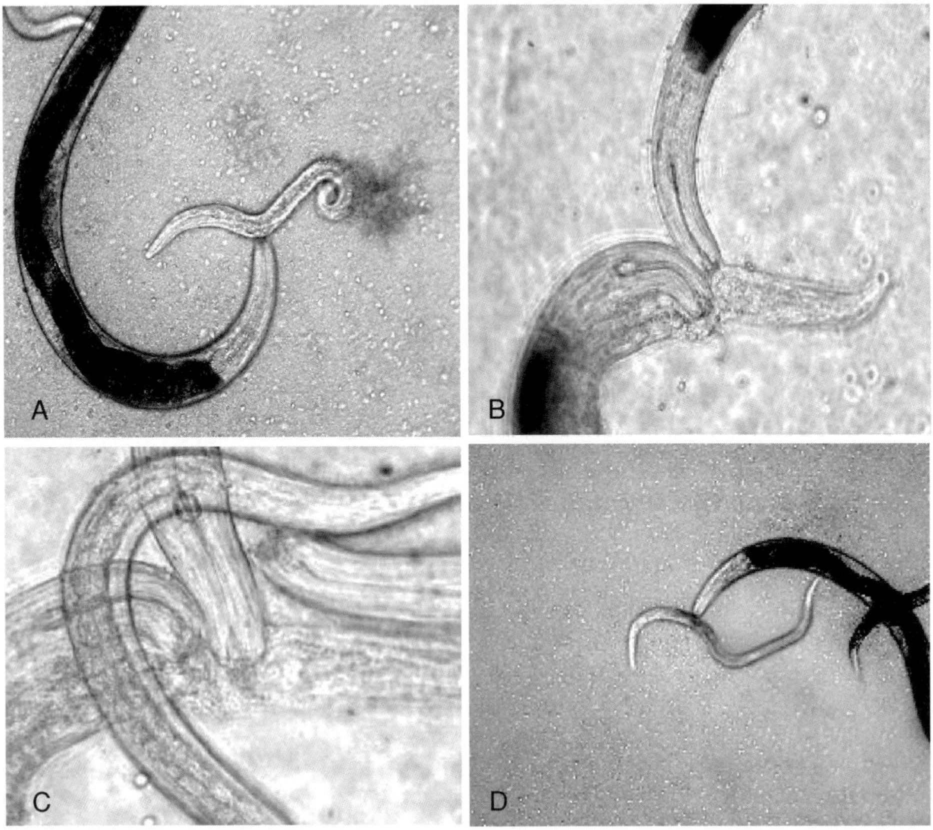

Fig. 26.2. (A) *Mononchoides* feeding on the infective juvenile (IJ) of *Steinernema carpocapsae*. (B) Two individuals of *Mononchoides* feeding together on the IJ of *Heterorhabditis bacteriophora*. (C) Three individuals of *Mononchoides longicaudatus* feeding together on *H. bacteriophora*. (D) *M. longicaudatus* feeding on IJ of *S. carpocapsae*.

biocontrol potential depends on their rate of predation, prey-searching abilities, strike rate, resistance to environmental conditions, etc.

26.4.1 Mononch predators

Webster (1972) stated that there is 'no possibility of using predatory nematodes in biological control'. Since then there has been little change in this perception over the intervening 30 years. That is, the conventional wisdom continues to be that nematode predators have far too many limitations to be seriously considered for biocontrol. This premise, however, is based on nematodes from the Order Mononchida. All mononchs are predaceous, and advocacy in using them for controlling phytoparasitic species goes back more than 80 years (Cobb, 1917).

Despite pessimism in the use of predatory mononchs as nematode biocontrol agents, studies have shown that they have reduced *Tylenchulus semipenetrans*, *Globodera rostochiensis* and *Meloidogyne incognita* populations in pots (Cohn and Mordechai, 1974; Small, 1979). They also decreased *Trichodorus* and *Hemicriconemoides* populations under field conditions (Ahmad and Jairajpuri, 1982). Studies on the predation abilities, food preference, strike rate of predators and resistance and susceptibility of prey nematodes to predation, cannibalism, predator–predator interactions, range of prey and factors influencing predation were all examined to evaluate predatory potential of mononchs against plant-parasitic and free-living nematode species.

Mononchs feed on a range of plant-parasitic nematodes (Table 26.1), rotifers and other soil microorganisms. They swallow free-living nematodes (75%) more than tylenchs (45%) or dorylaims (41%) (Bilgrami *et al.*, 1986). It is difficult to conclude whether the presence of more free-living nematodes in the intestine of mononchs is due to any feeding preference or the consequence of abundance of free-living nematodes in a localized area, since these observations were made on mounted specimens. Arpin (1979) also found significant correlation between mononch predators and free-living nematodes, which Nelmes and McCulloch (1975) did not observe during their study.

Mononchs have some flaws from a practical biocontrol standpoint, which led inevitably to Webster's (1972) conclusion, most notably the inability to culture them *in vitro*, but also a long life cycle, large size, modest fecundity, high cannibalism and lack of a resistant stage tolerant to environmental extremes. Thorne (1927) judged mononchs populations as too unstable to be useful in biocontrol programmes. Moreover, as extreme generalist predators, the prospects of using mononchs effectively to control a specific pest species seemed remote (Jones, 1974).

26.4.2. Dorylaim and nygolaim predators

Feeding by *Eudorylaimus obtusicaudatus* on *Heterodera schachtii* eggs inside the cyst suggested more than a casual role of stylet-bearing predators in nematode biocontrol under natural conditions and generated optimism in using these predators as biocontrol agents (Boosalis and Mankau, 1965). Dorylaim and nygolaims are most ubiquitous and occur in all types of soils, climates and habitats. Due to their widespread natural occurrence in fields they have advantages over other predator types. The presence of two, three or more genera in one field is quite common (Bilgrami *et al.*, 2002, 2003). These predators may be maintained and established under field conditions by making slight modifications in their environment. In addition to nematodes, the dorylaim and nygolaim predators also feed on algae and fungi (Ferris, 1968). The efficient rate of predation (Khan *et al.*, 1991), moderate to high strike rates (Bilgrami, 1992), positive correlation between predation and prey trophic groups (Bilgrami, 1993, 1995), prey preference (Khan *et al.*, 1995b), efficient prey searching (Bilgrami and Pervez, 2000), feeding, attraction and aggregation activities at the feeding site (Bilgrami *et al.*, 2000, 2001) and wide range of predation on plant-parasitic nematode species (Table 26.2) have established their credentials as an efficient biocontrol agent.

26.4.3. Diplogasterid predators

The diplogasterid nematodes have parasitic (Poinar *et al.*, 1976) and predaceous members (Bilgrami, 1997), although few studies have examined diplogasterid predation on plant-parasitic nematodes (Table 26.3). Yeates (1969) evaluated predation ability of *Diplenteron colobocercus*, which feed on bacteria in the absence of prey nematodes, whereas Grootaert *et al.* (1977) examined feeding habits of *Butlerius degrissei*. Both studies suggested the future use of diplogasterid predators in plant-parasitic nematode management. The real importance of diplogasterid predators was revealed when their predatory abilities were evaluated (Small and Grootaert, 1983; Bilgrami and Jairajpuri, 1989b), where they were found to be attracted towards prey and bacterial colonies (Bilgrami and Jairajpuri, 1988a).

Biocontrol potential of diplogasterid predators lies in the ease of their *in vitro* culture (Jairajpuri and Bilgrami, 1990), wide prey range (Grootaert *et al.*, 1977; Bilgrami and Jairajpuri, 1989b), short life cycle (6–8 days) (Tahseen *et al.*, 1990), high rate

Table 26.1. List of plant-parasitic nematodes recorded as prey of predatory mononchs.

Predators	Prey nematodes	References
Anatonchus amiciae	*Tylenchus*, *Xiphinema*	Coomans and Lima, 1965
A. ginglymodontus	*Meloidogyne hapla* (juvenile)	Szczygiel, 1966, 1971
A. tridentatus	*Paratylenchus macrophallus*, *Aphelenchus*, *Longidorus*, *Pratylenchus*	Mulvey, 1961; Banage, 1963
Clarkus mulveyi	*Tylenchorhynchus nudus*, *Helicotylenchus multicinctus*, *Rotylenchus reniformis*, *Meloidogyne incognita* (juvenile)	Mohandas and Prabhoo, 1980
C. papillatus	*Tylenchus*, *Tylenchulus semipenetrans*, *Tylocephalus auriculatus*, *Heterodera schachtii* (juvenile), *Hemicriconemoides*, *Aphelenchoides*, *M. hapla* (juvenile)	Cobb, 1917; Menzel, 1920; Steiner and Heinley, 1922
C. sheri	*Tylenchorhynchus*, *Aphelenchus*	Bilgrami *et al.*, 1986
Coomansus indicus	*Pratylenchus*, *Tylenchorhynchus*, *Hemicriconemoides*, *Xiphinema*	Bilgrami *et al.*, 1986
Iotonchus acutus	*Trichodorus obtusus*, *R. robustus*, *Xiphinema americanum*	Cobb, 1917; Thorne, 1932
I. amphigonicus	*H. schachtii* (juvenile)	Thorne, 1924
I. antidontus	*Tylenchorhynchus*	Bilgrami *et al.*, 1986
I. basidontus	*Tylenchorhynchus*	Bilgrami *et al.*, 1986
I. brachylaimus	*Rhadopholus similis*, *T. semipenetrans*	Cassidy, 1931; Mankau, 1982
I. indicus	*Tylenchorhynchus*	Bilgrami *et al.*, 1986
I. kherai	*T. nudus*, *Hirschmanniella oryzae*, *Scutellonema curvata*, *H. multicinctus*, *R. reniformis*, *M. incognita* (juvenile), *Xiphinema elongatum*	Mohandas and Prabhoo, 1980
I. longicaudatus	*Hoplolaimus*, *Hirschmanniella*	Bilgrami *et al.*, 1986
I. monhystera	*T. nudus*, *H. oryzae*, *H. multicinctus*, *R. reniformis*, *M. incognita* (juvenile)	Azmi, 1983; Bilgrami *et al.*, 1986
I. nayari	*X. elongatum*, *H. oryzae*, *H. multicinctus*, *R. reniformis*, *M. incognita* (juvenile), *T. nudus*	Mohandas and Prabhoo, 1980
I. parabasidontus	*Hirschmanniella*	Bilgrami *et al.*, 1986
I. prabhooi	*R. reniformis*, *M. incognita* (juvenile)	Mohandas and Prabhoo, 1980; Bilgrami *et al.*, 1986
I. risoceiae	*Pratylenchus*	Bilgrami *et al.*, 1986
I. shafi	*Hoplolaimus*	Bilgrami *et al.*, 1986
I. trichurus	*Pratylenchus*, *Hoplolaimus*, *Tylenchorhynchus*, *Xiphinema*	Bilgrami *et al.*, 1986
I. vulvapapillatus	*Tylenchorhynchus*	Andrassy, 1964
Miconchus aquaticus	*Helicotylenchus*, *Xiphinema*, *Hemicycliophora*	Bilgrami *et al.*, 1986
M. citri	*Pratylenchus*, *Tylenchorhynchus*	Bilgrami *et al.*, 1986
M. dalhousiensis	*Aphelenchoides*	Bilgrami *et al.*, 1986
Mononchus aquaticus	*Tylenchorhynchus mashoodi*, *H. oryzae*, *Hoplolaimus indicus*, *Helicotylenchusindicus*, *X. americanum*, *Longidorus*, *Paralongidorus citri*, *Paratrichodorus*, *Anguina tritici* (juvenile), *M. incognita* (juvenile), *Meloidogyne naasi* (juvenile), *Heterodera mothi* (juvenile), *Rotylenchus fallorobustus*, *Globodera rostochiensis* (juvenile)	Grootaert and Maertens, 1976; Grootaert *et al.*, 1977; Grootaert and Wyss, 1979; Small and Grootaert, 1983; Bilgrami, 1992; Bilgrami and Jairajpuri, 1984; Bilgrami *et al.*, 1986
M. truncates	*H. schachtii*	Thorne, 1927

M. tunbridgensis	*Aphelenchus avenae*, *T. semipenetrans*, *Hoplolaimus*, *Tylenchorhynchus*, *Hemicriconemoides*	Mankau, 1980, 1982; Bilgrami *et al.*, 1986
Mylonchulus agilis	*Helicotylenchus vulgaris*, *R. fallorobustus*, *Longidorus caespiticola*	Doucet, 1980
M. brachyuris	*Subanguina radicicola*, *R. similis*	Cassidy, 1931
M. dentatus	*A. avenae*, *Helicotylenchus indicus*, *Hoplolaimus indicus*, *T. mashhoodi*, *M. incognita* (juvenile), *H. mothi* (juvenile), *T. semipenetrans*, *H. oryzae*, *Basiria*, *Xiphinema*, *Longidorus*, *P. citri*	Jairajpuri and Azmi, 1978; Bilgrami and Kulshreshtha, 1994
M. hawaiiensis	*T. nudus*, *H. oryzae*, *R. reniformis*, *M. incognita* (juvenile)	Mohandas and Prabhoo, 1980
M. minor	*A. tritici* (juvenile), *M. incognita* (juvenile), *T. semipenetrans*, *X. americanum*, *R. reniformis*	Kulshreshtha *et al.*, 1993; Choudhary and Sivakumar, 2000
M. parabrachuris	*H. schachtii* (juvenile)	Thorne, 1927
M. sigmaturus	*H. schachtii* (juvenile), *R. similis*, *T. semipenetrans*, *Meloidogyne javanica* (juvenile), *Subanguina radicicola*	Thorne, 1927; Cassidy, 1931; Cohn and Mordechai, 1973,1974; Mankau, 1982
Prionchulus muscorum	*Aphelenchus*, *Hoplolaimus*, *Tylenchorhynchus*, *Hemicriconemoides*, *Aphelenchus*	Szczygiel, 1971; Arpin, 1976; Bilgrami *et al.*, 1986
P. punctatus	*A. avenae*, *M. naasi* (juvenile), *G. rosctochiensis* (juvenile), *R. fallorobustus*, *Helicotylenchus*, *A. tritici* (juvenile)	Nelmes, 1974; Maertens, 1975; Grootaert *et al.*, 1977; Small and Grootaert, 1983; Small, 1979
Sporonchulus ibitiensis	*Aphelenchus*, *Aphelenchoides*	Carvalho, 1953
S. vagabundus	*Aphelenchoides*, *Hemicycliophora*, *Trichodorus*	Bilgrami *et al.*, 1986

of predation (Bilgrami and Jairajpuri, 1989b), ability to detect and respond to prey chemoattractants (Yeates, 1969; Bilgrami and Jairajpuri, 1988a), ability to switch to bacteria in the absence of prey (Bilgrami and Jairajpuri, 1989b), aggregation at the feeding site (Bilgrami and Jairajpuri, 1988b), high reproductive rate, predatory post-embryonic stages, rare cannibalism and environmentally tolerant resting stage (Bilgrami, 1997). Nevertheless, exploitation of the biocontrol potential of diplogasterids has been slow due to the lack of studies.

26.5. Resistance and Susceptibility of Prey Nematodes to Predation

Successful biocontrol could be achieved if predators possess high strike rate and prey nematodes are highly susceptible. Cohn and Mordechai (1974), Grootaert *et al.* (1977), Small and Grootaert (1983) and Bilgrami and Jairajpuri (1989a) differentiated prey nematodes depending on their abilities to resist predation.

The ability of prey nematodes to defend themselves against a predator's onslaught vary from species to species and individual to individual. Resistance to predation is due to the thick or double body cuticles (e.g. *Hoplolaimus*), coarse body annulations (e.g. *Hemicriconemoides*), gelatinous matrix or toxic body repellents (e.g. *Helicotylenchus*) and rapid undulatory body movements (e.g. *Rhabditis*). Bilgrami and Jairajpuri (1989a) have proposed the following equations to determine strike rate of predators and resistance and susceptibility of prey nematodes:

$$\text{Strike rate of predators (SR\%)} = \text{EA/E} \times 100$$

$$\text{Prey resistance (PR\%)} = \text{EA} - \text{AW/EA} \times 100$$

$$\text{Prey susceptibility (PS\%)} = 100 - \text{PR},$$

Table 26.2. List of plant-parasitic nematodes recorded as prey of predatory dorylaim, nygolaim and tylenchid predators.

Predators	Prey nematodes	References
Allodorylaimus americanus	*Meloidogyne incognita* (juvenile), *Anguina tritici* (juvenile), *Xiphinema basiri*, *Longidorus*, *Tylenchorhynchus mashhoodi*, *Hirschmanniella oryzae*, *Helicotylenchus indicus*, *Aphelenchoides*, *Basiria*, *Aphelenchus avenae*, *Tylenchulus semipenetrans*, *Trichodorus*	Khan *et al.*, 1995a,b
A. amylovorus	*T. semipenetrans*	Mankau, 1982
A. obscurus	*Heterodora schachtii* (juvenile)	Thorne and Swanger, 1936
A. obtusicaudatus	*H. schachtii* (juvenile)	Marinari *et al.*, 1982
A. nivalis	*M. incognita* (juvenile), *Heterodora mothi* (juvenile), *X. basiri*, *Longidorus*, *T. mashoodi*, *H. oryzae*, *H. indicus*, *Aphelenchoides*, *Basiria*, *A. avenae*, *T. semipenetrans*, *Trichodorus*	Bilgrami, 1993; Khan *et al.*, 1991
Discolaimus arenicolus	*M. incognita* (juvenile)	Yeates, 1993
D. silvicolus	*M. incognita* (juvenile), *H. mothi* (juvenile), *A. tritici* (juvenile), *X. basiri*, *Longidorus*, *T. mashhoodi*, *H. oryzae*, *H. indicus*, *Aphelenchoides*, *Basiria*, *A. avenae*, *T. semipenetrans*, *Trichodorus*	Khan *et al.*, 1995a
Dorylaimus obscurus	*H. schachtii* eggs	Thorne and Swanger, 1936
D. obtusicaudatus	*H. schachtii* eggs	Cobb, 1929
D. stagnalis	*T. mashhoodi*, *H. oryzae*, *H. indicus*, *X. americanum*, *Longidorus*, *Paralongidorus citri*, *A. tritici* (juvenile), *M. incognita* (juvenile), *H. mothi* (juvenile)	Bilgrami, 1992; Shafqat *et al.*, 1987
Eudorylaimus obtusicaudatus	*H. schachtii*	Esser, 1987
Labronema vulvapapillatum	*A. avenae*, *A. tritici* (juvenile), *Meloidogyne naasi* (juvenile), *Globodera rostochiensis* (juvenile)	Wyss and Grootaert, 1977; Grootaert and Small, 1982; Small and Grootaert, 1983; Esser, 1987
Mesodorylaimus bastiani	*M. incognita* (juvenile), *H. mothi* (juvenile), *X. basiri*, *X. indicus*, *X. americanum*, *X. insigne*, *Longidorus*, *T. mashoodi*, *H. oryzae*, *H. indicus*, *Aphelenchoides*, *Basiria*, *A. avenae*, *T. semipenetrans*, *Trichodorus*, *Paratrichodorus*, *A. tritici* (juvenile), *Longidorus*, *T. mashhoodi*	Bilgrami, 1992
Pungentus monohystera	*T. semipenetrans*	Mankau, 1982
Seinura celeris	*A. avenae*	Hechler and Taylor, 1966
S. demani	*A. bicaudatus*, *A. avenae*	Wood, 1974
S. oliveirae	*A. avenae*	Hechler and Taylor, 1966
S. oxura	*A. avenae*, *Ditylenchus myceliophagus*	Hechler and Taylor, 1966; Cayrol, 1970
S. steineri	*A. avenae*	Hechler and Taylor, 1966

S. tenuicaudata	*M. marioni* (juvenile), *Pratylenchus pratensis*, *Aphelenchus avenae*, *A. parietinus*, *D. dipsaci*, *Heterodera trifoli* (juvenile), *Me. hapla* (juvenile), *Neotylenchus linfordi*	Linford 1937; Linford and Oliviera, 1937; Hechler, 1963

where SR = strike rate of predators; PR = prey resistance; PS = prey susceptibility; EA = number of encounters of predators with prey resulting into attack; AW = number of attacks by predators resulting into prey wounding; E = total number of encounters with the prey.

26.6. Prey Specificity

The host range of predatory nematodes is known from chance observations in Petri plates or from gut contents of preserved material. Mononchs generally lack prey specificity. Essentially, they engulf any organism that can be swallowed completely, including all types of nematodes, rotifers, protozoa, oligocheates and other invertebrates (Bilgrami *et al.*, 1986). They are rapacious, with reports of a single individual mononch killing up to 83 cyst nematode (*Heterodera*) prey per day and another ingesting 1332 prey over its lifespan (Steiner and Heinly, 1922).

Where mononchs are broad in prey specificity, diplogasterids are moderate and resemble entomopathogenic nematodes (EPNs) (e.g. *Steinernema carpocapsae*), with a specificity that is neither excessively wide nor narrow (Gaugler *et al.*, 1997). Thus, diplogasterid predators are not environmentally risky biocides. For example, Chitambar and Noffsinger (1989) reported a 'high degree of specificity' in *Odontopharynx longicaudata* with 6 of 17 nematode prey species readily consumed, but little or no feeding on the remaining 11 species. High prey specificity was also shown by differences in prey population consumed. For example, consumption of *Trichodorus*, *Pratylenchus vulnus* and *Anguina amsinckia* was moderate at 70–78%, whereas it was 100% for *Acrobeloides* and *Anguina pacificae*. Prey selectivity was similarly reported in *Butlerius* and *Mononchoides* (Grootaert *et al.*, 1977; Bilgrami and Jairajpuri, 1989b), with endoparasitic being preferred over ectoparasitic species. Small and Grootaert (1983) examined predation parameters in five species and

Table 26.3. List of plant-parasitic nematodes recorded as prey of diplogasterid predators.

Predators	Prey nematodes	References
Butlerius degrissei	*Aphelenchus avenae*, *A. fragariae*, *Pratylenchus*, *Globodera rostochiensis* (juvenile), *Rotylenchus robustus*	Grootaert *et al.*, 1977; Grootaert and Jaques, 1979; Small and Grootaert, 1983
B. micans	*A. avenae*	Pillai and Taylor, 1968
Fictor anchicoprophaga	*A. avenae*	Pillai and Taylor, 1968
Mononchoides bollingeri	*A. avenae*	Goodrich *et al.*, 1968
M. changi	*A. avenae*	Goodrich *et al.*, 1968
M. longicaudatus	*Meloidogyne incognita* (juvenile), *Anguina tritici* (juvenile), *Tylenchorhynchus mashhoodi*, *Xiphinema americanum*, *Helicotylenchus indicus*, *Longidorus*, *Trichodorus*	Bilgrami and Jairajpuri, 1988a, 1989b
M. fortidens	*M. incognita* (juvenile), *A. tritici* (juvenile), *T. mashhoodi*, *X. americanum*, *H. indicus*, *Longidorus*, *Trichodorus*	Bilgrami and Jairajpuri, 1988a, 1989b

concluded, 'the range of prey nematodes available to predatory nematodes may be more limited than previously thought'. Thus, the available evidence indicates that the earlier argument of Webster (1972) and Jones (1974) that predatory nematodes are unlikely to control plant-parasitic nematodes due to their non-specificity may well be invalid for diplogasterid predators. Our own studies with *Mononchoides* (A.L. Bilgrami, 2002, unpublished data) lend further support to this conclusion.

Because all previous work has been conducted in laboratory or greenhouse assays where predator–prey contact was assured, it may be safe to assume that prey specificity in the field is even greater. Conversely, Yeates (1969) reported that *D. colobocercus* was not selective, but this conclusion was based on only three closely related prey species. This predator is reported to have consumed 180–380 *Bursilla littoralis* prey in 24 h and could ingest five prey in as little as 7 min.

26.7. Cannibalism

Unlike dorylaim, nygolaim and diplogasterid predators, cannibalism is common in mononchs (e.g. *Mylonchulus*, *Mononchus*, *Iotonchus*, *Coomansus*, *Sporonchulus*) (Bilgrami *et al.*, 1986). *P. punctatus* is an exception (Nelmes, 1974). Of the total number of mononchs observed 20% showed cannibalistic tendency (Bilgrami *et al.*, 1986). Thus, the tendency of feeding on their own members is the reason why mononchs have poor biocontrol prospects (Webster, 1972; Bilgrami, 1997). Diplogasterid predators rarely resort to cannibalism in the presence of prey nematodes. However, they do so when prey density is less or prey is not available (Bilgrami, 1997). Yeates (1969) observed only a single case of cannibalism during his extensive study of *D. colobocercus*.

26.8. Efficacy

Interest in predatory nematodes for biocontrol is long-standing, yet remarkably few efficacy experiments have been reported while these have been restricted to Petri dish or soil pot studies. Boosalis and Mankau (1965) found *Thornia* sp. did not reduce populations of the citrus pest *T. semipenetrans* in pots, even after 29 months. Cohn and Mordechai (1974) found *Mylonchulus* suppressed the population of the reniform nematode in pot experiments. Small (1979) subsequently reported a significant reduction in the populations of *G. rostochiensis* and *M. incognita* in the presence of *P. punctatus*, also in pots, and indicated that further study was clearly justified. Root-knot development caused by *M. incognita* on chilli (*Capsicum annuum*) cv. Jawala significantly declined in the presence of *M. aquaticus*. The severity of root-knot infection was reduced when chopped leaves of neem (*Azadirachta indica*) and castor (*Ricinus communis*) were incorporated into the soil. Improvement in plant growth was positively correlated with the level of nematode control (Akhtar and Mahmood, 1993).

The efficacy of predatory diplogasterids has been suggested from a handful of small, short-term laboratory tests. Studies by Yeates (1969) and Bilgrami and Jairajpuri (1989b) were highly encouraging in demonstrating that species of *Diplenteron* and *Mononchoides* showed prey density dependence, a highly desirable trait in any biocontrol agent. Fauzia *et al.* (1998) found that *M. longicaudatus* significantly reduced galling by root-knot nematodes, resulting in improved vegetative growth and increased root-mass. Osman (1988) showed *Diplogaster* reduced populations of *Meloidogyne javanica* and *T. semipenetrans* juveniles in pots. The rate of predation on parasitic nematodes in laboratory tests was independent of prey species but dependent on prey density.

26.9. Life History

Diplogasterid life history has been well studied. Most of the species are small (~1 mm as adults) yet they have a high fecundity and a short life cycle of approximately 1 week (Tahseen *et al.*, 1990). Pillai and Taylor (1968) recorded progeny production of five females of *Paroigolaimella bernensis* and *Fictor anchicoprophaga* to average 18,000 after 10 days at 20°C. Higher temperatures, however, greatly accelerated growth and development with an astonishing fast generation time of 40–44 h at 35°C without significant loss of reproductive capacity. Reproduction is amphimictic but can be parthenogenic in the absence of males. Yeates (1969) doubted whether copulation with males provided any advantage. Optimum hatch and growth are achieved near 30°C. All post-embryonic stages are predatory.

26.10. Ecology

Diplogasterids differ from entomopathogenic species in one fundamental way: under natural conditions EPNs feed on specific symbiotic bacteria only within the host cadaver, whereas diplogasterids also feed on bacteria besides prey nematodes (Pillai and Taylor, 1968; Yeates, 1969; Bilgrami, 1990; Yeates *et al.*, 1993). Yeates (1969) suggested that detritus–bacterial complex has greater food value for diplogasterid nematodes. It appears that diplogasterid predators can also sustain themselves on nonspecific bacteria in the soil environment when the prey population is at low density. Hassell (1978) observed that switching behaviour buffers predator populations and thereby serves as a 'powerful stabilizing mechanism'. Thus, the significance of the ability of diplogasterids to 'switch' between predator and microbivore feeding modes rests in the anticipated ability to survive periods of low prey densities.

Environmental stress tends to induce the formation of the 'dauer juvenile' (DJ) that enhances the tolerance to extremes of moisture, temperature and chemicals. These DJs are metabolically active and motile, nonageing but developmentally arrested. Only predatory diplogasterids, the cutting and sucking type of predators, have such a resting stage that shares strong similarities with that of EPNs in being induced when conditions are unfavourable and in possessing enhanced survival abilities. Most other differences remain uncertain, as in sharp contrast to the DJs of entomopathogenic species, dauers of diplogasterids have received little attention. Although investigations are awaited, it may be hypothesized that diplogasterid DJs possess some degree of tolerance to anhydrobiotic conditions.

26.11. Culture

The dearth of efficacy studies for predatory nematodes in large part reflects the lack of *in vitro* production methodology. With a few exceptions, predators are reared using *in vivo* methods, which require maintaining concurrent prey cultures, thereby greatly reducing efficiency. The ability to mass-rear EPNs was the catalyst driving their development (Gaugler and Han, 2002). Ease of culture here is due to the ability of entomopathogenic species to feed on symbiotic bacteria, leading ultimately to rearing in 80,000-l bioreactors (Georgis, 2002).

Mononchs were never considered as a good choice of biocontrol agent although they possess significant potential to reduce parasitic nematode populations under field conditions. Predatory mononchs are fastidious to culture due to their localized distribution in the field, long life cycles and low rate of fecundity. In contrast, dorylaim, nygolaim and other stylet-bearing predators provide better candidates for biocontrol since they are widely distributed in fields, occur naturally at high densities and are relatively easy to culture. Their long life cycle is presumed to be the main hurdle for any practical application. Diplogasterids can be reared on either prey nematodes or various bacteria both by *in vivo* or *in vitro* methods since they are facultative and biphasic. *D. colobocercus*, *B. degrissei*, *M. fortidens*, *M. longicaudatus* have been successfully maintained on nematodes like *Caenorhabditis*, *Rhabditis*, *Panagrellus*, *Cephalobus*, bacteria or on a combination of prey nematodes such as bacteria for multiple generations over a period of several months. Pillai and Taylor (1968) cultured diplogasterids on a dixenic culture of bacteria and *Aphelenchus avenae*. Prey nematodes and bacteria have supported growth and development of diplogasterid predators, although some appeared to provide better nematode reproduction than others. In a study on reproductive capacity of *Mononchoides* cultures with 25 adult female nematodes/5.5-cm agar Petri dish was initiated with *Escherichia coli*. After 20 days at 30°C culture plates averaged an impressive 10,376 individuals; rate of oviposition was 8–10 eggs/day/female (A.L. Bilgrami, 2002, unpublished data).

26.12. Conservation

The conservation of predatory nematodes under natural environment is another important aspect that could make their practical utilization possible. The conservation of predatory nematodes appears to be simple and cost-effective in comparison to insects and other beneficial organisms. With relatively little effort their activity may be stimulated and density elevated to counter populations of target pest nematodes. Studies are, however, lacking, justifying pessimism to developing methods to conserve predatory nematodes under their natural habitats. Few studies on soil amendments with neem (*A. indica*) products such as leaf powder, sawdust and oilseed cake showed encouraging results in maintaining and conserving predatory nematode populations in the field (Akhtar, 1995). Winter mulching could be another option to improve predatory nematode preservation in the field. This technique has been found to be effective in stabilizing the populations of *Iotonchus tenuicaudatus* that feed on *T. semipenetrans* and *Helicotylenchus dihystera* in orange orchards (Rama and Dasgupta, 1998). More studies are needed on the role of organic soil amendments and nitrogenous compounds in predatory nematode conservation.

26.13. Future Prospects

Predatory nematodes represent a small amount of the available biomass in the soil, but their presence across so many trophic levels, e.g. plant, fungal- and bacterial-feeding nematodes and carrion feeders besides other microorganisms, is vitally important in soil ecosystem processes (Barker and Koenning, 1998). Their future role in nematode management depends greatly on advances made on other control methods, their effectiveness and the resources provided to establish research programmes.

Biocontrol potential and efficacy of predatory nematodes vary with their type. Mononchs because of their long life cycle, low rate of fecundity, susceptibility to changing environmental conditions, difficulty of mass-production and cannibalism do not fulfil requirements of an efficient biocontrol agent. On the other hand, the advantage of dorylaim, nygolaim and other stylet-bearing predators lies in their widespread natural occurrence (200–500 million/acre (Thorne, 1930)) and ease of mass culturing due to their polyphagous feeding habits. Also, their populations may be elevated in the field by adding organic nutrients. However, more field studies are needed to test this hypothesis. Their long life cycle and low rate of fecundity are also causes of concern.

The real possibility of using predatory nematodes in nematode management programmes lies in the diplogasterid predators, due to their biphasic feeding, high rates of predation and fecundity, short life cycle, ability to search for prey and the presence of resistant DJs. Diplogasterid predators, rarely resort to cannibalism due to their bacteriophagous feeding habits. Despite remarkable similarities with the attributes of EPN species, diplogasterids should not be considered as unilateral inundative agents (i.e. repeated applications for short-term control). The flexible biphasic feeding behaviour of diplogasterids should endow them with superior persistence; i.e. when prey become scarce they should switch to feeding on soil bacteria to maintain themselves. Nematode predators are likely to offer the most promise as augmentative agents in colonization efforts in combination with cultural control tactics, such as rotation, cover cropping, green manuring, organic amendments and plant resistance.

References

Ahmad, N. and Jairajpuri, M.S. (1982) Population fluctuations of the predatory nematode *Parahadronchus shakili* (Jairajpuri, 1969) (Mononchida). *Proceedings of the Symposium on Animal Population, India* 3, 1–12.

Akhtar, M. (1995) Biological control of the root-knot nematode *Meloidogyne incognita* in tomato by the predatory nematode *Mononchus aquaticus. International Pest Control* 37, 18–19.

Akhtar, M. and Mahmood, I. (1993) Effect of *Mononchus aquaticus* and organic amendments on *Meloidogyne incognita* development on chilli. *Nematologia Mediterranea* 21, 251–262.

Andrassy, I. (1964) Subwasserbematoden aus groben *Gebrigsgegenden ostafrikas. Acta Zoologica Academiae Scientiarum Hungaricae* 10, 1–59.

Arpin, P. (1976) Etude et discussion sur un milieu de culture pour Mononchidae (Nematoda). *Revue de Ecologie et de Biologie du Sol* 13, 629–634.

Arpin, P. (1979) Ecologie et systematique des nematodes Mononchidae des zones foresteries et harpaeces zones climate, temperate humid I. Types de sol et groupements specifiques. *Revue de Nematologie* 4, 131–143.

Azmi, M.I. (1983) Predatory behaviour of nematodes: biological control of *Helicotylenchus dihystera* through the predaceous nematode, *Iotonchus monhystera. Indian Journal of Nematology* 13, 1–8.

Banage, W.B. (1963) The ecological importance of free living soil nematodes with special reference to those of moorland soil. *Journal of Animal Ecology* 32, 133–140.

Barker, K.R. and Koenning, S. (1998) Developing sustainable systems for nematode management. *Annual Review of Phytopathology* 36, 165–205.

Bilgrami, A.L. (1990) Diplogasterid predators. In: *Nematode Bio-control: Aspects and Prospects.* Jairajpuri, Alam and Ahmad (eds). CBS Publishers and Distributors, New Delhi, India, pp. 135–142.

Bilgrami, A.L. (1992) Resistance and susceptibility of prey nematodes to predation and strike rate of the predators, *Mononchus aquaticus, Dorylaimus stagnalis* and *Aquatides thornei. Fundamental and Applied Nematology* 15, 265–270.

Bilgrami, A.L. (1993) Analyses of relationships between predation by *Aporcelaimellus nivalis* and different prey trophic categories. *Nematologica* 39, 356–365.

Bilgrami, A.L. (1995) Numerical analysis of the relationship between the predation by *Mesodorylaimus bastiani* (Nematoda: Dorylaimida) and different prey trophic categories. *Nematologia Mediterranea* 23, 81–88.

Bilgrami, A.L. (1997) *Nematode Biopesticides.* Aligarh University Press, Aligarh, India, 262 pp.

Bilgrami, A.L. and Jairajpuri, M.S. (1984) Cannibalism in *Mononchus aquaticus. Indian Journal of Nematology* 14, 202–203.

Bilgrami, A.L. and Jairajpuri, M.S. (1988a) Attraction of *Mononchoides longicaudatus* and *M. fortidens* (Nematoda: Diplogasterida) towards prey and factors influencing attraction. *Revue de Nematologie* 11, 195–202.

Bilgrami, A.L. and Jairajpuri, M.S. (1988b) Aggregation of *Mononchoides longicaudatus* and *M. fortidens* (Nematoda: Diplogasterida) at feeding sites. *Nematologica* 34, 119–121.

Bilgrami, A.L. and Jairajpuri, M.S. (1989a) Resistance of prey to predation and strike rate of the predators *Mononchoides longicaudatus* and *M. fortidens* (Nematoda: Diplogasterida). *Revue de Nematologie* 12, 45–49.

Bilgrami, A.L. and Jairajpuri, M.S. (1989b) Predatory abilities of *Mononchoides longicaudatus* and *M. fortidens* (Nematoda: Diplogasterida) and factors influencing predation. *Nematologica* 35, 475–488.

Bilgrami, A.L. and Kulshreshtha, R. (1994) Evaluation of predation abilities of *Mylonchulus dentatus. Indian Journal of Nematology* 23, 191–198.

Bilgrami, A.L. and Pervez, R. (2000) Prey searching and attraction behaviours of *Mesodorylaimus bastiani* and *Aquatides thornei* (Nematoda: Dorylaimida). *International Journal of Nematology* 10, 199–206.

Bilgrami, A.L., Ahmad, I. and Jairajpuri, M.S. (1986) A study on the intestinal contents of some mononchs. *Revue de Nematologie* 9, 191–194.

Bilgrami, A.L., Pervez, R., Yoshiga, T. and Kondo, E. (2000) Feeding, attraction and aggregation behaviour of *Mesodorylaimus bastiani* and *Aquatides thornei* at the feeding site using *Hirschmanniella oryzae* as prey. *International Journal of Nematology* 10, 207–214.

Bilgrami, A.L., Pervez, R., Kondo, E. and Yoshiga, T. (2001) Attraction and aggregation behaviour of *Mesodorylaimus bastiani* and *Aquatides thornei* (Nematoda: Dorylaimida). *Applied Entomology and Zoology* 36, 243–249.

Bilgrami, A.L., Khan, Z., Kondo, E. and Yoshiga, T. (2002) Generic diversity and community dynamics of nematodes with particular reference to predaceous nematodes at deciduous forest of Henukuma, Saga Prefecture, Japan. *International Journal of Nematology* 12, 46–54.

Bilgrami, A.L. Wenju, L., Wang, P. and Qi Li. (2003) Generic diversity, population structure and community ecology of plant and soil nematodes. *International Journal of Nematology* 13, 104–117.

Boosalis, M. and Mankau, R. (1965) Parasitism and predation of soil microorganisms. In: Baker, F. and Synder, W. (eds) *Ecology of Soil-Borne Plant Pathogens*. University of California Press, Berkeley, California, 374–391.

Carvalho, J.C. (1953) *Mononchus*-um predator voraz. *Revista do Instituto Adolfo Lutz* 13, 75–81.

Cassidy, G.H. (1931) Some mononchs of Hawaii. *Hawaiian Planters Records* 35, 305–339.

Cayrol, J.C. (1970) Action des autres composants de la biocenose du champignon de couche sur le nematode mycophage, *Ditylenchus myceliophagus*. J.B. Goodey, 1958 et etude de son anabiose: Forme de survie en conditions defavorables. *Review of Ecological and Biological Sciences* 7, 409–440.

Chitambar, J.J. and Noffsinger, M. (1989) Predaceous behavior and life history of *Odontopharynx longicaudata* (Diplogasterida). *Journal of Nematology* 21, 284–291.

Choudhary, B.N and Sivakumar, C.V. (2000) Biocontrol potential of *Mylonchulus minor* against some plant parasitic nematodes. *Annals of Plant Protection Sciences* 8, 53–57.

Cobb, N.A. (1917) The *Mononchus*: a genus of free living predatory nematodes. *Soil Science* 3, 431–486.

Cobb, N.A. (1929) Nemas of the genus *Dorylaimus* attacking the edges of mites. *Journal of Parasitology* 15, 284 (Abstr.).

Cohn, E. and Mordechai, M. (1973) Biological control of the citrus nematode. *Phytoparasitica* 1, 32 (Abstr.).

Cohn, E. and Mordechai, M. (1974) Experiments in suppressing citrus nematode populations by use of marigold and predaceous nematode. *Nematologia Mediterranea* 2, 43–53.

Coomans, A. and Lima, B. (1965) Description of *Anatonchus amiciae* n. sp. (Nematoda: Mononchidae) with observations on its juvenile stages and anatomy. *Nematologica* 11, 413–431.

Doucet, M.E. (1980) Description d'une nourelle espece de genere *Mylonchulus* (Nematoda: Dorylaimida). *Nematologia Mediterranea* 8, 37–42.

Esser, E. (1987) Biological control of nematodes by nematodes I. Dorylaims (Nematoda: Dorylaimida). *Nematology Circular* No. 144, 4.

Fauzia, M., Jairajpuri, M.S. and Khan, Z. (1998) Biocontrol potential of *Mononchoides longicaudatus* on *Meloidogyne incognita* on tomato plants. *International Journal of Nematology* 8, 89–91.

Ferris, V.R. (1968) Biometric analysis in the genus *Labornema* (Nematoda: Dorylaimida) with a redescription of *L. thornei* n. sp. *Nematologica* 14, 276–284.

Gaugler, R. and Han, R. (2002) Mass production technology. In: Gaugler, R. (ed.) *Entomopathogenic Nematology*. CAB International, Wallingford, UK, pp. 289–310.

Gaugler, R., Lewis, E.E. and Stuart, R.J. (1997) Ecology in the service of biological control. *Oecologica* 109, 483–489.

Georgis, R. (2002). The Biosys experiment: an insider's perspective. In: Gaugler, R. (ed.) *Entomopathogenic Nematology*. CAB International, Wallingford, UK, pp. 357–372.

Goodrich, M., Hechler, H.C. and Taylor, D.B. (1968) *Mononchoides changi* n. sp., and *M. bollingeri* n. sp. (Nematoda: Diplogasteridae) from a waste treatment plant. *Nematologica* 14, 25–36.

Grewal, P.S., Siddiqi, M.R. and Atkey, P.T. (1991) *Aphelenchoides richardsoni* sp. nov. and *Seinura paynei* sp. nov. from mushrooms in the British Isles and *S. obscura* sp. nov. from India (Nematoda: Aphelenchida). *Afro-Asian Journal of Nematology* 1, 204–211.

Grootaert, P. and Jaques, A. (1979) Description of *Butlerius degrissei* n. sp. (Nematoda: Diplogasteridae). *Nematologica* 25, 203–214.

Grootaert, P. and Maertens, D. (1976) Cultivation and life cycle of *Mononchus aquaticus*. *Nematologica* 22, 173–181.

Grootaert, P. and Wyss, U. (1979) Aspects of the biology of *Labronema Vulvapapillatum* (Meyl) (Nematoda, Dorylaimidae) in laboratory culture. *Biologisch Jaarboek Dodonaea* 50, 135–148.

Grootaert, P., Jaques, A. and Small, R.W. (1977) Prey selection in *Butlerius* sp. (Rhabditida: Diplogasteridae). *Mededelingen van de Faculteit Landbouwwetenschappen Rijksuniversiteit Gent* 42, 1559–1563.

Hassell, M.P. (1978) *The Dynamics of Arthropod Predator-Prey Systems*. Monographs in Population Biology 13, Princeton University Press, Princeton, New Jersey, 25 pp.

Hechler, H.C. (1963) Description, developmental biology and feeding habits of *Seinura tenuicaudata* (de Man), a nematode predator. *Proceedings of the Helminthological Society of Washington* 30, 183–195.

Hechler, H.C. and Taylor, D.P. (1966) The life cycle histories of *Seinura celeries, S. olivierae, S. oxura* and *S. steineri* (Nematoda: Aphelenchoididae). *Proceedings of the Helminthological Society of Washington* 33, 71–83.

Jairajpuri, M.S. and Azmi, M.I. (1978) Some studies on the predatory behaviour of *Mylonchulus dentatus*. *Nematologia Mediterranea* 6, 205–212.

Jairajpuri, M.S. and Bilgrami, A.L. (1990) Predatory nematodes. In: Jairajpuri, M.S., Alam, M.M. and Ahmad, I. (eds) *Nematode Bio-Control: Aspects and Prospects*. CBS Publishers, New Delhi, pp. 95–125.

Jones, F.G.W. (1974) Control of nematode pests, background and outlook for biological control. In: Price-Jones, D. and Solomon, M.E. (eds) *Biology in Pest and Disease Control*. Oxford Press, The British Ecological Society, UK, pp. 249–268.

Kerry, B. (2000) Rhizosphere interactions and the exploitation of microbial agents for the biological control of plant-parasitic nematodes. *Annual Review of Phytopathology* 38, 423–441.

Khan, Z., Bilgrami, A.L. and Jairajpuri, M.S. (1991) Some studies on the predation abilities of *Aporcelaimellus nivalis* (Nematoda: Dorylaimida). *Nematologica* 37, 333–342.

Khan, Z., Bilgrami, A.L. and Jairajpuri, M.S. (1995a) Observations on the predation abilities of *Neoactinolaimus agilis* (Dorylaimida: Actinolaimoidea). *Indian Journal of Nematology* 25, 129–135.

Khan, Z., Bilgrami, A.L. and Jairajpuri, M.S. (1995b) A comparative study on the predation by *Allodorylaimus americanus* n. sp. and *Discolaimus silvicolus* (Nematoda: Dorylaimida) on different species of plant parasitic nematodes. *Fundamental and Applied Nematology* 18, 99–108.

Kulshreshtha, R., Bilgrami, A.L. and Khan, Z. (1993) Predation abilities of *Mylonchulus minor* and factors influencing predation. *Annals of Plant Protection Sciences* 1, 79–84.

Linford, M.B. (1937) The feeding of some hollow stylet bearing nematodes. *Proceedings of the Helminthological Society of Washington* 4, 41–46.

Linford, M.B. and Oliviera, J.M. (1937) The feeding of hollow spear nematodes on other nematodes. *Science* 85, 295–297.

Maertens, D. (1975) Observations on the life cycle of *Prionchulus punctatus* (Cobb, 1917) and culture conditions. *Biologische Jaarboek Dodonaea* 43, 197–218.

Mankau, R. (1980) Biological control of nematode pests by natural enemies. *Annual Review of Phytopathology* 18, 415–440.

Mankau, R. (1982) Biological control of nematode pests by natural enemies. *Annual Review of Phytopathology* 18, 415–440.

Marinari, A., Vinciguerra, M.T., Vovlus, V. and Zullini, A. (1982) Nematodi delle dune costiere d'Italia. In: struttura delle zoocenosi terrestri 3. *Anbiente Mediterranei Le Coste Sabbiose*, 27–50.

Menzel, R. (1920) Uber die Nahrung der freilebenden nematoden und die Art ihrer Aufnahme. Ein Beitrag zur Kenntnis der Krnahrung der wurmer. *Verhandle. Nturf. Ges. Basel* 31, 153–188.

Mohandas, C. and Prabhoo, N.R. (1980) The feeding behaviour and food preferences of predatory nematodes (Mononchida) from the soils of Kerala (India). *Review of Ecology and Biology of Soil* 17, 53–60.

Mulvey, R.H. (1961) The Mononchidae: a family of predaceous nematodes. I. Genus *Mylonchulus* (Enoplida: Nematoda). *Canadian Journal of Zoology* 39, 665–696.

Nelmes, A.J. (1974) Evaluation of the feeding behaviour of *Prionchulus punctatus* (Cobb) a nematode predator. *Journal of Animal Ecology* 43, 553–565.

Nelmes, A.J. and McCulloch, J.S. (1975) Number of mononchid nematodes in soil sown to cereals and grasses. *Annals of Applied Biology* 79, 231–242.

Osman, G.Y. (1988) Studies on the potential use of the predator, *Diplogaster* sp. (Nematoda, Diplogasteridae) on certain root-parasitic nematodes. *Anzeiger fur Schadlingskunde Pflanzenschutz Umweltschutz* 61, 70–73.

Pillai, J. and Taylor, D. (1968) Biology of *Paroigolaimella bernensis* and *Fictor anchicoprophaga* (Diplogasteridae) in laboratory culture. *Nematologica* 14, 159–170.

Poinar, G.O., Triggiani, O. and Merritt, R. (1976). Life history of *Eudiplogaster aphodii*, a facultative parasite of *Aphodius fimet arius*. *Nematologica* 22, 79–86.

Rama, K. and Dasgupta, M.K. (1998) Biocontrol of nematodes associated with mandarin orange decline by the promotion of predatory nematode *Iotonchus tenuicaudatus* (Kreis, 1924). *Indian Journal of Nematology* 28, 118–124.

Shafqat, S., Bilgrami, A.L. and Jairajpuri, M.S. (1987) Evaluation of the predatory behaviour of *Dorylaimus stagnalis* (Nematoda: Dorylaimida). *Revue de Nematologie* 10, 455–461.

Small, R.W. (1979) The effects of predatory nematodes on populations of plant parasitic nematodes in pots. *Nematologica* 25, 94–103.

Small, R.W. and Grootaert, P. (1983) Observations on the predatory abilities of some soil dwelling predatory nematodes. *Nematologica* 29, 109–118.

Steiner, G. and Heinley, H. (1922) The possibility of control of *Heterodera radicolola* and other plant injurious nematodes by means of predatory nemas, especially *Mononchus papillatus*. *Journal of Washington Academy of Science* 12, 367–385.

Szczygiel, A. (1966) Studies on the fauna and population dynamics of nematodes occurring on strawberry plantations. *Ekologia Polaska* 14, 651–709.

Szcyzygiel, A. (1971) Wystepowanie drapieznych nicieni zrodzniy Mononchidae w glebach uprawnych w polsce. *Zeszyty Problemowe Postepow Nauk Rolniczych* 121, 145–158.

Tahseen, Q., Jairajpuri, M.S. and Ahmad, I. (1990) The life cycle of *Mononchoides fortidens* with emphasis on gonad development. *Nematologica* 36, 440–447.

Thorne, G. (1924) Utah nematodes of the genus *Mononchus*. *Trans American Microscope Society* 43, 157–171.

Thorne, G. (1927) The life history, habits and economic importance of some mononchs. *Journal of Agricultural Research* 34, 265–286.

Thorne, G. (1930) Predaceous nemas of the genus *Nygolaimus* and new genus *Sectonema*. *Journal of Agricultural Research* 44, 445–446.

Thorne, G. (1932) Specimens of *Mononchus acutus* found to contain *Trichodorus obtusus, Tylenchulus robustus* and *Xiphinema americanum*. *Journal of Parasitology* 19, 90.

Thorne, G. and Swanger, H.H. (1936) A monograph of the nematode genera *Dorylaimus, Aporcelaimus, Dorylaimoides* and *Pungentus*. *Capita Zoologica* 6, 1–233.

van der Heiden. (1974) The structure of the anterior feeding apparatus in members of the Ironidae (Nematoda: Enoplida). *Nematologica* 20, 419–436.

Webster, J.M. (1972) Nematodes and biocontrol. In: Webster, J.M. (ed.) *Economic Nematology*. Academic Press, London, pp. 469–496.

Wood, F.H. (1974) Biology of *Seinura demani* (Nematoda: Aphelenchoididae). *Nematologica* 20, 347–353.

Wyss, U. (1971) Der Mechanismus der Nahrungsaufnahme bei *Trichodorus similes*. *Nematologica* 7, 508–518.

Wyss, U. and Grootaert, P. (1977) Feeding mechanisms of *Labornema* sp. *Mededelingen van de Faculteit Landbouwwetenschappen Rijksuniversiteit Gent* 42, 1521–1527.

Yeates, G.W. (1969) Predation by *Mononchoides potohikus* in laboratory culture. *Nematologica* 15, 1–9.

Yeates, G.W., Bongers, T., De Goede, R.G.M., Freckman, D.W. and Georgieva, S.S. (1993) Feeding habits in soil nematode families and genera – an outline for soil ecologist. *Journal of Nematology* 25, 315–331.

Part VI

Fungal-feeding Nematodes

27 Potential of Fungal-feeding Nematodes for the Control of Soil-borne Plant Pathogens

N. Ishibashi
Saga University, 1090–3, Kinryu-Chifu, Saga, 849-0905, Japan

27.1. Introduction

Nematode species including *Aphelenchus* spp., *Filenchus* spp., *Tylenchus* spp. and *Iotonchium* spp. have been reported to feed on fungi (Wood, 1973; Grewal, 1991; Mizukubo, 1993; Brzeski, 1997; Karegar and Geraert, 1998; Chen and Ferris, 2000; Tsuda and Futai, 2000; Okada, 2001; Okada *et al.*, 2002). Although fungivorous nematodes have been viewed to play an important role in organic matter decomposition, only a few studies have focused on their role as biocontrol agents for fungal pathogens of plants (Barker, 1964; Evans, 1970; Barnes *et al.*, 1981; Caubel *et al.*, 1981; Beagle-Ristanio and Paravizas, 1985; Choo and Estey, 1985; Choi *et al.*, 1988; Choi and Ishibashi, 1989; Ishibashi and Choi, 1991). Most of these studies have concentrated on one species, *Aphelenchus avenae* Bastian 1865, which is ubiquitous in the temperate zone and has been cultured experimentally on 76 species of fungi (Townshend, 1964; N. Ishibashi, 2003, unpublished data). This chapter describes recent research on *A. avenae* and demonstrates its potential for the control of soil-borne fungal diseases of plants.

27.2. Mass-production

27.2.1. Solid substrate culture

Agricultural and food industry wastes such as lees from sugarcane (bagasse), fruit juice and distillery wastes, chaff, bran, brewer's grains, beet pulp, potato chip waste and used green tea leaves and grounds can be used as substrates to cultivate fungi and fungal-feeding nematodes, provided temperature and humidity can be controlled. Nematode yield is dependent on the suitability of the fungus species cultivated (Ishibashi *et al.*, 2000). Although most fungi can grow on the above vegetable substrates, multiplication efficiency may differ with fungal species. The quality of the media can be improved by mixing more than one substrate. For example, beet pulp mixed with a tea-based substrate (green or oolong) can be beneficial because mixing the plurality of substrates buffers against any increase in pH during nematode cultivation as the used green or oolong tea leaves or tea grounds absorb the generated ammonia, effectively retarding the pH increase (Ishibashi *et al.*, 2000). Solid substrates must have a high degree of porosity to allow the fungi and nematodes to penetrate into the medium, otherwise reproduction only occurs on the surface and yield is reduced. Bagasse, beet pulp, chaff or potato chip wastes have adequate porosity and can be used as base substrates. When bagasse is used as the base substrate, a preferred formulation (on a dry weight basis) consists of equal proportions of bagasse, used green tea leaves or tea grounds and brewer's grains or potato chip wastes. The water content is usually 60% in any substrate. During the cultivation period (approximately 30 days), the pH range should be kept at 5.5–5.8, preferably 5.5 with Sörensen phosphate buffer ($NaHPO_4$ 8.91 g;KH_2PO_4 0.53 g/l H_2O).

Once the media is autoclaved (1 kg (wet weight) in a 5-l jar), the host fungi and nematodes are inoculated. The inoculum (nematodes plus *Rhizoctonia solani* or *Botrytis cinerea*) is prepared on agar plates. The hyphae of *R. solani* AG-4 grow rapidly forming a fungal mat on potato dextrose agar (PDA), and the seed nematodes (pellet containing *c.* 10,000 nematodes on a Millipore filter membrane) are added to the Petri dish at the same time as the fungi. *B. cinerea* grows more slowly than *R. solani*; thus the nematodes may be added 3–4 days after the fungus. If the nematodes are collected from soil, they should be surface-sterilized in a 1000 ppm streptomycin solution for 30 min, centrifugally washed with sterilized water three times and inoculated on to a fungal mat on PDA in a Petri dish. After about 30 days, multiplied nematodes can be inoculated along with agar substrate (diameter *c.* 3–5 cm) on to the above-mentioned media in 5-l jars. From 20 g of solid medium (water content 60%, dry weight *c.* 7 g) in a 100-ml culture vessel, about 4×10^6 nematodes were harvested after 3 weeks with a feed of *B. cinerea* (Ishibashi *et al.*, 2000). On *R. solani*, in a 5-l jar with 1 kg substrate (wet weight; 1 part of potato chip waste, 1 part of beet pulp and 1 part of tea grounds with a water content of 60%), 57×10^7 nematodes are usually produced. One merit of the solid waste substrate method may be the recycling of used media (after nematode harvest) as organic fertilizers; moreover, inadvertent establishment of nematode populations in the field might be expected because the used substrates still contain many nematodes that may become established in the soil. Although this approach is intriguing, the 5-l jars occupy a great deal of space, which severely limits the efficiency of the process. More efficient processes are needed in terms of economy of scale.

27.2.2. Semi-solid substrate culture

This method aims at mass production of the nematodes using a fermenter. The entire substrate cannot be a liquid because the nematodes attach their stylets perpendicularly to the hyphal tissue to feed. Therefore, a solid substrate is impregnated with a liquid medium for propagation of the fungus (Ishibashi *et al.*, 2000). Spongy materials, such as multiporous resins (e.g. formed polyurethane, microporous polyvinyl formal resins), or fibrous aggregates (e.g. organic or inorganic fibre linters, woven or non-woven fabrics), animal or vegetable microporous materials and solid materials having desirable water absorbency and water retention, are applicable. These materials are shredded using a blender and aggregated into an appropriate size, sufficiently washed with water and dried before use. The standard PDA or potato sucrose agar (PSA) medium with the agar removed (thus termed potato-dextrose solution), a phosphate-buffered saline used in lieu of distilled water, and additional starch, dextrose or sucrose, serves as the liquid medium. The liquid medium is preferred to have a pH in the range of 5.6–5.8.

Yields of *A. avenae* on the polyester–polyurethane sponge (PPS) substrate permeated with potato-dextrose solution with various quantities of potato and dextrose are shown in Table 27.1. On the sponge substrate yield also depends on the compositions and amount of potato-dextrose solution added per unit weight of sponge. The highest population (108.3×10^4) of *A. avenae* was achieved 30 days after inoculation on 20 g PPS substrate permeated with potato-dextrose solution made with 800 g potato soup and 80 g dextrose/l at the rate of 16 ml/g PPS.

Bulk quantities of prepared PPS impregnated with the above liquid medium can be used to grow *A. avenae* in a fermenter. A 5-l vessel will hold 2 kg (wet weight) of PPS impregnated with liquid nutrients. The prepared PPS is placed in the fermenter and subjected to sterilization at 120°C. A piece of host fungal mat and the nematodes (dried pellet with *c.* 10,000 individuals) are inoculated on the medium in the fermenter. However, since the fungal growth on the liquid medium is not so fast as on the solid medium, the host fungus (even *R. solani*) should be placed on the medium before the nematode inoculation. The incubation is conducted with addition of sterilized air. The amount of aeration may be the minimum amount recommended for the particular container size used. Additionally, the aeration may be stopped occasionally, e.g. aeration may not be necessary for about 1 week after the inoculation of the fungus. When nets of nematodes are visible on the inside surface glass wall (after about 30 days), the nematodes are washed out of the container and sifted on 100- and 400-mesh sieves. The expected yield of nematodes is at least 100 million. In addition to high yields, the culture on PPS with artificial

Table 27.1. Propagation of *Aphelenchus avenae* on sponge impregnated with various quantities of potato and dextrose solution. (From Ishibashi *et al.*, 2000.)

Potato:dextrose	Number of nematodes ($\times 10^4$) at days after inoculation		
(g/l)	10	20	30
200:20	5.2a	40.8c	44.0c
400:20	4.2a	53.8d	65.1c
800:20	0.8a	33.3bc	69.6c
200:40	4.9a	26.8b	15.2b
200:80	4.9a	16.3b	10.6b
400:40	6.2a	67.7c	76.5c
800:80	0.7a	68.5c	108.6e

Data are the means of five replicates. Values followed by the same letter(s) are not significantly different according to *t*-test at 5% level. Two hundred *A. avenae* (mostly fourth stage juvenile and adults) were introduced to 100 ml flasks containing 20 g of potato dextrose solution at the rate of 16 ml/g sponge. *Rhizoctonia solani* AG-4 was inoculated simultaneously with the nematodes and incubated at 25°C.

liquid medium is advantageous compared with the aforementioned solid substrates composed of vegetable wastes, due to efficiency of space requirements.

27.3. Formulations

27.3.1. Drying in used solid medium

Nematodes collected from the culture vessel are injected into a block (about $2 \times 2 \times 2$ cm) of used substrates and subjected to 97% RH with saturated K_2SO_4 for 2–3 days at 25°C, then 85% RH with saturated KCl for 4–6 days, finally when the blocks become almost dried; blocks holding 10,000–50,000 nematodes each can be kept at 30–50% RH in a desiccation chamber until use. Each block can be applied to a nursery pot or planting spot.

27.3.2. Drying on filter papers

A nematode suspension in 0.01% formalin containing *c.* 10,000 nematodes is filtered by an aspirator on to a filter paper in a separable suction funnel. Nematodes on a filter paper are exposed to 97% RH for 1–2 days, then at 88% RH (with saturated $ZnSO_4{:}7H_2O$) at 25°C for 2–3 days. The desiccated nematodes can then be stored in a refrigerator at 5°C. The filter paper with nematode pellet can be directly placed on a fungal mat as an inoculum.

27.3.3. Drying with inert materials

Concentrated nematodes (several million) collected by suction funnel are mixed with 100 g of dried fine vermiculite, talc or perlite. The moisture content will be 30–40%. The mixture is gradually dried for a week. When used, the mixture will be blended with soil and applied to furrows before planting.

27.4. Efficacy

27.4.1. Control of fungal diseases in sterilized soil

The fungi *R. solani* AG-4, *Fusarium oxysporum* f. sp. *lagenariae*, *Pythium* sp. and *Phytophthora nicotianae* var. *parasitica* were cultured for 2 weeks on wheat bran media. One gram of medium containing the cultured fungi was inoculated with 10,000, 50,000 or 100,000 *A. avenae* into polystyrol pots with mixed soil (300 ml autoclaved mixed soil, water content 50%, of 15 parts of sandy loam and 1 part of vermiculite). The soil in the pot was blended, and the pots were closed and placed in the dark at 25°C for a week. Then, four grains of surface-sterilized cucumber seeds were sown per pot. Pots without fungi or nematodes were used as controls. The experiment contained ten replicates. Percentage plant survival was determined 5 and 14 days after sowing. In this test, because the sterilized soil was used, survival was generally lower in the nematode-alone treatment and decreased as nematode rate increased (Table 27.2). However, when nematodes and fungi were inoculated together, substantial plant protection was achieved. The best results were observed with 10,000 nematodes (70–80% plant survival). No seeds germinated in any of the fungus-alone treatments.

27.4.2. Control of soil fungal diseases in non-sterilized soil

R. solani AG-4 III-A was cultured in liquid potato dextrose benomyl medium. Hyphae were collected and divided into four groups: 0.375, 0.75, 1.5 and 3 g/l of distilled water, then mixed with 15 l non-sterilized loamy soil, corresponding to 20, 50, 100 and 200 mg/l soil, respectively. These corresponded to 25, 50, 100, 200 mg a.i./100 ml soil. Suspensions containing 1.25×10^2, 1.25×10^3, 1.25×10^4, 2.5×10^4 and

Table 27.2. Effects of different concentrations of *Aphelenchus avenae* on the survival of cucumber plants inoculated with different soil-borne fungal pathogens 14 days after sowing.

	% survival of plants			
	Number of *A. avenae*/300 ml soil			
Pathogens	0	10,000	50,000	100,000
None	100.0a	79.3b	67.5b	43.2c
Rhizoctonia solani AG-4	0d	82.5b	84.1b	51.0c
Fusarium oxysporum f. sp. *lagenaria*	0d	72.2b	77.6b	52.4c
Pythium sp.	0d	76.7b	80.0b	49.0c
Phytophthora nicotianae var. *parasitica*	0d	76.5b	81.2b	54.7c

The numerals with same letter(s) are not significantly different by Turkey's test at 5% level.

5×10^4 nematodes/100 ml soil were prepared. The soil was naturally contaminated with *Pythium* sp. As a chemical control, the fungicide, Flutolanil, was used at a recommended rate of 25 mg/l soil to control *R. solani*. Two days after treatment, ten sprouting cucumber seeds per pot were sown with five replicates. Plant survival was recorded 10 days after sowing and the results are shown in Table 27.3. At the 25–50 mg/l soil *R. solani* density, which is within the fungus's natural range, the nematodes caused substantial control of *R. solani* even with low nematode densities (1.25×10^3–1.25×10^4/l soil). Plant survival was higher in the *A. avenae* treatments than Flutolanil at the recommended dosage. This may be attributed to the resurgence of *Pythium* sp. in the chemical treatment, since Flutolanil is specific to *R. solani*, not to *Pythium* spp. That is, the application of *A. avenae* controlled *Pythium* sp. as well as *R. solani*, while Flutolanil controlled only *R. solani*, so that *Pythium* sp. increased in lieu of *R. solani*.

27.5. Inhibition of Plant-parasitic Nematode Root Penetration

The attraction of *A. avenae* to plant roots impedes colonization of plant-parasitic nematodes (Matsunaga *et al.*, 1996). Table 27.4 shows the penetration of *Pratylenchus coffeae* and *Meloidogyne incognita* in cucumber transformed hairy roots on agar plate in the presence of *A. avenae*. The reduction in the penetration rate of plant-parasitic nematodes is also induced by the

Table 27.3. Effects of different concentrations of *Aphelenchus avenae* on the survival of cucumber plants inoculated with different densities of *Rhizoctonia solani* 10 days after sowing.

	% control			
	Hyphal density (*R. solani*) mg/l soil			
Treatments	200	100	50	25
A. avenae 5.0×10^5/l soil	64.0c	93.6b	—	—
2.5×10^5/l soil	46.0d	87.2c	—	—
1.25×10^5/l soil	6.0e	42.6d	97.6b	100.0a
1.25×10^4/l soil	—	—	95.8b	100.0a
1.25×10^3/l soil	—	—	95.8b	87.0c
Flutolanil 25 mg/l soil	96.0b	97.7b	60.4c	76.1c

Numerals with the same letter(s) are not significantly different according to Tukey's test at 5% level; — = not tested.

Table 27.4. Penetration rate of *Pratylenchus coffeae* (Pc) and *Meloidogyne incognita* (Mi) to cucumber roots as affected by *Aphelenchus avenae* (Aa) when applied simultaneously.

	Number of penetrated nematodes ± SE	
Number of nematodes inoculated	*P. coffeae*	*M. incognita*
100 each without Aa	30.8 ± 2.6 (100)a	34.7 ± 3.7 (100)a
(Pc or Mi) + Aa 100	7.3 ± 1.9 (14.8)b	11.6 ± 2.1 (35.3)b
(Pc or Mi) + Aa 1,000	2.6 ± 1.6 (7.1)c	5.6 ± 1.4 (16.1)c
(Pc or Mi) + Aa 5,000	1.2 ± 0.6 (3.1)d	2.3 ± 0.9 (6.3)d
(Pc or Mi) + Aa 10,000	0.6 ± 0.6 (2.0)e	1.3 ± 0.6 (3.5)e

Numerals in parentheses indicate percentage of control.

application of entomopathogenic nematodes (EPNs) (Bird and Bird, 1986; Choi *et al.*, 1988; Ishibashi and Choi, 1991; Matsunaga *et al.*, 1996; Grewal *et al.*, 1999). However, the mixed application of entomopathogenic and fungivorous nematodes does not give a synergistic effect, but is rather less effective (Ishibashi and Choi, 1991).

27.6. Integrated Control of Soil-borne Fungal Pathogens and Insect Pests

27.6.1. Simultaneous applications with entomopathogenic nematodes (EPNs)

When *A. avenae* nematodes are inoculated in combination with the EPNs *Steinernema* spp., plant survival generally improves (Ishibashi and Choi, 1991). For example, cucumber germinates earlier and the germination is higher than when *A. avenae* is inoculated without EPNs. The mechanism is not known; cucumber seeds may have been stimulated by the EPNs. In order to confirm the efficacy of EPNs against insect pests when applied together with *A. avenae*, several tests were conducted. Table 27.5 indicates that high application rates of *A. avenae* caused a decrease in steinernematid virulence to the turnip moth (which was not very susceptible without addition of *A. avenae*). In contrast, *A. avenae* did not adversely affect control of the more susceptible common cutworm. With a view to the future, it can be expected that the fungivorous nematodes will be for the most part compatible with beneficial nematodes such as steinernematids (Ishibashi and Choi, 1991; Ishibashi, 1993). Certainly,

Table 27.5. Effect of mixed application of fungivorous nematode, *Aphelenchus avenae*, and entomopathogenic nematode (EPN), *Steinernema carpocapsae*, on the mortality of common cutworm, *Spodoptera litura*, and turnip moth, *Agrotis segetum*. (From Ishibashi and Choi, 1991.)

Mix ratio	% mortality of insects						
	S. litura			*A. segetum*			
S. carpocapsae + *A. avenae*	48 h	72 h	96 h	48 h	72 h	96 h	120 h
0 + 0	0a	0a	0a	0a	0a	0a	0a
0 + 5000	0a	0a	0a	0a	0a	0a	0a
50 + 0	85c	95c	100d	50b	80c	90c	90c
50 + 50	95c	100d	100d	40b	60b	80b	90c
50 + 500	95c	100d	—	10a	40b	60b	60b
50 + 5000	95c	100d	—	0a	20b	40b	50b

The mortality suffixed with the same letter is not significantly different at 5% by χ^2 test; — = not tested.

there were no adverse effects on the efficacy of *A. avenae* in combination with EPNs (Ishibashi, 1998).

27.6.2. Simultaneous applications with microbial control agents

The fungivorous nematode, *A. avenae*, is also compatible with *Bacillus thuringiensis*. The exotoxin from the bacteria was previously reported to be detrimental to this nematode as well as to plant nematodes (Ignoffo and Dropkin, 1976). Our experiments, however, showed no adverse effect on *A. avenae* when mixed with *B. thuringiensis*. Likewise, none of the strains or isolates of *Pasteuria penetrans* that we have tested attached to the cuticle of *A. avenae* (Ishibashi, 1998). Root-knot nematodes or root-lesion nematodes and fungal pathogens have been suggested to work synergistically to cause plant disease even at low pathogen population densities (Saeed *et al.*, 1999). The combination of *A. avenae* and *P. penetrans* may offer the possibility of conquering these disease complexes in soils (Ishibashi, 1998). Thus, it is conceivable that mixed applications of *A. avenae* with other beneficial biocontrol agents can be compatible for simultaneous control of soil pests, soil fungal diseases and plant parasitic nematodes.

27.7. Conclusions

The fungivorous nematode, *A. avenae*, can be mass-produced and a long-term preservation method has been established. A number of advantages are associated with the use of *A. avenae* for biocontrol of soil-borne plant diseases. Industrial wastes can be effectively utilized in mass production of the nematodes, and the repeated application of the used substrates (which still contain the nematodes) leads to additional suppression of plant nematodes and soil fungal pathogens such as *Rhizoctonia*, *Pythium*, *Phytophthora*, *Verticillium* or *Fusarium*. Additionally, condensed nematode cakes can be prepared to reduce transportation costs. The biocontrol of soil-borne diseases by *A. avenae* and mixed application with other beneficial nematodes (Ishibashi, 1998) can be effectively utilized in agriculture. Conclusively, this beneficial nematode, *A. avenae*, cannot be used to treat disease-suffering plants, but can be applied as a prophylactic treatment. Applications should be made at the recommended dose because higher application rates may decrease efficacy.

However, when the nematodes are serially subcultured with the same fungus, even if the fungus is the best for the propagation of the nematodes, the fecundity of the nematodes gradually declines (Mankau and Mankau, 1962), eventually rendering the population extinct. This deterioration of the nematodes owing to subculturing can be prevented by storing dried nematodes or by subculturing on several different host fungi. When large numbers of nematodes (more than 1×10^4/100 ml soil) are applied to sterilized soil together with seeds, seed germination may be hindered (Kobayashi and Chikuo, 1993). For example, in small cup experiments with sterilized soil, germination of cucumber and spinach seeds was hindered when planting was concurrent with large numbers of *A. avenae* (N. Ishibashi, 2003, unpublished data). However, detrimental effects on germination may not occur in natural soils due to the presence of other sources of food for the nematodes as well as nematode antagonists. There are many *A. avenae* strains or isolates from various locations, which differ in host preference (Pill and Taylor, 1967; Evans and Fisher, 1970a,b; Ali *et al.*, 1999a,b,c). Differences in virulence among strains may account for substantial discrepancies observed between efficacy studies. A good strain collection is also lacking. In some experiments, nematodes were not applied at an appropriate time and thus the efficacy of nematodes was unclear (Gupta, 1986).

References

Ali, M.R., Amin, B., Adachi, T. and Ishibashi, N. (1999a) Host and temperature preference, male occurrence and morphometrics of fungivorous nematode, *Aphelenchus avenae* isolates from Japan. *Japanese Journal of Nematology* 29, 1–11.

Ali, M.R., Yamaguchi, Y. and Ishibashi, N. (1999b) RAPD and PCR-RFLP analysis on genetic diversity of *Aphelenchus avenae* isolates collected from Kyushu and some other districts of Japan. *Japanese Journal of Nematology* 29, 24–34.

Ali, M.R., Yamaguchi, Y. and Ishibashi, N. (1999c) Esterase isozymes phenotypes of 15 isolates of fungivorous nematode, *Aphelenchus avenae* Bastian 1865, from Kyushu and other districts of Japan. *Japanese Journal of Nematology* 29, 35–41.

Barker, K.P. (1964) On the disease reduction and production of the nematode *Aphelenchus avenae* on isolates of *Rhizoctonia solani*. *Plant Disease Reporter* 48, 428–432.

Barnes, G.L., Russel, C.C., Foster, W.D. and McNew, R.W. (1981) *Aphelenchus avenae*, a potential biological control for root rot fungi. *Plant Diseases* 65, 423–424.

Beagle-Ristanio, J.E. and Paravizas, G.C. (1985) Biological control of *Rhizoctonia* stem canker and black scurf of potato. *Phytopathology* 75, 560–564.

Bird, A.F. and Bird, J. (1986) Observations on the use of insect parasitic nematodes as a means of biological control of root-knot nematodes. *International Journal of Parasitology* 16, 511–516.

Brzeski, M.W (1997) Redescription of some species of the genus *Filenchus* Andrássy, 1954 (Nematoda, Tylenchidae). *Midcel lània Zoologica* 20, 45–64.

Caubel, G., Jouan, B., Queneherve, P. and Radwan, J. (1981) Lutte contre *Rhizoctonia solani* Kuhn, parasite du contonnier par le nematode *Aphelenchus avenae* Bastian. *Review de Nematology* 4, 93–98.

Chen, J. and Ferris, H. (2000) Growth and nitrogen mineralization of selected fungi and fungal-feeding nematodes on sand amended with organic matter. *Plant and Soil* 218, 91–101.

Choi, D.R. and Ishibashi, N. (1989) Propagation of *Aphelenchus avenae* isolates on six species of fungi and five substrates. *Japanese Journal of Nematology* 19, 13–17.

Choi, D.R., Ishibashi, N. and Tanaka, K. (1988) Possible integrated control of soil insect pests, soil-borne diseases, and plant nematodes by mixed application of fungivorous and entomogenous nematodes. *Bulletin of the Faculty of Agriculture*, Saga University No. 65, 27–35.

Choo, P.H. and Estey, H. (1985) Control of the damping-off disease of pea by *Aphelenchus avenae*. *Indian Journal of Nematology* 15, 1–4.

Evans, A.A.F. (1970) Mass culture of mycophagous nematodes. *Journal of Nematology* 2, 99–100.

Evans, A.A.F. and Fisher, J.M. (1970a) Some factors affecting the number and size of nematodes in population of *Aphelenchus avenae*. *Nematologica* 16, 295–304.

Evans, A.A.F. and Fisher, J.M. (1970b) Ecology difference between four Australian isolates of *Aphelenchus avenae* Bastian, 1985. *Australian Journal of Biological Sciences* 23, 507–509.

Grewal, P.S. (1991) Description of *Filenchus recisus* sp. n. and observations on *F. thornei* 'England Population' and *F. ditissimus* (Nematoda: Tylenchidae) from mushroom casing material. *Nematologica* 37, 243–251.

Grewal, P.S., Lewis, E.E. and Venkatachari, S. (1999) Allelopathy: a possible mechanism of suppression of plant-parasitic nematodes by entomopathogenic nematodes. *Nematology* 1, 735–743.

Gupta, M.C. (1986) Biological control of *Fusarium moniliformae* Sheldon and *Pythium butleri* Subramanium by *Aphelenchus avenae* Bastian, 1865 in chitin and cellulose amended soils. *Soil Biology and Biochemistry* 18, 327–329.

Ignoffo, C.M. and Dropkin, V.H. (1976) Deleterious effects of the thermostable toxin of *Bacillus thuringiensis* on species of soil-inhabiting, myceliophagous, and plant-parasitic nematodes. *Journal of Kansas Entomological Society* 50, 394–398.

Ishibashi, N. (1993) Integrated control of insect pests by *Steinernema carpocapsae*. In: Bedding, R., Akhurst, R., and Kaya, H.K. (eds) *Nematodes and the Biological Control of Insect Pests.* CSIRO Publications, East Melbourne, Australia, pp. 105–113.

Ishibashi, N. (1998) Possible integration of beneficial nematodes and microbial agents for biocontrol of soil pests. *Japanese Journal of Nematology* 28 (Special Issue), 22–28.

Ishibashi, N. and Choi, D.R. (1991) Biological control of soil pests by mixed application of entomopathogenic and fungivorous nematodes. *Journal of Nematology* 23, 175–181.

Ishibashi, N., Ali, M.D. and Saramoto, M. (2000) Mass-production of fungivorous nematode, *Aphelenchus avenae* Bastian 1865, on industrial vegetable/animal wastes. *Japanese Journal of Nematology* 30, 8–17.

Karegar, A. and Geraert, E. (1998) Descriptions of *Filenchus paravesiculosus* sp. n. and three other species of the *Filenchus* Andrássy, 1954 (Nematoda: Tylenchidae) from Iran. *Nematologica* 44, 225–239.

Kobayashi, N. and Chikuo, Y. (1993) Biological control of cucumber *Fusarium* wilt by the fungivorous nematode, *Aphelenchus avenae.* In: Ishibashi, N. (ed.) *Detection of Beneficial Nematodes in Japan and Systematized Nematode Application.* Technical Bulletin Grant-in-Aid, Ministry of Education, Japan Grant No. 02506001, Saga University, Saga, Japan, pp. 116–122.

Mankau, R. and Mankau, S.K. (1962) Multiplication of *Aphelenchus avenae* on phytopathogenic soil fungi. *Phytopathology* 52, 741.

Matsunaga, T., Wang, X.D. and Ishibashi, N. (1996) Interference by entomopathogenic and fungivorous nematodes of root invasion by plant-parasitic nematodes. *Japanese Journal of Nematology* 26, 1–11.

Mizukubo, T. (1993) Description of *Filenchus discolabialis* sp. and additional records of *F. helenae* (Szczyfgiel). *Nematologica* 39, 466–475.

Okada, H. (2001) A fungal-feeding nematode of the family Tylenchidae from decomposing rice straw. *Phytopathology* 91 (Suppl.), 140 (Abstr.).

Okada, H., Tsukiboshi, T. and Kadota, I. (2002) Mycetophagy in *Filenchus misellus* (Andrássy, 1958) Lownsbery & Lownsbery, 1985 (Nematoda: Tylenchidae), with notes on its morphology. *Nematology* 4, 795–801.

Pill, J.K. and Taylor, D.P. (1967) Influence of fungi on host preference, host suitability, and morphometrics of five mycophagous nematodes. *Nematologica* 13, 529–540.

Saeed, I.A.M., MacGuidwin, A.E., Rouse, D.I. and Sharkey, T.D. (1999) Limitation to photosynthesis in *Pratylenchus penetrans*- and *Verticillium dahliae*-infected potato. *Crop Science* 39, 1340–1346.

Townshend, J.L. (1964) Fungus hosts of *Aphelenchus avenae* Bastian, 1865, and *Bursaphelenchus fungivorous* Franklin and Hooper, 1962 and their attractiveness to these nematode species. *Canadian Journal of Microbiology* 10, 727–737.

Tsuda, K. and Futai, K. (2000) The insect-parasitic stage and life cycle of *Iotonchium unglatum* (Tylenchida: Iotonchidae) the causal agent of gill-knot disease of the oyster nysgroom. *Japanese Journal of Nematology* 30, 1–7.

Wood, F.H. (1973) Life cycle and host–parasite relationships of *Aphelenchus costanus* (de Man, 1921) Meyl, 1961 (Nematoda: Tylenchidae). *New Zealand Journal of Agricultural Research* 16, 373–380.

Part VII

Conclusions and Future Directions

28 Critical Issues and Research Needs for Expanding the Use of Nematodes in Biocontrol

P.S. Grewal,[1] R.-U. Ehlers[2] and D.I. Shapiro-Ilan[3]
[1]*Department of Entomology, Ohio State University, Wooster, OH 44691, USA;*
[2]*E-nema GmbH, Klausdorfer Straße 28-36, D-24223 Raisdorf, Germany;*
[3]*USDA-ARS, Southeastern Fruit and Tree Nut Research Laboratory, Byron, GA 31008, USA*

28.1. Introduction

Nematodes are important biocontrol agents of pest insects, molluscs, plant-parasitic nematodes and soil-borne fungal pathogens of plants. Entomopathogenic nematodes (EPNs) (Rhabditida: Steinernematidae and Heterorhabditidae) in particular have emerged as excellent biocontrol agents of soil-dwelling insect pests. They are now used in citrus groves, strawberry plantations, cranberry bogs, production nurseries, greenhouses and turfgrass for the management of important insect pests (Table 28.1). Currently, over 19,000 ha of citrus are treated annually with nematodes for the control of root feeding *Diaprepes* weevil in Florida (see Shapiro-Ilan *et al.*, Chapter 11, this volume), and nematodes have replaced aldrin and carbofuran as a control measure for the black vine weevil in the greenhouse and nursery industries in Europe (see van Tol *et al.*, Chapter 9, this volume). They are also used for the control of the hunting billbug on golf courses in Japan (see Grewal *et al.*, Chapter 7, this volume) and against black vine weevil and cranberry girdler in cranberry bogs in North America (see Cowles *et al.*, Chapter 12, this volume). Similarly, slug-parasitic nematodes (Rhabditida: Rhabditidae) have

Table 28.1. List of commercially produced biocontrol nematodes and their current target pests.

Nematode species	Strain	Major target pest
Steinernema carpocapsae	All	Artichoke plume moth, armyworms, webworms, cutworms, billbugs, cat flea, mint flea beetle, mint root borer, cranberry girdler
S. feltiae	SN	Fungus gnats, sciarid flies, European crane fly
S. glaseri	NJ43	White grubs
S. riobrave	RGV	Citrus root weevils
S. scapterisci	Uruguay	Mole crickets
Heterorhabditis bacteriophora	HP88	White grubs, black vine weevil, stem borers
H. indica	LN2	Citrus root weevils
H. marelata	Oregon	Black vine weevil
H. megidis	UK	Black vine weevil
H. zealandica	X1	White grubs
Phasmarhabditis hermaphrodita	UK	Slugs and snails
Beddingia siricidicola[a]	Australia	Wood wasps

[a]Used in classical biocontrol.

proven effective against pest slugs and snails (see Wilson and Grewal, Chapter 24, this volume) and are now commercially used in agriculture and horticulture in Europe (see Ester and Wilson, Chapter 25, this volume). However, scores of field tests demonstrate efficacy of nematodes against dozens of other pests, and many more pests have been shown to be susceptible to nematodes under laboratory and greenhouse conditions. Also, entomophilic nematodes including *Beddingia siricidicola*, *Thripinema* spp. and mermithid nematodes are proving useful in classical biocontrol of insect pests (see Chapters 21, 22, 23, this volume). The explorations of the potential of predatory nematodes to manage plant-parasitic nematodes (see Chapter 26, this volume) and of the fungal feeding nematodes to control fungal pathogens of plants (see Ishibashi, Chapter 27, this volume) have just begun. It is our assessment that nematodes are underutilized in pest control programmes and their biocontrol potential can be further expanded substantially. In this chapter we explore the reasons for under-utilization of nematodes, identify critical issues and discuss the research needs that must be addressed to expand the use of nematodes in biocontrol.

28.2. Factors Limiting the Expansion of Nematode Use

End-users choose a pest control product using one or more of the following four main criteria: product cost, availability, ease-of-use and efficacy. A small number of end-users, especially in urban areas, use biocontrol products due to their concerns about the health and safety of their children and pets. It is also possible that some end-users may lack confidence in the use of biocontrol agents due to lack of experience, or may simply perceive them to be less effective. We discuss below four key factors that predominantly limit the expansion of nematode use against susceptible pests.

28.2.1. High product cost

Nematode products are generally more expensive than the standard chemical pesticides. This is due to the high cost of nematode production, formulation, storage and transport. The nematode production process is long, taking 10–14 days for the completion of a single run. Controlled temperature and aeration requirements further add to the cost of nematode production. EPNs are now mass-produced in over 18 countries by either *in vivo* or *in vitro* methods. During the last decade, a distinct cottage industry has emerged in the USA that utilizes the *in vivo* process for nematode mass production for sale, especially in the home and garden markets. The wax moth, *Galleria mellonella*, larvae are most commonly used for rearing the nematodes *in vivo* because of their commercial availability. Using this *in vivo* process, yields between 0.5×10^5 and 4×10^5 infective juveniles (IJs)/larva, depending on the nematode species, have been obtained. The *in vivo* process, however, lacks economy of scale; the labour, equipment and material (insect) costs increase as a linear function of production capacity. Perhaps even more important is the lack of improved quality while increasing scale. Also *in vivo* nematode production is increasingly sensitive to biological variations and catastrophes as scale increases, thus increasing the cost of production. Conceivably, modifications of the *in vivo* process could increase efficiency and profitability (Gaugler *et al.*, 2002).

The first successful commercial scale *in vitro* culture was established by Bedding (1981), which is now known as the 'solid' culture. In this method, nematodes are cultured on a crumbed polyether polyurethane sponge impregnated with emulsified beef-fat and pig's kidneys or in vegetable protein sources, along with symbiotic bacteria. Using this method, nematode yields of approximately $6–10 \times 10^5$ IJs/g of medium were achieved. Since then, this method has been commercially used in Australia, China, Poland, Sweden, Switzerland, the Netherlands and the USA. In a scale-up model, Friedman (1990) reported that the solid culture method is economically feasible up to a production level of approximately 10×10^{12} nematodes/month. Labour costs increase significantly for nematode production beyond this level, making a less expensive method of large-scale production a necessity.

In the 1980s, a liquid fermentation technique was developed for large-scale production of EPNs. Although initial investment is high, production costs rapidly decline up to a capacity of approximately 50×10^{12} IJs/month, in the liquid culture method (Friedman, 1990). This method allows consistent production of steinernematids in up to 80,000-l fermenters. Improvements in nematode fermentation and media formulation processes further improved nematode quality and yields, resulting in a production cost of approximately US$0.10 per million nematodes (without the formulation). The current yields of *Steinernema carpocapsae* in the liquid culture average at about 2.5×10^5 IJs/ml; those of *H. indica* can reach 5.0×10^5/ml. In addition to *S. carpocapsae* and *Heterorhabditis indica*, *S. riobrave*, *S. scapterisci*, *S. feltiae*, *S. glaseri*, *H. bacteriophora* and *H. megidis* have been produced successfully in large-scale liquid cultures.

Formulation, storage and transport costs further add to the price of nematode products. Due to the poor storage stability of concentrated nematodes in water (bulk storage), they are formulated soon after production. The nematodes, whether formulated or not, need to be stored at low temperatures (2–10°C) at all times, which also adds a significant expense. Almost all commercially available formulations require cold storage and are shipped on ice, often via overnight or priority mail, further increasing nematode costs by several times.

28.2.2. Limited product availability

Another major factor limiting the use of nematodes is their lack of availability in the retail market. Due to the lack of a significant room temperature shelf-life, nematode products are not available in the lawn and garden stores, farmer cooperatives and typical pesticide distribution outlets. The only formulation, the water-dispersible granules (WDGs), that possesses 5- to 6-month room temperature shelf-life was taken out of the retail market due to fungal contamination in 1994. Currently, all nematode products require refrigeration and are shipped out of cold storage facilities after the receipt of a mail order. Thus, end-users have to specifically search for nematode products at unusual places, such as the Internet, and have to wait for the arrival of the product for application.

Although EPNs are easily mass-produced, their supply depends upon production capacity and financial capability of the producers to build inventory in anticipation of demand. Limited production capacity and poor storage stability coupled with seasonal nature of demand seriously limit further expansion in the use of nematodes.

28.2.3. Suboptimum ease-of-use

Although progress has been made so that some nematode products are as easy to apply as the standard chemical pesticides, they still require a more educated user for optimum results. All nematode products have to be applied as soon as they are purchased, due to the lack of shelf life. Applications need to be made in the late evening to avoid exposure of nematodes to lethal sunlight. When applying nematodes, a post-application irrigation is always required and a pre-application irrigation is often needed. Humidity should be high for foliar application, and both cold and warm temperatures limit nematode efficacy. Some nematode products are quite tedious to apply and are thus not suitable for large area application. For example, the nematodes held on sponges need to be hand-squeezed into water before application. Due to the labour-intensive application and constant refrigeration requirements, these formulations are only applicable in the home lawn and garden situations.

28.2.4. Suboptimum efficacy

Many insect pests, e.g. the black vine weevil, can be better controlled with nematodes than with chemicals. However, in other systems, efficacy of nematode products may be lower than the standard pesticides. Many reasons can contribute to low efficacy. Poor efficacy can result from an inappropriate application method, the use of a poor quality product or suboptimum application conditions. Quality of commercially produced nematodes is essentially self-regulated (see Grewal and Peters, Chapter 4, this volume). Poor quality of commercially produced nematodes may result from factors unknown to the producer, such as deterioration during transport and distribution. Experience with commercial distribution shows that in most cases in which nematodes failed to control the target pests, the reasons were either misuse (other pest insect than indicated, mistakes in application or storage techniques) or exposure to high temperature during transportation. Low efficacy may also result from the recommendation of inappropriate nematode species or strain and the lack of proper instructions on handling and application. Suboptimum application conditions such as high or low moisture and temperature, exposure to sunlight, lack of or delay in postapplication irrigation, impervious soil surface (e.g. presence of thatch in turfgrass) and high clay content in soil can result in low efficacy of nematodes.

28.3. Research Needs for Enhancing the Use of Nematodes in Biocontrol

28.3.1. Lowering product cost

In the major markets (citrus, ornamentals, mushrooms, turf) nematodes are sold at reasonable costs to bring benefit to the farmer. However, when it comes to larger-scale outdoor markets, the costs are still too high. Several possibilities exist to lower product costs. Increasing production efficiency through improvement of media constituents and optimizing process conditions to reduce process time are the major factors to reduce production costs. These factors, however, have a biological limit and some producers have optimized their production processes for these factors. Significant potential for cost reduction exists in the economies of scale. Producing nematodes at a larger scale can substantially reduce production costs, which can bring down the cost to approximately US$0.01/million IJs. Larger production facilities, however, require larger markets. Thus, reduction in nematode production costs is directly related to the market size. Storage at the factory and inventory in anticipation of demand add significant costs to nematode products. Cheaper and longer storability of nematodes will also bring the costs down.

28.3.2. Increasing product availability

Availability of nematode products can be enhanced by increasing production capacity and by enhancing storage stability. Since the demise of Biosys in 1997, several small companies have increased nematode production capacity to meet demand. At present, the largest fermenter used for nematode production is only 30,000 l. It is likely that nematode producers will further expand production capacity to meet the expanding demand for nematode products due to the ban on the use of chemical pesticides in several urban municipalities and the emergence of new markets for nematode products. Nematode production capacity can be expanded without any major technical hurdles, as demonstrated by the production of up to 80,000 l by Biosys (Grewal and Georgis, 1999). At this production scale, a single fermenter can produce enough nematodes to treat 6000–8000 ha, depending upon the nematode species, at an application rate of 2.5 billion nematodes/ha.

The availability of nematode products in the market may also be limited by their poor storage stability. In fact, poor storage stability is the major factor limiting widespread availability of nematodes. Due to the lack of cold storage facilities at pesticide distribution outlets, nematode inventory at the distribution site is not possible. An attractive solution to this problem can be the building of cold storage capacity at distribution outlets. It should be noted that distributor education about proper storage and handling of the nematodes is also important.

Poor stability results from short lifespan and rapid inactivation of IJs due to environmental extremes including temperature and desiccation. The availability of EPNs with 1- to 2-year shelf-life and increased resistance to environmental extremes will make a significant difference in their use pattern in agricultural and horticultural markets. Increase in storage stability has been achieved through improved formulations (Grewal, 2000a,b) and by the discovery of nematode strains with prolonged IJ longevity (Grewal *et al.*, 2002), but these approaches cannot meet the 1- to 2-year room temperature shelf-life requirement. Thus, transgenic approaches will be needed to achieve this goal.

Another major obstacle limiting nematode availability is the lack of professional marketing. The largest market is the citrus market; due to the well-developed introduction of the system and satisfying control results for over a decade, this market is relatively stable. Another large market is sciarid control in glasshouse ornamentals. In this market, biocontrol with invertebrate biocontrol agents is well established and nematodes are just ‘another product’ among many biocontrol agents. The sales logistics can be developed with insects and mites and in this system nematodes are the easier products regarding shelf-life. However, only very small amounts are necessary to meet the demand. Other markets are less well developed because of the smaller market sizes or the lack of products for the control of all the pests, as growers expect to use complete control systems like those supplied by the chemical industry. For biocontrol agents complete product portfolios have only been developed for the glasshouse market. Limited amounts of biocontrol products for other crops usually cannot justify the build-up of a sales force. Thus, many markets lack marketing activities and the sales are low.

28.3.3. Enhancing product ease-of-use

28.3.3.1. Improved formulation

Development of improved 'easy-to-use' formulations will facilitate expanded use of EPNs and could increase efficacy as well. Certainly, extension of shelf-life is one of the critical limiting factors that can be addressed through improved formulations. A breakthrough in formulation technology was cited in the introduction of WDGs, in which the steinernematids enter a partially anhydrobiotic state, allowing them to survive up to 6 months at 4–25°C (substantially longer than previous formulations) (Grewal and Georgis, 1999). Yet this formulation does not appear to be equally suitable for all EPNs. Low viability in WDGs can be an issue for some species and detract from consumer acceptance (McCoy *et al.*, 2000; Shapiro-Ilan *et al.*, 2002). In response to the problem, an easy-to-use liquid formulation containing concentrated non-desiccated nematodes (that maintain high viability) was developed, but it must be applied within 48–72 h of receipt (Grewal, 1999). Better adapted to the requirements of transport stability are formulations in which EPNs are desiccated to induce immobility, which will conserve energy resources of the nematodes (Grewal, 2002). In these formulations nematodes survive detrimental conditions during transport much better than in non-desiccated formulations and regain activity soon after mixing with water. Clearly, additional research is needed to develop better formulations with extended shelf-life.

Another issue that can be addressed through improved formulation is enhancing EPN survival upon exposure to environmental extremes. This can have a profound impact on the potential for above-ground application with EPNs, which is a recent subject of renewed activity. For example, use of new polymeric formulations combined with low rates, better placement of nematodes and adapted application equipment have been utilized to control foliar pests such as *Spodoptera exigua* and leafminers in the genus *Liriomyza* (Piggott *et al.*, 2003). In addition to formulation, enhancing EPN tolerance to environmental extremes can be addressed through strain discovery and improvement (Grewal *et al.*, 2002; Strauch *et al.*, 2004; see Section 28.3.4.5).

28.3.3.2. Improved delivery

Development of improved methodology for EPN delivery can result in improved efficacy through superior preservation of nematode quality and more accurate or efficient application to the target site. Additionally, improved application methods that are easier to use will decrease application costs and increase the attractiveness of nematode products to consumers.

Some of the novel approaches to nematode delivery may be quite simple yet effective, e.g. application in infected hosts, 'nema bags' and root dips. The potential to effectively apply nematodes to the target site in infected host cadavers is one example of a relatively simple technology that has recently received renewed attention. Laboratory experiments indicted greater nematode dispersal (Shapiro and Glazer, 1996), infectivity (Shapiro and Lewis, 1999) and persistence in soil (Perez *et al.*, 2003) when the nematodes were applied in host cadavers compared with aqueous application. To facilitate storage of nematode-infected cadavers, and avoid rupture or sticking together during application, the cadavers can be coated with a protective formulation (Shapiro-Ilan *et al.*, 2001), or hard-bodied insects (e.g. *Tenebrio molitor*) could be used (Shapiro-Ilan *et al.*, 2003). Effective pest suppression has been reported in field trials when nematodes were applied in infected hosts using this method (Jansson *et al.*, 1993; Parkman *et al.*, 1993). Furthermore, greenhouse trials indicated superior pest suppression through application of nematode in cadavers relative to aqueous suspension (Shapiro-Ilan *et al.*, 2003). The superior pest suppression observed in cadaver applications may have been due to reduced physical damage relative to spray applications, or to metabolites present in the cadavers that enhance dispersal or infection (Shapiro and Lewis, 1999; Shapiro *et al.*, 2000). Application of EPNs in host cadavers could reduce costs to *in vivo* producers because labour-intensive steps in the production process, such as harvesting and concentration of IJs, would be eliminated. However, the practical application of host cadavers has yet to occur.

An additional advantage to application of nematodes in infected host cadavers relative to application in aqueous suspension may be that the nematodes emerge from infected hosts over a period of time, thus creating a 'slow-release' effect and possibly extending the period of control. This slow-release concept was the primary motivation in developing another approach to nematode delivery, the 'nema bag'. In this approach, nematodes are applied in tea bags containing superabsorbent gel (Menzler-Hokkanen and Hokkanen, 2003). Field suppression of oilseed rape pests was achieved using nema bags (Menzler-Hokkanen and Hokkanen, 2003).

Another simple approach to nematode delivery that can be easily incorporated into certain grower practices is prophylactic treatment before planting. This approach has yielded promising preliminary results when strawberry roots were dipped in a nematode suspension before planting (A. Peters, A. Susurluk and R.-U. Ehlers, unpublished data). Conceivably, a similar approach could be applied for prophylactic (or curative) protection of various ornamentals in the nursery industry; potted plants or trees in burlap root balls could be dipped in nematode suspension before shipping or planting. In addition to these simple approaches to improved delivery (cadaver application, nema bags, prophylactic treatments), more advanced improvements in application equipment and precision will certainly facilitate ease-of-use and expand utility of EPNs (see Section 28.3.4.2).

28.3.4. Enhancing product efficacy

28.3.4.1. Consumer education

Consumer education is extremely important for optimum use of EPNs to obtain the expected high efficacy against the target pests. Provision of clear instructions on how to handle and apply nematode products is critical for obtaining successful control of the target pest. An instructional video on EPNs was made in 1999 and was widely distributed throughout the world. A website housing information on EPNs (http://www.oardc.ohio-state.edu/nematodes) was established in 1999 and has been extensively accessed by users worldwide. However, more extension materials such as fact sheets, bulletins and instructional videos are needed to disseminate the latest information to growers. Also, training of crop consultants, and golf course agronomists, in the optimum use of EPNs should be a high priority. In Japan, supervision of nematode applications at golf courses by the consultants resulted in rapid adoption of nematode products for the control of the hunting billbug (Satoshi Yamanaka, personal communication, 1998). Appropriate participatory learning models need to be developed to further enhance the adoption and optimum use of biocontrol nematodes.

28.3.4.2. Optimum application

Ensuring appropriate application conditions can result in improved efficacy of nematodes. For example, pretreatment irrigation may be needed to increase soil moisture and moderate soil temperature before nematode application. Although the importance of post-application irrigation is often emphasized, its timing, amount and frequency are extremely important. A delay of only a few minutes in post-application irrigation on a bright sunny day may result in desiccation and death of nematodes stuck to the exposed plant surfaces. A study has revealed that the total amount of water (irrigation and precipitation) applied during the 3–4-week post-application period determines the level of white grub control achieved by EPNs in turfgrass (Grewal *et al.*, 2004).

Enhanced nematode delivery may also be brought about through a more in-depth understanding of the effects of application equipment on nematode physiology and efficacy. Until recently relatively little attention has been given to effects of application equipment on EPNs (Fife *et al.*, 2003). Recommendations have tended to be overly general, for example, that nematodes can be applied using various spray equipment as long as the nozzle sizes are 50 μm or greater and pressure remains below 2070 kPa (300 psi) (Georgis, 1990). However, Fife *et al.* (2003, 2004) reported that various parameters in spray equipment such as nozzle and pumping system type can affect nematode fitness, and resistance to pressure can vary among nematode species. Additional research on application technology is required such that, eventually, all operating conditions within each application system are defined and optimized. Furthermore, technological advances in the application equipment itself, such as use of precision-based methodology being developed for application of chemical pesticides (Miller and Paice, 1995; Bongiovanni and Lowenberg-Deboer, 2004), will surely increase pest control efficacy and acceptance by growers.

28.3.4.3. Correct nematode species, strain and application rate

Matching of the appropriate nematode species with the target pest is another requirement for achieving successful control. It is now well documented that nematode species differ in their host-finding behaviour that narrows their field host range (Gaugler *et al.*, 1997; see Chapter 2, this volume). Some species like *S. carpocapsae* use an ambush type host-finding strategy and are therefore not suitable for use against less mobile hosts that feed below ground, such as white grubs. Other species like *H. bacteriophora* use a cruising type host-finding behaviour and are therefore less appropriate for use against highly mobile surface feeding insects, such as cutworms and armyworms. More recent studies have documented large differences in virulence among different strains of nematode species (Grewal *et al.*, 2002, 2004; also see chapters 7, 9 and 12, this volume). Thus, in addition to the appropriate nematode species, attention needs to be paid to the strain of the nematodes.

Recommendations on the application rate of the same nematode species for the control of the same pest species vary widely by the product. As application rate is a major economic driver, companies may recommend lower rates of nematode application. There needs to be more research on the evaluation of commercial products for validation of their efficacy, a role typically played by the University Extension.

28.3.4.4. Synergistic combinations

In certain systems, EPNs will not provide adequate suppression of a given target pest, even when all biotic and abiotic requirements are met. In these cases, where use of EPNs as a 'stand-alone' product is not feasible, it may still be beneficial to apply nematodes in combination with other pest control agents. Application of EPNs is compatible with concurrent application of a number of chemical and biotic agents and several chemical or biotic agents interact synergistically with EPNs (see Chapter 20, this volume). Thus, a combination of multiple tactics may be advantageous due to an overall increase in pest suppression, reduced rate requirements or reduction in the quantity of chemical pesticides in the environment. Hence the use of EPNs can be expanded by incorporating the tactic into integrated management strategies that result in economic and environmental benefits.

28.3.4.5. Genetic improvement

There is a huge untapped natural genetic diversity in EPN populations around the world. Only a fraction of this genetic diversity has been exploited for biocontrol. Emphasis needs to be placed on conducting more comprehensive surveys, and preservation of the isolated strains. One of the problems of the surveys conducted so far has been that many of the nematode strains collected during the surveys have already been lost due to the inappropriate storage protocols and the lack of resources for long-term preservation of the live material. In fact, many of the collected strains were lost even before they were correctly identified let alone their biocontrol potential assessed. Therefore, the strains collected during any future surveys should be carefully preserved.

Further improvements in important traits such as host finding, infectivity, virulence, desiccation tolerance, heat tolerance and cold activity can result in enhanced use of biocontrol nematodes. Selective breeding, mutagenesis and genetic engineering are potential techniques that offer promise for the improvement of biocontrol nematodes. The utility of selective breeding or hybridization for the improvement of host finding (Gaugler *et al.*, 1989), temperature tolerance (Grewal *et al.*, 1996a,b; Shapiro *et al.*, 1997), pesticide tolerance (Glazer *et al.*, 1997) and infectivity (Tomalak, 1994) has been demonstrated. Isolation of desiccation-tolerant mutants has also been reported (O'Leary and Burnell, 1997). However, much needs to be learned about the basic physiological, biochemical, genetic and molecular mechanisms controlling desirable traits of biocontrol nematodes (see Grewal *et al.*, 2004) before genetic engineering techniques can be applied. Cloning of genes involved in heat tolerance, desiccation tolerance and chemoreception has already begun. The next biggest challenge is to determine the functions of these and many more yet-to-be-isolated genes. In this regard, the establishment of functional genomic techniques such as gene knockout using site-directed mutagenesis, transposons and RNA interference are being developed. The genome of the bacterial symbiont of *H. bacteriophora*, *Photorhabdus luminescens*, has already been sequenced and an international consortium to sequence the genome of *H. bacteriophora* has been established.

Genetic deterioration of nematodes due to inadvertent selection during laboratory culture and mass production may occur and should be prevented. Recent research has documented that deterioration in important nematode traits such as heat and desiccation tolerance and IJ longevity can occur in just 3–6 passages through the laboratory host *G. mellonella* larvae (Wang and Grewal, 2002). Such trait deterioration can, however, be prevented by preserving master stocks in liquid nitrogen (Wang and Grewal, 2002). Beneficial traits can also be stabilized by establishing inbred lines (Bai *et al.*, 2004). Stringent quality control assessment on the part of commercial nematode producers, batch-coding of products to track product age and quality and measures to prevent/minimize trait deterioration due to inadvertent selection during mass production will ensure quality improvement and high field efficacy.

28.3.5. Enhancing the carryover effect

Most biocontrol nematodes, particularly the EPNs, are currently used as inundative biocontrol agents for immediate pest control. However, the documentation of their widespread natural occurrence in diverse ecosystems, their role in natural epizootics in pest populations, persistence beyond immediate pest control in some cases and some successful inoculative release experiments now indicate that they can be managed as a sustainable biocontrol mechanism (see Chapter 18, this volume). The development of a sound conservation approach requires basic information on the population biology, genetics and ecology of the endemic nematode populations. Although considerable advances have been made in our understanding of the behaviour, physiology and ecology of EPNs, much of this research has occurred on laboratory populations. Thus, there is a huge gap in our understanding of the biology and ecology of biocontrol nematodes in the field. For instance, it has long been recognized that inadvertent laboratory selection may lead to reduced virulence and fitness, but no tests have been conducted to demonstrate the effects of inbreeding depression on the persistence of EPNs following application. A recent study (Wang and Grewal, 2002) has demonstrated that environmental fitness traits such as IJ longevity, heat tolerance and resistance to ultraviolet (UV) and desiccation in *H. bacteriophora* can deteriorate very rapidly following recovery from the wild (within 6–12 generations), while virulence against the laboratory host may actually increase. Thus, the rapid deterioration in environmental fitness during laboratory culture and handling may be a cause for the poor persistence of EPNs noted in many field trials. Much research is needed to develop practical conservation approached to use biocontrol nematodes.

References

Bai, C., Shapiro-Ilan, D.I., Gaugler, R. and Hopper, K.R. (2004) Stabilization of beneficial traits in *Heterorhabditis bacteriophora* through creation of inbred lines. *Biological Control* 32, 220–227.

Bedding, R.A. (1981) Low-cost in vitro mass-production of *Neoaplectana* and *Heterorhabditis* species (nematoda) for field control of insect pests. *Nematologica* 27, 109–114.

Bongiovanni, R.A. and Lowenberg-Deboer, J.B. (2004) Precision agriculture and sustainability. *Precision Agriculture* 5, 359–387.

Fife, J.P., Derksen, R.C., Ozkan, H.E. and Grewal, P.S. (2003) Effects of pressure differentials on the viability and infectivity of entomopathogenic nematodes. *Biological Control* 27, 65–72.

Fife, J.P., Derksen, R.C., Ozkan, H.E., Grewal, P.S., Chalmers, J.J. and Krause, R. (2004) Evaluation of a contraction flow field on hydrodynamic damage to entomopathogenic nematodes – a biological pest control agent. *Biotechnology and Bioengineering* 86, 96–107.

Friedman, M.J. (1990) Commercial production and development. In: Gaugler, R. and Kaya, H.K. (eds) *Entomopathogenic Nematodes in Biological Control.* CRC Press, Boca Raton, Florida, pp. 153–172.

Gaugler, R., Campbell, J.F. and McGuire, T.R. (1989) Selection for host-finding in *Steinernema feltiae. Journal of Invertebrate Pathology* 54, 363–372.

Gaugler, R., Wilson, M. and Shearer, P. (1997) Field release and environmental fate of a transgenic entomopathogenic nematode. *Biological Control* 9, 75–80.

Gaugler, R., Brown, I., Shapiro-Ilan, D. and Atwa, A. (2002) Automated technology for *in vivo* mass production of entomopathogenic nematodes. *Biological Control* 24, 199–206.

Georgis, R. (1990) Formulation and application technology. In: Gaugler, R. and Kaya, H.K. (eds) *Entomopathogenic Nematodes in Biological Control.* CRC Press, Boca Raton, Florida, pp. 173–191.

Glazer, I., Salame, L. and Segal, D. (1997) Genetic enhancement of nematicide resistance in entomopathogenic nematodes. *Biocontrol Science and Technology* 7, 499–512.

Grewal, P.S. (1999) Factors in the success and failure of microbial control in turfgrass. *Integrated Pest Management Reviews* 4, 287–294.

Grewal, P.S. (2000a) Enhanced ambient storage stability of an entomopathogenic nematode through anhydrobiosis. *Pest Management Science* 56, 401–406.

Grewal, P.S. (2000b) Anhydrobiotic potential and long-term storage stability of entomopathogenic nematodes (Rhabditida: Steinernematidae). *International Journal for Parasitology* 30, 995–1000.

Grewal, P.S. (2002) Formulation and application technology. In: Gaugler, R. (ed.) *Entomopathogenic Nematology.* CAB International, Wallingford, UK, pp. 265–287.

Grewal, P.S. and Georgis, R. (1999) Entomopathogenic nematodes. In: Hall, F.R. and Menn, J. (eds) *Methods in Biotechnology: Biopesticides: Use and Delivery.* Humana Press, Totowa, New Jersey, pp. 271–299.

Grewal, P.S., Gaugler, R., and Wang, Y. (1996a) Enhanced cold tolerance of the entomopathogenic nematode *Steinernema feltiae* through genetic selection. *Annals of Applied Biology* 129, 335–341.

Grewal, P.S., Gaugler, R. and Shupe, C. (1996b) Rapid evolution of thermal sensitivity of entomopathogenic nematodes in response to selection at temperature extremes. *Journal of Invertebrate Pathology* 68, 65–73.

Grewal, P.S., Wang, X. and Taylor, R.A.J. (2002) Dauer juvenile longevity and stress tolerance in natural populations of entomopathogenic nematodes. *International Journal for Parasitology* 32, 717–725.

Grewal, P.S., Power, K.T., Grewal, S.K., Suggars, A. and Haupricht, S. (2004) Enhanced predictability in biological control of white grubs with new strains of entomopathogenic nematodes. *Biological Control* 30, 73–82.

Jansson, R.K., Lecrone, S.H. and Gaugler, R. (1993) Field efficacy and persistence of entomopathogenic nematodes (*Rhabditida: Steinernematidae, Heterorhabditidae*) as biological control agents of the sweet potato weevil (*Coleoptera: Apionidae*) in southern Florida. *Journal of Economic Entomology* 86, 1055–1063.

McCoy, C.W., Shapiro, D.I., Duncan, L.W. and Nguyen, K. (2000) Entomopathogenic nematodes and other natural enemies as mortality factors for larvae of *Diaprepes abbreviatus. Biological Control* 19, 182–190.

Menzler-Hokkanen, I. and Hokkanen, H.M.T. (2003) Entomopathogenic nematode delivery systems for biological control of pests on major outdoor crops: the case of oilseed rape. *The Society of Invertebrate Pathology Program and Abstracts for the 36th Annual Meeting,* Burlington, Vermont, p. 60 (Abstr.).

Miller, P.C.H. and Paice, M.E.R. (1995) Applying the patch spraying concept to reduce total herbicide input, Silsoe Link, February 22, 1995: *Agrochemicals: Effective use and the Environment,* Silsoe, UK.

O'Leary, S.A. and Burnell, A.M. (1997) The isolation of mutants of *Heterorhabditis megidis* (Strain UK211) with increased desiccation tolerance. *Fundamental and Applied Nematology* 20, 197–205.

Parkman, J.P., Hudson, W.G., Frank, J.H., Nguyen, K.B. and Smart, G.C. (1993) Establishment and persistence of *Steinernema scapterisci* (Rhabditida: Steinernematidae) in field populations of *Scapteriscus* spp. mole crickets (Orthoptera: Gryllotalpidae). *Journal of Entomological Science* 28, 182–190.

Perez, E.E., Lewis, E.E. and Shapiro-Ilan, D.I. (2003) Impact of the host cadaver on survival and infectivity of entomopathogenic nematodes (Rhabditida: Steinernematidae and Heterorhabditidae) under desiccating conditions. *Journal of Invertebrate Pathology* 82, 111–118.

Piggott, S.J., Clayton, R., Matthews, G.A. and Wright, D.J. (2003) Development of a new application apparatus for entomopathogenic nematodes. *Pest Management Science* 59, 1344–1348.

Shapiro, D.I. and Glazer, I. (1996) Comparison of entomopathogenic nematode dispersal from infected hosts versus aqueous suspension. *Environmental Entomology* 25, 1455–1461.

Shapiro, D.I. and Lewis, E.E. (1999) Comparison of entomopathogenic nematode infectivity from infected hosts versus aqueous suspension. *Environmental Entomology* 28, 907–911.

Shapiro, D.I., Glazer, I. and Segal, D. (1997) Genetic improvement of heat tolerance in *Heterorhabditis bacteriophora* through hybridization. *Biological Control* 8, 153–159.

Shapiro, D.I., Lewis, E.E., Paramasivam, S. and McCoy, C.W. (2000) Nitrogen partitioning in *Heterorhabditis bacteriophora*-infected hosts and the effects of nitrogen on attraction/repulsion. *Journal of Invertebrate Pathology* 76, 43–48.

Shapiro-Ilan, D.I., Lewis, E.E., Behle, R.W. and McGuire, M.R. (2001) Formulation of entomopathogenic nematodes-infected cadavers. *Journal of Invertebrate Pathology* 78, 17–23.

Shapiro-Ilan, D.I., Gouge, D.H. and Koppenhöfer, A.M. (2002) Factors affecting commercial success: case studies in cotton, turf and citrus. In: Gaugler, R. (ed.) *Entomopathogenic Nematology.* CAB International, Wallingford, UK, pp. 333–355.

Shapiro-Ilan, D.I., Lewis, E.E., Tedders, W.L. and Son, Y. (2003) Superior efficacy observed in entomopathogenic nematodes applied in infected-host cadavers compared with application in aqueous suspension. *Journal of Invertebrate Pathology* 83, 270–272.

Strauch, O., Oestergaard, J., Hollmer, S., Ehlers, R.U. (2004) Genetic improvement of the desiccation tolerance of the entomopathogenic nematode *Heterorhabditis bacteriophora* through selective breeding. *Biological Control* 31, 218–226.

Tomalak, M. (1994) Selective breeding of *Steinernema feltiae* (Filipjev) (Nematoda: Steinernematidae) for improved efficacy in control of a mushroom fly *Lycoriella solani* Winnertz (Diptera: Sciaridae). *Biocontrol Science and Technology* 4, 187–198.

Wang, X. and Grewal, P.S. (2002) Rapid deterioration of environmental tolerance and reproductive potential of an entomopathogenic nematode during laboratory maintenance. *Biological Control* 23, 71–78.

Index

Figures in bold indicate major references.
Figures in italic refer to diagrams, photographs and tables.